Probability and Statistics

The Science of Uncertainty

Michael J. Evans and Jeffrey S. Rosenthal

University of Toronto

W. H. Freeman and Company

New York

Senior Acquisitions Editor: Patrick Farace
Development Editor: Danielle Swearengin
New Media/Supplements Editor: Brian Donnellan
Marketing Manager: Jeffrey Rucker
Project Editor: Vivien Weiss
Text Designer: Blake Logan
Cover Image: Joan Spavins, Doorways, acrylic and watercolor on paper, 2002
Illustrations and Composition: Michael J. Evans, Jeffrey S. Rosenthal
Director of Production: Ellen Cash
Printing and Binding: R. R. Donnelley

Library of Congress Control Number: 2003108117

W. H. Freeman and Company
41 Madison Avenue
New York, NY 10010
Houndmills, Basingstoke RG21 6XS, England

www.whfreeman.com

Contents

Preface

This book is an introductory text on probability and statistics. The book is targeted at students who have studied one year of calculus at the university level and are seeking an introduction to probability and statistics that has mathematical content. Where possible, we provide mathematical details, and it is expected that students are seeking to gain some mastery over these, as well as learn how to conduct data analyses. All of the usual methodologies covered in a typical introductory course are introduced, as well as some of the theory that serves as their justification.

The text can be used with or without a statistical computer package. It is our opinion that students should see the importance of various computational techniques in applications, and the book attempts to do this. Accordingly, we feel that computational aspects of the subject, such as Monte Carlo, should be covered, even if a statistical package is not used. All of the computations in this text were carried out using Minitab. Minitab is a suitable computational platform to accompany the text, but others could be used. There is a **Computations** appendix that contains the Minitab code for those computations that are slightly involved (for example, if looping is required); these can be used by students as templates for their own calculations. If a software package like Minitab is used with the course, then no programming is required by the students to do problems.

We have organized the exercises in the book into groups, as an aid to users. **Exercises** are suitable for all students and are there to give practice in applying the concepts discussed in a particular section. **Problems** require greater understanding, and a student can expect to spend more thinking time on these. If a problem is marked (MV), then it will require some facility with multi-variable calculus beyond the first calculus course, although these problems are not necessarily hard. **Challenges** are problems that most students will find difficult. The **Challenges** are only for students who have no difficulty with the **Exercises** and the **Problems**. There are also **Computer Exercises** and **Computer Problems**, where it is expected that students will make use of a statistical package in deriving solutions.

We have also included a number of **Discussion Topics** that are designed to promote critical thinking in students. Throughout the book we try to point students beyond the mastery of technicalities to think of the subject in a larger frame of reference. It is important that students acquire a sound mathematical foundation in the basic techniques of probability and statistics. We believe that this book will help students accomplish this. Ultimately, however, these subjects are applied in real-world contexts, so it is equally important that students understand how to go about their application and understand what issues arise. Often there are no right answers to **Discussion Topics**. Their purpose is to get a student thinking about the subject matter. If these were to be used for evaluation, then they would be answered in essay format and graded on the maturity the student showed with respect to the issues involved. **Discussion Topics** are probably most suitable for smaller classes, but there will also be benefit to students if these are simply read over and thought about.

Some sections of the book are labelled **Advanced**. This material is aimed at students who are more mathematically mature (for example, they are taking, or have taken, a second course in calculus). All of the **Advanced** material can be skipped, with no loss of continuity, by an instructor who wishes to do so. In particular, the final chapter of the text is labelled **Advanced** and would only be taught in a high-level introductory course aimed at specialists. Also, many proofs are put in a final section of each chapter, labelled **Further Proofs (Advanced)**. An instructor can choose which (if any) of these proofs they wish to present to their students. As such, we feel that the material in the text is presented in a flexible way that allows the instructor to find an appropriate level for the students they are teaching. There is a **Mathematical Background** appendix that reviews some mathematical concepts that students may be rusty on from a first course in calculus, as well as brief introductions to partial derivatives, double integrals, etc.

Chapter 1 introduces the probability model and provides motivation for the study of probability. The basic properties of a probability measure are developed.

Chapter 2 deals with discrete, continuous, joint distributions, and the effects of a change of variable. The multivariate change of variable is developed in an Advanced section. The topic of simulating from a probability distribution is introduced in this chapter.

Chapter 3 introduces expectation. The probability-generating function is introduced as well as the moments and the moment-generating function of a random variable. This chapter develops some of the major inequalities used in probability. There is a section available on characteristic functions as an Advanced topic.

Chapter 4 deals with sampling distributions and limits. Convergence in probability, convergence with probability 1, the weak and strong laws of large numbers, convergence in distribution, and the central limit theorem are all introduced along with various applications such as Monte Carlo. The normal distribution theory, necessary for many statistical applications, is also dealt with here.

As mentioned, Chapters 1 through 4 include material on Monte Carlo techniques. Simulation is a key aspect of the application of probability theory, and it is our view that its teaching should be integrated with the theory right from the start. This reveals the power of probability to solve real-world problems and helps convince students that it is far more than just an interesting mathematical theory. No practitioner divorces himself from the theory when using the computer for computations or vice versa. We believe this is a more modern way of teaching the subject. This material can be skipped, however, if an instructor doesn't agree with this, or feels they do not have enough time to cover it effectively.

Chapter 5 is an introduction to statistical inference. For the most part this is concerned with laying the groundwork for the development of more formal methodology in later chapters. So practical issues — such as proper data collection, presenting data via graphical techniques, and informal inference methods like descriptive statistics — are discussed here.

Chapter 6 deals with many of the standard methods of inference for one-sample problems. The theoretical justification for these methods is developed primarily through the likelihood function, but the treatment is still fairly informal. Basic methods of inference, such as the standard error of an estimate, confidence intervals, and P-values, are introduced. There is also a section devoted to distribution-free (nonparametric) methods like the bootstrap.

Chapter 7 involves many of the same problems discussed in Chapter 6 but now from a Bayesian perspective. The point of view adopted here is not that Bayesian methods are better or, for that matter, worse than those of Chapter 6. Rather, we take the view that Bayesian methods arise naturally when the statistician adds another ingredient — the prior — to the model. The appropriateness of this, or the sampling model for the data, is resolved through the model-checking methods of Chapter 9. It is not our intention to have students adopt a particular philosophy. Rather, the text introduces students to a broad spectrum of statistical thinking.

Subsequent chapters deal with both frequentist and Bayesian approaches to the various problems discussed. The Bayesian material is in clearly labelled sections and can be skipped with no loss of continuity, if so desired. It has become apparent in recent years, however, that Bayesian methodology is widely used in applications. As such, we feel that it is important for students to be exposed to this, as well as to the frequentist approaches, early in their statistical education.

Chapter 8 deals with the traditional optimality justifications offered for some statistical inferences. In particular, some aspects of optimal unbiased estimation and the Neyman-Pearson theorem are discussed in this chapter. There is also a brief introduction to decision theory. This chapter is more formal and mathematical than Chapters 5, 6, and 7, and it can be skipped, with no loss of continuity, if an instructor wants to emphasize methods and applications.

Chapter 9 is on model checking. We placed model checking in a separate chapter to emphasize its importance in applications. In practice, model checking is the way statisticians justify the methods of inference they use. So this is a very important topic.

Chapter 10 is concerned with the statistical analysis of relationships among variables. This includes material on simple linear and multiple regression, ANOVA, the design of experiments, and contingency tables. The emphasis in this chapter is on applications.

Chapter 11 is concerned with stochastic processes. In particular, Markov chains and Markov chain Monte Carlo are covered in this chapter, as are Brownian motion and its relevance to finance. Fairly sophisticated topics are introduced, but the treatment is entirely elementary. Chapter 11 depends only on the material in Chapters 1 through 4.

A one-semester course on probability would cover Chapters 1–4 and perhaps some of Chapter 11. A one-semester, follow-up course on statistics would cover Chapters 5–7 and 9–10. Chapter 8 is not necessary, but some parts, such as the theory of unbiased estimation and optimal testing, are suitable for a more theoretical course.

A basic two-semester course in probability and statistics would cover Chapters 1–6 and 9–10. Such a course covers all the traditional topics, including basic probability theory, basic statistical inference concepts, and the usual introductory applied statistics topics. To cover the entire book would take three semesters, which could be organized in a variety of ways.

The Advanced sections can be skipped or included, depending on the level of the students, with no loss of continuity. A similar comment applies to Chapters 7, 8, and 11.

Students who have already taken an introductory noncalculus-based, applied statistics course will also benefit from a course based on this text. While similar topics are covered, they are presented with more depth and rigor here. For example, *Introduction to the Practice of Statistics*, Fourth Edition, by D. Moore and G. McCabe (W. H. Freeman, 2003) is an excellent text, and we feel that this book will serve as the basis for a good follow-up course.

Many thanks to the reviewers and class testers for their comments: Michelle Baillargeon (McMaster University), Lisa A. Bloomer (Middle Tennessee State University), Eugene Demidenko (Dartmouth College), Robert P. Dobrow (Carleton College), John Ferdinands (Calvin College), Soledad A. Fernandez (The Ohio State University), Dr. Paramjit Gill (Okanagan University College), Ellen Gundlach (Purdue University), Paul Gustafson (University of British Columbia), Jan Hannig (Colorado State University), Susan Herring (Sonoma State University), George F. Hilton, Ph.D., (Pacific Union College), Paul Joyce (University of Idaho), Hubert Lilliefors (George Washington University), Phil McDonnough (University of Toronto), Julia Morton (Nipissing University), Randall H. Rieger (West Chester University), Robert L. Schaefer (Miami University), Osnat Stramer (University of Iowa), Tim B. Swartz (Simon Fraser University), Glen Takahara (Queen's University), Robert D. Thompson (Hunter College), Dr. David C. Vaughan (Wilfrid Laurier University), Joseph J. Walker (Georgia State University), Dongfeng Wu (Mississippi State University), Yuehua Wu (York University), Nicholas Zaino (University of Rochester).

The authors would also like to thank many who have assisted in the development of this project. In particular our colleagues and students at the University of Toronto have been very supportive. Hadas Moshonov, Aysha Hashim, and Natalia Cheredeko of the University of Toronto helped in many ways. A number of the data sets in Chapter 10 have been used in courses at the University of Toronto for many years and were, we believe, compiled through the work of the late Professor Daniel B. DeLury. Professor David Moore of Purdue University was of assistance in providing several of the tables at the back of the text. Patrick Farace, Chris Spavins, and Danielle Swearengin of W. H. Freeman provided much support and encouragement. Our families helped us with their patience and care while we worked at what seemed at times an unending task; many thanks to Rosemary and Heather Evans and Margaret Fulford.

Michael Evans and Jeffrey Rosenthal
Toronto, 2003

Chapter 1

Probability Models

This chapter introduces the basic concept of the entire course, namely probability. We discuss why probability was introduced as a scientific concept and how it has been formalized mathematically in terms of a probability model. Following this we develop some of the basic mathematical results associated with the probability model.

1.1 Probability: A Measure of Uncertainty

Often in life we are confronted by our own ignorance. Whether we are pondering tonight's traffic jam, tomorrow's weather, next week's stock prices, an upcoming election, or where we left our hat, often we do not know an outcome with certainty. Instead, we are forced to guess, to estimate, to hedge our bets.

Probability is the science of uncertainty. It provides precise mathematical rules for understanding and analyzing our own ignorance. It does not tell us tomorrow's weather or next week's stock prices; rather, it gives us a framework for working with our limited knowledge and for making sensible decisions based on what we do and do not know.

To say there is a 40% chance of rain tomorrow is not to know tomorrow's weather. Rather, it is to know what we do not know about tomorrow's weather.

In this text, we will develop a more precise understanding of what it means to say there is a 40% chance of rain tomorrow. We will learn how to work with ideas of randomness, probability, expected value, prediction, estimation, etc., in ways that are sensible and mathematically clear.

There are also other sources of randomness besides uncertainty. For example, computers often use *pseudorandom numbers* to make games fun, simulations accurate, and searches efficient. Also, according to the modern theory of quantum mechanics, the makeup of atomic matter is in some sense *truly* random. All such sources of randomness can be studied using the techniques of this text.

Another way of thinking about probability is in terms of *relative frequency*. For example, to say a coin has a 50% chance of coming up heads can be interpreted as saying that, if we flipped the coin many, many times, then approximately half of the time it would come up heads. This interpretation has some limitations. In many cases (such as tomorrow's weather or next week's stock prices), it is impossible to repeat the experiment many, many times. Furthermore, what precisely does "approximately" mean in this case? However, despite these limitations, the relative frequency interpretation is a useful way to think of probabilities and to develop intuition about them.

Uncertainty has been with us forever, of course, but the mathematical theory of probability originated in the seventeenth century. In 1654, the Paris gambler Le Chevalier de Méré asked Blaise Pascal about certain probabilities that arose in gambling (such as, if a game of chance is interrupted in the middle, what is the probability that each player would have won had the game continued). Pascal was intrigued, and corresponded with the great mathematician and lawyer Pierre de Fermat about these questions. Pascal later wrote the book *Traité du Triangle Arithmetique*, discussing binomial coefficients (Pascal's triangle) and the binomial probability distribution.

At the beginning of the twentieth century, Russians such as Andrei Andreyevich Markov, Andrey Nikolaevich Kolmogorov, and Pafnuty L. Chebyshev (and American Norbert Wiener) developed a more formal mathematical theory of probability. In the 1950s, Americans William Feller and Joe Doob wrote important books about the mathematics of probability theory. They popularized the subject in the western world, both as an important area of pure mathematics and as having important applications in physics, chemistry, and later in computer science, economics, and finance.

1.1.1   Why Do We Need Probability Theory?

Probability theory comes up very often in our daily lives. We offer a few examples here.

Suppose you are considering buying a "Lotto 6/49" lottery ticket. In this lottery, you are to pick six distinct integers between 1 and 49. Another six distinct integers between 1 and 49 are then selected at random by the lottery company. If the two sets of six integers are identical, then you win the jackpot.

After mastering Section 1.4, you will know how to calculate that the probability of the two sets matching is equal to one chance in 13,983,816. That is, it is about 14 million times more likely that you will not win the jackpot than that you will. (These are not very good odds!)

Suppose the lottery tickets cost $1 each. After mastering expected values in Chapter 3, you will know that you should not even *consider* buying a lottery ticket unless the jackpot is more than 14 million dollars (which it usually is not). Furthermore, if the jackpot ever is more than 14 million dollars, then likely many other people will buy lottery tickets that week, leading to a larger probability that you will have to *share* the jackpot with other winners even if you do win — so it is probably not in your favor to buy a lottery ticket even then.

Suppose instead that a "friend" offers you a bet. He has three cards, one red on both sides, one black on both sides, and one red on one side and black on the other. He mixes the three cards in a hat, picks one at random, and places it flat on the table with only one side showing. Suppose that one side is red. He then offers to bet his $4 against your $3 that the other side of the card is also red.

At first you might think that the probability that the other side is also red is 50%, so that this is a good bet. However, after mastering conditional probability (Section 1.5), you will know that conditional on one side being red, the conditional probability that the other side is also red is equal to 2/3. So, by the theory of expected values (Chapter 3), you will know that you should not accept your "friend's" bet.

Finally, suppose your "friend" suggests that you flip a coin one thousand times. Your "friend" says that if the coin comes up heads at least six hundred times, then he will pay you $100; otherwise, you have to pay him just $1.

At first you might think that, while 500 heads is the most likely, there is still a *reasonable* chance that 600 heads will appear, at least good enough to justify accepting your friend's $100 to $1 bet. However, after mastering the laws of large numbers (Chapter 4), you will know that as the number of coin flips gets large, it becomes more and more likely that the number of heads is very close to half of the total number of coin flips. In fact, in this case, there is less than one chance in ten billion of getting more than 600 heads! So, you should not accept this bet either.

As these examples show, a good understanding of probability theory will allow you to correctly assess probabilities in everyday situations. This will allow you to make wiser decisions. It might even save you money!

Probability theory also plays a key role in many important applications of science and technology. For example, the design of a nuclear reactor must be such that the escape of radioactivity into the environment is an extremely rare event. Of course, we would like to say that it is categorically impossible for this to ever happen, but reactors are complicated systems, built up from many interconnected subsystems, each of which we know will fail to function properly at some time. Further, we can never definitely say that a natural event like an earthquake cannot occur that would damage the reactor sufficiently to allow an emission. The best we can do is to try and quantify our uncertainty concerning the failures of reactor components or the occurrence of natural events that would lead to such an event. This is where probability enters the picture. Using probability as a tool to deal with the uncertainties, the reactor can be designed to ensure that an unacceptable emission has an extremely small probability, say once in a billion years, of occurring.

The gambling and nuclear reactor examples deal essentially with the concept of *risk* — the risk of losing money, the risk of being exposed to an injurious level of radioactivity, etc. In fact, we are exposed to risk all the time. When we ride in a car, or take an airplane flight, or even walk down the street, we are exposed to risk. We know that the risk of injury in such circumstances is never zero, yet still we engage in these activities. This is because we intuitively realize that the probability of an accident occurring is extremely low.

So, we are using probability every day in our lives to assess risk. As the problems we face, individually or collectively, become more complicated, we need to refine and develop our rough, intuitive ideas about probability to form a clear and precise approach. This is why probability theory has been developed as a subject. In fact, the insurance industry has been developed to help us cope with risk. Probability is the tool used to determine what you pay to reduce your risk or to compensate you or your family in case of a personal injury.

Summary of Section 1.1

- Probability theory provides us with a precise understanding of uncertainty.

- This understanding can help to make predictions, make better decisions, assess risk, and even make money.

Discussion Topics

1.1.1 Do you think that tomorrow's weather and next week's stock prices are "really" random, or is this just a convenient way to discuss and analyze them?

1.1.2 Do you think it is possible for probabilities to depend on who is observing them, or at what time?

1.1.3 Do you find it surprising that probability theory was not discussed as a mathematical subject until the seventeenth century? Why or why not?

1.1.4 In what ways is probability important for such subjects as physics, computer science, and finance? Explain.

1.1.5 What are examples from your own life where thinking about probabilities did save — or could have saved — you money or helped you to make a better decision? (List as many as you can.)

1.1.6 Probabilities are often depicted in popular movies and television programs. List as many examples as you can. Do you think the probabilities were portrayed there in a "reasonable" way?

1.2   Probability Models

A formal definition of probability begins with a *sample space*, often written S. This sample space is any set that lists all possible *outcomes* (or, *responses*) of some unknown experiment or situation. For example, perhaps

$$S = \{\text{rain, snow, clear}\}$$

when predicting tomorrow's weather. Or perhaps S is the set of all positive real numbers, when predicting next week's stock price. The point is, S can be any set at all, even an infinite set. We usually write s for an element of S, so that

$s \in S$. Note that S describes only those things that we are interested in; if we are studying weather, then rain and snow are in S, but tomorrow's stock prices are not.

A probability model also requires a collection of *events*, which are subsets of S to which probabilities can be assigned. For the above weather example, the subsets {rain}, {snow}, {rain, snow}, {rain, clear}, {rain, snow, clear}, and even the empty set $\emptyset = \{\ \}$, are all examples of subsets of S which could be events. Note that here the comma means "or"; thus, {rain, snow} is the event that it will rain *or* snow. We will generally assume that *all* subsets of S are events. (In fact, in complicated situations there are some technical restrictions on what subsets can or cannot be events, according to the mathematical subject of measure theory. But we will not concern ourselves with such technicalities here.)

Finally, and most importantly, a probability model requires a *probability measure*, usually written P. This probability measure must assign, to each event A, a probability $P(A)$. We require the following properties:

1. $P(A)$ is always a nonnegative real number, between 0 and 1 inclusive.

2. $P(\emptyset) = 0$, i.e., if A is the empty set $\emptyset$, then $P(A) = 0$.

3. $P(S) = 1$, i.e., if A is the entire sample space S, then $P(A) = 1$.

4. P is *(countably) additive*, meaning that if $A_1, A_2, \ldots$ is a finite or countable sequence of disjoint events, then

$$P(A_1 \cup A_2 \cup \cdots) = P(A_1) + P(A_2) + \cdots . \tag{1.2.1}$$

The first of these properties says that we shall measure all probabilities on a scale from 0 to 1, where 0 means impossible and 1 (or 100%) means certain. The second property says the probability that *nothing* happens is 0; in other words, it is impossible that *no* outcome will occur. The third property says the probability that *something* happens is 1; in other words, it is certain that *some* outcome must occur.

The fourth property is the most subtle. It says that we can calculate probabilities of complicated events by adding up the probabilities of smaller events, provided those smaller events are *disjoint* and together contain the entire complicated event. Note that events are *disjoint* if they contain no outcomes in common. For example, {rain} and {snow, clear} are disjoint, while {rain} and {rain, clear} are not. (We are assuming for simplicity that it cannot both rain *and* snow tomorrow.) Thus, we should have $P(\{\text{rain}\}) + P(\{\text{snow, clear}\}) = P(\{\text{rain, snow, clear}\})$, but do *not* expect to have $P(\{\text{rain}\}) + P(\{\text{rain, clear}\}) = P(\{\text{rain, rain, clear}\})$ (the latter being the same as $P(\{\text{rain, clear}\})$).

We now formalize the definition of a probability model.

Definition 1.1 A *probability model* consists of a nonempty set called the sample space S; a collection of events which are subsets of S; and a probability measure

P assigning a probability between 0 and 1 to each event, with $P(\emptyset) = 0$ and $P(S) = 1$, and with P additive as in (1.2.1).

Example 1.2.1

Consider again the weather example, with $S = \{$rain, snow, clear$\}$. Suppose that the probability of rain is 40%, the probability of snow is 15%, and the probability of a clear day is 45%. We can express this as $P(\{$rain$\}) = 0.40$, $P(\{$snow$\}) = 0.15$, and $P(\{$clear$\}) = 0.45$.

For this example, of course $P(\emptyset) = 0$, i.e., it is impossible that *nothing* will happen tomorrow. Also $P(\{$rain, snow, clear$\}) = 1$, because we are assuming that exactly *one* of rain, snow, or clear must occur tomorrow. (To be more realistic, we might say that we are predicting the weather at exactly 11:00 A.M. tomorrow.) Now, what is the probability that it will rain *or* snow tomorrow? Well, by the additivity property, we see that

$$P(\{\text{rain}, \text{snow}\}) = P(\{\text{rain}\}) + P(\{\text{snow}\}) = 0.40 + 0.15 = 0.55.$$

We thus conclude that, as expected, there is a 55% chance of rain *or* snow tomorrow. ∎

Example 1.2.2

Suppose your candidate has a 60% chance of winning an election in progress. Then $S = \{$win, lose$\}$, with $P(\text{win}) = 0.6$ and $P(\text{lose}) = 0.4$. Note that $P(\text{win}) + P(\text{lose}) = 1$. ∎

Example 1.2.3

Suppose we flip a fair coin, which can come up either heads (H) or tails (T) with equal probability. Then $S = \{H, T\}$, with $P(H) = P(T) = 0.5$. Of course, $P(H) + P(T) = 1$. ∎

Example 1.2.4

Suppose we flip three fair coins in a row, and keep track of the sequence of heads and tails that results. Then

$$S = \{HHH, HHT, HTH, HTT, THH, THT, TTH, TTT\}.$$

Furthermore, each of these eight outcomes is equally likely. Thus, $P(HHH) = 1/8$, $P(TTT) = 1/8$, etc. Also, the probability that the first coin is heads, *and* the second coin is tails, but the third coin can be anything, is equal to the sum of the probabilities of the events HTH and HTT, i.e., is equal to $P(HTH) + P(HTT) = 1/8 + 1/8 = 1/4$. ∎

Example 1.2.5

Suppose we flip three fair coins in a row, but care only about the number of heads that result. Then $S = \{0, 1, 2, 3\}$. However, the probabilities of these four outcomes are *not* all equally likely; we will see later that in fact $P(0) = P(3) = 1/8$, while $P(1) = P(2) = 3/8$. ∎

We note that it is possible to define probability models on more complicated (e.g., uncountably infinite) sample spaces, as well.

Example 1.2.6

Suppose that $S = [0, 1]$ is the unit interval. We can define a probability measure P on S by saying that

$$P([a, b]) = b - a, \qquad \text{whenever } 0 \le a \le b \le 1. \qquad (1.2.2)$$

In words, for any[1] subinterval $[a, b]$ of $[0, 1]$, the probability of the interval is simply the *length* of that interval. This example is called the *uniform distribution on* $[0, 1]$. The uniform distribution is just the first of many distributions on uncountable state spaces. Many further examples will be given in Chapter 2. ∎

1.2.1 Venn Diagrams and Subsets

Venn diagrams provide a very useful graphical method for depicting the sample space S and subsets of it. For example, in Figure 1.2.1 we have a Venn diagram showing the subset $A \subset S$ and the *complement*

$$A^c = \{s : s \notin A\}$$

of A. The rectangle denotes the entire sample space S. The circle (and its interior) denotes the subset A, and the region outside the circle, but inside S, denotes A^c.

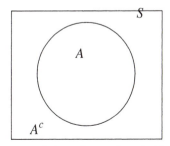

Figure 1.2.1: Venn diagram of the subsets A and A^c of the sample space S.

Two subsets $A \subset S$ and $B \subset S$ are depicted as two circles, as in Figure 1.2.2. The *intersection*

$$A \cap B = \{s : s \in A \text{ and } s \in B\}$$

of the subsets A and B is the set of elements common to both sets and is depicted by the region where the two circles overlap. The set

$$A \cap B^c = \{s : s \in A \text{ and } s \in B\}$$

---

[1] For the uniform distribution on $[0, 1]$, it turns out that not all subsets of $[0, 1]$ can properly be regarded as *events* for this model. However, this is merely a technical property, and any subset that we can explicitly write down will always be an event. See more advanced probability books, e.g., page 3 of *A First Look at Rigorous Probability Theory*, by J. S. Rosenthal (World Scientific Publishing, Singapore, 2000).

is called the *complement of B in A* and is depicted as the region inside the A circle, but not inside the B circle. This is the set of elements in A but not in B. Similarly, we have the complement of A in B, namely, $A^c \cap B$. Observe that the sets $A \cap B, A \cap B^c$, and $A^c \cap B$ are mutually disjoint.

The *union*

$$A \cup B = \{s : s \in A \text{ or } s \in B\}$$

of the sets A and B is the set of elements that are in either A or B. In Figure 1.2.2, it is depicted by the region covered by both circles. Notice that $A \cup B = (A \cap B^c) \cup (A \cap B) \cup (A^c \cap B)$.

There is one further region in Figure 1.2.2. This is the complement of $A \cup B$, namely the set of elements that are in neither A nor B. So we immediately have

$$(A \cup B)^c = A^c \cap B^c.$$

Similarly, we can show that

$$(A \cap B)^c = A^c \cup B^c,$$

namely, the subset of elements that are not in both A and B is given by the set of elements not in A or not in B.

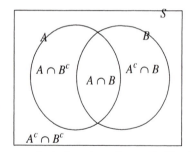

Figure 1.2.2: Venn diagram depicting the subsets A, B, $A \cap B$, $A \cap B^c$, $A^c \cap B$, $A^c \cap B^c$, and $A \cup B$.

Finally, we note that if A and B are disjoint subsets, then it makes sense to depict these as drawn in Figure 1.2.3, i.e., as two nonoverlapping circles, as they have no elements in common.

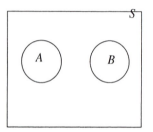

Figure 1.2.3: Venn diagram of the disjoint subsets A and B.

Summary of Section 1.2

- A probability model consists of a sample space S and a probability measure P assigning probabilities to each event.

- All different sorts of sets can arise as sample spaces.

- Venn diagrams provide a convenient method for representing sets and the relationships among them.

Exercises

1.2.1 Suppose $S = \{1, 2, 3\}$, with $P(\{1\}) = 1/2$, $P(\{2\}) = 1/3$, and $P(\{3\}) = 1/6$.
(a) What is $P(\{1, 2\})$?
(b) What is $P(\{1, 2, 3\})$?
(c) List all events A such that $P(A) = 1/2$.

1.2.2 Suppose $S = \{1, 2, 3, 4, 5, 6, 7, 8\}$, with $P(\{s\}) = 1/8$ for $1 \le s \le 8$.
(a) What is $P(\{1, 2\})$?
(b) What is $P(\{1, 2, 3\})$?
(c) How many events A are there such that $P(A) = 1/2$?

1.2.3 Suppose $S = \{1, 2, 3\}$, with $P(\{1\}) = 1/2$ and $P(\{1, 2\}) = 2/3$. What must $P(\{2\})$ be?

1.2.4 Suppose $S = \{1, 2, 3\}$, and we try to define P by $P(\{1, 2, 3\}) = 1$, $P(\{1, 2\}) = 0.7$, $P(\{1, 3\}) = 0.5$, $P(\{2, 3\}) = 0.7$, $P(\{1\}) = 0.2$, $P(\{2\}) = 0.5$, $P(\{3\}) = 0.3$. Is P a valid probability measure? Why or why not?

1.2.5 Consider the uniform distribution on $[0, 1]$. Let $s \in [0, 1]$ be any outcome. What is $P(\{s\})$? Do you find this result surprising?

1.2.6 Label the subregions in the Venn diagram in Figure 1.2.4 using the sets $A, B,$ and C and their complements (just as we did in Figure 1.2.2).

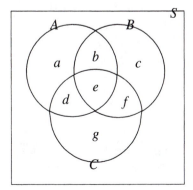

Figure 1.2.4: Venn diagram of subsets $A, B,$ and C.

1.2.7 On a Venn diagram, depict the set of elements that are in subsets A or B but *not* in both. Also write this as a subset involving unions and intersections of A, B, and their complements.

Problems

1.2.8 Consider again the uniform distribution on $[0, 1]$. Is it true that

$$P([0, 1]) = \sum_{s \in [0,1]} P(\{s\})?$$

How does this relate to the additivity property of probability measures?

1.2.9 Suppose S is a finite or countable set. Is it possible that $P(\{s\}) = 0$ for every single $s \in S$? Why or why not?

1.2.10 Suppose S is an uncountable set. Is it possible that $P(\{s\}) = 0$ for every single $s \in S$? Why or why not?

Discussion Topics

1.2.11 Does the additivity property make sense intuitively? Why or why not?

1.2.12 Is it important that we always have $P(S) = 1$? How would probability theory change if this were not the case?

1.3   Properties of Probability Models

The additivity property of probability measures automatically implies certain basic properties. These are true for *any* probability model at all.

   If A is any event, we write A^c (read "A complement") for the event that A does *not* occur. In the weather example, if $A = \{rain\}$, then $A^c = \{snow, clear\}$. In the coin examples, if A is the event that the first coin is heads, then A^c is the event that the first coin is tails.

   Now, A and A^c are always disjoint. Furthermore, their union is always the entire sample space: $A \cup A^c = S$. Hence, by the additivity property, we must have $P(A) + P(A^c) = P(S)$. But we always have $P(S) = 1$. Hence, $P(A) + P(A^c) = 1$, or

$$P(A^c) = 1 - P(A). \tag{1.3.1}$$

In words, the probability that any event does *not* occur is equal to one minus the probability that it *does* occur. This is a very helpful fact that we shall use often.

   Now suppose that $A_1, A_2, \ldots$ are events that form a *partition* of the sample space S. This means that $A_1, A_2, \ldots$ are disjoint, and furthermore their union is equal to S, i.e., $A_1 \cup A_2 \cup \cdots = S$. We have the following basic theorem that allows us to decompose the calculation of the probability of B into the sum of the probabilities of the sets $A_i \cap B$. Often these are easier to compute.

Theorem 1.3.1 (*Law of total probability, unconditioned version*) Let $A_1, A_2, \ldots$ be events that form a partition of the sample space S. Let B be any event. Then

$$P(B) = P(A_1 \cap B) + P(A_2 \cap B) + \cdots$$

Proof: The events $(A_1 \cap B), (A_2 \cap B), \ldots$ are disjoint, and their union is B. Hence, the result follows immediately from the additivity property (1.2.1). ∎

A somewhat more useful version of the law of total probability, and applications of its use, are provided in Section 1.5.

Suppose now that A and B are two events such that A *contains* B (in symbols, $A \supseteq B$). In words, all outcomes in B are also in A. Intuitively, A is a "larger" event than B, so we would expect its probability to be larger. We have the following result.

Theorem 1.3.2 Let A and B be two events with $A \supseteq B$. Then

$$P(A) = P(B) + P(A \cap B^c). \tag{1.3.2}$$

Proof: We can write $A = B \cup (A \cap B^c)$, where B and $A \cap B^c$ are disjoint. Hence, $P(A) = P(B) + P(A \cap B^c)$ by additivity. ∎

Because we always have $P(A \cap B^c) \geq 0$, we conclude the following.

Corollary 1.3.1 (*Monotonicity property*) Let A and B be two events, with $A \supseteq B$. Then

$$P(A) \geq P(B).$$

On the other hand, rearranging (1.3.2), we obtain the following.

Corollary 1.3.2 Let A and B be two events, with $A \supseteq B$. Then

$$P(A \cap B^c) = P(A) - P(B). \tag{1.3.3}$$

More generally, even if we do not have $A \supseteq B$, we have the following property.

Theorem 1.3.3 (*Principle of inclusion–exclusion, two-event version*) Let A and B be two events. Then

$$P(A \cup B) = P(A) + P(B) - P(A \cap B). \tag{1.3.4}$$

Proof: We can write $A \cup B = (A \cap B^c) \cup (B \cap A^c) \cup (A \cap B)$, where $A \cap B^c$, $B \cap A^c$, and $A \cap B$ are disjoint. By additivity, we have

$$P(A \cup B) = P(A \cap B^c) + P(B \cap A^c) + P(A \cap B). \tag{1.3.5}$$

On the other hand, using Corollary 1.3.2 (with B replaced by $A \cap B$), we have

$$P(A \cap B^c) = P(A \cap (A \cap B)^c) = P(A) - P(A \cap B) \tag{1.3.6}$$

and similarly,

$$P(B \cap A^c) = P(B) - P(A \cap B). \tag{1.3.7}$$

Substituting (1.3.6) and (1.3.7) into (1.3.5), the result follows. ∎

A more general version of the principle of inclusion-exclusion is developed in Challenge 1.3.6.

Sometimes we do not need to evaluate the probability content of a union, and need only know it is bounded above by the sum of the probabilities of the individual events. This is called subadditivity.

Theorem 1.3.4 (*Subadditivity*) Let $A_1, A_2, \ldots$ be a finite or countably infinite sequence of events, not necessarily disjoint. Then

$$P(A_1 \cup A_2 \cup \cdots) \leq P(A_1) + P(A_2) + \cdots$$

Proof: See Section 1.7 for the proof of this result. ∎

We note that some properties in the definition of a probability model actually follow from other properties. For example, once we know the probability P is additive and that $P(S) = 1$, it follows that we *must* have $P(\emptyset) = 0$. Indeed, because S and $\emptyset$ are disjoint, $P(S \cup \emptyset) = P(S) + P(\emptyset)$. But of course, $P(S \cup \emptyset) = P(S) = 1$, so we must have $P(\emptyset) = 0$.

Similarly, once we know P is additive on countably infinite sequences of disjoint events, it follows that P must be additive on finite sequences of disjoint events, too. Indeed, given a finite disjoint sequence $A_1, \ldots, A_n$, we can just set $A_i = \emptyset$ for all $i > n$, to get a countably infinite disjoint sequence with the same union and the same sum of probabilities.

Summary of Section 1.3

- The probability of the complement of an event equals one minus the probability of the event.

- Probabilities always satisfy the basic properties of total probability, subadditivity, and monotonicity.

- The principle of inclusion-exclusion allows for the computation of $P(A \cup B)$ in terms of simpler events.

Exercises

1.3.1 Suppose $S = \{1, 2, \ldots, 100\}$. Suppose further that $P(\{1\}) = 0.1$.
(a) What is the probability $P(\{2, 3, 4, \ldots, 100\})$?
(b) What is the smallest possible value of $P(\{1, 2, 3\})$?

1.3.2 Suppose that Al watches the six o'clock news 2/3 of the time, watches the eleven o'clock news 1/2 of the time, and watches both the six o'clock and eleven o'clock news 1/3 of the time. For a randomly selected day, what is the probability that Al watches only the six o'clock news? For a randomly selected day, what is the probability that Al watches neither news?

1.3.3 Suppose that an employee arrives late 10% of the time, leaves early 20% of the time, and both arrives late *and* leaves early 5% of the time. What is the probability that on a given day that employee will either arrive late *or* leave early (or both)?

1.3.4 Suppose your right knee is sore 15% of the time, and your left knee is sore 10% of the time. What is the largest possible percentage of time that at least one of your knees is sore? What is the smallest possible percentage of time that at least one of your knees is sore?

Problems

1.3.5 Suppose we choose a positive integer at random, according to some unknown probability distribution. Suppose we know that $P(\{1, 2, 3, 4, 5\}) = 0.3$, that $P(\{4, 5, 6\}) = 0.4$, and that $P(\{1\}) = 0.1$. What are the largest and smallest possible values of $P(\{2\})$?

Challenges

1.3.6 Generalize the principle of inclusion–exclusion, as follows.
(a) Suppose there are three events A, B, and C. Prove that

$$P(A \cup B \cup C) = P(A) + P(B) + P(C) - P(A \cap B) - P(A \cap C)$$
$$- P(B \cap C) + P(A \cap B \cap C).$$

(b) Suppose there are n events $A_1, A_2, \ldots, A_n$. Prove that

$$P(A_1 \cup \cdots \cup A_n) = \sum_{i=1}^{n} P(A_i) - \sum_{\substack{i,j=1 \\ i<j}}^{n} P(A_i \cap A_j) + \sum_{\substack{i,j,k=1 \\ i<j<k}}^{n} P(A_i \cap A_j \cap A_k)$$
$$- \cdots \pm P(A_1 \cap \cdots \cap A_n).$$

(Hint: Use induction.)

Discussion Topics

1.3.7 Of the various theorems presented in this section, which ones do you think are the most important? Which ones do you think are the least important? Explain the reasons for your choices.

1.4 Uniform Probability on Finite Spaces

If the sample space S is finite, then one possible probability measure on S is the *uniform* probability measure, which assigns probability $1/|S|$ to each outcome. Here $|S|$ is the number of elements in the sample space S. By additivity, it then follows that for any event A we have

$$P(A) = \frac{|A|}{|S|}. \tag{1.4.1}$$

Example 1.4.1
Suppose we roll a six-sided die. The possible outcomes are $S = \{1, 2, 3, 4, 5, 6\}$, so that $|S| = 6$. If the die is fair, then we believe each outcome is equally likely. We thus set $P(\{i\}) = 1/6$ for each $i \in S$ so that $P(\{3\}) = 1/6$, $P(\{4\}) = 1/6$, etc. It follows from (1.4.1) that, for example, $P(\{3, 4\}) = 2/6 = 1/3$, $P(\{1, 5, 6\}) = 3/6 = 1/2$, etc. This is a good model of rolling a fair six-sided die once. ∎

Example 1.4.2
For a second example, suppose we flip a fair coin once. Then $S = \{\text{heads, tails}\}$, so that $|S| = 2$, and $P(\{\text{heads}\}) = P(\{\text{tails}\}) = 1/2$. ∎

Example 1.4.3
Suppose now that we flip *three different* fair coins. The outcome can be written as a sequence of three letters, with each letter being H (for heads) or T (for tails). Thus,

$$S = \{HHH, HHT, HTH, HTT, THH, THT, TTH, TTT\}.$$

Here $|S| = 8$, and each of the events is equally likely. Hence $P(\{HHH\}) = 1/8$, $P(\{HHH, TTT\}) = 2/8 = 1/4$, etc. Note also that, by additivity, we have, for example, that $P(\text{exactly two heads}) = P(\{HHT, HTH, THH\}) = 1/8 + 1/8 + 1/8 = 3/8$, etc. ∎

Example 1.4.4
For a final example, suppose we roll a fair six-sided die *and* flip a fair coin. Then we can write

$$S = \{1H, 2H, 3H, 4H, 5H, 6H, 1T, 2T, 3T, 4T, 5T, 6T\}.$$

Hence, $|S| = 12$ in this case, and $P(s) = 1/12$ for each $s \in S$. ∎

1.4.1   Combinatorial Principles

Because of (1.4.1), problems involving uniform distributions on finite sample spaces often come down to being able to compute the sizes $|A|$ and $|S|$ of the sets involved. That is, we need to be good at *counting* the number of elements in various sets. The science of counting is called *combinatorics*, and some aspects of it are very sophisticated. In the remainder of this section, we consider a few simple combinatorial rules and their application in probability theory when the uniform distribution is appropriate.

Example 1.4.5 *Counting Sequences: The Multiplication Principle*
Suppose we flip three fair coins and roll two fair six-sided dice. What is the probability that all three coins come up heads, and both dice come up 6? Each coin has two possible outcomes (heads and tails), and each die has six possible outcomes $\{1, 2, 3, 4, 5, 6\}$. The total number of possible outcomes of the three coins and two dice is thus given by *multiplying* three 2's and two 6's, i.e.,

$2 \times 2 \times 2 \times 6 \times 6 = 288$. This is sometimes referred to as the *multiplication principle*. There are thus 288 possible outcomes of our experiment (e.g., HHH66, HTH24, TTH15, etc.). Of these outcomes, only one (namely, HHH66) counts as a success. Thus, the probability that all three coins come up heads and both dice come up 6 is equal to $1/288$.

Notice that we can obtain this result in an alternative way. The chance that any one of the coins comes up heads is $1/2$, and the chance that any one die comes up 6 is $1/6$. Furthermore, these events are all *independent* (see the next section). Then, under independence, the probability that they *all* occur is given by the product of their individual probabilities, namely

$$(1/2)(1/2)(1/2)(1/6)(1/6) = 1/288.$$

More generally, suppose we have k finite sets $S_1, \ldots, S_k$ and we want to count the number of sequences of length k where the ith element comes from S_i, i.e., count the number of elements in

$$S = \{(s_1, \ldots, s_k) : s_i \in S_i\} = S_1 \times \cdots \times S_k.$$

The multiplication principle says that the number of such sequences is obtained by multiplying together the number of elements in each set S_i, i.e.,

$$|S| = |S_1| \cdots |S_k|. \quad \blacksquare$$

Example 1.4.6
Suppose we roll two fair six-sided dice. What is the probability that the sum of the numbers showing is equal to 10? By the above multiplication principle, the total number of possible outcomes is equal to $6 \times 6 = 36$. Of these outcomes, there are three that sum to 10, namely $(4, 6)$, $(5, 5)$, and $(6, 4)$. Thus, the probability that the sum is 10 is equal to $3/36$, or $1/12$. $\blacksquare$

Example 1.4.7 *Counting Permutations*
Suppose four friends go to a restaurant, and each checks his or her coat. At the end of the meal, the four coats are *randomly* returned to the four people. What is the probability that each of the four people gets his or her own coat? Here the total number of different ways the coats can be returned is equal to $4 \times 3 \times 2 \times 1$, or 4! (i.e., four factorial). This is because the first coat can be returned to any of the four friends, the second coat to any of the three remaining friends, and so on. Only one of these assignments is correct. Hence, the probability that each of the four people gets his or her own coat is equal to $1/4!$, or $1/24$.

Here we are counting permutations, or sequences of elements from a set where no element appears more than once. We can use the multiplication principle to count permutations more generally. For example, suppose $|S| = n$ and we want to count the number of permutations of length $k \leq n$ obtained from S, i.e., we want to count the number of elements of the set

$$\{(s_1, \ldots, s_k) : s_i \in S, s_i \neq s_j \text{ when } i \neq j\}.$$

Then we have n choices for the first element s_1, $n-1$ choices for the second element, and finally $n-(k-1)=n-k+1$ choices for the last element. So there are $n(n-1)\cdots(n-k+1)$ permutations of length k from a set of n elements. This can also be written as $n!/(n-k)!$. Notice that when $k=n$, there are

$$n! = n(n-1)\cdots 2\cdot 1$$

permutations of length n. ∎

Example 1.4.8 *Counting Subsets*

Suppose 10 fair coins are flipped. What is the probability that exactly seven of them are heads? Here each possible sequence of ten heads or tails (e.g., $HHHTTTHTTT, THTTTTHHHT$, etc.) is equally likely, and by the multiplication principle the total number of possible outcomes is equal to 2 multiplied by itself 10 times, or $2^{10} = 1024$. Hence, the probability of any particular sequence occurring is $1/1024$. But of these sequences, how many have exactly seven heads?

To answer this, notice that we may specify such a sequence by giving the positions of the seven heads, which involves choosing a subset of size 7 from the set of possible indices $\{1, \ldots, 10\}$. There are $10!/3! = 10\cdot 9\cdots 5\cdot 4$ different permutations of length 7 from $\{1, \ldots, 10\}$ and each such permutation specifies a sequence of seven heads and three tails. But we can permute the indices specifying where the heads go in 7! different ways without changing the sequence of heads and tails. So, the total number of outcomes with exactly seven heads is equal to $10!/3!7! = 120$. The probability that exactly seven of the ten coins are heads is therefore equal to $120/1024$, or just under 12%.

In general, if we have a set S of n elements, then the number of different subsets of size k that we can construct by choosing elements from S is

$$\binom{n}{k} = \frac{n!}{k!\,(n-k)!},$$

which is called the *binomial coefficient*. This follows by the same argument, namely, there are $n!/(n-k)!$ permutations of length k obtained from the set and each such permutation, and the $k!$ permutations obtained by permuting it, specify a unique subset of S. ∎

It follows, for example, that the probability of obtaining exactly k heads when flipping a total of n fair coins is given by

$$\binom{n}{k}2^{-n} = \frac{n!}{k!\,(n-k)!}2^{-n}.$$

This is because there are $\binom{n}{k}$ different patterns of k heads and $n-k$ tails, and a total of 2^n different sequences of n heads and tails.

More generally, if each coin has probability θ of being heads (and probability $1-\theta$ of being tails), where $0 \le \theta \le 1$, then the probability of obtaining exactly k heads when flipping a total of n such coins is given by

$$\binom{n}{k}\theta^k(1-\theta)^{n-k} = \frac{n!}{k!\,(n-k)!}\theta^k(1-\theta)^{n-k}, \tag{1.4.2}$$

because each of the $\binom{n}{k}$ different patterns of k heads and $n - k$ tails has probability $\theta^k(1 - \theta)^{n-k}$ (this follows from the discussion of independence in Section 1.5.2) of occurring. If $\theta = 1/2$, then this reduces to the previous formula.

Example 1.4.9 *Counting Sequences of Subsets and Partitions*
Suppose we have a set S of n elements and we want to count the number of elements of

$$\{(S_1, S_2, \ldots, S_l) : S_i \subset S, |S_i| = k_i, S_i \cap S_j = \emptyset \text{ when } i \neq j\},$$

namely, we want to count the number of sequences of l subsets of a set where no two subsets have any elements in common and the ith subset has k_i elements. By the multiplication principle this equals

$$\binom{n}{k_1}\binom{n - k_1}{k_2} \cdots \binom{n - k_1 - \cdots - k_{l-1}}{k_l}$$

$$= \frac{n!}{k_1! \cdots k_{l-1}! k_l! (n - k_1 - \cdots - k_l)!}, \qquad (1.4.3)$$

because we can choose the elements of S_1 in $\binom{n}{k_1}$ ways, choose the elements of S_2 in $\binom{n-k_1}{k_2}$ ways, etc.

When we have that $S = S_1 \cup S_2 \cup \cdots \cup S_l$, in addition to the individual sets being mutually disjoint, then we are counting the number of *ordered partitions* of a set of n elements with k_1 elements in the first set, k_2 elements in the second set, etc. In this case, (1.4.3) equals

$$\binom{n}{k_1 \, k_2 \, \ldots \, k_l} = \frac{n!}{k_1! k_{21}! \cdots k_l!}, \qquad (1.4.4)$$

which is called the *multinomial coefficient*. ∎

For example, how many different bridge hands are there? By this we mean how many different ways can a deck of 52 cards be divided up into four hands of 13 cards each, with the hands labelled North, East, South, and West, respectively. By (1.4.4) this equals

$$\binom{52}{13 \, 13 \, 13 \, 13} = \frac{52!}{13! \, 13! \, 13! \, 13!} \approx 5.364474 \times 10^{28},$$

which is a very large number.

Summary of Section 1.4

- The uniform probability distribution on a finite sample space S satisfies $P(A) = |A| / |S|$.

- Computing $P(A)$ in this case requires computing the *sizes* of the sets A and S. This may require combinatorial principles such as the multiplication principle, factorials, and binomial/multinomial coefficients.

Exercises

1.4.1 Suppose we roll eight fair six-sided dice.
(a) What is the probability that all eight dice show a 6?
(b) What is the probability that all eight dice show the same number?
(c) What is the probability that the sum of the eight dice is equal to 9?

1.4.2 Suppose we roll ten fair six-sided dice. What is the probability that there are exactly two 2's showing?

1.4.3 Suppose we flip 100 fair independent coins. What is the probability that at least three of them are heads? (Hint: You may wish to use (1.3.1).)

1.4.4 Suppose we are dealt five cards from an ordinary 52-card deck. What is the probability that
(a) we get all four aces, plus the king of spades?
(b) all five cards are spades?
(c) we get no pairs (i.e., all five cards are different values)?
(d) we get a full house (i.e., three cards are the same value, and the other two cards are the same value)?

1.4.5 Suppose we deal four 13-card bridge hands from an ordinary 52-card deck. What is the probability that
(a) all 13 spades end up in the same hand?
(b) all four aces end up in the same hand?

1.4.6 Suppose we pick two cards at random from an ordinary 52-card deck. What is the probability that the sum of the values of the two cards (where we count jacks, queens, and kings as ten, and count aces as one) is at least four?

1.4.7 Suppose we keep dealing cards from an ordinary 52-card deck until the first jack appears. What is the probability that at least ten cards go by *before* the first jack?

1.4.8 In a well-shuffled ordinary 52-card deck, what is the probability that the ace of spades and the ace of clubs are adjacent to each other?

1.4.9 Suppose we repeatedly roll two fair six-sided dice, considering the *sum* of the two values showing each time. What is the probability that the first time the sum is exactly seven is on the third roll?

1.4.10 Suppose we roll three fair six-sided dice. What is the probability that two of them show the same value, but the third one does not?

1.4.11 Consider two urns, labelled urn #1 and urn #2. Suppose urn #1 has 5 red and 7 blue balls. Suppose urn #2 has 6 red and 12 blue balls. Suppose we pick three balls uniformly at random from each of the two urns. What is the probability that all six chosen balls are the same color?

Problems

1.4.12 Show that a probability measure defined by (1.4.1) is always additive in the sense of (1.2.1).

1.4.13 Suppose we roll eight fair six-sided dice. What is the probability that the sum of the eight dice is equal to 9? What is the probability that the sum of the eight dice is equal to 10? What is the probability that the sum of the eight dice is equal to 11?

1.4.14 Suppose we roll one fair six-sided die, and flip six coins. What is the probability that the number of heads is equal to the number showing on the die?

1.4.15 Suppose we roll ten fair six-sided dice. What is the probability that there are exactly two 2's showing, *and* exactly three 3's showing?

1.4.16 Suppose we deal four 13-card bridge hands from an ordinary 52-card deck. What is the probability that the North and East hands each have exactly the same number of spades?

1.4.17 Suppose we pick a card at random from an ordinary 52-card deck, and also flip ten fair coins. What is the probability that the number of heads equals the value of the card (where we count jacks, queens, and kings as ten, and count aces as one)?

Challenges

1.4.18 Suppose we roll two fair six-sided dice and flip 12 coins. What is the probability that the number of heads is equal to the sum of the numbers showing on the two dice?

1.4.19 *(The birthday problem)* Suppose there are C people, each of whose birthdays (month and day only) are equally likely to fall on any of the 365 days of a normal (i.e., non-leap) year.
(a) Suppose $C = 2$. What is the probability that the two people have the same exact birthday?
(b) Suppose $C \geq 2$. What is the probability that all C people have the same exact birthday?
(c) Suppose $C \geq 2$. What is the probability that *some pair* of the C people have the same exact birthday? (Hint: You may wish to use (1.3.1).)
(d) What is the smallest value of C such that the probability in part (c) is more than 0.5? Do you find this result surprising?

1.5 Conditional Probability and Independence

Consider again the three-coin example as in Example 1.4, where we flip three different fair coins, and

$$S = \{HHH, HHT, HTH, HTT, THH, THT, TTH, TTT\},$$

with $P(s) = 1/8$ for each $s \in S$. What is the probability that the first coin comes up heads? Well, of course, this should be 1/2. We can see this more formally by saying that $P(\text{first coin heads}) = P(\{HHH, HHT, HTH, HTT\}) = 4/8 = 1/2$, as it should.

$$\left(\frac{1}{6}\right)^2 \times \left(\frac{5}{6}\right)^8 \binom{10}{2}$$

But suppose now that an informant tells us that exactly two of the three coins came up heads. *Now* what is the probability that the first coin was heads?

The point is that this informant has changed our available information, i.e., changed our level of ignorance. It follows that our corresponding probabilities should also change. Indeed, if we know that exactly two of the coins were heads, then we know that the outcome was one of HHT, HTH, and THH. Because those three outcomes should (in this case) still all be equally likely, and because only the first two correspond to the first coin being heads, we conclude the following: If we know that exactly two of the three coins are heads, *then* the probability that the first coin is heads is 2/3.

More precisely, we have computed a *conditional probability*. That is, we have determined that, *conditional* on knowing that exactly two coins came up heads, the *conditional probability* of the first coin being heads is 2/3. We write this in mathematical notation as

$$P(\text{first coin heads} \mid \text{two coins heads}) = 2/3.$$

Here the vertical bar | stands for "conditional on," or "given that."

1.5.1 Conditional Probability

In general, given two events A and B with $P(B) > 0$, the *conditional probability* of A given B, written $P(A \mid B)$, stands for the fraction of the time that A occurs once we *know* that B occurs. It is computed as the ratio of the probability that A and B *both* occur, divided by the probability that B occurs, as follows.

Definition 1.5.1 Given two events A and B, with $P(B) > 0$, the conditional probability of A given B is equal to

$$P(A \mid B) = \frac{P(A \cap B)}{P(B)}. \tag{1.5.1}$$

The motivation for (1.5.1) is as follows. The event B will occur a fraction $P(B)$ of the time. Also, *both* A and B will occur a fraction $P(A \cap B)$ of the time. The ratio $P(A \cap B)/P(B)$ thus gives the *proportion* of the times when B occurs, that A *also* occurs. That is, if we *ignore* all the times that B does not occur, and consider only those times that B does occur, then the ratio $P(A \cap B)/P(B)$ equals the fraction of the time that A will also occur. This is precisely what is meant by the conditional probability of A given B.

In the example just computed, A is the event that the first coin is heads, while B is the event that exactly two coins were heads. Hence, in mathematical terms, $A = \{HHH, HHT, HTH, HTT\}$ and $B = \{HHT, HTH, THH\}$. It follows that $A \cap B = \{HHT, HTH\}$. Therefore,

$$P(A \mid B) = \frac{P(A \cap B)}{P(B)} = \frac{P(\{HHT, HTH\})}{P(\{HHT, HTH, THH\})} = \frac{2/8}{3/8} = 2/3,$$

as already computed.

On the other hand, we similarly compute that

$$P(\text{first coin tails} \mid \text{two coins heads}) = 1/3.$$

We thus see that conditioning on some event (such as "two coins heads") can make probabilities either increase (as for the event "first coin heads") or decrease (as for the event "first coin tails").

The definition of $P(B \mid A)$ immediately leads to the *multiplication formula*

$$P(A \cap B) = P(A)P(B \mid A). \tag{1.5.2}$$

This allows us to compute the joint probability of A and B when we are given the probability of A and the conditional probability of B given A.

Conditional probability allows us to express Theorem 1.3.1, the law of total probability, in a different and sometimes more helpful way.

Theorem 1.5.1 (*Law of total probability, conditioned version*) Let $A_1, A_2, \ldots$ be events that form a partition of the sample space S, each of positive probability. Let B be any event. Then

$$P(B) = P(A_1)P(B \mid A_1) + P(A_2)P(B \mid A_2) + \cdots$$

Proof: The multiplication formula (1.5.2) gives that $P(A_i \cap B) = P(A_i)P(A_i \mid B)$. The result then follows immediately from Theorem 1.3.1. ∎

Example 1.5.1
Suppose a class contains 60% girls and 40% boys. Suppose that 30% of the girls have long hair, and 20% of the boys have long hair. A student is chosen uniformly at random from the class. What is the probability that the chosen student will have long hair?

To answer this, we let A_1 be the set of girls, and A_2 be the set of boys. Then $\{A_1, A_2\}$ is a partition of the class. We further let B be the set of all students with long hair.

We are interested in $P(B)$. We compute this by Theorem 1.5.1 as

$$P(B) = P(A_1)P(B \mid A_1) + P(A_2)P(B \mid A_2) = (0.6)(0.3) + (0.4)(0.2) = 0.26,$$

so there is a 26% chance that the randomly-chosen student has long hair. ∎

Suppose now that A and B are two events, each of positive probability. In some applications we are given the values of $P(A)$, $P(B)$, and $P(B \mid A)$ and want to compute $P(A \mid B)$. The following result establishes a simple relationship among these quantities.

Theorem 1.5.2 (*Bayes' theorem*) Let A and B be two events, each of positive probability. Then

$$P(A \mid B) = \frac{P(A)}{P(B)} P(B \mid A).$$

Proof: We compute that

$$\frac{P(A)}{P(B)}P(B\,|\,A) = \frac{P(A)}{P(B)}\frac{P(A\cap B)}{P(A)} = \frac{P(A\cap B)}{P(B)} = P(A\,|\,B).$$

This gives the result. ∎

Standard applications of the multiplication formula, the law of total probabilities, and Bayes' theorem occur with *two-stage systems*. The response for such systems can be thought of as occurring in two steps or stages. Typically we are given the probabilities for the first stage and the conditional probabilities for the second stage. The multiplication formula is then used to calculate joint probabilities for what happens at both stages; the law of total probability is used to compute the probabilities for what happens at the second stage; and Bayes' theorem is used to calculate the conditional probabilities for the first stage given what has occurred at the second stage. We illustrate this via an example.

Example 1.5.2

Suppose urn #1 has 3 red and 2 blue balls, and urn #2 has 4 red and 7 blue balls. Suppose one of the two urns is selected with probability 1/2 each, and then one of the balls within that urn is picked uniformly at random.

Then we might ask, what is the probability that urn #2 is selected at the first stage (event A) and a blue ball is selected at the second stage (event B)? Then the multiplication formula provides the correct way to compute this probability as

$$P\,(A\cap B) = P(A)P(B\,|\,A) = \frac{1}{2}\frac{7}{11} = \frac{7}{22}.$$

Suppose instead we want to compute the probability that a blue ball is obtained. Then using the law of total probability (Theorem 1.5.1), we have that

$$P(B) = P(A)P(B\,|\,A) + P(A^c)P(B\,|\,A^c) = \frac{1}{2}\frac{2}{5} + \frac{1}{2}\frac{7}{11}.$$

Now suppose that we are given the information that the ball picked is blue. Then, using Bayes' theorem, the conditional probability that we had selected urn #2 is given by

$$\begin{aligned}
P(A\,|\,B) &= \frac{P(A)}{P(B)}P(B\,|\,A) = \left(\frac{1/2}{(1/2)(2/5) + (1/2)(7/11)}\right)\frac{7}{11} \\
&= 35/57 = 0.614.
\end{aligned}$$

Note that, without the information that a blue ball occurred at the second stage, we have that

$$P(\text{urn \#2 selected}) = 1/2.$$

We see that knowing the ball was blue significantly *increases* the probability that urn #2 was selected. ∎

We can represent a two-stage system using a tree, as in Figure 1.5.1. It can be helpful to draw such a figure when carrying out probability computations for such systems. There are two possible outcomes at the first stage and three possible outcomes at the second stage.

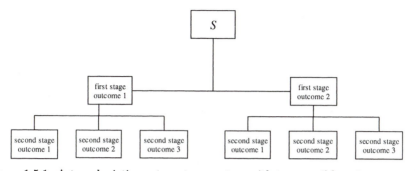

Figure 1.5.1: A tree depicting a two-stage system with two possible outcomes at the first stage and three possible outcomes at the second stage.

1.5.2 Independence of Events

Consider now Example 1.4.4, where we roll one fair die and flip one fair coin, so that

$$S = \{1H, 2H, 3H, 4H, 5H, 6H, 1T, 2T, 3T, 4T, 5T, 6T\}$$

and $P(\{s\}) = 1/12$ for each $s \in S$. Here the probability that the die comes up 5 is equal to $P(\{5H, 5T\}) = 2/12 = 1/6$, as it should be.

But now, what is the probability that the die comes up 5, *conditional* on knowing that the coin came up tails? Well, we can compute that probability as

$$
\begin{aligned}
P(\text{die} = 5 \mid \text{coin} = \text{tails}) &= \frac{P(\text{die} = 5 \text{ and coin} = \text{tails})}{P(\text{coin} = \text{tails})} \\
&= \frac{P(\{5T\})}{P(\{1T, 2T, 3T, 4T, 5T, 6T\})} \\
&= \frac{1/12}{6/12} = 1/6.
\end{aligned}
$$

This is the same as the unconditional probability, $P(\text{die} = 5)$. It seems that knowing that the coin was tails had no effect whatsoever on the probability that the coin came up 5. This property is called *independence*. We say that the coin and the die are *independent* in this example, to indicate that the occurrence of one does not have any influence on the probability of the other occurring.

More formally, we make the following definition.

Definition 1.5.2 Two events A and B are *independent* if

$$P(A \cap B) = P(A)\, P(B).$$

Now, because $P(A \mid B) = P(A \cap B)/P(B)$, we see that A and B are independent if and only if $P(A \mid B) = P(A)$ or $P(B \mid A) = P(B)$, provided that $P(A) > 0$ and $P(B) > 0$. Definition 1.5.2 has the advantage that it remains valid even if $P(B) = 0$ or $P(A) = 0$, respectively. Intuitively, events A and B are independent if neither one has any impact on the probability of the other.

Example 1.5.3
In Example 1.4.4, if A is the event that the die was 5, and B is the event that the coin was tails, then $P(A) = P(\{5H, 5T\}) = 2/12 = 1/6$, and $P(B) = P(\{1T, 2T, 3T, 4T, 5T, 6T\}) = 6/12 = 1/2$. Also $P(A \cap B) = P(\{5T\}) = 1/12$, which is indeed equal to $(1/6)(1/2)$. Hence, A and B are independent in this case. ∎

For multiple events, the definition of independence is somewhat more involved.

Definition 1.5.3 A collection of events $A_1, A_2, A_3, \ldots$ are *independent* if

$$P(A_{i_1} \cap \cdots \cap A_{i_j}) = P(A_{i_1}) \cdots P(A_{i_j}),$$

for *any* finite subcollection $A_{i_1}, \ldots, A_{i_j}$ of distinct events.

Example 1.5.4
According to Definition 1.5.3, three events A, B, and C are independent if *all* of the following equations hold:

$$\begin{aligned}
P(A \cap B) &= P(A)P(B), \\
P(A \cap C) &= P(A)P(C), \\
P(B \cap C) &= P(B)P(C),
\end{aligned} \qquad (1.5.3)$$

and

$$P(A \cap B \cap C) = P(A)P(B)P(C). \qquad (1.5.4)$$

It is not sufficient to check just *some* of these conditions to verify independence. For example, suppose that $S = \{1, 2, 3, 4\}$, with $P(\{1\}) = P(\{2\}) = P(\{3\}) = P(\{4\}) = 1/4$. Let $A = \{1, 2\}$, $B = \{1, 3\}$, and $C = \{1, 4\}$. Then each of the three equations (1.5.3) holds, but equation (1.5.4) does not hold. Here, the events A, B, and C are called *pairwise independent*, but they are not independent. ∎

Summary of Section 1.5

- Conditional probability measures the probability that A occurs given that B occurs; it is given by $P(A \mid B) = P(A \cap B) / P(B)$.

- Conditional probability satisfies its own law of total probability.

- Events are independent if they have no effect on each other's probabilities. Formally, this means that $P(A \cap B) = P(A)P(B)$.

- If A and B are independent, and $P(A) > 0$ and $P(B) > 0$, then $P(A \mid B) = P(A)$ and $P(B \mid A) = P(B)$.

Exercises

1.5.1 Suppose that we roll four fair six-sided dice.
(a) What is the conditional probability that the first die shows 2, conditional on the event that exactly three dice show 2?
(b) What is the conditional probability that the first die shows 2, conditional on the event that *at least* three dice show 2?

1.5.2 Suppose we flip two fair coins, and roll one fair six-sided die.
(a) What is the probability that the number of heads equals the number showing on the die?
(b) What is the conditional probability that the number of heads equals the number showing on the die, conditional on knowing that the die showed 1?
(c) Is the answer for part (b) larger or smaller than the answer for part (a)? Explain intuitively why this is so.

1.5.3 Suppose we flip three fair coins.
(a) What is the probability that all three coins are heads?
(b) What is the conditional probability that all three coins are heads, conditional on knowing that the number of heads is odd?
(c) What is the conditional probability that all three coins are heads, given that the number of heads is even?

1.5.4 Suppose we deal five cards from an ordinary 52-card deck. What is the conditional probability that all five cards are spades, given that at least four of them are spades?

1.5.5 Suppose we deal five cards from an ordinary 52-card deck. What is the conditional probability that the hand contains all four aces, given that the hand contains at least four aces?

1.5.6 Suppose we deal five cards from an ordinary 52-card deck. What is the conditional probability that the hand contains no pairs, given that it contains no spades?

1.5.7 Suppose a baseball pitcher throws fast balls 80% of the time and curve balls 20% of the time. Suppose a batter hits a home run on 8% of all fastball pitches, and on 5% of all curve ball pitches. What is the probability that this batter will hit a home run on this pitcher's next pitch?

1.5.8 Suppose the probability of snow is 20%, and the probability of a traffic accident is 10%. Suppose further that the *conditional* probability of an accident, given that it snows, is 40%. What is the conditional probability that it snows, given that there is an accident?

1.5.9 Suppose we roll two fair six-sided dice, one red and one blue. Let A be the event that the two dice show the same value. Let B be the event that the

sum of the two dice is equal to 12. Let C be the event that the red die shows 4. Let D be the event that the blue die shows 4.
(a) Are A and B independent?
(b) Are A and C independent?
(c) Are A and D independent?
(d) Are C and D independent?
(e) Are A, C, and D all independent?

1.5.10 Consider two urns, labelled urn #1 and urn #2. Suppose, as in Exercise 1.4.11, that urn #1 has 5 red and 7 blue balls, that urn #2 has 6 red and 12 blue balls, and that we pick three balls uniformly at random from each of the two urns. Conditional on the fact that all six chosen balls are the same color, what is the conditional probability that this color is red?

Problems

1.5.11 Consider three cards, as follows: One is red on both sides, one is black on both sides, and one is red on one side and black on the other. Suppose the cards are placed in a hat, and one is chosen at random. Suppose further that this card is placed flat on the table, so we can see one side only.
(a) What is the probability that this one side is red?
(b) Conditional on this one side being red, what is the probability that the card showing is the one which is red on both sides? (Hint: The answer is somewhat surprising.)
(c) Suppose you wanted to verify the answer in part (b), using an actual, physical experiment. Explain how you could do this.

1.5.12 Prove that A and B are independent if and only if A^C and B are independent.

1.5.13 Let A and B be events of positive probability. Prove that $P(A \mid B) > P(A)$ if and only if $P(B \mid A) > P(B)$.

Challenges

1.5.14 Suppose we roll three fair six-sided dice. Compute the conditional probability that the first die shows 4, given that the sum of the three numbers showing is 12.

1.5.15 *(The game of craps)* The game of craps is played by rolling two fair, six-sided dice. On the first roll, if the sum of the two numbers showing equals 2, 3, or 12 then the player immediately loses. If the sum equals 7 or 11 then the player immediately wins. If the sum equals any other value, then this value becomes the player's "point." The player then repeatedly rolls the two dice, until such time as he or she either rolls the point value again (in which case he or she wins), or rolls a 7 (in which case he or she loses).
(a) Suppose the player's point is equal to 4. Conditional on this, what is the conditional probability that he or she will win (i.e., will roll another 4 before rolling a 7)? (Hint: The final roll will be either a 4 or 7; what is the conditional probability that it is a 4?)

(b) For $2 \leq i \leq 12$, let p_i be the conditional probability that the player will win, conditional on having rolled i on the first roll. Compute p_i for all i with $2 \leq i \leq 12$. (Hint: You've already done this for $i = 4$ in part (b). Also, the cases $i = 2, 3, 7, 11, 12$ are trivial. The other cases are similar to the $i = 4$ case.)
(c) Compute the overall probability that a player will win at craps. (Hint: Use part (b) and Theorem 1.5.1.)

1.5.16 *(The Monty Hall problem)* Suppose there are three doors, labeled A, B, and C. A new car is behind one of the three doors, but you don't know which. You select one of the doors, say door A. The host then opens one of doors B or C, as follows: If the car is behind B, then they open C; if the car is behind C, then they open B; if the car is behind A, then they open either B or C with probability $1/2$ each. (In any case, the door opened by the host will *not* have the car behind it.) The host then gives you the *option* of either sticking with your original door choice (i.e., A), or switching to the remaining unopened door (i.e., whichever of B or C the host did not open). You then win (i.e., get to keep the car) if and only if the car is behind your final door selection. (Source: *Parade Magazine*, "Ask Marilyn" column, September 9, 1990.) Suppose for definiteness that the host opens door B.
(a) If you stick with your original choice (i.e., door A), conditional on the host having opened door B, then what is your probability of winning? (Hint: First condition on the true location of the car. Then use Theorem 1.5.2.)
(b) If you switch to the remaining door (i.e., door C), conditional on the host having opened door B, then what is your probability of winning?
(c) Do you find the result of parts (a) and (b) surprising? How could you design a physical experiment to verify the result?
(d) Suppose we change the rules so that, if you originally chose A and the car was indeed behind A, then the host always opens door B. How would the answers to parts (a) and (b) change in this case?
(e) Suppose we change the rules so that, if you originally chose A, then the host always opens door B *no matter where the car is*. We then *condition* on the fact that door B happened *not* to have a car behind it. How would the answers to parts (a) and (b) change in this case?

Discussion Topics

1.5.17 Suppose two people simultaneously each flip a fair coin. Will the results of the two flips usually be independent? Under what sorts of circumstances might they not be independent? (List as many such circumstances as you can.)

1.5.18 Suppose you are able to repeat an experiment many times, and you wish to check whether or not two events are independent. How might you go about this?

1.5.19 The Monty Hall problem (Challenge 1.5.16) was originally presented by Marilyn von Savant, writing in the "Ask Marilyn" column of *Parade Magazine*. She gave the correct answer. However, many people (including some well-known mathematicians, plus many lay people) wrote in to complain that her answer

was incorrect. The controversy dragged on for months, with many letters and very strong language written by both sides (in the end von Savant was vindicated). Part of the confusion lay in the assumptions being made, e.g., some people misinterpreted her question as that of the modified version of part (e) of Challenge 1.5.16. However, a lot of the confusion was simply due to mathematical errors and misunderstandings. (Source: *Parade Magazine*, "Ask Marilyn" column, September 9, 1990; December 2, 1990; February 17, 1991; July 7, 1991.)
(a) Does it surprise you that so many people, including well-known mathematicians, made errors in solving this problem? Why or why not?
(b) Does it surprise you that so many people, including many lay people, cared so strongly about the answer to this problem? Why or why not?

1.6   Continuity of P

Suppose $A_1, A_2, \ldots$ is a sequence of events that are getting "closer" (in some sense) to another event, A. Then we might expect that the probabilities $P(A_1)$, $P(A_2), \ldots$ are getting close to $P(A)$, i.e., that $\lim_{n \to \infty} P(A_n) = P(A)$. But can we be sure about this?

Properties like this, which say that $P(A_n)$ is close to $P(A)$ whenever A_n is "close" to A, are called *continuity properties*. The above question can thus be translated, roughly, as asking whether or not probability measures P are "continuous." It turns out that P is indeed continuous in some sense.

Specifically, let us write $\{A_n\} \nearrow A$, and say that the sequence $\{A_n\}$ *increases* to A, if $A_1 \subseteq A_2 \subseteq A_3 \subseteq \cdots$, and also $\bigcup_{n=1}^{\infty} A_n = A$. That is, the sequence of events is an *increasing* sequence, and furthermore its union is equal to A. For example, if $A_n = (1/n, n]$, then $A_1 \subseteq A_2 \subseteq \cdots$ and $\bigcup_{n=1}^{\infty} A_n = (0, \infty)$. Hence $\{(1/n, n]\} \nearrow (0, \infty)$. Figure 1.6.1 depicts an increasing sequence of subsets.

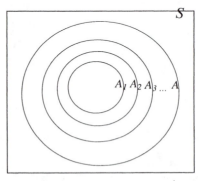

Figure 1.6.1: An increasing sequence of subsets $A_1 \subseteq A_2 \subseteq A_3 \subseteq \ldots$.

Similarly, let us write $\{A_n\} \searrow A$ and say that the sequence $\{A_n\}$ *decreases* to A, if $A_1 \supseteq A_2 \supseteq A_3 \supseteq \cdots$, and also $\bigcap_{n=1}^{\infty} A_n = A$. That is, the sequence of events is a *decreasing* sequence, and furthermore its intersection is equal to A. For example, if $A_n = (-1/n, 1/n]$, then $A_1 \supseteq A_2 \supseteq \cdots$ and $\bigcap_{n=1}^{\infty} A_n = \{0\}$.

Hence $\{(-1/n, 1/n]\} \searrow \{0\}$. Figure 1.6.2 depicts a decreasing sequence of subsets.

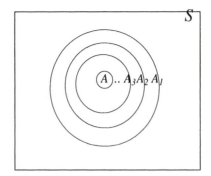

Figure 1.6.2: A decreasing sequence of subsets $A_1 \supseteq A_2 \supseteq A_3 \supseteq \ldots$.

We will consider such sequences of sets at several points in the text. For this we need the following result.

Theorem 1.6.1 Let $A, A_1, A_2, \ldots$ be events, and suppose that either $\{A_n\} \nearrow A$ or $\{A_n\} \searrow A$. Then

$$\lim_{n \to \infty} P(A_n) = P(A).$$

Proof: See Section 1.7 for the proof of this theorem. ∎

Example 1.6.1
Suppose S is the set of all positive integers, with $P(s) = 2^{-s}$ for all $s \in S$. Then what is $P(\{5, 6, 7, 8, \ldots\})$?

We begin by noting that the events $A_n = \{5, 6, 7, 8, \ldots, n\}$ increase to $A = \{5, 6, 7, 8, \ldots\}$, i.e., $\{A_n\} \nearrow A$. Hence, using continuity of probabilities, we must have

$$
\begin{aligned}
P(\{5, 6, 7, 8, \ldots\}) &= \lim_{n \to \infty} P(\{5, 6, 7, 8, \ldots, n\}) \\
&= \lim_{n \to \infty} \left(P(5) + P(6) + \cdots + P(n) \right) \\
&= \lim_{n \to \infty} \left(2^{-5} + 2^{-6} + \cdots + 2^{-n} \right) = \lim_{n \to \infty} \left(\frac{2^{-5} - 2^{-n-1}}{1 - 2^{-1}} \right) \\
&= \lim_{n \to \infty} \left(2^{-4} - 2^{-n} \right) = 2^{-4} = 1/16.
\end{aligned}
$$

Alternatively, we could use countable additivity directly, to conclude that $P(\{5, 6, 7, 8, \ldots\}) = P(5) + P(6) + P(7) + \cdots$, which amounts to the same thing. ∎

Example 1.6.2
Let P be *some* probability measure on the space $S = R^1$. Suppose

$$P\left((3, 5 + 1/n)\right) \geq \delta$$

for all n, where $\delta > 0$. Let $A_n = (3, 5 + 1/n)$. Then $\{A_n\} \searrow A$ where $A = (3, 5]$. Hence, we must have $P(A) = P((3, 5]) \geq \delta$ as well.

Note, however, that we could still have $P((3, 5)) = 0$. For example, perhaps $P(\{5\}) = \delta$, but $P((3, 5)) = 0$. ∎

Summary of Section 1.6

- If $\{A_n\} \nearrow A$ or $\{A_n\} \searrow A$, then $\lim_{n\to\infty} P(A_n) = P(A)$.

- This allows us to compute or bound various probabilities that otherwise could not be understood.

Exercises

1.6.1 Suppose that $S = \{1, 2, 3, \ldots\}$ is the set of all positive integers, and that $P(\{s\}) = 2^{-s}$ for all $s \in S$. Compute $P(A)$ where $A = \{2, 4, 6, \ldots\}$ is the set of all *even* positive integers. Do this in two ways — by using continuity of P (together with finite additivity) and by using countable additivity.

1.6.2 Consider the uniform distribution on $[0, 1]$. Compute (with proof)

$$\lim_{n\to\infty} P([1/4, \ 1 - e^{-n}]).$$

1.6.3 Suppose that $S = \{1, 2, 3, \ldots\}$ is the set of all positive integers, and that P is *some* probability measure on S. Prove that we must have

$$\lim_{n\to\infty} P(\{1, 2, \ldots, n\}) = 1.$$

1.6.4 Let P be *some* probability measure on *some* space S.
(a) Prove that we must have $\lim_{n\to\infty} P((0, 1/n)) = 0$.
(b) Show by example that we might have $\lim_{n\to\infty} P([0, 1/n)) > 0$.

Challenge

1.6.5 Suppose we know that P is *finitely* additive, but we do *not* know that it is *countably* additive. In other words, we know that $P(A_1 \cup \cdots \cup A_n) = P(A_1) + \cdots + P(A_n)$ for any finite collection of disjoint events $\{A_1, \ldots, A_n\}$, but we do not know about $P(A_1 \cup A_2 \cup \cdots)$ for infinite collections of disjoint events. Suppose further that we know that P is continuous in the sense of Theorem 1.6.1. Using this, give a *proof* that P must be countably additive. (In effect, you are proving that continuity of P is *equivalent* to countable additivity of P, at least once we know that P is finitely additive.)

1.7   Further Proofs (Advanced)

Theorem 1.3.4 (*Subadditivity*) Let $A_1, A_2, \ldots$ be a finite or countably infinite sequence of events, not necessarily disjoint. Then

$$P(A_1 \cup A_2 \cup \cdots) \leq P(A_1) + P(A_2) + \cdots$$

Proof: Let $B_1 = A_1$, and for $n \geq 2$, let $B_n = A_n \cap (A_1 \cup \cdots \cup A_{n-1})^c$. Then $B_1, B_2, \ldots$ are disjoint, $B_1 \cup B_2 \cup \cdots = A_1 \cup A_2 \cup \cdots$ and, by additivity,

$$P(A_1 \cup A_2 \cup \cdots) = P(B_1 \cup B_2 \cup \cdots) = P(B_1) + P(B_2) + \cdots. \qquad (1.7.1)$$

Furthermore, $A_n \supseteq B_n$, so by monotonicity, we have $P(A_n) \geq P(B_n)$. It follows from (1.7.1) that

$$P(A_1 \cup A_2 \cup \cdots) = P(B_1) + P(B_2) + \cdots \leq P(A_1) + P(A_2) + \cdots$$

as claimed. ∎

Theorem 1.6.1 Let $A, A_1, A_2, \ldots$ be events, and suppose that either $\{A_n\} \nearrow A$ or $\{A_n\} \searrow A$. Then $\lim_{n \to \infty} P(A_n) = P(A)$.

Proof: Suppose first that $\{A_n\} \nearrow A$. Then we can write

$$A = A_1 \cup (A_2 \cap A_1^c) \cup (A_3 \cap A_2^c) \cup \cdots$$

where the union is disjoint. Hence, by additivity,

$$P(A) = P(A_1) + P(A_2 \cap A_1^c) + P(A_3 \cap A_2^c) + \cdots$$

Now, by definition, writing this infinite sum is the same thing as writing

$$P(A) = \lim_{n \to \infty} \left(P(A_1) + P(A_2 \cap A_1^c) + \cdots + P(A_n \cap A_{n-1}^c) \right). \qquad (1.7.2)$$

However, again by additivity, we see that

$$P(A_1) + P(A_2 \cap A_1^c) + P(A_3 \cap A_2^c) + \cdots + P(A_n \cap A_{n-1}^c) = P(A_n).$$

Substituting this information into (1.7.2), we obtain $P(A) = \lim_{n \to \infty} P(A_n)$, which was to be proved.

Suppose now that $\{A_n\} \searrow A$. Let $B_n = A_n^c$, and let $B = A^c$. Then we see that $\{B_n\} \nearrow B$ (why?). Hence, by what we just proved, we must have $P(B) = \lim_{n \to \infty} P(B_n)$. But then, using (1.3.1), we have

$$1 - P(A) = \lim_{n \to \infty} \left\{ 1 - P(A_n) \right\},$$

from which it follows that $P(A) = \lim_{n \to \infty} P(A_n)$. This completes the proof. ∎

Chapter 2

Random Variables and Distributions

In Chapter 1 we discussed the probability model as the central object of study in the theory of probability. This required defining a probability measure P on a class of subsets of the sample space S. It turns out that there are simpler ways of presenting a particular probability assignment than this — ways that are much more convenient to work with than P. This chapter is concerned with the definitions of random variables, distribution functions, probability functions, density functions, and the development of the concepts necessary for carrying out calculations for a probability model using these entities. Also this chapter discusses the concept of the conditional distribution of one random variable, given the values of others. Conditional distributions of random variables provide the framework for discussing what it means to say that variables are related, and this is important in many applications of probability and statistics.

2.1 Random Variables

The previous chapter explained how to construct probability models, including a sample space S and a probability measure P. Once we have a probability model, we may define *random variables* for that probability model.

Intuitively, a random variable assigns a numerical value to each possible outcome in the sample space. For example, if the sample space is {rain, snow, clear}, then we might define a random variable X such that $X = 3$ if it rains, $X = 6$ if it snows, and $X = -2.7$ if it is clear.

More formally, we have the following definition.

Definition 2.1.1 A *random variable* is a function from the sample space S to the set R^1 of all real numbers.

Figure 2.1.1 provides a graphical representation of a random variable X taking a response value $s \in S$ into a real number $X(s) \in R^1$.

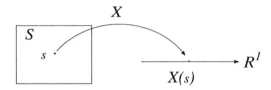

Figure 2.1.1: A random variable X as a function on the sample space S and taking values in R^1.

Example 2.1.1 *A Very Simple Random Variable*
The random variable described above could be written formally as X : {rain, snow, clear} $\to R^1$ by $X(\text{rain}) = 3$, $X(\text{snow}) = 6$, and $X(\text{clear}) = -2.7$. We will return to this example below. ∎

We now present several further examples. The point is, we can define random variables any way we like, as long as they are functions from the sample space to R^1.

Example 2.1.2
For the case $S = \{\text{rain}, \text{snow}, \text{clear}\}$, we might define a second random variable Y by saying that $Y = 0$ if it rains, $Y = -1/2$ if it snows, and $Y = 7/8$ if it is clear. That is $Y(\text{rain}) = 0$, $Y(\text{snow}) = 1/2$, and $Y(\text{rain}) = 7/8$. ∎

Example 2.1.3
If the sample space corresponds to flipping three different coins, then we could let X be the total number of heads showing, let Y be the total number of tails showing, let $Z = 0$ if there is exactly one head, and otherwise $Z = 17$, etc. ∎

Example 2.1.4
If the sample space corresponds to rolling two fair dice, then we could let X be the square of the number showing on the first die, let Y be the square of the number showing on the second die, let Z be the sum of the two numbers showing, let W be the square of the sum of the two numbers showing, let R be the sum of the squares of the two numbers showing, etc. ∎

Example 2.1.5 *Constants as Random Variables*
As a special case, every *constant* value c is also a random variable, by saying that $c(s) = c$ for all $s \in S$. Thus, 5 is a random variable, as is 3 or -21.6. ∎

Example 2.1.6 *Indicator Functions*
One special kind of random variable is worth mentioning. If A is any event, then we can define the *indicator function* of A, written I_A, to be the random variable

$$I_A(s) = \begin{cases} 1 & s \in A \\ 0 & s \notin A, \end{cases}$$

which is equal to 1 on A, and is equal to 0 on A^C. ∎

Given random variables X and Y, we can perform the usual arithmetic operations on them. Thus, for example, $Z = X^2$ is another random variable, defined by $Z(s) = X^2(s) = (X(s))^2 = X(s) \times X(s)$. Similarly, if $W = XY^3$, then $W(s) = X(s) \times Y(s) \times Y(s) \times Y(s)$, etc. Also, if $Z = X + Y$, then $Z(s) = X(s) + Y(s)$, and so on.

Example 2.1.7
Consider rolling a fair six-sided die, so that $S = \{1, 2, 3, 4, 5, 6\}$. Let X be the number showing, so that $X(s) = s$ for $s \in S$. Let Y be three more than the number showing, so that $Y(s) = s + 3$. Let $Z = X^2 + Y$. Then $Z(s) = X(s)^2 + Y(s) = s^2 + s + 3$. So $Z(1) = 5$, $Z(2) = 9$, etc. ∎

We write $X = Y$ to mean that $X(s) = Y(s)$ for all $s \in S$. Similarly, we write $X \leq Y$ to mean that $X(s) \leq Y(s)$ for all $s \in S$, and $X \geq Y$ to mean that $X(s) \geq Y(s)$ for all $s \in S$. For example, we write $X \leq c$ to mean that $X(s) \leq c$ for all $s \in S$.

Example 2.1.8
Again consider rolling a fair six-sided die, with $S = \{1, 2, 3, 4, 5, 6\}$. For $s \in S$, let $X(s) = s$, and let $Y = X + I_{\{6\}}$. This means that

$$Y(s) = X(s) + I_{\{6\}}(s) = \begin{cases} s & s \leq 5 \\ 7 & s = 6. \end{cases}$$

Hence, $Y(s) = X(s)$ for $1 \leq s \leq 5$. But it is not true that $Y = X$, because $Y(6) \neq X(6)$. On the other hand, it is true that $Y \geq X$. ∎

Example 2.1.9
For the random variable of Example 2.1.1 above, it is not true that $X \geq 0$, nor is it true that $X \leq 0$. However, it is true that $X \geq -2.7$ and that $X \leq 6$. It is also true that $X \geq -10$ and $X \leq 100$. ∎

If S is infinite, then a random variable X can take on infinitely many different values.

Example 2.1.10
If $S = \{1, 2, 3, \ldots\}$, with $P\{s\} = 2^{-s}$ for all $s \in S$, and if X is defined by $X(s) = s^2$, then we always have $X \geq 1$. But there is no largest value of $X(s)$, because the value $X(s)$ increases without bound as $s \to \infty$. We shall call such a random variable an *unbounded* random variable. ∎

Finally, suppose X is a random variable. We know that different states s occur with different probabilities. It follows that also $X(s)$ takes different values with different probabilities. These probabilities are called the *distribution* of X; we consider them next.

Summary of Section 2.1

- A random variable is a function from the state space to the set of real numbers.

- The function could be constant, or correspond to counting some random quantity that arises, or any other sort of function.

Exercises

2.1.1 Let $S = \{1, 2, 3, \ldots\}$, and let $X(s) = s^2$ and $Y(s) = 1/s$ for $s \in S$. For each of the following quantities, determine (with explanation) whether or not it exists. If it does exist then give its value.
(a) $\min_{s \in S} X(s)$
(b) $\max_{s \in S} X(s)$
(c) $\min_{s \in S} Y(s)$
(d) $\max_{s \in S} Y(s)$

2.1.2 Let $S = \{\text{high, middle, low}\}$. Define random variables X, Y, and Z by $X(\text{high}) = -12$, $X(\text{middle}) = -2$, $X(\text{low}) = 3$, $Y(\text{high}) = 0$, $Y(\text{middle}) = 0$, $Y(\text{low}) = 1$, $Z(\text{high}) = 6$, $Z(\text{middle}) = 0$, $Z(\text{low}) = 4$. Determine whether each of the following relations is true or false:
(a) $X < Y$
(b) $X \leq Y$
(c) $Y < Z$
(d) $Y \leq Z$
(e) $XY < Z$
(f) $XY \leq Z$

2.1.3 Let $S = \{1, 2, 3, 4, 5\}$.
(a) Define two different (i.e., nonequal) nonconstant random variables, X and Y, on S.
(b) For the random variables X and Y that you have chosen, let $Z = X + Y^2$. Compute $Z(s)$ for all $s \in S$.

2.1.4 Consider rolling a fair six-sided die, so that $S = \{1, 2, 3, 4, 5, 6\}$. Let $X(s) = s$, and $Y(s) = s^3 + 2$. Let $Z = XY$. Compute $Z(s)$ for all $s \in S$.

2.1.5 Let A and B be events, and let $X = I_A \cdot I_B$. Is X an indicator function? If yes, then of what event?

Problems

2.1.6 Let X be a random variable.
(a) Is it necessarily true that $X \geq 0$?
(b) Is it necessarily true that there is some real number c such that $X + c \geq 0$?
(c) Suppose the sample space S is finite. Then is it necessarily true that there is some real number c such that $X + c \geq 0$?

2.1.7 Suppose the sample space S is finite. Is it possible to define an unbounded random variable on S? Why or why not?

2.1.8 Suppose X is a random variable that takes only the values 0 or 1. Must X be an indicator function? Explain.

2.1.9 Suppose the sample space S is finite, of size m. How many different indicator functions can be defined on S?

2.1.10 Suppose X is a random variable. Let $Y = \sqrt{X}$. Must Y be a random variable? Explain.

Discussion Topics

2.1.11 Mathematical probability theory was introduced to the English-speaking world largely by two American mathematicians, William Feller and Joe Doob, writing in the early 1950s. According to Professor Doob, the two of them had an argument about whether random variables should be called "random variables" or "chance variables." They decided by flipping a coin – and "random variables" won. (Source: *Statistical Science* **12** (1997), No. 4, page 307.) Which name do *you* think would have been a better choice?

2.2 Distributions of Random Variables

Because random variables are defined to be functions of the outcome s, and because the outcome s is assumed to be random (i.e., to take on different values with different probabilities), it follows that the value of a random variable will itself be random (as the name implies).

Specifically, if X is a random variable, then what is the probability that X will equal some particular value x? Well, $X = x$ precisely when the outcome s is chosen such that $X(s) = x$.

Example 2.2.1
Let us again consider the random variable of Example 2.1.1, where $S = \{$rain, snow, clear$\}$, and X is defined by $X(\text{rain}) = 3$, $X(\text{snow}) = 6$, and $X(\text{clear}) = -2.7$. Suppose further that the probability measure P is such that $P(\text{rain}) = 0.4$, $P(\text{snow}) = 0.15$, and $P(\text{clear}) = 0.45$. Then clearly $X = 3$ only when it rains, $X = 6$ only when it snows, and $X = -2.7$ only when it is clear. Thus, $P(X = 3) = P(\text{rain}) = 0.4$, $P(X = 6) = P(\text{snow}) = 0.15$, and $P(X = -2.7) = P(\text{clear}) = 0.45$. Also, $P(X = 17) = 0$, and in fact $P(X = x) = P(\emptyset) = 0$ for all $x \notin \{3, 6, -2.7\}$. We can also compute that

$$P(X \in \{3,6\}) = P(X = 3) + P(X = 6) = 0.4 + 0.15 = 0.55,$$

while

$$P(X < 5) = P(X = 3) + P(X = -2.7) = 0.4 + 0.45 = 0.85,$$

etc. ∎

We see from this example that, if B is any subset of the real numbers, then $P(X \in B) = P(\{s \in S : X(s) \in B\})$. Furthermore, to understand X well requires knowing the probabilities $P(X \in B)$ for different subsets B. That is the motivation for the following definition.

Definition 2.2.1 If X is a random variable, then the *distribution* of X is the collection of probabilities $P(X \in B)$, for all[1] subsets B of the real numbers.

In Figure 2.2.1 we provide a graphical representation of how we compute the distribution of a random variable X. For a set B, we must find the elements in $s \in S$ such that $X(s) \in B$. These elements are given by the set $\{s \in S : X(s) \in B\}$. Then we evaluate the probability $P(\{s \in S : X(s) \in B\})$. We must do this for every subset $B \subset R^1$.

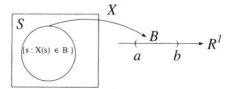

Figure 2.2.1: If $B = (a, b) \subset R^1$, then $\{s \in S : X(s) \in B\}$ is the set of elements such that $a < X(s) < b$.

Example 2.2.2 *A Very Simple Distribution*
Consider once again the above random variable, where $S = \{\text{rain, snow, clear}\}$, and X is defined by $X(\text{rain}) = 3$, $X(\text{snow}) = 6$, and $X(\text{clear}) = -2.7$, and $P(\text{rain}) = 0.4$, $P(\text{snow}) = 0.15$, and $P(\text{clear}) = 0.45$. What is the distribution of X? Well, if B is any subset of the real numbers, then $P(X \in B)$ should count 0.4 if $3 \in B$, plus 0.15 if $6 \in B$, plus 0.45 if $-2.7 \in B$. We can formally write all this information at once by saying that

$$P(X \in B) = 0.4\, I_B(3) + 0.15\, I_B(6) + 0.45\, I_B(-2.7),$$

where again $I_B(x) = 1$ if $x \in B$, and $I_B(x) = 0$ if $x \notin B$. ∎

Example 2.2.3 *An Almost-As-Simple Distribution*
Consider once again the above setting, with $S = \{\text{rain, snow, clear}\}$, and $P(\text{rain}) = 0.4$, $P(\text{snow}) = 0.15$, and $P(\text{clear}) = 0.45$. Consider a random variable Y defined by $Y(\text{rain}) = 5$, $Y(\text{snow}) = 7$, and $Y(\text{clear}) = 5$.

What is the distribution of Y? Well, clearly $Y = 7$ only when it snows, so that $P(Y = 7) = P(\text{snow}) = 0.15$, However, here $Y = 5$ if it rains *or* if it is clear. Hence, $P(Y = 5) = P(\{\text{rain, clear}\}) = 0.4 + 0.45 = 0.85$. Therefore, if B is any subset of the real numbers, then

$$P(Y \in B) = 0.15\, I_B(7) + 0.85\, I_B(5). \quad ∎$$

While the above examples show that it is possible to keep track of $P(X \in B)$ for all subsets B of the real numbers, they also indicate that it is rather cumbersome to do so. Fortunately, there are simpler functions available to help us keep track of probability distributions, including cumulative distribution functions, probability functions, and density functions. We discuss these next.

---

[1] Strictly speaking, it is required that B be a "Borel" subset, which is a technical restriction from measure theory that need not concern us here. Any subset that we could ever write down is a Borel subset.

Summary of Section 2.2

- The distribution of a random variable X is the collection of probabilities $P(X \in B)$ of X belonging to various sets.

- The probability $P(X \in B)$ is calculated by calculating the probability of the set of response values s such that $X(s) \in B$, i.e., $P(X \in B) = P(\{s \in S : X(s) \in B\})$.

Exercises

2.2.1 Consider flipping two independent fair coins. Let X be the number of heads that appear. Compute $P(X = x)$ for all real numbers x.

2.2.2 Suppose we flip three fair coins, and let X be the number of heads showing.
(a) Compute $P(X = x)$ for every real number x.
(b) Write a formula for $P(X \in B)$, for any subset B of the real numbers.

2.2.3 Suppose we roll two fair six-sided dice, and let Y be the sum of the two numbers showing.
(a) Compute $P(Y = y)$ for every real number y.
(b) Write a formula for $P(Y \in B)$, for any subset B of the real numbers.

2.2.4 Suppose we roll one fair six-sided die, and let Z be the number showing. Let $W = Z^3 + 4$, and let $V = \sqrt{Z}$.
(a) Compute $P(W = w)$ for every real number w.
(b) Compute $P(V = v)$ for every real number v.
(c) Compute $P(ZW = x)$ for every real number x.
(d) Compute $P(VW = y)$ for every real number y.
(e) Compute $P(V + W = r)$ for every real number r.

2.2.5 Suppose that a bowl contains 100 chips with 30 chips labelled 1, 20 chips labelled 2, and 50 chips labelled 3. The chips are thoroughly mixed, a chip is drawn, and the number X on the chip is noted.
(a) Compute $P(X = x)$ for every real number x.
(b) Suppose the first chip is replaced and then a second chip is drawn and the number Y on the chip is noted. Compute $P(Y = y)$ for every real number y.
(c) Compute $P(W = w)$ for every real number w when $W = X + Y$.

2.2.6 Suppose a standard deck of 52 playing cards is thoroughly shuffled and a single card is drawn. Suppose an ace has value 1, a jack has value 11, a queen has value 12, and a king has value 13.
(a) Compute $P(X = x)$ for every real number x, when X is the value of the card drawn.
(b) Suppose that $Y = 1, 2, 3,$ or 4 when a diamond, heart, club, or spade is drawn. Compute $P(Y = y)$ for every real number y.
(c) Compute $P(W = w)$ for every real number w when $W = X + Y$.

2.2.7 Suppose a university is comprised of 55% female students and 45% male students. A student is selected to complete a questionnaire. There are 25 questions on the questionnaire administered to a male student and 30 questions on the questionnaire administered to a female student. If X denotes the number of questions answered by a randomly selected student, then compute $P(X = x)$ for every real number x.

2.2.8 Suppose that a bowl contains ten chips, each uniquely numbered one of $0, 1, \ldots, 9$. The chips are thoroughly mixed, one is drawn and the number on it, X_1 is noted. This chip is then replaced in the bowl. A second chip is drawn and the number on it, X_2, is noted. Compute $P(W = w)$ for every real number w when $W = X_1 + 10X_2$.

Problem

2.2.9 Suppose that a bowl contains ten chips each uniquely numbered one of $0, 1, \ldots, 9$. The chips are thoroughly mixed, one is drawn and the number on it, X_1, is noted. This chip is *not* replaced in the bowl. A second chip is drawn and the number on it, X_2, is noted. Compute $P(W = w)$ for every real number w when $W = X_1 + 10X_2$.

Challenge

2.2.10 Suppose Alice flips three fair coins, and let X be the number of heads showing. Suppose Barbara flips five fair coins, and let Y be the number of heads showing. Let $Z = X - Y$. Compute $P(Z = z)$ for every real number z.

2.3   Discrete Distributions

For many random variables X, we have $P(X = x) > 0$ for certain x values. This means there is positive probability that the variable will be equal to certain particular values.

If

$$\sum_{x \in R^1} P(X = x) = 1,$$

then that means that *all* of the probability associated with the random variable X can be found from the probability that X will be equal to certain particular values. This prompts the following definition.

Definition 2.3.1 A random variable X is *discrete* if

$$\sum_{x \in R^1} P(X = x) = 1. \tag{2.3.1}$$

At first glance one might expect (2.3.1) to be true for *any* random variable. However, (2.3.1) does not hold for the uniform distribution on $[0, 1]$ or for other *continuous* distributions, as we shall see in the next section.

Random variables satisfying (2.3.1) are simple in some sense, because we can understand them completely just by understanding their probabilities of being equal to particular values x. Indeed, by simply listing out all the possible values x such that $P(X = x) > 0$, we obtain a second, equivalent definition, as follows.

Definition 2.3.2 A random variable X is *discrete* if there is a finite or countable sequence $x_1, x_2, \ldots$ of distinct real numbers, and a corresponding sequence $p_1, p_2, \ldots$ of nonnegative real numbers, such that $P(X = x_i) = p_i$ for all i, and $\sum_i p_i = 1$.

This second definition also suggests how to keep track of discrete distributions. It prompts the following definition.

Definition 2.3.3 For a discrete random variable X, its *probability function* is the function $p_X : R^1 \to [0, 1]$ defined by

$$p_X(x) = P(X = x).$$

Hence, if $x_1, x_2, \ldots$ are the distinct values such that $P(X = x_i) = p_i$ for all i with $\sum_i p_i = 1$, then

$$p_X(x) = \begin{cases} p_i & x = x_i \text{ for some } i \\ 0 & \text{otherwise.} \end{cases}$$

Clearly, all the information about the distribution of X is contained in its probability function, but only if we *know* that X is a discrete random variable.

Finally, we note that Theorem 1.5.1 immediately implies the following.

Theorem 2.3.1 (*Law of total probability, discrete random variable version*) Let X be a discrete random variable, and let A be some event. Then

$$P(A) = \sum_{x \in R^1} P(X = x) \, P(A \,|\, X = x).$$

2.3.1 Important Discrete Distributions

Certain particular discrete distributions are so important that we list them here.

Example 2.3.1 *Degenerate Distributions*
Let c be some fixed real number. Then, as already discussed, c is also a random variable (in fact, c is a *constant random variable*). In this case, clearly c is discrete, with probability function p_c satisfying that $p_c(c) = 1$, and $p_c(x) = 0$ for $x \neq c$. Because c is always equal to a particular value (namely, c) with probability 1, the distribution of c is sometimes called a *point mass* or *point distribution* or *degenerate distribution*. ∎

Example 2.3.2 *The Bernoulli Distribution*
Consider flipping a coin that has probability θ of coming up heads and probability $1 - \theta$ of coming up tails, where $0 < \theta < 1$. Let $X = 1$ if the coin is heads, while $X = 0$ if the coin is tails. Then $p_X(1) = P(X = 1) = \theta$, while $p_X(0) = P(X = 0) = 1 - \theta$. The random variable X is said to have the Bernoulli(θ) distribution; we write this as $X \sim$ Bernoulli(θ).

Bernoulli distributions arise anytime we have a response variable that takes only two possible values, and we label one of these outcomes by 1 and the other by 0. For example, 1 could correspond to success and 0 to failure of some quality test applied to an item produced in a manufacturing process. In this case, θ is the proportion of manufactured items that will pass the test. Alternatively, we could be randomly selecting an individual from a population and recording a 1 when the individual is female and a 0 if the individual is a male. In this case, θ is the proportion of females in the population. ∎

Example 2.3.3 *The Binomial Distribution*
Consider flipping n coins, each of which has (independent) probability θ of coming up heads, and probability $1 - \theta$ of coming up tails. (Again $0 < \theta < 1$.) Let X be the total number of heads showing. Then by (1.4.2), we see that for $x = 0, 1, 2, \ldots, n$

$$p_X(x) = P(X = x) = \binom{n}{x} \theta^x (1 - \theta)^{n-x} = \frac{n!}{x!\,(n-x)!} \, \theta^x (1 - \theta)^{n-x}.$$

The random variable X is said to have the Binomial(n, θ) distribution; we write this as $X \sim$ Binomial(n, θ). The Bernoulli(θ) distribution corresponds to the special case of the Binomial(n, θ) distribution when $n = 1$, namely Bernoulli$(\theta) =$ Binomial$(1, \theta)$. Figure 2.3.1 contains the plots of several Binomial$(20, \theta)$ probability functions.

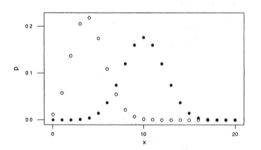

Figure 2.3.1: Plot of the Binomial$(20, .5)$ (● ● ●) and the Binomial$(20, .2)$ (○ ○ ○) probability functions.

The binomial distribution is applicable to any situation involving n independent performances of a random system and, for each performance, we are recording whether a particular event has occurred, called a *success*, or has not occurred, called a *failure*. If we denote the event in question by A and put $\theta = P(A)$, we have that the number of successes in the n performances is distributed Binomial(n, θ). For example, we could be testing light bulbs produced by a manufacturer, and θ is the probability that a bulb works when we test it. Then the number of bulbs that work in a batch of n is distributed Binomial(n, θ). If a baseball player has probability θ of getting a hit when at bat, then the number of hits obtained in n at-bats is distributed Binomial(n, θ).

There is another way of expressing the binomial distribution that is sometimes useful. For example, if $X_1, X_2, \ldots, X_n$ are chosen independently and each has the Bernoulli(θ) distribution, and $Y = X_1 + \cdots + X_n$, then Y will have the Binomial(n, θ) distribution. ∎

Example 2.3.4 *The Geometric Distribution*

Consider repeatedly flipping a coin that has probability θ of coming up heads, and probability $1 - \theta$ of coming up tails, where again $0 < \theta < 1$. Let X be the number of tails that appear before the first head. Then for $k \geq 0$, $X = k$ if and only if the coin shows exactly k tails followed by a head. The probability of this is equal to $(1 - \theta)^k \theta$. (In particular, the probability of getting an *infinite* number of tails before the first head is equal to $(1 - \theta)^\infty \theta = 0$, so X is never equal to infinity.) Hence, $p_X(k) = (1 - \theta)^k \theta$, for $k = 0, 1, 2, 3, \ldots$. The random variable X is said to have the Geometric(θ) distribution; we write this as $X \sim$ Geometric(θ). Figure 2.3.2 contains the plots of several Geometric(θ) probability functions.

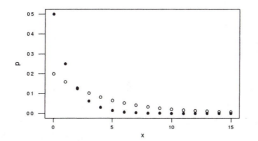

Figure 2.3.2: Plot of the Geometric$(1/2)$ ($\bullet\,\bullet\,\bullet$) and the Geometric$(1/5)$ ($\circ\,\circ\,\circ$) probability functions at the values $0, 1, \ldots, 15$.

The geometric distribution applies whenever we are counting the number of failures until the first success for independent performances of a random system where the occurrence of some event is considered a success. For example, the number of light bulbs tested that work until the first bulb that does not (a working bulb is considered a "failure" for the test) and the number of at-bats until the first hit for the baseball player both follow the geometric distribution.

We note that some books instead define the geometric distribution to be the number of coin flips up to *and including* the first head, which is simply equal to one plus the random variable defined here. ∎

Example 2.3.5 *The Negative-Binomial Distribution*

Generalizing the previous example, consider again repeatedly flipping a coin that has probability θ of coming up heads, and probability $1 - \theta$ of coming up tails. Let r be a positive integer, and let Y be the number of tails that appear before the rth head. Then for $k \geq 0$, $Y = k$ if and only if the coin shows exactly $r - 1$ heads (and k tails) on the first $r - 1 + k$ flips, and then shows a head on the $(r + k)$th flip. The probability of this is equal to

$$p_Y(k) = \binom{r - 1 + k}{r - 1} \theta^{r-1}(1 - \theta)^k \theta = \binom{r - 1 + k}{k} \theta^r (1 - \theta)^k,$$

for $k = 0, 1, 2, 3, \ldots$.

The random variable Y is said to have the Negative-Binomial(r, θ) distribution; we write this as $Y \sim$ Negative-Binomial(r, θ). Of course, the special case

$r = 1$ corresponds to the Geometric(θ) distribution. So, in terms of our notation we have that Negative-Binomial($1, \theta$) = Geometric(θ). Figure 2.3.3 contains the plots of several Negative-Binomial(r, θ) probability functions.

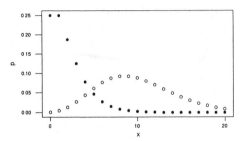

Figure 2.3.3: Plot of the Negative-Binomial($2, 1/2$) ($\bullet \bullet \bullet$) and the Negative-Binomial($10, 1/2$) ($\circ \circ \circ$) probability functions at the values $0, 1, \ldots, 20$.

The Negative-Binomial(r, θ) distribution applies whenever we are counting the number of failures until the rth success for independent performances of a random system where the occurrence of some event is considered a success. For example, the number of light bulbs tested that work until the third bulb that does not and the number of at-bats until the fifth hit for the baseball player both follow the negative-binomial distribution. ∎

Example 2.3.6 *The Poisson Distribution*
We say that a random variable Y has the Poisson(λ) distribution, and write $Y \sim$ Poisson(λ), if

$$p_Y(y) = P(Y = y) = \frac{\lambda^y}{y!} e^{-\lambda}$$

for $y = 0, 1, 2, 3, \ldots$. We note that since (from calculus), $\sum_{y=0}^{\infty} \lambda^y / y! = e^{\lambda}$ it is indeed true (as it must be) that $\sum_{y=0}^{\infty} P(Y = y) = 1$. Figure 2.3.4 contains the plots of several Poisson(λ) probability functions.

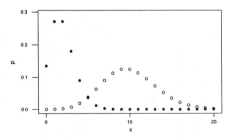

Figure 2.3.4: Plot of the Poisson(2) ($\bullet \bullet \bullet$) and the Poisson(10) ($\circ \circ \circ$) probability functions at the values $0, 1, \ldots, 20$.

We motivate the Poisson distribution as follows. Suppose $X \sim$ Binomial(n, θ), i.e., X has the Binomial(n, θ) distribution as in Example 2.3.3. Then for

$0 \leq x \leq n$,

$$P(X = x) = \binom{n}{x} \theta^x (1 - \theta)^{n-x}.$$

If we set $\theta = \lambda / n$ for some $\lambda > 0$, then this becomes

$$
\begin{aligned}
P(X = x) &= \binom{n}{x} \left(\frac{\lambda}{n}\right)^x \left(1 - \frac{\lambda}{n}\right)^{n-x} \\
&= \frac{n(n-1)\cdots(n-x+1)}{x!} \left(\frac{\lambda}{n}\right)^x \left(1 - \frac{\lambda}{n}\right)^{n-x}. \quad (2.3.2)
\end{aligned}
$$

Let us now consider what happens if we let $n \to \infty$ in (2.3.2), while keeping x fixed at some nonnegative integer. In that case,

$$\frac{n(n-1)(n-2)\cdots(n-x+1)}{n^x} = 1 \left(1 - \frac{1}{n}\right) \left(1 - \frac{2}{n}\right) \cdots \left(1 - \frac{x+1}{n}\right)$$

converges to 1 while (since from calculus $(1 + (c/n))^n \to e^c$ for any c)

$$\left(1 - \frac{\lambda}{n}\right)^{n-x} = \left(1 - \frac{\lambda}{n}\right)^n \left(1 - \frac{\lambda}{n}\right)^{-x} \to e^{-\lambda} \cdot 1 = e^{-\lambda}.$$

Substituting these limits into (2.3.2), we see that

$$\lim_{n \to \infty} P(X = x) = \frac{\lambda^x}{x!} e^{-\lambda}$$

for $x = 0, 1, 2, 3, \ldots$.

Intuitively, we can phrase this result as follows. If we flip a *very large* number of coins n, and each coin has a *very small* probability $\theta = \lambda/n$ of coming up heads, then the probability that the total number of heads will be x is approximately given by $\lambda^x e^{-\lambda}/x!$. Figure 2.3.5 displays the accuracy of this approximation when we are approximating the Binomial$(100, 1/10)$ distribution by the Poisson(λ) distribution where $\lambda = n\theta = 100(1/10) = 10$.

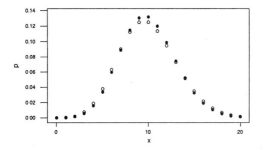

Figure 2.3.5: Plot of the Binomial$(100, 1/10)$ ($\bullet \bullet \bullet$) and the Poisson(10) ($\circ \circ \circ$) probability functions at the values $0, 1, \ldots, 20$.

The Poisson distribution is a good model for counting random occurrences of an event, when there are many possible occurrences, but each occurrence has very small probability. Examples include the number of house fires in a city on a given day; the number of radioactive events recorded by a Geiger counter; the number of phone calls arriving at a switchboard; the number of hits on a popular World Wide Web page on a given day; etc. ∎

Example 2.3.7 *The Hypergeometric Distribution*
Suppose that an urn contains M white balls and $N - M$ black balls. Suppose we draw $n \leq N$ balls from the urn in such a fashion that each subset of n balls has the same probability of being drawn. Because there are $\binom{N}{n}$ such subsets, this probability is $1/\binom{N}{n}$.

One way of accomplishing this is to thoroughly mix the balls in the urn and then draw a first ball. Accordingly each ball has probability $1/N$ of being drawn. Then, without replacing the first ball, we thoroughly mix the balls in the urn and draw a second ball. So each ball in the urn has probability $1/(N-1)$ of being drawn. We then have that any two balls, say the ith and jth balls, have probability

$P(\text{ball } i \text{ and } j \text{ are drawn})$

$= P(\text{ball } i \text{ is drawn first})P(\text{ball } j \text{ is drawn second} \,|\, \text{ball } i \text{ is drawn first})$

$\quad + P(\text{ball } j \text{ is drawn first})P(\text{ball } i \text{ is drawn second} \,|\, \text{ball } j \text{ is drawn first})$

$= \dfrac{1}{N}\dfrac{1}{N-1} + \dfrac{1}{N}\dfrac{1}{N-1} = 1/\binom{N}{2}$

of being drawn in the first two draws. Continuing in this fashion for n draws, we obtain that the probability of any particular set of n balls being drawn is $1/\binom{N}{n}$. This type of sampling is called *sampling with replacement*.

Given that we take a sample of n, let X denote the number of white balls obtained. Note that we must have $X \geq 0$ and $X \geq n - (N - M)$, because at most $N - M$ of the balls could be black. Hence, $X \geq \max(0, n + M - N)$. Further, $X \leq n$, and $X \leq M$, because there are only M white balls. Hence $X \leq \min(n, M)$.

So suppose $\max(0, n + M - N) \leq x \leq \min(n, M)$. What is the probability that x white balls are obtained? In other words, what is $P(X = x)$? To evaluate this, we know that we need to count the number of subsets of n balls that contain x white balls. Using the combinatorial principles of Section 1.4.1, we see that this number is given by $\binom{M}{x}\binom{N-M}{n-x}$. Therefore,

$$P(X = x) = \binom{M}{x}\binom{N-M}{n-x} \bigg/ \binom{N}{n}$$

for $\max(0, n + M - N) \leq x \leq \min(n, M)$. The random variable X is said to have the Hypergeometric(N, M, n) distribution. In Figure 2.3.6, we have plotted some hypergeometric probability functions. The Hypergeometric$(20, 10, 10)$ probability function is 0 for $x > 10$, while the Hypergeometric$(20, 10, 5)$ probability function is 0 for $x > 5$.

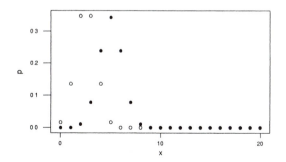

Figure 2.3.6: Plot of Hypergeometric$(20, 10, 10)$ ($\bullet \bullet \bullet$) and Hypergeometric$(20, 10, 5)$ ($\circ \circ \circ$) probability functions.

Obviously, the hypergeometric distribution will apply in any context where we are *sampling without replacement* from a finite set of N elements, and where each element of the set either has a characteristic or does not. For example, if we randomly select people to participate in an opinion poll so that each set of n individuals in a population of N has the same probability of being selected, then the number of people who respond yes to a particular question is distributed Hypergeometric(N, M, n), where M is the number of people in the entire population who would respond yes. We will see the relevance of this to statistics in Section 5.4.2. ∎

Suppose in Example 2.3.7 we had instead *replaced* the drawn ball before drawing the next ball. This is called *sampling with replacement*. It is then clear, from Example 2.3.3, that the number of white balls obtained in n draws is distributed Binomial$(n, M/N)$.

Summary of Section 2.3

- A random variable X is discrete if $\sum_x P(X = x) = 1$, i.e., if all its probability comes from being equal to particular values.

- A discrete random variable X only takes on a finite or countable number of distinct values.

- Important discrete distributions include the degenerate, Bernoulli, binomial, geometric, negative-binomial, Poisson, and hypergeometric distributions.

Exercises

2.3.1 Consider rolling two fair six-sided dice. Let Y be the sum of the numbers showing. What is the probability function of Y?

2.3.2 Consider flipping a fair coin. Let $Z = 1$ if the coin is heads, and $Z = 3$ if the coin is tails. Let $W = Z^2 + Z$.

(a) What is the probability function of Z?
(b) What is the probability function of W?

2.3.3 Consider flipping two fair coins. Let $X = 1$ if the first coin is heads, and $X = 0$ if the first coin is tails. Let $Y = 1$ if the second coin is heads, and $Y = 5$ if the second coin is tails. Let $Z = XY$. What is the probability function of Z?

2.3.4 Consider flipping two fair coins. Let $X = 1$ if the first coin is heads, and $X = 0$ if the first coin is tails. Let $Y = 1$ if the two coins show the *same* thing (i.e., both heads or both tails), with $Y = 0$ otherwise. Let $Z = X + Y$, and $W = XY$.
(a) What is the probability function of Z?
(b) What is the probability function of W?

2.3.5 Consider rolling two fair six-sided dice. Let W be the product of the numbers showing. What is the probability function of W?

2.3.6 Let $Z \sim$ Geometric(θ). Compute $P(5 \leq Z \leq 9)$.

2.3.7 Let $X \sim$ Binomial$(12, \theta)$. For what value of θ is $P(X = 11)$ maximized?

2.3.8 Let $W \sim$ Poisson(λ). For what value of λ is $P(W = 11)$ maximized?

2.3.9 Let $Z \sim$ Negative-Binomial$(3, 1/4)$. Compute $P(Z \leq 2)$.

2.3.10 Let $X \sim$ Geometric$(1/5)$. Compute $P(X^2 \leq 15)$.

2.3.11 Let $Y \sim$ Binomial$(10, \theta)$. Compute $P(Y = 10)$.

2.3.12 Let $X \sim$ Poisson(λ). Let $Y = X - 7$. What is the probability function of Y?

2.3.13 Let $X \sim$ Hypergeometric$(20, 7, 8)$. What is the probability that $X = 3$? What is the probability that $X = 8$?

2.3.14 Suppose that a symmetrical die is rolled 20 independent times, and each time we record whether or not the event $\{2, 3, 5, 6\}$ has occurred.
(a) What is the distribution of the number of times this event occurs in 20 rolls?
(b) Calculate the probability that the event occurs five times.

2.3.15 Suppose that a basketball player sinks a basket from a certain position on the court with probability 0.35.
(a) What is the probability that the player sinks three baskets in ten independent throws?
(b) What is the probability that the player throws ten times before obtaining the first basket?
(c) What is the probability that the player throws ten times before obtaining two baskets?

2.3.16 An urn contains 4 black balls and 5 white balls. After a thorough mixing a ball is drawn from the urn, its color is noted, and the ball is returned to the urn.
(a) What is the probability that 5 black balls are observed in 15 such draws?

(b) What is the probability that 15 draws are required until the first black ball is observed?

(c) What is the probability that 15 draws are required until the fifth black ball is observed?

2.3.17 An urn contains 4 black balls and 5 white balls. After a thorough mixing, a ball is drawn from the urn, its color is noted, and the ball is set aside. The remaining balls are then mixed and a second ball is drawn.

(a) What is the probability distribution of the number of black balls observed?

(b) What is the probability distribution of the number of white balls observed?

2.3.18 (*Poisson processes and queues*) Consider a situation involving a server, e.g., a cashier at a fast-food restaurant, an automatic bank teller machine, a telephone exchange, etc. Units typically arrive for service in a random fashion and form a queue when the server is busy. It is often the case that the number of arrivals at the server, for some specific unit of time t, can be modeled by a Poisson(λt) distribution and is such that the number of arrivals in nonoverlapping periods are independent. In Chapter 3 we will show that λt is the average number of arrivals during a time period of length t, and so λ is the rate of arrivals per unit of time.

Suppose telephone calls arrive at a help line at the rate of two per minute. A Poisson process provides a good model.

(a) What is the probability that five calls arrive in the next 2 minutes?

(b) What is the probability that five calls arrive in the next 2 minutes and then five more calls arrive in the following 2 minutes?

(c) What is the probability that no calls will arrive during a 10-minute period?

2.3.19 Suppose that an urn contains 1000 balls — one of these is black, and the other 999 are white. Suppose that 100 balls are randomly drawn from the urn with replacement. Use the appropriate Poisson distribution to approximate the probability that five black balls are observed.

2.3.20 Suppose that there is a loop in a computer program and that the test to exit the loop depends on the value of a random variable X. The program exits the loop whenever $X \in A$, and this occurs with probability $1/3$. If the loop is executed at least once, what is the probability that the loop is executed five times before exiting?

Computer Exercises

2.3.21 Tabulate and plot the Hypergeometric$(20, 8, 10)$ probability function.

2.3.22 Tabulate and plot the Binomial$(30, 0.3)$ probability function. Tabulate and plot the Binomial$(30, 0.7)$ probability function. Explain why the Binomial$(30, 0.3)$ probability function at x agrees with the Binomial$(30, 0.7)$ probability function at $n - x$.

Problems

2.3.23 Let X be a discrete random variable with probability function $p_X(x) = 2^{-x}$ for $x = 1, 2, 3, \ldots$, with $p_X(x) = 0$ otherwise.

(a) Let $Y = X^2$. What is the probability function p_Y of Y?

(b) Let $Z = X - 1$. What is the distribution of Z? (Identify the distribution by name, and specify all parameter values.)

2.3.24 Let $X \sim \text{Binomial}(n_1, \theta)$ and $Y \sim \text{Binomial}(n_2, \theta)$, with X and Y chosen independently. Let $Z = X + Y$. What will be the distribution of Z? (Explain your reasoning.) (Hint: See the end of Example 2.3.3.)

2.3.25 Let $X \sim \text{Geometric}(\theta)$ and $Y \sim \text{Geometric}(\theta)$, with X and Y chosen independently. Let $Z = X + Y$. What will be the distribution of Z? Generalize this to r coins. (Explain your reasoning.)

2.3.26 Let $X \sim \text{Geometric}(\theta_1)$ and $Y \sim \text{Geometric}(\theta_2)$, with X and Y chosen independently. Compute $P(X \leq Y)$. Explain what this probability is in terms of coin tossing.

2.3.27 Suppose that $X \sim \text{Geometric}(\lambda/n)$. Compute $\lim_{n \to \infty} P(X \leq n)$.

2.3.28 Let $X \sim \text{Negative-Binomial}(r, \theta)$ and $Y \sim \text{Negative-Binomial}(s, \theta)$, with X and Y chosen independently. Let $Z = X + Y$. What will be the distribution of Z? (Explain your reasoning.)

2.3.29 (*Generalized hypergeometric distribution*) Suppose that a set contains N objects, M_1 of which are labelled 1, M_2 of which are labelled 2, and the remainder of which are labelled 3. Suppose we select a sample of $n \leq N$ objects from the set using sampling without replacement, as described in Example 2.3.7. Determine the probability that we obtain the counts (f_1, f_2, f_3) where f_i is the number of objects labelled i in the sample.

2.3.30 Suppose that units arrive at a server according to a Poisson process at rate λ (see Exercise 2.3.18). Let T be the amount of time until the first call. Calculate $P(T > t)$.

2.4   Continuous Distributions

In the previous section, we considered discrete random variables X for which $P(X = x) > 0$ for certain values of x. However, for some random variables X, such as one having the uniform distribution, we have $P(X = x) = 0$ for all x. This prompts the following definition.

Definition 2.4.1 A random variable X is *continuous* if

$$P(X = x) = 0, \tag{2.4.1}$$

for all $x \in R^1$.

Example 2.4.1 *The Uniform[0, 1] Distribution*

Consider a random variable whose distribution is the uniform distribution on $[0, 1]$, as presented in (1.2.2). That is,

$$P(a \leq X \leq b) = b - a, \tag{2.4.2}$$

whenever $0 \leq a \leq b \leq 1$, with $P(X < 0) = P(X > 1) = 0$. The random variable X is said to have the Uniform$[0, 1]$ distribution; we write this as $X \sim$ Uniform$[0, 1]$. For example,

$$P\left(\frac{1}{2} \leq X \leq \frac{3}{4}\right) = \frac{3}{4} - \frac{1}{2} = \frac{1}{4}.$$

Also,

$$P\left(X \geq \frac{2}{3}\right) = P\left(\frac{2}{3} \leq X \leq 1\right) + P(X > 1)) = \left(1 - \frac{2}{3}\right) + 0 = \frac{1}{3}.$$

In fact, for any $x \in [0, 1]$,

$$P(X \leq x) = P(X < 0) + P(0 \leq X \leq x) = 0 + (x - 0) = x.$$

Note that setting $a = b = x$ in (2.4.2), we see in particular that $P(X = x) = x - x = 0$ for every $x \in R^1$. Thus, the uniform distribution is an example of a continuous distribution. In fact, it is one of the most important examples! ∎

The Uniform$[0, 1]$ distribution is fairly easy to work with. However, in general, continuous distributions are very difficult to work with. Because $P(X = x) = 0$ for all x, we cannot simply add up probabilities like we can for discrete random variables. Thus, how can we keep track of all the probabilities?

A possible solution is suggested by rewriting (2.4.2), as follows. For $x \in R^1$, let

$$f(x) = \begin{cases} 1 & 0 \leq x \leq 1 \\ 0 & \text{otherwise.} \end{cases} \tag{2.4.3}$$

Then (2.4.2) may be rewritten as

$$P(a \leq X \leq b) = \int_a^b f(x)\, dx, \tag{2.4.4}$$

whenever $a \leq b$.

One might wonder about the wisdom of converting the simple equation (2.4.2) into the complicated integral equation (2.4.4). However, the advantage of (2.4.4) is that, by modifying the function f, we can obtain many other continuous distributions besides the uniform distribution. To explore this, we make the following definitions.

Definition 2.4.2 Let $f : R^1 \to R^1$ be a function. Then f is a *density function* if $f(x) \geq 0$ for all $x \in R^1$, and $\int_{-\infty}^{\infty} f(x)\, dx = 1$.

Definition 2.4.3 A random variable X is *absolutely continuous* if there is a density function f, such that

$$P(a \leq X \leq b) = \int_a^b f(x)\, dx, \tag{2.4.5}$$

whenever $a \leq b$, as in (2.4.4).

In particular, if $b = a + \delta$ with δ a small positive number, and if f is continuous at a, then we see that

$$P(a \leq X \leq a + \delta) = \int_a^{a+\delta} f(x)\,dx \approx \delta\, f(a).$$

Thus, a density function evaluated at a may be thought of as measuring the probability of a random variable being in a small interval about a.

To better understand absolutely continuous random variables, we note the following.

Theorem 2.4.1 Let X be an absolutely continuous random variable. Then X is a continuous random variable, i.e., $P(X = a) = 0$ for all $a \in R^1$.

Proof: Let a be any real number. Then $P(X = a) = P(a \leq X \leq a)$. On the other hand, setting $a = b$ in (2.4.5), we see that $P(a \leq X \leq a) = \int_a^a f(x)\,dx = 0$. Hence, $P(X = a) = 0$ for all a, as required. ∎

It turns out that the converse to Theorem 2.4.1 is false. That is, not all continuous distributions are absolutely continuous.[2] However, most of the continuous distributions that arise in statistics are absolutely continuous. Furthermore, absolutely continuous distributions are much easier to work with than are other kinds of continuous distributions. Hence, we restrict our discussion to absolutely continuous distributions here. In fact, statisticians sometimes say that X is *continuous* as shorthand for saying that X is absolutely continuous.

2.4.1 Important Absolutely Continuous Distributions

Certain absolutely continuous distributions are so important that we list them here.

Example 2.4.2 *The Uniform*[0, 1] *Distribution*
Clearly, the uniform distribution is absolutely continuous, with the density function given by (2.4.3). We will see, in Section 2.10, that the Uniform[0, 1] distribution has an important relationship with every absolutely continuous distribution. ∎

Example 2.4.3 *The Uniform*[L, R] *Distribution*
Let L and R be any two real numbers with $L < R$. Consider a random variable X such that

$$P(a \leq X \leq b) = \frac{b-a}{R-L} \tag{2.4.6}$$

whenever $L \leq a \leq b \leq R$, with $P(X < L) = P(X > R) = 0$. The random variable X is said to have the Uniform[L, R] distribution; we write this as $X \sim \text{Uniform}[L, R]$. (If $L = 0$ and $R = 1$, then this definition coincides with the previous definition of the Uniform[0, 1] distribution.) Note that $X \sim$

---

[2] For examples of this, see more advanced probability books, e.g., page 118 of *A First Look at Rigorous Probability Theory* by J. S. Rosenthal, World Scientific Publishing, Singapore, 2000.

Uniform[L, R] has the same probability of being in any two subintervals of [L, R] that have the same length.

Note that the Uniform[L, R] distribution is also absolutely continuous, with density given by

$$f(x) = \begin{cases} \frac{1}{R-L} & L \le x \le R \\ 0 & \text{otherwise.} \end{cases}$$

In Figure 2.4.1 we have plotted a Uniform[2, 4] density. ∎

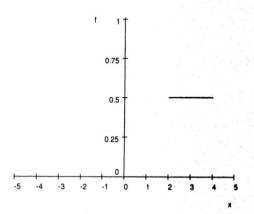

Figure 2.4.1: A Uniform[2, 4] density function.

Example 2.4.4 *The Exponential*(1) *Distribution*
Define a function $f : R^1 \to R^1$ by

$$f(x) = \begin{cases} e^{-x} & x \ge 0 \\ 0 & x < 0. \end{cases}$$

Then clearly $f(x) \ge 0$ for all x. Also,

$$\int_{-\infty}^{\infty} f(x)\, dx = \int_0^{\infty} e^{-x}\, dx = -e^{-x} \Big|_0^{\infty} = (-0) - (-1) = 1.$$

Hence, f is a density function. See Figure 2.4.2 for a plot of this density.

Consider now a random variable X having this density function f. If $0 \le a \le b < \infty$, then

$$P(a \le X \le b) = \int_a^b f(x)\, dx = \int_a^b e^{-x}\, dx = (-e^{-b}) - (-e^{-a}) = e^{-a} - e^{-b}.$$

The random variable X is said to have the Exponential(1) distribution, which we write as $X \sim$ Exponential(1). The exponential distribution has many important properties, which we will explore in the coming sections. ∎

Example 2.4.5 *The Exponential(λ) Distribution*
Let $\lambda > 0$ be a fixed constant. Define a function $f : R^1 \to R^1$ by

$$f(x) = \begin{cases} \lambda e^{-\lambda x} & x \geq 0 \\ 0 & x < 0. \end{cases}$$

Then clearly $f(x) \geq 0$ for all x. Also,

$$\int_{-\infty}^{\infty} f(x)\, dx = \int_0^{\infty} \lambda e^{-\lambda x}\, dx = -e^{-\lambda x} \Big|_0^{\infty} = (-0) - (-1) = 1.$$

Hence, f is again a density function. (If $\lambda = 1$, then this corresponds to the Exponential(1) density.)

If X is a random variable having this density function f, then

$$P(a \leq X \leq b) = \int_a^b \lambda e^{-\lambda x}\, dx = (-e^{-\lambda b}) - (-e^{-\lambda a}) = e^{-\lambda a} - e^{-\lambda b}$$

for $0 \leq a \leq b < \infty$. The random variable X is said to have the Exponential(λ) distribution; we write this as $X \sim$ Exponential(λ). Note that some books and software packages instead replace λ by $1/\lambda$ in the definition of the Exponential(λ) distribution — always check this when using another book or when using software.

An exponential distribution can often be used to model lifelengths. For example, a certain type of light bulb produced by a manufacturer might follow an Exponential(λ) distribution for an appropriate choice of λ. By this we mean that the lifelength X of a randomly selected light bulb from those produced by this manufacturer has probability

$$P(X \geq x) = \int_x^{\infty} \lambda e^{-\lambda z}\, dz = e^{-\lambda x}$$

of lasting longer than x, in whatever units of time are being used. We will see in Chapter 3 that, in a specific application, the value $1/\lambda$ will correspond to the average lifelength of the light bulbs.

As another application of this distribution, consider a situation involving a server, e.g., a cashier at a fast-food restaurant, an automatic bank teller machine, a telephone exchange, etc. Units arrive for service in a random fashion and form a queue when the server is busy. It is often the case that the number of arrivals at the server, for some specific unit of time t, can be modeled by a Poisson(λt) distribution. Now let T_1 be the time until the first arrival. Then we have

$$P(T_1 > t) = P(\text{no arrivals in } (0, t]) = \frac{(\lambda t)^0}{0!} e^{-\lambda t} = e^{-\lambda t}$$

and T_1 has density given by

$$f(t) = -\frac{d}{dt} \int_t^{\infty} f(z)\, dz = -\frac{d}{dt} P(T_1 > t) = \lambda e^{-\lambda t}.$$

So $T_1 \sim$ Exponential(λ). ∎

Example 2.4.6 *The Gamma(α, λ) Distribution*
The *gamma function* is defined by

$$\Gamma(\alpha) = \int_0^\infty t^{\alpha-1} e^{-t}\, dt, \qquad \alpha > 0.$$

It turns out (Problem 2.4.7) that

$$\Gamma(\alpha + 1) = \alpha\Gamma(\alpha) \tag{2.4.7}$$

and that if n is a positive integer, then $\Gamma(n) = (n-1)!$, while $\Gamma(1/2) = \sqrt{\pi}$.

We can use the gamma function to define the density of the Gamma(α, λ) distribution, as follows. Let $\alpha > 0$ and $\lambda > 0$, and define a function f by

$$f(x) = \frac{\lambda^\alpha x^{\alpha-1}}{\Gamma(\alpha)}\, e^{-\lambda x} \tag{2.4.8}$$

when $x > 0$, with $f(x) = 0$ for $x \le 0$. Then clearly $f \ge 0$. Furthermore, it is not hard to verify (Problem 2.4.9) that $\int_0^\infty f(x)\, dx = 1$. Hence, f is a density function.

A random variable X having density function f given by (2.4.8) is said to have the Gamma(α, λ) distribution; we write this as $X \sim$ Gamma(α, λ). Note that some books and software packages instead replace λ by $1/\lambda$ in the definition of the Gamma(α, λ) distribution — always check this when using another book or when using software.

The case $\alpha = 1$ corresponds (because $\Gamma(1) = 0! = 1$) to the Exponential(λ) distribution: Gamma($1, \lambda$) = Exponential(λ). In Figure 2.4.2, we have plotted several Gamma(α, λ) density functions.

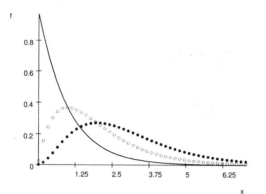

Figure 2.4.2: Graph of an Exponential(1) (—), a Gamma(2, 1) (∘ ∘ ∘), and a Gamma(3, 1) (● ● ●) density.

A gamma distribution can also be used to model lifelengths. As Figure 2.4.2 shows, the gamma family gives a much greater variety of shapes to choose among than the exponential family does. ∎

We now define a function $\phi : R^1 \to R^1$ by

$$\phi(x) = \frac{1}{\sqrt{2\pi}} e^{-x^2/2}. \qquad (2.4.9)$$

This function ϕ is the famous "bell-shaped curve", because its graph is in the shape of a bell, as shown in Figure 2.4.3.

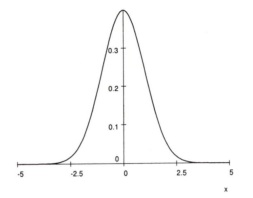

Figure 2.4.3: Plot of the function ϕ in (2.4.9).

We have the following result for ϕ.

Theorem 2.4.2 The function ϕ given by (2.4.9) is a density function.
Proof: See Section 2.11 for the proof of this result.

This leads to the following important distributions.

Example 2.4.7 *The $N(0,1)$ Distribution*
Let X be a random variable having the density function ϕ given by (2.4.9). This means that for $-\infty < a \le b < \infty$,

$$P(a \le X \le b) = \int_a^b \phi(x)\, dx = \int_a^b \frac{1}{\sqrt{2\pi}} e^{-x^2/2}\, dx.$$

The random variable X is said to have the $N(0,1)$ distribution (or, the *standard normal distribution*); we write this as $X \sim N(0,1)$. ∎

Example 2.4.8 *The $N(\mu, \sigma^2)$ Distribution*
Let $\mu \in R^1$, and let $\sigma > 0$. Let f be the function defined by

$$f(x) = \frac{1}{\sigma} \phi\left(\frac{x - \mu}{\sigma}\right) = \frac{1}{\sigma\sqrt{2\pi}} e^{-(x-\mu)^2/2\sigma^2}.$$

(If $\mu = 0$ and $\sigma = 1$, then this corresponds with the previous example.) Clearly, $f \ge 0$. Also, letting $y = (x - \mu)/\sigma$, we have

$$\int_{-\infty}^{\infty} f(x)\, dx = \int_{-\infty}^{\infty} \sigma^{-1}\phi((x-\mu)/\sigma)\, dx = \int_{-\infty}^{\infty} \sigma^{-1}\phi(y)\sigma\, dy = \int_{-\infty}^{\infty} \phi(y)\, dy = 1.$$

Hence, f is a density function.

Let X be a random variable having this density function f. The random variable X is said to have the $N(\mu, \sigma^2)$ distribution; we write this as $X \sim N(\mu, \sigma^2)$ or $X \sim N(\mu, \sigma^2)$. In Figure 2.4.4, we have plotted the $N(0, 1)$ and the $N(1, 1)$ densities. Note that changes in μ simply shift the density without changing its shape. In Figure 2.4.5, we have plotted the $N(0, 1)$ and the $N(0, 4)$ densities. Note that both densities are centered on 0 but the $N(0, 4)$ density is much more spread out. The value of σ^2 controls the amount of spread.

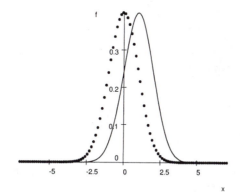

Figure 2.4.4: Graph of the $N(1, 1)$ density (—) and the $N(0, 1)$ (• • •) density.

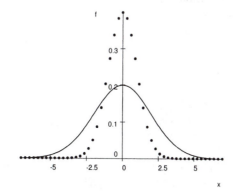

Figure 2.4.5: Graph of a $N(0, 4)$ (—) density and a $N(0, 1)$ (• • •) density.

The $N(\mu, \sigma^2)$ distribution, for some choice of μ and σ^2, arises quite often in applications. Part of the reason for this is an important result known as the central limit theorem which we will discuss in Section 4.4. In particular, this result leads to using a normal distribution to approximate other distributions, just as we used the Poisson distribution to approximate the binomial distribution in Example 2.3.6.

In a large human population, it is not uncommon for various body measurements to be normally distributed (at least to a reasonable degree of approximation). For example, let us suppose that heights (measured in feet) of students at a particular university are distributed $N(\mu, \sigma^2)$ for some choice of μ and σ^2. Then the probability that a randomly selected student has height between a and b feet, with $a > b$, is given by

$$\int_a^b \frac{1}{\sigma\sqrt{2\pi}} e^{-(x-\mu)^2/2\sigma^2}.$$

In Section 2.5 we will discuss how to evaluate such an integral. Later in this text we will discuss how to select an appropriate value for μ and σ^2 and to assess whether or not any normal distribution is appropriate to model the distribution of a variable defined on a particular population. ∎

Given an absolutely continuous random variable X, we will write its density as f_X, or as f if no confusion arises. Absolutely continuous random variables will be used extensively in later chapters of this book.

Remark 2.4.1 Finally, we note that density functions are not unique. Indeed, if f is a density function, and we change its value at a finite number of points, then the value of $\int_a^b f(x)\,dx$ will remain unchanged. Hence, the changed function will also qualify as a density corresponding to the same distribution. On the other hand, often a particular "best" choice of density function is clear. For example, if the density function can be chosen to be continuous, or even piecewise continuous, then this is to be preferred over some other version of the density function.

To take a specific example, for the Uniform$[0, 1]$ distribution, we could replace the density f of (2.4.3) by

$$g(x) = \begin{cases} 1 & 0 < x < 1 \\ 0 & \text{otherwise,} \end{cases}$$

or even by

$$h(x) = \begin{cases} 1 & 0 < x < 3/4 \\ 17 & x = 3/4 \\ 1 & 3/4 < x < 1 \\ 0 & \text{otherwise.} \end{cases}$$

Either of these new densities would again define the Uniform$[0, 1]$ distribution, because we would have $\int_a^b f(x)\,dx = \int_a^b g(x)\,dx = \int_a^b h(x)\,dx$ for any $a < b$.

On the other hand, the densities f and g are both piecewise continuous and are therefore natural choices for the density function, while h is an unnecessarily complicated choice. Hence, when dealing with density functions, we shall always assume that they are as continuous as possible, such as f and g, rather than having removable discontinuities such as h. This will be particularly important when discussing likelihood methods in Chapter 6.

Summary of Section 2.4

- A random variable X is continuous if $P(X = x) = 0$ for all x, i.e., if none of its probability comes from being equal to particular values.

- X is absolutely continuous if there exists a density function f_X with $P(a \le X \le b) = \int_a^b f_X(x)\,dx$ for all $a < b$.

- Important absolutely continuous distributions include the uniform, exponential, gamma, and normal.

Exercises

2.4.1 Let $U \sim \text{Uniform}[0, 1]$. Compute each of the following.
(a) $P(U \le 0)$
(b) $P(U = 1/2)$
(c) $P(U < -1/3)$
(d) $P(U \le 2/3)$
(e) $P(U < 2/3)$
(f) $P(U < 1)$
(g) $P(U \le 17)$

2.4.2 Let $W \sim \text{Uniform}[1, 4]$. Compute each of the following.
(a) $P(W \ge 5)$
(b) $P(W \ge 2)$
(c) $P(W^2 \le 9)$ (Hint: If $W^2 \le 9$, what must W be?)
(d) $P(W^2 \le 2)$

2.4.3 Let $Z \sim \text{Exponential}(4)$. Compute each of the following.
(a) $P(Z \ge 5)$
(b) $P(Z \ge -5)$
(c) $P(Z^2 \ge 9)$
(d) $P(Z^4 - 17 \ge 9)$

2.4.4 Establish for which constants c the following functions are densities.
(a) $f(x) = cx$ on $(0, 1)$ and 0 otherwise
(b) $f(x) = cx^n$ on $(0, 1)$ and 0 otherwise, for n a nonnegative integer
(c) $f(x) = cx^{1/2}$ on $(0, 2)$ and 0 otherwise
(d) $f(x) = c \sin x$ on $(0, \pi/2)$ and 0 otherwise

2.4.5 Is the function defined by $f(x) = x/3$ for $-1 < x < 2$ and 0 otherwise, a density? Why or why not?

Problems

2.4.6 Let $Y \sim \text{Exponential}(\lambda)$ for some $\lambda > 0$. Let $y, h \ge 0$. Prove that $P(Y - h \ge y \mid Y \ge h) = P(Y \ge y)$. That is, conditional on knowing that $Y \ge h$, the random variable $Y - h$ has the same distribution as Y did originally. This is called the *memoryless* property of the exponential distributions; it says that they immediately "forget" their past behavior.

2.4.7 Consider the gamma function $\Gamma(\alpha) = \int_0^\infty t^{\alpha-1} e^{-t} dt$, for $\alpha > 0$.
(a) Prove that $\Gamma(\alpha + 1) = \alpha \Gamma(\alpha)$. (Hint: Use integration by parts.)
(b) Prove that $\Gamma(1) = 1$.
(c) Use parts (a) and (b) to show that $\Gamma(n) = (n - 1)!$ if n is a positive integer.

2.4.8 Use the fact that $\Gamma(1/2) = \sqrt{\pi}$ to give an alternate proof that $\int_{-\infty}^{\infty} \phi(x)\, dx = 1$ (as in Theorem 2.4.2). (Hint: Make the substitution $t = x^2/2$.)

2.4.9 Let f be the density of the Gamma(α, λ) distribution, as in (2.4.8). Prove that $\int_0^{\infty} f(x)\, dx = 1$. (Hint: Let $t = \lambda x$.)

2.4.10 (*Logistic distribution*) Consider the function given by $f(x) = e^{-x} (1 + e^{-x})^{-2}$ for $-\infty < x < \infty$. Prove that f is a density function.

2.4.11 (*Weibull*(α) *distribution*) Consider, for $\alpha > 0$ fixed, the function given by $f(x) = \alpha x^{\alpha - 1} e^{-x^{\alpha}}$ for $0 < x < \infty$ and 0 otherwise. Prove that f is a density function.

2.4.12 (*Pareto*(α) *distribution*) Consider, for $\alpha > 0$ fixed, the function given by $f(x) = \alpha (1 + x)^{-\alpha - 1}$ for $0 < x < \infty$ and 0 otherwise. Prove that f is a density function.

2.4.13 (*Cauchy distribution*) Consider the function given by

$$f(x) = \frac{1}{\pi} \frac{1}{1 + x^2}$$

for $-\infty < x < \infty$. Prove that f is a density function. (Hint: Recall the derivative of $\arctan(x)$.)

2.4.14 (*Laplace distribution*) Consider the function given by $f(x) = e^{-|x|}/2$ for $-\infty < x < \infty$ and 0 otherwise. Prove that f is a density function.

2.4.15 (*Extreme value distribution*) Consider the function given by $f(x) = e^{-x} \exp\{-e^{-x}\}$ for $-\infty < x < \infty$ and 0 otherwise. Prove that f is a density function.

2.4.16 (*Beta*(α, β) *distribution*) The *beta function* is the function $B : (0, \infty)^2 \to R^1$ given by

$$B(a, b) = \int_0^1 x^{a-1} (1 - x)^{b-1}\, dx.$$

It can be proved (see Challenge 2.4.17) that

$$B(a, b) = \frac{\Gamma(a)\Gamma(b)}{\Gamma(a + b)} \tag{2.4.10}$$

(a) Prove that the function f given by $f(x) = B^{-1}(a, b) x^{a-1} (1 - x)^{b-1}$, for $0 < x < 1$ and 0 otherwise, is a density function.
(b) Determine and plot the density when $a = 1, b = 1$. Can you name this distribution?
(c) Determine and plot the density when $a = 2, b = 1$.
(d) Determine and plot the density when $a = 1, b = 2$.
(e) Determine and plot the density when $a = 2, b = 2$.

Challenge

2.4.17 Prove (2.4.10). (Hint: Use $\Gamma(a)\Gamma(b) = \int_0^{\infty} \int_0^{\infty} x^{a-1} y^{b-1} e^{-x-y}\, dx\, dy$ and make the change of variable $u = x = y, v = x/u$.)

2.5 Cumulative Distribution Functions

If X is a random variable, then its distribution consists of the values of $P(X \in B)$ for all subsets B of the real numbers. However, there are certain special subsets B that are convenient to work with. Specifically, if $B = (-\infty, x]$ for some real number x, then $P(X \in B) = P(X \leq x)$. It turns out (see **Theorem 2.5.1**) that it is sufficient to keep track of $P(X \leq x)$ for all real numbers x.

This motivates the following definition.

Definition 2.5.1 Given a random variable X, its *cumulative distribution function* (or *distribution function* or *cdf* for short) is the function $F_X : R^1 \to [0, 1]$, defined by $F_X(x) = P(X \leq x)$. (Where there is no confusion, we sometimes write $F(x)$ for $F_X(x)$.)

The reason for calling F_X the "distribution function" is that the full distribution of X can be determined directly from F_X. We demonstrate this for some events of particular importance.

First, suppose that $B = (a, b]$ is a left-open interval. Then using (1.3.3),

$$P(X \in B) = P(a < X \leq b) = P(X \leq b) - P(X \leq a) = F_X(b) - F_X(a).$$

Now, suppose that $B = [a, b]$ is a closed interval. Then using the continuity of probability (Theorem 1.6.1), we have

$$
\begin{aligned}
P(X \in B) &= P(a \leq X \leq b) = \lim_{n \to \infty} P(a - 1/n < X \leq b) \\
&= \lim_{n \to \infty} (F_X(b) - F_X(a - 1/n)) = F_X(b) - \lim_{n \to \infty} F_X(a - 1/n).
\end{aligned}
$$

We sometimes write $\lim_{n \to \infty} F_X(a - 1/n)$ as $F_X(a^-)$, so that $P(X \in [a, b]) = F_X(b) - F_X(a^-)$. In the special case where $a = b$, we have

$$P(X = a) = F_X(a) - F_X(a^-). \tag{2.5.1}$$

Similarly, if $B = (a, b)$ is an open interval, then

$$P(X \in B) = P(a < X < b) = \lim_{n \to \infty} F_X(b - 1/n) - F_X(a) = F_X(b^-) - F_X(a),$$

while if $B = [a, b)$ is a right-open interval, then

$$
\begin{aligned}
P(X \in B) &= P(a \leq X < b) = \lim_{n \to \infty} F_X(b - 1/n) - \lim_{n \to \infty} F_X(a - 1/n) \\
&= F_X(b^-) - F_X(a^-).
\end{aligned}
$$

We conclude that we can determine $P(X \in B)$ from F_X whenever B is any kind of interval.

Now, if B is instead a *union* of intervals, then we can use additivity to again compute $P(X \in B)$ from F_X. For example, if

$$B = (a_1, b_1] \cup (a_2, b_2] \cup \cdots \cup (a_k, b_k],$$

with $a_1 < b_1 < a_2 < b_2 < \cdots < a_k < b_k$, then by additivity,

$$
\begin{aligned}
P(X \in B) &= P(X \in (a_1, b_1]) + \cdots + P(X \in (a_k, b_k]) \\
&= F_X(b_1) - F_X(a_1) + \cdots + F_X(b_k) - F_X(a_k).
\end{aligned}
$$

Hence, we can still compute $P(X \in B)$ solely from the values of $F_X(x)$.

Theorem 2.5.1 Let X be any random variable, with cumulative distribution function F_X. Let B be any subset of the real numbers. Then $P(X \in B)$ can be determined solely from the values of $F_X(x)$.

Outline of Proof: It turns out that all relevant subsets B can be obtained by applying limiting operations to unions of intervals. Hence, because F_X determines $P(X \in B)$ when B is a union of intervals, it follows that F_X determines $P(X \in B)$ for all relevant subsets B. ∎

2.5.1 Properties of Distribution Functions

In light of Theorem 2.5.1, we see that cumulative distribution functions F_X are very useful. Thus, we note a few of their basic properties here.

Theorem 2.5.2 Let F_X be the cumulative distribution function of a random variable X. Then

(a) $0 \leq F_X(x) \leq 1$ for all x,
(b) $F_X(x) \leq F_X(y)$ whenever $x \leq y$ (i.e., F_X is increasing),
(c) $\lim_{x \to +\infty} F_X(x) = 1$,
(d) $\lim_{x \to -\infty} F_X(x) = 0$.

Proof: (a) Because $F_X(x) = P(X \leq x)$ is a probability, it is between 0 and 1.

(b) Let $A = \{X \leq x\}$ and $B = \{X \leq y\}$. Then if $x \leq y$, then $A \subseteq B$, so that $P(A) \leq P(B)$. But $P(A) = F_X(x)$ and $P(B) = F_X(y)$, so the result follows.

(c) Let $A_n = \{X \leq n\}$. Because X must take on *some* value and hence $X \leq n$ for sufficiently large n, we see that $\{A_n\}$ increases to S, i.e., $\{A_n\} \nearrow S$ (see Section 1.6). Hence, by continuity of P (Theorem 1.6.1), $\lim_{n \to \infty} P(A_n) = P(S) = 1$. But $P(A_n) = P(X \leq n) = F_X(n)$, so the result follows.

(d) Let $B_n = \{X \leq -n\}$. Because $X \geq -n$ for sufficiently large n, $\{B_n\}$ decreases to the empty set, i.e., $\{B_n\} \searrow \emptyset$. Hence, again by continuity of P, $\lim_{n \to \infty} P(B_n) = P(\emptyset) = 0$. But $P(B_n) = P(X \leq -n) = F_X(-n)$, so the result follows. ∎

If F_X is a cumulative distribution function, then F_X is also *right continuous*; see Problem 2.5.8. It turns out that if a function $F : R^1 \to R^1$ satisfies properties (a) through (d) and is right continuous, then there is a unique probability measure P on R^1 such that F is the cdf of P. We will not establish this result here.

2.5.2 Cdfs of Discrete Distributions

We can compute the cumulative distribution function F_X of a discrete random variable from its probability function p_X, as follows.

Theorem 2.5.3 Let X be a discrete random variable with probability function p_X. Then its cumulative distribution function F_X satisfies $F_X(x) = \sum_{y \le x} p_X(y)$.

Proof: Let $x_1, x_2, \ldots$ be the possible values of X. Then $F_X(x) = P(X \le x) = \sum_{x_i \le x} P(X = x_i) = \sum_{y \le x} P(X = y) = \sum_{y \le x} p_X(y)$, as claimed. ∎

Hence, if X is a discrete random variable, then by Theorem 2.5.3, F_X is piecewise constant, with a jump of size $p_X(x_i)$ at each value x_i. A plot of such a distribution looks like that depicted in Figure 2.5.1.

We consider an example of a distribution function of a discrete random variable.

Example 2.5.1
Consider flipping one fair six-sided die, so that $S = \{1, 2, 3, 4, 5, 6\}$, with $P(s) = 1/6$ for each $s \in S$. Let X be the number showing on the die divided by 6, so that $X(s) = s/6$ for $s \in S$. What is $F_X(x)$? We have that

$$F_X(x) = P(X \le x) = \sum_{s \in S, s \le x} P(s) = \sum_{s \in S, s \le x} \frac{1}{6} = \frac{1}{6} |\{s \in S : s \le x\}|.$$

That is, to compute $F_X(x)$, we count how many elements $s \in S$ satisfy $s/6 \le x$, and multiply that number by $1/6$. Therefore

$$F_X(x) = \begin{cases} 0 & x < 1/6 \\ 1/6 & 1/6 \le x < 2/6 \\ 2/6 & 2/6 \le x < 3/6 \\ 3/6 & 3/6 \le x < 4/6 \\ 4/6 & 4/6 \le x < 5/6 \\ 5/6 & 5/6 \le x < 1 \\ 6/6 & 1 \le x. \end{cases}$$

In Figure 2.5.1, we present a graph of the function F_X and note that this is a step function. Note (Exercise 2.5.1) that the properties of Theorem 2.5.2 are indeed satisfied by the function F_X. ∎

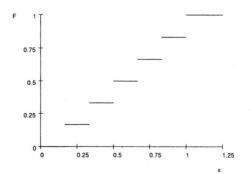

Figure 2.5.1: Graph of the cdf F_X in Example 2.5.1.

2.5.3   Cdfs of Absolutely Continuous Distributions

Once we know the density f_X of X, then it is easy to compute the cumulative distribution function of X, as follows.

Theorem 2.5.4 Let X be an absolutely continuous random variable, with density function f_X. Then the cumulative distribution function F_X of X satisfies

$$F_X(x) = \int_{-\infty}^x f_X(t)\, dt$$

for $x \in R^1$.

Proof: This follows immediately from (2.4.5), by setting $b = x$ and letting $a \to -\infty$. ∎

From the fundamental theorem of calculus, we see that it is also possible to compute a density f_X once we know the cumulative distribution function F_X.

Corollary 2.5.1 Let X be an absolutely continuous random variable, with cumulative distribution function F_X. Let

$$f_X(x) = \frac{d}{dx} F_X(x) = F_X'(x).$$

Then f_X is a density function for X.

We note that F_X might not be differentiable everywhere, so that the function f_X of the corollary might not be defined at certain isolated points. The density function may take any value at such points.

Consider again the $N(0,1)$ distribution, with density ϕ given by (2.4.9). According to Theorem 2.5.4, the cumulative distribution function F of this distribution is given by

$$F(x) = \int_{-\infty}^x \phi(t)\, dt = \int_{-\infty}^x \frac{1}{\sqrt{2\pi}} e^{-t^2/2}\, dt.$$

It turns out that it is provably impossible to evaluate this integral exactly, except for certain specific values of x (e.g., $x = -\infty$, $x = 0$, or $x = \infty$). Nevertheless, the cumulative distribution function of the $N(0,1)$ distribution is so important that it is assigned a special symbol. Furthermore, this is tabulated in Table D.2 of Appendix D for certain values of x.

Definition 2.5.2 The symbol Φ stands for the cumulative distribution function of a standard normal distribution, defined by

$$\Phi(x) = \int_{-\infty}^x \phi(t)\, dt = \int_{-\infty}^x \frac{1}{\sqrt{2\pi}} e^{-t^2/2}\, dt, \qquad (2.5.2)$$

for $x \in R^1$.

Example 2.5.2 *Normal Probability Calculations*

Suppose that $X \sim N(0, 1)$, and we want to calculate

$$P(-0.63 \le X \le 2.0) = P(X \le 2.0) - P(X \le -0.63).$$

Then $P(X \le 2) = \Phi(2)$, while $P(X \le -0.63) = \Phi(-0.63)$. Unfortunately, $\Phi(2)$ and $\Phi(-0.63)$ cannot be computed exactly, but they can be computed approximately using a computer to numerically calculate the integral (2.5.2). Virtually all statistical software packages will provide such approximations, but there are also available many tabulations such as Table D.2. Using this table we obtain $\Phi(2) = 0.9772$ while $\Phi(-0.63) = 0.2643$. This implies that

$$P(-0.63 \le X \le 2.0) = \Phi(2.0) - \Phi(-0.63) = 0.9772 - 0.2643 = 0.7129.$$

Now suppose that $X \sim N(\mu, \sigma^2)$, and we want to calculate $P(a \le X \le b)$. Letting f denote the density of X and following Example 2.4.8, we have

$$P(a \le X \le b) = \int_a^b f(x)\, dx = \int_a^b \frac{1}{\sigma} \phi\left(\frac{x - \mu}{\sigma}\right) dx.$$

Then, again following Example 2.4.8, we make the substitution $y = (x - \mu)/\sigma$ in the above integral to obtain

$$P(a \le X \le b) = \int_{\frac{a-\mu}{\sigma}}^{\frac{b-\mu}{\sigma}} \phi(x)\, dx = \Phi\left(\frac{b - \mu}{\sigma}\right) - \Phi\left(\frac{a - \mu}{\sigma}\right).$$

Therefore, general normal probabilities can be computed using the function Φ.

Suppose now that $a = -0.63, b = 2.0, \mu = 1.3$ and $\sigma^2 = 4$. We obtain

$$
\begin{aligned}
P(-0.63 \le X \le 2.0) &= \Phi\left(\frac{2.0 - 1.3}{2}\right) - \Phi\left(\frac{-0.63 - 1.3}{2}\right) \\
&= \Phi(0.35) - \Phi(-0.965) = 0.6368 - 0.16725 \\
&= 0.46955
\end{aligned}
$$

because, using Table D.2, $\Phi(0.35) = 0.6368$ and we approximate $\Phi(-0.965)$ by the linear interpolation between the values $\Phi(-0.96) = 0.1685, \Phi(-0.97) = 0.1660$ given by

$$
\begin{aligned}
\Phi(-0.965) &\approx \Phi(-0.96) + \frac{\Phi(-0.97) - \Phi(-0.96)}{-0.97 - (-0.96)}(-0.965 - (-0.96)) \\
&= 0.1685 + \frac{0.1660 - 0.1685}{-0.97 - (-0.96)}(-0.965 - (-0.96)) = 0.16725. \ \blacksquare
\end{aligned}
$$

Example 2.5.3

Let X be a random variable with cumulative distribution function given by

$$
F_X(x) = \begin{cases} 0 & x < 2 \\ (x - 2)^4/16 & 2 \le x < 4 \\ 1 & 4 \le x. \end{cases}
$$

In Figure 2.5.2 we present a graph of F_X.

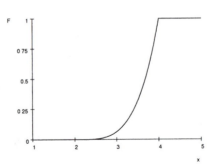

Figure 2.5.2: Graph of the cdf F_X in Example 2.5.3.

Suppose for this random variable X we want to compute $P(X \le 3)$, $P(X < 3)$, $P(X > 2.5)$ and $P(1.2 < X \le 3.4)$. We can compute all of these probabilities directly from F_X. We have that

$$
\begin{aligned}
P(X \le 3) &= F_X(3) = (3 - 2)^4/16 = 1/16, \\
P(X < 3) &= F_X(3^-) = \lim_{n \to \infty} (3 - (1/n) - 2)^4/16 = 1/16, \\
P(X > 2.5) &= 1 - P(X \le 2.5) = 1 - F_X(2.5) \\
&= 1 - (2.5 - 2)^4/16 = 1 - 0.0625/16 = 0.996, \\
P(1.2 < X \le 3.4) &= F_X(3.4) - F_X(1.2) = (3.4 - 2)^4/16 - 0 = 0.2401. \quad \blacksquare
\end{aligned}
$$

2.5.4   Mixture Distributions

Suppose now that $F_1, F_2, \ldots, F_k$ are cumulative distribution functions, corresponding to various distributions. Also let $p_1, p_2, \ldots, p_k$ be positive real numbers with $\sum_{i=1}^k p_i = 1$ (so these values form a probability distribution). Then we can define a new function G by

$$
G(x) = p_1 F_1(x) + p_2 F_2(x) + \cdots + p_k F_k(x). \tag{2.5.3}
$$

It is easily verified (Exercise 2.5.6) that the function G given by (2.5.3) will satisfy properties (a) through (d) of Theorem 2.5.2 and is right continuous. Hence, G is also a cdf.

The distribution whose cdf is given by (2.5.3) is called a *mixture distribution*, because it *mixes* the various distributions with cdfs $F_1, \ldots, F_k$ according to the probability distribution given by the $p_1, p_2, \ldots, p_k$.

To see how a mixture distribution arises in applications, consider a two-stage system as discussed in Section 1.5.1. Let Z be a random variable describing the outcome of the first stage and such that $P(Z = i) = p_i$ for $i = 1, 2, \ldots, k$. Suppose that for the second stage, we observe a random variable Y where the distribution of Y depends on the outcome of the first stage, so that Y has cdf F_i when $Z = i$. In effect F_i is the conditional distribution of Y given that $Z = i$

(see Section 2.8). Then, by the law of total probability (Theorem 1.5.1), the distribution function of Y is given by

$$P(Y \le y) = \sum_{i=1}^{k} P(Y \le y \,|\, Z = i) P(Z = i) = \sum_{i=1}^{k} p_i F_i(y) = G(y).$$

Therefore the distribution function of Y is given by a mixture of the F_i.

Consider the following example of this.

Example 2.5.4

Suppose we have two bowls containing chips. Bowl #1 contains one chip labelled 0, two chips labelled 3, and one chip labelled 5. Bowl #2 contains one chip labelled 2, one chip labelled 4, and one chip labelled 5. Now let X_i be the random variable corresponding to randomly drawing a chip from bowl #i. Therefore $P(X_1 = 0) = 1/4$, $P(X_1 = 3) = 1/2$, and $P(X_1 = 5) = 1/4$, while $P(X_2 = 2) = P(X_2 = 4) = P(X_2 = 5) = 1/3$. Then X_1 has distribution function given by

$$F_1(x) = \begin{cases} 0 & x < 0 \\ 1/4 & 0 \le x < 3 \\ 3/4 & 3 \le x < 5 \\ 1 & x \ge 5 \end{cases}$$

and X_2 has distribution function given by

$$F_2(x) = \begin{cases} 0 & x < 2 \\ 1/3 & 2 \le x < 4 \\ 2/3 & 4 \le x < 5 \\ 1 & x \ge 5. \end{cases}$$

Now suppose that we choose a bowl by randomly selecting a card from a deck of five cards where one card is labelled 1 and four cards are labelled 2. Let Z denote the value on the card obtained, so that $P(Z = 1) = 1/5$ and $P(Z = 2) = 4/5$. Then, having obtained the value $Z = i$, we observe Y by randomly drawing a chip from bowl #i. We see immediately that the cdf of Y is given by

$$G(x) = (1/5) F_1(x) + (4/5) F_2(x),$$

and this is a mixture of the cdfs F_1 and F_2. ∎

As the following examples illustrate, it is also possible to have *infinite* mixtures of distributions.

Example 2.5.5 *Location and Scale Mixtures*

Suppose F is some cumulative distribution function. Then for any real number y, the function F_y defined by $F_y(x) = F(x - y)$ is also a cumulative distribution function. In fact, F_y is just a "shifted" version of F. An example of this is depicted in Figure 2.5.3.

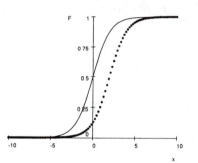

Figure 2.5.3: Plot of the distribution functions F (—) and F_2 (• • •) in Example 2.5.5, where $F(x) = e^x / (e^x + 1)$ for $x \in R^1$.

If $p_i \geq 0$ with $\sum_i p_i = 1$ (so the p_i form a probability distribution), and $y_1, y_2, \ldots$ are real numbers, then we can define a *discrete location mixture* by

$$H(x) = \sum_i p_i F_{y_i}(x) = \sum_i p_i F(x - y_i).$$

Indeed, the shift $F_y(x) = F(x - y)$ itself corresponds to a special case of a discrete location mixture, with $p_1 = 1$ and $y_1 = y$.

Further, if g is some nonnegative function with $\int_{-\infty}^{\infty} g(t)\, dt = 1$ (so g is a density function), then we can define

$$H(x) = \int_{-\infty}^{\infty} F_y(x)\, g(y)\, dy = \int_{-\infty}^{\infty} F(x - y)\, g(y)\, dy.$$

Then it is not hard to see that H is also a cumulative distribution function. The distribution function H is called a *continuous location mixture* of F. The idea is that H corresponds to a *mixture* of different shifted distributions F_y, with the density g giving the distribution of the mixing coefficient y.

We can also define a *discrete scale mixture* by

$$K(x) = \sum_i p_i F(x/y_i),$$

whenever $y_i > 0$, $p_i \geq 0$, and $\sum_i p_i = 1$. Similarly, if $\int_0^{\infty} g(t)\, dt = 1$, then we can put

$$K(x) = \int_0^{\infty} F(x/y) g(y)\, dy.$$

Then K is also a cumulative distribution function, called a *continuous scale mixture* of F. ∎

You might wonder at this point if a mixture distribution is discrete or continuous. This depends on the distributions being mixed and the mixing distribution. For example, discrete location mixtures of discrete distributions are discrete and discrete location mixtures of continuous distributions are continuous.

There is nothing restricting us, however, to mixing only discrete distributions or only continuous distributions. Other kinds of distribution are considered in the following section.

2.5.5 Distributions Neither Discrete Nor Continuous

There are some distributions that are neither discrete nor continuous, as the following example shows.

Example 2.5.6

Suppose that $X_1 \sim \text{Poisson}(3)$ is discrete with cdf F_1, while $X_2 \sim N(0,1)$ is continuous with cdf F_2 and Y has the mixture distribution given by $F_Y(y) = (1/5)\,F_1(y) + (4/5)\,F_2(y)$. Using (2.5.1), we have

$$
\begin{aligned}
P(Y = y) &= F_Y(y) - F_Y(y^-) \\
&= (1/5)F_1(y) + (4/5)\,F_2(y) - (1/5)F_1(y^-) - (4/5)F_2(y^-) \\
&= (1/5)\left(F_1(y) - F_1(y^-)\right) + (4/5)\left(F_2(y) - F_2(y^-)\right) \\
&= \frac{1}{5}P(X_1 = y) + \frac{4}{5}P(X_2 = y).
\end{aligned}
$$

Therefore,

$$
P(Y = y) = \begin{cases} \frac{1}{5}\frac{3^y}{y!}e^{-3} & y \text{ a nonnegative integer} \\ 0 & \text{otherwise.} \end{cases}
$$

Because $P(Y = y) > 0$ for nonnegative integers y, the random variable Y is not continuous. On the other hand, we have

$$
\sum_y P(Y = y) = \sum_{y=0}^{\infty} \frac{1}{5}\frac{3^y}{y!}e^{-3} = \frac{1}{5} < 1.
$$

Hence, Y is not discrete either.

In fact, Y is neither discrete nor continuous. Rather, Y is a mixture of a discrete and a continuous distribution. ∎

For the most part in this book, we shall treat discrete and continuous distributions separately. However, it is important to keep in mind that actual distributions may not be discrete or continuous but may be a mixture of the two. In most applications, however, the distributions we deal with are either continuous or discrete.

Recall that a continuous distribution need not be absolutely continuous, i.e., have a density. Hence, a distribution which is a mixture of a discrete and a continuous distribution might not be a mixture of a discrete and an absolutely continuous distribution.

Summary of Section 2.5

- The cumulative distribution function (cdf) of X is $F_X(x) = P(X \leq x)$.

- All probabilities associated with X can be determined from F_X.

- As x increases from $-\infty$ to ∞, $F_X(x)$ increases from 0 to 1.

- If X is discrete, then $F_X(x) = \sum_{y \leq x} P(X = y)$.

- If X is absolutely continuous, then $F_X(x) = \int_{-\infty}^{x} f_X(t)\, dt$, and $f_X(x) = F_X'(x)$.

- We write $\Phi(x)$ for the cdf of the standard normal distribution evaluated at x.

- A mixture distribution has a cdf which is a linear combination of other cdfs. Two special cases are location and scale mixtures.

- Some mixture distributions are neither discrete nor continuous.

Exercises

2.5.1 Verify explicitly that properties (a) through (d) of Theorem 2.5.2 are indeed satisfied by the function F_X in Example 2.5.1.

2.5.2 Consider flipping one fair six-sided die, so that $S = \{1, 2, 3, 4, 5, 6\}$, and $P(s) = 1/6$ for all $s \in S$. Let X be the number showing on the die, so that $X(s) = s$ for $s \in S$. Let $Y = X^2$. Compute the cumulative distribution function $F_Y(y) = P(Y \leq y)$, for all $y \in R^1$. Verify explicitly that properties (a) through (d) of Theorem 2.5.2 are satisfied by this function F_Y.

2.5.3 For each of the following functions F, determine whether or not F is a valid cumulative distribution function, i.e., whether or not F satisfies properties (a) through (d) of Theorem 2.5.2.
(a) $F(x) = x$ for all $x \in R^1$
(b)

$$F(x) = \begin{cases} 0 & x < 0 \\ x & 0 \leq x \leq 1 \\ 1 & x > 1 \end{cases}$$

(c)

$$F(x) = \begin{cases} 0 & x < 0 \\ x^2 & 0 \leq x \leq 1 \\ 1 & x > 1 \end{cases}$$

(d)

$$F(x) = \begin{cases} 0 & x < 0 \\ x^2 & 0 \leq x \leq 3 \\ 1 & x > 3 \end{cases}$$

(e)

$$F(x) = \begin{cases} 0 & x < 0 \\ x^2/9 & 0 \le x \le 3 \\ 1 & x > 3 \end{cases}$$

(f)

$$F(x) = \begin{cases} 0 & x < 1 \\ x^2/9 & 1 \le x \le 3 \\ 1 & x > 3 \end{cases}$$

(g)

$$F(x) = \begin{cases} 0 & x < -1 \\ x^2/9 & -1 \le x \le 3 \\ 1 & x > 3 \end{cases}$$

2.5.4 Let $X \sim N(0,1)$. Compute each of the following in terms of the function Φ of Definition 2.5.2 and use Table D.2 (or software) to evaluate these probabilities numerically.
(a) $P(X \le -5)$
(b) $P(-2 \le X \le 7)$
(c) $P(X \ge 3)$

2.5.5 Let $Y \sim N(-8,4)$. Compute each of the following, in terms of the function Φ of Definition 2.5.2 and use Table D.2 (or software) to evaluate these probabilities numerically.
(a) $P(Y \le -5)$
(b) $P(-2 \le Y \le 7)$
(c) $P(Y \ge 3)$

2.5.6 Verify that the function G given by (2.5.3) satisfies properties (a) through (d) of Theorem 2.5.2.

Problems

2.5.7 Let F be a cumulative distribution function. Compute (with explanation) the value of $\lim_{n \to \infty}[F(2n) - F(n)]$.

2.5.8 Let F be a cumulative distribution function. For $x \in R^1$, we could define $F(x^+)$ by $F(x^+) = \lim_{n \to \infty} F(x + \frac{1}{n})$. Prove that F is *right-continuous*, meaning that for each $x \in R^1$, we have $F(x^+) = F(x)$. (Hint: You will need to use continuity of P (Theorem 1.6.1).)

2.5.9 Let X be a random variable, with cumulative distribution function F_X. Prove that $P(X = a) = 0$ if and only if the function F_X is continuous at a. (Hint: Use (2.5.1) and the previous exercise.)

2.5.10 Let Φ be as in Definition 2.5.4. Derive a formula for $\Phi(-x)$ in terms of $\Phi(x)$. (Hint: Let $s = -t$ in (2.5.2), and don't forget Theorem 2.5.2.)

2.5.11 Determine the distribution function for the logistic distribution of Problem 2.4.10.

2.5.12 Determine the distribution function for the Weibull(α) distribution of Problem 2.4.11.

2.5.13 Determine the distribution function for the Pareto(α) distribution of Problem 2.4.12.

2.5.14 Determine the distribution function for the Cauchy distribution of Problem 2.4.13.

2.5.15 Determine the distribution function for the Laplace distribution of Problem 2.4.14.

2.5.16 Determine the distribution function for the extreme value distribution of Problem 2.4.15.

2.5.17 Determine the distribution function for the beta distributions of Problem 2.4.16 for parts (b) through (e).

Discussion Topics

2.5.18 Does it surprise you that all information about the distribution of a random variable X can be stored by a single function F_X? Why or why not? What other examples can you think of where lots of different information is stored by a single function?

2.6   One-Dimensional Change of Variable

Let X be a random variable with a known distribution. Suppose that $Y = h(X)$, where $h : R^1 \rightarrow R^1$ is some function. (Recall that this really means that $Y(s) = h(X(s))$, for all $s \in S$.) Then what is the distribution of Y?

2.6.1   The Discrete Case

If X is a *discrete* random variable, this is quite straightforward. To compute the probability that $Y = y$, we need to compute the probability of the set consisting of all the x values satisfying $h(x) = y$, namely, compute $P(X \in \{x : h(x) = y\})$. This is depicted graphically in Figure 2.6.1.

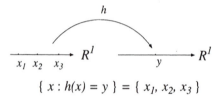

$$\{\, x : h(x) = y \,\} = \{\, x_1, x_2, x_3 \,\}$$

Figure 2.6.1: An example where the set of x values that satisfy $h(x) = y$ consists of three points x_1, x_2, and x_3.

We now establish the basic result.

Theorem 2.6.1 Let X be a discrete random variable, with probability function p_X. Let $Y = h(X)$, where $h : R^1 \to R^1$ is some function. Then Y is also discrete, and its probability function p_Y satisfies $p_Y(y) = \sum_{x \in h^{-1}\{y\}} p_X(x)$, where $h^{-1}\{y\}$ is the set of all real numbers x with $h(x) = y$.

Proof: We compute that $p_Y(y) = P(h(X) = y) = \sum_{x \in h^{-1}\{y\}} P(X = x) = \sum_{x \in h^{-1}\{y\}} p_X(x)$, as claimed. ∎

Example 2.6.1
Let X be the number of heads when flipping three fair coins. Let $Y = 1$ if $X \geq 1$, with $Y = 0$ if $X = 0$. Then $Y = h(X)$ where $h(0) = 0$ and $h(1) = h(2) = h(3) = 1$. Hence, $h^{-1}\{0\} = \{0\}$, so $P(Y = 0) = P(X = 0) = 1/8$. On the other hand, $h^{-1}\{1\} = \{1, 2, 3\}$, so $P(Y = 1) = P(X = 1) + P(X = 2) + P(X = 3) = 3/8 + 3/8 + 1/8 = 7/8$. ∎

Example 2.6.2
Let X be the number showing on a fair six-sided die, so that $P(X = x) = 1/6$ for $x = 1, 2, 3, 4, 5$, and 6. Let $Y = X^2 - 3X + 2$. Then $Y = h(X)$ where $h(x) = x^2 - 3x + 2$. Note that $h(x) = 0$ if and only if $x = 1$ or $x = 2$. Hence, $h^{-1}\{0\} = \{1, 2\}$, and

$$P(Y = 0) = p_X(1) + p_X(2) = \frac{1}{6} + \frac{1}{6} = \frac{1}{3}. \quad ∎$$

2.6.2 The Continuous Case

If X is *continuous* and $Y = h(X)$, then the situation is more complicated. Indeed, Y might not be continuous at all, as the following example shows.

Example 2.6.3
Let X have the uniform distribution on $[0, 1]$, i.e., $X \sim \text{Uniform}[0, 1]$, as in Example 2.4.2. Let $Y = h(X)$, where

$$h(x) = \begin{cases} 7 & x \leq 3/4 \\ 5 & x > 3/4 \end{cases}$$

Here $Y = 7$ if and only if $X \leq 3/4$ (which happens with probability 3/4), while $Y = 5$ if and only if $X > 3/4$ (which happens with probability 1/4). Hence, Y is discrete, with probability function p_Y satisfying $p_Y(7) = 3/4, p_Y(5) = 1/4$, and $p_Y(y) = 0$ when $y \neq 5, 7$. ∎

On the other hand, if X is absolutely continuous, and the function h is *strictly increasing*, then the situation is considerably simpler, as the following theorem shows.

Theorem 2.6.2 Let X be an absolutely continuous random variable, with density function f_X. Let $Y = h(X)$, where $h : R^1 \to R^1$ is a function that is differentiable and strictly increasing. Then Y is also absolutely continuous, and its density function f_Y is given by

$$f_Y(y) = f_X(h^{-1}(y)) \, / \, |h'(h^{-1}(y))|, \qquad (2.6.1)$$

where h' is the derivative of h, and where $h^{-1}(y)$ is the unique number x such that $h(x) = y$.

Proof: See Section 2.11 for the proof of this result. ∎

Example 2.6.4
Let $X \sim \text{Uniform}[0, 1]$, and let $Y = 3X$. What is the distribution of Y?
  Here X has density f_X given by $f_X(x) = 1$ if $0 \le x \le 1$, and $f_X(x) = 0$ otherwise. Also, $Y = h(X)$, where h is defined by $h(x) = 3x$. Note that h is strictly increasing, because if $x < y$, then $3x < 3y$, i.e., $h(x) < h(y)$. Hence, we may apply Theorem 2.6.2.
  We note first that $h'(x) = 3$ and that $h^{-1}(y) = y/3$. Then, according to Theorem 2.6.2, Y is absolutely continuous with density

$$
\begin{aligned}
f_Y(y) &= f_X(h^{-1}(y))/|h'(h^{-1}(y))| = \frac{1}{3} f_X(y/3) \\
&= \begin{cases} 1/3 & 0 \le y/3 \le 1 \\ 0 & \text{otherwise} \end{cases} = \begin{cases} 1/3 & 0 \le y \le 3 \\ 0 & \text{otherwise.} \end{cases}
\end{aligned}
$$

  By comparison with Example 2.4.3, we see that $Y \sim \text{Uniform}[0, 3]$, i.e., that Y has the Uniform$[L, R]$ distribution with $L = 0$ and $R = 3$. ∎

Example 2.6.5
Let $X \sim N(0, 1)$, and let $Y = 2X + 5$. What is the distribution of Y?
  Here X has density f_X given by

$$f_X(x) = \phi(x) = \frac{1}{\sqrt{2\pi}} e^{-x^2/2}.$$

Also, $Y = h(X)$, where h is defined by $h(x) = 2x + 5$. Note that again h is strictly increasing, because if $x < y$, then $2x + 5 < 2y + 5$, i.e., $h(x) < h(y)$. Hence, we may again apply Theorem 2.6.2.
  We note first that $h'(x) = 2$ and that $h^{-1}(y) = (y - 5)/2$. Then, according to Theorem 2.6.2, Y is absolutely continuous with density

$$f_Y(y) = f_X(h^{-1}(y))/|h'(h^{-1}(y))| = f_X((y-5)/2)/2 = \frac{1}{2\sqrt{2\pi}} e^{(y-5)^2/8}.$$

  By comparison with Example 2.4.8, we see that $Y \sim N(5, 4)$, i.e., that Y has the $N(\mu, \sigma^2)$ distribution with $\mu = 5$ and $\sigma^2 = 4$. ∎

  If instead the function h is strictly *decreasing*, then a similar result holds.

Theorem 2.6.3 Let X be an absolutely continuous random variable, with density function f_X. Let $Y = h(X)$, where $h : R^1 \to R^1$ is a function that is differentiable and strictly decreasing. Then Y is also absolutely continuous, and its density function f_Y may again be defined by (2.6.1).

Proof: See Section 2.11 for the proof of this result. ∎

Example 2.6.6

Let $X \sim \text{Uniform}[0, 1]$, and let $Y = \ln(1/X)$. What is the distribution of Y?

Here X has density f_X given by $f_X(x) = 1$ for $0 \leq x \leq 1$, and $f_X(x) = 0$ otherwise. Also, $Y = h(X)$, where h is defined by $h(x) = \ln(1/x)$. Note that here h is strictly decreasing, because if $x < y$, then $1/x > 1/y$, so $\ln(1/x) > \ln(1/y)$, i.e., $h(x) > h(y)$. Hence, we may apply Theorem 2.6.3.

We note first that $h'(x) = -1/x$ and that $h^{-1}(y) = e^{-y}$. Then, by Theorem 2.6.3, Y is absolutely continuous with density

$$
\begin{aligned}
f_Y(y) &= f_X(h^{-1}(y))/|h'(h^{-1}(y))| = e^{-y} f_X(e^{-y}) \\
&= \begin{cases} e^{-y} & 0 \leq e^{-y} \leq 1 \\ 0 & \text{otherwise} \end{cases} = \begin{cases} e^{-y} & y \geq 0 \\ 0 & \text{otherwise.} \end{cases}
\end{aligned}
$$

By comparison with Example 2.4.4, we see that $Y \sim \text{Exponential}(1)$, i.e., that Y has the Exponential(1) distribution. ∎

Finally, we note the following.

Theorem 2.6.4 Theorem 2.6.2 (and 2.6.3) remains true assuming only that h is strictly increasing (or decreasing) at places for which $f_X(x) > 0$. If $f_X(x) = 0$ for an interval of x values, then it does not matter how the function h behaves in that interval (or even if it is well-defined there).

Example 2.6.7

If $X \sim \text{Exponential}(\lambda)$, then $f_X(x) = 0$ for $x < 0$. Therefore, it is required that h be strictly increasing (or decreasing) only for $x \geq 0$. Thus, functions such as $h(x) = x^2$, $h(x) = x^8$, and $h(x) = \sqrt{x}$ could still be used with Theorem 2.6.2, while functions such as $h(x) = -x^2$, $h(x) = -x^8$, and $h(x) = -\sqrt{x}$ could still be used with Theorem 2.6.3, even though such functions may not necessarily be strictly increasing (or decreasing) and well-defined on the entire real line. ∎

Summary of Section 2.6

- If X is discrete, and $Y = h(X)$, then $P(Y = y) = \sum_{x:\, h(x)=y} P(X = x)$.

- If X is absolutely continuous, and $Y = h(X)$ with h strictly increasing or strictly decreasing, then the density of Y is given by $f_Y(y) = f_X(h^{-1}(y))\,/\,|h'(h^{-1}(y))|$.

- This allows us to compute the distribution of a function of a random variable.

Exercises

2.6.1 Let $X \sim \text{Uniform}[L, R]$. Let $Y = cX + d$, where $c > 0$. Prove that $Y \sim \text{Uniform}[cL + d, cR + d]$. (This generalizes Example 2.6.4.)

2.6.2 Let $X \sim \text{Uniform}[L, R]$. Let $Y = cX + d$, where $c < 0$. Prove that $Y \sim \text{Uniform}[cR + d, cL + d]$. (In particular, if $L = 0$ and $R = 1$ and $c = -1$ and $d = 1$, then $X \sim \text{Uniform}[0, 1]$ and also $Y = 1 - X \sim \text{Uniform}[0, 1]$.)

2.6.3 Let $X \sim N(\mu, \sigma^2)$. Let $Y = cX + d$, where $c > 0$. Prove that $Y \sim N(c\mu + d, c^2\sigma^2)$. (This generalizes Example 2.6.5.)

2.6.4 Let $X \sim \text{Exponential}(\lambda)$. Let $Y = cX$, where $c > 0$. Prove that $Y \sim \text{Exponential}(\lambda/c)$.

2.6.5 Let $X \sim \text{Exponential}(\lambda)$. Let $Y = X^3$. Compute the density f_Y of Y.

2.6.6 Let $X \sim \text{Exponential}(\lambda)$. Let $Y = X^{1/4}$. Compute the density f_Y of Y. (Hint: Use Theorem 2.6.4.)

2.6.7 Let $X \sim \text{Uniform}[0, 3]$. Let $Y = X^2$. Compute the density function f_Y of Y.

2.6.8 Let X have a density such that $f_X(x - \mu) = f_X(\mu - x)$, i.e., it is symmetric about μ. Let $Y = 2\mu - X$. Show that the density of Y is given by f_X. Use this to determine the distribution of Y when $X \sim N(\mu, \sigma^2)$.

Problems

2.6.9 Let $X \sim \text{Uniform}[2, 7]$, $Y = X^3$ and $Z = \sqrt{Y}$. Compute the density f_Z of Z, in two ways.
(a) Apply Theorem 2.6.2 first to obtain the density of Y, and then apply Theorem 2.6.2 *again* to obtain the density of Z.
(b) Observe that $Z = \sqrt{Y} = \sqrt{X^3} = X^{3/2}$, and apply Theorem 2.6.2 just *once*.

2.6.10 Let $X \sim \text{Uniform}[L, R]$, and let $Y = h(X)$ where $h(x) = (x - c)^6$. According to Theorem 2.6.4, under what conditions on L, R, and c can we apply Theorem 2.6.2 or Theorem 2.6.3 to this choice of X and Y?

2.6.11 Let $X \sim N(\mu, \sigma^2)$. Let $Y = cX + d$, where $c < 0$. Prove that again $Y \sim N(c\mu + d, c^2\sigma^2)$, just like in Exercise 2.6.3.

2.6.12 (*Log-normal(τ) distribution*) Suppose that $X \sim N(0, \tau^2)$. Prove that $Y = e^X$ has density

$$f_\tau(y) = \frac{1}{\sqrt{2\pi}\tau} \exp\left(-\frac{(\ln y)^2}{2\tau^2}\right) \frac{1}{y}$$

for $y > 0$ and where $\tau > 0$ is unknown. We say that $Y \sim \text{Log-normal}(\tau)$.

2.6.13 Suppose that $X \sim \text{Weibull}(\alpha)$ (Problem 2.4.11). Determine the distribution of $Y = X^\beta$.

2.6.14 Suppose that $X \sim \text{Pareto}(\alpha)$ (Problem 2.4.12). Determine the distribution of $Y = (1 + X)^{\beta} - 1$.

2.6.15 Suppose that X has the extreme value distribution (Problem 2.4.15). Determine the distribution of $Y = e^{-X}$.

Challenge

2.6.16 Theorems 2.6.2 and 2.6.3 require that h be an increasing or decreasing function, at least at places where the density of X is positive (see Theorem 2.6.4). Suppose now that $X \sim N(0, 1)$ and $Y = h(X)$, where $h(x) = x^2$. Then $f_X(x) > 0$ for all x, while h is increasing only for $x > 0$ and decreasing only for $x < 0$, hence, Theorems 2.6.2 and 2.6.3 do not directly apply. Compute $f_Y(y)$ anyway. (Hint: $P(a \leq Y \leq b) = P(a \leq Y \leq b, \ X > 0) + P(a \leq Y \leq b, \ X < 0)$.)

2.7 Joint Distributions

Suppose X and Y are two random variables. Even if we know the distributions of X and Y exactly, this still does not tell us anything about the *relationship* between X and Y.

Example 2.7.1
Let $X \sim \text{Bernoulli}(1/2)$, so that $P(X = 0) = P(X = 1) = 1/2$. Let $Y_1 = X$, and let $Y_2 = 1 - X$. Then we clearly have $Y_1 \sim \text{Bernoulli}(1/2)$ and $Y_2 \sim \text{Bernoulli}(1/2)$ as well.

On the other hand, the *relationship* between X and Y_1 is very different from the relationship between X and Y_2. For example, if we know that $X = 1$, then we also must have $Y_1 = 1$, but $Y_2 = 0$. Hence, merely knowing that X, Y_1, and Y_2 all have the distribution Bernoulli$(1/2)$, does not give us complete information about the relationships among these random variables. ∎

A formal definition of joint distribution is as follows.

Definition 2.7.1 If X and Y are random variables, then the *joint distribution* of X and Y is the collection of probabilities $P((X, Y) \in B)$, for all subsets $B \subseteq R^2$ of pairs of real numbers.

Joint distributions, like other distributions, are so complicated that we use various functions to describe them, including joint cumulative distribution functions, joint probability functions, and joint density functions, as we now discuss.

2.7.1 Joint Cumulative Distribution Functions

Definition 2.7.2 Let X and Y be random variables. Then their *joint cumulative distribution function* is the function $F_{X,Y} : R^2 \to R^1$ defined by

$$F_{X,Y}(x, y) = P(X \leq x, \ Y \leq y).$$

(Recall that the comma means "and" here, so that $F_{X,Y}(x, y)$ is the probability that $X \leq x$ *and* $Y \leq y$.)

Example 2.7.2 (*Example 2.7.1 continued*)
Again let $X \sim \text{Bernoulli}(1/2)$, $Y_1 = X$, and $Y_2 = 1 - X$. Then we compute that

$$F_{X,Y_1}(x,y) = P(X \le x, Y_1 \le y) = \begin{cases} 0 & \min(x,y) < 0 \\ 1/2 & 0 \le \min(x,y) < 1 \\ 1 & \min(x,y) \ge 1. \end{cases}$$

On the other hand,

$$F_{X,Y_2}(x,y) = P(X \le x, Y_2 \le y) = \begin{cases} 0 & \min(x,y) < 0 \text{ or } \max(x,y) < 1 \\ 1/2 & 0 \le \min(x,y) < 1 \le \max(x,y) \\ 1 & \min(x,y) \ge 1. \end{cases}$$

We thus see that F_{X,Y_1} is quite a different function from F_{X,Y_2}. This reflects the fact that, even though Y_1 and Y_2 each have the same distribution, their relationship with X is quite different. On the other hand, the functions F_{X,Y_1} and F_{X,Y_2} are rather cumbersome and awkward to work with. ∎

We see from this example that joint cumulative distribution functions (or joint cdfs) do indeed keep track of the relationship between X and Y. Indeed, joint cdfs tell us everything about the joint probabilities of X and Y, as the following theorem (an analog of Theorem 2.3.1) shows.

Theorem 2.7.1 Let X and Y be any random variables, with joint cumulative distribution function $F_{X,Y}$. Let B be a subset of R^2. Then $P((X,Y) \in B)$ can be determined solely from the values of $F_{X,Y}(x,y)$.

We shall not give a proof of Theorem 2.7.1, although it is similar to the proof of Theorem 2.5.1. However, the following theorem indicates why Theorem 2.7.1 is true, and it also provides a useful computational fact.

Theorem 2.7.2 Let X and Y be any random variables, with joint cumulative distribution function $F_{X,Y}$. Suppose $a \le b$ and $c \le d$. Then

$$P(a < X \le b, \ c < Y \le d) = F_{X,Y}(b,d) - F_{X,Y}(a,d) - F_{X,Y}(b,c) + F_{X,Y}(a,c).$$

Proof: According to (1.3.3),

$$P(a < X \le b, \ c < Y \le d)$$
$$= P(X \le b, \ Y \le d) - P(X \le b, \ Y \le d, \text{and either } X \le a \text{ or } Y \le c).$$

But by the principle of inclusion-exclusion (1.3.4),

$$P(X \le b, \ Y \le d, \text{ and either } X \le a \text{ or } Y \le c)$$
$$= P(X \le b, \ Y \le c) + P(X \le a, \ Y \le d) - P(X \le a, \ Y \le c).$$

Combining these two equations, we see that

$$P(a < X \le b, c < Y \le d)$$
$$= P(X \le b, Y \le d) - P(X \le b, Y \le c) - P(X \le a, Y \le d) + P(X \le a, Y \le c)$$

and from this we obtain

$$P(a < X \le b, c < Y \le d) = F_{X,Y}(b,d) - F_{X,Y}(b,c) - F_{X,Y}(a,d) + F_{X,Y}(a,c),$$

as claimed. ∎

Joint cdfs are not easy to work with. Thus, in this section we shall also consider other functions, which are more convenient for pairs of discrete or absolutely continuous random variables.

2.7.2   Marginal Distributions

We have seen how a joint cumulative distribution function $F_{X,Y}$ tells us about the relationship between X and Y. However, the function $F_{X,Y}$ also tells us everything about each of X and Y separately, as the following theorem shows.

Theorem 2.7.3 Let X and Y be two random variables, with joint cumulative distribution function $F_{X,Y}$. Then the cumulative distribution function F_X of X satisfies

$$F_X(x) = \lim_{y \to \infty} F_{X,Y}(x,y),$$

for all $x \in R^1$. Similarly, the cumulative distribution function F_Y of Y satisfies

$$F_Y(y) = \lim_{x \to \infty} F_{X,Y}(x,y),$$

for all $y \in R^1$

Proof: Note that we *always* have $Y \le \infty$. Hence, using continuity of P, we have

$$
\begin{aligned}
F_X(x) &= P(X \le x) = P(X \le x,\ Y \le \infty) \\
&= \lim_{y \to \infty} P(X \le x,\ Y \le y) = \lim_{y \to \infty} F_{X,Y}(x,y),
\end{aligned}
$$

as claimed. Similarly,

$$
\begin{aligned}
F_Y(y) &= P(Y \le y) = P(X \le \infty,\ Y \le y) \\
&= \lim_{x \to \infty} P(X \le x,\ Y \le y) = \lim_{x \to \infty} F_{X,Y}(x,y),
\end{aligned}
$$

completing the proof. ∎

In the context of Theorem 2.7.3, F_X is called the *marginal cumulative distribution function* of X, and the distribution of X is called the *marginal distribution* of X. (Similarly, F_Y is called the marginal cumulative distribution function of Y, and the distribution of Y is called the marginal distribution of Y.) Intuitively, if we think of $F_{X,Y}$ as being a function of a pair (x,y), then F_X and F_Y are functions of x and y, respectively, which could be written into the "margins" of a graph of $F_{X,Y}$.

Example 2.7.3

In Figure 2.7.1 we have plotted the joint distribution function

$$F_{X,Y}(x,y) = \begin{cases} 0 & x < 0 \text{ or } y < 0 \\ xy^2 & 0 \leq x \leq 1, 0 \leq y \leq 1 \\ x & 0 \leq x \leq 1, y \geq 1 \\ y^2 & x \geq 1, 0 \leq y \leq 1 \\ 1 & x > 1 \text{ and } y > 1. \end{cases}$$

It is easy to see that $F_X(x) = F_{X,Y}(x,1) = x$ for $0 \leq x \leq 1$ and $F_Y(y) = F_{X,Y}(1,y) = y^2$ for $0 \leq y \leq 1$. The graphs of these functions are given by the outermost edges of the surface depicted in Figure 2.7.1. ∎

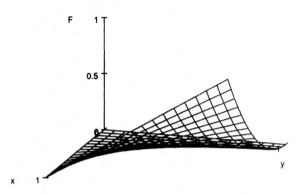

Figure 2.7.1: Graph of the joint distribution function $F_{X,Y}(x,y) = xy^2$ for $0 \leq x \leq 1$ and $0 \leq y \leq 1$ in Example 2.7.3.

Theorem 2.7.3 thus tells us that the joint cdf $F_{X,Y}$ is very useful indeed. Not only does it tell us about the relationship of X to Y, but it also contains all the information about the marginal distributions of X and of Y.

We will see in the next subsections that joint probability functions, and joint density functions, similarly contain information about both the relationship of X and Y and the marginal distributions of X and Y.

2.7.3   Joint Probability Functions

Suppose X and Y are both *discrete* random variables. Then we can define a joint probability function for X and Y, as follows.

Definition 2.7.3 Let X and Y be discrete random variables. Then their joint probability function, $p_{X,Y}$, is a function from R^2 to R^1, defined by

$$p_{X,Y}(x,y) = P(X = x, Y = y).$$

Consider the following example.

Example 2.7.4 *(Examples 2.7.1 and 2.7.2 continued)*
Again let $X \sim \text{Bernoulli}(1/2)$, $Y_1 = X$, and $Y_2 = 1 - X$. Then we see that

$$p_{X,Y_1}(x,y) = P(X = x,\ Y_1 = y) = \begin{cases} 1/2 & x = y = 1 \\ 1/2 & x = y = 0 \\ 0 & \text{otherwise.} \end{cases}$$

On the other hand,

$$p_{X,Y_2}(x,y) = P(X = x,\ Y_2 = y) = \begin{cases} 1/2 & x = 1,\ y = 0 \\ 1/2 & x = 0,\ y = 1 \\ 0 & \text{otherwise.} \end{cases}$$

We thus see that p_{X,Y_1} and p_{X,Y_2} are two simple functions that are easy to work with and that clearly describe the relationships between X and Y_1, and between X and Y_2. Hence, for pairs of discrete random variables, joint probability functions are usually the best way of describing their relationships. ∎

Once we know the joint probability function $p_{X,Y}$, the marginal probability functions of X and Y are easily obtained.

Theorem 2.7.4 Let X and Y be two discrete random variables, with joint probability function $p_{X,Y}$. Then the probability function p_X of X may be computed as

$$p_X(x) = \sum_y p_{X,Y}(x,y).$$

Similarly, the probability function p_Y of Y may be computed as

$$p_Y(y) = \sum_x p_{X,Y}(x,y).$$

Proof: Using additivity of P, we have that

$$p_X(x) = P(X = x) = \sum_y P(X = x,\ Y = y) = \sum_y p_{X,Y}(x,y),$$

as claimed. Similarly,

$$p_Y(y) = P(Y = y) = \sum_x P(X = x,\ Y = y) = \sum_x p_{X,Y}(x,y). \quad \blacksquare$$

Example 2.7.5
Suppose the joint probability function of X and Y is given by

$$p_{X,Y}(x,y) = \begin{cases} 1/7 & x = 5,\ y = 0 \\ 1/7 & x = 5,\ y = 3 \\ 1/7 & x = 5,\ y = 4 \\ 3/7 & x = 8,\ y = 0 \\ 1/7 & x = 8,\ y = 4 \\ 0 & \text{otherwise.} \end{cases}$$

Then

$$p_X(5) = \sum_y p_{X,Y}(5,y) = p_{X,Y}(5,0) + p_{X,Y}(5,3) + p_{X,Y}(5,4)$$

$$= \frac{1}{7} + \frac{1}{7} + \frac{1}{7} = \frac{3}{7},$$

while

$$p_X(8) = \sum_y p_{X,Y}(8,y) = p_{X,Y}(8,0) + p_{X,Y}(8,4) = \frac{3}{7} + \frac{1}{7} = \frac{4}{7}.$$

Similarly,

$$p_Y(4) = \sum_x p_{X,Y}(x,4) = p_{X,Y}(5,4) + p_{X,Y}(8,4) = \frac{1}{7} + \frac{1}{7} = \frac{2}{7},$$

etc.

Note that in such a simple context it is possible to tabulate the joint probability function in a table. The following table illustrates this for $p_{X,Y}, p_X$, and p_Y of this example.

	$Y = 0$	$Y = 3$	$Y = 4$	
$X = 5$	1/7	1/7	1/7	3/7
$X = 8$	3/7	0	1/7	4/7
	4/7	1/7	2/7	

Summing the rows and columns and placing the totals in the margins, gives the marginal distributions of X and Y. ∎

2.7.4 Joint Density Functions

If X and Y are *continuous* random variables, then clearly $p_{X,Y}(x,y) = 0$ for all x and y. Hence, joint probability functions are not useful in this case. On the other hand, we shall see here that if X and Y are *jointly absolutely continuous*, then their relationship may be usefully described by a joint density function.

Definition 2.7.4 Let $f : R^2 \to R^1$ be a function. Then f is a *joint density function* if $f(x,y) \geq 0$ for all x and y, and $\int_{-\infty}^{\infty} \int_{-\infty}^{\infty} f(x,y)\, dx\, dy = 1$.

Definition 2.7.5 Let X and Y be random variables. Then X and Y are *jointly absolutely continuous* if there is a joint density function f, such that

$$P(a \leq X \leq b,\ c \leq Y \leq d) = \int_c^d \int_a^b f(x,y)\, dx\, dy,$$

for all $a \leq b, c \leq d$.

Consider the following example.

Example 2.7.6

Let X and Y be jointly absolutely continuous, with joint density function f given by

$$f(x,y) = \begin{cases} 4x^2y + 2y^5 & 0 \le x \le 1,\ 0 \le y \le 1 \\ 0 & \text{otherwise.} \end{cases}$$

We first verify that f is indeed a density function. Clearly $f(x,y) \ge 0$ for all x and y. Also,

$$\int_{-\infty}^{\infty} \int_{-\infty}^{\infty} f(x,y)\,dx\,dy = \int_0^1 \int_0^1 (4x^2y + 2y^5)\,dx\,dy = \int_0^1 \left(\frac{4}{3}y + 2y^5\right)dy$$

$$= \frac{4}{3}\frac{1}{2} + 2\frac{1}{6} = \frac{2}{3} + \frac{1}{3} = 1.$$

Hence, f is a joint density function. In Figure 2.7.2, we have plotted the function f, which gives a surface over the unit square.

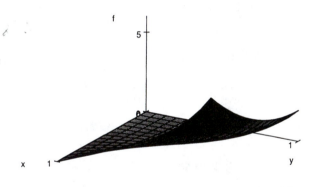

Figure 2.7.2: A plot of the density f in Example 2.7.6.

We next compute $P(0.5 \le X \le 0.7,\ 0.2 \le Y \le 0.9)$. Indeed, we have

$$P(0.5 \le X \le 0.7,\ 0.2 \le Y \le 0.9)$$

$$= \int_{0.2}^{0.9} \int_{0.5}^{0.7} (4x^2y + 2y^5)\,dx\,dy$$

$$= \int_{0.2}^{0.9} \left(\frac{4}{3}((0.7)^3 - (0.5)^3)\,y + 2y^5(0.7 - 0.5)\right)dy$$

$$= \frac{4}{3}\left((0.7)^3 - (0.5)^3\right)\frac{1}{2}((0.9)^2 - (0.2)^2) + \frac{2}{6}((0.9)^6 - (0.2)^6)(0.7 - 0.5))$$

$$= \frac{2}{3}((0.7)^3 - (0.5)^3)((0.9)^2 - (0.2)^2) + \frac{1}{3}((0.9)^6 - (0.2)^6)(0.7 - 0.5)) \doteq 0.147.$$

Other probabilities can be computed similarly. ∎

Once we know a joint density $f_{X,Y}$, then computing the marginal densities of X and Y is very easy, as the following theorem shows.

Theorem 2.7.5 Let X and Y be jointly absolutely continuous random variables, with joint density function $f_{X,Y}$. Then the (marginal) density f_X of X satisfies

$$f_X(x) = \int_{-\infty}^{\infty} f_{X,Y}(x,y)\, dy,$$

for all $x \in R^1$. Similarly, the (marginal) density f_Y of Y satisfies

$$f_Y(y) = \int_{-\infty}^{\infty} f_{X,Y}(x,y)\, dx,$$

for all $y \in R^1$.

Proof: We need to show that, for $a \le b$, $P(a \le X \le b) = \int_a^b f_X(x)\, dx = \int_a^b \int_{-\infty}^{\infty} f_{X,Y}(x,y)\, dy\, dx$. Now, we always have $-\infty < Y < \infty$. Hence, using continuity of P, we have that $P(a \le X \le b) = P(a \le X \le b, -\infty < Y < \infty)$ and

$$P(a \le X \le b, -\infty < Y < \infty)$$

$$= \lim_{\substack{c \to -\infty \\ d \to \infty}} P(a \le X \le b, c \le Y \le d) = \lim_{\substack{c \to -\infty \\ d \to \infty}} \int_c^d \int_a^b f(x,y)\, dx\, dy$$

$$= \lim_{\substack{c \to -\infty \\ d \to \infty}} \int_a^b \int_c^d f(x,y)\, dy\, dx = \int_a^b \int_{-\infty}^{\infty} f_{X,Y}(x,y)\, dy\, dx,$$

as claimed. The result for f_Y follows similarly. ∎

Example 2.7.7 (*Example 2.7.6 continued*)
Let X and Y again have joint density

$$f_{X,Y}(x,y) = \begin{cases} 4x^2y + 2y^5 & 0 \le x \le 1,\ 0 \le y \le 1 \\ 0 & \text{otherwise.} \end{cases}$$

Then by Theorem 2.7.5, for $0 \le x \le 1$,

$$f_X(x) = \int_{-\infty}^{\infty} f_{X,Y}(x,y)\, dy = \int_0^1 (4x^2y + 2y^5)\, dy = 2x^2 + (1/3),$$

while for $x < 0$ or $x > 1$,

$$f_X(x) = \int_{-\infty}^{\infty} f_{X,Y}(x,y)\, dy = \int_{-\infty}^{\infty} 0\, dy = 0.$$

Similarly, for $0 \le y \le 1$,

$$f_Y(y) = \int_{-\infty}^{\infty} f_{X,Y}(x,y)\, dx = \int_0^1 (4x^2y + 2y^5)\, dx = \frac{4}{3}y + 2y^5,$$

while for $y < 0$ or $y > 1$, $f_Y(y) = 0$. ∎

Example 2.7.8

Suppose X and Y are jointly absolutely continuous, with joint density

$$f_{X,Y}(x,y) = \begin{cases} 120x^3y & x \geq 0,\ y \geq 0,\ x+y \leq 1 \\ 0 & \text{otherwise.} \end{cases}$$

Then the region where $f_{X,Y}(x,y) > 0$ is a *triangle*, as depicted in Figure 2.7.3.

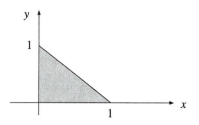

Figure 2.7.3: Region of the plane where the density $f_{X,Y}$ in Example 2.7.8 is positive.

We check that

$$\int_{-\infty}^{\infty} \int_{-\infty}^{\infty} f_{X,Y}(x,y)\,dx\,dy = \int_0^1 \int_0^{1-x} 120x^3y\,dy\,dx = \int_0^1 120x^3 \frac{(1-x)^2}{2}\,dx$$

$$= \int_0^1 60(x^3 - 2x^4 + x^5)\,dx = 60(\frac{1}{4} - 2\frac{1}{5} + \frac{1}{6})$$

$$= 15 - 2(12) + 10 = 1,$$

so that $f_{X,Y}$ is indeed a joint density function. We then compute that, for example,

$$f_X(x) = \int_0^{1-x} 120x^3y\,dy = 120x^3 \frac{(1-x)^2}{2} = 60(x^3 - 2x^4 + x^5)$$

for $0 \leq x \leq 1$ (with $f_X(x) = 0$ for $x < 0$ or $x > 1$). ∎

Example 2.7.9 *Bivariate Normal*$(\mu_1, \mu_2, \sigma_1, \sigma_2, \rho)$ *Distribution*
Let $\mu_1, \mu_2, \sigma_1, \sigma_2$, and ρ be real numbers, with $\sigma_1, \sigma_2 > 0$ and $-1 \leq \rho \leq 1$. Let X and Y have joint density given by

$$f_{X,Y}(x,y) = \frac{1}{2\pi\sigma_1\sigma_2\sqrt{1-\rho^2}} \exp\left\{ -\frac{1}{2(1-\rho^2)} \left[\left(\frac{x-\mu_1}{\sigma_1}\right)^2 + \left(\frac{y-\mu_2}{\sigma_2}\right)^2 - 2\rho\left(\frac{x-\mu_1}{\sigma_1}\right)\left(\frac{y-\mu_2}{\sigma_2}\right) \right] \right\}$$

for $x \in R^1, y \in R^1$. Then X and Y are said to have the *Bivariate Normal*$(\mu_1, \mu_2, \sigma_1, \sigma_2, \rho)$ *distribution*.

It can be shown (Problem 2.7.8) that $X \sim N(\mu_1, \sigma_1^2)$ and $Y \sim N(\mu_2, \sigma_2^2)$. Hence, X and Y are each normally distributed. The parameter ρ measures the degree of the *relationship* that exists between X and Y (see Problem 3.3.11) and is called the *correlation*. In particular, X and Y are independent (see Section 2.8.3), and so unrelated, if and only if $\rho = 0$ (Problem 2.8.17).

Figure 2.7.4 is a plot of the *standard bivariate normal* density, given by setting $\mu_1 = 0, \mu_2 = 0, \sigma_1 = 1, \sigma_2 = 1$, and $\rho = 0$. This is a bell-shaped surface in R^3 with its peak at the point $(0,0)$ in the xy-plane. The graph of the general Bivariate Normal$(\mu_1, \mu_2, \sigma_1, \sigma_2, \rho)$ distribution is also a bell-shaped surface, but the peak is at the point (μ_1, μ_2) in the xy-plane and the shape of the bell is controlled by σ_1, σ_2, and ρ.

Figure 2.7.4: A plot of the standard bivariate normal density function.

It can be shown (Problem 2.9.13) that, when Z_1, Z_2 are independent random variables, both distributed $N(0,1)$, and we put

$$X = \mu_1 + \sigma_1 Z_1, \quad Y = \mu_2 + \sigma_2 \left(\rho Z_1 + \left(1 - \rho^2\right)^{1/2} Z_2 \right), \qquad (2.7.1)$$

then $(X, Y) \sim$ Bivariate Normal$(\mu_1, \mu_2, \sigma_1, \sigma_2, \rho)$. This relationship can be quite useful in establishing various properties of this distribution. We can also write an analogous version $Y = \mu_2 + \sigma_2 Z_1, X = \mu_1 + \sigma_1(\rho Z_1 + \left(1 - \rho^2\right)^{1/2} Z_2)$ and obtain the same distributional result.

The bivariate normal distribution is one of the most commonly used bivariate distributions in applications. For example, if we randomly select an individual from a population and measure his weight X and height Y, then a bivariate normal distribution will often provide a reasonable description of the joint distribution of these variables. ∎

Joint densities can also be used to compute probabilities of more general regions, as the following result shows. (We omit the proof. The special case $B = [a, b] \times [c, d]$ corresponds directly to the definition of $f_{X,Y}$.)

Theorem 2.7.6 Let X and Y be jointly absolutely continuous random variables, with joint density $f_{X,Y}$, and let $B \subseteq R^2$ be any region. Then

$$P((X, Y) \in B) = \int_B \int f(x, y) \, dx \, dy.$$

The previous discussion has centered around having just *two* random variables, X and Y. More generally, we may consider n random variables $X_1, \ldots, X_n$. If the random variables are all discrete, then we can further define a joint probability function $p_{X_1,\ldots,X_n} : R^n \to [0,1]$ by $p_{X_1,\ldots,X_n}(x_1, \ldots, x_n) = P(X_1 = x_1, \ldots, X_n = x_n)$. If the random variables are jointly absolutely continuous, then we can define a joint density function $f_{X_1,\ldots,X_n} : R^n \to [0,1]$ so that

$$P(a_1 \leq X_1 \leq b_1, \ldots, a_n \leq X_n \leq b_n)$$
$$= \int_{a_n}^{b_n} \cdots \int_{a_1}^{b_1} f_{X_1,\ldots,X_n}(x_1, \ldots, x_n) \, dx_1 \cdots dx_n,$$

whenever $a_i \leq b_i$ for all i.

Summary of Section 2.7

- It is often important to keep track of the joint probabilities of two random variables, X and Y.

- Their joint cumulative distribution function is given by $F_{X,Y}(x,y) = P(X \leq x, \, Y \leq y)$.

- If X and Y are discrete, then their joint probability function is given by $p_{X,Y}(x,y) = P(X = x, \, Y = y)$.

- If X and Y are absolutely continuous, then their joint density function $f_{X,Y}(x,y)$ is such that $P(a \leq X \leq b, \, c \leq Y \leq d) = \int_c^d \int_a^b f_{X,Y}(x,y) \, dx \, dy$.

- The marginal density of X and Y can be computed from any of $F_{X,Y}$, or $p_{X,Y}$, or $f_{X,Y}$.

- An important example of a joint distribution is the bivariate normal distribution.

Exercises

2.7.1 Let $X \sim$ Bernoulli($1/3$), and let $Y = 4X - 2$. Compute the joint cdf $F_{X,Y}$.

2.7.2 Let $X \sim$ Bernoulli($1/4$), and let $Y = -7X$. Compute the joint cdf $F_{X,Y}$.

2.7.3 Suppose

$$p_{X,Y}(x,y) = \begin{cases} 1/5 & x = 2, \ y = 3 \\ 1/5 & x = 3, \ y = 2 \\ 1/5 & x = -3, \ y = -2 \\ 1/5 & x = -2, \ y = -3 \\ 1/5 & x = 17, \ y = 19 \\ 0 & \text{otherwise.} \end{cases}$$

(a) Compute p_X.

(b) Compute p_Y.

(c) Compute $P(Y > X)$.

(d) Compute $P(Y = X)$.

(e) Compute $P(XY < 0)$.

2.7.4 For each of the following joint density functions $f_{X,Y}$, find the value of C, and compute $f_X(x), f_Y(y)$, and $P(X \le 0.8, Y \le 0.6)$.

(a)
$$f_{X,Y}(x,y) = \begin{cases} 2x^2y + Cy^5 & 0 \le x \le 1, \ 0 \le y \le 1 \\ 0 & \text{otherwise} \end{cases}$$

(b)
$$f_{X,Y}(x,y) = \begin{cases} C(xy + x^5y^5) & 0 \le x \le 1, \ 0 \le y \le 1 \\ 0 & \text{otherwise} \end{cases}$$

(c)
$$f_{X,Y}(x,y) = \begin{cases} C(xy + x^5y^5) & 0 \le x \le 4, \ 0 \le y \le 10 \\ 0 & \text{otherwise} \end{cases}$$

(d)
$$f_{X,Y}(x,y) = \begin{cases} Cx^5y^5 & 0 \le x \le 4, \ 0 \le y \le 10 \\ 0 & \text{otherwise} \end{cases}$$

2.7.5 Prove that $F_{X,Y}(x,y) \le \min(F_X(x), F_Y(y))$.

Problems

2.7.6 Let $X \sim$ Exponential(λ), and let $Y = X^3$. Compute the joint cdf, $F_{X,Y}(x,y)$.

2.7.7 Let $F_{X,Y}$ be a joint cdf. Prove that for all $y \in R^1$, $\lim_{x \to -\infty} F_{X,Y}(x,y) = 0$.

2.7.8 Let X and Y have the Bivariate Normal($\mu_1, \mu_2, \sigma_1, \sigma_2, \rho$) distribution, as in Example 2.7.9. Prove that $X \sim N(\mu_1, \sigma_1^2)$, by proving that

$$\int_{-\infty}^{\infty} f_{X,Y}(x,y) \, dy = \frac{1}{\sigma_1\sqrt{2\pi}} \exp\left\{-\frac{(x-\mu_1)^2}{2\sigma_1^2}\right\}.$$

2.7.9 Suppose that the joint density $f_{X,Y}$ is given by $f_{X,Y}(x,y) = Cye^{-xy}$ for $0 < x < 1, 0 < y < 1$ and is 0 otherwise.

(a) Determine C so that $f_{X,Y}$ is a density.

(b) Compute $P(1/2 < X < 1, 1/2 < Y < 1)$.

(c) Compute the marginal densities of X and Y.

2.7.10 Suppose that the joint density $f_{X,Y}$ is given by $f_{X,Y}(x,y) = Cye^{-xy}$ for $0 < x < y < 1$ and is 0 otherwise.

(a) Determine C so that $f_{X,Y}$ is a density.

(b) Compute $P(1/2 < X < 1, 1/2 < Y < 1)$.
(c) Compute the marginal densities of X and Y.

2.7.11 Suppose that the joint density $f_{X,Y}$ is given by $f_{X,Y}(x,y) = Ce^{-(x+y)}$ for $0 < x < y < \infty$ and is 0 otherwise.
(a) Determine C so that $f_{X,Y}$ is a density.
(b) Compute the marginal densities of X and Y.

2.7.12 (*Dirichlet$(\alpha_1, \alpha_2, \alpha_3)$ distribution*) Let (X_1, X_2) have the joint density

$$f_{X_1, X_2}(x_1, x_2) = \frac{\Gamma(\alpha_1 + \alpha_2 + \alpha_3)}{\Gamma(\alpha_1)\Gamma(\alpha_2)\Gamma(\alpha_3)} x_1^{\alpha_1 - 1} x_2^{\alpha_2 - 1} (1 - x_1 - x_2)^{\alpha_3 - 1}$$

for $x_1 \geq 0, x_2 \geq 0$, and $0 \leq x_1 + x_2 \leq 1$. A Dirichlet distribution is often applicable when X_1, X_2, and $1 - X_1 - X_2$ correspond to random proportions.
(a) Prove that f_{X_1, X_2} is a density. (Hint: Sketch the region where f_{X_1, X_2} is nonnegative, integrate out x_1 first by making the transformation $u = x_1/(1-x_2)$ in this integral, and use (2.4.10) from Problem 2.4.16.)
(b) Prove that $X_1 \sim \text{Beta}(\alpha_1, \alpha_2 + \alpha_3)$ and $X_2 \sim \text{Beta}(\alpha_2, \alpha_1 + \alpha_3)$.

2.7.13 (*Dirichlet$(\alpha_1, \ldots, \alpha_{k+1})$ distribution*) Let $(X_1, \ldots, X_k)$ have the joint density

$$\begin{aligned} &f_{X_1, \ldots, X_k}(x_1, \ldots, x_k) \\ &= \frac{\Gamma(\alpha_1 + \cdots + \alpha_{k+1})}{\Gamma(\alpha_1) \cdots \Gamma(\alpha_{k+1})} x_1^{\alpha_1 - 1} \cdots x_k^{\alpha_k - 1} (1 - x_1 - \cdots - x_k)^{\alpha_{k+1} - 1} \end{aligned}$$

for $x_i \geq 0, i = 1, \ldots, k$, and $0 \leq x_1 + \cdots + x_k \leq 1$. Prove that $f_{X_1, \ldots, X_k}$ is a density. (Hint: Problem 2.7.12.)

Challenge

2.7.14 Find an example of two random variables X and Y, and a function $h : R^1 \to R^1$, such that $F_X(x) > 0$ and $F_Y(x) > 0$ for all $x \in R^1$, but $\lim_{x \to \infty} F_{X,Y}(x, h(x)) = 0$.

Discussion Topic

2.7.15 What are examples of pairs of real-life random quantities that have interesting relationships? (List as many as you can, and describe each relationship as well as you can.)

2.8 Conditioning and Independence

Let X and Y be two random variables. Suppose we know that $X = 5$. What does that tell us about Y? Depending on the relationship between X and Y, that may tell us everything about Y (e.g., if $Y = X$), or nothing about Y. Usually the situation will be between these two extremes, and the knowledge that $X = 5$ will change the probabilities for Y somewhat.

2.8.1 Conditioning on Discrete Random Variables

Suppose X is a discrete random variable, with $P(X = 5) > 0$. Let $a < b$, and suppose we are interested in the conditional probability $P(a < Y \leq b \,|\, X = 5)$. Well, we already know how to compute such conditional probabilities. Indeed, by (1.5.1),

$$P(a < Y \leq b \,|\, X = 5) = \frac{P(a < Y \leq b,\ X = 5)}{P(X = 5)},$$

provided that $P(X = 5) > 0$. This prompts the following definition.

Definition 2.8.1 Let X and Y be random variables, and suppose that $P(X = x) > 0$. The *conditional distribution* of Y, given that $X = x$, is the probability distribution assigning probability

$$\frac{P(Y \in B, X = x)}{P(X = x)}$$

to each event $Y \in B$. In particular, it assigns probability

$$\frac{P(a < Y \leq b, X = x)}{P(X = x)}$$

to the event that $a < Y \leq b$.

Example 2.8.1
Suppose as in Example 2.7.5 that X and Y have joint probability function

$$p_{X,Y}(x, y) = \begin{cases} 1/7 & x = 5,\ y = 0 \\ 1/7 & x = 5,\ y = 3 \\ 1/7 & x = 5,\ y = 4 \\ 3/7 & x = 8,\ y = 0 \\ 1/7 & x = 8,\ y = 4 \\ 0 & \text{otherwise.} \end{cases}$$

Let us compute $P(Y = 4 \,|\, X = 8)$. We compute this as

$$P(Y = 4 \,|\, X = 8) = \frac{P(Y = 4,\ X = 8)}{P(X = 8)} = \frac{1/7}{(3/7) + (1/7)} = \frac{1/7}{4/7} = 1/4.$$

On the other hand,

$$P(Y = 4 \,|\, X = 5) = \frac{P(Y = 4,\ X = 5)}{P(X = 5)} = \frac{1/7}{(1/7) + (1/7) + (1/7)} = \frac{1/7}{3/7} = 1/3.$$

We thus see that, depending on the value of X, we obtain different probabilities for Y. ∎

Generalizing from the above example, we see that if X and Y are discrete, then

$$P(Y = y \,|\, X = x) = \frac{P(Y = y,\ X = x)}{P(X = x)} = \frac{p_{X,Y}(x, y)}{p_X(x)} = \frac{p_{X,Y}(x, y)}{\sum_z p_{X,Y}(x, z)}.$$

This prompts the following definition.

Definition 2.8.2 Suppose X and Y are two discrete random variables. Then the *conditional probability function* of Y, given X, is the function $p_{Y|X}$ defined by

$$p_{Y|X}(y \,|\, x) = \frac{p_{X,Y}(x,y)}{\sum_z p_{X,Y}(x,z)} = \frac{p_{X,Y}(x,y)}{p_X(x)},$$

defined for all $y \in R^1$ and all x with $p_X(x) > 0$.

2.8.2 Conditioning on Continuous Random Variables

If X is continuous, then we will have $P(X = x) = 0$. In this case, Definitions 2.8.1 and 2.8.2 cannot be used, because we cannot divide by 0. So, how can we condition on $X = x$ in this case?

One approach is suggested by instead conditioning on $x - \epsilon < X \le x + \epsilon$, where $\epsilon > 0$ is a very small number. Even if X is continuous, we might still have $P(x - \epsilon \le X \le x + \epsilon) > 0$. On the other hand, if ϵ is very small and $x - \epsilon \le X \le x + \epsilon$, then X must be very close to x.

Indeed, suppose that X and Y are jointly absolutely continuous, with joint density function $f_{X,Y}$. Then

$$
\begin{aligned}
P(a \le Y \le b \,|\, x - \epsilon \le X \le x + \epsilon) &= \frac{P(a \le Y \le b, x - \epsilon \le X \le x + \epsilon)}{P(x - \epsilon \le X \le x + \epsilon)} \\
&= \frac{\int_a^b \int_{x-\epsilon}^{x+\epsilon} f_{X,Y}(t,y)\, dt\, dy}{\int_{-\infty}^{\infty} \int_{x-\epsilon}^{x+\epsilon} f_{X,Y}(t,y)\, dt\, dy}.
\end{aligned}
$$

In Figure 2.8.1, we have plotted the region $\{(x,y) : a \le y \le b,\ x - \epsilon < x \le x + \epsilon\}$ for (X,Y).

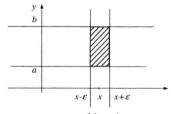

Figure 2.8.1: The shaded region is the set $\{(x,y) : a \le y \le b,\ x - \epsilon \le x \le x + \epsilon\}$.

Now, if ϵ is very small, then in the above integrals we will always have t very close to x. If $f_{X,Y}$ is a continuous function, then this implies that $f_{X,Y}(t,y)$ will be very close to $f_{X,Y}(x,y)$. We conclude that, if ϵ is very small, then

$$
P(a \le Y \le b \,|\, x - \epsilon \le X \le x + \epsilon) \approx \frac{\int_a^b \int_{x-\epsilon}^{x+\epsilon} f_{X,Y}(x,y)\, dt\, dy}{\int_{-\infty}^{\infty} \int_{x-\epsilon}^{x+\epsilon} f_{X,Y}(x,y)\, dt\, dy}
$$

$$
= \frac{\int_a^b 2\epsilon\, f_{X,Y}(x,y)\, dy}{\int_{-\infty}^{\infty} 2\epsilon\, f_{X,Y}(x,y)\, dy} = \int_a^b \frac{f_{X,Y}(x,y)}{\int_{-\infty}^{\infty} f_{X,Y}(x,z)\, dz}\, dy.
$$

This suggests that the quantity

$$\frac{f_{X,Y}(x,y)}{\int_{-\infty}^{\infty} f_{X,Y}(x,z)\,dz} = \frac{f_{X,Y}(x,y)}{f_X(x)}$$

plays the role of a density, for the conditional distribution of Y given that $X = x$. This prompts the following definitions.

Definition 2.8.3 Let X and Y be jointly absolutely continuous, with joint density function $f_{X,Y}$. The *conditional density* of Y given $X = x$, is the function $f_{Y|X}(y\,|\,x)$, defined by

$$f_{Y|X}(y\,|\,x) = \frac{f_{X,Y}(x,y)}{f_X(x)},$$

valid for all $y \in R^1$, and for all x such that $f_X(x) > 0$.

Definition 2.8.4 Let X and Y be jointly absolutely continuous, with joint density function $f_{X,Y}$. The *conditional distribution* of Y, given $X = x$, is defined by saying that

$$P(a \leq Y \leq b\,|\,X = x) = \int_a^b f_{Y|X}(y\,|\,x)\,dy,$$

when $a \leq b$, with $f_{Y|X}$ as in Definition 2.8.3, valid for all x such that $f_X(x) > 0$.

Example 2.8.2

Let X and Y have joint density

$$f_{X,Y}(x,y) = \begin{cases} 4x^2y + 2y^5 & 0 \leq x \leq 1,\ 0 \leq y \leq 1 \\ 0 & \text{otherwise,} \end{cases}$$

as in Examples 2.7.6 and 2.7.7.

We know from Example 2.7.7 that

$$f_X(x) = \begin{cases} 2x^2 + (1/3) & 0 \leq x \leq 1 \\ 0 & \text{otherwise,} \end{cases}$$

while

$$f_Y(y) = \begin{cases} \frac{4}{3}y + 2y^5 & 0 \leq y \leq 1 \\ 0 & \text{otherwise.} \end{cases}$$

Let us now compute $P(0.2 \leq Y \leq 0.3\,|\,X = 0.8)$. Using Definitions 2.8.4 and 2.8.3, we have

$$P(0.2 \leq Y \leq 0.3\,|\,X = 0.8)$$

$$= \int_{0.2}^{0.3} f_{Y|X}(y\,|\,0.8)\,dy = \frac{\int_{0.2}^{0.3} f_{X,Y}(0.8,\,y)\,dy}{f_X(0.8)} = \frac{\int_{0.2}^{0.3} \left(4\,(0.8)^2\,y + 2y^5\right)\,dy}{2\,(0.8)^2 + \frac{1}{3}}$$

$$= \frac{\frac{4}{2}\,(0.8)^2\,\left((0.3)^2 - (0.2)^2\right) + \frac{2}{6}((0.3)^6 - (0.2)^6)}{2\,(0.8)^2 + \frac{1}{3}} = 0.0398.$$

By contrast, if we compute the *unconditioned* (i.e., usual) probability that $0.2 \leq Y \leq 0.3$, we see that

$$
\begin{aligned}
P(0.2 \leq Y \leq 0.3) &= \int_{0.2}^{0.3} f_Y(y)\,dy = \int_{0.2}^{0.3} (\tfrac{4}{3}y + 2y^5)\,dy \\
&= \frac{4}{3}\frac{1}{2}((0.3)^2 - (0.2)^2) + \frac{2}{6}((0.3)^6 - (0.2)^6) = 0.0336.
\end{aligned}
$$

We thus see that conditioning on $X = 0.8$ decreases the probability that $0.2 \leq Y \leq 0.3$, from about 0.0398 to about 0.0336. ∎

By analogy with Theorem 2.4.2, we have the following.

Theorem 2.8.1 (*Law of total probability, absolutely continuous random variable version*) Let X and Y be jointly absolutely continuous random variables, and let $a \leq b$ and $c \leq d$. Then

$$
P(a \leq X \leq b,\ c \leq Y \leq d) = \int_c^d \int_a^b f_X(x)\,f_{Y|X}(y\,|\,x)\,dx\,dy.
$$

More generally, if $B \subseteq R^2$ is any region, then

$$
P\big((X,Y) \in B\big) = \int_B \int f_X(x)\,f_{Y|X}(y\,|\,x)\,dx\,dy.
$$

Proof: By Definition 2.8.3,

$$
f_X(x)\,f_{Y|X}(y\,|\,x) = f_{X,Y}(x,y).
$$

Hence, the result follows immediately from Definition 2.7.4 and Theorem 2.7.6. ∎

2.8.3 Independence of Random Variables

Recall from Definition 1.5.2 that two events A and B are *independent* if $P(A \cap B) = P(A)\,P(B)$. We wish to have a corresponding definition of independence for random variables X and Y. Intuitively, independence of X and Y means that X and Y have no influence on each other, i.e., that the values of X make no change to the probabilities for Y (and vice versa).

The idea of the formal definition is that X and Y give rise to events, of the form "$a < X \leq b$" or "$Y \in B$", and we want all such events involving X to be independent of all such events involving Y. Specifically, our definition is the following.

Definition 2.8.5 Let X and Y be two random variables. Then X and Y are *independent* if, for all subsets B_1 and B_2 of the real numbers,

$$
P(X \in B_1, Y \in B_2) = P(X \in B_1)P(Y \in B_2).
$$

That is, the events "$X \in B_1$" and "$Y \in B_2$" are independent events.

Intuitively, X and Y are independent if they have no influence upon each other, as we shall see.

Now, Definition 2.8.5 is very difficult to work with. Fortunately, there is a much simpler characterization of independence.

Theorem 2.8.2 Let X and Y be two random variables. Then X and Y are independent if and only if

$$P(a \leq X \leq b, c \leq Y \leq d) = P(a \leq X \leq b)P(c \leq Y \leq d) \qquad (2.8.1)$$

whenever $a \leq b$ and $c \leq d$.

That is, X and Y are independent if and only if the events "$a \leq X \leq b$" and "$c \leq Y \leq d$" are independent events whenever $a \leq b$ and $c \leq d$.

We shall not prove Theorem 2.8.2 here, although it is similar in spirit to the proof of Theorem 2.5.1. However, we shall sometimes use (2.8.1) to check for the independence of X and Y.

Still, even (2.8.1) is not so easy to check directly. For discrete and for absolutely continuous distributions, easier conditions are available, as follows.

Theorem 2.8.3 Let X and Y be two random variables.
(a) If X and Y are discrete, then X and Y are independent if and only if their joint probability function $p_{X,Y}$ satisfies

$$p_{X,Y}(x, y) = p_X(x)\, p_Y(y)$$

for all $x, y \in R^1$.
(b) If X and Y are jointly absolutely continuous, then X and Y are independent if and only if their joint density function $f_{X,Y}$ can be chosen to satisfy

$$f_{X,Y}(x, y) = f_X(x)\, f_Y(y)$$

for all $x, y \in R^1$.

Proof: (a) If X and Y are independent, then setting $a = b = x$ and $c = d = y$ in (2.8.1), we see that $P(X = x, Y = y) = P(X = x)P(Y = y)$. Hence, $p_{X,Y}(x, y) = p_X(x)\, p_Y(y)$.

Conversely, if $p_{X,Y}(x, y) = p_X(x)\, p_Y(y)$ for all x and y, then

$$P(a \leq X \leq b, c \leq Y \leq d)$$
$$= \sum_{a \leq x \leq b\, c \leq y \leq d} p_{X,Y}(x, y) = \sum_{a \leq x \leq b\, c \leq y \leq d} p_X(x)\, p_Y(y)$$
$$= \left(\sum_{a \leq x \leq b} p_X(x) \right) \left(\sum_{c \leq y \leq d} p_Y(y) \right) = P(a \leq X \leq b)\, P(c \leq Y \leq d).$$

This completes the proof of (a).

(b) If $f_{X,Y}(x,y) = f_X(x) f_Y(y)$ for all x and y, then

$$P(a \le X \le b, c \le Y \le d)$$

$$= \int_a^b \int_c^d f_{X,Y}(x,y) \, dy \, dx = \int_a^b \int_c^d f_X(x) f_Y(y) \, dy \, dx$$

$$= \left(\int_a^b f_X(x) \, dx \right) \left(\int_c^d f_Y(y) \, dy \right) = P(a \le X \le b) \, P(c \le Y \le d).$$

This completes the proof of the "if" part of (b). The proof of the "only if" part of (b) is more technical, and we do not include it here. ∎

Example 2.8.3
Let X and Y have, as in Example 2.7.6, joint density

$$f_{X,Y}(x,y) = \begin{cases} 4x^2y + 2y^5 & 0 \le x \le 1, \ 0 \le y \le 1 \\ 0 & \text{otherwise} \end{cases}$$

and so, as derived in as in Example 2.7.7, marginal densities

$$f_X(x) = \begin{cases} 2x^2 + (1/3) & 0 \le x \le 1 \\ 0 & \text{otherwise} \end{cases}$$

and

$$f_Y(y) = \begin{cases} \frac{4}{3}y + 2y^5 & 0 \le y \le 1 \\ 0 & \text{otherwise.} \end{cases}$$

Then we compute that

$$f_X(x) f_Y(y) = \begin{cases} (2x^2 + (1/3))(\frac{4}{3}y + 2y^5) & 0 \le x \le 1, \ 0 \le y \le 1 \\ 0 & \text{otherwise.} \end{cases}$$

We therefore see that $f_X(x) f_Y(y) \ne f_{X,Y}(x,y)$. Hence, X and Y are *not* independent. ∎

Example 2.8.4
Let X and Y have joint density

$$f_{X,Y}(x,y) = \begin{cases} \frac{1}{8080}(12xy^2 + 6x + 4y^2 + 2) & 0 \le x \le 6, \ 3 \le y \le 5 \\ 0 & \text{otherwise.} \end{cases}$$

We compute the marginal densities as

$$f_X(x) = \int_{-\infty}^{\infty} f_{X,Y}(x,y) \, dy = \begin{cases} \frac{1}{60} + \frac{1}{20}x & 0 \le x \le 6 \\ 0 & \text{otherwise,} \end{cases}$$

and

$$f_Y(y) = \int_{-\infty}^{\infty} f_{X,Y}(x,y) \, dx = \begin{cases} \frac{3}{202} + \frac{3}{101}y^2 & 3 \le y \le 5 \\ 0 & \text{otherwise.} \end{cases}$$

Then we compute that

$$f_X(x)f_Y(y) = \begin{cases} (\frac{1}{60} + \frac{1}{20}x)(\frac{3}{202} + \frac{3}{101}y^2) & 0 \le x \le 6,\ 3 \le y \le 5 \\ 0 & \text{otherwise.} \end{cases}$$

Multiplying this out, we see that $f_X(x)\,f_Y(y) = f_{X,Y}(x,y)$. Hence, X and Y *are* independent in this case. ∎

Combining Theorem 2.8.3 with Definitions 2.8.2 and 2.8.3, we immediately obtain the following result about independence. It says that independence of random variables is the same as saying that conditioning on one has no effect on the other, which corresponds to an intuitive notion of independence.

Theorem 2.8.4 Let X and Y be two random variables.
(a) If X and Y are discrete, then X and Y are independent if and only if $p_{Y|X}(y\,|\,x) = p_Y(y)$, for every $x, y \in R^1$.
(b) If X and Y are jointly absolutely continuous, then X and Y are independent if and only if $f_{Y|X}(y\,|\,x) = f_Y(y)$, for every $x, y \in R^1$.

While Definition 2.8.5 is quite difficult to work with, it does provide the easiest way of proving one very important property of independence, as follows.

Theorem 2.8.5 Let X and Y be independent random variables. Let $f, g : R^1 \to R^1$ be any two functions. Then the random variables $f(X)$ and $g(Y)$ are also independent.

Proof: Using Definition 2.8.5, we compute that

$$\begin{aligned} P(f(X) \in B_1,\ g(Y) \in B_2) &= P\left(X \in f^{-1}(B_1),\ Y \in g^{-1}(B_2)\right) \\ &= P\left(X \in f^{-1}(B_1)\right)\ P\left(Y \in g^{-1}(B_2)\right) \\ &= P(f(X) \in B_1)\ P(g(Y) \in B_2). \end{aligned}$$

(Here $f^{-1}(B_1) = \{x \in R^1 : f(x) \in B_1\}$ and $g^{-1}(B_2) = \{y \in R^1 : g(y) \in B_2\}$.) Because this is true for any B_1 and B_2, we see that $f(X)$ and $g(Y)$ are independent. ∎

Suppose now that we have n random variables $X_1, \ldots, X_n$. The random variables are *independent* if and only if the collection of events $\{a_i \le X_i \le b_i\}$ are independent, whenever $a_i \le b_i$, for all $i = 1, 2, \ldots, n$. Generalizing Theorem 2.8.3, we have the following result.

Theorem 2.8.6 Let $X_1, \ldots, X_n$ be a collection of random variables.
(a) If $X_1, \ldots, X_n$ are discrete, then $X_1, \ldots, X_n$ are independent if and only if their joint probability function $p_{X_1,\ldots,X_n}$ satisfies

$$p_{X_1,\ldots,X_n}(x_1, \ldots, x_n) = p_{X_1}(x_1) \cdots p_{X_n}(x_n)$$

for all $x_1, \ldots, x_n \in R^1$.
(b) If $X_1, \ldots, X_n$ are jointly absolutely continuous, then $X_1, \ldots, X_n$ are independent if and only if their joint density function $f_{X_1,\ldots,X_n}$ can be chosen to satisfy

$$f_{X_1,\ldots,X_n}(x, y) = f_{X_1}(x_1) \cdots f_{X_n}(x_n)$$

for all $x_1, \ldots, x_n \in R^1$.

A particularly common case in statistics is the following.

Definition 2.8.6 A collection $X_1, \ldots, X_n$ of random variables is *independent and identically distributed* (or *i.i.d.*) if the collection is independent, and if furthermore each of the n variables has the same distribution. The i.i.d. sequence $X_1, \ldots, X_n$ is also referred to as a *sample* from the common distribution.

In particular, if a collection $X_1, \ldots, X_n$ of random variables is i.i.d. and *discrete*, then each of the probability functions p_{X_i} is the same, so that $p_{X_1}(x) = p_{X_2}(x) = \cdots = p_{X_n}(x) \equiv p(x)$, for all $x \in R^1$. Furthermore, from Theorem 2.8.6(a), it follows that

$$p_{X_1, \ldots, X_n}(x_1, \ldots, x_n) = p_{X_1}(x_1)p_{X_2}(x_2) \cdots p_{X_n}(x_n) = p(x_1)p(x_2) \cdots p(x_n)$$

for all $x_1, \ldots, x_n \in R^1$.

Similarly, if a collection $X_1, \ldots, X_n$ of random variables is i.i.d. and *jointly absolutely continuous*, then each of the density functions f_{X_i} is the same, so that $f_{X_1}(x) = f_{X_2}(x) = \cdots = f_{X_n}(x) \equiv f(x)$, for all $x \in R^1$. Furthermore, from Theorem 2.8.6(b), it follows that

$$f_{X_1, \ldots, X_n}(x_1, \ldots, x_n) = f_{X_1}(x_1) f_{X_2}(x_2) \cdots f_{X_n}(x_n) = f(x_1) f(x_2) \cdots f(x_n)$$

for all $x_1, \ldots, x_n \in R^1$.

We now consider an important family of discrete distributions that arise via sampling.

Example 2.8.5 *Multinomial Distributions*
Suppose we have a response s that can take three possible values — for convenience labelled $1, 2$, and 3 — with the probability distribution

$$P(s = 1) = \theta_1, P(s = 2) = \theta_2, P(s = 3) = \theta_3$$

so that each $\theta_i \geq 0$ and $\theta_1 + \theta_2 + \theta_3 = 1$. As a simple example, consider a bowl of chips of which a proportion θ_i of the chips are labelled i (for $i = 1, 2, 3$). If we randomly draw a chip from the bowl and observe its label s, then $P(s = i) = \theta_i$. Alternatively, consider a population of students at a university of which a proportion θ_1 live on campus (denoted by $s = 1$), a proportion θ_2 live off-campus with their parents (denoted by $s = 2$), and a proportion θ_3 live off-campus independently (denoted by $s = 3$). If we randomly draw a student from this population and determine s for that student, then $P(s = i) = \theta_i$.

We can also write

$$P(s = i) = \theta_1^{I_{\{1\}}(i)} \theta_2^{I_{\{2\}}(i)} \theta_3^{I_{\{3\}}(i)}$$

for $i \in \{1, 2, 3\}$, where $I_{\{j\}}$ is the indicator function for $\{j\}$. Therefore, if $(s_1, \ldots, s_n)$ is a sample from the distribution on $\{1, 2, 3\}$ given by the θ_i, Theorem 2.8.6(a) implies that the joint probability function for the sample equals

$$P(s_1 = k_1, \ldots, s_n = k_n) = \prod_{j=1}^{n} \theta_1^{I_{\{1\}}(k_j)} \theta_2^{I_{\{2\}}(k_j)} \theta_3^{I_{\{3\}}(k_j)} = \theta_1^{x_1} \theta_2^{x_2} \theta_3^{x_3} \quad (2.8.2)$$

where $x_i = \sum_{j=1}^{n} I_{\{i\}}(k_j)$ is equal to the number of i's in $(k_1, \ldots, k_n)$.

Now, based on the sample $(s_1, \ldots, s_n)$, define the random variables

$$X_i = \sum_{j=1}^{n} I_{\{i\}}(s_j)$$

for $i = 1, 2,$ and 3. Clearly X_i is the number of i's observed in the sample and we always have $X_i \in \{0, 1, \ldots, n\}$ and $X_1 + X_2 + X_3 = n$. We refer to the X_i as the *counts* formed from the sample.

For (x_1, x_2, x_3) satisfying $x_i \in \{0, 1, \ldots, n\}$ and $x_1 + x_2 + x_3 = n$, (2.8.2) implies that the joint probability function for (X_1, X_2, X_3) is given by

$$
\begin{aligned}
p_{(X_1, X_2, X_3)}(x_1, x_2, x_3) &= P(X_1 = x_1, X_2 = x_2, X_3 = x_3) \\
&= C(x_1, x_2, x_3) \theta_1^{x_1} \theta_2^{x_2} \theta_3^{x_3}
\end{aligned}
$$

where $C(x_1, x_2, x_3)$ equals the number of samples $(s_1, \ldots, s_n)$ with x_1 of its elements equal to 1, x_2 of its elements equal to 2, and x_3 of its elements equal to 3. To calculate $C(x_1, x_2, x_3)$, we note that there are $\binom{n}{x_1}$ choices for the places of the 1's in the sample sequence, $\binom{n-x_1}{x_2}$ choices for the places of the 2's in the sequence, and finally $\binom{n-x_1-x_2}{x_3} = 1$ choices for the places of the 3's in the sequence (recall the multinomial coefficient defined in (1.4.4)). Therefore the probability function for the counts (X_1, X_2, X_3) is equal to

$$
\begin{aligned}
p_{(X_1, X_2, X_3)}(x_1, x_2, x_3) &= \binom{n}{x_1}\binom{n-x_1}{x_2}\binom{n-x_1-x_2}{x_3} \theta_1^{x_1} \theta_2^{x_2} \theta_3^{x_3} \\
&= \binom{n}{x_1\ x_2\ x_3} \theta_1^{x_1} \theta_2^{x_2} \theta_3^{x_3}.
\end{aligned}
$$

We say that

$$(X_1, X_2, X_3) \sim \text{Multinomial}(n, \theta_1, \theta_2, \theta_3).$$

Notice that the Multinomial$(n, \theta_1, \theta_2, \theta_3)$ generalizes the Binomial(n, θ) distribution as we are now counting the number of response values in three possible categories rather than two. Also, it is immediate that

$$X_i \sim \text{Binomial}(n, \theta_i)$$

because X_i equals the number of occurrences of i in the n independent response values, and i occurs for an individual response with probability equal to θ_i (also see Problem 2.8.18).

As a simple example, suppose that we have an urn containing 10 red balls, 20 white balls, and 30 black balls. If we randomly draw 10 balls from the urn with replacement, what is the probability that we will obtain 3 red, 4 white, and 3 black balls? Because we are drawing with replacement, the draws are i.i.d. and so the counts are distributed Multinomial$(10, 10/60, 20/60, 30/60)$. The required probability equals

$$\binom{10}{3\ 4\ 3}\left(\frac{10}{60}\right)^3 \left(\frac{20}{60}\right)^4 \left(\frac{30}{60}\right)^3 = 3.0007 \times 10^{-2}.$$

Note that if we had drawn without replacement, then the draws are not i.i.d., and the counts do not follow a multinomial distribution but rather a generalization of the hypergeometric distribution, as discussed in Problem 2.3.29.

Now suppose we have a response s that takes k possible values — for convenience labelled $1, 2, \ldots, k$ — with the probability distribution given by $P(s = i) = \theta_i$. For a sample $(s_1, \ldots, s_n)$, define the counts $X_i = \sum_{j=1}^{n} I_{\{i\}}(s_j)$ for $i = 1, \ldots k$. Then, arguing as above, and recalling the development of (1.4.4), we have

$$p_{(X_1,\ldots,X_k)}(x_1, \ldots, x_k) = \binom{n}{x_1 \ldots x_k} \theta_1^{x_1} \cdots \theta_k^{x_k}$$

whenever each $x_i \in \{0, \ldots, n\}$ and $x_1 + \cdots + x_k = n$. In this case we write

$$(X_1, \ldots, X_k) \sim \text{Multinomial}(n, \theta_1, \ldots, \theta_k). \quad \blacksquare$$

2.8.4 Order Statistics

Suppose now that $(X_1, \ldots, X_n)$ is a sample. In many applications of statistics we will have n data values where the assumption that these arise as an i.i.d. sequence makes sense. It is often of interest then to order these from smallest to largest to obtain the *order statistics*

$$X_{(1)}, \ldots, X_{(n)}.$$

Here $X_{(i)}$ is equal to the ith smallest value in the sample $X_1, \ldots, X_n$. So, for example, if $n = 5$ and

$$X_1 = 2.3, \ X_2 = 4.5, \ X_3 = -1.2, \ X_4 = 2.2, \ X_5 = 4.3$$

then

$$X_{(1)} = -1.2, \ X_{(2)} = 2.2, \ X_{(3)} = 2.3, \ X_{(4)} = 4., \ X_{(5)} = 34.5.$$

Of considerable interest in many situations are the distributions of the order statistics. Consider the following examples.

Example 2.8.6 *Distribution of the Sample Maximum*
Suppose $X_1, X_2, \ldots, X_n$ are i.i.d. so that $F_{X_1}(x) = F_{X_2}(x) = \cdots = F_{X_n}(x)$. Then the *largest-order statistic* $X_{(n)} = \max(X_1, X_2, \ldots, X_n)$ is the *maximum* of these n random variables.

Now $X_{(n)}$ is another random variable. What is its cumulative distribution function? We see that $X_{(n)} \leq x$ if and only if $X_i \leq x$ for all i. Hence,

$$
\begin{aligned}
F_{X_{(n)}}(x) &= P(X_{(n)} \leq x) = P(X_1 \leq x, \ X_2 \leq x, \ \ldots, X_n \leq x) \\
&= P(X_1 \leq x)P(X_2 \leq x) \cdots P(X_n \leq x) = F_{X_1}(x)F_{X_2}(x) \cdots F_{X_n}(x) \\
&= (F_{X_1}(x))^n.
\end{aligned}
$$

If F_{X_1} corresponds to an absolutely continuous distribution, then we can differentiate this expression to obtain the density of $X_{(n)}$. $\blacksquare$

Example 2.8.7

As a special case of Example 2.8.6, suppose that $X_1, X_2, \ldots, X_n$ are identically and independently distributed Uniform$[0, 1]$. From the above, for $0 \leq x \leq 1$, we have $F_{X_{(n)}}(x) = (F_{X_1}(x))^n = x^n$. It then follows from Corollary 2.5.1 that the density $f_{X_{(n)}}$ of $X_{(n)}$ equals $f_{X_{(n)}}(x) = F'_{X_{(n)}}(x) = nx^{n-1}$ for $0 \leq n \leq 1$, with (of course) $f_{X_{(n)}}(x) = 0$ for $x < 0$ and $x > 1$. Note that, from Problem 2.4.16, we can write $X_{(n)} \sim \text{Beta}(n, 1)$. ∎

Example 2.8.8 *Distribution of the Sample Minimum*

Following Example 2.8.6, we can also obtain the distribution function of the sample minimum, or *smallest-order statistic*, $X_{(1)} = \min(X_1, X_2, \ldots, X_n)$. We have

$$
\begin{aligned}
F_{X_{(1)}}(x) &= P(X_{(1)} \leq x) \\
&= 1 - P(X_{(1)} > x) \\
&= 1 - P(X_1 > x, \; X_2 > x, \; \ldots, X_n > x) \\
&= 1 - P(X_1 > x)\, P(X_2 > x) \cdots P(X_n > x) \\
&= 1 - (1 - F_{X_1}(x))\,(1 - F_{X_2}(x)) \cdots (1 - F_{X_n}(x)) \\
&= 1 - (1 - F_{X_1}(x))^n.
\end{aligned}
$$

Again, if F_{X_1} corresponds to an absolutely continuous distribution, we can differentiate this expression to obtain the density of $X_{(1)}$. ∎

Example 2.8.9

Let $X_1, \ldots, X_n$ be i.i.d. Uniform$[0, 1]$. Hence, for $0 \leq x \leq 1$,

$$
F_{X_{(1)}}(x) = P(X_{(1)} \leq x) = 1 - P(X_{(1)} > x) = 1 - (1 - x)^n.
$$

It then follows from Corollary 2.5.1 that the density $f_{X_{(1)}}$ of $X_{(1)}$ satisfies $f_{X_{(1)}}(x) = F'_{X_{(1)}}(x) = n(1 - x)^{n-1}$ for $0 \leq x \leq 1$, with (of course) $f_{X_{(1)}}(x) = 0$ for $x < 0$ and $x > 1$. Note that, from Problem 2.4.16, we can write $X_{(1)} \sim \text{Beta}(1, n)$. ∎

The sample median and sample quartiles are defined in terms of order statistics and used in statistical applications. These quantities, and their uses, are discussed in Section 5.5.

Summary of Section 2.8

- If X and Y are discrete, then the conditional probability function of Y, given X, equals $p_{Y|X}(y \,|\, x) = p_{X,Y}(x, y) / p_X(x)$.

- If X and Y are absolutely continuous, then the conditional density function of Y, given X, equals $f_{Y|X}(y \,|\, x) = f_{X,Y}(x, y) / f_X(x)$.

- X and Y are independent if $P(X \in B_1, Y \in B_2) = P(X \in B_1)P(Y \in B_2)$ for all $B_1, B_2 \subseteq R^1$.

- Discrete X and Y are independent if and only if $p_{X,Y}(x,y) = p_X(x)p_Y(y)$ for all $x, y \in R^1$ or, equivalently, $p_{Y|X}(y\,|\,x) = p_Y(y)$.

- Absolutely continuous X and Y are independent if and only if $f_{X,Y}(x,y) = f_X(x)f_Y(y)$ for all $x, y \in R^1$ or, equivalently, $f_{Y|X}(y\,|\,x) = f_Y(y)$.

- A sequence $X_1, X_2, \ldots, X_n$ is i.i.d. if the random variables are independent, and each X_i has the same distribution.

Exercises

2.8.1 Suppose X and Y have joint probability function

$$p_{X,Y}(x,y) = \begin{cases} 1/6 & x = -2,\ y = 3 \\ 1/12 & x = -2,\ y = 5 \\ 1/6 & x = 9,\ y = 3 \\ 1/12 & x = 9,\ y = 5 \\ 1/3 & x = 13,\ y = 3 \\ 1/6 & x = 13,\ y = 5 \\ 0 & \text{otherwise.} \end{cases}$$

(a) Compute $p_X(x)$ for all $x \in R^1$.
(b) Compute $p_Y(y)$ for all $y \in R^1$.
(c) Determine whether or not X and Y are independent.

2.8.2 Suppose X and Y have joint probability function

$$p_{X,Y}(x,y) = \begin{cases} 1/16 & x = -2,\ y = 3 \\ 1/4 & x = -2,\ y = 5 \\ 1/2 & x = 9,\ y = 3 \\ 1/16 & x = 9,\ y = 5 \\ 1/16 & x = 13,\ y = 3 \\ 1/16 & x = 13,\ y = 5 \\ 0 & \text{otherwise.} \end{cases}$$

(a) Compute $p_X(x)$ for all $x \in R^1$.
(b) Compute $p_Y(y)$ for all $y \in R^1$.
(c) Determine whether or not X and Y are independent.

2.8.3 Suppose X and Y have joint density function

$$f_{X,Y}(x,y) = \begin{cases} \frac{12}{49}\left(2 + x + xy + 4y^2\right) & 0 \le x \le 1,\ 0 \le y \le 1 \\ 0 & \text{otherwise.} \end{cases}$$

(a) Compute $f_X(x)$ for all $x \in R^1$.
(b) Compute $f_Y(y)$ for all $y \in R^1$.
(c) Determine whether or not X and Y are independent.

2.8.4 Suppose X and Y have joint density function

$$f_{X,Y}(x,y) = \begin{cases} \frac{2}{5(2+e)} \left(3 + e^x + 3y + 3ye^y + ye^x + ye^{x+y}\right) & \begin{array}{l} 0 \le x \le 1, \\ 0 \le y \le 1 \end{array} \\ 0 & \text{otherwise.} \end{cases}$$

(a) Compute $f_X(x)$ for all $x \in R^1$.
(b) Compute $f_Y(y)$ for all $y \in R^1$.
(c) Determine whether or not X and Y are independent.

2.8.5 Suppose X and Y have joint probability function

$$p_{X,Y}(x,y) = \begin{cases} 1/9 & x = -4, \ y = -2 \\ 2/9 & x = 5, \ y = -2 \\ 3/9 & x = 9, \ y = -2 \\ 2/9 & x = 9, \ y = 0 \\ 1/9 & x = 9, \ y = 4 \\ 0 & \text{otherwise.} \end{cases}$$

(a) Compute $P(Y = 4 \,|\, X = 9)$.
(b) Compute $P(Y = -2 \,|\, X = 9)$.
(c) Compute $P(Y = 0 \,|\, X = -4)$.
(d) Compute $P(Y = -2 \,|\, X = 5)$.
(e) Compute $P(X = 5 \,|\, Y = -2)$.

2.8.6 Let $X \sim$ Bernoulli(θ) and $Y \sim$ Geometric(θ), with X and Y independent. Let $Z = X + Y$. What is the probability function of Z?

2.8.7 For each of the following joint density functions $f_{X,Y}$ (taken from Exercise 2.7.4), compute the conditional density $f_{Y|X}(y \,|\, x)$, and determine whether or not X and Y are independent.
(a)

$$f_{X,Y}(x,y) = \begin{cases} 2x^2y + Cy^5 & 0 \le x \le 1, \ 0 \le y \le 1 \\ 0 & \text{otherwise} \end{cases}$$

(b)

$$f_{X,Y}(x,y) = \begin{cases} C(xy + x^5y^5) & 0 \le x \le 1, \ 0 \le y \le 1 \\ 0 & \text{otherwise} \end{cases}$$

(c)

$$f_{X,Y}(x,y) = \begin{cases} C(xy + x^5y^5) & 0 \le x \le 4, \ 0 \le y \le 10 \\ 0 & \text{otherwise} \end{cases}$$

(d)

$$f_{X,Y}(x,y) = \begin{cases} Cx^5y^5 & 0 \le x \le 4, \ 0 \le y \le 10 \\ 0 & \text{otherwise} \end{cases}$$

2.8.8 Let X and Y be jointly absolutely continuous random variables. Suppose $X \sim \text{Exponential}(2)$, and that $P(Y > 5 \,|\, X = x) = e^{-3x}$. Compute $P(Y > 5)$.

2.8.9 Give an example of two random variables X and Y each taking values in the set $\{1, 2, 3\}$, such that $P(X = 1, Y = 1) = P(X = 1) \, P(Y = 1)$, but X and Y are *not* independent.

2.8.10 Let $X \sim \text{Bernoulli}(\theta)$ and $Y \sim \text{Bernoulli}(\psi)$, where $0 < \theta < 1$ and $0 < \psi < 1$. Suppose $P(X = 1, Y = 1) = P(X = 1) \, P(Y = 1)$. Prove that X and Y must be independent.

2.8.11 Suppose that X is a *constant* random variable and that Y is any random variable. Prove that X and Y must be independent.

2.8.12 Suppose $X \sim \text{Bernoulli}(1/3)$ and $Y \sim \text{Poisson}(\lambda)$, with X and Y independent, and with $\lambda > 0$. Compute $P(X = 1 \,|\, Y = 5)$.

Problems

2.8.13 Let X and Y be jointly absolutely continuous random variables, having joint density of the form

$$f_{X,Y}(x, y) = \begin{cases} C_1(2x^2 y + C_2 y^5) & 0 \le x \le 1, \ 0 \le y \le 1 \\ 0 & \text{otherwise.} \end{cases}$$

Determine values of C_1 and C_2, such that $f_{X,Y}$ is a valid joint density function, and X and Y are independent.

2.8.14 Let X and Y be discrete random variables. Suppose $p_{X,Y}(x, y) = g(x) \, h(y)$, for some functions g and h. Prove that X and Y are independent. (Hint: Use Theorem 2.8.3(a) and Theorem 2.7.4.)

2.8.15 Let X and Y be jointly absolutely continuous random variables. Suppose $f_{X,Y}(x, y) = g(x) \, h(y)$, for some functions g and h. Prove that X and Y are independent. (Hint: Use Theorem 2.8.3(b) and Theorem 2.7.5.)

2.8.16 Let X and Y be discrete random variables, with $P(X = 1) > 0$ and $P(X = 2) > 0$. Suppose $P(Y = 1 \,|\, X = 1) = 3/4$ and $P(Y = 2 \,|\, X = 2) = 3/4$. Prove that X and Y cannot be independent.

2.8.17 Let X and Y have the bivariate normal distribution, as in Example 2.7.9. Prove that X and Y are independent if and only if $\rho = 0$.

2.8.18 Suppose that $(X_1, X_2, X_3) \sim \text{Multinomial}(n, \theta_1, \theta_2, \theta_3)$. Prove, by summing the joint probability function, that $X_1 \sim \text{Binomial}(n, \theta_1)$.

2.8.19 Suppose that $(X_1, X_2, X_3) \sim \text{Multinomial}(n, \theta_1, \theta_2, \theta_3)$. Find the conditional distribution of X_2 given that $X_1 = x_1$.

2.8.20 Suppose that $X_1, \ldots, X_n$ is a sample from the Exponential(λ) distribution. Find the densities $f_{X_{(1)}}$ and $f_{X_{(n)}}$.

2.8.21 Suppose that $X_1, \ldots, X_n$ is a sample from a distribution with cdf F. Prove that

$$F_{X_{(i)}}(x) = \sum_{j=i}^{n} \binom{n}{j} F^j(x)(1 - F(x))^{n-j}.$$

(Hint: Note that $X_{(i)} \leq i$ if and only if at least i of $X_1, \ldots, X_n$ are less than or equal to x.)

2.8.22 Suppose that $X_1, \ldots, X_5$ is a sample from the Uniform$[0, 1]$ distribution. If we define the sample median to be $X_{(3)}$, find the density of the sample median. Can you identify this distribution? (Hint: Use Problem 2.8.21.)

2.8.23 Suppose that $(X, Y) \sim$ Bivariate Normal$(\mu_1, \mu_2, \sigma_1, \sigma_2, \rho)$. Prove that Y given $X = x$ is distributed $N(\mu_2 + \rho\sigma_2(x - \mu_1)/\sigma_1, (1 - \rho^2)\sigma_2^2)$. Establish the analogous result for the conditional distribution of X given $Y = y$. (Hint: Use (2.7.1) for Y given $X = x$ and its analog for X given $Y = y$.)

Challenge

2.8.24 Let X and Y be random variables.
(a) Suppose X and Y are both discrete. Prove that X and Y are independent if and only if $P(Y = y \mid X = x) = P(Y = y)$ for all x and y such that $P(X = x) > 0$.
(b) Suppose X and Y are jointly absolutely continuous. Prove that X and Y are independent if and only if $P(a \leq Y \leq b \mid X = x) = P(a \leq Y \leq b)$ for all x and y such that $f_X(x) > 0$.

2.9 Multidimensional Change of Variable

Let X and Y be random variables with known joint distribution. Suppose that $Z = h_1(X, Y)$ and $W = h_2(X, Y)$, where $h_1, h_2 : R^2 \to R^1$ are two functions. What is the joint distribution of Z and W?

This is similar to the problem considered in Section 2.6, except that we have moved from a one-dimensional to a two-dimensional setting. The two-dimensional setting is more complicated; however, the results remain essentially the same, as we shall see.

2.9.1 The Discrete Case

If X and Y are *discrete* random variables, then the distribution of Z and W is essentially straightforward.

Theorem 2.9.1 Let X and Y be discrete random variables, with joint probability function $p_{X,Y}$. Let $Z = h_1(X, Y)$ and $W = h_2(X, Y)$, where $h_1, h_2 : R^2 \to R^1$ are some functions. Then Z and W are also discrete, and their joint probability function $p_{Z,W}$ satisfies

$$p_{Z,W}(z, w) = \sum_{\substack{x, y \\ h_1(x,y)=z,\ h_2(x,y)=w}} p_{X,Y}(x, y).$$

Here, the sum is taken over all pairs (x, y) such that $h_1(x, y) = z$, and $h_2(x, y) = w$.

Proof: We compute that $p_{Z,W}(z, w) = P(Z = z, W = w) = P(h_1(X, Y) = z, h_2(X, Y) = w)$, and this equals

$$\sum_{\substack{x,y \\ h_1(x,y)=z, \, h_2(x,y)=w}} P(X = x, Y = y) = \sum_{\substack{x,y \\ h_1(x,y)=z, \, h_2(x,y)=w}} p_{X,Y}(x, y),$$

as claimed. ∎

As a special case, we note the following.

Corollary 2.9.1 Suppose in the context of Theorem 2.9.1 that the joint function $h = (h_1, h_2) : R^2 \to R^2$ defined by $h(x, y) = \big(h_1(x, y), \, h_2(x, y)\big)$ is one-to-one, i.e., if $h_1(x_1, y_1) = h_1(x_2, y_2)$ and $h_2(x_1, y_1) = h_2(x_2, y_2)$, then $x_1 = x_2$ and $y_1 = y_2$. Then

$$p_{Z,W}(z, w) = p_{X,Y}(h^{-1}(z, w)),$$

where $h^{-1}(z, w)$ is the unique pair (x, y) such that $h(x, y) = (z, w)$.

Example 2.9.1
Suppose X and Y have joint density function

$$p_{X,Y}(x, y) = \begin{cases} 1/6 & x = 2, y = 6 \\ 1/12 & x = -2, y = -6 \\ 1/4 & x = -3, y = 11 \\ 1/2 & x = 3, y = -8 \\ 0 & \text{otherwise.} \end{cases}$$

Let $Z = X + Y$, and $W = Y - X^2$. Then $p_{Z,W}(8, 2) = P(Z = 8, W = 2) = P(X = 2, Y = 6) + P(X = -3, Y = 11) = 1/6 + 1/4 = 5/12$. On the other hand, $p_{Z,W}(-5, -17) = P(Z = -5, W = -17) = P(X = 3, Y = -8) = \frac{1}{2}$. ∎

2.9.2 The Continuous Case (Advanced)

If X and Y are *continuous*, and the function $h = (h_1, h_2)$ is *one to one*, then it is again possible to compute a formula for the joint density of Z and W, as the following theorem shows. To state it, recall from multivariable calculus that, if $h = (h_1, h_2) : R^2 \to R^2$ is a differentiable function, then its *Jacobian derivative* J is defined by

$$J(x, y) = \det \begin{pmatrix} \frac{\partial h_1}{\partial x} & \frac{\partial h_2}{\partial x} \\[2mm] \frac{\partial h_1}{\partial y} & \frac{\partial h_2}{\partial y} \end{pmatrix} = \frac{\partial h_1}{\partial x} \frac{\partial h_2}{\partial y} - \frac{\partial h_2}{\partial x} \frac{\partial h_1}{\partial y}.$$

Theorem 2.9.2 Let X and Y be jointly absolutely continuous, with joint density function $f_{X,Y}$. Let $Z = h_1(X, Y)$ and $W = h_2(X, Y)$, where $h_1, h_2 : R^2 \to R^1$ are differentiable functions. Define the joint function $h = (h_1, h_2) : R^2 \to R^2$ by

$$h(x, y) = (h_1(x, y), h_2(x, y)).$$

Assume that h is one-to-one, at least on the region $\{(x,y) : f(x, 0 > 0\}$, i.e., if $h_1(x_1, y_1) = h_1(x_2, y_2)$ and $h_2(x_1, y_1) = h_2(x_2, y_2)$, then $x_1 = x_2$ and $y_1 = y_2$. Then Z and W are also jointly absolutely continuous, with joint density function $f_{Z,W}$ given by

$$f_{Z,W}(z, w) = f_{X,Y}(h^{-1}(z, w))\,/\,|J(h^{-1}(z, w))|,$$

where J is the Jacobian derivative of h, and where $h^{-1}(z, w)$ is the unique pair (x, y) such that $h(x, y) = (z, w)$.

Proof: See Section 2.11 for the proof of this result. ∎

Example 2.9.2

Let X and Y be jointly absolutely continuous, with joint density function $f_{X,Y}$ given by

$$f_{X,Y}(x, y) = \begin{cases} 4x^2y + 2y^5 & 0 \le x \le 1, 0 \le y \le 1 \\ 0 & \text{otherwise,} \end{cases}$$

as in Example 2.7.6. Let $Z = X + Y^2$ and $W = X - Y^2$. What is the joint density of Z and W?

We first note that $Z = h_1(X, Y)$ and $W = h_2(X, Y)$, where $h_1(x, y) = x + y^2$ and $h_2(x, y) = x - y^2$. Hence,

$$J(x, y) = \frac{\partial h_1}{\partial x}\frac{\partial h_2}{\partial y} - \frac{\partial h_2}{\partial x}\frac{\partial h_1}{\partial y} = (1)(-2y) - (1)(2y) = -4y.$$

We may *invert* the relationship h by solving for X and Y, to obtain that

$$X = \frac{1}{2}(Z + W) \text{ and } Y = \sqrt{\frac{Z - W}{2}}.$$

This means that $h = (h_1, h_2)$ is invertible, with

$$h^{-1}(z, w) = \left(\frac{1}{2}(z + w),\ \sqrt{\frac{z - w}{2}}\right).$$

Hence, using Theorem 2.9.2, we see that

$$
\begin{aligned}
&f_{Z,W}(z, w) \\
=\ & f_{X,Y}(h^{-1}(z, w))\,/\,|J(h^{-1}(z, w))| \\
=\ & f_{X,Y}\left(\frac{1}{2}(z + w),\ \sqrt{\frac{z - w}{2}}\right)\,/\,|J(h^{-1}(z, w))| \\
=\ & \begin{cases} \left\{4(\frac{1}{2}(z + w))^2\sqrt{\frac{z-w}{2}} + 2\left(\sqrt{\frac{z-w}{2}}\right)^5\right\}/4\sqrt{\frac{z-w}{2}} & \begin{array}{l} 0 \le \frac{1}{2}(z + w) \le 1, \\ 0 \le \sqrt{\frac{z-w}{2}} \le 1 \end{array} \\ 0 & \text{otherwise} \end{cases} \\
=\ & \begin{cases} (\frac{z+w}{2})^2 + \frac{1}{2}\left(\frac{z-w}{2}\right)^2 & 0 \le z + w \le 2,\ 0 \le z - w \le 2 \\ 0 & \text{otherwise.} \end{cases}
\end{aligned}
$$

We have thus obtained the joint density function for Z and W. ∎

Example 2.9.3

Let U_1 and U_2 be independent, each having the Uniform$[0,1]$ distribution. (We could write this as U_1, U_2 are i.i.d. Uniform$[0,1]$.) Thus,

$$f_{U_1,U_2}(u_1, u_2) = \begin{cases} 1 & 0 \leq u_1 \leq 1, \ 0 \leq u_2 \leq 1 \\ 0 & \text{otherwise.} \end{cases}$$

Then define X and Y by

$$X = \sqrt{2 \log(1/U_1)} \cos(2\pi U_2), \quad Y = \sqrt{2 \log(1/U_1)} \sin(2\pi U_2).$$

What is the joint density of X and Y?

We see that here $X = h_1(U_1, U_2)$ and $Y = h_2(U_1, U_2)$, where

$$h_1(u_1, u_2) = \sqrt{2 \log(1/u_1)} \cos(2\pi u_2), \quad h_2(u_1, u_2) = \sqrt{2 \log(1/u_1)} \sin(2\pi u_2).$$

Therefore

$$\frac{\partial h_1}{\partial u_1}(u_1, u_2) = \frac{1}{2}(2 \log(1/u_1))^{-1/2}(2u_1(-1/u_1^2)) \cos(2\pi u_2).$$

Continuing in this way, we eventually compute (Exercise 2.9.1) that

$$J(u_1, u_2) = \frac{\partial h_1}{\partial u_1}\frac{\partial h_2}{\partial u_2} - \frac{\partial h_2}{\partial u_1}\frac{\partial h_1}{\partial u_2} = -\frac{2\pi}{u_1}\left(\cos^2(2\pi u_2) + \sin^2(2\pi u_2)\right) = -\frac{2\pi}{u_1}.$$

On the other hand, inverting the relationship h, we compute that

$$U_1 = e^{-(X^2+Y^2)/2}, \quad U_2 = \arctan(Y/X)\,/\,2\pi.$$

Hence, using Theorem 2.9.2, we see that

$$\begin{aligned} f_{X,Y}(x, y) &= f_{U_1,U_2}(h^{-1}(x,y))\,/\,|J(h^{-1}(x,y))| \\ &= f_{U_1,U_2}\left(e^{-(x^2+y^2)/2}, \ \arctan(y/x)\,/\,2\pi\right) \\ &\quad \times |J(e^{-(x^2+y^2)/2}, \ \arctan(y/x)\,/\,2\pi))|^{-1} \\ &= \begin{cases} 1\,/\,|-2\pi\,/\,e^{-(x^2+y^2)/2}| & 0 \leq e^{-(x^2+y^2)/2} \leq 1, \\ & 0 \leq \arctan(y/x)\,/\,2\pi \leq 1 \\ 0 & \text{otherwise} \end{cases} \\ &= \frac{1}{2\pi}e^{-(x^2+y^2)/2} \end{aligned}$$

where the last expression is valid for all x and y, because we *always* have $0 \leq e^{-(x^2+y^2)/2} \leq 1$ and $0 \leq \arctan(y/x)\,/\,2\pi \leq 1$.

We conclude that

$$f_{X,Y}(x, y) = \left(\frac{1}{\sqrt{2\pi}}e^{-x^2/2}\right)\left(\frac{1}{\sqrt{2\pi}}e^{-y^2/2}\right).$$

We recognize this as a product of two standard normal densities. We thus conclude that $X \sim N(0,1)$ and $Y \sim N(0,1)$, and furthermore X and Y are independent. ∎

2.9.3   Convolution

Suppose now that X and Y are independent, with known distributions, and that $Z = X + Y$. What is the distribution of Z? In this case, the distribution of Z is called the *convolution* of the distributions of X and of Y. Fortunately, the convolution is often reasonably straightforward to compute.

Theorem 2.9.3 Let X and Y be independent, and let $Z = X + Y$.
(a) If X and Y are both discrete, with probability functions p_X and p_Y, then Z is also discrete, with probability function p_Z given by

$$p_Z(z) = \sum_x p_X(x) p_Y(z - x).$$

(b) If X and Y are jointly absolutely continuous, with density functions f_X and f_Y, then Z is also absolutely continuous, with density function f_Z given by

$$f_Z(z) = \int_{-\infty}^{\infty} f_X(x) f_Y(z - x)\, dx.$$

Proof: (a) We let $W = Y$, and consider the two-dimensional transformation from (X, Y) to $(Z, W) = (X + Y, Y)$.

In the discrete case, by Corollary 2.9.1, $p_{Z,W}(z, w) = p_{X,Y}(z - w, w)$. Then from Theorem 2.7.4, $p_Z(z) = \sum_w p_{Z,W}(z, w) = \sum_w p_{X,Y}(z - w, w)$. But because X and Y are independent, $p_{X,Y}(x, y) = p_X(x)\, p_Y(y)$, so $p_{X,Y}(z - w, w) = p_X(z - w)\, p_Y(w)$. This proves part (a).

(b) In the continuous case, we must compute the Jacobian derivative $J(x, y)$ of the transformation from (X, Y) to $(Z, W) = (X + Y, Y)$. Fortunately, this is very easy, as we obtain

$$J(x, y) = \frac{\partial(x + y)}{\partial x}\frac{\partial y}{\partial y} - \frac{\partial y}{\partial x}\frac{\partial(x + y)}{\partial y} = (1)(1) - (0)(1) = 1.$$

Hence, from Theorem 2.9.2, $f_{Z,W}(z, w) = f_{X,Y}(z - w, w)/|1| = f_{X,Y}(z - w, w)$ and from Theorem 2.7.5,

$$f_Z(z) = \int_{-\infty}^{\infty} f_{Z,W}(z, w)\, dw = \int_{-\infty}^{\infty} f_{X,Y}(z - w, w)\, dw.$$

But because X and Y are independent, we may take $f_{X,Y}(x, y) = f_X(x)\, f_Y(y)$, so $f_{X,Y}(z - w, w) = f_X(z - w)\, f_Y(w)$. This proves part (b). ∎

Example 2.9.4
Let $X \sim \text{Binomial}(4, 1/5)$ and $Y \sim \text{Bernoulli}(1/4)$, with X and Y independent. Let $Z = X + Y$. Then

$$
\begin{aligned}
p_Z(3) &= P(X + Y = 3) = P(X = 3, Y = 0) + P(X = 2, Y = 1) \\
&= \binom{4}{3}(1/5)^3(4/5)^1\,(3/4) + \binom{4}{2}(1/5)^2(4/5)^2\,(1/4) \\
&= 4(1/5)^3(4/5)^1\,(3/4) + 6(1/5)^2(4/5)^2\,(1/4) \doteq 0.0576. \quad \blacksquare
\end{aligned}
$$

Example 2.9.5

Let $X \sim \text{Uniform}[3, 7]$ and $Y \sim \text{Exponential}(6)$, with X and Y independent. Let $Z = X + Y$. Then

$$
\begin{aligned}
f_Z(5) &= \int_{-\infty}^{\infty} f_X(x)\, f_Y(5-x)\, dx = \int_{3}^{5} (1/4)\, 6\, e^{-6(5-x)}\, dx \\
&= -(1/4)e^{-6(5-x)} \Big|_{x=3}^{x=5} = -(1/4)e^{-12} + (1/4)e^0 \doteq 0.2499985.
\end{aligned}
$$

Note that here the limits of integration go from 3 to 5 only, because $f_X(x) = 0$ for $x < 3$, while $f_Y(5-x) = 0$ for $x > 5$. ∎

Summary of Section 2.9

- If X and Y are discrete, and $Z = h_1(X, Y)$ and $W = h_2(X, Y)$, then

$$
p_{Z,W}(z, w) = \sum_{\{(x,y):\, h_1(x,y)=z,\, h_2(x,y)=w\}} p_{X,Y}(x, y).
$$

- If X and Y are absolutely continuous, and $Z = h_1(X, Y)$ and $W = h_2(X, Y)$, and $h = (h_1, h_2) : R^2 \to R^2$ is one-to-one with Jacobian $J(x, y)$, then $f_{Z,W}(z, w) = f_{X,Y}(h^{-1}(z, w))/|J(h^{-1}(z, w))|$.

- This allows us to compute the joint distribution of functions of pairs of random variables.

Exercises

2.9.1 Verify explicitly in Example 2.9.3 that $J(u_1, u_2) = -2\pi/u_1$.

2.9.2 Let $X \sim \text{Exponential}(3)$ and $Y \sim \text{Uniform}[1, 4]$, with X and Y independent. Let $Z = X + Y$ and $W = X - Y$.
(a) Write down the joint density $f_{X,Y}(x, y)$ of X and Y. (Be sure to consider the ranges of valid x and y values.)
(b) Find a two-dimensional function h such that $(Z, W) = h(X, Y)$.
(c) Find a two-dimensional function h^{-1} such that $(X, Y) = h^{-1}(Z, W)$.
(d) Compute the joint density $f_{Z,W}(z, w)$ of Z and W. (Again, be sure to consider the ranges of valid z and w values.)

2.9.3 Repeat parts (b) through (d) of Exercise 2.9.2, for the same random variables X and Y, if instead $Z = X^2 + Y^2$ and $W = X^2 - Y^2$.

2.9.4 Repeat parts (b) through (d) of Exercise 2.9.2, for the same random variables X and Y, if instead $Z = X + 4$ and $W = Y - 3$.

2.9.5 Repeat parts (b) through (d) of Exercise 2.9.2, for the same random variables X and Y, if instead $Z = Y^4$ and $W = X^4$.

2.9.6 Suppose the joint probability function of X and Y is given by

$$p_{X,Y}(x,y) = \begin{cases} 1/7 & x = 5, y = 0 \\ 1/7 & x = 5, y = 3 \\ 1/7 & x = 5, y = 4 \\ 3/7 & x = 8, y = 0 \\ 1/7 & x = 8, y = 4 \\ 0 & \text{otherwise.} \end{cases}$$

Let $Z = X + Y$, $W = X - Y$, $A = X^2 + Y^2$, and $B = 2X - 3Y^2$.
(a) Compute the joint probability function $p_{Z,W}(z,w)$.
(b) Compute the joint probability function $p_{A,B}(a,b)$.
(c) Compute the joint probability function $p_{Z,A}(z,a)$.
(d) Compute the joint probability function $p_{W,B}(w,b)$.

2.9.7 Let X have probability function

$$p_X(x) = \begin{cases} 1/3 & x = 0 \\ 1/2 & x = 2 \\ 1/6 & x = 3 \\ 0 & \text{otherwise,} \end{cases}$$

and let Y have probability function

$$p_Y(y) = \begin{cases} 1/6 & y = 2 \\ 1/12 & y = 5 \\ 3/4 & y = 9 \\ 0 & \text{otherwise.} \end{cases}$$

Suppose X and Y are independent. Let $Z = X + Y$. Compute $p_Z(z)$ for all $z \in R^1$.

2.9.8 Let $X \sim \text{Geometric}(1/4)$, and let Y have probability function

$$p_Y(y) = \begin{cases} 1/6 & y = 2 \\ 1/12 & y = 5 \\ 3/4 & y = 9 \\ 0 & \text{otherwise.} \end{cases}$$

Let $W = X + Y$. Suppose X and Y are independent. Compute $p_W(w)$ for all $w \in R^1$.

Problems

2.9.9 Let X and Y be independent, with $X \sim \text{Binomial}(n_1, \theta)$ and $Y \sim \text{Binomial}(n_2, \theta)$. Let $Z = X + Y$. Use Theorem 2.9.3(a) to prove that $Z \sim \text{Binomial}(n_1 + n_2, \theta)$.

2.9.10 Let X and Y be independent, with $X \sim \text{Negative-Binomial}(r_1, \theta)$ and $Y \sim \text{Negative-Binomial}(r_2, \theta)$. Let $Z = X + Y$. Use Theorem 2.9.3(a) to prove that $Z \sim \text{Negative-Binomial}(r_1 + r_2, \theta)$.

2.9.11 Let X and Y be independent, with $X \sim N(\mu_1, \sigma_1^2)$ and $Y \sim N(\mu_2, \sigma_2^2)$. Let $Z = X + Y$. Use Theorem 2.9.3(b) to prove that $Z \sim N(\mu_1 + \mu_2, \sigma_1^2 + \sigma_2^2)$.

2.9.12 Let X and Y be independent, with $X \sim \text{Gamma}(\alpha_1, \lambda)$ and $Y \sim \text{Gamma}(\alpha_2, \lambda)$. Let $Z = X + Y$. Use Theorem 2.9.3(b) to prove that $Z \sim \text{Gamma}(\alpha_1 + \alpha_2, \lambda)$.

2.9.13 (MV) Show that when Z_1, Z_2 are i.i.d. $N(0, 1)$ and X, Y are given by (2.7.1), then $(X, Y) \sim \text{Bivariate Normal}(\mu_1, \mu_2, \sigma_1, \sigma_2, \rho)$.

2.10 Simulating Probability Distributions

So far, we have been concerned primarily with mathematical theory and manipulations of probabilities and random variables. However, modern high-speed computers can be used to simulate probabilities and random variables *numerically*. Such simulations have many applications, including:

- To approximate quantities that are too difficult to compute mathematically;

- To graphically simulate complicated physical or biological systems;

- To randomly sample from large data sets to search for errors or illegal activities, etc.

- To implement complicated algorithms to sharpen pictures, recognize speech and handwriting, etc.

- To simulate intelligent behavior;

- To encrypt data or generate passwords;

- To solve puzzles or break codes by trying lots of random solutions;

- To generate random choices for on-line quizzes, computer games, etc.

Indeed, as computers become faster and more widespread, probabilistic simulations are becoming more and more common in software applications, scientific research, quality control, marketing, law enforcement, etc.

In most applications of probabilistic simulation, the first step is to simulate random variables having certain distributions. That is, a certain probability distribution will be specified, and it will be desired to generate one or more random variables having that distribution.

Now, nearly all modern computer languages come with a *pseudorandom number generator*, which is a device for generating a sequence $U_1, U_2, \ldots$ of random values that are approximately independent and have approximately the uniform distribution on $[0, 1]$. Now, in fact, the U_i are usually generated from some sort of deterministic iterative procedure, which is designed to "appear" random. So, the U_i are, in fact, not random, but rather *pseudorandom*.

Nevertheless, we shall ignore any concerns about pseudorandomness, and shall simply assume that

$$U_1, U_2, U_3, \ldots \sim \text{Uniform}[0, 1], \qquad (2.10.1)$$

i.e., the U_i are i.i.d. Uniform$[0, 1]$.

Hence, if all we ever need are Uniform$[0, 1]$ random variables, then according to (2.10.1), we are all set. However, in most applications, other kinds of randomness are also required. We therefore consider how to use the uniform random variables of (2.10.1) to generate random variables having other distributions.

Example 2.10.1 *The Uniform$[L, R]$ Distribution*
Suppose we want to generate $X \sim \text{Uniform}[L, R]$. According to Exercise 2.6.1, we can simply set

$$X = (R - L)U_1 + L,$$

to ensure that $X \sim \text{Uniform}[L, R]$. ∎

2.10.1   Simulating Discrete Distributions

We now consider the question of how to simulate from discrete distributions.

Example 2.10.2 *The Bernoulli(θ) Distribution*
Suppose we want to generate $X \sim \text{Bernoulli}(\theta)$, where $0 < \theta < 1$. We can simply set

$$X = \begin{cases} 1 & U_1 \leq \theta \\ 0 & U_1 > p \end{cases}$$

Then clearly we always have either $X = 0$ or $X = 1$. Furthermore, $P(X = 1) = P(U_1 \leq \theta) = \theta$, because $U_1 \sim \text{Uniform}[0, 1]$. Hence, we see that $X \sim \text{Bernoulli}(\theta)$. ∎

Example 2.10.3 *The Binomial(n, θ) Distribution*
Suppose we want to generate $Y \sim \text{Binomial}(n, \theta)$, where $0 < \theta < 1$ and $n \geq 1$. There are two natural methods for doing this.

For the first method, we can simply define Y as follows

$$Y = \min\{j : \sum_{k=0}^{j} \binom{n}{k} \theta^k (1 - \theta)^{n-k} \geq U_1\}.$$

That is, we let Y be the largest value of j such that the sum of the binomial probabilities up to $j - 1$ is still no more than U_1. In that case,

$$
\begin{aligned}
P(Y = y) &= P\left(
\begin{array}{c}
\sum_{k=0}^{y-1} \binom{n}{k} \theta^k (1-\theta)^{n-k} < U_1 \\
\text{and } \sum_{k=0}^{y} \binom{n}{k} \theta^k (1-\theta)^{n-k} \geq U_1
\end{array}
\right) \\
&= P\left(\sum_{k=0}^{y-1} \binom{n}{k} \theta^k (1-\theta)^{n-k} < U_1 \leq \sum_{k=0}^{y} \binom{n}{k} \theta^k (1-\theta)^{n-k} \right) \\
&= \sum_{k=0}^{y} \binom{n}{k} \theta^k (1-\theta)^{n-k} - \sum_{k=0}^{y-1} \binom{n}{k} \theta^k (1-\theta)^{n-k} \\
&= \binom{n}{y} \theta^y (1-\theta)^{n-y}.
\end{aligned}
$$

Hence, we have $Y \sim \text{Binomial}(n, \theta)$, as desired.

Alternatively, we can set

$$
X_i = \begin{cases} 1 & U_i \leq p \\ 0 & U_i > p \end{cases}
$$

for $i = 1, 2, 3, \ldots$. Then, by Example 2.10.2, we have $X_i \sim \text{Bernoulli}(\theta)$ for each i, with the $\{X_i\}$ independent because the $\{U_i\}$ are independent. Hence, by the observation at the end of Example 2.3.3, if we set $Y = X_1 + \cdots + X_n$, then we will again have $Y \sim \text{Binomial}(n, \theta)$. ∎

In Example 2.10.3, the second method is more elegant and is also simpler computationally (as it does not require computing any binomial coefficients). On the other hand, the first method of Example 2.10.3 is more general, as the following theorem shows.

Theorem 2.10.1 Let p be a probability function for a discrete probability distribution. Let $x_1 < x_2 < x_3 < \cdots$ be all the values for which $p(x_i) > 0$. Let $U_1 \sim \text{Uniform}[0, 1]$. Define Y by

$$
Y = \min\{x_j : \sum_{k=1}^{j} p(x_k) \geq U_1\}.
$$

Then Y is a discrete random variable, having probability function p.

Proof: We have

$$
\begin{aligned}
P(Y = x_i) &= P\left(\sum_{k=1}^{i-1} p(x_k) < U_1, \text{ and } \sum_{k=1}^{i} p(x_k) \geq U_1 \right) \\
&= P\left(\sum_{k=1}^{i-1} p(x_k) < U_1 \leq \sum_{k=1}^{i} p(x_k) \right) \\
&= \sum_{k=1}^{i} p(x_k) - \sum_{k=1}^{i-1} p(x_k) = p(x_i).
\end{aligned}
$$

Also, clearly $P(Y = y) = 0$ if $y \notin \{x_1, x_2, \ldots\}$. Hence, for all $y \in R^1$, we have $P(Y = y) = p(y)$, as desired. ∎

Example 2.10.4 *The Geometric(θ) Distribution*
To simulate $Y \sim$ Geometric(θ), we again have two choices. Using Theorem 2.10.1, we can let $U_1 \sim$ Uniform$[0, 1]$ and then set

$$
\begin{aligned}
Y &= \min\{j : \sum_{k=0}^{j} \theta(1 - \theta)^k \geq U_1\} = \min\{j : 1 - (1 - \theta)^{j+1} \geq U_1\} \\
&= \min\{j : j \geq \frac{\log(1 - U_1)}{\log(1 - \theta)} - 1\} = \left\lfloor \frac{\log(1 - U_1)}{\log(1 - \theta)} \right\rfloor,
\end{aligned}
$$

where $\lfloor r \rfloor$ means to round down r to the next integer value, i.e., $\lfloor r \rfloor$ is the *greatest integer* not exceeding r (sometimes called the *floor* of r).

Alternatively, using the definition of Geometric(θ) from Example 2.3.4, we can set

$$
X_i = \begin{cases} 1 & U_i \leq \theta \\ 0 & U_i > \theta \end{cases}
$$

for $i = 1, 2, 3, \ldots$ (where $U_i \sim$ Uniform$[0, 1]$), and then let $Y = \min\{i : X_i = 1\}$. Either way, we have $Y \sim$ Geometric(θ), as desired. ∎

2.10.2 Simulating Continuous Distributions

We next turn to the subject of simulating absolutely continuous distributions. In general, this is not an easy problem. However, for certain particular continuous distributions, it is not difficult, as we now demonstrate.

Example 2.10.5 *The* Uniform$[L, R]$ *Distribution*
We have already seen in Example 2.10.1 that if $U_1 \sim$ Uniform$[0, 1]$, and we set

$$
X = (R - L)U_1 + L,
$$

then $X \sim$ Uniform$[L, R]$. Thus, simulating from any uniform distribution is straightforward. ∎

Example 2.10.6 *The Exponential(λ) Distribution*
We have also seen, in Example 2.6.6, that if $U_1 \sim$ Uniform$[0, 1]$, and we set

$$
Y = \ln(1/U_1),
$$

then $Y \sim$ Exponential(1). Thus, simulating from the Exponential(1) distribution is straightforward.

Furthermore, we know from Exercise 2.6.4 that once $Y \sim$ Exponential(1), then if $\lambda > 0$, and we set

$$
Z = Y / \lambda = \ln(1/U_1)/\lambda,
$$

then $Z \sim$ Exponential(λ). Thus, simulating from any Exponential(λ) distribution is also straightforward. ∎

Example 2.10.7 *The $N(\mu, \sigma^2)$ Distribution*
Simulating from the standard normal distribution, $N(0,1)$, may appear to be more difficult. However, by Example 2.9.3, if $U_1 \sim$ Uniform$[0,1]$ and $U_2 \sim$ Uniform$[0,1]$, with U_1 and U_2 independent, and we set

$$X = \sqrt{2\log(1/U_1)} \cos(2\pi U_2), \quad Y = \sqrt{2\log(1/U_1)} \sin(2\pi U_2), \qquad (2.10.2)$$

then $X \sim N(0,1)$ and $Y \sim N(0,1)$ (and furthermore X and Y are independent). So, using this trick, the standard normal distribution can be easily simulated as well.

It then follows from Exercise 2.6.3 that, once we have $X \sim N(0,1)$, if we set $Z = \sigma X + \mu$, then $Z \sim N(\mu, \sigma^2)$. Hence, it is straightforward to sample from any normal distribution. ∎

These examples illustrate that, for certain special continuous distributions, sampling from them is straightforward. To provide a *general* method of sampling from a continuous distribution, we first state the following definition.

Definition 2.10.1 Let X be a random variable, with cumulative distribution function F. Then the *inverse cdf* (or *quantile function*) of X is the function F^{-1} defined by

$$F^{-1}(t) = \min\{x : F(x) \geq t\},$$

for $0 < t < 1$.

In Figure 2.10.1, we have provided a plot of the inverse cdf of an $N(0,1)$ distribution. Note that this function goes to $-\infty$ as the argument goes to 0, and goes to ∞ as the argument goes to 1.

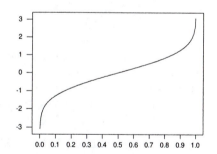

Figure 2.10.1: The inverse cdf of the $N(0,1)$ distribution.

Using the inverse cdf, we obtain a general method of sampling from a continuous distribution, as follows.

Theorem 2.10.2 (*Inversion method for generating random variables*) Let F be any cumulative distribution function, and let $U \sim$ Uniform$[0,1]$. Define a random variable Y by $Y = F^{-1}(U)$. Then $P(Y \leq y) = F(y)$, i.e., Y has cumulative distribution function given by F.

Proof: We begin by noting that $P(Y \leq y) = P(F^{-1}(U) \leq y)$. But $F^{-1}(U)$ is the smallest value x such that $F(x) \geq U$. Hence, $F^{-1}(U) \leq y$ if and only if $F(y) \geq U$, i.e., $U \leq F(y)$. Hence,

$$P(Y \leq y) = P(F^{-1}(U) \leq y) = P(U \leq F(y)).$$

But $0 \leq F(y) \leq 1$, and $U \sim \text{Uniform}[0, 1]$, so $P(U \leq F(y)) = F(y)$. Hence,

$$P(Y \leq y) = P(U \leq F(y)) = F(y).$$

It follows that F is the cdf of Y, as claimed. ∎

We note that Theorem 2.10.2 is valid for *any* cumulative distribution function, whether it corresponds to a continuous distribution, a discrete distribution, or a mixture of the two as in Section 2.5.4. In fact, this was proved for discrete distributions in Theorem 2.10.1.

Example 2.10.8 *Generating from an Exponential Distribution*
Let F be the cdf of an Exponential(1) random variable. Then

$$F(x) = \int_0^x e^{-t}\, dt = 1 - e^{-x}.$$

It then follows that

$$
\begin{aligned}
F^{-1}(t) &= \min\{x : F(x) \geq t\} = \min\{x : 1 - e^{-x} \geq t\} \\
&= \min\{x : x \geq -\ln(1-t)\} = -\ln(1-t) = \ln(1/(1-t)).
\end{aligned}
$$

Therefore, by Theorem 2.10.2, if $U \sim \text{Uniform}[0, 1]$, and we set

$$Y = F^{-1}(U) = \ln(1/(1-U)), \tag{2.10.3}$$

then $Y \sim \text{Exponential}(1)$.

Now, we have already seen from Example 2.6.6 that, if $U \sim \text{Uniform}[0, 1]$, and we set $Y = \ln(1/U)$, then $Y \sim \text{Exponential}(1)$. This is essentially the same as (2.10.3), except that we have replaced U by $1 - U$. On the other hand, this is not surprising, because we already know by Exercise 2.6.2 that, if $U \sim \text{Uniform}[0, 1]$, then also $1 - U \sim \text{Uniform}[0, 1]$. ∎

Example 2.10.9 *Generating from the Standard Normal Distribution*
Let Φ be the cdf of a $N(0, 1)$ random variable, as in Definition 2.5.4. Then

$$\Phi^{-1}(t) = \min\{x : \Phi(x) \geq t\},$$

and there is no simpler formula for $\Phi^{-1}(t)$. By Theorem 2.10.2, if $U \sim \text{Uniform}[0, 1]$, and we set

$$Y = \Phi^{-1}(U), \tag{2.10.4}$$

then $Y \sim N(0, 1)$.

On the other hand, due to the difficulties of computing with Φ and Φ^{-1}, the method (2.10.4) is not very practical. It is far better to use the method of (2.10.2), to simulate a normal random variable. $\blacksquare$

For distributions that are too complicated to sample using the inversion method of Theorem 2.10.2, and for which no simple trick is available, it may still be possible to do sampling using *Markov chain methods,* which we will discuss in later chapters, or by *rejection sampling* (see Challenge 2.10.13).

Summary of Section 2.10

- It is important to be able to simulate probability distributions.

- If X is discrete, taking the value x_i with probability p_i, where $x_1 < x_2 < \cdots$, and $U \sim \text{Uniform}[0, 1]$, and $Y = \min\left\{x_j : \sum_{k=1}^{j} p_k \geq U\right\}$, then Y has the same distribution as X. This allows us to simulate virtually any discrete distribution.

- If F is any cumulative distribution with inverse cdf F^{-1}, $U \sim \text{Uniform}[0, 1]$, and $Y = F^{-1}(U)$, then Y has cumulative distribution function F. This allows us to simulate virtually any distribution.

- There are simple methods of simulating many standard distributions, including the binomial, uniform, exponential, and normal.

Exercises

2.10.1 Let Y be a discrete random variable with $P(Y = -7) = 1/2$, $P(Y = -2) = 1/3$, and $P(Y = 5) = 1/6$. Find a formula for Z in terms of U, such that if $U \sim \text{Uniform}[0, 1]$, then Z has the same distribution as Y.

2.10.2 For each of the following cumulative distribution functions F, find a formula for X in terms of U, such that if $U \sim \text{Uniform}[0, 1]$, then X has cumulative distribution function F.

(a)
$$F(x) = \begin{cases} 0 & x < 0 \\ x & 0 \leq x \leq 1 \\ 1 & x > 1 \end{cases}$$

(b)
$$F(x) = \begin{cases} 0 & x < 0 \\ x^2 & 0 \leq x \leq 1 \\ 1 & x > 1 \end{cases}$$

(c)
$$F(x) = \begin{cases} 0 & x < 0 \\ x^2/9 & 0 \leq x \leq 3 \\ 1 & x > 3 \end{cases}$$

(d)

$$F(x) = \begin{cases} 0 & x < 1 \\ x^2/9 & 1 \leq x \leq 3 \\ 1 & x > 3 \end{cases}$$

(e)

$$F(x) = \begin{cases} 0 & x < 0 \\ x^5/32 & 0 \leq x \leq 2 \\ 1 & x > 2 \end{cases}$$

(f)

$$F(x) = \begin{cases} 0 & x < 0 \\ 1/3 & 0 \leq x < 7 \\ 3/4 & 7 \leq x < 11 \\ 1 & x \geq 11 \end{cases}$$

Computer Exercises

2.10.3 For each of the following distributions, use the computer (you can use any algorithms available to you as part of a software package) to simulate $X_1, X_2, \ldots, X_N$ i.i.d. having the given distribution. (Take $N = 1000$ at least, with $N = 10,000$ or $N = 100,000$ if possible.) Then compute $\bar{X} = (1/N) \sum_{i=1}^{N} X_i$ and $(1/N) \sum_{i=1}^{N} (X_i - \bar{X})^2$.

(a) Uniform$[0, 1]$
(b) Uniform$[5, 8]$
(c) Bernoulli$(1/3)$
(d) Binomial$(12, 1/3)$
(e) Geometric$(1/5)$
(f) Exponential(1)
(g) Exponential(13)
(h) $N(0, 1)$
(i) $N(5, 9)$

Problems

2.10.4 Let $G(x) = p_1 F_1(x) + p_2 F_2(x) + \cdots + p_k F_k(x)$, where $p_i \geq 0$, $\sum_i p_i = 1$, and F_i are cdfs, as in (2.5.3). Suppose we can generate X_i to have cdf F_i, for $i = 1, 2, \ldots, k$. Describe a procedure for generating a random variable Y that has cdf G.

2.10.5 Let X be an absolutely continuous random variable, with density given by $f_X(x) = x^{-2}$ for $x \geq 1$, with $f_X(x) = 0$ otherwise. Find a formula for Z in terms of U, such that if $U \sim \text{Uniform}[0, 1]$, then Z has the same distribution as X.

2.10.6 Find the inverse cdf of the logistic distribution of Problem 2.4.10. (Hint: See Problem 2.5.11.)

2.10.7 Find the inverse cdf of the Weibull(α) distribution of Problem 2.4.11. (Hint: See Problem 2.5.12.)

2.10.8 Find the inverse cdf of the Pareto(α) distribution of Problem 2.4.12. (Hint: See Problem 2.5.13.)

2.10.9 Find the inverse cdf of the Cauchy distribution of Problem 2.4.13. (Hint: See Problem 2.5.14.)

2.10.10 Find the inverse cdf of the Laplace distribution of Problem 2.4.14. (Hint: See Problem 2.5.15.)

2.10.11 Find the inverse cdf of the extreme value distribution of Problem 2.4.15. (Hint: See Problem 2.5.16.)

2.10.12 Find the inverse cdfs of the beta distributions in Problem 2.4.16 (b) through (d). (Hint: See Problem 2.5.17.)

2.10.13 *(Method of composition)* If we generate $X \sim f_X$ obtaining x, and then generate Y from $f_{Y|X}(\cdot | x)$, prove that $Y \sim f_Y$.

Challenge

2.10.14 *(Rejection sampling)* Suppose f is a complicated density function. Suppose g is a density function from which it is easy to sample (e.g., the density of a uniform or exponential or normal distribution). Suppose we know a value of c such that $f(x) \leq cg(x)$ for all $x \in R^1$. The following provides a method, called rejection sampling, for sampling from a complicated density f by using a simpler density g, provided only that we know $f(x) \leq cg(x)$ for all $x \in R^1$.
(a) Suppose Y has density g. Let $U \sim \text{Uniform}[0, c]$, with U and Y independent. Prove that

$$P(a \leq Y \leq b \mid f(Y) \geq Ucg(Y)) = \int_a^b f(x) \, dx.$$

(Hint: Use Theorem 2.8.1 to show that $P(a \leq Y \leq b, \ f(Y) \geq cUg(Y)) = \int_a^b g(y)P(f(Y) \geq cUg(Y) \mid Y = y) \, dy$.)
(b) Suppose that $Y_1, Y_2, \ldots$ are i.i.d., each with density g, and independently $U_1, U_2, \ldots$ are i.i.d. Uniform$[0, c]$. Let $i_0 = 0$, and for $n \geq 1$, let $i_n = \min\{j > i_{n-1} : U_j f(Y_j) \geq cg(Y_j)\}$. Prove that $X_{i_1}, X_{i_2}, \ldots$ are i.i.d., each with density f. (Hint: Prove this for X_{i_1}, X_{i_2}.)

2.11 Further Proofs (Advanced)

Theorem 2.4.2 The function ϕ given by (2.4.9) is a density function.

Proof: Clearly $\phi(x) \geq 0$ for all x. However, to prove that $\int_{-\infty}^{\infty} \phi(x) \, dx = 1$ is more difficult.

To proceed, we set $I = \int_{-\infty}^{\infty} \phi(x)\, dx$. Then, using multivariable calculus,

$$
\begin{aligned}
I^2 &= \left(\int_{-\infty}^{\infty} \phi(x)\, dx \right)^2 = \left(\int_{-\infty}^{\infty} \phi(x)\, dx \right) \left(\int_{-\infty}^{\infty} \phi(y)\, dy \right) \\
&= \int_{-\infty}^{\infty} \int_{-\infty}^{\infty} \phi(x)\, \phi(y)\, dx\, dy = \int_{-\infty}^{\infty} \int_{-\infty}^{\infty} \frac{1}{2\pi} e^{-(x^2+y^2)/2}\, dx\, dy.
\end{aligned}
$$

We now switch to polar coordinates (r, θ), so that $x = r \cos \theta$ and $y = r \sin \theta$, where $r > 0$ and $0 \leq \theta \leq 2\pi$. Then $x^2 + y^2 = r^2$, and by the multivariable change of variable theorem from calculus, $dx\, dy = r\, dr\, d\theta$. Hence,

$$
\begin{aligned}
I^2 &= \int_0^{2\pi} \int_0^{\infty} \frac{1}{2\pi} e^{-r^2/2} r\, dr\, d\theta = \int_0^{\infty} e^{-r^2/2} r\, dr \\
&= -e^{-r^2/2} \Big|_{r=0}^{r=\infty} = (-0) - (-1) = 1,
\end{aligned}
$$

and we have $I^2 = 1$. But clearly $I \geq 0$ (because $\phi \geq 0$), so we must have $I = 1$, as claimed. ∎

Theorem 2.6.2 Let X be an absolutely continuous random variable, with density function f_X. Let $Y = h(X)$, where $h : R^1 \to R^1$ is a function that is differentiable and strictly increasing. Then Y is also absolutely continuous, and its density function f_Y is given by

$$
f_Y(y) = f_X(h^{-1}(y)) \,/\, |h'(h^{-1}(y))|, \tag{2.11.1}
$$

where h' is the derivative of h, and where $h^{-1}(y)$ is the unique number x such that $h(x) = y$.

Proof: We must show that whenever $a \leq b$, we have

$$
P(a \leq Y \leq b) = \int_a^b f_Y(y)\, dy,
$$

where f_Y is given by (2.11.1). To that end, we note that, because h is strictly increasing, so is h^{-1}. Hence, applying h^{-1} preserves inequalities, so that

$$
\begin{aligned}
P(a \leq Y \leq b) &= P(h^{-1}(a) \leq h^{-1}(Y) \leq h^{-1}(b)) = P(h^{-1}(a) \leq X \leq h^{-1}(b)) \\
&= \int_{h^{-1}(a)}^{h^{-1}(b)} f_X(x)\, dx.
\end{aligned}
$$

We then make the substitution $y = h(x)$, so that $x = h^{-1}(y)$, and

$$
dx = \left| \frac{d}{dy} h^{-1}(y) \right| dy.
$$

But by the inverse function theorem from calculus, $\frac{d}{dy} h^{-1}(y) = 1/h'(h^{-1}(y))$. Furthermore, as x goes from $h^{-1}(a)$ to $h^{-1}(b)$, we see that $y = h(x)$ goes from *a* to *b*. We conclude that

$$P(a \le Y \le b) = \int_{h^{-1}(a)}^{h^{-1}(b)} f_X(x)\,dx = \int_a^b f_X(h^{-1}(y))(1/|h'(h^{-1}(y))|)\,dy$$
$$= \int_a^b f_Y(y)\,dy,$$

as required. ∎

Theorem 2.6.3 Let X be an absolutely continuous random variable, with density function f_X. Let $Y = h(X)$, where $h : R^1 \to R^1$ is a function that is differentiable and strictly decreasing. Then Y is also absolutely continuous, and its density function f_Y may again be defined by (2.11.1).

Proof: We note that, because h is strictly decreasing, so is h^{-1}. Hence, applying h^{-1} reverses the inequalities, so that

$$P(a \le Y \le b) = P(h^{-1}(b) \le h^{-1}(Y) \le h^{-1}(a)) = P(h^{-1}(b) \le X \le h^{-1}(a))$$
$$= \int_{h^{-1}(b)}^{h^{-1}(a)} f_X(x)\,dx.$$

We then make the substitution $y = h(x)$, so that $x = h^{-1}(y)$, and

$$dx = \left| \frac{d}{dy} h^{-1}(y) \right| dy.$$

But by the inverse function theorem from calculus,

$$\frac{d}{dy} h^{-1}(y) = \frac{1}{h'(h^{-1}(y))}.$$

Furthermore, as x goes from $h^{-1}(b)$ to $h^{-1}(a)$, we see that $y = h(x)$ goes from a to b. We conclude that

$$P(a \le Y \le b) = \int_{h^{-1}(b)}^{h^{-1}(a)} f_X(x)\,dx = \int_a^b f_X(h^{-1}(y))\,(1/|h'(h^{-1}(y))|)\,dy$$
$$= \int_a^b f_Y(y)\,dy,$$

as required. ∎

Theorem 2.9.2 Let X and Y be jointly absolutely continuous, with joint density function $f_{X,Y}$. Let $Z = h_1(X,Y)$ and $W = h_2(X,Y)$, where $h_1, h_2 : R^2 \to R^1$ are differentiable functions. Define the joint function $h = (h_1, h_2) : R^2 \to R^2$ by

$$h(x,y) = (h_1(x,y), h_2(x,y)).$$

Assume that h is one-to-one, at least on the region $\{(x,y) : f(x,0 > 0\}$, i.e., if $h_1(x_1,y_1) = h_1(x_2,y_2)$ and $h_2(x_1,y_1) = h_2(x_2,y_2)$, then $x_1 = x_2$ and $y_1 = y_2$.

Then Z and W are also jointly absolutely continuous, with joint density function $f_{Z,W}$ given by

$$f_{Z,W}(z,w) = f_{X,Y}(h^{-1}(z,w)) / |J(h^{-1}(z,w))|,$$

where J is the Jacobian derivative of h, and where $h^{-1}(z,w)$ is the unique pair (x,y) such that $h(x,y) = (z,w)$.

Proof: We must show that whenever $a \leq b$ and $c \leq d$, we have

$$P(a \leq Z \leq b, \ c \leq W \leq d) = \int_c^d \int_a^b f_{Z,W}(z,w) \, dw \, dz.$$

If we let $S = [a,b] \times [c,d]$ be the two-dimensional rectangle, then we can rewrite this as

$$P((Z,W) \in S) = \int_S \int f_{Z,W}(z,w) \, dz \, dw.$$

Now, using the theory of multivariable calculus, and making the substitution $(x,y) = h^{-1}(z,w)$ (which is permissible because h is one-to-one), we have

$$\int_S \int f_{Z,W}(z,w) \, dz \, dw$$

$$= \int_S \int \left(f_{X,Y}(h^{-1}(z,w)) / |J(h^{-1}(z,w))| \right) \, dz \, dw$$

$$= \int_{h^{-1}(S)} \int \left(f_{X,Y}(x,y) / |J(x,y)| \right) |J(x,y)| \, dx \, dy$$

$$= \int_{h^{-1}(S)} \int f_{X,Y}(x,y) \, dx \, dy = P((X,Y) \in h^{-1}(S))$$

$$= P(h^{-1}(Z,W) \in h^{-1}(S)) \ = P((Z,W) \in S),$$

as required. ∎

Chapter 3

Expectation

In the first two chapters we learned about probability models, random variables, and distributions. There is one more concept that is fundamental to all of probability theory, that of expected value.

Intuitively, the expected value of a random variable is the average value that the random variable takes on. For example, if half the time $X = 0$, and the other half of the time $X = 10$, then the average value of X is 5. We shall write this as $E(X) = 5$. Similarly, if one-third of the time $Y = 6$ while two-thirds of the time $Y = 15$, then $E(Y) = 12$.

Another interpretation of expected value is in terms of fair gambling. Suppose someone offers you a ticket (e.g., a lottery ticket) worth a certain random amount X. How much would you be willing to pay to buy the ticket? It seems reasonable that you would be willing to pay the expected value $E(X)$ of the ticket, but no more. However, this interpretation does have certain limitations; see Example 3.1.12.

To understand expected value more precisely, we consider discrete and absolutely continuous random variables separately.

3.1   The Discrete Case

We begin with a definition.

Definition 3.1.1 Let X be a discrete random variable. Then the *expected value* (or *mean value* or *mean*) of X, written $E(X)$ (or μ_X), is defined by

$$E(X) = \sum_{x \in R^1} x\, P(X = x) = \sum_{x \in R^1} x\, p_X(x).$$

We will have $P(X = x) = 0$ except for those values x that are possible values of X. Hence, an equivalent definition is the following.

Definition 3.1.2 Let X be a discrete random variable, taking on distinct values $x_1, x_2, \ldots,$ with $p_i = P(X = x_i)$. Then the *expected value* of X is given by

$$E(X) = \sum_i x_i\, p_i.$$

The definition (in either form) is best understood through examples.

Example 3.1.1
Suppose, as above, that $P(X = 0) = P(X = 10) = 1/2$. Then

$$E(X) = (0)(1/2) + (10)(1/2) = 5,$$

as predicted. ∎

Example 3.1.2
Suppose, as above, that $P(Y = 6) = 1/3$, and $P(Y = 15) = 2/3$. Then

$$E(Y) = (6)(1/3) + (15)(2/3) = 2 + 10 = 12,$$

again as predicted. ∎

Example 3.1.3
Suppose that $P(Z = -3) = 0.2$, and $P(Z = 11) = 0.7$, and $P(Z = 31) = 0.1$. Then

$$E(Z) = (-3)(0.2) + (11)(0.7) + (31)(0.1) = -0.6 + 7.7 + 3.1 = 10.2. \;∎$$

Example 3.1.4
Suppose that $P(W = -3) = 0.2$, and $P(W = -11) = 0.7$, and $P(W = 31) = 0.1$. Then

$$E(W) = (-3)(0.2) + (-11)(0.7) + (31)(0.1) = -0.6 - 7.7 + 3.1 = -5.2.$$

In this case, the expected value of W is *negative*. ∎

We thus see that, for a discrete random variable X, once we know the probabilities that $X = x$ (or equivalently, once we know the probability function p_X), it is straightforward (at least in simple cases) to compute the expected value of X.

   We now consider some of the common discrete distributions introduced in Section 2.3.

Example 3.1.5 *Degenerate Distributions*
If $X \equiv c$ is a constant, then $P(X = c) = 1$, so

$$E(X) = (c)(1) = c,$$

as it should. ∎

Example 3.1.6 *The Bernoulli(θ) Distribution and Indicator Functions*
If $X \sim$ Bernoulli(θ), then $P(X = 1) = \theta$ and $P(X = 0) = 1 - \theta$, so

$$E(X) = (1)(\theta) + (0)(1 - \theta) = \theta.$$

As a particular application of this, suppose that we have a response s taking values in a sample S and $A \subset S$. Letting $X(s) = I_A(s)$, we have that X is the indicator function of the set A and so takes the values 0 and 1. Then we have that $P(X = 1) = P(A)$, and so $X \sim$ Bernoulli($P(A)$). This implies that

$$E(X) = E(I_A) = P(A).$$

Therefore, we have shown that the expectation of the indicator function of the set A is equal to the probability of A. ∎

Example 3.1.7 *The Binomial(n, θ) Distribution*
If $Y \sim$ Binomial(n, θ), then

$$P(Y = k) = \binom{n}{k} \theta^k (1 - \theta)^{n-k}$$

for $k = 0, 1, \ldots, n$. Hence,

$$
\begin{aligned}
E(Y) &= \sum_{k=0}^{n} k\, P(Y = k) = \sum_{k=0}^{n} k \binom{n}{k} \theta^k (1 - \theta)^{n-k} \\
&= \sum_{k=0}^{n} k \frac{n!}{k!\,(n-k)!} \theta^k (1 - \theta)^{n-k} = \sum_{k=1}^{n} \frac{n!}{(k-1)!\,(n-k)!} \theta^k (1 - \theta)^{n-k} \\
&= \sum_{k=1}^{n} \frac{n\,(n-1)!}{(k-1)!\,(n-k)!} \theta^k (1 - \theta)^{n-k} = \sum_{k=1}^{n} n \binom{n-1}{k-1} \theta^k (1 - \theta)^{n-k}.
\end{aligned}
$$

Now, the *binomial theorem* says that for any a and b, and any positive integer m,

$$(a + b)^m = \sum_{j=0}^{m} \binom{m}{j} a^j b^{m-j}.$$

Using this, and setting $j = k - 1$, we see that

$$
\begin{aligned}
E(Y) &= \sum_{k=1}^{n} n \binom{n-1}{k-1} \theta^k (1 - \theta)^{n-k} = \sum_{j=0}^{n-1} n \binom{n-1}{j} \theta^{j+1} (1 - \theta)^{n-j-1} \\
&= n\theta \sum_{j=0}^{n-1} \binom{n-1}{j} \theta^j (1 - \theta)^{n-j-1} = n\theta \left((\theta + 1 - \theta)^{n-1} \right) = n\theta.
\end{aligned}
$$

Hence, the expected value of Y is $n\theta$. Note that this is precisely n times the expected value of X, where $X \sim$ Bernoulli(θ) as in Example 3.1.6. We shall see in Example 3.1.15 that this is not a coincidence. ∎

Example 3.1.8 *The Geometric(θ) Distribution*
If $Z \sim$ Geometric(θ), then $P(Z = k) = (1 - \theta)^k \theta$ for $k = 0, 1, 2, \ldots$. Hence,

$$E(Z) = \sum_{k=0}^{\infty} k(1 - \theta)^k \theta. \tag{3.1.1}$$

Therefore, we can write

$$(1 - \theta)E(Z) = \sum_{\ell=0}^{\infty} \ell(1 - \theta)^{\ell+1} \theta.$$

Using the substitution $k = \ell + 1$, we compute that

$$(1 - \theta)E(Z) = \sum_{k=1}^{\infty} (k - 1)(1 - \theta)^k \theta. \tag{3.1.2}$$

Subtracting (3.1.2) from (3.1.1), we see that

$$\theta E(Z) = (E(Z)) - ((1 - \theta)E(Z)) = \sum_{k=1}^{\infty} (k - (k - 1))(1 - \theta)^k \theta$$

$$= \sum_{k=1}^{\infty} (1 - \theta)^k \theta = \frac{1 - \theta}{1 - (1 - \theta)} \theta = 1 - \theta.$$

Hence, $\theta E(Z) = 1 - \theta$, and we obtain $E(Z) = (1 - \theta)/\theta$. ∎

Example 3.1.9 *The Poisson(λ) Distribution*
If $X \sim$ Poisson(λ), then $P(X = k) = e^{-\lambda} \lambda^k / k!$ for $k = 0, 1, 2, \ldots$. Hence, setting $\ell = k - 1$,

$$E(X) = \sum_{k=0}^{\infty} k e^{-\lambda} \frac{\lambda^k}{k!} = \sum_{k=1}^{\infty} e^{-\lambda} \frac{\lambda^k}{(k-1)!} = \lambda e^{-\lambda} \sum_{k=1}^{\infty} \frac{\lambda^{k-1}}{(k-1)!}$$

$$= \lambda e^{-\lambda} \sum_{\ell=0}^{\infty} \frac{\lambda^\ell}{\ell!} = \lambda e^{-\lambda} e^{\lambda} = \lambda,$$

and we conclude that $E(X) = \lambda$. ∎

It should be noted that expected values can sometimes be infinite, as the following example demonstrates.

Example 3.1.10
Let X be a discrete random variable, with probability function p_X given by

$$p_X(2^k) = 2^{-k}$$

for $k = 1, 2, 3, \ldots$, with $p_X(x) = 0$ for other values of x. That is, $p_X(2) = 1/2$, $p_X(4) = 1/4$, $p_X(8) = 1/8$, etc., while $p_X(1) = p_X(3) = p_X(5) = p_X(6) = \cdots = 0$.

Then it is easily checked that p_X is indeed a valid probability function (i.e., $p_X(x) \geq 0$ for all x, with $\sum_x p_X(x) = 1$). On the other hand, we compute that

$$E(X) = \sum_{k=1}^{\infty} (2^k)(2^{-k}) = \sum_{k=1}^{\infty} (1) = \infty.$$

We therefore say that $E(X) = \infty$, i.e., that the expected value of X is infinite. ∎

Sometimes the expected value simply does not exist, as in the following example.

Example 3.1.11
Let Y be a discrete random variable, with probability function p_Y given by

$$p_Y(y) = \begin{cases} 1/2y & y = 2, 4, 8, 16, \ldots \\ 1/2|y| & y = -2, -4, -8, -16, \ldots \\ 0 & \text{otherwise} \end{cases}$$

That is, $p_Y(2) = p_Y(-2) = 1/4$, $p_Y(4) = p_Y(-4) = 1/8$, $p_Y(8) = p_Y(-8) = 1/16$, etc. Then it is easily checked that p_Y is indeed a valid probability function (i.e., $p_Y(y) \geq 0$ for all y, with $\sum_y p_Y(y) = 1$).
On the other hand, we compute that

$$\begin{aligned} E(Y) &= \sum_y y\, p_Y(y) = \sum_{k=1}^{\infty} (2^k)(1/2\, 2^k) + \sum_{k=1}^{\infty} (-2^k)(1/2\, 2^k) \\ &= \sum_{k=1}^{\infty} (1/2) - \sum_{k=1}^{\infty} (1/2) = \infty - \infty, \end{aligned}$$

which is *undefined*. We therefore say that $E(Y)$ does not exist, i.e., that the expected value of Y is *undefined* in this case. ∎

Example 3.1.12 *The St. Petersburg Paradox*
Suppose someone makes you the following deal. You will repeatedly flip a fair coin. You will receive an award of 2^Z pennies, where Z is the number of tails that appear before the first head. How much would you be willing to pay for this deal?

Well, the probability that the award will be 2^z pennies is equal to the probability that you will flip z tails and then one head, which is equal to $1/2^{z+1}$. Hence, the expected value of the award (in pennies) is equal to

$$\sum_{z=0}^{\infty} (2^z)(1/2^{z+1}) = \sum_{z=0}^{\infty} 1/2 = \infty.$$

In words, the average amount of the award is infinite!

Hence, according to the "fair gambling" interpretation of expected value, as discussed at the beginning of this chapter, it seems that you should be willing to pay an infinite amount (or, at least, any finite amount no matter how large) to get the award promised by this deal! How much do you think you should *really* be willing to pay for it?[1] ∎

Example 3.1.13 *The St. Petersburg Paradox, Truncated*
Suppose in the St. Petersburg paradox (Example 3.1.12), it is agreed that the award will be truncated at 2^{30} cents (which is just over 10 million dollars!). That is, the award will be the same as for the original deal, except the award will be frozen once it exceeds 2^{30} cents. Formally, the award is now equal to $2^{\min(30,Z)}$ pennies, where Z is as before.

How much would you be willing to pay for this new award? Well, the expected value of the new award (in cents) is equal to

$$\sum_{z=1}^{\infty}(2^{\min(30,z)})(1/2^{z+1}) = \sum_{z=1}^{30}(2^z)(1/2^{z+1}) + \sum_{z=31}^{\infty}(2^{30})(1/2^{z+1})$$

$$= \sum_{z=1}^{30}(1/2) + (2^{30})(1/2^{31})$$

$$= (30/2) + (1/2) = 31/2 = 15.5.$$

That is, truncating the award at just over 10 million dollars changes its expected value enormously, from infinity to less than 16 cents! ∎

In *utility theory*, it is often assumed that each person has a utility function U such that, if they win x cents, their amount of "utility" (i.e., benefit or joy or pleasure) is equal to $U(x)$. In this context, the truncation of Example 3.1.13 may be thought of not as changing the rules of the game but as corresponding to a utility function of the form $U(x) = \min(x, 2^{30})$. In words, this says that your utility is equal to the amount of money you get, until you reach 2^{30} cents (approximately 10 million dollars), after which point you don't care about money[2] anymore. The result of Example 3.1.13 then says that, with this utility function, the St. Petersburg paradox is only worth 15.5 cents to you – even though its expected value is infinite.

We often need to compute expected values of *functions* of random variables. Fortunately, this is not too difficult, as the following theorem shows.

Theorem 3.1.1
(a) Let X be a discrete random variable, and let $g : R^1 \to R^1$ be some function such that the expectation of the random variable $g(X)$ exists. Then

$$E(g(X)) = \sum_{x} g(x) P(X = x).$$

(b) Let X and Y be discrete random variables, and let $h : R^2 \to R^1$ be some

---

[1] When one of the authors first heard about this deal, he decided to try it, and agreed to pay one dollar. In fact, he got four tails before the first head, so his award was 16 cents, but he still lost 84 cents overall.

[2] Or, perhaps, you think it is unlikely you will be able to *collect* the money!

function such that the expectation of the random variable $h(X, Y)$ exists. Then

$$E(h(X,Y)) = \sum_{x,y} h(x,y) P(X = x, \ Y = y).$$

Proof: We prove part (b) here. Part (a) then follows by simply setting $h(x, y) = g(x)$ and noting that

$$\sum_{x,y} g(x) P(X = x, \ Y = y) = \sum_{x} g(x) P(X = x).$$

Let $Z = h(X, Y)$. We have that

$$
\begin{aligned}
E(Z) &= \sum_{z} z\, P(Z = z) = \sum_{z} z\, P\left(h(X,Y) = z\right) \\
&= \sum_{z} z \sum_{\substack{x,y \\ h(x,y)=z}} P(X = x, \ Y = y) = \sum_{x,y} \sum_{\substack{z \\ z=h(x,y)}} z\, P(X = x, \ Y = y) \\
&= \sum_{x,y} h(x,y)\, P(X = x, \ Y = y),
\end{aligned}
$$

as claimed. ∎

One of the most important properties of expected value is that it is *linear*, as follows.

Theorem 3.1.2 (*Linearity of expected values*) Let X and Y be discrete random variables, let a and b be real numbers, and put $Z = aX + bY$. Then $E(Z) = aE(X) + bE(Y)$.

Proof: Let $p_{X,Y}$ be the joint probability function of X and Y. Then using Theorem 3.1.1,

$$
\begin{aligned}
E(Z) &= \sum_{x,y} (ax + by)\, p_{X,Y}(x,y) = a \sum_{x,y} x\, p_{X,Y}(x,y) + b \sum_{x,y} y\, p_{X,Y}(x,y) \\
&= a \sum_{x} x \sum_{y} p_{X,Y}(x,y) + b \sum_{y} y \sum_{x} p_{X,Y}(x,y).
\end{aligned}
$$

Because $\sum_{y} p_{X,Y}(x,y) = p_X(x)$ and $\sum_{x} p_{X,Y}(x,y) = p_Y(y)$, we have that

$$E(Z) = a \sum_{x} x\, p_X(x) + b \sum_{y} y\, p_Y(y) = aE(X) + bE(Y),$$

as claimed. ∎

Example 3.1.14
Let $X \sim \text{Binomial}(n, \theta_1)$, and let $Y \sim \text{Geometric}(\theta_2)$. What is $E(3X - 2Y)$?

We already know (Examples 3.1.6 and 3.1.7) that $E(X) = n\theta_1$ and $E(Y) = (1 - \theta_2)/\theta_2$. Hence, by Theorem 3.1.2, $E(3X - 2Y) = 3E(X) - 2E(Y) = 3n\theta_1 - 2(1 - \theta_2)/\theta_2$. ∎

Example 3.1.15

Let $Y \sim \text{Binomial}(n, \theta)$. Then we know (cf. Example 2.4.3) that we can think of $Y = X_1 + \cdots + X_n$, where each $X_i \sim \text{Bernoulli}(\theta)$ (in fact, $X_i = 1$ if the ith coin is heads, otherwise $X_i = 0$). Because $E(X_i) = \theta$ for each i, it follows immediately from Theorem 3.1.2 that

$$E(Y) = E(X_1) + \cdots + E(X_n) = \theta + \cdots + \theta = n\theta.$$

This gives the same answer as Example 3.1.7, but much more easily. ∎

Suppose that X is a random variable and $Y = c$ is a constant. Then from Theorem 3.1.2, we have that $E(X + c) = E(X) + c$. From this we see that the mean value μ_X of X is a measure of the *location* of the probability distribution of X. For example, if X takes the value x with probability p and the value y with probability $1 - p$, then the mean of X is $\mu_X = px + (1 - p)y$ and this is a value between x and y. For a constant c, the probability distribution of $X + c$ is concentrated on the points $x + c$ and $y + c$ with probabilities p and $1 - p$, respectively. The mean of $X + c$ is $\mu_X + c$, which is between the points $x + c$ and $y + c$, i.e., the mean shifts with the probability distribution. It is also true that if X is concentrated on the finite set of points $x_1 < x_2 < \cdots < x_k$, then $x_1 \leq \mu_X \leq x_k$, and the mean shifts exactly as we shift the distribution. This is depicted in Figure 3.1.1 for a distribution concentrated on $k = 4$ points. Using the results of Section 2.6.1, we have that $p_{X+c}(x) = p_X(x - c)$.

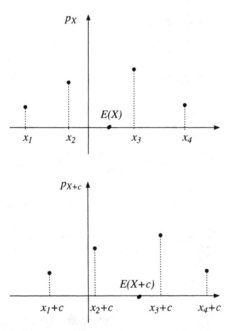

Figure 3.1.1: The probability functions and means of discrete random variables X, and $X + c$.

Theorem 3.1.2 says, in particular, that $E(X+Y) = E(X) + E(Y)$, i.e., that expectation preserves sums. It is reasonable to ask if the same property holds for products. That is, do we necessarily have $E(XY) = E(X)E(Y)$? In general, the answer is no, as the following example shows.

Example 3.1.16
Let X and Y be discrete random variables, with joint probability function given by

$$p_{X,Y}(x,y) = \begin{cases} 1/2 & x = 3,\ y = 5 \\ 1/6 & x = 3,\ y = 9 \\ 1/6 & x = 6,\ y = 5 \\ 1/6 & x = 6,\ y = 9 \\ 0 & \text{otherwise.} \end{cases}$$

Then

$$E(X) = \sum_x xP(X = x) = (3)(1/2 + 1/6) + (6)(1/6 + 1/6) = 4$$

and

$$E(Y) = \sum_y y\,P(Y = y) = (5)(1/2 + 1/6) + (9)(1/6 + 1/6) = 19/3,$$

while

$$\begin{aligned} E(XY) &= \sum_z z\,P(XY = z) \\ &= (3)(5)(1/2) + (3)(9)(1/6) + (6)(5)(1/6) + (6)(9)(1/6) \\ &= 26. \end{aligned}$$

Because $(4)(19/3) \neq 26$, we see that $E(X)\,E(Y) \neq E(XY)$ in this case. ∎

On the other hand, if X and Y are *independent*, then we do have $E(X)E(Y) = E(XY)$.

Theorem 3.1.3 Let X and Y be discrete random variables that are independent. Then $E(XY) = E(X)E(Y)$.

Proof: Independence implies (Theorem 2.8.3) that $P(X = x, Y = y) = P(X = x)\,P(Y = y)$. Using this, we compute by Theorem 3.1.1 that

$$\begin{aligned} E(XY) &= \sum_{x,y} xy\,P(X = x, Y = y) = \sum_{x,y} xy\,P(X = x)\,P(Y = y) \\ &= \left(\sum_x x\,P(X = x)\right)\left(\sum_y y\,P(Y = y)\right) = E(X)\,E(Y), \end{aligned}$$

as claimed. ∎

Theorem 3.1.3 will be used often in subsequent chapters, as will the following important property.

Theorem 3.1.4 (*Monotonicity*) Let X and Y be discrete random variables, and suppose that $X \leq Y$. (Remember that this means $X(s) \leq Y(s)$ for all $s \in S$.) Then $E(X) \leq E(Y)$.

Proof: Let $Z = Y - X$. Then Z is also discrete. Furthermore, because $X \leq Y$, we have $Z \geq 0$, so that all possible values of Z are nonnegative. Hence, if we list the possible values of Z as $z_1, z_2, \ldots$, then $z_i \geq 0$ for all i, so that

$$E(Z) = \sum_i z_i P(Z = z_i) \geq 0.$$

But by Theorem 3.1.2, $E(Z) = E(Y) - E(X)$. Hence, $E(Y) - E(X) \geq 0$, so that $E(Y) \geq E(X)$. ∎

Summary of Section 3.1

- The expected value $E(X)$ of a random variable X represents the long-run average value that it takes on.

- If X is discrete, then $E(X) = \sum_x x\, P(X = x)$.

- The expected values of the Bernoulli, binomial, geometric, and Poisson distributions were computed.

- Expected value has an interpretation in terms of fair gambling, but such interpretations require utility theory to accurately reflect human behavior.

- Expected values of functions of one or two random variables can also be computed by summing the function values times the probabilities.

- Expectation is linear and monotone.

- If X and Y are independent, then $E(XY) = E(X)\, E(Y)$. But without independence, this property may fail.

Exercises

3.1.1 Compute $E(X)$ when the probability function of X is given by each of the following.

(a)
$$p_X(x) = \begin{cases} 1/7 & x = -4 \\ 2/7 & x = 0 \\ 4/7 & x = 3 \\ 0 & \text{otherwise} \end{cases}$$

(b)
$$p_X(x) = \begin{cases} 2^{-x-1} & x = 0, 1, 2, \ldots \\ 0 & \text{otherwise} \end{cases}$$

(c)
$$p_X(x) = \begin{cases} 2^{-x+6} & x = 7, 8, 9, \ldots \\ 0 & \text{otherwise} \end{cases}$$

3.1.2 Let X and Y have joint probability function given by

$$p_{X,Y}(x, y) = \begin{cases} 1/7 & x = 5, \ y = 0 \\ 1/7 & x = 5, \ y = 3 \\ 1/7 & x = 5, \ y = 4 \\ 3/7 & x = 8, \ y = 0 \\ 1/7 & x = 8, \ y = 4 \\ 0 & \text{otherwise,} \end{cases}$$

as in Example 2.7.5. Compute each of the following.
(a) $E(X)$
(b) $E(Y)$
(c) $E(3X + 7Y)$
(d) $E(X^2)$
(e) $E(Y^2)$
(f) $E(XY)$
(g) $E(XY + 14)$

3.1.3 Let X and Y have joint probability function given by

$$p_{X,Y}(x, y) = \begin{cases} 1/2 & x = 2, \ y = 10 \\ 1/6 & x = -7, \ y = 10 \\ 1/12 & x = 2, \ y = 12 \\ 1/12 & x = -7, \ y = 12 \\ 1/12 & x = 2, \ y = 14 \\ 1/12 & x = -7, \ y = 14 \\ 0 & \text{otherwise.} \end{cases}$$

Compute each of the following.
(a) $E(X)$
(b) $E(Y)$
(c) $E(X^2)$
(d) $E(Y^2)$
(e) $E(X^2 + Y^2)$
(f) $E(XY - 4Y)$

3.1.4 Let $X \sim \text{Bernoulli}(\theta_1)$ and $Y \sim \text{Binomial}(n, \theta_2)$. Compute $E(4X - 3Y)$.

3.1.5 Let $X \sim \text{Geometric}(\theta)$ and $Y \sim \text{Poisson}(\lambda)$. Compute $E(8X - Y + 12)$.

3.1.6 Let $Y \sim \text{Binomial}(100, 0.3)$, and $Z \sim \text{Poisson}(7)$. Compute $E(Y + Z)$.

3.1.7 Let $X \sim \text{Binomial}(80, 1/4)$ and let $Y \sim \text{Poisson}(3/2)$. Assume X and Y are independent. Compute $E(XY)$.

3.1.8 Starting with one penny, suppose you roll one fair six-sided die and get paid an additional number of pennies equal to three times the number showing on the die. Let X be the total number of pennies you have at the end. Compute $E(X)$.

3.1.9 Suppose you start with eight pennies, and flip one fair coin. If the coin comes up heads, you get to keep all your pennies, but if the coin comes up tails, you have to give half of them back. Let X be the total number of pennies you have at the end. Compute $E(X)$.

Problems

3.1.10 Suppose you start with one penny, and repeatedly flip a fair coin. Each time you get heads, *before* the first time you get tails, you get two more pennies. Let X be the total number of pennies you have at the end. Compute $E(X)$.

3.1.11 Suppose you start with one penny, and repeatedly flip a fair coin. Each time you get heads, *before* the first time you get tails, your number of pennies is *doubled.* Let X be the total number of pennies you have at the end. Compute $E(X)$.

3.1.12 Let $X \sim \text{Geometric}(\theta)$ and let $Y = \min(X, 100)$.
(a) Compute $E(Y)$.
(b) Compute $E(Y - X)$.

3.1.13 Give an example of a random variable X such that $E(\min(X, 100)) = E(X)$.

3.1.14 Give an example of a random variable X such that $E(\min(X, 100)) = E(X)/2$.

3.1.15 Give an example of a joint probability function $p_{X,Y}$ for random variables X and Y, such that $X \sim \text{Bernoulli}(1/4)$ and $Y \sim \text{Bernoulli}(1/2)$, but $E(XY) \neq 1/8$.

3.1.16 For $X \sim \text{Hypergeometric}(N, M, n)$ prove that $E(X) = nM/N$.

3.1.17 For $X \sim \text{Negative-Binomial}(r, \theta)$ prove that $E(X) = r(1 - \theta)/\theta$. (Hint: Argue that if $X_1, \ldots, X_r$ are independent and identically distributed $\text{Geometric}(\theta)$, then $X = X_1 + \cdots + X_r \sim \text{Negative-Binomial}(r, \theta)$.)

3.1.18 Suppose that $(X_1, X_2, X_3) \sim \text{Multinomial}(n, \theta_1, \theta_2, \theta_3)$. Prove that $E(X_i) = n\theta_i$.

Challenges

3.1.19 Let $X \sim \text{Geometric}(\theta)$. Compute $E(X^2)$.

3.1.20 Suppose X is a discrete random variable, such that $E(\min(X, M)) = E(X)$. Prove that $P(X > M) = 0$.

Discussion Topics

3.1.21 How much would *you* be willing to pay for the deal corresponding to the St. Petersburg paradox (Example 3.1.12)? Justify your answer.

3.1.22 What utility function U (as in the text following Example 3.1.13) best describes your own personal attitude toward money? Why?

3.2 The Absolutely Continuous Case

Suppose now that X is absolutely continuous, with density function f_X. How can we compute $E(X)$ then? By analogy with the discrete case, we might try computing $\sum_x x P(X = x)$, but because $P(X = x)$ is always zero, this sum is always zero as well.

On the other hand, if ϵ is a small positive number, then we could try approximating $E(X)$ by

$$E(X) \approx \sum_i i\epsilon \, P(i\epsilon \leq X < (i+1)\epsilon),$$

where the sum is over all integers i. This makes sense because, if ϵ is small and $i\epsilon \leq X < (i+1)\epsilon$, then $X \approx i\epsilon$.

Now, we know that

$$P(i\epsilon \leq X < (i+1)\epsilon) = \int_{i\epsilon}^{(i+1)\epsilon} f_X(x) \, dx.$$

This tells us that

$$E(X) \approx \sum_i \int_{i\epsilon}^{(i+1)\epsilon} i\epsilon \, f_X(x) \, dx.$$

Furthermore, in this integral, $i\epsilon \leq x < (i+1)\epsilon$. Hence, $i\epsilon \approx x$. We therefore see that

$$E(X) \approx \sum_i \int_{i\epsilon}^{(i+1)\epsilon} x \, f_X(x) \, dx = \int_{-\infty}^{\infty} x \, f_X(x) \, dx.$$

This prompts the following definition.

Definition 3.2.1 Let X be an absolutely continuous random variable, with density function f_X. Then the *expected value* of X is given by

$$E(X) = \int_{-\infty}^{\infty} x \, f_X(x) \, dx.$$

From this definition, it is not too difficult to compute the expected values of many of the standard absolutely continuous distributions.

Example 3.2.1 *The Uniform[0, 1] Distribution*

Let $X \sim$ Uniform[0, 1] so that the density of X is given by

$$f_X(x) = \begin{cases} 1 & 0 \le x \le 1 \\ 0 & \text{otherwise.} \end{cases}$$

Hence,

$$E(X) = \int_{-\infty}^{\infty} x\, f_X(x)\, dx = \int_0^1 x\, dx = \frac{x^2}{2}\Big|_{x=0}^{x=1} = 1/2,$$

as one would expect. ∎

Example 3.2.2 *The Uniform[L, R] Distribution*

Let $X \sim$ Uniform[L, R] so that the density of X is given by

$$f_X(x) = \begin{cases} 1/(R-L) & L \le x \le R \\ 0 & \text{otherwise.} \end{cases}$$

Hence,

$$\begin{aligned} E(X) &= \int_{-\infty}^{\infty} x\, f_X(x)\, dx = \int_L^R x\, \frac{1}{R-L}\, dx = \frac{x^2}{2(R-L)}\Big|_{x=L}^{x=R} \\ &= \frac{R^2 - L^2}{2(R-L)} = \frac{(R-L)(R+L)}{2(R-L)} = \frac{R+L}{2}, \end{aligned}$$

again as one would expect. ∎

Example 3.2.3 *The Exponential(λ) Distribution*

Let $Y \sim$ Exponential(λ) so that the density of Y is given by

$$f_Y(y) = \begin{cases} \lambda e^{-\lambda y} & y \ge 0 \\ 0 & y < 0. \end{cases}$$

Hence, integration by parts, with $u = y$ and $dv = \lambda e^{-\lambda y}$ (so $du = dy, v = -e^{-\lambda y}$), leads to

$$\begin{aligned} E(Y) &= \int_{-\infty}^{\infty} y\, f_Y(y)\, dy = \int_0^{\infty} y\, \lambda e^{-\lambda y}\, dy = -\left. y e^{-\lambda y}\right|_0^{\infty} + \int_0^{\infty} e^{-\lambda y}\, dy \\ &= \int_0^{\infty} e^{-\lambda y}\, dy = -\left. \frac{e^{-\lambda y}}{\lambda}\right|_0^{\infty} = -\frac{0-1}{\lambda} = \frac{1}{\lambda}. \end{aligned}$$

In particular, if $\lambda = 1$, then $Y \sim$ Exponential(1), and $E(Y) = 1$. ∎

Example 3.2.4 *The N(0, 1) Distribution*

Let $Z \sim N(0, 1)$ so that the density of Z is given by

$$f_Z(z) = \phi(z) = \frac{1}{\sqrt{2\pi}} e^{-z^2/2}.$$

Hence,

$$
\begin{aligned}
E(Z) &= \int_{-\infty}^{\infty} z\, f_Z(z)\, dz \\
&= \int_{-\infty}^{\infty} z\, \frac{1}{\sqrt{2\pi}} e^{-z^2/2}\, dz \\
&= \int_{-\infty}^{0} z\, \frac{1}{\sqrt{2\pi}} e^{-z^2/2}\, dz + \int_{0}^{\infty} z\, \frac{1}{\sqrt{2\pi}} e^{-z^2/2}\, dz.
\end{aligned}
$$

But using the substitution $w = -z$, we see that

$$
\int_{-\infty}^{0} z\, \frac{1}{\sqrt{2\pi}} e^{-z^2/2}\, dz = \int_{0}^{\infty} (-w)\, \frac{1}{\sqrt{2\pi}} e^{-w^2/2}\, dw.
$$

Then the two integrals in (3.2.1) cancel each other out, and we are left with $E(Z) = 0$. ∎

As with discrete variables, means of absolutely continuous random variables can also be infinite or undefined.

Example 3.2.5
Let X have density function given by

$$
f_X(x) = \begin{cases} 1/x^2 & x \ge 1 \\ 0 & \text{otherwise.} \end{cases}
$$

Then

$$
E(X) = \int_{-\infty}^{\infty} x\, f_X(x)\, dx = \int_{1}^{\infty} x\, (1/x^2)\, dx = \int_{1}^{\infty} (1/x)\, dx = \log x \Big|_{x=1}^{x=\infty} = \infty.
$$

Hence, the expected value of X is infinite. ∎

Example 3.2.6
Let Y have density function given by

$$
f_Y(y) = \begin{cases} 1/2y^2 & y \ge 1 \\ 1/2y^2 & y \le -1 \\ 0 & \text{otherwise.} \end{cases}
$$

Then

$$
\begin{aligned}
E(Y) &= \int_{-\infty}^{\infty} y\, f_Y(y)\, dy = \int_{-\infty}^{-1} y(1/y^2)\, dy + \int_{1}^{\infty} y(1/y^2)\, dy \\
&= -\int_{1}^{\infty} (1/y)\, dy + \int_{1}^{\infty} (1/y)\, dy = -\infty + \infty,
\end{aligned}
$$

which is undefined. Hence, the expected value of Y is undefined (i.e., does not exist) in this case. ∎

Theorem 3.1.1 remains true in the continuous case, as follows.

Theorem 3.2.1
(a) Let X be an absolutely continuous random variable, with density function f_X, and let $g : R^1 \to R^1$ be some function. Then when the expectation of $g(X)$ exists,

$$E\left(g(X)\right) = \int_{-\infty}^{\infty} g(x) f_X(x) \, dx.$$

(b) Let X and Y be jointly absolutely continuous random variables, with joint density function $f_{X,Y}$, and let $h : R^2 \to R^1$ be some function. Then when the expectation of $h(X,Y)$ exists,

$$E\left(h(X,Y)\right) = \int_{-\infty}^{\infty} \int_{-\infty}^{\infty} h(x,y) f_{X,Y}(x,y) \, dx \, dy.$$

We do not prove Theorem 3.2.1 here; however, we shall use it often. For a first use of this result, we prove that expected values for absolutely continuous random variables are still linear.

Theorem 3.2.2 (*Linearity of expected values*) Let X and Y be jointly absolutely continuous random variables, and let a and b be real numbers. Then $E(aX + bY) = aE(X) + bE(Y)$.

Proof: Let $f_{X,Y}$ be the joint density function of X and Y. Then using Theorem 3.2.1, we compute that

$$
\begin{aligned}
E(Z) &= \int_{-\infty}^{\infty} \int_{-\infty}^{\infty} (ax + by) \, f_{X,Y}(x,y) \, dx \, dy \\
&= a \int_{-\infty}^{\infty} \int_{-\infty}^{\infty} x f_{X,Y}(x,y) \, dx \, dy + b \int_{-\infty}^{\infty} \int_{-\infty}^{\infty} y f_{X,Y}(x,y) \, dx \, dy \\
&= a \int_{-\infty}^{\infty} x \left(\int_{-\infty}^{\infty} f_{X,Y}(x,y) \, dy \right) dx \\
&\quad + b \int_{-\infty}^{\infty} y \left(\int_{-\infty}^{\infty} f_{X,Y}(x,y) \, dx \right) dy
\end{aligned}
$$

But $\int_{-\infty}^{\infty} f_{X,Y}(x,y) \, dy = f_X(x)$ and $\int_{-\infty}^{\infty} f_{X,Y}(x,y) \, dx = f_Y(y)$, so

$$E(Z) = a \int_{-\infty}^{\infty} x f_X(x) \, dx + b \int_{-\infty}^{\infty} y f_Y(y) \, dy = aE(X) + bE(Y),$$

as claimed. ∎

Just as in the discrete case, we have that $E(X + c) = E(X) + c$ for an absolutely continuous random variable X. Note, however, that this is not implied by Theorem 3.2.2, because the constant c is a discrete, not absolutely continuous, random variable. In fact, we need a more general treatment of expectation to obtain this result (see Section 3.7). In any case, the result is true and we again have that the mean of a random variable serves as a measure of the location of the probability distribution of X. In Figure 3.2.1, we have plotted the densities and means of the absolutely continuous random variables X and $X+c$. The change of variable results from Section 2.6.2 give $f_{X+c}(x) = f_X(x - c)$.

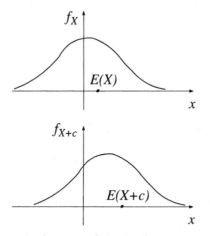

Figure 3.2.1: The densities and means of absolutely continuous random variables X and $X + c$.

If X and Y are *independent*, then the following results show that we again have $E(XY) = E(X)\,E(Y)$.

Example 3.2.7 *The $N(\mu, \sigma^2)$ Distribution*
Let $X \sim N(\mu, \sigma^2)$. Then we know (cf. Exercise 2.6.3) that if $Z = (X - \mu)/\sigma$, then $Z \sim N(0, 1)$. Hence, we can write $X = \mu + \sigma Z$, where $Z \sim N(0, 1)$. But we know (Example 3.2.4) that $E(Z) = 0$, and (Example 3.1.5) that $E(\mu) = \mu$. Hence, using Theorem 3.2.2,

$$E(X) = E(\mu + \sigma Z) = E(\mu) + \sigma E(Z) = \mu + \sigma(0) = \mu. \ \blacksquare$$

Theorem 3.2.3 Let X and Y be jointly absolutely continuous random variables that are independent. Then $E(XY) = E(X)E(Y)$.

Proof: Independence implies (Theorem 2.8.3) that $f_{X,Y}(x, y) = f_X(x)\,f_Y(y)$. Using this, we compute, using Theorem 3.2.1, that

$$
\begin{aligned}
E(XY) &= \int_{-\infty}^{\infty} \int_{-\infty}^{\infty} xy\, f_{X,Y}(x, y)\, dx\, dy = \int_{-\infty}^{\infty} \int_{-\infty}^{\infty} xy\, f_X(x) f_Y(y)\, dx\, dy \\
&= \left(\int_{-\infty}^{\infty} x\, f_X(x)\, dx \right) \left(\int_{-\infty}^{\infty} y\, f_Y(y)\, dy \right) = E(X)E(Y),
\end{aligned}
$$

as claimed. $\blacksquare$

The monotonicity property (Theorem 3.1.4) still holds as well.

Theorem 3.2.4 (*Monotonicity*) Let X and Y be jointly continuous random variables, and suppose that $X \leq Y$. Then $E(X) \leq E(Y)$.

Proof: Let $f_{X,Y}$ be the joint density function of X and Y. Because $X \leq Y$, the density $f_{X,Y}$ can be chosen so that $f_{X,Y}(x, y) = 0$ whenever $x > y$. Now let $Z = Y - X$. Then by Theorem 3.2.1 (b),

$$E(Z) = \int_{-\infty}^{\infty} \int_{-\infty}^{\infty} (y - x) \, f_{X,Y}(x, y) \, dx \, dy .$$

Because $f_{X,Y}(x, y) = 0$ whenever $x > y$, this implies that $E(Z) \geq 0$. But by Theorem 3.2.2, $E(Z) = E(Y) - E(X)$. Hence, $E(Y) - E(X) \geq 0$, so that $E(Y) \geq E(X)$. ∎

Summary of Section 3.2

- If X is absolutely continuous, then $E(X) = \int x \, f_X(x) \, dx$.

- The expected values of the uniform, exponential, and normal distributions were computed.

- Expectation for absolutely continuous random variables is still linear and monotone.

- If X and Y are independent, then we still have $E(XY) = E(X) \, E(Y)$.

Exercises

3.2.1 Compute C and $E(X)$ when the density function of X is given by each of the following.

(a)
$$f_X(x) = \begin{cases} C & 5 \leq x \leq 9 \\ 0 & \text{otherwise} \end{cases}$$

(b)
$$f_X(x) = \begin{cases} C(x + 1) & 6 \leq x \leq 8 \\ 0 & \text{otherwise} \end{cases}$$

(c)
$$f_X(x) = \begin{cases} Cx^4 & -5 \leq x \leq -2 \\ 0 & \text{otherwise} \end{cases}$$

3.2.2 Let X and Y have joint density

$$f_{X,Y}(x, y) = \begin{cases} 4x^2y + 2y^5 & 0 \leq x \leq 1, \ 0 \leq y \leq 1 \\ 0 & \text{otherwise,} \end{cases}$$

as in Examples 2.7.6 and 2.7.7. Compute each of the following.
(a) $E(X)$
(b) $E(Y)$
(c) $E(3X + 7Y)$

(d) $E(X^2)$
(e) $E(Y^2)$
(f) $E(XY)$
(g) $E(XY + 14)$

3.2.3 Let X and Y have joint density

$$f_{X,Y}(x, y) = \begin{cases} (4xy + 3x^2y^2)/18 & 0 \le x \le 1, 0 \le y \le 3 \\ 0 & \text{otherwise.} \end{cases}$$

Compute each of the following.
(a) $E(X)$
(b) $E(Y)$
(c) $E(X^2)$
(d) $E(Y^2)$
(e) $E(Y^4)$
(f) $E(X^2Y^3)$

3.2.4 Let X and Y have joint density

$$f_{X,Y}(x, y) = \begin{cases} 6xy + (9/2)x^2y^2 & 0 \le y \le x \le 1 \\ 0 & \text{otherwise.} \end{cases}$$

Compute each of the following.
(a) $E(X)$
(b) $E(Y)$
(c) $E(X^2)$
(d) $E(Y^2)$
(e) $E(Y^4)$
(f) $E(X^2Y^3)$

3.2.5 Let $X \sim \text{Uniform}[3, 7]$ and $Y \sim \text{Exponential}(9)$. Compute $E(-5X - 6Y)$.

3.2.6 Let $X \sim \text{Uniform}[-12, -9]$ and $Y \sim N(-8, 9)$. Compute $E(11X + 14Y + 3)$.

3.2.7 Let $Y \sim \text{Exponential}(9)$, and $Z \sim \text{Exponential}(8)$. Compute $E(Y + Z)$.

3.2.8 Let $Y \sim \text{Exponential}(9)$, and $Z \sim \text{Gamma}(5, 4)$. Compute $E(Y + Z)$. (You may use Problem 3.2.9 below.)

Problems

3.2.9 Let $\alpha > 0$ and $\lambda > 0$, and let $X \sim \text{Gamma}(\alpha, \lambda)$. Prove that $E(X) = \alpha/\lambda$. (Hint: The computations are somewhat similar to those of Problem 2.4.7. You will also need property (2.4.7) of the gamma function.)

3.2.10 Suppose that X follows the logistic distribution (Problem 2.4.10). Prove that $E(X) = 0$.

3.2.11 Suppose that X follows the Weibull(α) distribution (Problem 2.4.11). Prove that $E(X) = \Gamma\left(\alpha^{-1} + 1\right)$.

3.2.12 Suppose that X follows the Pareto(α) distribution (Problem 2.4.12) for $\alpha > 1$. Prove that $E(X) = 1/(\alpha - 1)$. What is $E(X)$ when $0 < \alpha \le 1$?

3.2.13 Suppose that X follows the Cauchy distribution (Problem 2.4.13). Argue that $E(X)$ does not exist. (Hint: Compute the integral in two parts, where the integrand is positive and where the integrand is negative.)

3.2.14 Suppose that X follows the Laplace distribution (Problem 2.4.14). Prove that $E(X) = 0$.

3.2.15 Suppose that X follows the Beta(a, b) distribution (Problem 2.4.16). Prove that $E(X) = a/(a + b)$.

3.2.16 Suppose that $(X_1, X_2) \sim$ Dirichlet$(\alpha_1, \alpha_2, \alpha_3)$ (Problem 2.7.12). Prove that $E(X_i) = \alpha_i/(\alpha_1 + \alpha_2 + \alpha_3)$.

3.3   Variance, Covariance, and Correlation

Now that we understand expected value, we can use it to define various other quantities of interest. The numerical values of these quantities provide us with information about the distribution of random variables.

Given a random variable X, we know that the average value of X will be $E(X)$. However, this tells us nothing about how far X tends to be from $E(X)$. For that, we have the following definition.

Definition 3.3.1 The *variance* of a random variable X is the quantity

$$\sigma_x^2 = \mathrm{Var}(X) = E\left((X - \mu_X)^2\right), \qquad (3.3.1)$$

where $\mu_X = E(X)$ is the mean of X.

We note that it is also possible to write (3.3.1) as $\mathrm{Var}(X) = E\left((X - E(X))^2\right)$; however, the multiple uses of "E" may be confusing. Also, because $(X - \mu_X)^2$ is always nonnegative, its expectation is always defined, and so the variance of X is always defined.

Intuitively, the variance $\mathrm{Var}(X)$ is a measure of how *spread out* the distribution of X is, or how *random* X is, or how much X *varies*, as the following example illustrates.

Example 3.3.1
Let X and Y be two discrete random variables, with probability functions

$$p_X(x) = \begin{cases} 1 & x = 10 \\ 0 & \text{otherwise} \end{cases}$$

and

$$p_Y(y) = \begin{cases} 1/2 & y = 5 \\ 1/2 & y = 15 \\ 0 & \text{otherwise,} \end{cases}$$

respectively.

Then $E(X) = E(Y) = 10$. However,

$$\text{Var}(X) = (10 - 10)^2(1) = 0,$$

while
$$\text{Var}(Y) = (5 - 10)^2(1/2) + (15 - 10)^2(1/2) = 25.$$

We thus see that, while X and Y have the same expected value, the variance of Y is much greater than that of X. This corresponds to the fact that Y is more random than X; that is, it varies more than X does. ∎

Example 3.3.2
Let X have probability function given by

$$p_X(x) = \begin{cases} 1/2 & x = 2 \\ 1/6 & x = 3 \\ 1/6 & x = 4 \\ 1/6 & x = 5 \\ 0 & \text{otherwise.} \end{cases}$$

Then $E(X) = (2)(1/2) + (3)(1/6) + (4)(1/6) + (5)(1/6) = 3$. Hence,

$$\text{Var}(X) = ((2 - 3)^2)\frac{1}{2} + ((3 - 3)^2)\frac{1}{6} + ((4 - 3)^2)\frac{1}{6} + ((5 - 3)^2)\frac{1}{6} = 4/3. \blacksquare$$

Example 3.3.3
Let $Y \sim \text{Bernoulli}(\theta)$. Then $E(Y) = \theta$. Hence,

$$\begin{aligned} \text{Var}(Y) &= E((Y - \theta)^2) = ((1 - \theta)^2)(\theta) + ((0 - \theta)^2)(1 - \theta) \\ &= \theta - 2\theta^2 + \theta^3 + \theta^2 - \theta^3 = \theta - \theta^2 = \theta(1 - \theta). \blacksquare \end{aligned}$$

The square in (3.3.1) implies that the "scale" of $\text{Var}(X)$ is different from the scale of X. For example, if X were measuring a distance in meters (m), then $\text{Var}(X)$ would be measuring in meters squared (m^2). If we then switched from meters to feet, we would have to multiply X by about 3.28084, but would have to multiply $\text{Var}(X)$ by about $(3.28084)^2$.

To correct for this "scale" problem, we can simply take the square root, as follows.

Definition 3.3.2 The *standard deviation* of a random variable X is the quantity

$$\sigma_X = \text{Sd}(X) = \sqrt{\text{Var}(X)} = \sqrt{E((X - \mu_X)^2)}.$$

It is reasonable to ask why, in (3.3.1), we need the square at all. Now, if we simply *omitted* the square, and considered $E((X - \mu_X))$, we would always get zero (because $\mu_X = E(X)$), which is useless. On the other hand, we could instead use $E(|X - \mu_X|)$. This would, like (3.3.1), be a valid measure of the average distance of X from μ_X. Furthermore, it would not have the "scale problem" that $\text{Var}(X)$ does. However, we shall see that $\text{Var}(X)$ has many convenient properties. By contrast, $E(|X - \mu_X|)$ is very difficult to work with.

Thus, it is purely for *convenience* that we define variance by $E\left((X - \mu_X)^2\right)$ instead of $E\left(|X - \mu_X|\right)$.

Variance will be very important throughout the remainder of this book. Thus, we pause to present some important properties of Var.

Theorem 3.3.1 Let X be any random variable, with expected value $\mu_X = E(X)$, and variance Var(X). Then
(a) Var$(X) \geq 0$.
(b) if a and b are real numbers, Var$(aX + b) = a^2$ Var(X).
(c) Var$(X) = E(X^2) - (\mu_X)^2 = E(X^2) - E(X)^2$. (That is, variance is equal to the second moment minus the square of the first moment.)
(d) Var$(X) \leq E(X^2)$.
Proof: (a) This is immediate, because we always have $(X - \mu_X)^2 \geq 0$.

(b) We note that $\mu_{aX+b} \equiv E(aX + b) = aE(X) + b = a\mu_X + b$, by linearity. Hence, again using linearity,

$$
\begin{aligned}
\text{Var}(aX + b) &= E\left((aX + b - \mu_{aX+b})^2\right) = E\left((aX + b - a\mu_X + b)^2\right) \\
&= a^2 E\left((X - \mu_X)^2\right) = a^2 \text{Var}(X).
\end{aligned}
$$

(c) Again using linearity,

$$
\begin{aligned}
\text{Var}(X) &= E\left((X - \mu_X)^2\right) = E\left(X^2 - 2X\mu_X + (\mu_X)^2\right) \\
&= E(X^2) - 2E(X)\mu_X + (\mu_X)^2 = E(X^2) - 2(\mu_X)^2 + (\mu_X)^2 \\
&= E(X^2) - (\mu_X)^2.
\end{aligned}
$$

(d) This follows immediately from part (c), because we have $-(\mu_X)^2 \leq 0$. ∎

Theorem 3.3.1 often provides easier ways of computing variance, as in the following examples.

Example 3.3.4 *Variance of the Exponential(λ) Distribution*
Let $W \sim$ Exponential(λ), so that $f_W(w) = \lambda e^{-\lambda w}$. Then $E(W) = 1/\lambda$. Also, using integration by parts,

$$
\begin{aligned}
E(W^2) &= \int_0^\infty w^2 \lambda e^{-\lambda w} dw = \int_0^\infty 2w e^{-\lambda w} dw \\
&= (2/\lambda) \int_0^\infty w\lambda e^{-\lambda w} dw = (2/\lambda)E(W) = 2/\lambda^2.
\end{aligned}
$$

Hence, by part (c) of Theorem 3.3.1,

$$
\text{Var}(W) = E(W^2) - (E(W))^2 = (2/\lambda^2) - (1/\lambda)^2 = 1/\lambda^2. \quad\blacksquare
$$

Example 3.3.5
Let $W \sim$ Exponential(λ), and let $Y \sim 5W + 3$. Then from the above example, Var$(W) = 1/\lambda^2$. Then, using part (b) of Theorem 3.3.1,

$$
\text{Var}(Y) = \text{Var}(5W + 3) = 25\,\text{Var}(W) = 25/\lambda^2. \quad\blacksquare
$$

Because $\sqrt{a^2} = |a|$, part (b) of Theorem 3.3.1 immediately implies a corresponding fact about standard deviation.

Corollary 3.3.1 Let X be any random variable, with standard deviation $\text{Sd}(X)$, and let a be any real number. Then $\text{Sd}(aX) = |a|\,\text{Sd}(X)$.

Example 3.3.6
Let $W \sim \text{Exponential}(\lambda)$, and let $Y \sim 5W + 3$. Then using the above examples, we see that $\text{Sd}(W) = (\text{Var}(W))^{1/2} = (1/\lambda^2)^{1/2} = 1/\lambda$. Also, $\text{Sd}(Y) = (\text{Var}(Y))^{1/2} = (25/\lambda^2)^{1/2} = 5/\lambda$. This agrees with Corollary 3.4.1, since $\text{Sd}(Y) = |5|\,\text{Sd}(W)$. ∎

Example 3.3.7 *Variance and Standard deviation of the $N(\mu, \sigma^2)$ Distribution*
Suppose that $X \sim N(\mu, \sigma^2)$. In Example 3.2.7 we established that $E(X) = \mu$. Now we compute $\text{Var}(X)$.

First consider $Z \sim N(0, 1)$. Then from Theorem 3.3.1(c) we have that

$$\text{Var}(Z) = E(Z^2) = \int_{-\infty}^{\infty} z^2 \frac{1}{\sqrt{2\pi}} \exp\left\{-\frac{z^2}{2}\right\} dz.$$

Then, putting $u = z, dv = z \exp\left\{-z^2/2\right\}$ (so $du = 1, v = -\exp\left\{-z^2/2\right\}$) and using integration by parts, we obtain

$$\text{Var}(Z) = -z \exp\left\{-z^2/2\right\}\Big|_{-\infty}^{\infty} + \int_{-\infty}^{\infty} \frac{1}{\sqrt{2\pi}} \exp\left\{-\frac{z^2}{2}\right\} dz = 1$$

and $\text{Sd}(Z) = 1$.

Now, for $\sigma > 0$, put $X = \mu + \sigma Z$. We then have $X \sim N(\mu, \sigma^2)$. From Theorem 3.3.1(b) we have that

$$\text{Var}(X) = \text{Var}(\mu + \sigma Z) = \sigma^2 \text{Var}(Z) = \sigma^2$$

and $\text{Sd}(X) = \sigma$. This establishes the variance of the $N(\mu, \sigma^2)$ distribution as σ^2 and the standard deviation as σ.

In Figure 3.3.1, we have plotted three normal distributions all with mean 0 but different variances.

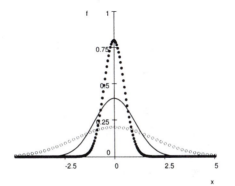

Figure 3.3.1: Plots of the the $N(0, 1)$ (—), the $N(0, 1/4)$ (• • •) and the $N(0, 4)$ (∘ ∘ ∘) density functions.

The effect of the variance on the amount of spread of the distribution about the mean is quite clear from these plots. As σ^2 increases the distribution becomes more diffuse; as it decreases, it becomes more concentrated about the mean 0. ∎

So far we have considered the variance of one random variable at a time. However, there is a related concept, covariance, which measures the *relationship* between two random variables.

Definition 3.3.3 The *covariance* of two random variables X and Y is given by

$$\text{Cov}(X, Y) = E\left((X - \mu_X)(Y - \mu_Y)\right),$$

where $\mu_X = E(X)$ and $\mu_Y = E(Y)$.

Example 3.3.8
Let X and Y be a discrete random variable, with joint probability function $p_{X,Y}$ given by

$$p_{X,Y}(x, y) = \begin{cases} 1/2 & x = 3, y = 4 \\ 1/3 & x = 3, y = 6 \\ 1/6 & x = 5, y = 6 \\ 0 & \text{otherwise.} \end{cases}$$

Then $E(X) = (3)(1/2) + (3)(1/3) + (5)(1/6) = 10/3$, and $E(Y) = (4)(1/2) + (6)(1/3) + (6)(1/6) = 5$. Hence,

$$\text{Cov}(X, Y) = E((X - 10/3)(Y - 5))$$
$$= (3 - 10/3)(4 - 5)/2 + (3 - 10/3)(6 - 5)/3 + (5 - 10/3)(6 - 5)/6$$
$$= 1/3. ∎$$

Example 3.3.9
Let X be any random variable with $\text{Var}(X) > 0$. Let $Y = 3X$, and let $Z = -4X$. Then $\mu_Y = 3\mu_X$ and $\mu_Z = -4\mu_X$. Hence,

$$\begin{aligned} \text{Cov}(X, Y) &= E\left((X - \mu_X)(Y - \mu_Y)\right) = E\left((X - \mu_X)(3X - 3\mu_X)\right) \\ &= 3E\left((X - \mu_X)^2\right) = 3\,\text{Var}(X), \end{aligned}$$

while

$$\begin{aligned} \text{Cov}(X, Z) &= E\left((X - \mu_X)(Z - \mu_Z)\right) = E\left((X - \mu_X)((-4)X - (-4)\mu_X)\right) \\ &= (-4)E\left((X - \mu_X)^2\right) = -4\,\text{Var}(X). \end{aligned}$$

Note in particular that $\text{Cov}(X, Y) > 0$, while $\text{Cov}(X, Z) < 0$. Intuitively, this says that Y increases when X increases, while Z decreases when X increases. ∎

We begin with some simple facts about covariance. Obviously, we always have $\text{Cov}(X, Y) = \text{Cov}(Y, X)$. We also have the following result.

Theorem 3.3.2 (*Linearity of covariance*) Let X, Y, and Z be three random variables. Let a and b be real numbers. Then

$$\text{Cov}(aX + bY, Z) = a\,\text{Cov}(X, Z) + b\,\text{Cov}(Y, Z).$$

Proof: Note that by linearity, $\mu_{aX+bY} \equiv E(aX + bY) = aE(X) + bE(Y) \equiv a\mu_X + b\mu_Y$. Hence,

$$
\begin{aligned}
\mathrm{Cov}(aX + bY, Z) &= E\left((aX + bY - \mu_{aX+bY})(Z - \mu_Z)\right) \\
&= E\left((aX + bY - a\mu_X - b\mu_Y)(Z - \mu_Z)\right) \\
&= E\left((aX - a\mu_X + bY - b\mu_Y)(Z - \mu_Z)\right) \\
&= aE\left((X - \mu_X)(Z - \mu_Z)\right) + bE\left((Y - \mu_Y)(Z - \mu_Z)\right) \\
&= a\,\mathrm{Cov}(X, Z) + b\,\mathrm{Cov}(Y, Z),
\end{aligned}
$$

and the result is established. ∎

We also have the following identity, which is similar to Theorem 3.3.1(c).

Theorem 3.3.3 Let X and Y be two random variables. Then

$$\mathrm{Cov}(X, Y) = E(XY) - E(X)E(Y).$$

Proof: Using linearity, we have

$$
\begin{aligned}
\mathrm{Cov}(X, Y) &= E\left((X - \mu_X)(Y - \mu_Y)\right) = E\left(XY - \mu_X Y - X\mu_Y + \mu_X \mu_Y\right) \\
&= E(XY) - \mu_X E(Y) - E(X)\mu_Y + \mu_X \mu_Y \\
&= E(XY) - \mu_X \mu_Y - \mu_X \mu_Y + \mu_X \mu_Y = E(XY) - \mu_X \mu_Y. \;\blacksquare
\end{aligned}
$$

Corollary 3.3.2 If X and Y are independent, then $\mathrm{Cov}(X, Y) = 0$.

Proof: Because X and Y are independent, we know (Theorems 3.1.3 and 3.2.3) that $E(XY) = E(X)\,E(Y)$. Hence, the result follows immediately from Theorem 3.3.3. ∎

We note that the converse to Corollary 3.4.2 is false, as the following example shows.

Example 3.3.10 *Covariance 0 Does Not Imply Independence.*
Let X and Y be discrete random variables, with joint probability function $p_{X,Y}$ given by

$$
p_{X,Y}(x, y) = \begin{cases}
1/4 & x = 3, \ y = 5 \\
1/4 & x = 4, \ y = 9 \\
1/4 & x = 7, \ y = 5 \\
1/4 & x = 6, \ y = 9 \\
0 & \text{otherwise.}
\end{cases}
$$

Then

$$E(X) = (3)(1/4) + (4)(1/4) + (7)(1/4) + (6)(1/4) = 5,$$

$$E(Y) = (5)(1/4) + (9)(1/4) + (5)(1/4) + (9)(1/4) = 7,$$

and

$$E(XY) = (3)(5)(1/4) + (4)(9)(1/4) + (7)(5)(1/4) + (6)(9)(1/4) = 35.$$

We obtain $\text{Cov}(X,Y) = E(XY) - E(X)\,E(Y) = 35 - (5)(7) = 0$.

On the other hand, X and Y are clearly not independent. For example, $P(X = 4) > 0$ and $P(Y = 5) > 0$, but $P(X = 4,\ Y = 5) = 0$, so $P(X = 4,\ Y = 5) \neq P(X = 4)\,P(Y = 5)$. ∎

There is also an important relationship between variance and covariance.

Theorem 3.3.4
(a) For any random variables X and Y,

$$\text{Var}(X + Y) = \text{Var}(X) + \text{Var}(Y) + 2\,\text{Cov}(X,Y).$$

(b) More generally, for any random variables $X_1, \ldots, X_n$,

$$\text{Var}\left(\sum_i X_i\right) = \sum_i \text{Var}(X_i) + 2\sum_{i<j} \text{Cov}(X_i, X_j).$$

Proof: We prove part (b) here; part (a) then follows as the special case $n = 2$. Note that by linearity,

$$\mu_{\sum_i X_i} \equiv E\left(\sum_i X_i\right) = \sum_i E(X_i) \equiv \sum_i \mu_{X_i}.$$

Therefore we have that

$$\text{Var}\left(\sum_i X_i\right)$$

$$= E\left(\left(\sum_i X_i - \mu_{\sum_i X_i}\right)^2\right) = E\left(\left(\sum_i X_i - \sum_i \mu_i\right)^2\right)$$

$$= E\left(\left(\sum_i (X_i - \mu_i)\right)^2\right) = E\left(\left(\sum_i (X_i - \mu_i)\right)\left(\sum_j (X_j - \mu_j)\right)\right)$$

$$= E\left(\sum_{i,j}(X_i - \mu_i)(X_j - \mu_j)\right) = \sum_{i,j} E\left((X_i - \mu_i)(X_j - \mu_j)\right)$$

$$= \sum_{i=j} E\left((X_i - \mu_i)(X_j - \mu_j)\right) + 2\sum_{i<j} E\left((X_i - \mu_i)(X_j - \mu_j)\right)$$

$$= \sum_i \text{Var}(X_i) + 2\sum_{i<j} \text{Cov}(X_i, X_j). \quad ∎$$

Combining Theorem 3.3.4 with Corollary 3.3.2, we immediately obtain the following.

Corollary 3.3.3
(a) If X and Y are independent, then $\text{Var}(X + Y) = \text{Var}(X) + \text{Var}(Y)$.

(b) If $X_1, \ldots, X_n$ are independent, then $\mathrm{Var}(\sum_{i=1}^{n} X_i) = \sum_{i=1}^{n} \mathrm{Var}(X_i)$.

One use of Corollary 3.4.3 is the following.

Example 3.3.11
Let $Y \sim \mathrm{Binomial}(n, \theta)$. What is $\mathrm{Var}(Y)$? Recall that we can write

$$Y = X_1 + X_2 + \cdots + X_n,$$

where the X_i are independent, with $X_i \sim \mathrm{Bernoulli}(\theta)$. We have already seen that $\mathrm{Var}(X_i) = \theta(1 - \theta)$. Hence, from Corollary 3.3.3,

$$\begin{aligned}
\mathrm{Var}(Y) &= \mathrm{Var}(X_1) + \mathrm{Var}(X_2) + \cdots + \mathrm{Var}(X_n) \\
&= \theta(1 - \theta) + \theta(1 - \theta) + \cdots + \theta(1 - \theta) = n\theta(1 - \theta). \ \blacksquare
\end{aligned}$$

There is another concept very closely related to covariance, that of correlation.

Definition 3.3.4 The *correlation* of two random variables X and Y is given by

$$\mathrm{Corr}(X, Y) = \frac{\mathrm{Cov}(X, Y)}{\mathrm{Sd}(X)\,\mathrm{Sd}(Y)} = \frac{\mathrm{Cov}(X, Y)}{\sqrt{\mathrm{Var}(X)\,\mathrm{Var}(Y)}}$$

provided $\mathrm{Var}(X) < \infty$ and $\mathrm{Var}(Y) < \infty$.

Example 3.3.12
As in Example 3.3.2, let X be any random variable with $\mathrm{Var}(X) > 0$. let $Y = 3X$, and let $Z = -4X$. Then $\mathrm{Cov}(X, Y) = 3\,\mathrm{Var}(X)$ and $\mathrm{Cov}(X, Z) = -4\,\mathrm{Var}(X)$. But by Corollary 3.3.1, $\mathrm{Sd}(Y) = 3\,\mathrm{Sd}(X)$ and $\mathrm{Sd}(Z) = 4\,\mathrm{Sd}(X)$. Hence,

$$\mathrm{Corr}(X, Y) = \frac{\mathrm{Cov}(X, Y)}{\mathrm{Sd}(X)\,\mathrm{Sd}(Y)} = \frac{3\,\mathrm{Var}(X)}{\mathrm{Sd}(X)3\,\mathrm{Sd}(X)} = \frac{\mathrm{Var}(X)}{\mathrm{Sd}(X)^2} = 1,$$

because $\mathrm{Sd}(X)^2 = \mathrm{Var}(X)$. Also, we have that

$$\mathrm{Corr}(X, Z) = \frac{\mathrm{Cov}(X, Z)}{\mathrm{Sd}(X)\,\mathrm{Sd}(Z)} = \frac{-4\,\mathrm{Var}(X)}{\mathrm{Sd}(X)4\,\mathrm{Sd}(X)} = -\frac{\mathrm{Var}(X)}{\mathrm{Sd}(X)^2} = -1.$$

Intuitively, this again says that Y increases when X increases, while Z decreases when X increases. However, note that the scale factors 3 and -4 have cancelled out; only their signs were important. $\blacksquare$

We shall see later, in Section 3.6, that we always have $-1 \le \mathrm{Corr}(X, Y) \le 1$, for any random variables X and Y. Hence, in Example 3.3.12, Y has the largest possible correlation with X (which makes sense because Y increases whenever X does, without exception), while Z has the smallest possible correlation with X (which makes sense because Z decreases whenever X does). We will also see that $\mathrm{Corr}(X, Y)$ is a measure of the extent to which a linear relationship exists between X and Y.

Example 3.3.13 *The Bivariate Normal($\mu_1, \mu_2, \sigma_1, \sigma_2, \rho$) Distribution*
We defined this distribution in Example 2.7.9. It turns out that when (X, Y)
follows this joint distribution then from Problem 2.7.8 $X \sim N(\mu_1, \sigma_1^2)$ and
$Y \sim N(\mu_2, \sigma_2^2)$. Further using (Problem 3.3.11) $\text{Corr}(X, Y) = \rho$. In the following
graphs we have plotted samples of $n = 1000$ values of (X, Y) from bivariate
normal distributions with $\mu_1 = \mu_2 = 0, \sigma_1^2 = \sigma_2^2 = 1$, and various values of ρ.
Note that we used (2.7.1) to generate these samples.

From these plots we can see the effect of ρ on the joint distribution. Figure
3.3.2 shows that when $\rho = 0$, the point cloud is roughly circular, becoming
elliptical in Figure 3.3.3 with $\rho = 0.5$, and more tightly concentrated about a
line in Figure 3.3.4 with $\rho = 0.9$. As we will see in Section 3.6, the points will
lie exactly on a line when $\rho = 1$.

Figure 3.3.5 demonstrates the effect of a negative correlation. With positive
correlations, the value of Y tends to increase with X, as reflected in the upward
slope of the point cloud. With negative correlations, Y tends to decrease with
X, as reflected in the negative slope of the point cloud. ∎

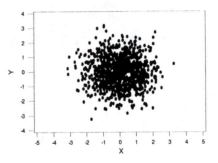

Figure 3.3.2: A sample of $n = 1000$ values (X, Y) from the Bivariate Normal
$(0, 0, 1, 1, 0)$ distribution.

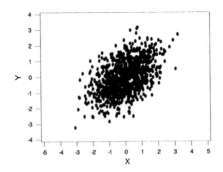

Figure 3.3.3: A sample of $n = 1000$ values (X, Y) from the Bivariate Normal
$(0, 0, 1, 1, 0.5)$ distribution.

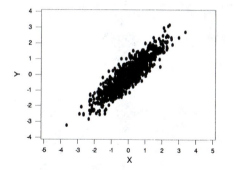

Figure 3.3.4: A sample of $n = 1000$ values (X, Y) from the Bivariate Normal $(0, 0, 1, 1, 0.9)$ distribution.

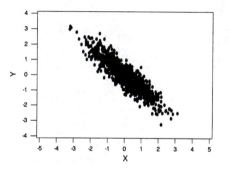

Figure 3.3.5: A sample of $n = 1000$ values (X, Y) from the Bivariate Normal $(0, 0, 1, 1, -0.9)$ distribution.

Summary of Section 3.3

- The variance of a random variable X measures how far it tends to be from its mean and is given by $\text{Var}(X) = E((X - \mu_X)^2) = E(X^2) - (E(X))^2$.

- The variances of many standard distributions were computed.

- The standard deviation of X equals $\text{Sd}(X) = \sqrt{\text{Var}(X)}$.

- $\text{Var}(X) \geq 0$, and $\text{Var}(aX + b) = a^2 \text{Var}(X)$; also $\text{Sd}(aX + b) = |a| \text{Sd}(X)$.

- The covariance of random variables X and Y measures how they are related and is given by $\text{Cov}(X, Y) = E((X - \mu_X)(Y - \mu_y)) = E(XY) - E(X) E(Y)$.

- If X and Y are independent, then $\text{Cov}(X, Y) = 0$.

- $\mathrm{Var}(X+Y) = \mathrm{Var}(X) + \mathrm{Var}(Y) + 2\,\mathrm{Cov}(X,Y)$. If X and Y are independent, this equals $\mathrm{Var}(X) + \mathrm{Var}(Y)$.

- The correlation of X and Y is $\mathrm{Corr}(X,Y) = \mathrm{Cov}(X,Y)/(\mathrm{Sd}(X)\,\mathrm{Sd}(Y))$.

Exercises

3.3.1 Suppose the joint probability function of X and Y is given by

$$p_{X,Y}(x,y) = \begin{cases} 1/2 & x=3,\ y=5 \\ 1/6 & x=3,\ y=9 \\ 1/6 & x=6,\ y=5 \\ 1/6 & x=6,\ y=9 \\ 0 & \text{otherwise,} \end{cases}$$

with $E(X)=4$, $E(Y)=19/3$, and $E(XY)=26$, as in Example 3.1.14.
(a) Compute $\mathrm{Cov}(X,Y)$.
(b) Compute $\mathrm{Var}(X)$ and $\mathrm{Var}(Y)$.
(c) Compute $\mathrm{Corr}(X,Y)$.

3.3.2 Suppose the joint probability function of X and Y is given by

$$p_{X,Y}(x,y) = \begin{cases} 1/7 & x=5,\ y=0 \\ 1/7 & x=5,\ y=3 \\ 1/7 & x=5,\ y=4 \\ 3/7 & x=8,\ y=0 \\ 1/7 & x=8,\ y=4 \\ 0 & \text{otherwise,} \end{cases}$$

as in Example 2.7.5.
(a) Compute $E(X)$ and $E(Y)$.
(b) Compute $\mathrm{Cov}(X,Y)$.
(c) Compute $\mathrm{Var}(X)$ and $\mathrm{Var}(Y)$.
(d) Compute $\mathrm{Corr}(X,Y)$.

3.3.3 Let X and Y have joint density

$$f_{X,Y}(x,y) = \begin{cases} 4x^2y + 2y^5 & 0 \le x \le 1,\ 0 \le y \le 1 \\ 0 & \text{otherwise,} \end{cases}$$

as in Exercise 3.2.2. Compute $\mathrm{Corr}(X,Y)$.

3.3.4 Let X and Y have joint density

$$f_{X,Y}(x,y) = \begin{cases} 15x^3y^4 + 6x^2y^7 & 0 \le x \le 1,\ 0 \le y \le 1 \\ 0 & \text{otherwise.} \end{cases}$$

Compute $E(X)$, $E(Y)$, $\mathrm{Var}(X)$, $\mathrm{Var}(Y)$, $\mathrm{Cov}(X,Y)$, and $\mathrm{Corr}(X,Y)$.

3.3.5 Let Y and Z be two independent random variables, each with positive variance. Prove that $\mathrm{Corr}(Y, Z) = 0$.

3.3.6 Let X, Y, and Z be three random variables, and suppose that X and Z are independent. Prove that $\mathrm{Cov}(X + Y, Z) = \mathrm{Cov}(Y, Z)$.

3.3.7 Let $X \sim$ Exponential(3) and $Y \sim$ Poisson(5). Assume X and Y are independent. Let $Z = X + Y$.
(a) Compute $\mathrm{Cov}(X, Z)$.
(b) Compute $\mathrm{Corr}(X, Z)$.

3.3.8 Prove that the variance of the Uniform$[L, R]$ distribution is given by $(R - L)^2/12$.

3.3.9 Prove that $\mathrm{Var}(X) = E\left(X\left(X - 1\right)\right) - E\left(X\right)E\left(X - 1\right)$. Use this to compute directly from the probability function that when $X \sim$ Binomial(n, θ), then $\mathrm{Var}(X) = n\theta\left(1 - \theta\right)$.

Problems

3.3.10 Let $X \sim N(0, 1)$, and let $Y = cX$.
(a) Compute $\lim_{c \searrow 0} \mathrm{Cov}(X, Y)$.
(b) Compute $\lim_{c \nearrow 0} \mathrm{Cov}(X, Y)$.
(c) Compute $\lim_{c \searrow 0} \mathrm{Corr}(X, Y)$.
(d) Compute $\lim_{c \nearrow 0} \mathrm{Corr}(X, Y)$.
(e) Explain why the answers in parts (c) and (d) are not the same.

3.3.11 Let X and Y have the bivariate normal distribution, as in Example 2.7.9. Prove that $\mathrm{Corr}(X, Y) = \rho$. (Hint: Use (2.7.1).)

3.3.12 Prove that the variance of the Geometric(θ) distribution is given by $(1 - \theta)/\theta^2$. (Hint: Use Exercise 3.3.9 and $((1 - \theta)^x)'' = x\left(x - 1\right)\left(1 - \theta\right)^{x-2}$.)

3.3.13 Prove that the variance of the Negative-Binomial(r, θ) distribution is given by $r\left(1 - \theta\right)/\theta^2$. (Hint: Use Problem 3.3.12.)

3.3.14 Let $\alpha > 0$ and $\lambda > 0$, and let $X \sim$ Gamma(α, λ). Prove that $\mathrm{Var}(X) = \alpha/\lambda^2$. (Hint: Recall Problem 3.2.9.)

3.3.15 Suppose that $X \sim$ Weibull(α) distribution (Problem 2.4.11). Prove that $\mathrm{Var}(X) = \Gamma\left(2/\alpha + 1\right) - \Gamma^2\left(1/\alpha + 1\right)$. (Hint: Recall Problem 3.2.11.)

3.3.16 Suppose that $X \sim$ Pareto(α) (Problem 2.4.12) for $\alpha > 2$. Prove that $\mathrm{Var}(X) = \alpha/((\alpha - 1)^2\left(\alpha - 2\right))$. (Hint: Recall Problem 3.2.12.)

3.3.17 Suppose that X follows the Laplace distribution (Problem 2.4.14). Prove that $\mathrm{Var}(X) = 2$. (Hint: Recall Problem 3.2.14.)

3.3.18 Suppose that $X \sim$ Beta(a, b) (Problem 2.4.16). Prove that $\mathrm{Var}(X) = ab/((a + b)^2(a + b + 1))$. (Hint: Recall Problem 3.2.15.)

3.3.19 Suppose that $(X_1, X_2, X_3) \sim$ Multinomial$(n, \theta_1, \theta_2, \theta_3)$. Prove that

$$\mathrm{Var}(X_i) = n\theta_i\left(1 - \theta_i\right), \quad \mathrm{Cov}\left(X_i, X_j\right) = -n\theta_i\theta_j \text{ when } i \neq j.$$

(Hint: Recall Problem 3.1.18.)

3.3.20 Suppose that $(X_1, X_2) \sim$ Dirichlet$(\alpha_1, \alpha_2, \alpha_3)$ (Problem 2.7.12). Prove that

$$\text{Var}(X_i) = \frac{\alpha_i (\alpha_1 + \alpha_2 + \alpha_3 - \alpha_i)}{(\alpha_1 + \alpha_2 + \alpha_3)^2 (\alpha_1 + \alpha_2 + \alpha_3 + 1)},$$

$$\text{Cov}(X_1, X_2) = \frac{-\alpha_1 \alpha_2}{(\alpha_1 + \alpha_2 + \alpha_3)^2 (\alpha_1 + \alpha_2 + \alpha_3 + 1)}.$$

(Hint: Recall Problem 3.2.16.)

3.3.21 Suppose that $X \sim$ Hypergeometric(N, M, n). Prove that

$$\text{Var}(X) = n\frac{M}{N} \left(1 - \frac{M}{N}\right) \frac{N - n}{N - 1}.$$

(Hint: Recall Problem 3.1.16 and use Exercise 3.3.9.)

Challenges

3.3.22 Let Y be a nonnegative random variable. Prove that $E(Y) = 0$ if and only if $P(Y = 0) = 1$. (You may assume for simplicity that Y is discrete, but the result is true for any Y.)

3.3.23 Prove that $\text{Var}(X) = 0$ if and only if there is a real number c with $P(X = c) = 1$. (You may use the result of Challenge 3.3.22.)

3.3.24 Give an example of a random variable X, such that $E(X) = 5$, and $\text{Var}(X) = \infty$.

3.4   Generating Functions

Let X be a random variable. Recall that the cumulative distribution function of X, defined by $F_X(x) = P(X \leq x)$, contains all the information about the distribution of X (see Theorem 2.5.1). It turns out that there are other functions, the probability-generating function and the moment-generating function, which also provide information (sometimes all the information) about X and its expected values.

Definition 3.4.1 Let X be a random variable (usually discrete). Then we define its *probability-generating function*, r_X, by $r_X(t) = E(t^X)$ for $t \in R^1$.

Consider the following examples of probability-generating functions.

Example 3.4.1 *The Binomial(n, θ) Distribution*
If $X \sim$ Binomial(n, θ), then

$$r_X(t) = E(t^X) = \sum_{i=0}^{n} P(X = i)t^i = \sum_{i=0}^{n} \binom{n}{i} \theta^i (1 - \theta)^{n-i} t^i$$

$$= \sum_{i=0}^{n} \binom{n}{i} (t\theta)^i (1 - \theta)^{n-i} = (t\theta + 1 - \theta)^n,$$

using the binomial theorem. ∎

Example 3.4.2 *The Poisson(λ) Distribution*
If $Y \sim$ Poisson(λ), then

$$
\begin{aligned}
r_Y(t) &= E(t^Y) = \sum_{i=0}^{\infty} P(Y = i)t^i = \sum_{i=0}^{\infty} e^{-\lambda} \frac{\lambda^i}{i!} t^i \\
&= \sum_{i=0}^{\infty} e^{-\lambda} \frac{(\lambda t)^i}{i!} = e^{-\lambda} e^{\lambda t} = e^{\lambda(t-1)}. \quad \blacksquare
\end{aligned}
$$

The following theorem tells us that once we know the probability-generating function $r_X(t)$, then we can compute all the probabilities $P(X = 0)$, $P(X = 1)$, $P(X = 2)$, etc.

Theorem 3.4.1 Let X be a discrete random variable, whose possible values are all nonnegative integers. Assume that $r_X(t_0) < \infty$ for some $t_0 > 0$. Then

$$
\begin{aligned}
r_X(0) &= P(X = 0), \\
r'_X(0) &= P(X = 1), \\
r''_X(0) &= 2 P(X = 2),
\end{aligned}
$$

etc. In general,

$$
r_X^{(k)}(0) = k! \, P(X = k),
$$

where $r_X^{(k)}$ is the kth derivative of r_X.

Proof: Because the possible values are all nonnegative integers of the form $i = 0, 1, 2, \ldots$, we have

$$
\begin{aligned}
r_X(t) &= E(t^X) = \sum_x t^x P(X = x) = \sum_{i=0}^{\infty} t^i P(X = i) \\
&= t^0 P(X = 0) + t^1 P(X = 1) + t^2 P(X = 2) + t^3 P(X = 3) + \cdots,
\end{aligned}
$$

so that

$$
r_X(t) = 1P(X = 0) + t^1 P(X = 1) + t^2 P(X = 2) + t^3 P(X = 3) + \cdots. \quad (3.4.1)
$$

Substituting $t = 0$ into (3.4.1) every term vanishes except the first one, and we obtain $r_X(0) = P(X = 0)$. Taking derivatives of both sides of (3.4.1), we obtain

$$
r'_X(t) = 1P(X = 1) + 2t^1 P(X = 2) + 3t^2 P(X = 3) + \cdots,
$$

and setting $t = 0$ gives $r'_X(0) = P(X = 1)$. Taking another derivative of both sides gives

$$
r''_X(t) = 2P(X = 2) + 3 \cdot 2t^1 P(X = 3) + \cdots
$$

and setting $t = 0$ gives $r''_X(0) = 2 P(X = 2)$. Continuing in this way, we obtain the general formula. ∎

We now apply Theorem 3.4.1 to the binomial and Poisson distributions.

Example 3.4.3 *The Binomial(n, θ) Distribution*
From Example 3.4.1 we have that

$$
\begin{aligned}
r_X(0) &= (1 - \theta)^n \\
r'_X(0) &= n(t\theta + 1 - \theta)^{n-1}(\theta)\Big|_{t=0} = n(1 - \theta)^{n-1}\theta \\
r''_X(0) &= n(n-1)(t\theta + 1 - \theta)^{n-2}(\theta)(\theta)\Big|_{t=0} = n(n-1)(1 - \theta)^{n-2}\theta^2,
\end{aligned}
$$

etc. It is thus verified directly that

$$
\begin{aligned}
P(X = 0) &= r_X(0) \\
P(X = 1) &= r'_X(0) \\
2\,P(X = 2) &= r''_X(0),
\end{aligned}
$$

etc. ∎

Example 3.4.4 *The Poisson(λ) Distribution*
From Example 3.4.2, we have that

$$
\begin{aligned}
r_X(0) &= e^{-\lambda} \\
r'_X(0) &= \lambda e^{-\lambda} \\
r''_X(0) &= \lambda^2 e^{-\lambda},
\end{aligned}
$$

etc. It is again verified that

$$
\begin{aligned}
P(X = 0) &= r_X(0) \\
P(X = 1) &= r'_X(0) \\
2\,P(X = 2) &= r''_X(0),
\end{aligned}
$$

etc. ∎

From Theorem 3.4.1, we can see why r_X is called the probability-generating function. For, at least in the discrete case with the distribution concentrated on the nonnegative integers, we can indeed generate the probabilities for X from r_X. From this we see immediately that for a random variable X that takes values only in $\{0, 1, 2, \ldots\}$, then r_X is unique. By this we mean that if X and Y are concentrated on $\{0, 1, 2, \ldots\}$ and $r_X = r_Y$, then X and Y have the same distribution. This uniqueness property of the probability-generating function can be very useful when we are trying to determine the distribution of a random variable that takes only values in $\{0, 1, 2, \ldots\}$.

It is clear that the probability-generating function tells us a lot — in fact everything — about the distribution of random variables concentrated on the nonnegative integers. But what about other random variables? It turns out that there are other quantities, called moments, associated with random variables that are quite informative about their distributions.

Definition 3.4.2 Let X be a random variable, and let k be a positive integer. Then the kth *moment* of X is the quantity $E(X^k)$, provided this expectation exists.

Note that if $E(X^k)$ exists and is finite, it can be shown that $E(X^l)$ exists and is finite when $0 \leq l < k$.

The first moment is just the mean of the random variable, and this can be taken as a measure of where the central mass of probability for X lies in the real line, at least when this distribution is unimodal (has a single peak) and is not too highly skewed. The second moment $E(X^2)$, together with the first moment, gives us the variance through $\text{Var}(X) = E(X^2) - (E(X))^2$. Therefore, the first two moments of the distribution tell us about the location of the distribution and the spread, or degree of concentration, of that distribution about the mean. In fact, the higher moments also provide us with information about the distribution.

Many of the most important distributions of probability and statistics have all of their moments finite, and in fact, have what is called a moment-generating function.

Definition 3.4.3 Let X be any random variable. Then its *moment-generating function* m_X is defined by $m_X(s) = E(e^{sX})$ at $s \in R^1$.

The following example computes the moment generating function of a well-known distribution.

Example 3.4.5 *The Exponential(λ) Distribution*
Let $X \sim \text{Exponential}(\lambda)$. Then for $s < \lambda$,

$$
\begin{aligned}
m_X(s) &= E(e^{sX}) = \int_{-\infty}^{\infty} e^{sx} f_X(x)\, dx = \int_0^{\infty} e^{sx} \lambda e^{-\lambda x}\, dx \\
&= \int_0^{\infty} \lambda e^{(s-\lambda)x}\, dx = \frac{\lambda e^{(s-\lambda)x}}{s-\lambda} \bigg|_{x=0}^{x=\infty} = -\frac{\lambda e^{(s-\lambda)0}}{s-\lambda} \\
&= -\frac{\lambda}{s-\lambda} = \lambda(\lambda-s)^{-1}. \quad \blacksquare
\end{aligned}
$$

A comparison of Definitions 3.4.1 and 3.4.3 immediately gives the following.

Theorem 3.4.2 Let X be any random variable. Then $m_X(s) = r_X(e^s)$.

This result can obviously help us in evaluating some moment-generating functions when we have r_X already.

Example 3.4.6
Let $Y \sim \text{Binomial}(n, \theta)$. Then we know that $r_Y(t) = (t\theta + 1 - \theta)^n$. Hence, $m_Y(s) = r_Y(e^s) = (e^s\theta + 1 - \theta)^n$. $\quad \blacksquare$

Example 3.4.7
Let $Z \sim \text{Poisson}(\lambda)$. Then we know that $r_Z(t) = e^{\lambda(t-1)}$. Hence, $m_Z(s) = r_Z(e^s) = e^{\lambda(e^s-1)}$. $\quad \blacksquare$

The following theorem tells us that once we know the moment-generating function $m_X(t)$, we can compute all the moments $E(X)$, $E(X^2)$, $E(X^3)$, etc.

Theorem 3.4.3 Let X be any random variable. Suppose that for some $s_0 > 0$, it is true that $m_X(s) < \infty$ whenever $s \in (-s_0, s_0)$. Then

$$
\begin{aligned}
m_X(0) &= 1 \\
m_X'(0) &= E(X) \\
m_X''(0) &= E(X^2),
\end{aligned}
$$

etc. In general,

$$m_X^{(k)}(0) = E(X^k),$$

where $m_X^{(k)}$ is the kth derivative of m_X.

Proof: We know that $m_X(s) = E(e^{sX})$. We have

$$m_X(0) = E(e^{0X}) = E(e^0) = E(1) = 1.$$

Also, taking derivatives, we see[3] that $m_X'(s) = E(X\,e^{sX})$, so

$$m_X'(0) = E(X\,e^{0X}) = E(Xe^0) = E(X).$$

Taking derivatives again, we see that $m_X''(s) = E(X^2 e^{sX})$, so

$$m_X''(0) = E(X^2\,e^{0X}) = E(X^2 e^0) = E(X^2).$$

Continuing in this way, we obtain the general formula. ∎

We now consider an application of Theorem 3.4.3.

Example 3.4.8 *The Mean and Variance of the Exponential(λ) Distribution*
Using the moment-generating function computed in Example 3.4.5, we have

$$m_X'(s) = (-1)\lambda(\lambda - s)^{-2}(-1) = \lambda(\lambda - s)^{-2}.$$

Therefore,

$$E(X) = m_X'(0) = \lambda(\lambda - 0)^{-2} = \lambda/\lambda^2 = 1/\lambda,$$

as it should. Also,

$$E(X^2) = m_X''(0) = (-2)\lambda(\lambda - 0)^{-3}(-1) = 2\lambda/\lambda^3 = 2/\lambda^2$$

so we have

$$\mathrm{Var}(X) = E(X^2) - (E(X)) = (2/\lambda^2) - (1/\lambda)^2 = 1/\lambda^2.$$

This provides an easy way of computing the variance of X. ∎

---

[3] Strictly speaking, interchanging the order of derivative and expectation is justified by analytic function theory, and requires that $m_X(s) < \infty$ whenever $|s| < s_0$.

Example 3.4.9 *The Mean and Variance of the Poisson(λ) Distribution*
In Example 3.4.7, we obtained $m_Z(s) = \exp\left(\lambda(e^s - 1)\right).$ So we have

$$
\begin{aligned}
E(X) &= m'_X(0) = \lambda e^0 \exp\left(\lambda(e^0 - 1)\right) = \lambda \\
E(X^2) &= m''_X(0) = \lambda e^0 \exp\left(\lambda(e^0 - 1)\right) + \left(\lambda e^0\right)^2 \exp\left(\lambda(e^0 - 1)\right) = \lambda + \lambda^2.
\end{aligned}
$$

Therefore, $\mathrm{Var}(X) = E(X^2) - (E(X))^2 = \lambda + \lambda^2 - \lambda^2 = \lambda.$ ∎

Computing the moment-generating function of a normal distribution is also important, but is somewhat more difficult.

Theorem 3.4.4 If $X \sim N(0, 1)$, then $m_X(s) = e^{s^2/2}$.

Proof: Because X has density $\phi(x) = (2\pi)^{-1/2} e^{-x^2/2}$, we have that

$$
\begin{aligned}
m_X(s) &= E(e^{sX}) = \int_{-\infty}^{\infty} e^{sx} \phi(x)\, dx = \int_{-\infty}^{\infty} e^{sx} \frac{1}{\sqrt{2\pi}} e^{-x^2/2}\, dx \\
&= \frac{1}{\sqrt{2\pi}} \int_{-\infty}^{\infty} e^{sx - (x^2/2)}\, dx = \frac{1}{\sqrt{2\pi}} \int_{-\infty}^{\infty} e^{-(x-s)^2/2 + (s^2/2)}\, dx \\
&= e^{s^2/2} \frac{1}{\sqrt{2\pi}} \int_{-\infty}^{\infty} e^{-(x-s)^2/2}\, dx.
\end{aligned}
$$

Setting $y = x - s$ (so that $dy = dx$), this becomes (using Theorem 2.4.2)

$$
m_X(s) = e^{s^2/2} \frac{1}{\sqrt{2\pi}} \int_{-\infty}^{\infty} e^{-y^2/2}\, dy = e^{s^2/2} \int_{-\infty}^{\infty} \phi(y)\, dy = e^{s^2/2},
$$

as claimed. ∎

One useful property of both probability-generating and moment-generating functions is the following.

Theorem 3.4.5 Let X and Y be random variables that are independent. Then we have
(a) $r_{X+Y}(t) = r_X(t)\, r_Y(t)$.
(b) $m_{X+Y}(t) = m_X(t)\, m_Y(t)$.

Proof: Because X and Y are independent, so are t^X and t^Y (by Theorem 2.8.5). Hence, we know (by Theorems 3.1.3 and 3.2.3) that $E(t^X t^Y) = E(t^X)\, E(t^Y)$. Using this, we have

$$
r_{X+Y}(t) = E\left(t^{X+Y}\right) = E\left(t^X t^Y\right) = E(t^X)E(t^Y) = r_X(t) r_Y(t).
$$

Similarly,

$$
m_{X+Y}(t) = E\left(e^{t(X+Y)}\right) = E\left(e^{tX} e^{tY}\right) = E(e^{tX})\, E(e^{tY}) = m_X(t) m_Y(t). ∎
$$

Example 3.4.10
Let $Y \sim \mathrm{Binomial}(n, \theta)$. Then as in Example 3.1.13, we can write

$$
Y = X_1 + \cdots + X_n,
$$

where the $\{X_i\}$ are i.i.d. with $X_i \sim$ Bernoulli(θ). Hence, Theorem 3.4.5 says we must have $r_Y(t) = r_{X_1}(t)\, r_{X_2}(t) \cdots r_{X_n}(t)$. But for any i,

$$r_{X_i}(t) = \sum_x t^x P(X = x) = t^1 \theta + t^0 (1 - \theta) = \theta t + 1 - \theta.$$

Hence, we must have

$$r_Y(t) = (\theta t + 1 - \theta)(\theta t + 1 - \theta) \cdots (\theta t + 1 - \theta) = (\theta t + 1 - \theta)^n,$$

as already verified in Example 3.4.1. ∎

Moment-generating functions, when defined in a neighborhood of 0, completely define a distribution in the following sense. (We omit the proof, which is advanced.)

Theorem 3.4.6 (*Uniqueness theorem*) Let X be a random variable, such that for some $s_0 > 0$, we have $m_X(s) < \infty$ whenever $s \in (-s_0, s_0)$. Then if Y is some other random variable with $m_Y(s) = m_X(s)$ whenever $s \in (-s_0, s_0)$, then X and Y have the same distribution.

Theorems 3.4.1 and 3.4.6 provide a powerful technique for identifying distributions. For example, if we determine that the moment-generating function of X is $m_X(t) = \exp\left(s^2/2\right)$, then we know, from Theorems 3.4.4 and 3.4.6, that $X \sim N(0, 1)$. We can use this approach to determine the distributions of some complicated random variables.

Example 3.4.11
Suppose that $X_i \sim N(\mu_i, \sigma_i^2)$ for $i = 1, \ldots, n$ and that these random variables are independent. Consider the distribution of $Y = \sum_{i=1}^n X_i$.
When $n = 1$ we have (from Problem 3.4.15)

$$m_Y(s) = \exp\left\{\mu_1 s + \frac{\sigma_1^2 s^2}{2}\right\}.$$

Then, using Theorem 3.4.5, we have that

$$
\begin{aligned}
m_Y(s) &= \prod_{i=1}^n m_{X_i}(s) = \prod_{i=1}^n \exp\left\{\mu_i s + \frac{\sigma_i^2 s^2}{2}\right\} \\
&= \exp\left\{\left(\sum_{i=1}^n \mu_i\right) s + \frac{\left(\sum_{i=1}^n \sigma_i^2\right) s^2}{2}\right\}.
\end{aligned}
$$

From Problem 3.4.15, and applying Theorem 3.4.6, we have that

$$Y \sim N\left(\sum_{i=1}^n \mu_i, \sum_{i=1}^n \sigma_i^2\right). \quad \blacksquare$$

Generating functions can also help us with compound distributions, defined as follows.

Definition 3.4.4 Let $X_1, X_2, \ldots$ be i.i.d., and let N be a nonnegative, integer-valued random variable which is independent of the $\{X_i\}$. Let

$$S = \sum_{i=1}^{N} X_i. \qquad (3.4.2)$$

Then S is said to have a *compound distribution*.

A compound distribution is obtained from a sum of i.i.d. random variables, where the number of terms in the sum is randomly distributed independently of the terms in the sum. Note that $S = 0$ when $N = 0$. Such distributions have applications in areas like insurance where the $X_1, X_2, \ldots$ are claims and N is the number of claims presented to an insurance company during a period. Therefore, S represents the total amount claimed against the insurance company during the period. Obviously, the insurance company wants to study the distribution of S, as this will help determine what it has to charge for insurance to ensure a profit.

The following theorem is important in the study of compound distributions.

Theorem 3.4.7 If S has a compound distribution as in (3.4.2), then
(a) $E(S) = E(X_1)E(N)$.
(b) $m_S(s) = r_N(m_{X_1}(s))$.
Proof: See Section 3.8 for the proof of this result. ∎

3.4.1 Characteristic Functions (Advanced)

One problem with moment-generating functions is that they can be *infinite* in any open interval about $s = 0$. Consider the following example.

Example 3.4.12
Let X be a random variable having density

$$f_X(x) = \begin{cases} 1/x^2 & x \geq 1 \\ 0 & \text{otherwise.} \end{cases}$$

Then

$$m_X(s) = E(e^{sX}) = \int_1^\infty e^{sx}(1/x^2)\, dx.$$

For any $s > 0$, we know that e^{sx} grows faster than x^2, so that $\lim_{x \to \infty} e^{sx}/x^2 = \infty$. Hence, $m_X(s) = \infty$ whenever $s > 0$.

Does X have any finite moments? We have that

$$E(X) = \int_1^\infty x\,(1/x^2)\, dx = \int_1^\infty (1/x)\, dx = \ln x \big|_{x=1}^{x=\infty} = \infty,$$

so, in fact, the first moment does not exist. From this we conclude that X does not have any moments. ∎

The random variable X in the above example does *not* satisfy the condition of Theorem 3.4.3 that $m_X(s) < \infty$ whenever $|s| < s_0$, for some $s_0 > 0$. Hence, Theorem 3.4.3 (like most other theorems that make use of moment-generating functions) does not apply. There is, however, a similarly defined function that does not suffer from this defect and is given by the following definition.

Definition 3.4.5 Let X be any random variable. Then we define its *characteristic function, c_X,* by

$$c_X(s) = E(e^{isX}) \tag{3.4.3}$$

for $s \in R^1$.

So the definition of c_X is just like the definition of m_X, except for the introduction of the imaginary number $i = \sqrt{-1}$. Using properties of complex numbers, we see that (3.4.3) can also be written as $c_X(s) = E(\cos(sX)) + i\,E(\sin(sX))$ for $s \in R^1$.

Consider the following examples.

Example 3.4.13 *The Bernoulli Distribution*
Let $X \sim$ Bernoulli(θ). Then

$$
\begin{aligned}
c_X(s) &= E(e^{isX}) = (e^{is0})(1-\theta) + (e^{is1})(\theta) \\
&= (1)(1-\theta) + e^{is}(\theta) = 1 - \theta + \theta e^{is} \\
&= 1 - \theta + \theta \cos s + i\theta \sin s. \ \blacksquare
\end{aligned}
$$

Example 3.4.14
Let X have probability function given by

$$
p_X(x) = \begin{cases}
1/6 & x = 2 \\
1/3 & x = 3 \\
1/2 & x = 4 \\
0 & \text{otherwise.}
\end{cases}
$$

Then

$$
\begin{aligned}
c_X(s) &= E(e^{isX}) = (e^{is2})(1/6) + (e^{is3})(1/3) + (e^{is4})(1/2) \\
&= (1/6)\cos 2s + (1/3)\cos 3s + (1/2)\cos 4s \\
&\quad + (1/6)i\sin 2s + (1/3)i\sin 3s + i(1/2)\sin 4s. \ \blacksquare
\end{aligned}
$$

Example 3.4.15
Let Z have probability function given by

$$
p_Z(z) = \begin{cases}
1/2 & z = 1 \\
1/2 & z = -1 \\
0 & \text{otherwise.}
\end{cases}
$$

Then

$$
\begin{aligned}
c_Z(s) &= E(e^{isZ}) = (e^{is})(1/2) + (e^{-is})(1/2) \\
&= (1/2)\cos(s) + (1/2)\cos(-s) + (1/2)\sin(s) + (1/2)\sin(-s) \\
&= (1/2)\cos s + (1/2)\cos s + (1/2)\sin s - (1/2)\sin s = \cos s.
\end{aligned}
$$

Hence, in this case, $c_Z(s)$ is a real (not complex) number for all $s \in R^1$. ∎

Once we overcome our "fear" of imaginary and complex numbers, we can see that the characteristic function is actually much better in some ways than the moment-generating function. The main advantage is that, because $e^{isX} = \cos(sX) + i \sin(sX)$ and $|e^{isX}| = 1$, the characteristic function (unlike the moment-generating function) is always finite (although it could be a complex number).

Theorem 3.4.8 Let X be any random variable, and let s be any real number. Then $c_X(s)$ is finite.

The characteristic function has many nice properties similar to the moment generating function. In particular we have the following. (The proof is just like the proof of Theorem 3.4.3.)

Theorem 3.4.9 Let X be any random variable with its first k moments finite. Then $c_X(0) = 1$, $c_X'(0) = iE(X)$, $c_X''(0) = i^2 E(X^2) = -E(X^2)$, etc. In general, $c_X^{(k)}(0) = i^k E(X^k)$, where $i = \sqrt{-1}$, and where $c_X^{(k)}$ is the kth derivative of c_X.

We also have the following. (The proof is just like the proof of Theorem 3.4.5.)

Theorem 3.4.10 Let X and Y be random variables which are *independent*. Then $c_{X+Y}(s) = c_X(s) \, c_Y(s)$.

For simplicity, we shall generally not use characteristic functions in this book. However, it is worth keeping in mind that whenever we do anything with moment-generating functions, we could usually do the same thing in greater generality using characteristic functions.

Summary of Section 3.4

- The probability-generating function of a random variable X is $r_X(t) = E(t^X)$.

- If X is discrete, then the derivatives of r_X satisfy $r_X^{(k)}(0) = k! \, P(X = k)$.

- The kth moment of a random variable X is $E(X^k)$.

- The moment-generating function of a random variable X is $m_X(s) = E(e^{sX}) = r_X(e^s)$.

- The derivatives of m_X satisfy $m_X^{(k)}(0) = E(X^k)$, for $k = 0, 1, 2, \ldots$.

- If X and Y are independent, then $r_{X+Y}(t) = r_X(t) \, r_Y(y)$ and $m_{X+Y}(s) = m_X(s) \, m_Y(s)$.

- If $m_X(s)$ is finite in a neighborhood of $s = 0$, then it uniquely characterizes the distribution of X.

- The characteristic function $c_X(s) = E(e^{isX})$ can be used in place of $m_X(s)$ to avoid infinities.

Exercises

3.4.1 Let Z be a discrete random variable with $P(Z = z) = 1/2^z$ for $z = 1, 2, 3, \ldots$.
(a) Compute $r_Z(t)$. Verify that $r_Z'(0) = P(Z = 1)$ and $r_Z''(0) = 2\,P(Z = 2)$.
(b) Compute $m_Z(t)$. Verify that $m_Z'(0) = E(Z)$ and $m_Z''(0) = E(Z^2)$.

3.4.2 Let $X \sim \text{Binomial}(n, \theta)$. Use m_X to prove that $\text{Var}(X) = n\theta(1 - \theta)$.

3.4.3 Let $Y \sim \text{Poisson}(\lambda)$. Use m_Y to compute the mean and variance of Y.

3.4.4 Let $Y = 3X + 4$. Compute $r_Y(t)$ in terms of r_X.

3.4.5 Let $Y = 3X + 4$. Compute $m_Y(s)$ in terms of m_X.

3.4.6 Let $X \sim \text{Binomial}(n, \theta)$. Compute $E(X^3)$, the third moment of X.

3.4.7 Let $Y \sim \text{Poisson}(\lambda)$. Compute $E(Y^3)$, the third moment of Y.

3.4.8 Suppose $P(X = 2) = 1/2$, $P(X = 5) = 1/3$, and $P(X = 7) = 1/6$.
(a) Compute $r_X(t)$ for $t \in R^1$.
(b) Verify that $r_X'(0) = P(X = 1)$ and $r_X''(0) = 2P(X = 2)$.
(c) Compute $m_X(s)$ for $s \in R^1$.
(d) Verify that $m_X'(0) = E(X)$ and $m_X''(0) = E(X^2)$.

Problems

3.4.9 Suppose $f_X(x) = 1/10$ for $0 < x < 10$, with $f_X(x) = 0$ otherwise.
(a) Compute $m_X(s)$ for $s \in R^1$.
(b) Verify that $m_X'(0) = E(X)$. (Hint: L'Hospital's rule.)

3.4.10 Let $X \sim \text{Geometric}(\theta)$. Compute $r_X(t)$ and $r_X''(0)/2$.

3.4.11 Let $X \sim \text{Negative-Binomial}(r, \theta)$. Compute $r_X(t)$ and $r_X''(0)/2$.

3.4.12 Let $X \sim \text{Geometric}(\theta)$.
(a) Compute $m_X(s)$.
(b) Use m_X to compute the mean of X.
(c) Use m_X to compute the variance of X.

3.4.13 Let $X \sim \text{Negative-Binomial}(r, \theta)$.
(a) Compute $m_X(s)$.
(b) Use m_X to compute the mean of X.
(c) Use m_X to compute the variance of X.

3.4.14 If $Y = a + bX$, where a and b are constants then show that $r_Y(t) = t r_X(t^b)$ and $m_Y(t) = e^{at} m_X(bt)$.

3.4.15 Let $Z \sim N(\mu, \sigma^2)$. Show that

$$m_Z(s) = \exp\left\{\mu s + \frac{\sigma^2 s^2}{2}\right\}.$$

(Hint: Write $Z = \mu + \sigma X$ where $X \sim N(0,1)$, and use Theorem 3.4.4.)

3.4.16 Let Y be distributed according to the Laplace distribution (Problem 2.4.14).
(a) Compute $m_Y(s)$. (Hint: Break the integral up into two pieces.)
(b) Use m_Y to compute the mean of Y.
(c) Use m_Y to compute the variance of Y.

3.4.17 Compute the kth moment of the Weibull(α) distribution in terms of Γ (Problem 2.4.11).

3.4.18 Compute the kth moment of the Pareto(α) distribution (Problem 2.4.12). (Hint: Make the transformation $u = (1+x)^{-1}$ and recall the beta distribution.)

3.4.19 Compute the kth moment of the Log-normal(τ) distribution (Problem 2.6.12). (Hint: Make the transformation $z = \ln x$ and use Problem 3.4.15.)

3.4.20 Prove that the moment-generating function of the Gamma(α, λ) distribution is given by $\lambda^\alpha / (\lambda - t)^\alpha$ when $t < \lambda$.

3.4.21 Suppose that $X_i \sim$ Poisson(λ_i) and $X_1, \ldots, X_n$ are independent. Using moment-generating functions, determine the distribution of $Y = \sum_{i=1}^n X_i$.

3.4.22 Suppose that $X_i \sim$ Negative-Binomial(r_i, θ) and $X_1, \ldots, X_n$ are independent. Using moment-generating functions, determine the distribution of $Y = \sum_{i=1}^n X_i$.

3.4.23 Suppose that $X_i \sim$ Gamma(α_i, λ) and $X_1, \ldots, X_n$ are independent. Using moment-generating functions, determine the distribution of $Y = \sum_{i=1}^n X_i$.

3.4.24 Suppose $X_1, X_2, \ldots$ is i.i.d. Exponential(λ) and $N \sim$ Poisson(λ) independent of the $\{X_i\}$. Determine the moment-generating function of S_N. Determine the first moment of this distribution by differentiating this function.

3.4.25 Suppose $X_1, X_2, \ldots$ are i.i.d. Exponential(λ) random variables and $N \sim$ Geometric(θ), independent of the $\{X_i\}$. Determine the moment-generating function of S_N. Determine the first moment of this distribution by differentiating this function.

3.4.26 Let $X \sim$ Bernoulli(θ). Use $c_X(s)$ to compute the mean of X.

3.4.27 Let $Y \sim$ Binomial(n, θ).
(a) Compute the characteristic function $c_Y(s)$. (Hint: Make use of $c_X(s)$ in Problem 3.4.26.)
(b) Use $c_Y(s)$ to compute the mean of Y.

3.4.28 The characteristic function of the Cauchy distribution (Problem 2.4.13) is given by $c(t) = e^{-|t|}$. Use this to determine the characteristic function of the sample mean

$$\bar{X} = \frac{1}{n} \sum_{i=1}^n X_i$$

based on a sample of n from the Cauchy distribution. Explain why this implies that the sample mean is also Cauchy distributed. What do you find surprising about this result?

3.4.29 The *kth cumulant* (when it exists) of a random variable X is obtained by calculating the kth derivative of $\ln c_X(s)$ with respect to s, evaluating this at $s = 0$, and dividing by i^k. Evaluate $c_X(s)$ and all the cumulants of the $N(\mu, \sigma^2)$ distribution.

3.5 Conditional Expectation

We have seen in Sections 1.5 and 2.8 that *conditioning* on some event, or some random variable, can change various probabilities. Now, because expectations are defined in terms of probabilities, it seems reasonable that expectations should also change when conditioning on some event or random variable. Such modified expectations are called *conditional expectations,* as we now discuss.

3.5.1 Discrete Case

The simplest case is when X is a discrete random variable, and A is some event of positive probability. We have the following.

Definition 3.5.1 Let X be a discrete random variable, and let A be some event with $P(A) > 0$. Then the *conditional expectation* of X, given A, is equal to

$$E(X \mid A) = \sum_{x \in R^1} x\, P(X = x \mid A) = \sum_{x \in R^1} x\, \frac{P(X = x, \ A)}{P(A)}.$$

Example 3.5.1
Consider rolling a fair six-sided die, so that $S = \{1, 2, 3, 4, 5, 6\}$. Let X be the number showing, so that $X(s) = s$ for $s \in S$. Let $A = \{3, 5, 6\}$ be the event that the die shows 3, 5, or 6. What is $E(X \mid A)$?
    Here we know that

$$P(X = 3 \mid A) = P(X = 3 \mid X = 3, 5, \text{ or } 6) = 1/3,$$

and, similarly, $P(X = 5 \mid A) = P(X = 6 \mid A) = 1/3$. Hence,

$$
\begin{aligned}
E(X \mid A) &= \sum_{x \in R^1} x\, P(X = x \mid A) \\
&= 3\, P(X = 3 \mid A) + 5\, P(X = 5 \mid A) + 6\, P(X = 6 \mid A) \\
&= 3(1/3) + 5(1/3) + 6(1/3) = 14/3. \ \blacksquare
\end{aligned}
$$

    Often we wish to condition on the value of some other random variable. If the other random variable is also discrete, and if the conditioned value has positive probability, then this works as above.

Definition 3.5.2 Let X and Y be discrete random variables, with $P(Y = y) > 0$. Then the *conditional expectation* of X, given $Y = y$, is equal to

$$E(X \mid Y = y) = \sum_{x \in R^1} x P(X = x \mid Y = y) = \sum_{x \in R^1} x \frac{p_{X,Y}(x, y)}{p_Y(y)}.$$

Example 3.5.2
Suppose the joint probability function of X and Y is given by

$$p_{X,Y}(x, y) = \begin{cases} 1/7 & x = 5, y = 0 \\ 1/7 & x = 5, y = 3 \\ 1/7 & x = 5, y = 4 \\ 3/7 & x = 8, y = 0 \\ 1/7 & x = 8, y = 4 \\ 0 & \text{otherwise.} \end{cases}$$

Then

$$
\begin{aligned}
E(X \mid Y = 0) &= \sum_{x \in R^1} x\, P(X = x \mid Y = 0) \\
&= 5 P(X = 5 \mid Y = 0) + 8\, P(X = 8 \mid Y = 0) \\
&= 5 \frac{P(X = 5,\ Y = 0)}{P(Y = 0)} + 8 \frac{P(X = 8,\ Y = 0)}{P(Y = 0)} \\
&= 5 \frac{1/7}{1/7 + 3/7} + 8 \frac{3/7}{1/7 + 3/7} = \frac{29}{4}.
\end{aligned}
$$

Similarly

$$
\begin{aligned}
E(X \mid Y = 4) &= \sum_{x \in R^1} x P(X = x \mid Y = 4) \\
&= 5 P(X = 5 \mid Y = 4) + 8\, P(X = 8 \mid Y = 4) \\
&= 5 \frac{1/7}{1/7 + 1/7} + 8 \frac{1/7}{1/7 + 1/7} = 13/2.
\end{aligned}
$$

Also,

$$
\begin{aligned}
E(X \mid Y = 3) &= \sum_{x \in R^1} x\, P(X = x \mid Y = 3) = 5 P(X = 5 \mid Y = 3) \\
&= 5 \frac{1/7}{1/7} = 5. \quad \blacksquare
\end{aligned}
$$

Sometimes we wish to condition on a random variable Y, without specifying in advance what value of Y we are conditioning on. In this case, the conditional expectation $E(X \mid Y)$ is itself a random variable — namely, it depends on the (random) value of Y that occurs.

Definition 3.5.3 Let X and Y be discrete random variables. Then the *conditional expectation* of X, given Y, is the random variable $E(X \mid Y)$, which is equal to $E(X \mid Y = y)$ when $Y = y$. In particular, $E(X \mid Y)$ is a random variable that depends on the random value of Y.

Example 3.5.3
Suppose again that the joint probability function of X and Y is given by

$$p_{X,Y}(x, y) = \begin{cases} 1/7 & x = 5, y = 0 \\ 1/7 & x = 5, y = 3 \\ 1/7 & x = 5, y = 4 \\ 3/7 & x = 8, y = 0 \\ 1/7 & x = 8 \ \ y = 4 \\ 0 & \text{otherwise.} \end{cases}$$

We have already computed that $E(X \mid Y = 0) = 29/4$, $E(X \mid Y = 4) = 13/2$, and $E(X \mid Y = 3) = 5$. We can express these results together by saying that

$$E(X \mid Y) = \begin{cases} 29/4 & Y = 0 \\ 5 & Y = 3 \\ 13/2 & Y = 4. \end{cases}$$

That is, $E(X \mid Y)$ is a random variable, which depends on the value of Y. Note that, because $P(Y = y) = 0$ for $y \neq 0, 3, 4$, the random variable $E(X \mid Y)$ is *undefined* in that case; but this is not a problem, because that case will never occur. ∎

Finally, we note that just like for regular expectation, conditional expectation is linear.

Theorem 3.5.1 Let X_1, X_2 and Y be random variables; let A be an event; let a, b and y be real numbers; and let $Z = aX_1 + bX_2$. Then
(a) $E(Z \mid A) = aE(X_1 \mid A) + bE(X_2 \mid A)$.
(b) $E(Z \mid Y = y) = aE(X_1 \mid Y = y) + bE(X_2 \mid Y = y)$.
(c) $E(Z \mid Y) = aE(X_1 \mid Y) + bE(X_2 \mid Y)$.

3.5.2 Absolutely Continuous Case

Suppose now that X and Y are jointly absolutely continuous. Then conditioning on $Y = y$, for some particular value of y, seems problematic, because $P(Y = y) = 0$. However, we have already seen in Section 2.8.2 that we can define a *conditional density* $f_{X|Y}(x|y)$ that gives us a density function for X, conditional on $Y = y$. And because density functions give rise to expectations, similarly conditional density functions give rise to conditional expectations, as follows.

Definition 3.5.4 Let X and Y be jointly absolutely continuous random variables, with joint density function $f_{X,Y}(x, y)$. Then the *conditional expectation* of X, given $Y = y$, is equal to

$$E(X \mid Y = y) = \int_{x \in R^1} x \, f_{X|Y}(x \mid y) \, dx = \int_{x \in R^1} x \, \frac{f_{X,Y}(x, y)}{f_Y(y)} \, dx.$$

Example 3.5.4

Let X and Y be jointly absolutely continuous, with joint density function $f_{X,Y}$ given by

$$f_{X,Y}(x,y) = \begin{cases} 4x^2y + 2y^5 & 0 \le x \le 1,\ 0 \le y \le 1 \\ 0 & \text{otherwise.} \end{cases}$$

Then for $0 < y < 1$,

$$f_Y(y) = \int_{-\infty}^{\infty} f_{X,Y}(x,y)\,dx = \int_0^1 (4x^2y + 2y^5)\,dx = 4y/3 + 2y^5.$$

Hence,

$$\begin{aligned} E(X\,|\,Y=y) &= \int_{x \in R^1} x\,\frac{f_{X,Y}(x,y)}{f_Y(y)}\,dx = \int_0^1 x\,\frac{4x^2y + 2y^5}{4y/3 + 2y^5}\,dx \\ &= \frac{y + y^5}{4y/3 + 2y^5} = \frac{1 + y^4}{4/3 + 2y^4}.\ \blacksquare \end{aligned}$$

As in the discrete case, we often wish to condition on a random variable without specifying in advance the value of that variable. Thus, $E(X\,|\,Y)$ is again a random variable, depending on the random value of Y.

Definition 3.5.5 Let X and Y be jointly absolutely continuous random variables. Then the *conditional expectation* of X, given Y, is the random variable $E(X\,|\,Y)$, which is equal to $E(X\,|\,Y=y)$ when $Y = y$. Thus, $E(X\,|\,Y)$ is a random variable that depends on the random value of Y.

Example 3.5.5

Let X and Y again have joint density

$$f_{X,Y}(x,y) = \begin{cases} 4x^2y + 2y^5 & 0 \le x \le 1, 0 \le y \le 1 \\ 0 & \text{otherwise.} \end{cases}$$

We already know that $E(X\,|\,Y=y) = \left(1 + y^4\right)/\left(4/3 + 2y^4\right)$. Because this formula is valid for any y between 0 and 1, we conclude that $E(X\,|\,Y) = \left(1 + Y^4\right)/\left(4/3 + 2Y^4\right)$. Note that in this last formula, Y is a random variable, so $E(X\,|\,Y)$ is also a random variable. $\blacksquare$

Finally, we note that in the absolutely continuous case, conditional expectation is still linear, i.e., Theorem 3.5.1 continues to hold.

3.5.3 Double Expectations

Because the conditional expectation $E(X\,|\,Y)$ is itself a random variable (as a function of Y), it makes sense to take its expectation, $E\left(E(X\,|\,Y)\right)$. This is a *double expectation*. One of the key results about conditional expectation is that this expectation is always equal to $E(X)$.

Theorem 3.5.2 (*Theorem of total expectation*) If X and Y are random variables, then $E\left(E(X\,|\,Y)\right) = E(X)$.

This theorem follows as a special case of Theorem 3.5.3 below. But it also makes sense intuitively. Indeed, conditioning on Y will change the conditional value of X in various ways, sometimes making it smaller and sometimes larger depending on the value of Y. However, if we then average over all possible values of Y, these various effects will cancel out, and we will be left with just $E(X)$.

Example 3.5.6

Suppose again that X and Y have joint probability function

$$p_{X,Y}(x,y) = \begin{cases} 1/7 & x = 5, y = 0 \\ 1/7 & x = 5, y = 3 \\ 1/7 & x = 5, y = 4 \\ 3/7 & x = 8, y = 0 \\ 1/7 & x = 8 \;\; y = 4 \\ 0 & \text{otherwise.} \end{cases}$$

Then we know that

$$E(X\,|\,Y = y) = \begin{cases} 29/4 & y = 0 \\ 5 & y = 3 \\ 13/2 & y = 4. \end{cases}$$

Also, $P(Y = 0) = 1/7 + 3/7 = 4/7$, $P(Y = 3) = 1/7$, and $P(Y = 4) = 1/7 + 1/7 = 2/7$. Hence,

$$E\left(E(X\,|\,Y)\right)$$
$$= \sum_{y \in R^1} E(X\,|\,Y = y)P(Y = y)$$
$$= E(X\,|\,Y = 0)P(Y = 0) + E(X\,|\,Y = 3)P(Y = 3) + E(X\,|\,Y = 4)P(Y = 4)$$
$$= (29/4)(4/7) + (5)(1/7) + (13/2)(2/7) = 47/7.$$

On the other hand, we compute directly that $E(X) = 5P(X = 5) + 8\,P(X = 8) = 5(3/7) + 8(4/7) = 47/7$. Hence, $E\left(E(X\,|\,Y)\right) = E(X)$, as claimed. ∎

Example 3.5.7

Let X and Y again have joint density

$$f_{X,Y}(x,y) = \begin{cases} 4x^2y + 2y^5 & 0 \le x \le 1,\ 0 \le y \le 1 \\ 0 & \text{otherwise.} \end{cases}$$

We already know that

$$E(X\,|\,Y) = \left(1 + Y^4\right) / \left(4/3 + 2Y^4\right),$$

and $f_Y(y) = 4y/3 + 2y^5$ for $0 \le y \le 1$. Hence,

$$E\left(E(X\,|\,Y)\right)$$

$$= E\left(\frac{1+Y^4}{4/3+2Y^4}\right) = \int_{-\infty}^{\infty} E(X\,|\,Y=y)\, f_Y(y)\, dy$$

$$= \int_0^1 \frac{1+y^4}{4/3+2y^4}\,(4y/3+2y^5)\, dy = \int_0^1 (y+y^5)\, dy = 1/2 + 1/6 = 2/3.$$

On the other hand,

$$E(X) = \int_{-\infty}^{\infty}\int_{-\infty}^{\infty} x\, f_{X,Y}(x,y)\, dy\, dx = \int_0^1\int_0^1 x\,(4x^2 y + 2y^5)\, dy\, dx$$

$$= \int_0^1 x\,(2x^2 + 2/6)\, dx = \int_0^1 (2x^3 + x/3)\, dx = 2/4 + 1/6 = 2/3.$$

Hence, $E\left(E(X\,|\,Y)\right) = E(X)$, as claimed. ∎

Theorem 3.5.2 is a special case (with $g(y) \equiv 1$) of the following more general result, which in fact *characterizes* conditional expectation.

Theorem 3.5.3 Let X and Y be random variables, and let $g : R^1 \to R^1$ be any function. Then $E\left(g(Y)E(X\,|\,Y)\right) = E\left(g(Y)X\right)$.

Proof: See Section 3.8 for the proof of this result.

We also note the following related result. It says that, when conditioning on Y, any function of Y can be factored out, since it is effectively a constant.

Theorem 3.5.4 Let X and Y be random variables, and let $g : R^1 \to R^1$ be any function. Then $E\left(g(Y)X\,|\,Y\right) = g(Y)E\left(X\,|\,Y\right)$.

Proof: See Section 3.8 for the proof of this result.

Finally, because conditioning *twice* on Y is the same as conditioning just once on Y, we immediately have the following.

Theorem 3.5.5 Let X and Y be random variables. Then $E\left(E(X\,|\,Y)\,|\,Y\right) = E\left(X\,|\,Y\right)$.

3.5.4 Conditional Variance (Advanced)

In addition to defining conditional expectation, we can define conditional variance. As usual, this involves the expected squared distance of a random variable to its mean. However, in this case the expectation is a conditional expectation. In addition, the mean is a conditional mean.

Definition 3.5.6 If X is a random variable, and A is an event with $P(A) > 0$, then the *conditional variance* of X, given A, is equal to

$$\text{Var}(X\,|\,A) = E\left((X - E(X\,|\,A))^2\,|\,A\right) = E(X^2\,|\,A) - (E(X\,|\,A))^2.$$

Similarly, if Y is another random variable, then

$$\begin{aligned}
\text{Var}(X \mid Y = y) &= E\left((X - E(X \mid Y = y))^2 \mid Y = y\right) \\
&= E(X^2 \mid Y = y) - (E(X \mid Y = y))^2,
\end{aligned}$$

and $\text{Var}(X \mid Y) = E\left((X - E(X \mid Y))^2 \mid Y\right) = E(X^2 \mid Y) - (E(X \mid Y))^2.$

Example 3.5.8

Consider again rolling a fair six-sided die, so that $S = \{1, 2, 3, 4, 5, 6\}$, with $P(s) = 1/6$ and $X(s) = s$ for $s \in S$, and with $A = \{3, 5, 6\}$. We have already computed that $P(X = s \mid A) = 1/3$ for $s \in A$, and that $E(X \mid A) = 14/3$. Hence,

$$\text{Var}(X \mid A) = E\left(X - E(X \mid A)\right)^2 \mid A)$$

$$= E\left((X - 14/3)^2 \mid A\right) = \sum_{s \in S}(s - 14/3)^2 P(X = s \mid A)$$

$$= (3 - 14/3)^2(1/3) + (5 - 14/3)^2(1/3) + (6 - 14/3)^2(1/3) = 14/9 \doteq 1.56.$$

By contrast, because $E(X) = 7/2$, we have

$$\text{Var}(X) = E\left((X - E(X))^2\right) = \sum_{x=1}^{6}(x - 7/2)^2(1/6) = 35/12 \doteq 2.92.$$

Hence, we see that the conditional variance $\text{Var}(X \mid A)$ is much smaller than the unconditional variance $\text{Var}(X)$. This indicates that, in this example, once we know that event A has occurred, we know more about the value of X than we did originally. ∎

Example 3.5.9

Suppose X and Y have joint density function

$$f_{X,Y}(x, y) = \begin{cases} 8xy & 0 < x < y < 1 \\ 0 & \text{otherwise.} \end{cases}$$

We have $f_Y(y) = 4y^3$, $f_{X \mid Y}(x \mid y) = 8xy/4y^3 = 2x/y^2$ for $0 < x < y$ and so

$$E(X \mid Y = y) = \int_0^y x \frac{2x}{y^2}\, dx = \int_0^y \frac{2x^2}{y^2}\, dx = \frac{2y^3}{3y^2} = \frac{2y}{3}.$$

Therefore,

$$\begin{aligned}
\text{Var}(X \mid Y = y) &= E\left((X - E(X \mid Y = y))^2 \mid Y = y\right) \\
&= \int_0^1 \left(x - \frac{2y}{3}\right)^2 \frac{2x}{y^2}\, dx = \frac{1}{2y^2} - \frac{8}{9y} + \frac{4}{9}. \ \blacksquare
\end{aligned}$$

Finally, we note that conditional expectation and conditional variance satisfy the following useful identity.

Theorem 3.5.6 For random variables X and Y,

$$\text{Var}(X) = \text{Var}\left(E(X \mid Y)\right) + E\left(\text{Var}(X \mid Y)\right).$$

Proof: See Section 3.8 for the proof of this result.

Summary of Section 3.5

- If X is discrete, then the conditional expectation of X, given an event A, is equal to $E(X \mid A) = \sum_{x \in R^1} x P(X = x \mid A)$.

- If X and Y are discrete random variables, then $E(X \mid Y)$ is itself a random variable, with $E(X \mid Y)$ equal to $E(X \mid Y = y)$ when $Y = y$.

- If X and Y are jointly absolutely continuous, then $E(X \mid Y = y) = \int x \, f_{X|Y}(x \mid y) \, dx$, and again $E(X \mid Y)$ is itself a random variable, with $E(X \mid Y)$ equal to $E(X \mid Y = y)$ when $Y = y$.

- Conditional expectation is linear.

- We always have $E(g(Y) E(X \mid Y)) = E(g(Y) X)$, and $E(E(X \mid Y) \mid Y) = E(X \mid Y)$.

- Conditional variance is given by $\mathrm{Var}(X \mid Y) = E(X^2 \mid Y) - (E(X \mid Y))^2$.

Exercises

3.5.1 Suppose X and Y are discrete, with

$$p_{X,Y}(x, y) = \begin{cases} 1/5 & x = 2, y = 3 \\ 1/5 & x = 3, y = 2 \\ 1/5 & x = 3, y = 3 \\ 1/5 & x = 2, y = 2 \\ 1/5 & x = 3, \ y = 17 \\ 0 & \text{otherwise.} \end{cases}$$

(a) Compute $E(X \mid Y = 3)$.
(b) Compute $E(Y \mid X = 3)$.
(c) Compute $E(X \mid Y)$.
(d) Compute $E(Y \mid X)$.

3.5.2 Suppose X and Y are jointly absolutely continuous, with

$$f_{X,Y}(x, y) = \begin{cases} 9(xy + x^5 y^5)/16000900 & 0 \le x \le 4, \ 0 \le y \le 5 \\ 0 & \text{otherwise.} \end{cases}$$

(a) Compute $f_X(x)$.
(b) Compute $f_Y(y)$.
(c) Compute $E(X \mid Y)$.
(d) Compute $E(Y \mid X)$.
(e) Compute $E(E(X \mid Y))$, and verify that it is equal to $E(X)$.

3.5.3 Suppose X and Y are discrete, with

$$p_{X,Y}(x,y) = \begin{cases} 1/11 & x = -4, y = 2 \\ 2/11 & x = -4, y = 3 \\ 4/11 & x = -4, y = 7 \\ 1/11 & x = 6, y = 2 \\ 1/11 & x = 6, y = 3 \\ 1/11 & x = 6, y = 7 \\ 1/11 & x = 6, y = 13 \\ 0 & \text{otherwise.} \end{cases}$$

(a) Compute $E(Y \mid X = 6)$.
(b) Compute $E(Y \mid X = -4)$.
(c) Compute $E(Y \mid X)$.

3.5.4 Let $p_{X,Y}$ be as in the previous exercise.
(a) Compute $E(X \mid Y = 2)$.
(b) Compute $E(X \mid Y = 3)$.
(c) Compute $E(X \mid Y = 7)$.
(d) Compute $E(X \mid Y = 13)$.
(e) Compute $E(X \mid Y)$.

3.5.5 Suppose that a student is considering one of two possible summer jobs to take. If it is not necessary to take a summer course, then a job as a waiter will produce earnings (rounded to the nearest $1000) with the following probability distribution.

$1000	$2000	$3000	$4000
0.1	0.3	0.4	0.2

If it is necessary to take a summer course, then a part-time job at a hotel will produce earnings (rounded to the nearest $1000) with the following probability distribution.

$1000	$2000	$3000	$4000
0.3	0.4	0.2	0.1

If the probability that the student will have to take the summer course is 0.6, then determine the student's expected summer earnings.

Problems

3.5.6 Suppose there are two urns. Urn I contains 100 chips, with 30 chips labelled 1, 40 chips labelled 2, and 30 chips labelled 3. Urn II contains 100 chips, with 20 chips labelled 1, 50 chips labelled 2, and 30 chips labelled 3. A coin is tossed and if head is observed, then a chip is randomly drawn from urn I, otherwise a chip is randomly drawn from urn II. The value Y on the chip is recorded. If an occurrence of head on the coin is denoted by $X = 1$, a tail by $X = 0$, and $X \sim \text{Bernoulli}(3/4)$, then determine $E(X \mid Y), E(Y \mid X), E(Y)$, and $E(X)$.

3.5.7 Suppose that five coins are each tossed until the first head is obtained on each coin and where each coin has probability θ of producing a head. If you are told that the total number of tails observed is $Y = 10$, then determine the expected number of tails observed on the first coin.

3.5.8 *(Simpson's paradox)* Suppose that the conditional distributions of Y, given X, are given in the following table. For example, $p_{Y|X}(1|i)$ could correspond to the probability that a randomly selected heart patient at hospital i has a successful treatment.

$p_{Y|X}(0|1)$	$p_{Y|X}(1|1)$
0.030	0.970
$p_{Y|X}(0|2)$	$p_{Y|X}(1|2)$
0.020	0.980

(a) Compute $E(Y|X)$.

(b) Now suppose that patients are additionally classified as being seriously ill ($Z = 1$), or not seriously ill ($Z = 0$). The conditional distributions of Y, given (X, Z) are given in the following tables. Compute $E(Y|X, Z)$.

$p_{Y|X,Z}(0|1,0)$	$p_{Y|X,Z}(1|1,0)$
0.010	0.990
$p_{Y|X,Z}(0|2,0)$	$p_{Y|X,Z}(1|2,0)$
0.013	0.987

$p_{Y|X,Z}(0|1,1)$	$p_{Y|X,Z}(1|1,1)$
0.038	0.962
$p_{Y|X,Z}(0|2,1)$	$p_{Y|X,Z}(1|2,1)$
0.040	0.960

(c) Explain why the conditional distributions in part (a) indicate that hospital 2 is the better hospital for a patient who needs to undergo this treatment, but all the conditional distributions in part (b) indicate that hospital 1 is the better hospital. This phenomenon is known as Simpson's paradox.

(d) Prove that, in general, $p_{Y|X}(y|x) = \sum_z p_{Y|X,Z}(y|x,z) p_{Z|X}(z|x)$ and $E(Y|X) = E(E(Y|X,Z)|X)$.

(e) If the conditional distributions $p_{Z|X}(\cdot|x)$, corresponding to the example discussed in parts (a) through (c) are given in the following table, verify the result in part (d) numerically and explain how this resolves Simpson's paradox.

$p_{Z|X}(0|1)$	$p_{Z|X}(1|1)$
0.286	0.714
$p_{Z|X}(0|2)$	$p_{Z|X}(1|2)$
0.750	0.250

3.5.9 Present an example of a random variable X, and an event A with $P(A) > 0$, such that $\text{Var}(X|A) > \text{Var}(X)$. (Hint: Suppose $S = \{1, 2, 3\}$ with $X(s) = s$, and $A = \{1, 3\}$.)

3.5.10 Suppose that X given $Y = y$ is distributed Gamma(α, y) and $1/Y \sim$ Exponential(λ). Determine $E(X)$.

3.5.11 Suppose that $(X, Y) \sim$ Bivariate Normal $(\mu_1, \mu_2, \sigma_1, \sigma_2, \rho)$. Use (2.7.1) (when given $Y = y$) and its analog (when given $X = x$) to determine $E(X \mid Y)$, $E(Y \mid X)$, $\text{Var}(X \mid Y)$, and $\text{Var}(Y \mid X)$.

3.5.12 Suppose that $(X_1, X_2, X_3) \sim$ Multinomial$(n, \theta_1, \theta_2, \theta_3)$. Determine $E(X_1 \mid X_2)$ and $\text{Var}(X_1 \mid X_2)$. (Hint: Show that X_1, given $X_2 = x_2$, has a binomial distribution.)

3.5.13 Suppose that $(X_1, X_2) \sim$ Dirichlet$(\alpha_1, \alpha_2, \alpha_3)$. Determine $E(X_1 \mid X_2)$ and $\text{Var}(X_1 \mid X_2)$. (Hint: First show that $X_1/(1 - x_2)$, given $X_2 = x_2$, has a beta distribution and then use Problem 3.3.18.)

3.5.14 Let $f_{X,Y}$ be as in Exercise 3.5.2.
(a) Compute $\text{Var}(X)$.
(b) Compute $\text{Var}(E(X \mid Y))$.
(c) Compute $\text{Var}(X \mid Y)$.
(d) Verify that $\text{Var}(X) = \text{Var}(E(X \mid Y)) + E(\text{Var}(X \mid Y))$.

3.5.15 Suppose we have three discrete random variables X, Y, and Z. We say that X and Y are *conditionally independent*, given Z, if

$$p_{X,Y|Z}(x, y \mid z) = p_{X|Z}(x \mid z) \, p_{Y|Z}(y \mid z)$$

for every x, y, and z such that $P(Z = z) > 0$. Prove that when X and Y are conditionally independent, given Z, then

$$E(g(X)h(Y) \mid Z) = E(g(X) \mid Z) \, E(h(Y) \mid Z).$$

3.6   Inequalities

Expectation and variance are closely related to the underlying distributions of random variables. This relationship allows us to prove certain inequalities that are often very useful. We begin with a classic result, Markov's inequality, which is very simple but also very useful and powerful.

Theorem 3.6.1 (*Markov's inequality*) If X is a nonnegative random variable, then for all $a > 0$,

$$P(X \geq a) \leq \frac{E(X)}{a}.$$

That is, the probability that X exceeds any given value a is no more than the mean of X divided by a.

Proof: Define a new random variable Z by

$$Z = \begin{cases} a & X \geq a \\ 0 & X < a. \end{cases}$$

Then clearly $Z \leq X$, so that $E(Z) \leq E(X)$ by monotonicity. On the other hand,

$$E(Z) = a \, P(Z = a) + 0 \, P(Z = 0) = a \, P(Z = a) = a \, P(X \geq a).$$

So, $E(X) \geq E(Z) = a\,P(X \geq a)$. Rearranging, $P(X \geq a) \leq E(X)/a$, as claimed. ∎

Intuitively, Markov's inequality says that if the expected value of X is small, then it is unlikely that X will be too large. We now consider some applications of Theorem 3.6.1.

Example 3.6.1
Suppose $P(X = 3) = 1/2$, $P(X = 4) = 1/3$, and $P(X = 7) = 1/6$. Then $E(X) = 3(1/2) + 4(1/3) + 7(1/6) = 4$. Hence, setting $a = 6$, Markov's inequality says that $P(X \geq 6) \leq 4/6 = 2/3$. In fact, $P(X \geq 6) = 1/6 < 2/3$.

Example 3.6.2
Suppose $P(X = 2) = P(X = 8) = 1/2$. Then $E(X) = 2(1/2) + 8(1/2) = 5$. Hence, setting $a = 8$, Markov's inequality says that $P(X \geq 8) \leq 5/8$. In fact, $P(X \geq 8) = 1/2 < 5/8$. ∎

Example 3.6.3
Suppose $P(X = 0) = P(X = 2) = 1/2$. Then $E(X) = 0(1/2) + 2(1/2) = 1$. Hence, setting $a = 2$, Markov's inequality says that $P(X \geq 2) \leq 1/2$. In fact, $P(X \geq 2) = 1/2$, so Markov's inequality is an *equality* in this case. ∎

Markov's inequality is also used to prove Chebychev's inequality, perhaps the most important inequality in all of probability theory.

Theorem 3.6.2 (*Chebychev's inequality*) Let Y be an arbitrary random variable, with finite mean μ_Y. Then for all $a > 0$,

$$P\left(|Y - \mu_Y| \geq a\right) \leq \frac{\text{Var}(Y)}{a^2}.$$

Proof: Set $X = (Y - \mu_Y)^2$. Then X is a nonnegative random variable. Thus, using Theorem 3.6.1, we have $P\left(|Y - \mu_Y| \geq a\right) = P\left(X \geq a^2\right) \leq E(X)/a^2 = \text{Var}(Y)/a^2$ and this establishes the result. ∎

Intuitively, Chebychev's inequality says that if the variance of Y is small, then it is unlikely that Y will be too far from its mean value μ_Y. We now consider some examples.

Example 3.6.4
Suppose again that $P(X = 3) = 1/2$, $P(X = 4) = 1/3$, and $P(X = 7) = 1/6$. Then $E(X) = 4$ as above. Also, $E(X^2) = 9(1/2) + 16(1/3) + 49(1/6) = 18$, so that $\text{Var}(X) = 18 - 4^2 = 2$. Hence, setting $a = 1$, Chebychev's inequality says that $P(|X - 4| \geq 1) \leq 2/1^2 = 2$, which tells us nothing because we always have $P(|X - 4| \geq 1) \leq 1$. On the other hand, setting $a = 3$, we get $P(|X - 4| \geq 3) \leq 2/3^2 = 2/9$, which is true, because in fact $P(|X - 4| \geq 3) = P(X = 7) = 1/6 < 2/9$. ∎

Example 3.6.5
Let $X \sim \text{Exponential}(3)$, and let $a = 5$. Then $E(X) = 1/3$ and $\text{Var}(X) = 1/9$. Hence, by Chebychev's inequality with $a = 1/2$, $P(|X - 1/3| \geq 1/2) \leq$

$(1/9)/(1/2)^2 = 4/9$. On the other hand, because $X \geq 0$, $P(|X - 1/3| \geq 1/2) = P(X \geq 5/6)$, and by Markov's inequality, $P(X \geq 5/6) \leq (1/3)/(5/6) = 2/5$. Because $2/5 < 4/9$, we actually get a better bound from Markov's inequality than from Chebychev's inequality in this case. ■

Example 3.6.6
Let $Z \sim N(0, 1)$, and $a = 5$. Then by Chebychev's inequality, $P(|Z| \geq 5) \leq 1/5$. ■

Example 3.6.7
Let X be a random variable having *very small* variance. Then Chebychev's inequality says that $P(|X - \mu_X| \geq a)$ is small whenever a is not too small. In other words, usually $|X - \mu_X|$ is very small, i.e., $X \approx \mu_X$. This makes sense, because if the variance of X is very small, then usually X is very close to its mean value μ_X. ■

Inequalities are also useful for covariances, as follows.

Theorem 3.6.3 (*Cauchy–Schwartz inequality*) Let X and Y be arbitrary random variables, each having finite, nonzero variance. Then

$$|\text{Cov}(X, Y)| \leq \sqrt{\text{Var}(X) \text{Var}(Y)}.$$

Furthermore, if $\text{Var}(Y) > 0$, then equality is attained if and only if $X - \mu_X = \lambda(Y - \mu_Y)$ where $\lambda = \text{Cov}(X, Y)/\text{Var}(Y)$.

Proof: See Section 3.8 for the proof. ■

The Cauchy-Schwartz inequality says that if the variance of X or Y is small, then the covariance of X and Y must also be small.

Example 3.6.8
Suppose $X = C$ is a constant. Then $\text{Var}(X) = 0$. It follows from the Cauchy-Schwartz inequality that, for *any* random variable Y, we must have $\text{Cov}(X, Y) \leq (\text{Var}(X) \text{Var}(Y))^{1/2} = (0 \text{Var}(Y))^{1/2} = 0$, so that $\text{Cov}(X, Y) = 0$. ■

Recalling that the *correlation* of X and Y is defined by

$$\text{Corr}(X, Y) = \frac{\text{Cov}(X, Y)}{\sqrt{\text{Var}(X) \text{Var}(Y)}},$$

we immediately obtain the following important result (which has already been referred to, back when correlation was first introduced).

Corollary 3.6.1 Let X and Y be a arbitrary random variables, having finite means and finite, nonzero variances. Then $|\text{Corr}(X, Y)| \leq 1$. Furthermore, $|\text{Corr}(X, Y)| = 1$ if and only if

$$X - \mu_X = \frac{\text{Cov}(X, Y)}{\text{Var}(Y)}(Y - \mu_Y).$$

So the correlation between two random variables is always between -1 and 1. We also see that X and Y are linearly related if and only if $|\text{Corr}(X, Y)| = 1$, and that this relationship is increasing (positive slope) when $\text{Corr}(X, Y) = 1$ and decreasing (negative slope) when $\text{Corr}(X, Y) = -1$.

3.6.1   Jensen's Inequality (Advanced)

Finally, we develop a more advanced inequality that is sometimes very useful. A function f is called *convex* if for every $x < y$, the line segment from $(x, f(x))$ to $(y, f(y))$ lies entirely *above* the graph of f, as depicted in Figure 3.6.1.

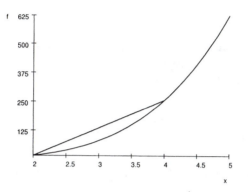

Figure 3.6.1: Plot of the convex function $f(x) = x^4$ and the line segment joining $(2, f(2))$ to $(4, f(4))$.

In symbols, we require that for every $x < y$ and every $0 < \lambda < 1$, we have $\lambda f(x) + (1 - \lambda)f(y) \geq f(\lambda x + (1 - \lambda)y)$. Examples of convex functions include $f(x) = x^2$, $f(x) = x^4$, and $f(x) = \max(x, C)$ for any real number C. We have the following.

Theorem 3.6.4 (*Jensen's inequality*) Let X be an arbitrary random variable, and let $f : R^1 \to R^1$ be a convex function such that $E(f(X))$ is finite. Then $f(E(X)) \leq E(f(X))$. Equality occurs if and only if $f(X) = a + bX$ for some a and b.

Proof: Because f is convex, we can find a linear function $g(x) = ax + b$ such that $g(E(X)) = f(E(X))$, and $g(x) \leq f(x)$ for all $x \in R^1$ (see, for example, Figure 3.6.2).

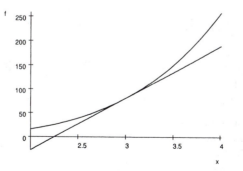

Figure 3.6.2: Plot of the convex function $f(x) = x^4$ and the function $g(x) = 81 + 108(x - 3)$, satisfying $g(x) \leq f(x)$ on the interval $(2, 4)$.

But then using monotonicity and linearity, we have $E(f(X)) \geq E(g(X)) = E(aX + b) = aE(X) + b = f(E(X))$, as claimed.

We have equality if and only if $0 = E(f(X) - g(X))$. Because $f(X) - g(X) \geq 0$, this occurs (using Challenge 3.3.22) if and only if $f(X) = g(X) = aX + b$ with probability 1. ∎

Example 3.6.9
Let X be a random variable with finite variance. Then setting $f(x) = x^2$, Jensen's inequality says that $E(X^2) \geq (E(X))^2$. Of course, we already knew this, because $E(X^2) - (E(X))^2 = \text{Var}(X) \geq 0$. ∎

Example 3.6.10
Let X be a random variable with finite fourth moment. Then setting $f(x) = x^4$, Jensen's inequality says that $E(X^4) \geq (E(X))^4$. ∎

Example 3.6.11
Let X be a random variable with finite mean, and let $M \in R^1$. Then setting $f(x) = \max(x, M)$, we have that $E(\max(X, M)) \geq \max(E(X), M)$ by Jensen's inequality. In fact, we could also have deduced this from the monotonicity property of expectation, using the two inequalities $\max(X, M) \geq X$ and $\max(X, M) \geq M$. ∎

Summary of Section 3.6

- For nonnegative X, Markov's inequality says $P(X \geq a) \leq E(X)/a$.

- Chebychev's inequality says $P(|Y - \mu_Y| \geq a) \leq \text{Var}(Y)/a^2$.

- The Cauchy–Schwartz inequality says $|\text{Cov}(X, Y)| \leq (\text{Var}(X)\,\text{Var}(Y))^{1/2}$, so that $|\text{Corr}(X, Y)| \leq 1$.

- Jensen's inequality says $f(E(X)) \leq E(f(X))$ whenever f is *convex*.

Exercises

3.6.1 Let $Z \sim \text{Poisson}(3)$. Use Markov's inequality to get an upper bound on $P(Z \geq 7)$.

3.6.2 Let $X \sim \text{Exponential}(5)$. Use Markov's inequality to get an upper bound on $P(X \geq 3)$ and compare it with the precise value.

3.6.3 Let $X \sim \text{Geometric}(1/2)$.
(a) Use Markov's inequality to get an upper bound on $P(X \geq 9)$.
(b) Use Markov's inequality to get an upper bound on $P(X \geq 2)$.
(c) Use Chebychev's inequality to get an upper bound on $P(|X - 1| \geq 1)$.
(d) Compare the answers obtained in parts (b) and (c).

3.6.4 Let $Z \sim N(5, 9)$. Use Chebychev's inequality to get an upper bound on $P(|Z - 5| \geq 30)$.

3.6.5 Let $W \sim$ Binomial(100, 1/2), as in the number of heads when flipping 100 fair coins. Use Chebychev's inequality to get an upper bound on $P(|W - 50| \geq 10)$.

3.6.6 Let $Y \sim N(0, 100)$, and let $Z \sim$ Binomial(80, 1/4). Determine (with explanation) the largest and smallest possible values of $\text{Cov}(Y, Z)$.

3.6.7 Let $X \sim$ Geometric(1/11). Use Jensen's inequality to determine a lower bound on $E(X^4)$, in two different ways:
(a) Apply Jensen's inequality to X with $f(x) = x^4$.
(b) Apply Jensen's inequality to X^2 with $f(x) = x^2$.

Problems

3.6.8 Prove that for any $\epsilon > 0$ and $\delta > 0$, there is a positive integer M, such that if X is the number of heads when flipping M fair coins, then $P(|(X/M) - (1/2)| \geq \delta) \leq \epsilon$.

3.6.9 Prove that for any μ and $\sigma^2 > 0$, there is $a > 0$ and a random variable X with $E(X) = \mu$ and $\text{Var}(X) = \sigma^2$, such that Chebychev's inequality holds with equality, i.e., such that $P(|X - \mu| \geq a) = \sigma^2/a^2$.

3.6.10 Suppose that (X, Y) is uniform on the set $\{(x_1, y_1), \ldots, (x_n, y_n)\}$ where the $x_1, \ldots, x_n$ are distinct values and the $y_1, \ldots, y_n$ are distinct values.
(a) Prove that X is uniformly distributed on $x_1, \ldots, x_n$, with mean given by $\bar{x} = n^{-1} \sum_{i=1}^{n} x_i$ and variance given by $\hat{s}_X^2 = n^{-1} \sum_{i=1}^{n} (x_i - \bar{x})^2$.
(b) Prove that the correlation coefficient between X and Y is given by

$$r_{XY} = \frac{\sum_{i=1}^{n} (x_i - \bar{x})(y_i - \bar{y})}{\sqrt{\sum_{i=1}^{n} (x_i - \bar{x})^2} \sqrt{\sum_{i=1}^{n} (y_i - \bar{y})^2}} = \frac{\hat{s}_{XY}}{\hat{s}_X \hat{s}_Y}$$

where $\hat{s}_{XY} = n^{-1} \sum_{i=1}^{n} (x_i - \bar{x})(y_i - \bar{y})$. The value $\hat{s}_{XY}$ is referred to as the *sample covariance* and r_{XY} is referred to as the *sample correlation coefficient* when the values $(x_1, y_1), \ldots, (x_n, y_n)$ are an observed sample from some bivariate distribution.
(c) Argue that r_{XY} is also the correlation coefficient between X and Y when we drop the assumption of distinctness for the x_i and y_i.
(d) Prove that $-1 \leq r_{XY} \leq 1$ and state the conditions under which $r_{XY} = \pm 1$.

3.6.11 Suppose that X is uniformly distributed on $\{x_1, \ldots, x_n\}$ and so has mean $\bar{x} = n^{-1} \sum_{i=1}^{n} x_i$ and variance $\hat{s}_X^2 = n^{-1} \sum_{i=1}^{n} (x_i - \bar{x})^2$ (see Problem 3.6.10(a)). What is the largest proportion of the values x_i that can lie outside the interval $(\bar{x} - 2\hat{s}_X, \bar{x} + 2\hat{s}_X)$?

3.6.12 Suppose that X is distributed with density given by $f_X(x) = 2/x^3$ for $x > 1$ and is 0 otherwise.
(a) Prove that f_X is a density.
(b) Calculate the mean of X.
(c) Compute $P(X \geq k)$ and compare this with the upper bound on this quantity given by Markov's inequality.

(d) What does Chebyshev's inequality say in this case?

3.6.13 Let $g(x) = \max(-x, -10)$.
(a) Verify that g is a convex function.
(b) Suppose $Z \sim$ Exponential(5). Use Jensen's inequality to obtain a lower bound on $E(g(Z))$.

3.6.14 It can be shown that a function f, with continuous second derivative, is convex on (a, b) if $f''(x) > 0$ for all $x \in (a, b)$.
(a) Use the above fact to show that $f(x) = x^p$ is convex on $(0, \infty)$ whenever $p \geq 1$.
(b) Use (a) to prove that $(E(|X|^p))^{1/p} \geq |E(X)|$ whenever $p \geq 1$.
(c) Prove that $\text{Var}(X) = 0$ if and only if X is degenerate at a constant c.

Challenge

3.6.15 Determine (with proof) all functions that are convex and whose *negatives* are also convex. (That is, find all functions f such that f is convex, and also $-f$ is convex.)

3.7 General Expectations (Advanced)

So far we have considered expected values separately for discrete and absolutely continuous random variables only. However, this separation into two different "cases" may seem unnatural. Furthermore, we know that some random variables are neither discrete nor continuous — for example, mixtures of discrete and continuous distributions.

Hence, it seems desirable to have a more general definition of expected value. Such generality is normally considered in the context of general measure theory, an advanced mathematical subject. However, it is also possible to give a general definition in elementary terms, as follows.

Definition 3.7.1 Let X be an arbitrary random variable (perhaps neither discrete nor continuous). Then the *expected value* of X is given by

$$E(X) = \int_0^\infty P(X > t)\, dt - \int_{-\infty}^0 P(X < t)\, dt,$$

provided either $\int_0^\infty P(X > t)\, dt < \infty$ or $\int_{-\infty}^0 P(X < t)\, dt < \infty$.

This definition appears to contradict our previous definitions of $E(X)$. However, in fact there is no contradiction, as the following theorem shows.

Theorem 3.7.1
(a) Let X be a discrete random variable with distinct possible values $x_1, x_2, \ldots$, and put $p_i = P(X = x_i)$. Then Definition 3.7.1 agrees with the previous definition of $E(X)$. That is,

$$\int_0^\infty P(X > t)\, dt - \int_{-\infty}^0 P(X < t)\, dt = \sum_i x_i p_i.$$

(b) Let X be an absolutely continuous random variable with density f_X. Then Definition 3.7.1 agrees with the previous definition of $E(X)$. That is,

$$\int_0^\infty P(X > t)\, dt - \int_{-\infty}^0 P(X < t)\, dt = \int_{-\infty}^\infty x f_X(x)\, dx.$$

Proof: The key to the proof is switching the order of the integration/summation.
(a) We have

$$\int_0^\infty P(X > t)\, dt = \int_0^\infty \sum_{i,x_i > t} p_i\, dt = \sum_i p_i \int_0^{x_i} dt = \sum_i p_i\, x_i,$$

as claimed.

(b) We have

$$\begin{aligned}
\int_0^\infty P(X > t)\, dt &= \int_0^\infty \left(\int_t^\infty f_X(x)\, dx \right) dt \\
&= \int_0^\infty \left(\int_0^x f_X(x)\, dt \right) dx = \int_0^\infty x\, f_X(x)\, dx.
\end{aligned}$$

Similarly,

$$\begin{aligned}
\int_{-\infty}^0 P(X < t)\, dt &= \int_{-\infty}^0 \left(\int_{-\infty}^t f_X(x)\, dx \right) dt \\
&= \int_{-\infty}^0 \left(\int_x^0 f_X(x)\, dt \right) dx = \int_{-\infty}^0 (-x)\, f_X(x)\, dx.
\end{aligned}$$

Hence,

$$\begin{aligned}
\int_0^\infty P(X > t)\, dt &- \int_{-\infty}^0 P(X < t)\, dt \\
&= \int_0^\infty x\, f_X(x)\, dx - \int_{-\infty}^0 (-x)\, f_X(x)\, dx = \int_{-\infty}^\infty x\, f_X(x)\, dx,
\end{aligned}$$

as claimed. ∎

In other words, Theorem 3.7.1 says that Definition 3.7.1 includes our previous definitions of expected value, for both discrete and absolutely continuous random variables, while working for *any* random variable at all. (Note that to apply Definition 3.7.1 we take an *integral*, not a *sum*, regardless of whether X is discrete or continuous!)

Furthermore, Definition 3.7.1 preserves the key properties of expected value, as the following theorem shows. (We omit the proof here, but see Challenge 3.7.5 for a proof of part (c).)

Theorem 3.7.2 Let X and Y be arbitrary random variables, perhaps neither discrete nor continuous, with expected values defined by Definition 3.7.1.

(a) (*Linearity*) If a and b are any real numbers, then $E(aX + bY) = aE(X) + bE(Y)$.

(b) If X and Y are independent, then $E(XY) = E(X)\,E(Y)$.

(c) (*Monotonicity*) If $X \leq Y$, then $E(X) \leq E(Y)$.

Definition 3.7.1 also tells us about expected values of mixture distributions, as follows.

Theorem 3.7.3 For $1 \leq i \leq k$, let Y_i be a random variable with cdf F_i. Let X be a random variable whose cdf corresponds to a finite mixture (as in Section 2.5.4) of the cdfs of the Y_i, so that $F_X(x) = \sum_i p_i F_i(x)$, where $p_i \geq 0$ and $\sum_i p_i = 1$. Then $E(X) = \sum_i p_i E(Y_i)$.

Proof: We compute that

$$
\begin{aligned}
P(X > t) &= 1 - F_X(t) = 1 - \sum_i p_i F_i(t) \\
&= \sum_i p_i (1 - F_i(t)) = \sum_i p_i P(Y_i > t).
\end{aligned}
$$

Similarly,

$$
P(X < t) = F_X(t^-) = \sum_i p_i F_i(t^-) = \sum_i p_i P(Y_i < t).
$$

Hence, from Definition 3.7.1,

$$
\begin{aligned}
E(X) &= \int_0^\infty P(X > t)\,dt - \int_{-\infty}^0 P(X < t)\,dt \\
&= \int_0^\infty \sum_i p_i P(Y_i > t)\,dt - \int_{-\infty}^0 \sum_i p_i P(Y_i < t)\,dt \\
&= \sum_i p_i \left(\int_0^\infty P(Y_i > t)\,dt - \int_{-\infty}^0 P(Y_i < t)\,dt \right) \\
&= \sum_i p_i E(Y_i),
\end{aligned}
$$

as claimed. ∎

Summary of Section 3.7

- For general random variables, we can define a general expected value by $E(X) = \int_0^\infty P(X > t)\,dt - \int_{-\infty}^0 P(X < t)\,dt$.

- This definition agrees with our previous one, for discrete or absolutely continuous random variables.

- General expectation is still linear and monotone.

Exercises

3.7.1 Let X_1, X_2, and Y be as in Example 2.5.6, so that Y is a mixture of X_1 and X_2. Compute $E(X_1)$, $E(X_2)$, and $E(Y)$.

3.7.2 Suppose we roll a fair six-sided die. If it comes up 1, then we roll the same die again and let X be the value showing. If it comes up anything other than 1, then we instead roll a fair eight-sided die (with the sides numbered 1 through 8), and let X be the value showing on the eight-sided die. Compute the expected value of X.

3.7.3 Let X be a positive constant random variable, so that $X = C$ for some constant $C > 0$. Prove directly from Definition 3.7.1 that $E(X) = C$.

3.7.4 Let Z be a general random variable (perhaps neither discrete nor continuous), and suppose that $P(Z \le 100) = 1$. Prove directly from Definition 3.7.1 that $E(Z) \le 100$.

Challenge

3.7.5 Prove part (c) of Theorem 3.7.2. (Hint: If $X \le Y$, then how does the event $\{X > t\}$ compare to the event $\{Y > t\}$? Hence, how does $P(X > t)$ compare to $P(Y > t)$? And what about $\{X < t\}$ and $\{Y < t\}$?)

3.8   Further Proofs (Advanced)

Theorem 3.4.7 If S has a compound distribution as in (3.4.2), then
(a) $E(S) = E(X_1)\,E(N)$.
(b) $m_S(s) = r_N(m_{X_1}(s))$.

Proof: (a) Because the $\{X_i\}$ are i.i.d., we have $E(X_i) = E(X_1)$ for all i. Define I_i by $I_i = I_{\{1,\ldots,N\}}(i)$. Then we can write $S = \sum_{i=1}^{\infty} X_i I_i$. Also note that $\sum_{i=1}^{\infty} I_i = N$.

Because N is independent of X_i, so is I_i, and we have

$$
\begin{aligned}
E(S) &= E\left(\sum_{i=1}^{\infty} X_i\, I_i\right) = \sum_{i=1}^{\infty} E\left(X_i I_i\right) \\
&= \sum_{i=1}^{\infty} E(X_i)E(I_i) = \sum_{i=1}^{\infty} E(X_1)E(I_i) \\
&= E(X_1) \sum_{i=1}^{\infty} E(I_i) = E(X_1)E\left(\sum_{i=1}^{\infty} I_i\right) \\
&= E(X_1)E(N).
\end{aligned}
$$

This proves part (a).

(b) Now, using an expectation version of the law of total probability (see Theorem 3.5.3), and recalling that $E(\exp\left(\sum_{i=1}^{n} s\,X_i\right)) = m_{X_1}(s)^n$ because the $\{X_i\}$ are i.i.d., we compute that

$$m_S(s)$$

$$= E\left(\exp\left(\sum_{i=1}^{n} sX_i\right)\right) = \sum_{n=0}^{\infty} P(N=n)E\left(\exp\left(\sum_{i=1}^{n} sX_i\right) \mid N=n\right)$$

$$= \sum_{n=0}^{\infty} P(N=n)E\left(\exp\left(\sum_{i=1}^{n} sX_i\right)\right) = \sum_{n=0}^{\infty} P(N=n)m_{X_1}(s)^n$$

$$= E(m_{X_1}(s)^N) = r_N(m_{X_1}(s)),$$

thus proving part (b). ∎

Theorem 3.5.3 Let X and Y be random variables, and let $g : R^1 \to R^1$ be any function. Then $E\left(g(Y)\,E(X\,|\,Y)\right) = E\left(g(Y)\,X\right).$

Proof: If X and Y are discrete, then

$$\begin{aligned}
E\left(g(Y)\,E(X\,|\,Y)\right) &= \sum_{y\in R^1} g(y)E(X\,|\,Y=y)\,P(Y=y) \\[2mm]
&= \sum_{y\in R^1} g(y)\left(\sum_{x\in R^1} x\,P(X=x\,|\,Y=y)\right)P(Y=y) \\[2mm]
&= \sum_{y\in R^1} g(y)\left(\sum_{x\in R^1} x\,\frac{P(X=x,\,Y=y)}{P(Y=y)}\right)P(Y=y) \\[2mm]
&= \sum_{x\in R^1}\sum_{y\in R^1} g(y)x P(X=x,\,Y=y) = E(g(Y)X),
\end{aligned}$$

as claimed.

Similarly, if X and Y are jointly absolutely continuous, then

$$\begin{aligned}
E\left(g(Y)\,E(X\,|\,Y)\right) &= \int_{-\infty}^{\infty} g(y)\,E(X\,|\,Y=y\,f_Y(y)\,dy \\[2mm]
&= \int_{-\infty}^{\infty} g(y)\left(\int_{-\infty}^{\infty} x\,f_{X|Y}(x|y)\,dx\right)f_Y(y)\,dy \\[2mm]
&= \int_{-\infty}^{\infty} g(y)\left(\int_{-\infty}^{\infty} x\,\frac{f_{X,Y}(x,y)}{f_Y(y)}\,dx\right)f_Y(y)\,dy \\[2mm]
&= \int_{-\infty}^{\infty}\int_{-\infty}^{\infty} g(y)\,x\,f_{X,Y}(x,y)\,dx\,dy = E(g(Y)\,X),
\end{aligned}$$

as claimed. ∎

Theorem 3.5.4 Let X and Y be random variables, and let $g : R^1 \to R^1$ be any function. Then $E\left(g(Y)\,X\,|\,Y\right) = g(Y)\,E\left(X\,|\,Y\right).$

Proof: For simplicity, we assume X and Y are discrete; the jointly absolutely continuous case is similar. Then for any y with $P(Y = y) > 0$,

$$
\begin{aligned}
E\left(g(Y)\,X\,|\,Y = y\right) &= \sum_{x \in R^1} \sum_{z \in R^1} g(z)\,x\,P(X = x, Y = z\,|\,Y = y) \\
&= \sum_{x \in R^1} g(y)\,x\,P(X = x\,|\,Y = y) \\
&= g(y) \sum_{x \in R^1} x\,P(X = x\,|\,Y = y) = g(y)\,E\left(X\,|\,Y = y\right).
\end{aligned}
$$

Because this is true for any y, we must have $E\left(g(Y)X\,|\,Y\right) = g(Y)\,E\left(X\,|\,Y\right)$, as claimed. ∎

Theorem 3.5.6 For random variables X and Y, $\mathrm{Var}(X) = \mathrm{Var}(E(X\,|\,Y)) + E\left(\mathrm{Var}(X\,|\,Y)\right)$.

Proof: Using Theorem 3.5.2, we have that

$$
\mathrm{Var}(X) = E((X - \mu_X)^2) = E\left(E\left((X - \mu_X)^2\,|\,Y\right)\right). \tag{3.8.1}
$$

Now,

$$
\begin{aligned}
(X - \mu_X)^2 &= (X - E(X\,|\,Y) + E(X\,|\,Y) - \mu_X)^2 \\
&= (X - E(X\,|\,Y))^2 + (E(X\,|\,Y) - \mu_X)^2 \\
&\quad + 2\,(X - E(X\,|\,Y))\,(E(X\,|\,Y) - \mu_X). \tag{3.8.2}
\end{aligned}
$$

But $E\left((X - E(X\,|\,Y))^2\,|\,Y\right) = \mathrm{Var}(X\,|\,Y)$.

Also, again using Theorem 3.5.2,

$$
E\left(E\left((E(X\,|\,Y) - \mu_X)^2\,|\,Y\right)\right) = E\left((E(X\,|\,Y) - \mu_X)^2\right) = \mathrm{Var}\left(E(X\,|\,Y)\right).
$$

Finally, using Theorem 3.5.4 and linearity (Theorem 3.5.1), we see that

$$
\begin{aligned}
&E\left((X - E(X\,|\,Y))\,(E(X\,|\,Y) - \mu_X)\,|\,Y\right) \\
&= (E(X\,|\,Y) - \mu_X)\,E\left(X - E(X\,|\,Y)\,|\,Y\right) \\
&= (E(X\,|\,Y) - \mu_X)\,(E(X\,|\,Y) - E\left(E(X\,|\,Y)\,|\,Y\right)) \\
&= (E(X\,|\,Y) - \mu_X)\,(E(X\,|\,Y) - E(X\,|\,Y)) = 0.
\end{aligned}
$$

From (3.8.1), (3.8.2), and linearity, we have that $\mathrm{Var}(X) = E\left(\mathrm{Var}(X\,|\,Y)\right) + \mathrm{Var}(E(X\,|\,Y)) + 0$, which completes the proof. ∎

Theorem 3.6.3 (*Cauchy–Schwartz inequality*) Let X and Y be arbitrary random variables, each having finite, nonzero variance. Then

$$
|\mathrm{Cov}(X, Y)| \leq \sqrt{\mathrm{Var}(X)\,\mathrm{Var}(Y)}.
$$

Furthermore, if $\mathrm{Var}(Y) > 0$, then equality is attained if and only if $X - \mu_X = \lambda(Y - \mu_Y)$ where $\lambda = \mathrm{Cov}(X, Y)/\mathrm{Var}(Y)$.

Proof: If $\text{Var}(Y) = 0$, then Challenge 3.3.23 implies that $Y = \mu_Y$ with probability 1 (because $\text{Var}(Y) = E((Y - \mu_Y)^2) \geq 0$). This implies that

$$\text{Cov}\,(X, Y) = E\left((X - \mu_X)(\mu_Y - \mu_Y)\right) = 0 = \sqrt{\text{Var}\,(X)\,\text{Var}\,(Y)},$$

and the Cauchy–Schwartz inequality holds.

If $\text{Var}(Y) \neq 0$, let $Z = X - \mu_X$ and $W = Y - \mu_Y$. Then for any real number λ, we compute, using linearity, that

$$
\begin{aligned}
E\left((Z - \lambda W)^2\right) &= E(Z^2) - 2\lambda E(ZW) + \lambda^2 E(W^2) \\
&= \text{Var}(X) - 2\lambda\,\text{Cov}(X, Y) + \lambda^2\,\text{Var}(Y) \\
&= a\lambda^2 + b\lambda + c,
\end{aligned}
$$

where $a = \text{Var}(Y) > 0$, $b = -2\,\text{Cov}(X, Y)$, and $c = \text{Var}(X)$. On the other hand, clearly $E\left((Z - \lambda W)^2\right) \geq 0$ for all λ. Hence, we have a quadratic equation that is always nonnegative, and so has at most one real root.

By the quadratic formula, any quadratic equation has *two* real roots provided that the discriminant $b^2 - 4ac > 0$. Because that is not the case here, we must have $b^2 - 4ac \leq 0$, i.e.,

$$4\,\text{Cov}(X, Y)^2 - 4\,\text{Var}(Y)\,\text{Var}(X) \leq 0.$$

Dividing by 4, rearranging, and taking square roots, we see that $|\text{Cov}(X, Y)| \leq (\text{Var}(X)\,\text{Var}(Y))^{1/2}$, as claimed.

Finally, $|\text{Cov}(X, Y)| = (\text{Var}(X)\,\text{Var}(Y))^{1/2}$ if and only if $b^2 - 4ac = 0$, which means the quadratic has one real root. Thus, there is some real number λ such that $E((Z - \lambda W)^2) = 0$. Since $(Z - \lambda W)^2 \geq 0$, it follows from Challenge 3.3.22 that this happens if and only if $Z - \lambda W = 0$ with probability 1, as claimed. When this is the case, then

$$\text{Cov}(X, Y) = E\left(ZW\right) = E\left(\lambda W^2\right) = \lambda E\left(W^2\right) = \lambda\,\text{Var}\,(Y)$$

and so $\lambda = \text{Cov}(X, Y)/\text{Var}(Y)$ when $\text{Var}(Y) \neq 0$. ∎

Chapter 4

Sampling Distributions and Limits

In many applications of probability theory, we will be faced with the following problem. Suppose that $X_1, X_2, \ldots, X_n$ is an identically and independently distributed (i.i.d.) sequence, i.e., $X_1, X_2, \ldots, X_n$ is a sample from some distribution, and we want to find the distribution of a new random variable $Y = h(X_1, X_2, \ldots, X_n)$ for some function h. In particular, we might want to compute the distribution function of Y or perhaps its mean and variance. The distribution of Y is sometimes referred to as its *sampling distribution*, as Y is based on a sample from some underlying distribution.

We will see that some of the methods developed in earlier chapters are useful in solving such problems — especially when it is possible to compute an exact solution, e.g., obtain an exact expression for the probability or density function of Y. Section 4.6 contains a number of exact distribution results for a variety of functions of normal random variables. These have important applications in statistics.

Quite often, however, exact results are impossible to obtain, as the problem is just too complex. In such a case, we must develop an approximation to the distribution of Y.

For many important problems, a version of Y is defined for each sample size n (e.g., a sample mean or sample variance), so that we can consider a sequence of random variables $Y_1, Y_2, \ldots$, etc. This leads us to consider the limiting distribution of such a sequence so that, when n is large, we can approximate the distribution of Y_n by the limit, which is often much simpler. This approach leads to a famous result, known as the central limit theorem, discussed in Section 4.4.

Sometimes we cannot even develop useful approximations for large n, due to the difficulty of the problem or perhaps because n is just too small in a particular application. Fortunately, however, we can then use the Monte Carlo approach where the power of the computer becomes available. This is discussed in Section 4.5.

In Chapter 5 we will see that, in statistical applications, we typically do not know much about the underlying distribution of the X_i from which we are sampling. We then collect a sample and a value, such as Y, which will serve as an estimate of a characteristic of the underlying distribution, e.g., the sample mean $\bar{X}$ will serve as an estimate of the mean of the distribution of the X_i. We then want to know what happens to these estimates as n grows. If we have chosen our estimates well, then the estimates will converge to the quantities we are estimating as n increases. Such an estimate is called *consistent*. In Sections 4.2 and 4.3 we will discuss the most important consistency theorems, namely, the weak and strong laws of large numbers.

4.1 Sampling Distributions

Let us consider a very simple example.

Example 4.1.1
Suppose we obtain a sample X_1, X_2 of size $n = 2$ from the discrete distribution with probability function given by

$$p_X(x) = \begin{cases} 1/2 & x = 1 \\ 1/4 & x = 2 \\ 1/4 & x = 3 \\ 0 & \text{otherwise.} \end{cases}$$

Let us take $Y_2 = (X_1 X_2)^{1/2}$. This is the *geometric mean* of the sample values (the geometric mean of n positive numbers $x_1, \ldots, x_n$ is defined as $(x_1 \cdots x_n)^{1/n}$).

To determine the distribution of Y_2, we first list the possible values for Y_2, the samples that give rise to these values, and their probabilities of occurrence. The values of these probabilities specify the sampling distribution of Y. We have the following table.

y	Sample	$p_{Y_2}(y)$
1	$\{(1,1)\}$	$\frac{1}{2}\frac{1}{2} = \frac{1}{4}$
$\sqrt{2}$	$\{(1,2),(2,1)\}$	$\frac{1}{2}\frac{1}{4} + \frac{1}{4}\frac{1}{2} = \frac{1}{4}$
$\sqrt{3}$	$\{(1,3),(1,3)\}$	$\frac{1}{2}\frac{1}{4} + \frac{1}{4}\frac{1}{2} = \frac{1}{4}$
2	$\{(2,2)\}$	$\frac{1}{4}\frac{1}{4} = \frac{1}{16}$
$\sqrt{6}$	$\{(2,3),(3,2)\}$	$\frac{1}{4}\frac{1}{4} + \frac{1}{4}\frac{1}{4} = \frac{1}{8}$
3	$\{(3,3)\}$	$\frac{1}{4}\frac{1}{4} = \frac{1}{16}$

Now suppose instead we have a sample $X_1, \ldots, X_{20}$ of size $n = 20$, and we want to find the distribution of $Y_{20} = (X_1 \cdots X_{20})^{1/20}$. Obviously, we can proceed as above, but this time the computations are much more complicated, as there are now $3^{20} = 3,486,784,401$ possible samples, as opposed to the $3^2 = 9$ samples used to form the previous table. Directly computing $p_{Y_{20}}$, as we have done for p_{Y_2}, would be onerous — even for a computer! So what can we do here?

One possibility is to look at the distribution of $Y_n = (X_1 \cdots X_n)^{1/n}$ when n is large and see if we can approximate this in some fashion. The results of Section 4.4.1 show that

$$\ln Y_n = \frac{1}{n} \sum_{i=1}^{n} \ln X_i$$

has an approximate normal distribution when n is large. In fact, the approximating normal distribution when $n = 20$, turns out to be an $N(0.447940, 0.105167)$ distribution, and we have plotted this density in Figure 4.1.1.

Another approach is to use the methods of Section 2.10 to generate N samples of size $n = 20$ from p_X, calculate $\ln Y_{20}$ for each ($\ln$ is a 1–1 transformation and we transform to avoid the potentially large values assumed by Y_{20}), and then use these N values to approximate the distribution of $\ln Y_{20}$. For example, in Figure 4.1.2 we have provided a plot of a density histogram (see Section 5.4.3 for more discussion of histograms) of $N = 10^4$ values of $\ln Y_{20}$ calculated from $N = 10^4$ samples of size $n = 20$ generated (using the computer) from p_X. The area of each rectangle corresponds to the proportion of values of $\ln Y_{20}$ that were in the interval given by the base of the rectangle. As we will see in Sections 4.2, 4.3, and 4.4, these areas approximate the actual probabilities that $\ln Y_{20}$ falls in these intervals. These approximations improve as we increase N.

Notice the similarity in the shapes of Figures 4.1.1 and 4.1.2. Figure 4.1.2 is not symmetrical about its center, however, as it is somewhat skewed. This is an indication that the normal approximation is not entirely adequate when $n = 20$. ∎

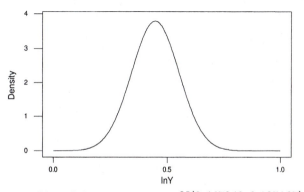

Figure 4.1.1: Plot of the approximating $N(0.447940, 0.105167)$ density to the distribution of $\ln Y_{20}$ in Example 4.1.1.

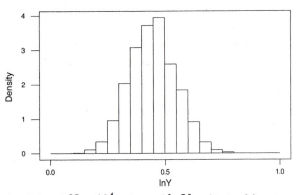

Figure 4.1.2: Plot of $N = 10^4$ values of $\ln Y_{20}$ obtained by generating $N = 10^4$ samples from p_X in Example 4.1.1.

Sometimes we are lucky and can work out the sampling distribution of $Y = h(X_1, X_2, \ldots, X_n)$ exactly in a form useful for computing probabilities and expectations for Y. In general, however, when we want to compute $P(Y \in B) = P_Y(B)$, we will have to determine the set of samples $(X_1, X_2, \ldots, X_n)$ such that $Y \in B$, as given by

$$h^{-1}B = \{(x_1, x_2, \ldots, x_n) : h(x_1, x_2, \ldots, x_n) \in B\},$$

and then compute $P\left((X_1, X_2, \ldots, X_n) \in h^{-1}B\right)$. This is typically an intractable problem and approximations or simulation (Monte Carlo) methods will be essential. Techniques for deriving such approximations will be discussed in subsequent sections of this chapter. In particular, we will develop an important approximation to the sampling distribution of the sample mean

$$\bar{X} = h(X_1, X_2, \ldots, X_n) = \frac{1}{n} \sum_{i=1}^{n} X_i.$$

Summary of Section 4.1

- A sampling distribution is the distribution of a random variable corresponding to a function of some i.i.d. sequence.

- Sampling distributions can sometimes be computed by direct computation or by approximations such as the central limit theorem.

Exercises

4.1.1 Suppose that X_1, X_2, X_3 are i.i.d. from p_X in Example 4.1.1. Determine the exact distribution of $Y_3 = (X_1 X_2 X_3)^{1/3}$.

4.1.2 Suppose that a fair six-sided die is tossed $n = 2$ independent times. Compute the exact distribution of the sample mean.

4.1.3 Suppose that an urn contains a proportion p of chips labelled 0 and proportion $1 - p$ of chips labelled 1. For a sample of $n = 2$, drawn with replacement, determine the distribution of the sample mean.

4.1.4 Suppose that an urn contains N chips labelled 0 and M chips labelled 1. For a sample of $n = 2$, drawn without replacement, determine the distribution of the sample mean.

4.1.5 Suppose that a symmetrical die is tossed $n = 20$ independent times. Work out the exact sampling distribution of the maximum of this sample.

Computer Exercises

4.1.6 Generate a sample of $N = 10^3$ values of Y_{50} in Example 4.1.1. Calculate the mean and standard deviation of this sample.

4.1.7 Suppose that $X_1, X_2, \ldots, X_{10}$ is an i.i.d. sequence from an $N(0, 1)$ distribution. Generate a sample of $N = 10^3$ values from the distribution of $\max(X_1, X_2, \ldots, X_{10})$. Calculate the mean and standard deviation of this sample.

Problems

4.1.8 Suppose that $X_1, X_2, \ldots, X_n$ is a sample from the Poisson(λ) distribution. Determine the exact sampling distribution of $Y = X_1 + X_2 + \cdots + X_n$. (Hint: Determine the moment-generating function of Y and use the uniqueness theorem.)

4.1.9 Suppose that X_1, X_2 is a sample from the Uniform(0,1) distribution. Determine the exact sampling distribution of $Y = X_1 + X_2$. (Hint: Determine the density of Y.)

4.1.10 Suppose that X_1, X_2 is a sample from the Uniform(0,1) distribution. Determine the exact sampling distribution of $Y = (X_1 X_2)^{1/2}$. (Hint: Determine the density of $\ln Y$ and then transform.)

4.2 Convergence in Probability

Notions of *convergence* are fundamental to much of mathematics. For example, if $a_n = 1 - 1/n$, then $a_1 = 0$, $a_2 = 1/2$, $a_3 = 2/3$, $a_4 = 3/4$, etc. We see that the values of a_n are getting "closer and closer" to 1, and indeed we know from calculus that $\lim_{n \to \infty} a_n = 1$ in this case.

For random variables, notions of convergence are more complicated. If the values themselves are random, then how can they "converge" to anything? On the other hand, we can consider various *probabilities* associated with the random variables, and see if *they* converge in some sense.

The simplest notion of convergence of random variables is convergence in probability, as follows. (Other notions of convergence will be developed in subsequent sections.)

Definition 4.2.1 Let $X_1, X_2, \ldots$ be an infinite sequence of random variables, and let Y be another random variable. Then the sequence $\{X_n\}$ *converges in probability* to Y, if for all $\epsilon > 0$, $\lim_{n \to \infty} P(|X_n - Y| \geq \epsilon) = 0$, and we write $X_n \xrightarrow{P} Y$.

In Figure 4.2.1, we have plotted the differences $X_n - Y$, for selected values of n, for 10 generated sequences $\{X_n - Y\}$ for a typical situation where the random variables X_n converge to a random variable Y in probability. Also we have plotted the horizontal lines at $\pm \epsilon$ for $\epsilon = 0.25$. From this we can see the increasing concentration of the distribution of $X_n - Y$ about 0, as n increases, as required by Definition 4.2.1. In fact, the 10 observed values of $X_{100} - Y$ all satisfy the inequality $|X_{100} - Y| < 0.25$.

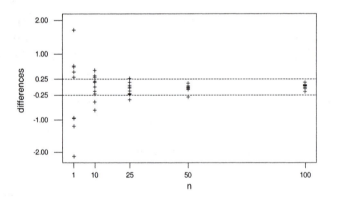

Figure 4.2.1: Plot of 10 replications of $\{X_n - Y\}$ illustrating the convergence in probability of X_n to Y.

We consider some applications of this definition.

Example 4.2.1
Let Y be any random variable, and let $X_1 = X_2 = X_3 = \cdots = Y$. (That is, the random variables are all *identical* to each other.) In that case, $|X_n - Y| = 0$, so of course

$$\lim_{n \to \infty} P(|X_n - Y| \geq \epsilon) = 0$$

for all $\epsilon > 0$. Hence, $X_n \xrightarrow{P} Y$. ∎

Example 4.2.2
Suppose $P(X_n = 1 - 1/n) = 1$ and $P(Y = 1) = 1$. Then $P(|X_n - Y| \geq \epsilon) = 0$ whenever $n > 1/\epsilon$. Hence, $P(|X_n - Y| \geq \epsilon) \to 0$ as $n \to \infty$ for all $\epsilon > 0$. Hence, the sequence $\{X_n\}$ converges in probability to Y. (Here, the distributions of X_n and Y are all *degenerate*.) ∎

Example 4.2.3

Let $U \sim \text{Uniform}[0,1]$. Define X_n by

$$X_n = \begin{cases} 3 & U \leq \frac{2}{3} - \frac{1}{n} \\ 8 & \text{otherwise,} \end{cases}$$

and define Y by

$$Y = \begin{cases} 3 & U \leq \frac{2}{3} \\ 8 & \text{otherwise.} \end{cases}$$

Then

$$P\left(|X_n - Y| \geq \epsilon\right) \leq P\left(X_n \neq Y\right) = P\left(\frac{2}{3} - \frac{1}{n} < U < \frac{2}{3}\right) = \frac{1}{n}.$$

Hence, $P\left(|X_n - Y| \geq \epsilon\right) \to 0$ as $n \to \infty$ for all $\epsilon > 0$, and the sequence $\{X_n\}$ converges in probability to Y. (This time, the distributions of X_n and Y are *not* degenerate.) ∎

A common case is where the distributions of the X_n are not degenerate, but Y is just a constant, as in the following example.

Example 4.2.4

Suppose $Z_n \sim \text{Exponential}(n)$ and let $Y = 0$. Then

$$P\left(|Z_n - Y| \geq \epsilon\right) = P(Z_n \geq \epsilon) = \int_\epsilon^\infty n e^{-nx} dx = e^{-n\epsilon}.$$

Hence, again $P\left(|Z_n - Y| \geq \epsilon\right) \to 0$ as $n \to \infty$ for all $\epsilon > 0$, so the sequence $\{Z_n\}$ converges in probability to Y. ∎

4.2.1 The Weak Law of Large Numbers

One of the most important applications of convergence in probability is the weak law of large numbers. Suppose $X_1, X_2, \ldots$ is a sequence of independent random variables that each have the same mean μ. For large n, what can we say about their average

$$M_n = \frac{1}{n}(X_1 + \cdots + X_n)?$$

We refer to M_n as the *sample average*, or *sample mean*, for $X_1, \ldots, X_n$. When the sample size n is fixed, we will often use $\bar{X}$ as a notation for sample mean instead of M_n.

For example, if we flip a sequence of fair coins, and if $X_i = 1$ or $X_i = 0$ as the ith coin comes up heads or tails, then M_n represents the *fraction* of the first n coins that came up heads. We might expect that for large n, this fraction will be close to $1/2$, i.e., to the expected value of the X_i.

The weak law of large numbers provides a precise sense in which average values M_n tend to get close to $E(X_i)$, for large n.

Theorem 4.2.1 (*Weak law of large numbers*) Let $X_1, X_2, \ldots$ be a sequence of independent random variables, each having the same mean μ and each having variance less than or equal to $v < \infty$. Then for all $\epsilon > 0$, $\lim_{n\to\infty} P(|M_n - \mu| \geq \epsilon) = 0$. That is, the averages converge in probability to the common mean μ or $M_n \overset{P}{\to} \mu$.

Proof: Using linearity of expected value, we see that $E(M_n) = \mu$. Also, using independence, we have

$$
\begin{aligned}
\text{Var}(M_n) &= \frac{1}{n^2}(\text{Var}(X_1) + \text{Var}(X_2) + \cdots + \text{Var}(X_n)) \\
&\leq \frac{1}{n^2}(v + v + \cdots + v) = \frac{1}{n^2}(nv) = v/n.
\end{aligned}
$$

Hence, by Chebychev's inequality (Theorem 3.6.2), we have

$$
P(|M_n - \mu| \geq \epsilon) \leq \text{Var}(M_n)/\epsilon^2 \leq v/\epsilon^2 n.
$$

This converges to 0 as $n \to \infty$, which proves the theorem. ∎

It is a fact that, in Theorem 4.2.1, if we require the X_i variables to be i.i.d. instead of merely independent, then we do not even need the X_i to have finite variance. But we will not discuss this result further here. Consider some applications of the weak law of large numbers.

Example 4.2.5
Consider flipping a sequence of identical fair coins. Let M_n be the fraction of the first n coins that are heads. Then $M_n = (X_1 + \cdots + X_n)/n$, where $X_i = 1$ if the ith coin is heads, otherwise $X_i = 0$. Hence, by the weak law of large numbers, we have

$$
\begin{aligned}
\lim_{n\to\infty} P(M_n < 0.49) &= \lim_{n\to\infty} P(M_n - .5 < -0.01) \\
&\leq \lim_{n\to\infty} P(M_n - .5 < -0.01 \text{ or } M_n - .5 > 0.01) \\
&= \lim_{n\to\infty} P(|M_n - .5| > 0.01) = 0
\end{aligned}
$$

and, similarly, $\lim_{n\to\infty} P(M_n > 0.51) = 0$. This illustrates that for large n, it is very likely that M_n is very close to 0.5. ∎

Example 4.2.6
Consider flipping a sequence of identical coins, each of which has probability p of coming up heads. Let M_n again be the fraction of the first n coins that are heads. Then by the weak law of large numbers, for any $\epsilon > 0$, $\lim_{n\to\infty} P(p - \epsilon < M_n < p + \epsilon) = 1$. We thus see that for large n, it is very likely that M_n is very close to p. (The previous example corresponds to the special case $p = 1/2$.) ∎

Example 4.2.7
Let $X_1, X_2, \ldots$ be i.i.d. with distribution $N(3, 5)$. Then $E(M_n) = 3$, and by the weak law of large numbers, $P(3 - \epsilon < M_n < 3 + \epsilon) \to 1$ as $n \to \infty$. Hence, for large n, the average value M_n is very close to 3. ∎

Example 4.2.8
Let $W_1, W_2, \ldots$ be i.i.d. with distribution Exponential(6). Then $E(M_n) = 1/6$, and by the weak law of large numbers, $P(1/6 - \epsilon < M_n < 1/6 + \epsilon) \to 1$ as $n \to \infty$. Hence, for large n, the average value M_n is very close to $1/6$. ∎

Summary of Section 4.2

- A sequence $\{X_n\}$ of random variables converges in probability to Y if $\lim_{n\to\infty} P(|X_n - Y| \geq \epsilon) = 0$.

- The weak law of large numbers says that if $\{X_n\}$ is i.i.d. (or is independent with constant mean and bounded variance), then the averages $M_n = (X_1 + \cdots + X_n)/n$ converge in probability to $E(X_i)$.

Exercises

4.2.1 Let $U \sim$ Uniform$[5, 10]$, and let $Z = I_{U \in [5, 7)}$ and $Z_n = I_{U \in [5, 7+1/n^2)}$. Prove that $Z_n \to Z$ in probability.

4.2.2 Let $Y \sim$ Uniform$[0, 1]$, and let $X_n = Y^n$. Prove that $X_n \to 0$ in probability.

4.2.3 Let $W_1, W_2, \ldots$ be i.i.d. with distribution Exponential(3). Prove that for some n, we have $P(W_1 + W_2 + \cdots + W_n < n/2) > 0.999$.

4.2.4 Let $Y_1, Y_2, \ldots$ be i.i.d. with distribution $N(2, 5)$. Prove that for some n, we have $P(Y_1 + Y_2 + \cdots + Y_n > n) > 0.999$.

4.2.5 Let $X_1, X_2, \ldots$ be i.i.d. with distribution Poisson(8). Prove that for some n, we have $P(X_1 + X_2 + \cdots + X_n > 9n) < 0.001$.

Computer Exercises

4.2.6 Generate i.i.d. $X_1, \ldots, X_n$ distributed Exponential(5) and compute M_n when $n = 20$. Repeat this N times, where N is large (if possible, take $N = 10^5$, otherwise as large as is feasible), and compute the proportion of values of M_n that lie between 0.19 and 0.21. Repeat this with $n = 50$. What property of convergence in probability do your results illustrate?

4.2.7 Generate i.i.d. $X_1, \ldots, X_n$ distributed Poisson(7) and compute M_n when $n = 20$. Repeat this N times, where N is large (if possible, take $N = 10^5$, otherwise as large as is feasible), and compute the proportion of values of M_n that lie between 6.99 and 7.01. Repeat this with $n = 100$. What property of convergence in probability do your results illustrate?

Problems

4.2.8 Give an example of random variables $X_1, X_2, \ldots$ such that $\{X_n\}$ converges to 0 in probability, but $E(X_n) = 1$ for all n. (Hint: Suppose $P(X_n = n) = 1/n$ and $P(X_n = 0) = 1 - 1/n$.)

4.2.9 Prove that $X_n \overset{P}{\to} 0$ if and only if $|X_n| \overset{P}{\to} 0$.

4.2.10 Prove or disprove that $X_n \overset{P}{\to} 5$ if and only if $|X_n| \overset{P}{\to} 5$.

4.3   Convergence with Probability 1

A notion of convergence for random variables that is closely associated with the convergence of a sequence of real numbers is provided by the concept of convergence with probability 1. This property is given in the following definition.

Definition 4.3.1 Let $X_1, X_2, \ldots$ be an infinite sequence of random variables. We shall say that the sequence $\{X_i\}$ *converges with probability 1* (or *converges almost surely*) to a random variable Y, if $P\left(\lim_{n\to\infty} X_n = Y\right) = 1$ and we write $X_n \overset{a.s.}{\to} Y$.

In Figure 4.3.1, we illustrate this convergence by graphing the sequence of differences $\{X_n - Y\}$ for a typical situation where the random variables X_n converge to a random variable Y with probability 1. We have also plotted the horizontal lines at $\pm\epsilon$ for $\epsilon = 0.1$. Notice that inevitably all the values $X_n - Y$ are in the interval $(-0.1, 0.1)$ or, in other words, the values of X_n are within 0.1 of the values of Y.

Definition 4.3.1 indicates that for any given $\epsilon > 0$, there will exist a value N_ϵ such that $|X_n - Y| < \epsilon$ for every $n \geq N_\epsilon$. The value of N_ϵ will vary depending on the observed value of the sequence $\{X_n - Y\}$, but it always exists. Contrast this with the situation depicted in Figure 4.2.1, which only says that the probability distribution $X_n - Y$ concentrates about 0 as n grows and not that the individual values of $X_n - Y$ will necessarily all be near 0 (also see Example 4.3.2).

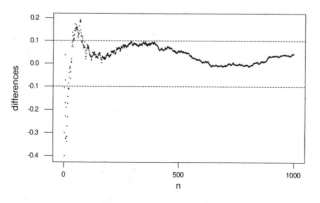

Figure 4.3.1: Plot of a single replication $\{X_n - Y\}$ illustrating the convergence with probability 1 of X_n to Y.

Consider an example of this.

Example 4.3.1
Consider again the setup of Example 4.2.2, where $U \sim \text{Uniform}[0, 1]$,

$$X_n = \left\{ \begin{array}{ll} 3 & U \leq \frac{2}{3} - \frac{1}{n} \\ 8 & \text{otherwise} \end{array} \right.$$

and

$$Y = \begin{cases} 3 & U \le \frac{2}{3} \\ 8 & \text{otherwise.} \end{cases}$$

If $U > 2/3$, then $Y = 8$ and also $X_n = 8$ for all n, so clearly $X_n \to Y$. If $U < 2/3$, then for large enough n we will also have

$$U \le \frac{2}{3} - \frac{1}{n},$$

so again $X_n \to Y$. On the other hand, if $U = 2/3$, then we will always have $X_n = 8$, even though $Y = 3$. Hence, $X_n \to Y$ except when $U = 2/3$. Because $P(U = 2/3) = 0$, we do have $X_n \to Y$ with probability 1. ∎

One might wonder what the relationship is between convergence in probability and convergence with probability 1. The following theorem provides an answer.

Theorem 4.3.1 Let $Z, Z_1, Z_2, \ldots$ be random variables. Suppose $Z_n \to Z$ with probability 1. Then $Z_n \to Z$ in probability. That is, if a sequence of random variables converges almost surely, then it converges in probability to the same limit.

Proof: See Section 4.7 for the proof of this result. ∎

On the other hand, the converse to Theorem 4.3.1 is false, as the following example shows.

Example 4.3.2
Let U have the uniform distribution on $[0, 1]$. We construct an infinite sequence of random variables $\{X_n\}$ by setting

$$\begin{aligned}
X_1 &= I_{[0,1/2)}(U), \ X_2 = I_{[1/2,1]}(U), \\
X_3 &= I_{[0,1/4)}(U), \ X_4 = I_{[1/4,1/2)}(U), \ X_5 = I_{[1/2,3/4)}(U), \ X_6 = I_{[3/4,1]}(U), \\
X_7 &= I_{[0,1/8)}(U), X_8 = I_{[1/8,1/4)}(U), \ldots
\end{aligned}$$

$$\vdots$$

where I_A is the *indicator function* of the event A, i.e., $I_A(s) = 1$ if $s \in A$, and $I_A(s) = 0$ if $s \notin A$.

Note that we first subdivided $[0, 1]$ into two equal-length subintervals and defined X_1 and X_2 as the indicator functions for the two subintervals. Next we subdivided $[0, 1]$ into four equal-length subintervals and defined X_3, X_4, X_5, and X_6 as the indicator functions for the four subintervals. We continued this process by next dividing $[0, 1]$ into eight equal-length subintervals, then 16 equal-length subintervals, etc., to obtain an infinite sequence of random variables.

Each of these random variables X_n takes the values 0 and 1 only and so must follow a Bernoulli distribution. In particular, $X_1 \sim$ Bernoulli$(1/2)$, $X_2 \sim$ Bernoulli$(1/2)$, $X_3 \sim$ Bernoulli$(1/4)$, etc.

Then for $0 < \epsilon < 1$, we have that $P(|X_n - 0| \geq \epsilon) = P(X_n = 1)$. Because the intervals for U that make $X_n \neq 0$ are getting smaller and smaller, we see that $P(X_n = 1)$ is converging to 0. Hence, X_n converges to 0 in probability.

On the other hand, X_n does *not* converge to 0 almost surely. Indeed, no matter what value U takes on, there will always be infinitely many different n for which $X_n = 1$. Hence, we will have $X_n = 1$ infinitely often, so that we will *not* have X_n converging to 0 for any particular value of U. Thus, $P(\lim_{n \to \infty} X_n \to 0) = 0$, and X_n does *not* converge to 0 with probability 1. ∎

Theorem 4.3.1 and Example 4.3.2 together show that convergence with probability 1 is a *stronger* notion than convergence in probability.

Now, the weak law of large numbers (above) concludes only that the averages M_n are converging in probability to $E(X_i)$. A stronger version of this result would instead conclude convergence with probability 1. We consider that now.

4.3.1   The Strong Law of Large Numbers

The following is a strengthening of the weak law of large numbers, because it concludes convergence with probability 1 instead of just convergence in probability.

Theorem 4.3.2 (*Strong law of large numbers*) Let $X_1, X_2, \ldots$ be a sequence of i.i.d. random variables, each having finite mean μ. Then

$$P\left(\lim_{n \to \infty} M_n = \mu\right) = 1.$$

That is, the averages converge with probability 1 to the common mean μ or $M_n \overset{a.s.}{\to} \mu$.

Proof: See *A First Look at Rigorous Probability Theory*, by J. S. Rosenthal (World Scientific Publishing Co., 2000) for a proof of this result. ∎

This result says that sample averages converge with probability 1 to μ.

Like Theorem 4.2.1, it says that for large n the averages M_n are usually close to $\mu = E(X_i)$ for large n. But it says in addition that if we wait long enough (i.e., if n is large enough), then eventually the averages will *all* be close to μ, for *all* sufficiently large n. In other words, the sample mean is consistent for μ.

Summary of Section 4.3

- A sequence $\{X_n\}$ of random variables converges with probability 1 (or converges almost surely) to Y if, $P(\lim_{n \to \infty} X_n = Y) = 1$.

- Convergence with probability 1 implies convergence in probability.

- The strong law of large numbers says that if $\{X_n\}$ is i.i.d., then the averages $M_n = (X_1 + \cdots + X_n)/n$ converge with probability 1 to $E(X_i)$.

Exercises

4.3.1 Let $U \sim$ Uniform$[5, 10]$, and let $Z = I_{[5, 7)}(U)$ (i.e., Z is the indicator function of $[5, 7)$) and $Z_n = I_{[5, 7+1/n^2)}(U)$. Prove that $Z_n \to Z$ with probability 1.

4.3.2 Let $Y \sim$ Uniform$[0, 1]$, and let $X_n = Y^n$. Prove that $X_n \to 0$ with probability 1.

4.3.3 Let $W_1, W_2, \ldots$ be i.i.d. with distribution Exponential(3). Prove that with probability 1, for some n, we have $W_1 + W_2 + \cdots + W_n < n/2$.

4.3.4 Let $Y_1, Y_2, \ldots$ be i.i.d. with distribution $N(2, 5)$. Prove that with probability 1, for some n, we have $Y_1 + Y_2 + \cdots + Y_n > n$.

4.3.5 Suppose $X_n \to X$ with probability 1, and also $Y_n \to Y$ with probability 1. Prove that $P(X_n \to X \text{ and } Y_n \to Y) = 1$.

4.3.6 Suppose $Z_1, Z_2, \ldots$ are i.i.d. with finite mean μ. Let $M_n = (Z_1 + \cdots + Z_n)/n$. Determine (with explanation) whether the following statements are true or false.
(a) With probability 1, $M_n = \mu$ for some n.
(b) With probability 1, $\mu - 0.01 < M_n < \mu + 0.01$ for some n.
(c) With probability 1, $\mu - 0.01 < M_n < \mu + 0.01$ for all but finitely many n.
(d) For any $x \in R^1$, with probability 1, $x - 0.01 < M_n < x + 0.01$ for some n.

Computer Exercises

4.3.7 Generate i.i.d. $X_1, \ldots, X_n$ distributed Exponential(5) with n large (take $n = 10^5$ if possible). Plot the values $M_1, M_2, \ldots, M_n$. To what value are they converging? How quickly?

4.3.8 Generate i.i.d. $X_1, \ldots, X_n$ distributed Poisson(7) with n large (take $n = 10^5$ if possible). Plot the values $M_1, M_2, \ldots, M_n$. To what value are they converging? How quickly?

4.3.9 Generate i.i.d. $X_1, X_2, \ldots, X_n$ distributed $N(-4, 3)$ with n large (take $n = 10^5$ if possible). Plot the values $M_1, M_2, \ldots, M_n$. To what value are they converging? How quickly?

Problems

4.3.10 Suppose for each positive integer k, there are random variables $W_k, X_{k,1}, X_{k,2}, \ldots$ such that $P(\lim_{n \to \infty} X_{n,k} = W_k) = 1$. Prove that

$$P(\lim_{n \to \infty} X_{n,k} = W_k \text{ for all } k) = 1.$$

4.3.11 Prove that $X_n \overset{a.s.}{\to} 0$ if and only if $|X_n| \overset{a.s.}{\to} 0$.

4.3.12 Prove or disprove that $X_n \overset{a.s.}{\to} 5$ if and only if $|X_n| \overset{a.s.}{\to} 5$.

Challenges

4.3.13 Suppose for each real number $r \in [0, 1]$, there are random variables

$W_r, X_{r,1}, X_{r,2}, \ldots$ such that $P(\lim_{n \to \infty} X_{n,r} = W_r) = 1$. Prove or disprove that we must have $P(\lim_{n \to \infty} X_{n,r} = W_r$ for all $r \in [0,1]) = 1$.

4.3.14 Give an example of random variables $X_1, X_2, \ldots$ such that $\{X_n\}$ converges to 0 with probability 1, but $E(X_n) = 1$ for all n.

4.4 Convergence in Distribution

There is yet another notion of convergence of a sequence of random variables that is important in applications of probability and statistics.

Definition 4.4.1 Let $X, X_1, X_2, \ldots$ be random variables. Then we say that the sequence $\{X_n\}$ *converges in distribution* to X, if for all $x \in R^1$ such that $P(X = x) = 0$ we have $\lim_{n \to \infty} P(X_n \le x) = P(X \le x)$, and we write $X_n \xrightarrow{D} X$.

Intuitively, $\{X_n\}$ converges in distribution to X if for large n, the distribution of X_n is close to that of X. The importance of this, as we will see, is that often the distribution of X_n is difficult to work with, while that of X is much simpler. With X_n converging in distribution to X, however, we can approximate the distribution of X_n by that of X.

Example 4.4.1
Suppose $P(X_n = 1) = 1/n$, and $P(X_n = 0) = 1 - 1/n$. Let $X = 0$ so that $P(X = 0) = 1$. Then $\{X_n\}$ converges in distribution to X. Intuitively, as $n \to \infty$, it is more and more likely that X_n will equal 0. ∎

Example 4.4.2
Suppose $P(X_n = 1) = 1/2 + 1/n$, and $P(X_n = 0) = 1/2 - 1/n$. Suppose further that $P(X = 0) = P(X = 1) = 1/2$. Then $\{X_n\}$ converges in distribution to X, because $P(X_n = 1) \to 1/2$ and $P(X_n = 0) \to 1/2$ as $n \to \infty$. ∎

Example 4.4.3
Let $X \sim \text{Uniform}[0,1]$, and let $P(X_n = i/n) = 1/n$ for $i = 1, 2, \ldots, n$. Then X is absolutely continuous, while X_n is discrete. On the other hand, for any $0 \le x \le 1$, we have $P(X \le x) = x$, and letting $\lfloor x \rfloor$ denote the greatest integer less than or equal to x, we have

$$P(X_n \le x) = \frac{\lfloor nx \rfloor}{n}.$$

Hence, $|P(X_n \le x) - P(X \le x)| \le 1/n$ for all n. Because $\lim_{n \to \infty} 1/n = 0$, we do indeed have $X_n \to X$ in distribution. ∎

Example 4.4.4
Suppose $X_1, X_2, \ldots$ are i.i.d. with finite mean μ, and $M_n = (X_1 + \cdots + X_n)/n$. Then the weak law of large numbers says that for any $\epsilon > 0$, we have

$$P(M_n \le \mu - \epsilon) \to 0 \text{ and } P(M_n \le \mu + \epsilon) \to 1$$

as $n \to \infty$. It follows that $\lim_{n \to \infty} P(M_n \le x) = P(M \le x)$ for any $x \ne \mu$, where M is the constant random variable $M = \mu$. Hence, $M_n \to M$ in distribution. Note that it is *not* necessarily the case that $P(M_n \leq \mu) \to P(M \leq \mu) = 1$. However, this does not contradict the definition of convergence in distribution, because $P(M = \mu) \neq 0$, so we do not need to worry about the case $x = \mu$. ∎

Example 4.4.5 *Poisson Approximation to the Binomial*
Suppose $X_n \sim$ Binomial$(n, \lambda/n)$ and $X \sim$ Poisson(λ). We have seen in Example 2.3.6 that,

$$P(X_n = j) = \binom{n}{j} \left(\frac{\lambda}{n}\right)^j (1 - \frac{\lambda}{n})^{n-j} \to e^{-\lambda} \frac{\lambda^j}{j!}$$

as $n \to \infty$. This implies that $F_{X_n}(x) \to F_X(x)$ at every point $x \notin \{0, 1, 2, \ldots\}$ and these are precisely the points for which $P(X = x) = 0$. Therefore, $\{X_n\}$ converges in distribution to X. (Indeed, this was our original motivation for the Poisson distribution.) ∎

Many more examples of convergence in distribution are given by the central limit theorem, discussed in the next section. We first pause to consider the relationship of convergence in distribution to our previous notions of convergence.

Theorem 4.4.1 If $X_n \xrightarrow{P} X$, then $X_n \xrightarrow{D} X$.

Proof: See Section 4.7 for the proof of this result. ∎

The converse to Theorem 4.4.1 is false. Indeed, the fact that X_n converges in distribution to X says nothing about the underlying *relationship* between X_n and X, it only says something about their distributions. The following example illustrates this.

Example 4.4.6
Suppose $X, X_1, X_2, \ldots$ are i.i.d., each equal to ± 1 with probability $\frac{1}{2}$ each. In this case, $P(X_n \leq x) = P(X \leq x)$ for all n and for all $x \in R^1$, so of course X_n converges in distribution to X. On the other hand, because X and X_n are independent,

$$P(|X - X_n| \geq 2) = \frac{1}{2}$$

for all n, which does *not* go to 0 as $n \to \infty$. Hence, X_n does *not* converge to X in probability (or with probability 1). So we can have convergence in distribution without having convergence in probability or convergence with probability 1. ∎

The following result, stated without proof, indicates how moment-generating functions can be used to check for convergence in distribution. (This generalizes Theorem 3.4.6.)

Theorem 4.4.2 Let X be a random variable, such that for some $s_0 > 0$, we have $m_X(s) < \infty$ whenever $s \in (-s_0, s_0)$. If $Z_1, Z_2, \ldots$ is a sequence of random variables with $m_{Z_n}(s) < \infty$ and $\lim_{n \to \infty} m_{Z_n}(s) = m_X(s)$ for all $s \in (-s_0, s_0)$, then $\{Z_n\}$ converges to X in distribution.

We will make use of this result to prove one of the most famous theorems of probability — the central limit theorem.

Finally, we note that combining Theorem 4.4.1 with Theorem 4.3.1 reveals the following.

Corollary 4.4.1 If $X_n \to X$ with probability 1, then $X_n \overset{D}{\to} X$.

4.4.1   The Central Limit Theorem

We now present the central limit theorem, one of the most important results in all of probability theory. Intuitively, it says that a large sum of i.i.d. random variables, properly normalized, will always have approximately a *normal* distribution. This shows that the normal distribution is extremely fundamental in probability and statistics — even though its density function is complicated and its cumulative distribution function is intractable.

Suppose $X_1, X_2, \ldots$ is an i.i.d. sequence of random variables each having finite mean μ and finite variance σ^2. Let $S_n = X_1 + \cdots + X_n$ be the sample sum and $M_n = S_n/n$ be the sample mean. The central limit theorem is concerned with the distribution of the random variable

$$Z_n = \frac{S_n - n\mu}{\sqrt{n}\sigma} = \frac{M_n - \mu}{\sigma/\sqrt{n}} = \sqrt{n}\left(\frac{M_n - \mu}{\sigma}\right)$$

where $\sigma = \sqrt{\sigma^2}$. We know $E(M_n) = \mu$ and $Var(M_n) = \sigma^2/n$ and this implies that $E(Z_n) = 0$ and $\text{Var}(Z_n) = 1$. The variable Z_n is thus obtained from the sample mean (or sample sum) by subtracting off its mean and dividing by its standard deviation. This transformation is referred to as *standardizing* a random variable, so that it has mean 0 and variance 1. Therefore, Z_n is the standardized version of the sample mean (sample sum).

Note that the distribution of Z_n shares two characteristics with the $N(0,1)$ distribution, namely, it has mean 0 and variance 1. The central limit theorem shows that there is an even stronger relationship.

Theorem 4.4.3 (*The central limit theorem*) Let $X_1, X_2, \ldots$ be i.i.d. with finite mean μ and finite variance σ^2. Let $Z \sim N(0,1)$. Then as $n \to \infty$, the sequence $\{Z_n\}$ converges in distribution to Z, i.e., $Z_n \overset{D}{\to} Z$.

Proof: See Section 4.7 for the proof of this result. ∎

The central limit theorem is so important that we shall restate its conclusions in several different ways.

Corollary 4.4.2 For each fixed $x \in R^1$, $\lim_{n\to\infty} P(Z_n \le x) = \Phi(x)$, where Φ is the cumulative distribution function for the standard normal distribution.

We can write this as follows.

Corollary 4.4.3 For each fixed $x \in R^1$,

$$\lim_{n\to\infty} P(S_n \le n\mu + x\sqrt{n}\sigma) = \Phi(x) \text{ and } \lim_{n\to\infty} P(M_n \le \mu + x\sigma/\sqrt{n}) = \Phi(x).$$

In particular, S_n is approximately equal to $n\mu$, with deviations from this value of order $\sqrt{n}$, and M_n is approximately equal to μ, with deviations from this value of order $1/\sqrt{n}$.

We note that it is not essential in the central limit theorem to divide by σ. Without doing so, the theorem asserts instead that $(S_n - n\mu)/\sqrt{n}$ (or $\sqrt{n}(M_n - n\mu)$) converges in distribution to the $N(0, \sigma^2)$ distribution. That is, the limiting distribution will still be normal but will have variance σ^2 instead of variance 1.

Similarly, instead of dividing by exactly σ, it suffices to divide by any quantity σ_n, provided $\sigma_n \overset{a.s.}{\to} \sigma$. A simple modification of the proof of Theorem 4.4.2 leads to the following result.

Corollary 4.4.4 If

$$Z_n^* = \frac{S_n - n\mu}{\sqrt{n}\sigma_n} = \frac{M_n - \mu}{\sigma_n/\sqrt{n}} = \sqrt{n}\left(\frac{M_n - \mu}{\sigma_n}\right).$$

and $\lim_{n\to\infty} \sigma_n \overset{a.s.}{\to} \sigma$, then $Z_n^* \overset{D}{\to} Z$ as $n \to \infty$.

To illustrate the central limit theorem, we consider a simulation experiment.

Example 4.4.7 *The Central Limit Theorem Illustrated in a Simulation*
Suppose we generate a sample $X_1, \ldots, X_n$ from the Uniform$(0, 1)$ density. Note that the Uniform$(0, 1)$ density is completely unlike a normal density. An easy calculation shows that when $X \sim$ Uniform$(0, 1)$, then $E(X) = 1/2$ and Var$(X) = 1/12$.

Now suppose we are interested in the distribution of the sample average $M_n = S_n/n = (X_1 + \cdots + X_n)/n$ for various choices of n. The central limit theorem tells us that

$$Z_n = \frac{S_n - n/2}{\sqrt{n/12}} = \sqrt{n}\left(\frac{M_n - 1/2}{\sqrt{1/12}}\right)$$

converges in distribution to an $N(0, 1)$ distribution. But how large does n have to be for this approximation to be accurate?

To assess this, we ran a Monte Carlo simulation experiment. In Figure 4.4.1, we have plotted a density histogram of $N = 10^5$ values from the $N(0, 1)$ distribution based on 800 subintervals of $(-4, 4)$, each of length $l = 0.01$. Density histograms are more extensively discussed in Section 5.4.3, but for now we note that above each interval we have plotted the proportion of sampled values that fell in the interval, divided by the length of the interval. As we increase N and decrease l, these histograms will look more and more like the density of the distribution we are sampling from. Indeed, Figure 4.4.1 looks very much like an $N(0, 1)$ density, as it should.

In Figure 4.4.2, we have plotted a density histogram (using the same values of N and l) of Z_1. Note that

$$Z_1 \sim \text{Uniform}\left(-\sqrt{12}/2, \sqrt{12}/2\right)$$

and indeed the histogram does look like a uniform density. Figure 4.4.3 presents
a density histogram of Z_2, and this still looks very nonnormal, but note that
the histogram of Z_3 in Figure 4.4.4 is beginning to look more like a normal
distribution. The histogram of Z_{10} in Figure 4.4.5 looks very normal. In fact,
the proportion of Z_{10} values in $(-\infty, 1.96]$, for this histogram, equals 0.9759,
while the exact proportion for an $N(0, 1)$ distribution is 0.9750.

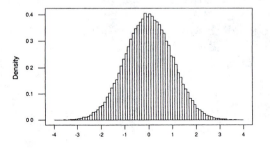

Figure 4.4.1: Density histogram of 10^5 standard normal values.

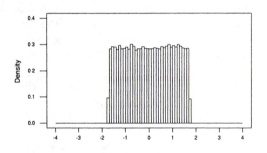

Figure 4.4.2: Density histogram for 10^5 values of Z_1 in Example 4.4.7.

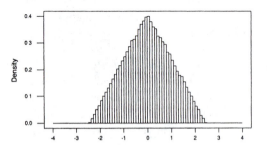

Figure 4.4.3: Density histogram for 10^5 values of Z_2 in Example 4.4.7.

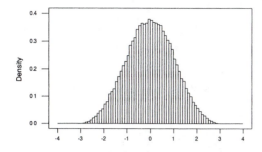

Figure 4.4.4: Density histogram for 10^5 values of Z_3 in Example 4.4.7.

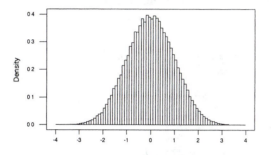

Figure 4.4.5: Density histogram for 10^5 values of Z_{10} in Example 4.4.7.

So in this example, the central limit theorem has taken effect very quickly, even though we are sampling from a very nonnormal distribution. As it turns out, it is primarily the tails of a distribution that determine how large n has to be for the central limit theorem approximation to be accurate. When a distribution has tails no longer than a normal distribution, we can expect the approximation to be quite accurate for relatively small sample sizes. ∎

We consider some further applications of the central limit theorem.

Example 4.4.8
For example, suppose $X_1, X_2, \ldots$ are i.i.d. random variables, each with the Poisson(5) distribution. Recall that this implies that $\mu = E(X_i) = 5$ and $\sigma^2 = \text{Var}(X_i) = 5$. Hence, for each fixed $x \in R^1$, we have

$$P\left(S_n \leq 5n + x\sqrt{5n}\right) \to \Phi(x)$$

as $n \to \infty$. ∎

Example 4.4.9 *Normal Approximation to the Binomial Distribution*
Suppose $X_1, X_2, \ldots$ are i.i.d. random variables, each with the Bernoulli(θ) distribution. Recall that this implies that $E(X_i) = \theta$ and $v = \text{Var}(X_i) = \theta(1 - \theta)$.

Hence, for each fixed $x \in R^1$, we have

$$P\left(S_n \leq n\theta + x\sqrt{n\theta(1-\theta)}\right) \rightarrow \Phi(x), \tag{4.4.1}$$

as $n \rightarrow \infty$.

But now note that we have previously shown that $Y_n = S_n \sim \text{Binomial}(n, \theta)$. So (4.4.1) implies that whenever we have a random variable $Y_n \sim \text{Binomial}(n, \theta)$, then

$$P\left(Y_n \leq y\right) = P\left(\frac{Y_n - n\theta}{\sqrt{n\theta(1-\theta)}} \leq \frac{y - n\theta}{\sqrt{n\theta(1-\theta)}}\right) \approx \Phi\left(\frac{y - n\theta}{\sqrt{n\theta(1-\theta)}}\right) \tag{4.4.2}$$

for large n.

Note that we are approximating a discrete distribution by a continuous distribution here. Reflecting this, a small improvement is often made to (4.4.2) when y is a nonnegative integer. Instead, we use

$$P\left(Y_n \leq y\right) \approx \Phi\left(\frac{y + 0.5 - n\theta}{\sqrt{n\theta(1-\theta)}}\right).$$

Adding 0.5 to y is called the *correction for continuity*. In effect this allocates all the relevant normal probability in the interval $(y - 0.5, y + 0.5)$ to the nonnegative integer y. This has been shown to improve the approximation (4.4.2). ∎

Example 4.4.10 *Approximating Probabilities Using the Central Limit Theorem* While there are tables for the binomial distribution (Table D.6) we often have to compute binomial probabilities for situations the tables do not cover. We can always use statistical software for this and, in fact, such software makes use of the normal approximation we derived from the central limit theorem.

For example, suppose that we have a biased coin, where the probability of getting a head on a single toss is $\theta = 0.6$, and we will toss the coin $n = 1000$ times, and we wish to calculate the probability of getting at least 550 heads and no more than 625 heads. If Y denotes the number of heads obtained in the 1000 tosses, we have that $Y \sim \text{Binomial}(1000, 0.6)$, so

$$E(Y) = 1000(0.6) = 600,$$
$$\text{Var}(Y) = 1000(0.6)(0.4) = 240.$$

Therefore, using the correction for continuity and Table D.2,

$$P\left(550 \leq Y \leq 625\right) = P\left(550 - 0.5 \leq Y \leq 625 + 0.5\right)$$
$$= P\left(\frac{549.5 - 600}{\sqrt{240}} \leq \frac{Y - 600}{\sqrt{240}} \leq \frac{625.5 - 600}{\sqrt{240}}\right)$$
$$= P\left(-3.2598 \leq \frac{Y - 600}{\sqrt{240}} \leq 1.646\right)$$
$$\approx \Phi(1.65) - \Phi(-3.26) = .9505 - .0006 = 0.9499.$$

Note that it would be impossible to compute this probability using the formulas for the binomial distribution. ∎

One of the most important uses of the central limit theorem is that it leads to a method for assessing the error in an average when this is estimating or approximating some quantity of interest. We now discuss this.

4.4.2 The Central Limit Theorem and Assessing Error

Suppose $X_1, X_2, \ldots$ is an i.i.d. sequence of random variables, each with finite mean μ and finite variance σ^2, and we are using the sample average M_n to approximate the mean μ. This situation arises commonly in many computational (see Section 4.5) and statistical (see Chapter 6) problems. In such a context, we can generate the X_i but we don't know the value of μ.

If we approximate μ by M_n, then a natural question to ask is: How much error is there in the approximation? The central limit theorem tells us that

$$
\begin{aligned}
\Phi(3) - \Phi(-3) &= \lim_{n \to \infty} P\left(-3 < \frac{M_n - \mu}{\sigma/\sqrt{n}} < 3\right) \\
&= \lim_{n \to \infty} P\left(M_n - 3\frac{\sigma}{\sqrt{n}} < \mu < M_n + 3\frac{\sigma}{\sqrt{n}}\right).
\end{aligned}
$$

Using Table D.2 (or statistical software), we have that $\Phi(3) - \Phi(-3) = 0.9987 - (1 - 0.9987) = 0.9974$. So, for large n, we have that the interval

$$
\left(M_n - 3\sigma/\sqrt{n}, M_n + 3\sigma/\sqrt{n}\right)
$$

contains the unknown value of μ with virtual certainty (actually with probability about .9974). Therefore, the half-length $3\sigma/\sqrt{n}$ of this interval gives us an assessment of the error in the approximation M_n. Note that $\text{Var}(M_n) = \sigma^2/n$ so the half-length of the interval equals three standard deviations of the estimate M_n.

Because we don't know μ, it is extremely unlikely that we will know σ (as its definition uses μ). But if we can find a consistent estimate σ_n of σ, then we can use Corollary 4.4.4 instead to construct such an interval.

As it turns out, the correct choice of σ_n depends on what we know about the sampling distribution we are sampling from (see Chapter 6 for more discussion of this). For example, if $X_1 \sim \text{Bernoulli}(\theta)$, then $\mu = \theta$ and $\sigma^2 = \text{Var}(X_1) = \theta(1 - \theta)$. By the strong law of large numbers (Theorem 4.3.2), $M_n \overset{a.s.}{\to} \mu = \theta$ and so

$$
\sigma_n = \sqrt{M_n(1 - M_n)} \overset{a.s.}{\to} \sqrt{\theta(1 - \theta)} = \sigma.
$$

Then, using the same argument as above, we have that, for large n, the interval

$$
\left(M_n - 3\sqrt{M_n(1 - M_n)/n}, M_n + 3\sqrt{M_n(1 - M_n)/n}\right) \tag{4.4.3}
$$

contains the true value of θ with virtual certainty (again with probability about .9974). The half-length of (4.4.3) is a measure of the accuracy of the estimate

M_n — notice that this can be computed from the values $X_1, \ldots, X_n$. We refer to the quantity $(M_n (1 - M_n)/n)^{1/2}$ as the *standard error* of the estimate M_n.

For a general random variable X_1, let

$$
\sigma_n^2 \;=\; \frac{1}{n-1} \sum_{i=1}^n (X_i - M_n)^2 = \frac{1}{n-1} \left(\sum_{i=1}^n X_i^2 - 2M_n \sum_{i=1}^n X_i + nM_n^2 \right)
$$

$$
\;=\; \frac{n}{n-1} \left(\frac{1}{n} \sum_{i=1}^n X_i^2 - 2M_n^2 + M_n^2 \right) = \frac{n}{n-1} \left(\frac{1}{n} \sum_{i=1}^n X_i^2 - M_n^2 \right).
$$

By the strong law of large numbers, we have that $M_n \overset{a.s.}{\to} \mu$, and

$$
\frac{1}{n} \sum_{i=1}^n X_i^2 \overset{a.s.}{\to} E\left(X_1^2\right) = \sigma^2 + \mu^2.
$$

Because $n/(n-1) \to 1$ and $M_n^2 \overset{a.s.}{\to} \mu^2$ as well, we conclude that $\sigma_n^2 \overset{a.s.}{\to} \sigma^2$. This implies that $\sigma_n \overset{a.s.}{\to} \sigma$ and so σ_n is consistent for σ. It is common to call σ_n^2 the *sample variance* of the sample $X_1, \ldots, X_n$. When the sample size n is fixed we will often denote this estimate of the variance by S^2.

Again, using the above argument we have that, for large n, the interval

$$
\left(M_n - 3\sigma_n/\sqrt{n}, M_n + 3\sigma_n/\sqrt{n}\right) = \left(M_n - 3S/\sqrt{n}, M_n + 3S/\sqrt{n}\right) \qquad (4.4.4)
$$

contains the true value of μ with virtual certainty (also with probability about .9974). Therefore, the half-length is a measure of the accuracy of the estimate M_n — notice that this can be computed from the values $X_1, \ldots, X_n$. The quantity $S/\sqrt{n}$ is referred to as the *standard error* of the estimate M_n.

We will make use of these estimates of the error in approximations in the following section.

Summary of Section 4.4

- A sequence $\{X_n\}$ of random variables converges in distribution to Y if, for all $y \in R^1$ with $P(Y = y) = 0$, we have $\lim_{n\to\infty} F_{X_n}(y) = F_Y(y)$, i.e., $\lim_{n\to\infty} P(X_n \le y) = P(Y \le y)$.

- If $\{X_n\}$ converges to Y in probability (or with probability 1), then $\{X_n\}$ converges to Y in distribution.

- The very important central limit theorem says that if $\{X_n\}$ are i.i.d. with finite mean μ and variance σ^2, then the random variables $Z_n = (S_n - n\mu)/\sqrt{n}\sigma$ converge in distribution to a standard normal distribution.

- The central limit theorem allows us to approximate various distributions by normal distributions, which is helpful in simulation experiments and in many other contexts. Table D.2 (or any statistical software package) provides values for the cumulative distribution function of a standard normal.

Exercises

4.4.1 Suppose $P(X_n = i) = (n+i)/(3n+6)$, for $i = 1, 2, 3$. Suppose also that $P(X = i) = 1/3$, for $i = 1, 2, 3$. Prove that $\{X_n\}$ converges in distribution to X.

4.4.2 Suppose $P(Y_n = k) = (1 - 2^{-n-1})^{-1}/2^{k+1}$ for $k = 0, 1, \ldots, n$. Let $Y \sim \text{Geometric}(1/2)$. Prove that $\{Y_n\}$ converges in distribution to Y.

4.4.3 Let Z_n have density $(n+1)x^n$ for $0 < x < 1$, and 0 otherwise. Let $Z = 1$. Prove that $\{Z_n\}$ converges in distribution to Z.

4.4.4 Let W_n have density

$$\frac{1 + x/n}{1 + 1/2n}$$

for $0 < x < 1$, and 0 otherwise. Let $W \sim \text{Uniform}[0, 1]$. Prove that $\{W_n\}$ converges in distribution to W.

4.4.5 Let $Y_1, Y_2, \ldots$ be i.i.d. with distribution Exponential(3). Use the central limit theorem and Table D.2 (or software) to estimate the probability $P(\sum_{i=1}^{1600} Y_i \leq 540)$.

4.4.6 Let $Z_1, Z_2, \ldots$ be i.i.d. with distribution Uniform$[-20, 10]$. Use the central limit theorem and Table D.2 (or software) to estimate the probability $P(\sum_{i=1}^{900} Z_i \geq -4470)$.

4.4.7 Let $X_1, X_2, \ldots$ be i.i.d. with distribution Geometric(1/4). Use the central limit theorem and Table D.2 (or software) to estimate the probability $P(\sum_{i=1}^{800} X_i \geq 2450)$.

4.4.8 Suppose $X_n \sim N(0, 1/n)$, i.e., X_n has a normal distribution with mean 0 and variance $1/n$. Does the sequence $\{X_n\}$ converge in distribution to some random variable? If yes, what is the distribution of the random variable?

Computer Exercises

4.4.9 Generate N samples $X_1, X_2, \ldots, X_{20} \sim \text{Exponential}(3)$ for N large ($N = 10^4$ if possible). Use these samples to estimate the probability $P(1/6 \leq M_{20} \leq 1/2)$. How does your answer compare to what the central limit theorem gives as an approximation?

4.4.10 Generate N samples $X_1, X_2, \ldots, X_{30} \sim \text{Uniform}[-20, 10]$ for N large ($N = 10^4$ if possible). Use these samples to estimate the probability $P(M_{30} \leq -5)$. How does your answer compare to what the central limit theorem gives as an approximation?

4.4.11 Generate N samples $X_1, X_2, \ldots, X_{20} \sim \text{Geometric}(1/4)$ for N large ($N = 10^4$ if possible). Use these samples to estimate the probability $P(2.5 \leq M_{20} \leq 3.3)$. How does your answer compare to what the central limit theorem gives as an approximation?

4.4.12 Generate N samples $X_1, X_2, \ldots, X_{20}$ from the distribution of $\log Z$ where $Z \sim \text{Gamma}(4,1)$ for N large ($N = 10^4$ if possible). Use these samples to construct a density histogram of the values of M_{20}. Comment on the shape of this graph.

4.4.13 Generate N samples $X_1, X_2, \ldots, X_{20}$ from the Binomial$(10, 0.01)$ distribution for N large ($N = 10^4$ if possible). Use these samples to construct a density histogram of the values of M_{20}. Comment on the shape of this graph.

Problems

4.4.14 Let $a_1, a_2, \ldots$ be any sequence of nonnegative real numbers with $\sum_i a_i = 1$. Suppose $P(X = i) = a_i$ for every positive integer i. Construct a sequence $\{X_n\}$ of *absolutely continuous* random variables, such that $X_n \to X$ in distribution.

4.4.15 Let $f : [0,1] \to (0, \infty)$ be a continuous positive function such that $\int_0^1 f(x)\, dx = 1$. Consider random variables X and $\{X_n\}$ such that $P(a \leq X \leq b) = \int_a^b f(x)\, dx$ for $a \leq b$, and

$$P\left(X_n = \frac{i}{n}\right) = \frac{f(i/n)}{\sum_{j=1}^n f(j/n)}$$

for $i = 1, 2, 3, \ldots, n$. Prove that $X_n \to X$ in distribution.

4.4.16 Suppose that $Y_i = X_i^3$ and that $X_1, \ldots, X_n$ is a sample from an $N(\mu, \sigma^2)$ distribution. Indicate how you would approximate the probability $P(M_n \leq m)$ where $M_n = (Y_1 + \cdots + Y_n)/n$.

4.4.17 Suppose that $Y_i = \cos(2\pi U_i)$ and that $U_1, \ldots, U_n$ is a sample from the Uniform$[0,1]$ distribution. Indicate how you would approximate the probability $P(M_n \leq m)$ where $M_n = (Y_1 + \cdots + Y_n)/n$.

Computer Problems

4.4.18 Suppose that $Y = X^3$ and $X \sim N(0,1)$. By generating a large sample ($n = 10^4$ if possible) from the distribution of Y, approximate the probability $P(Y \leq 1)$ and assess the error in your approximation. Compute this probability exactly and compare it with your approximation.

4.4.19 Suppose that $Y = X^3$ and $X \sim N(0,1)$. By generating a large sample ($n = 10^4$ if possible) from the distribution of Y, approximate the expectation $E(\cos(X^3))$ and assess the error in your approximation.

Challenge

4.4.20 Suppose $X_n \to C$ in distribution, where C is a constant. Prove that $X_n \to C$ in probability. (This proves that if X is *constant*, then the converse to Theorem 4.4.1 *does* hold, even though it does not hold for general X.)

4.5 Monte Carlo Approximations

The laws of large numbers say that if $X_1, X_2, \ldots$ is an i.i.d. sequence of random variables with mean μ, and

$$M_n = \frac{X_1 + \cdots + X_n}{n},$$

then for large n we will have $M_n \approx \mu$.

Suppose now that μ is *unknown*. Then, as discussed in Section 4.4.2, it is possible to change perspective and use M_n (for large n) as an *estimator* or approximation of μ. Any time we approximate or estimate a quantity, we must also say something about how much error there is in the estimate. Of course, we cannot say what this error is exactly, as that would require us knowing the exact value of μ. In Section 4.4.2, however, we showed how the central limit theorem leads to a very natural approach to assessing this error, using three times the standard error of the estimate. We consider some examples.

Example 4.5.1
Consider flipping a sequence of identical coins, each of which has probability θ of coming up heads, but where θ is unknown. Let M_n again be the fraction of the first n coins that are heads. Then we know that for large n, it is very likely that M_n is very close to θ. Hence, we can use M_n to *estimate* θ. Further the discussion in Section 4.4.2 indicates that (4.4.3) is the relevant interval to quote when assessing how accurate the estimate M_n is. ∎

Example 4.5.2
Suppose we believe a certain medicine lowers blood pressure, but we do not know by how much. We would like to know the mean amount μ, by which this medicine lowers blood pressure.

Suppose we observe n patients (chosen at random so they are i.i.d.), where patient i has blood pressure B_i before taking the medicine, and blood pressure A_i afterwards. Let $X_i = B_i - A_i$ and then

$$M_n = \frac{1}{n} \sum_{i=1}^{n} (B_i - A_i)$$

is the average amount of blood pressure decrease. (Note that $B_i - A_i$ may be negative for some patients, and it is important to also include those negative terms in the sum.) Then for large n, the value of M_n is a good *estimate* of $E(X_i) = \mu$. Further the discussion in Section 4.4.2 indicates that (4.4.4) is the relevant interval to quote when assessing how accurate the estimate M_n is. ∎

Such estimators can also be used to estimate purely mathematical quantities that do not involve any experimental data (such as coins or medical patients), but that are too difficult to compute directly. In this case, such estimators are called *Monte Carlo approximations* (named after the gambling casino in the principality of Monaco, because they introduce randomness to solve nonrandom problems).

Example 4.5.3

Suppose we wish to evaluate

$$I = \int_0^1 \cos(x^2) \sin(x^4)\, dx.$$

This integral cannot easily be solved exactly. But it can be approximately computed using a Monte Carlo approximation, as follows.

We note that

$$I = E(\cos(U^2) \sin(U^4)),$$

where $U \sim \text{Uniform}[0,1]$. Hence, for large n, the integral I is approximately equal to $M_n = (T_1 + \cdots + T_n)/n$, where $T_i = \cos(U_i^2) \sin(U_i^4)$, and where $U_1, U_2, \ldots$ are i.i.d. Uniform$[0,1]$.

Putting this all together, we obtain an algorithm for approximating the integral I, as follows.

1. Select a large positive integer n.

2. Obtain $U_i \sim \text{Uniform}[0,1]$, independently for $i = 1, 2, \ldots, n$.

3. Set $T_i = \cos(U_i^2) \sin(U_i^4)$, for $i = 1, 2, \ldots, n$.

4. Estimate I by $M_n = (T_1 + \cdots + T_n)/n$.

For large enough n, this algorithm will provide a good estimate of the integral I.

For example, the following table records the estimates M_n and the intervals (4.4.4) based on samples of Uniform[0,1] variables for various choices of n.

n	M_n	$M_n - 3S/\sqrt{n}$	$M_n + 3S/\sqrt{n}$
10^3	0.145294	0.130071	0.160518
10^4	0.138850	0.134105	0.143595
10^5	0.139484	0.137974	0.140993

From this we can see that the value of I is approximately 0.139484, and the true value is almost certainly in the interval $(0.137974, 0.140993)$. Notice how the lengths of the intervals decrease as we increase n. In fact, it can be shown that the exact value is $I = 0.139567$, so our approximation is excellent. ∎

Example 4.5.4

Suppose we want to evaluate the integral

$$I = \int_0^\infty 25x^2 \cos(x^2) e^{-25x}\, dx.$$

This integral cannot easily be solved exactly, but it can also be approximately computed using a Monte Carlo approximation, as follows.

We note first that $I = E(X^2 \cos(X^2))$ where $X \sim \text{Exponential}(25)$. Hence, for large n, the integral I is approximately equal to $M_n = (T_1 + \cdots + T_n)/n$, where $T_i = X_i^2 \cos(X_i^2)$, with $X_1, X_2, \ldots$ i.i.d. Exponential(25).

Now, we know from Section 2.10 that we can simulate $X \sim \text{Exponential}(25)$ by setting $X = -\ln(U)/25$ where $U \sim \text{Uniform}[0,1]$. Hence, putting this all together, we obtain an algorithm for approximating the integral I, as follows.

1. Select a large positive integer n.

2. Obtain $U_i \sim \text{Uniform}[0,1]$, independently for $i = 1, 2, \ldots, n$.

3. Set $X_i = -\ln(U_i)/25$, for $i = 1, 2, \ldots, n$.

4. Set $T_i = X_i^2 \cos(X_i^2)$, for $i = 1, 2, \ldots, n$.

5. Estimate I by $M_n = (T_1 + \cdots + T_n)/n$.

For large enough n, this algorithm will provide a good estimate of the integral I.

For example, the following table records the estimates M_n and the intervals (4.4.4) based on samples of Exponential(25) variables for various choices of n.

n	M_n	$M_n - 3S/\sqrt{n}$	$M_n + 3S/\sqrt{n}$
10^3	3.33846×10^{-3}	2.63370×10^{-3}	4.04321×10^{-3}
10^4	3.29933×10^{-3}	3.06646×10^{-3}	3.53220×10^{-3}
10^5	3.20629×10^{-3}	3.13759×10^{-3}	3.27499×10^{-3}

From this we can see that the value of I is approximately 3.20629×10^{-3} and the true value is almost certainly in the interval $\left(3.13759 \times 10^{-3}, 3.27499 \times 10^{-3}\right)$. ∎

Example 4.5.5
Suppose we want to evaluate the sum

$$S = \sum_{j=0}^{\infty} (j^2 + 3)^{-7} 5^{-j}.$$

This sum is very difficult to compute directly, but it can be approximately computed using a Monte Carlo approximation.

Let us rewrite the sum as

$$S = \left(\frac{5}{4}\right) \sum_{j=0}^{\infty} (j^2 + 3)^{-7} \left(\frac{4}{5}\right) \left(1 - \frac{4}{5}\right)^j.$$

We then see that $S = (5/4) E((X^2 + 3)^{-7})$ where $X \sim \text{Geometric}(4/5)$.

Now, we know from Section 2.10 that we can simulate $X \sim \text{Geometric}(4/5)$ by setting $X = \lfloor \ln(1-U)/\ln(1-4/5) \rfloor$ or, equivalently, $X = \lfloor \ln(U)/\ln(1-4/5) \rfloor$, where $U \sim \text{Uniform}[0,1]$ and where $\lfloor \cdot \rfloor$ means to round down to the next integer value. Hence, we obtain an algorithm for approximating the sum S, as follows.

1. Select a large positive integer n.

2. Obtain $U_i \sim \text{Uniform}[0,1]$, independently for $i = 1, 2, \ldots, n$.

3. Set $X_i = \lfloor \ln(U_i)/\ln(1 - 4/5) \rfloor$, for $i = 1, 2, \ldots, n$.

4. Set $T_i = (X_i^2 + 3)^{-7}$, for $i = 1, 2, \ldots, n$.

5. Estimate S by $M_n = (5/4)(T_1 + \cdots + T_n)/n$.

For large enough n, this algorithm will provide a good estimate of the sum S.

For example, the following table records the estimates M_n and the intervals (4.4.4) based on samples of Geometric(4/5) variables for various choices of n.

n	M_n	$M_n - 3S/\sqrt{n}$	$M_n + 3S/\sqrt{n}$
10^3	4.66773×10^{-4}	4.47078×10^{-4}	4.86468×10^{-4}
10^4	4.73538×10^{-4}	4.67490×10^{-4}	4.79586×10^{-4}
10^5	4.69377×10^{-4}	4.67436×10^{-4}	4.71318×10^{-4}

From this we can see that the value of S is approximately 4.69377×10^{-4} and the true value is almost certainly in the interval $\left(4.67436 \times 10^{-4}, 4.71318 \times 10^{-4}\right)$. ∎

Note that when using a Monte Carlo approximation, it is not necessary that the range of an integral or sum be the entire range of the corresponding random variable, as follows.

Example 4.5.6
Suppose we want to evaluate the integral

$$J = \int_0^\infty \sin(x) e^{-x^2/2} \, dx.$$

Again, this is extremely difficult to evaluate exactly.

Here

$$J = \sqrt{2\pi} \, E(\sin(X) I_{\{X > 0\}}),$$

where $X \sim N(0, 1)$, and $I_{\{X > 0\}}$ is the indicator function of the event $\{X > 0\}$. We know from Section 2.10 that we can simulate $X \sim N(0, 1)$ by setting

$$X = \sqrt{2 \log(1/U)} \, \cos(2\pi V),$$

where U and V are i.i.d. Uniform[0, 1]. Hence, we obtain the following algorithm for approximating the integral J.

1. Select a large positive integer n.

2. Obtain $U_i, V_i \sim \text{Uniform}[0, 1]$, independently for $i = 1, 2, \ldots, n$.

3. Set $X_i = \sqrt{2 \log(1/U_i)} \, \cos(2\pi V_i)$, for $i = 1, 2, \ldots, n$.

4. Set $T_i = \sin(X_i) I_{\{X_i > 0\}}$, for $i = 1, 2, \ldots, n$. (That is, set $T_i = \sin(X_i)$ if $X_i > 0$, otherwise set $T_i = 0$.)

5. Estimate J by $M_n = \sqrt{2\pi}(T_1 + \cdots + T_n)/n$.

For large enough n, this algorithm will again provide a good estimate of the integral I.

For example, the following table records the estimates M_n and the intervals (4.4.4) based on samples of $N(0, 1)$ variables for various choices of n.

n	M_n	$M_n - 3S/\sqrt{n}$	$M_n + 3S/\sqrt{n}$
10^3	0.744037	0.657294	0.830779
10^4	0.733945	0.706658	0.761233
10^5	0.722753	0.714108	0.731398

From this we can see that the value of J is approximately 0.722753, and the true value is almost certainly in the interval $(0.714108, 0.731398)$. ∎

Now we consider an important problem for statistical applications of probability theory.

Example 4.5.7 *Approximating Sampling Distributions Using Monte Carlo.*
Suppose that $X_1, X_2, \ldots, X_n$ is an i.i.d. sequence with distribution given by probability measure P and we want to find the distribution of a new random variable $Y = h(X_1, X_2, \ldots, X_n)$ for some function h. Provided we can generate from P, then Monte Carlo methods give us a method for approximating this distribution.

Denoting the cumulative distribution function of Y by F_Y, we have

$$F_Y(y) = P((-\infty, y]) = E_{P_Y}\left(I_{(-\infty, y]}(Y)\right) = E\left(I_{(-\infty, y]}(h(X_1, X_2, \ldots, X_n))\right).$$

So $F_Y(y)$ can be expressed as the expectation of the random variable

$$I_{(-\infty, y]}(h(X_1, X_2, \ldots, X_n))$$

based on sampling from P.

To estimate this, we generate N samples of size n

$$(X_{i1}, X_{i2}, \ldots, X_{in}),$$

for $i = 1, \ldots, N$ from P (note N is the Monte Carlo sample size and can be varied, while the sample size n is fixed here) and then calculate the proportion of values $h(X_{i1}, X_{i2}, \ldots, X_{in}) \in (-\infty, y]$. The estimate M_N is then given by

$$\hat{F}_Y(y) = \frac{1}{N} \sum_{i=1}^{N} I_{(-\infty, y]}(h(X_{i1}, X_{i2}, \ldots, X_{in})).$$

By the laws of large numbers, this converges to $F_Y(y)$ as $N \to \infty$. To evaluate the error in this approximation, we use (4.4.3), which now takes the form

$$\left(\hat{F}_Y(y) - 3\sqrt{\hat{F}_Y(y)\left(1 - \hat{F}_Y(y)\right)/n}, \hat{F}_Y(y) + 3\sqrt{\hat{F}_Y(y)\left(1 - \hat{F}_Y(y)\right)/n}\right).$$

We presented an application of this in Example 4.4.7. Note that if the base of a rectangle in the histogram of Figure 4.4.2 is given by $(a, b]$, then the height of this rectangle equals the proportion of values that fell in $(a, b]$ times $1/(b-a)$. This can be expressed as $(\hat{F}_Y(b) - \hat{F}_Y(a))/(b-a)$, which converges to $(F_Y(b) - F_Y(a))/(b-a)$ as $N \to \infty$. This proves that the areas of the rectangles in the histogram converge to $F_Y(b) - F_Y(a)$ as $N \to \infty$.

More generally, we can approximate an expectation $E(g(Y))$ using the average

$$\frac{1}{N} \sum_{i=1}^{N} g\left(h(X_{i1}, X_{i2}, \ldots, X_{in})\right).$$

By the laws of large numbers, this average converges to $E(g(Y))$ as $N \to \infty$. ∎

Typically, there is more than one possible Monte Carlo algorithm for estimating a quantity of interest. For example, suppose we want to approximate the integral $\int_a^b g(x)\,dx$ where we assume this integral is finite. Let f be a density on the interval (a, b), such that $f(x) > 0$ for every $x \in (a, b)$, and suppose that we have a convenient algorithm for generating $X_1, X_2, \ldots$ i.i.d. with distribution given by f. We have that

$$\int_a^b g(x)\,dx = \int_a^b \frac{g(x)}{f(x)} f(x)\,dx = E\left(\frac{g(X)}{f(X)}\right)$$

when X is distributed with density f. So then we can estimate $\int_a^b g(x)\,dx$ by

$$M_n = \frac{1}{n} \sum_{i=1}^{n} \frac{g(X_i)}{f(X_i)} = \frac{1}{n} \sum_{i=1}^{n} T_i$$

where $T_i = g(X_i)/f(X_i)$. In effect this is what we did in Example 4.5.3 (f is the Uniform$[0, 1]$ density), in Example 4.5.4 (f is the Exponential(25) density), and in Example 4.5.6 (f is the $N(0, 1)$ density). But note that there are many other possible choices. In Example 4.5.3 we could have taken f to be any beta density. In Example 4.5.4 we could have taken f to be any gamma density and similarly in Example 4.5.6. Most statistical computer packages have commands for generating from these distributions. In a given problem, what is the best one to use?

In such a case we would naturally choose to use the algorithm that was most *efficient*. For the algorithms we have been discussing here, this means that if, based on a sample of n, algorithm 1 leads to an estimate with standard error $\sigma_1/\sqrt{n}$, and algorithm 2 leads to an estimate with standard error $\sigma_2/\sqrt{n}$, then algorithm 1 is more efficient than algorithm 2 whenever $\sigma_1 < \sigma_2$. Naturally, we would prefer algorithm 1, because the intervals (4.4.3) or (4.4.4), will tend to be shorter for algorithm 1 for the same sample size. Actually, a more refined comparison of efficiency would also take into account the total amount of computer time used by each algorithm, but we will ignore this aspect of the problem here. See Problem 4.5.14 for more discussion of efficiency and the choice of algorithm in the context of the integration problem.

Summary of Section 4.5

- An unknown quantity can be approximately computed using a Monte Carlo approximation, whereby independent replications of a random experiment (usually on a computer) are averaged to estimate the quantity.

- Monte Carlo approximations can be used to approximate complicated sums, integrals, and sampling distributions, all by choosing the random experiment appropriately.

Exercises

4.5.1 Describe a Monte Carlo approximation of $\int_{-\infty}^{\infty} \cos^2(x)e^{-x^2/2}\,dx$.

4.5.2 Describe a Monte Carlo approximation of $\sum_{j=0}^{m} j^6 \binom{m}{j} 2^j 3^{-m}$. (Hint: Remember the Binomial$(m, 2/3)$ distribution.)

4.5.3 Describe a Monte Carlo approximation of $\int_0^{\infty} e^{-5x-14x^2}\,dx$. (Hint: Remember the Exponential(5) distribution.)

4.5.4 Suppose $X_1, X_2, \ldots$ are i.i.d. with distribution Poisson(λ), where λ is unknown. Consider $M_n = (X_1 + X_2 + \cdots + X_n)/n$ as an estimate of λ. Suppose we know that $\lambda \leq 10$. How large must n be to guarantee that M_n will be within 0.1 of the true value of λ with virtual certainty, i.e., when is 3 standard deviations smaller than 0.1?

4.5.5 Describe a Monte Carlo approximation of $\sum_{j=0}^{\infty} \sin(j^2) 5^j / j!$. Assume you have available an algorithm for generating from the Poisson(5) distribution.

4.5.6 Describe a Monte Carlo approximation of $\int_0^{10} e^{-x^4}\,dx$. (Hint: Remember the Uniform$[0, 10]$ distribution.)

Computer Exercises

4.5.7 Use a Monte Carlo algorithm to approximate $\int_0^1 \cos(x^3)\sin(x^4)\,dx$ based on a large sample (take $n = 10^5$ if possible). Assess the error in the approximation.

4.5.8 Use a Monte Carlo algorithm to approximate $\int_0^{\infty} 25\cos(x^4)e^{-25x}\,dx$ based on a large sample (take $n = 10^5$ if possible). Assess the error in the approximation.

4.5.9 Use a Monte Carlo algorithm to approximate $\sum_{j=0}^{\infty} (j^2+3)^{-5} 5^{-j}$ based on a large sample (take $n = 10^5$ if possible). Assess the error in the approximation.

4.5.10 Suppose $X \sim N(0,1)$. Use a Monte Carlo algorithm to approximate $P(X^2 - 3X + 2 \geq 0)$ based on a large sample (take $n = 10^5$ if possible). Assess the error in the approximation.

Problems

4.5.11 Suppose that $X_1, X_2, \ldots$ are i.i.d. Bernoulli(θ) where θ is unknown.

Determine a lower bound on n so that the probability that the estimate M_n will be within δ of the unknown value of θ, is about .9974. This allows us to run simulations with high confidence that the error in the approximation quoted, is less than some prescribed value δ. (Hint: Use the fact that $x(1-x) \le 1/4$ for all $x \in [0, 1]$.)

4.5.12 Suppose that $X_1, X_2, \ldots$ are i.i.d. with unknown mean μ, and unknown variance σ^2. Suppose we know, however, that $\sigma^2 \le \sigma_0^2$ where σ_0^2 is a known value. Determine a lower bound on n so that the probability that the estimate M_n will be within δ of the unknown value of μ, is about 0.9974. This allows us to run simulations with high confidence that the error in the approximation quoted, is less than some prescribed value δ.

4.5.13 Suppose $X_1, X_2, \ldots$ are i.i.d. with distribution Uniform$[0, \theta]$, where θ is unknown, and consider $Z_n = n^{-1}(n+1) X_{(n)}$ as an estimate of θ (see Section 2.8.4 on order statistics).
(a) Prove that $E(Z_n) = \theta$ and compute Var(Z_n).
(b) Use Chebyshev's inequality to show that Z_n converges in probability to θ.
(c) Show that $E(2M_n) = \theta$ and compare M_n and Z_n with respect to their efficiencies as estimators of θ. Which would you use to estimate θ and why ?

4.5.14 (*Importance sampling*) Suppose we want to approximate the integral $\int_a^b g(x)\,dx$ where we assume this integral is finite. Let f be a density on the interval (a, b), such that $f(x) > 0$ for every $x \in (a, b)$, and is such that we have a convenient algorithm for generating $X_1, X_2, \ldots$ i.i.d. with distribution given by f.
(a) Prove that

$$M_n(f) = \frac{1}{n} \sum_{i=1}^{n} \frac{g(X_i)}{f(X_i)} \overset{a.s.}{\to} \int_a^b g(x)\,dx.$$

(We refer to f as an *importance sampler* and note this shows that every f satisfying the above conditions, provides a consistent estimator $M_n(f)$ of $\int_a^b g(x)\,dx$.)
(b) Prove that

$$\text{Var}(M_n(f)) = \frac{1}{n} \left\{ \int_a^b \frac{g^2(x)}{f(x)}\,dx - \left(\int_a^b g(x)\,dx \right)^2 \right\}.$$

(c) Suppose that $g(x) = h(x)f(x)$ where f is as described above. Show that importance sampling with respect to f leads to the estimator

$$M_n(f) = \frac{1}{n} \sum_{i=1}^{n} h(X_i).$$

(d) Prove that Var$(M_n(f))$ is minimized by taking $f(x) = |g(x)| / \int_a^b |g(x)|\,dx$. Calculate the minimum variance and show that the minimum variance is 0 when $g(x) \ge 0$ for all $x \in (a, b)$. Why is this optimal importance sampler typically not feasible? (The optimal importance sampler does indicate, however, that in our search for an efficient importance sampler, we look for an f that is large when $|g|$ is large and small when $|g|$ is small.)

(e) Show that if there exists c such that $|g(x)| \leq cf(x)$ for all $x \in (a, b)$, then $\text{Var}(M_n(f)) < \infty$.

(f) Determine the standard error of $M_n(f)$ and indicate how you would use this to assess the error in the approximation $M_n(f)$ when $\text{Var}(M_n(f)) < \infty$.

Computer Problems

4.5.15 Use a Monte Carlo algorithm to approximate $P(X^3 + Y^3 \leq 3)$ where $X \sim N(1, 2)$ independently of $Y \sim \text{Gamma}(1, 1)$ based on a large sample (take $n = 10^5$ if possible). Assess the error in the approximation. How large does n have to be to guarantee the estimate is within 0.01 of the true value with virtual certainty? (Hint: Problem 4.5.11.)

4.5.16 Use a Monte Carlo algorithm to approximate $E(X^3 + Y^3)$ where $X \sim N(1, 2)$ independently of $Y \sim \text{Gamma}(1, 1)$ based on a large sample (take $n = 10^5$ if possible). Assess the error in the approximation.

4.5.17 For the integral of Exercise 4.5.3, compare the efficiencies of the algorithm based on generating from an Exponential(5) distribution with that based on generating from an $N(0, 1/7)$ distribution.

Challenge

4.5.18 (*Buffon's needle*) Suppose you drop a needle at random onto a large sheet of lined paper. Assume the distance between the lines is exactly equal to the length of the needle.

(a) Prove that the probability that the needle lands touching a line is equal to $2/\pi$. (Hint: Let D be the distance from the higher end of the needle to the line just below it, and let A be the angle the needle makes with that line. Then what are the distributions of D and A? Under what conditions on D and A will the needle be touching a line?)

(b) Explain how this experiment could be used to obtain a Monte Carlo approximation for the value of π.

Discussion Topics

4.5.19 An integral like $\int_0^\infty x^2 \cos(x^2) e^{-x}\, dx$ can be approximately computed using a numerical integration computer package (e.g., using Simpson's rule). What are some advantages and disadvantages of using a Monte Carlo approximation instead of a numerical integration package?

4.5.20 Carry out the Buffon's needle Monte Carlo experiment described in Challenge 4.5.18, by repeating the experiment at least 20 times. Present the estimate of π so obtained. How close is it to the true value of π? What could be done to make the estimate more accurate?

4.6 Normal Distribution Theory

Because of the central limit theorem (Theorem 4.4.3), the normal distribution plays an extremely important role in statistical theory. For this reason, we shall consider a number of important properties and distributions related to the normal distribution. These properties and distributions will be very important for the statistical theory in later chapters of this book.

We already know that if $X_1 \sim N(\mu_1, \sigma_1^2)$ and $X_2 \sim N(\mu_2, \sigma_2^2)$, then $cX_1 + d \sim N(c\mu_1 + d, c^2\sigma^2)$ (Exercise 2.6.3), and $X_1 + X_2 \sim N(\mu_1 + \mu_2, \sigma_1^2 + \sigma_2^2)$ (Problem 2.9.11). Combining these facts and using induction, we have the following result.

Theorem 4.6.1 Suppose $X_i \sim N(\mu_i, \sigma_i^2)$ for $i = 1, 2, \ldots, n$ and that they are independent random variables. Let $Y = (\sum_i a_i X_i) + b$ for some constants $\{a_i\}$ and b. Then

$$Y \sim N\left(\left(\sum_i a_i \mu_i\right) + b, \sum_i a_i^2 \sigma_i^2 \right).$$

This immediately implies the following.

Corollary 4.6.1 Suppose $X_i \sim N(\mu, \sigma^2)$ for $i = 1, 2, \ldots, n$ and that they are independent random variables. If $\bar{X} = (X_1 + \cdots + X_n)/n$ then $\bar{X} \sim N(\mu, \sigma^2/n)$.

A more subtle property of normal distributions is the following.

Theorem 4.6.2 Suppose $X_i \sim N(\mu_i, \sigma_i^2)$ for $i = 1, 2, \ldots, n$ and also that the $\{X_i\}$ are *independent*. Let $U = \sum_{i=1}^{n} a_i X_i$ and $V = \sum_{i=1}^{n} b_i X_i$, for some constants $\{a_i\}$ and $\{b_i\}$. Then $\mathrm{Cov}(U, V) = \sum_i a_i b_i \sigma_i^2$. Furthermore, $\mathrm{Cov}(U, V) = 0$ if and only if U and V are independent.

Proof: The formula for $\mathrm{Cov}(U, V)$ follows immediately from the linearity of covariance (Theorem 3.3.2), because we have

$$\mathrm{Cov}(U, V) = \mathrm{Cov}\left(\sum_{i=1}^{n} a_i X_i, \sum_{j=1}^{n} b_j X_j \right) = \sum_{i=1}^{n} \sum_{j=1}^{n} a_i b_j \, \mathrm{Cov}(X_i, X_j)$$

$$= \sum_{i=1}^{n} a_i b_i \, \mathrm{Cov}(X_i, X_i) = \sum_{i=1}^{n} a_i b_i \, \mathrm{Var}(X_i) = \sum_{i=1}^{n} a_i b_i \sigma_i^2$$

(note that $\mathrm{Cov}(X_i, X_j) = 0$ for $i \neq j$, by independence). Also, if U and V are independent, then we must have $\mathrm{Cov}(U, V) = 0$ by Corollary 3.4.3.

It remains to prove that, if $\mathrm{Cov}(U, V) = 0$, then U and V are independent. This involves a two-dimensional change of variable, as discussed in the advanced Section 2.9.2, so we refer the reader to Section 4.7 for this part of the proof. ∎

Theorem 4.6.2 says that, for the special case of linear combinations of independent normal distributions, if $\mathrm{Cov}(U, V) = 0$, then U and V are independent. However, it is important to remember that this property is *not* true in general, and there are random variables X and Y such that $\text{Cov}(X, Y) = 0$ even though X and Y are not independent (see Example 3.3.10). Furthermore, this property is not even true of *normal* distributions in general (see Problem 4.6.10).

Note that using linear algebra, we can write the equations $U = \sum_{i=1}^{n} a_i X_i$ and $V = \sum_{i=1}^{n} b_i X_i$ of Theorem 4.6.2 in matrix form as

$$\begin{pmatrix} U \\ \end{pmatrix} = A \begin{pmatrix} X_1 \\ X_2 \\ \vdots \\ X_n \end{pmatrix}, \qquad (4.6.1)$$

where

$$A = \begin{pmatrix} a_1 & a_2 & \cdots & a_n \\ b_1 & b_2 & \cdots & b_n \end{pmatrix}.$$

Furthermore, the rows of A are *orthogonal* if and only if $\sum_i a_i b_i = 0$. Now, in the case $\sigma_i = 1$ for all i, we have that $\text{Cov}(U, V) = \sum_i a_i b_i$. Hence, if $\sigma_i = 1$ for all i, then Theorem 4.6.2 can be interpreted as saying that if U and V are given by (4.6.1), then U and V are independent if and only if the rows of A are orthogonal. Linear algebra is used extensively in more advanced treatments of these ideas.

4.6.1 The Chi-Squared Distribution

We now introduce another distribution, related to the normal distribution.

Definition 4.6.1 The *chi-squared distribution* with n degrees of freedom (or chi-squared(n), or $\chi^2(n)$) is the distribution of the random variable

$$Z = X_1^2 + X_2^2 + \cdots + X_n^2,$$

where $X_1, \ldots, X_n$ are i.i.d., each with the standard normal distribution $N(0, 1)$.

Most statistical packages have built-in routines for the evaluation of chi-squared probabilities (also see Table D.4 in Appendix D).

One property of the chi-squared distribution is easy.

Theorem 4.6.3 If $Z \sim \chi^2(n)$ then $E(Z) = n$.

Proof: Write $Z = X_1^2 + X_2^2 + \cdots + X_n^2$, where $\{X_i\}$ are i.i.d. $\sim N(0, 1)$. Then $E((X_i)^2) = 1$. It follows by linearity that $E(Z) = 1 + \cdots + 1 = n$. ∎

The density function of the chi-squared distribution is a bit harder to obtain. We begin with the case $n = 1$.

Theorem 4.6.4 Let $Z \sim \chi^2(1)$. Then

$$f_Z(z) = \frac{1}{\sqrt{2\pi z}} e^{-z/2} = \frac{(1/2)^{-1/2}}{\Gamma(1/2)} z^{-1/2} e^{-z/2}$$

for $z > 0$, with $f_Z(z) = 0$ for $z < 0$. That is, $Z \sim \text{Gamma}(1/2, 1/2)$ (using $\Gamma(1/2) = \sqrt{\pi}$).

Proof: Because $Z \sim \chi^2(1)$, we can write $Z = X^2$ where $X \sim N(0,1)$. We then compute that, for $z > 0$,

$$\int_{-\infty}^{z} f_Z(s)\, ds = P(Z \le z) = P(X^2 \le z) = P(-\sqrt{z} \le X \le \sqrt{z}).$$

But because $X \sim N(0,1)$ with density function $\phi(s) = (2\pi)^{-1/2}\, e^{-s^2/2}$ we can rewrite this as

$$\int_{-\infty}^{\sqrt{z}} f_Z(s)\, ds = \int_{-\sqrt{z}}^{\sqrt{z}} \phi(s)\, ds = \int_{-\infty}^{\sqrt{z}} \phi(s)\, ds - \int_{-\infty}^{-\sqrt{z}} \phi(s)\, ds.$$

Because this is true for all $z > 0$, we can differentiate with respect to z to get (using the fundamental theorem of calculus and the chain rule) that

$$f_Z(z) = \frac{1}{2\sqrt{z}} \phi(\sqrt{z}) - \frac{-1}{2\sqrt{z}} \phi(-\sqrt{z}) = \frac{1}{\sqrt{z}} \phi(\sqrt{z}) = \frac{1}{\sqrt{2\pi z}} e^{-z/2},$$

as claimed. ∎

In Figure 4.6.1 we have plotted the $\chi^2(1)$ density. Note that the density becomes infinite at 0.

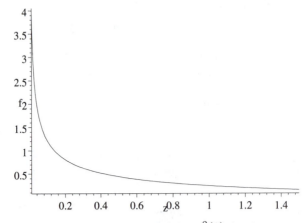

Figure 4.6.1: Plot of the $\chi^2(1)$ density.

Theorem 4.6.5 Let $Z \sim \chi^2(n)$. Then $Z \sim \text{Gamma}(n/2, 1/2)$. That is,

$$f_Z(z) = \frac{1}{2^{n/2}\Gamma(n/2)} z^{(n/2)-1} e^{-z/2}$$

for $z > 0$, with $f_Z(z) = 0$ for $z < 0$.

Proof: Because $Z \sim \chi^2(n)$, we can write $Z = X_1^2 + X_2^2 + \cdots + X_n^2$, where the X_i are i.i.d. $N(0,1)$. But this means that X_i^2 are i.i.d. $\chi^2(1)$. Hence, by

Theorem 4.6.4, we have X_i^2 i.i.d. Gamma$(1/2, 1/2)$ for $i = 1, 2, \ldots, n$. Therefore, Z is the sum of n independent random variables, each having distribution Gamma$(1/2, 1/2)$.

Now by Appendix C (see Problem 3.4.20), the moment-generating function of a Gamma(α, β) random variable is given by $m(s) = \beta^\alpha (\beta - s)^{-\alpha}$ for $s < \beta$. Putting $\alpha = 1/2$ and $\beta = 1/2$, and applying Theorem 3.4.5, the variable $Y = X_1^2 + X_2^2 + \cdots + X_n^2$ has moment-generating function given by

$$m_Y(s) = \prod_{i=1}^n m_{X_i^2}(s) = \prod_{i=1}^n \left(\frac{1}{2}\right)^{1/2} \left(\frac{1}{2} - s\right)^{-1/2} = \left(\frac{1}{2}\right)^{n/2} \left(\frac{1}{2} - s\right)^{-n/2}$$

for $s < 1/2$. We recognize this as the moment-generating function of the Gamma$(n/2, 1/2)$ distribution. Therefore, by Theorem 3.4.6 we have that $X_1^2 + X_2^2 + \cdots + X_n^2 \sim$ Gamma$(n/2, 1/2)$, as claimed.

This result can also be obtained using Problem 2.9.12 and induction. ∎

Note that the $\chi^2(2)$ density is the same as the Exponential(2) density. In Figure 4.6.2, we have plotted several χ^2 densities. Observe that the χ^2 are asymmetric and skewed to the right. As the degrees of freedom increases, the central mass of probability moves to the right.

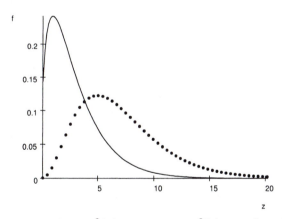

Figure 4.6.2: Plot of the $\chi^2(3)$ (—) and the $\chi^2(7)$ (• • •) density functions.

One application of the chi-squared distribution is the following.

Theorem 4.6.6 Let $X_1, \ldots, X_n$ be i.i.d. $N(\mu, \sigma^2)$. Put

$$\bar{X} = \frac{1}{n}(X_1 + \cdots + X_n) \text{ and } S^2 = \frac{1}{n-1} \sum_{i=1}^n (X_i - \bar{X})^2.$$

Then $(n-1) S^2/\sigma^2 \sim \chi^2(n-1)$, and furthermore, S^2 and $\bar{X}$ are independent.

Proof: See Section 4.7 for the proof of this result ∎

Because the $\chi^2(n-1)$ distribution has mean $n-1$, we obtain the following.

Corollary 4.6.2 $E(S^2) = \sigma^2$.

Proof: Theorems 4.6.6 and 4.6.3 imply that $E((n-1)S^2/\sigma^2) = n-1$ and $E(S^2) = \sigma^2$. ∎

Theorem 4.6.6 will find extensive use in Chapter 6. For example, this result together with Corollary 4.6.1, gives us the joint sampling distribution of the sample mean $\bar{X}$ and the sample variance S^2, when we are sampling from a $N(\mu, \sigma^2)$ distribution. If we do not know μ, then $\bar{X}$ is a natural estimator of this quantity and, similarly, S^2 is a natural estimator of σ^2, when it is unknown. Interestingly, we divide by $n-1$ in S^2, rather than n, precisely because we want $E(S^2) = \sigma^2$ to hold, as in Corollary 4.6.2. Actually this property does not depend on sampling from a normal distribution. It can be shown that anytime $X_1, \ldots, X_n$ is a sample from a distribution with variance σ^2, then $E(S^2) = \sigma^2$.

4.6.2 The t Distribution

The t distribution also has many statistical applications.

Definition 4.6.2 The t *distribution* with n degrees of freedom (or Student(n), or $t(n)$), is the distribution of the random variable

$$Z = \frac{X}{\sqrt{(X_1^2 + X_2^2 + \cdots + X_n^2)/n}},$$

where $X, X_1, \ldots, X_n$ are i.i.d., each with the standard normal distribution $N(0,1)$. (Equivalently, $Z = X/\sqrt{Y/n}$, where $Y \sim \chi^2(n)$.)

Most statistical packages have built-in routines for the evaluation of $t(n)$ probabilities (also see Table D.4 in Appendix D).

The density of the $t(n)$ distribution is given by the following result.

Theorem 4.6.7 Let $U \sim t(n)$. Then

$$f_U(u) = \frac{\Gamma\left(\frac{n+1}{2}\right)}{\sqrt{\pi}\,\Gamma\left(\frac{n}{2}\right)} \left(1 + \frac{u^2}{n}\right)^{-(n+1)/2} \frac{1}{\sqrt{n}}$$

for all $u \in R^1$.

Proof: For the proof of this result, see Section 4.7. ∎

The following result shows that, when n is large, the $t(n)$ distribution is very similar to the $N(0,1)$ distribution.

Theorem 4.6.8 As $n \to \infty$, the $t(n)$ distribution converges in distribution to a standard normal distribution.

Proof: Let $Z_1, \ldots, Z_n, Z$ be i.i.d. $N(0,1)$. As $n \to \infty$, by the strong law of large numbers, $(Z_1^2 + \cdots + Z_n^2)/n$ converges with probability 1 to the constant 1. Hence, the distribution of

$$\frac{Z}{\sqrt{(Z_1^2 + \cdots + Z_n^2)/n}} \tag{4.6.2}$$

converges to the distribution of Z, which is the standard normal distribution. By Definition 4.6.2, we have that (4.6.2) is distributed $t(n)$. ∎

In Figure 4.6.3, we have plotted several t densities. Notice that the densities of the t distributions are symmetric about 0 and look like the standard normal density.

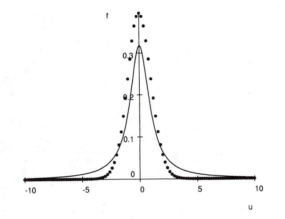

Figure 4.6.3: Plot of the $t(1)$ (—) and the $t(30)$ (• • •) density functions.

The $t(n)$ distribution has longer tails than the $N(0,1)$ distribution. For example, the $t(1)$ distribution (also known as the *Cauchy distribution*) has 0.9366 of its probability in the interval $(-10, 10)$, while the $N(0,1)$ distribution has all of its probability there (at least to four decimal places). The $t(30)$ and the $N(0,1)$ densities are very similar.

4.6.3 The F Distribution

Finally, we consider the F distribution.

Definition 4.6.3 The F *distribution* with m and n degrees of freedom (or $F(m, n)$) is the distribution of the random variable

$$Z = \frac{(X_1^2 + X_2^2 + \cdots + X_m^2)/m}{(Y_1^2 + Y_2^2 + \cdots + Y_n^2)/n},$$

where $X_1, \ldots, X_m, Y_1, \ldots, Y_n$ are i.i.d., each with the standard normal distribution. (Equivalently, $Z = (X/m)/(Y/n)$, where $X \sim \chi^2(m)$ and $Y \sim \chi^2(n)$.)

Most statistical packages have built-in routines for the evaluation of $F(m, n)$ probabilities (also see Table D.5 in Appendix D).

The density of the $F(m, n)$ distribution is given by the following result.

Theorem 4.6.9 Let $U \sim F(m, n)$. Then

$$f_U(u) = \frac{\Gamma\left(\frac{m+n}{2}\right)}{\Gamma\left(\frac{m}{2}\right)\Gamma\left(\frac{n}{2}\right)} \left(\frac{m}{n}u\right)^{(m/2)-1} \left(1 + \frac{m}{n}u\right)^{-(m+n)/2} \frac{m}{n}$$

for $u > 0$, with $f_U(u) = 0$ for $u < 0$.

Proof: For the proof of this result, see Section 4.7. ∎

In Figure 4.6.4, we have plotted several $F(m,n)$ densities. Notice that these densities are skewed to the right.

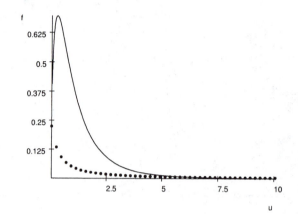

Figure 4.6.4: Plot of the $F(2,1)$ (• • •) and the $F(3,10)$ (—) density functions.

The following results are useful when it is necessary to carry out computations with the $F(m,n)$ distribution.

Theorem 4.6.10 If $Z \sim F(m,n)$, then $1/Z \sim F(n,m)$.

Proof: Using Definition 4.6.3, we have

$$\frac{1}{Z} = \frac{(Y_1^2 + Y_2^2 + \cdots + Y_n^2)/n}{(X_1^2 + X_2^2 + \cdots + X_m^2)/m}$$

and the result is immediate from the definition. ∎

Therefore, if $Z \sim F(m,n)$, then $P(Z \le z) = P(1/Z \ge 1/z) = 1 - P(1/Z \le 1/z)$ and $P(1/Z \le 1/z)$ is the cdf of the $F(n,m)$ distribution evaluated at $1/z$.

In many statistical applications, n can be very large. The following result then gives a useful approximation for that case.

Theorem 4.6.11 If $Z_n \sim F(m,n)$, then mZ_n converges in distribution to a $\chi^2(m)$ distribution as $n \to \infty$.

Proof: Using Definition 4.6.3, we have

$$mZ = \frac{X_1^2 + X_2^2 + \cdots + X_m^2}{(Y_1^2 + Y_2^2 + \cdots + Y_n^2)/n}.$$

By Definition 4.6.1, $X_1^2 + \cdots + X_m^2 \sim \chi^2(m)$. By Theorem 4.6.3, $E(Y_i^2) = 1$, and so the strong law of large numbers implies that $(Y_1^2 + Y_2^2 + \cdots + Y_n^2)/n$ converges almost surely to 1. This establishes the result. ∎

Finally, Definitions 4.6.2 and 4.6.3 immediately give the following result.

Theorem 4.6.12 If $Z \sim t(n)$, then $Z^2 \sim F(1,n)$.

Summary of Section 4.6

- Linear combinations of independent normal random variables are also normal, with appropriate mean and variance.

- Two linear combinations of the *same* collection of independent normal random variables are independent if and only if their covariance equals 0.

- The chi-squared distribution *with* n degrees of freedom is the distribution corresponding to a sum of squares of n i.i.d. standard normal random variables. It has mean n. It is equal to the Gamma$(n/2, 1/2)$ distribution.

- The t distribution with n degrees of freedom is the distribution corresponding to a standard normal random variable, divided by the square-root of $1/n$ times an independent chi-squared random variable with n degrees of freedom. Its density function was presented. As $n \to \infty$, it converges in distribution to a standard normal distribution.

- The F distribution with m and n degrees of freedom is the distribution corresponding to m/n times a chi-squared distribution with m degrees of freedom, divided by an independent chi-squared distribution with n degrees of freedom. Its density function was presented. If $m = 1$, then it is equal to a t distribution.

Exercises

4.6.1 Let $X_1 \sim N(3, 2^2)$ and $X_2 \sim N(-8, 5^2)$ be independent. Let $U = X_1 - 5X_2$, and $V = -6X_1 + CX_2$, where C is a constant.
(a) What are the distributions of U and V?
(b) What value of C makes U and V be independent?

4.6.2 Let $X \sim N(3, 5)$ and $Y \sim N(-7, 2)$ be independent.
(a) What is the distribution of $Z = 4X - Y/3$?
(b) What is the covariance of X and Z?

4.6.3 Let $X \sim N(3, 5)$ and $Y \sim N(-7, 2)$ be independent. Find values of C_1, C_2, C_3, C_4, C_5 so that $C_1(X + C_2)^2 + C_3(Y + C_4)^2 \sim \chi^2(C_5)$.

4.6.4 Let $X \sim \chi^2(n)$ and $Y \sim N(0, 1)$ be independent. Prove that $X + Y^2 \sim \chi^2(n+1)$.

4.6.5 Let $X \sim \chi^2(n)$ and $Y \sim \chi^2(m)$ be independent. Prove that $X + Y \sim \chi^2(n+m)$.

4.6.6 Let $X_1, X_2, \ldots, X_{4n}$ be i.i.d. with distribution $N(0, 1)$. Find a value of C such that
$$C \frac{X_1^2 + X_2^2 + \cdots + X_n^2}{X_{n+1}^2 + X_{n+2}^2 + \cdots + X_{4n}^2} \sim F(n, 3n).$$

4.6.7 Let $X_1, X_2, \ldots, X_{n+1}$ be i.i.d. with distribution $N(0,1)$. Find a value of C such that

$$C \frac{X_1}{\sqrt{X_2^2 + \cdots + X_n^2 + X_{n+1}^2}} \sim t(n).$$

4.6.8 Let $X \sim N(3,5)$ and $Y \sim N(-7,2)$ be independent. Find values of $C_1, C_2, C_3, C_4, C_5, C_6$ so that

$$\frac{C_1(X + C_2)^{C_3}}{(Y + C_4)^{C_5}} \sim t(C_6).$$

4.6.9 Let $X \sim N(3,5)$ and $Y \sim N(-7,2)$ be independent. Find values of $C_1, C_2, C_3, C_4, C_5, C_6, C_7$ so that

$$\frac{C_1(X + C_2)^{C_3}}{(Y + C_4)^{C_5}} \sim F(C_6, C_7).$$

Problems

4.6.10 Let $X \sim N(0,1)$, and let $P(Y = +1) = P(Y = -1) = 1/2$. Assume X and Y are independent. Let $Z = XY$.
(a) Prove that $Z \sim N(0,1)$.
(b) Prove that $\text{Cov}(X, Z) = 0$.
(c) Prove directly that X and Z are *not* independent.
(d) Why does this not contradict Theorem 4.6.2?

4.6.11 Let $Z \sim t(n)$. Prove that $P(Z < -x) = P(Z > x)$ for $x \in R^1$, namely, prove that the $t(n)$ distribution is symmetric about 0.

4.6.12 Let $X_n \sim F(n, 2n)$ for $n = 1, 2, 3, \ldots$ Prove that $X_n \to 1$ in probability and with probability 1.

4.6.13 *(The general Chi-squared distribution)* Prove that for $\alpha > 0$, the function

$$f(z) = \frac{1}{2^{\alpha/2}\Gamma(\alpha/2)} z^{(\alpha/2)-1} e^{-z/2}$$

defines a probability distribution on $(0, \infty)$. This distribution is known as the $\chi^2(\alpha)$ distribution, i.e., it generalizes the distribution in Section 4.6.2 by allowing the degrees of freedom to be an arbitrary positive real number. (Hint: The $\chi^2(\alpha)$ distribution is the same as a Gamma$(\alpha/2, 1/2)$ distribution.)

4.6.14 (MV) *(The general t distribution)* Prove that for $\alpha > 0$, the function

$$f(u) = \frac{\Gamma\left(\frac{\alpha+1}{2}\right)}{\sqrt{\pi}\,\Gamma\left(\frac{\alpha}{2}\right)} \left(1 + \frac{u^2}{\alpha}\right)^{-(\alpha+1)/2} \frac{1}{\sqrt{\alpha}}$$

defines a probability distribution on $(-\infty, \infty)$ by showing that the random variable

$$U = \frac{X}{\sqrt{Y/\alpha}}$$

has this density when $X \sim N(0,1)$ independent of $Y \sim \chi^2(\alpha)$ as in Problem 4.6.13. This distribution is known as the $t(\alpha)$ distribution, i.e., it generalizes the distribution in Section 4.6.3 by allowing the degrees of freedom to be an arbitrary positive real number. (Hint: The proof is virtually identical to that of Theorem 4.6.7.)

4.6.15 (MV) *(The general F distribution)* Prove that for $\alpha > 0, \beta > 0$, the function

$$f(u) = \frac{\Gamma\left(\frac{\alpha+\beta}{2}\right)}{\Gamma\left(\frac{\alpha}{2}\right)\Gamma\left(\frac{\beta}{2}\right)} \left(\frac{\alpha}{\beta}u\right)^{(\alpha/2)-1} \left(1 + \frac{\alpha}{\beta}u\right)^{-(\alpha+\beta)/2} \frac{\alpha}{\beta}$$

defines a probability distribution on $(0, \infty)$ by showing that the random variable

$$U = \frac{X/\alpha}{Y/\beta}$$

has this density whenever $X \sim \chi^2(\alpha)$ independent of $Y \sim \chi^2(\beta)$, as in Problem 4.6.13. This distribution is known as the $F(\alpha, \beta)$ distribution, i.e., it generalizes the distribution in Section 4.6.4 by allowing the numerator and denominator degrees of freedom to be arbitrary positive real numbers. (Hint: The proof is virtually identical to that of Theorem 4.6.9).

4.6.16 Prove that when $X \sim t(\alpha)$, as defined in Problem 4.6.14, and $\alpha > 1$, then $E(X) = 0$. Further prove that, when $\alpha > 2$, $\text{Var}(X) = \alpha/(\alpha - 2)$. You can assume the existence of these integrals — see Challenge 4.6.18. (Hint: To evaluate the second moment use $Y = X^2 \sim F(1, \alpha)$ as defined in Problem 4.6.15.)

4.6.17 Prove that when $X \sim F(\alpha, \beta)$, then $E(X) = \beta/(\beta - 2)$ when $\beta > 2$ and $\text{Var}(X) = 2\beta^2(\alpha + \beta - 2)/\alpha(\beta - 2)^2(\beta - 4)$ when $\beta > 4$.

Challenges

4.6.18 Following Problem 4.6.16, prove that the mean of X doesn't exist whenever $0 < \alpha \leq 1$. Further prove that the variance of X doesn't exist whenever $0 < \alpha \leq 1$ and is infinite when $1 < \alpha \leq 2$.

4.6.19 Prove the identity (4.7.1) in Section 4.7, which arises as part of the proof of Theorem 4.6.6.

4.7 Further Proofs (Advanced)

Theorem 4.3.1 Let $Z, Z_1, Z_2, \ldots$ be random variables. Suppose $Z_n \to Z$ with probability 1. Then $Z_n \to Z$ in probability. That is, if a sequence of random variables converges almost surely, then it converges in probability to the same limit.

Proof: Assume $P(Z_n \to Z) = 1$. Fix $\epsilon > 0$, and let $A_n = \{s : |Z_m - Z| \geq \epsilon$ for some $m \geq n\}$. Then $\{A_n\}$ is a decreasing sequence of events. Furthermore, if $s \in \cap_{n=1}^{\infty} A_n$, then $Z_n(s) \not\to Z(s)$ as $n \to \infty$. Hence,

$$P(\cap_{n=1}^{\infty} A_n) \leq P(Z_n \not\to Z) = 0.$$

By continuity of probabilities, we have $\lim_{n\to\infty} P(A_n) = P(\cap_{n=1}^{\infty} A_n) = 0$. Hence, $P(|Z_n - Z| \geq \epsilon) \leq P(A_n) \to 0$ as $n \to \infty$. Because this is true for any $\epsilon > 0$, we see that $Z_n \to Z$ in probability. ∎

Theorem 4.4.1 If $X_n \overset{P}{\to} X$, then $X_n \overset{D}{\to} X$.

Proof: Suppose $X_n \to X$ in probability, and that $P(X = x) = 0$. We wish to show that $\lim_{n\to\infty} P(X_n \leq x) = P(X \leq x)$.

Choose any $\epsilon > 0$. Now, if $X_n \leq x$ then we must have either $X \leq x + \epsilon$ or $|X - X_n| \geq \epsilon$. Hence, by subadditivity,

$$P(X_n \leq x) \leq P(X \leq x + \epsilon) + P(|X - X_n| \geq \epsilon).$$

Replacing x by $x - \epsilon$ in this equation, we see also that

$$P(X \leq x - \epsilon) \leq P(X_n \leq x) + P(|X - X_n| \geq \epsilon).$$

Rearranging and combining these two inequalities, we have

$$P(X \leq x - \epsilon) - P(|X - X_n| \geq \epsilon)$$
$$\leq P(X_n \leq x) \leq P(X \leq x + \epsilon) + P(|X - X_n| \geq \epsilon).$$

This is the key.

We next let $n \to \infty$. Because $X_n \to X$ in probability, we know that

$$\lim_{n\to\infty} P(|X - X_n| \geq \epsilon) = 0.$$

This means that $\lim_{n\to\infty} P(X_n \leq x)$ is "sandwiched" between $P(X \leq x - \epsilon)$ and $P(X \leq x + \epsilon)$.

We then let $\epsilon \searrow 0$. By continuity of probabilities,

$$\lim_{\epsilon \searrow 0} P(X \leq x + \epsilon) = P(X \leq x) \quad \text{and} \quad \lim_{\epsilon \searrow 0} P(X \leq x - \epsilon) = P(X < x).$$

This means that $\lim_{n\to\infty} P(X_n \leq x)$ is "sandwiched" between $P(X < x)$ and $P(X \leq x)$.

But because $P(X = x) = 0$, we must have $P(X < x) = P(X \leq x)$. Hence, $\lim_{n\to\infty} P(X_n \leq x) = P(X \leq x)$, as required. ∎

Theorem 4.4.3 (*The central limit theorem*) Let $X_1, X_2, \ldots$ be i.i.d. with finite mean μ and finite variance σ^2. Let $Z \sim N(0, 1)$. Set $S_n = X_1 + \cdots + X_n$, and

$$Z_n = \frac{S_n - n\mu}{\sqrt{n\sigma^2}}.$$

Then as $n \to \infty$, the sequence $\{Z_n\}$ converges in distribution to the Z, i.e., $Z_n \overset{D}{\to} Z$.

Proof: (Outline) Recall that the standard normal distribution has moment-generating function given by $m_Z(s) = \exp\left(s^2/2\right)$.

We shall now *assume* that $m_{Z_n}(s)$ is finite for $|s| < s_0$ for some $s_0 > 0$. (This assumption can be eliminated by using *characteristic functions* instead of moment-generating functions.) Assuming this, we will prove that for each real number s, we have $\lim_{n \to \infty} m_{Z_n}(s) = m_Z(s)$, where $m_{Z_n}(s)$ is the moment-generating function of Z_n. It then follows from Theorem 4.4.2 that Z_n converges to Z in distribution.

To proceed, let $Y_i = (X_i - \mu)/\sigma$. Then $E(Y_i) = 0$ and $E(Y_i^2) = \text{Var}(Y_i) = 1$. Also, we have

$$Z_n = \frac{1}{\sqrt{n}}(Y_1 + \cdots + Y_n).$$

Let $m_Y(s) = E(e^{sY_i})$ be the moment-generating function of Y_i (which is the same for all i, because they are i.i.d.). Then using independence, we compute that

$$
\begin{aligned}
\lim_{n \to \infty} m_{Z_n}(s) &= \lim_{n \to \infty} E\left(e^{sZ_n}\right) = \lim_{n \to \infty} E\left(e^{s(Y_1 + \cdots + Y_n)/\sqrt{n}}\right) \\
&= \lim_{n \to \infty} E\left(e^{sY_1/\sqrt{n}} e^{sY_2/\sqrt{n}} \cdots e^{sY_n/\sqrt{n}}\right) \\
&= \lim_{n \to \infty} E\left(e^{sY_1/\sqrt{n}}\right) E\left(e^{sY_2/\sqrt{n}}\right) \cdots E\left(e^{sY_n/\sqrt{n}}\right) \\
&= \lim_{n \to \infty} m_Y(s/\sqrt{n}) m_Y(s/\sqrt{n}) \cdots m_Y(s/\sqrt{n}) \\
&= \lim_{n \to \infty} m_Y(s/\sqrt{n})^n.
\end{aligned}
$$

Now, we know from Theorem 3.5.3 that $m_Y(0) = E(e^0) = 1$. Also, $m_Y'(0) = E(Y_i) = 0$, and $m_Y''(0) = E(Y_i^2) = 1$. But then expanding $m_Y(s)$ in a Taylor series around $s = 0$, we see that

$$m_Y(s) = 1 + 0s + \frac{1}{2!}s^2 + o(s^2) = 1 + s^2/2 + o(s^2),$$

where $o(s^2)$ stands for a quantity that, as $s \to 0$, goes to 0 faster than s^2 does — namely, $o(s^2)/s \to 0$ as $s \to 0$. This means that

$$m_Y(s/\sqrt{n}) = 1 + (s/\sqrt{n})^2/2 + o((s/\sqrt{n})^2) = 1 + s^2/2n + o(1/n),$$

where now $o(1/n)$ stands for a quantity that, as $n \to \infty$, goes to 0 faster than $1/n$ does.

Finally, we recall from calculus that, for any real number c, $\lim_{n \to \infty}(1 + c/n)^n = e^c$. It follows from this and the above that

$$\lim_{n \to \infty} (m_Y(s/2\sqrt{n}))^n = \lim_{n \to \infty} (1 + s^2/2n)^n = e^{s^2/2}.$$

That is, $\lim_{n \to \infty} m_{Z_n}(s) = e^{s^2/2}$, as claimed. ∎

Theorem 4.6.2 Suppose $X_i \sim N(\mu_i, \sigma_i^2)$ for $i = 1, 2, \ldots, n$, and also that the $\{X_i\}$ are *independent*. Let $U = \sum_{i=1}^{n} a_i X_i$ and $V = \sum_{i=1}^{n} b_i X_i$, for some constants $\{a_i\}$ and $\{b_i\}$. Then $\mathrm{Cov}(U, V) = \sum_i a_i b_i \sigma_i^2$. Furthermore, $\mathrm{Cov}(U, V) = 0$ if and only if U and V are independent.

Proof: It was proved in Section 4.6 that $\mathrm{Cov}(U, V) = \sum_i a_i b_i \sigma_i^2$ and that $\mathrm{Cov}(U, V) = 0$ if U and V are independent. It remains to prove that, if $\mathrm{Cov}(U, V) = 0$, then U and V are independent. For simplicity, we take $n = 2$ and $\mu_1 = \mu_2 = 0$ and $\sigma_1^2 = \sigma_2^2 = 1$; the general case is similar but messier. We therefore have

$$U = a_1 X_1 + a_2 X_2 \text{ and } V = b_1 X_1 + b_2 X_2.$$

The Jacobian derivative of this transformation is

$$J(x_1, x_2) = \frac{\partial U}{\partial X_1} \frac{\partial V}{\partial X_2} - \frac{\partial V}{\partial X_1} \frac{\partial U}{\partial X_2} = a_1 b_2 - b_1 a_2.$$

Inverting the transformation gives

$$X_1 = \frac{b_2 U - a_2 V}{a_1 b_2 - b_1 a_2} \text{ and } X_2 = \frac{a_1 V - b_1 U}{a_1 b_2 - b_1 a_2}.$$

Also,

$$f_{X_1, X_2}(x_1, x_2) = \frac{1}{2\pi} e^{-(x_1^2 + x_2^2)/2}.$$

Hence, from the multidimensional change of variable theorem (Theorem 2.9.2), we have

$$
\begin{aligned}
f_{U,V}(u, v) &= f_{X_1, X_2}(x_1, x_2) \left(\frac{b_2 u - a_2 v}{a_1 b_2 - b_1 a_2}, \frac{a_1 v - b_1 u}{a_1 b_2 - b_1 a_2} \right) |J(x_1, x_2)|^{-1} \\
&= \frac{1}{2\pi} \frac{\exp\left\{ -((b_2 u - a_2 v)^2 + (a_1 v - b_1 u)^2) / 2(a_1 b_2 - b_1 a_2)^2 \right\}}{|a_1 b_2 - b_1 a_2|}.
\end{aligned}
$$

But

$$(b_2 u - a_2 v)^2 + (a_1 v - b_1 u)^2 = (b_1^2 + b_2^2)u^2 + (a_1^2 + a_2^2)v^2 - 2(a_1 b_1 + a_2 b_2)uv,$$

and $\mathrm{Cov}(U, V) = a_1 b_1 + a_2 b_2$. Hence, if $\mathrm{Cov}(U, V) = 0$, then

$$(b_2 u - a_2 v)^2 + (a_1 v - b_1 u)^2 = (b_1^2 + b_2^2)u^2 + (a_1^2 + a_2^2)v^2,$$

and

$$
\begin{aligned}
f_{U,V}(u, v) &= \frac{\exp\left\{ -((b_1^2 + b_2^2)u^2 + (a_1^2 + a_2^2)v^2)/2(a_1 b_2 - b_1 a_2)^2 \right\}}{2\pi |a_1 b_2 - b_1 a_2|} \\
&= \frac{\exp\left\{ -(b_1^2 + b_2^2)u^2/2(a_1 b_2 - b_1 a_2)^2 \right\} \exp\left\{ -(a_1^2 + a_2^2)v^2/2(a_1 b_2 - b_1 a_2)^2 \right\}}{2\pi |a_1 b_2 - b_1 a_2|}.
\end{aligned}
$$

It follows that we can *factor* $f_{U,V}(u, v)$ as a function of u times a function of v. But this implies (Problem 2.8.15) that U and V are independent. $\blacksquare$

Theorem 4.6.6 Let $X_1, \ldots, X_n$ be i.i.d. $N(\mu, \sigma^2)$ and put

$$\bar{X} = \frac{1}{n}(X_1 + \cdots + X_n) \text{ and } S^2 = \frac{1}{n-1}\sum_{i=1}^{n}(X_i - \bar{X})^2.$$

Then $(n-1)S^2/\sigma^2 \sim \chi^2(n-1)$, and furthermore, S^2 and $\bar{X}$ are independent.

Proof: We have

$$\frac{n-1}{\sigma^2}S^2 = \sum_{i=1}^{n}\left(\frac{X_i - \bar{X}}{\sigma}\right)^2.$$

We rewrite this expression as (see Challenge 4.6.19)

$$\frac{n-1}{\sigma^2}S^2$$
$$= \left(\frac{X_1 - X_2}{\sigma\sqrt{2}}\right)^2 + \left(\frac{X_1 + X_2 - 2X_3}{\sigma\sqrt{2\cdot 3}}\right)^2 + \left(\frac{X_1 + X_2 + X_3 - 3X_4}{\sigma\sqrt{3\cdot 4}}\right)^2$$
$$+ \cdots + \left(\frac{X_1 + X_2 + \cdots + X_{n-1} - (n-1)X_n}{\sigma\sqrt{(n-1)n}}\right)^2. \tag{4.7.1}$$

Now, by Theorem 4.6.1, each of the $n-1$ expressions within brackets in (4.7.1) has the standard normal distribution. Furthermore, by Theorem 4.6.2, the expressions within brackets in (4.6.2) are all *independent* of one another and are also all independent of $\bar{X}$.

It follows that $(n-1)S^2/\sigma^2$ is independent of $\bar{X}$. It also follows, by the definition of the chi-squared distribution, that $(n-1)S^2/\sigma^2 \sim \chi^2(n-1)$. ∎

Theorem 4.6.7 Let $U \sim t(n)$. Then

$$f_U(u) = \frac{\Gamma\left(\frac{n+1}{2}\right)}{\sqrt{\pi}\,\Gamma\left(\frac{n}{2}\right)}\left(1 + \frac{u^2}{n}\right)^{-(n+1)/2}\frac{1}{\sqrt{n}}$$

for all $u \in R^1$.

Proof: Because $U \sim t(n)$, we can write $U = X/\sqrt{Y/n}$, where X and Y are independent with $X \sim N(0,1)$ and $Y \sim \chi^2(n)$. It follows that X and Y have joint density given by

$$f_{X,Y}(x, y) = \frac{e^{-x^2/2}y^{(n/2)-1}e^{-y/2}}{\sqrt{2\pi}2^{\frac{n}{2}}\Gamma\left(\frac{n}{2}\right)}$$

when $y > 0$ (with $f_{X,Y}(x, y) = 0$ for $y < 0$).

Let $V = Y$. We shall use the multivariate change of variables formula (Theorem 2.9.2) to compute the joint density $f_{U,V}(u, v)$ of U and V. Because $U = X/\sqrt{Y/n}$ and $V = Y$, it follows that $X = U\sqrt{V/n}$ and $Y = V$. We compute the Jacobian term as

$$J(x, y) = \det\begin{pmatrix} \frac{\partial u}{\partial x} & \frac{\partial v}{\partial x} \\ \frac{\partial u}{\partial y} & \frac{\partial v}{\partial y} \end{pmatrix} = \det\begin{pmatrix} \frac{1}{\sqrt{y/n}} & 0 \\ \frac{-x\sqrt{n}}{y^{3/2}} & 1 \end{pmatrix} = \frac{1}{\sqrt{y/n}}.$$

Hence,

$$f_{U,V}(u,v) = f_{X,Y}\left(u\sqrt{\frac{v}{n}}, v\right) J^{-1}\left(u\sqrt{\frac{v}{n}}, v\right)$$

$$= \frac{e^{-(u^2 v)/2n} v^{(n/2)-1} e^{-v/2}}{\sqrt{2\pi} 2^{\frac{n}{2}} \Gamma\left(\frac{n}{2}\right)} \sqrt{\frac{v}{n}}$$

$$= \frac{1}{\sqrt{\pi}\Gamma\left(\frac{n}{2}\right)} \frac{1}{2^{(n+1)/2}} \frac{1}{\sqrt{n}} v^{(n+1)/2-1} e^{-(v/2)\left(1+u^2/n\right)}$$

for $v > 0$ (with $f_{U,V}(u,v) = 0$ for $v < 0$).

Finally, we compute the marginal density of U:

$$\begin{aligned}
f_U(u) &= \int_{-\infty}^{\infty} f_{U,V}(u,v)\, dv \\
&= \frac{1}{\sqrt{\pi}\Gamma(n/2)} \frac{1}{2^{(n+1)/2}} \frac{1}{\sqrt{n}} \int_0^{\infty} v^{(n+1)/2-1} e^{-(v/2)\left(1+u^2/n\right)}\, dv \\
&= \frac{1}{\sqrt{\pi}\Gamma(n/2)} \left(1+\frac{u^2}{n}\right)^{-\frac{n+1}{2}} \frac{1}{\sqrt{n}} \int_0^{\infty} w^{(n+1)/2-1} e^{-w/2}\, dw \\
&= \frac{\Gamma\left(\frac{n+1}{2}\right)}{\sqrt{\pi}\Gamma(n/2)} \left(1+\frac{u^2}{n}\right)^{-\frac{n+1}{2}} \frac{1}{\sqrt{n}},
\end{aligned}$$

where we have made the substitution $w = \left(1+u^2/n\right) v/2$ to get the third equality and then used the definition of the gamma function to obtain the result. ∎

Theorem 4.6.9 Let $U \sim F(m,n)$. Then

$$f_U(u) = \frac{\Gamma\left(\frac{m+n}{2}\right)}{\Gamma\left(\frac{m}{2}\right)\Gamma\left(\frac{n}{2}\right)} \left(\frac{m}{n}u\right)^{(m/2)-1} \left(1+\frac{m}{n}u\right)^{-(m+n)/2} \frac{m}{n}$$

for $u > 0$, with $f_U(u) = 0$ for $u < 0$.

Proof: Because $U \sim F(n,m)$, we can write $U = (X/m)/(Y/n)$, where X and Y are independent with $X \sim \chi^2(m)$ and $Y \sim \chi^2(n)$. It follows that X and Y have joint density given by

$$f_{X,Y}(x,y) = \frac{x^{(m/2)-1} e^{-x/2} y^{(n/2)-1} e^{-y/2}}{2^{m/2}\Gamma\left(\frac{m}{2}\right) 2^{n/2}\Gamma\left(\frac{n}{2}\right)}$$

when $x, y > 0$ (with $f_{X,Y}(x,y) = 0$ for $x < 0$ or $y < 0$).

Let $V = Y$, and use the multivariate change of variables formula (Theorem 2.9.2) to compute the joint density $f_{U,V}(u,v)$ of U and V. Because $U = (X/m)/(Y/n)$ and $V = Y$, it follows that $X = (m/n)UV$ and $Y = V$. We compute the Jacobian term as

$$J(x,y) = \det \begin{pmatrix} \frac{\partial u}{\partial x} & \frac{\partial v}{\partial x} \\ \frac{\partial u}{\partial y} & \frac{\partial v}{\partial y} \end{pmatrix} = \det \begin{pmatrix} \frac{n}{my} & 0 \\ \frac{-nX}{mY^2} & 1 \end{pmatrix} = \frac{n}{my}.$$

Hence,

$$
\begin{aligned}
f_{U,V}(u,v) &= f_{X,Y}((m/n)uv,\, v)J^{-1}((m/n)uv,\, v) \\
&= \frac{\left(\frac{m}{n}uv\right)^{(m/2)-1} e^{-(m/n)(uv/2)} v^{(n/2)-1} e^{-(v/2)}}{2^{m/2}\Gamma\left(\frac{m}{2}\right) 2^{n/2}\Gamma\left(\frac{n}{2}\right)} \\
&= \frac{1}{\Gamma\left(\frac{m}{2}\right)\Gamma\left(\frac{n}{2}\right)} \left(\frac{m}{n}u\right)^{(m/2)-1} \frac{m}{n}\frac{1}{2^{m/2}} v^{(n/2)-1} e^{-(v/2)(1+mu/n)}
\end{aligned}
$$

for $u, v > 0$ (with $f_{U,V}(u,v) = 0$ for $u < 0$ or $v < 0$).

Finally, we compute the marginal density of U as

$$
\begin{aligned}
&f_U(u) \\
&= \int_{-\infty}^{\infty} f_{U,V}(u,v)\, dv \\
&= \frac{1}{\Gamma\left(\frac{m}{2}\right)\Gamma\left(\frac{n}{2}\right)} \left(\frac{m}{n}u\right)^{(m/2)-1} \frac{m}{n}\frac{1}{2^{m/2}} \int_0^{\infty} v^{(n/2)-1} e^{-(v/2)(1+mu/n)}\, dv \\
&= \frac{1}{\Gamma\left(\frac{m}{2}\right)\Gamma\left(\frac{n}{2}\right)} \left(\frac{m}{n}u\right)^{(m/2)-1} \left(1+\frac{m}{n}u\right)^{-(n+m)/2} \frac{m}{n} \int_0^{\infty} w^{(n/2)-1} e^{-w}\, dw \\
&= \frac{\Gamma\left(\frac{m+n}{2}\right)}{\Gamma\left(\frac{m}{2}\right)\Gamma\left(\frac{n}{2}\right)} \left(\frac{m}{n}u\right)^{(m/2)-1} \left(1+\frac{m}{n}u\right)^{-(n+m)/2}
\end{aligned}
$$

where we have used the substitution $w = (1 + mu/n)\, v/2$ to get the third equality and the final result follows from the definition of the gamma function. ∎

Chapter 5

Statistical Inference

In this chapter we begin our discussion of statistical inference. Probability theory is primarily concerned with calculating various quantities associated with a probability model. This requires that we *know* what the correct probability model is. In applications, this is often not the case, and the best we can say is that the correct probability measure to use is in a set of possible probability measures. We refer to this collection as the *statistical model*. So, in a sense, our uncertainty has increased; not only do we have the uncertainty associated with an outcome or response as described by a probability measure, but now we are uncertain about what the probability measure is as well.

Statistical inference is concerned with making statements or inferences about characteristics of the true underlying probability measure. Of course, these inferences must be based on some kind of information and the statistical model makes up part of it. Another important part of the information will be given by an observed outcome or response, which we refer to as the *data*. Inferences then take the form of various statements about the true underlying probability measure from which the data were obtained. These take a variety of forms, which we refer to as *types of inferences*.

The role of this chapter is to introduce the basic concepts and ideas of statistical inference. The most prominent approaches to inference are discussed in Chapters 6, 7, and 8. Likelihood methods require the least structure and are described in Chapter 6. Bayesian methods require some additional ingredients and are discussed in Chapter 7. Inference methods based on measures of performance and loss functions are described in Chapter 8.

5.1 Why Do We Need Statistics?

While we will spend much of our time discussing the theory of statistics, we should always remember that statistics is an applied subject. By this we mean that ultimately statistical theory will be applied to real-world situations to answer questions of practical importance.

What is it that characterizes those contexts in which statistical methods are useful? Perhaps the best way to answer this is to consider a practical example where statistical methodology plays an important role.

Example 5.1.1 *Stanford Heart Transplant Study*

In the paper by Turnbull, Brown, and Hu entitled *Survivorship of Heart Transplant Data* (*Journal of the American Statistical Association*, March 1974, Volume 69, 74–80), an analysis is conducted to determine whether or not a heart transplant program, instituted at Stanford University, is in fact producing the intended outcome. In this case, the intended outcome is an increased length of life, namely, a patient who receives a new heart should live longer than if one were not received.

It is obviously important that we be sure that a proposed medical treatment for a disease leads to an improvement in the condition. Clearly, we wouldn't want it to lead to a deterioration in the condition. Also, if it only produced a small improvement, it may not be worth carrying out if it is very expensive or causes additional suffering.

We can never know whether a particular patient who received a new heart has lived longer because of the transplant. So our only hope of determining if the treatment is working is to compare the lifelengths of patients who received new hearts with the lifelengths of patients who did not. There are many factors that influence a patient's lifelength, and many of these will have nothing to do with the condition of the patient's heart. For example, lifestyle and the existence of other pathologies will have a great influence, and these will vary greatly from patient to patient. So how can we make this comparison?

One approach to this problem is to imagine that there are probability distributions that describe the lifelengths of the two groups. Let these be given by the densities f_T and f_C, where T denotes transplant and C denotes no transplant. Here we have used C as our label, because this group is serving as a *control* in the study to provide something to compare the treatment (a heart transplant) to. Then we consider the lifelength of a patient who received a transplant as a random observation from f_T and the lifelength of a patient who did not receive a transplant as a random observation from f_C. We want to compare f_T and f_C, in some fashion, to determine whether or not the transplant treatment is working. For example, we might compute the mean lifelengths of each distribution and compare these. If the mean lifelength of f_T is greater than f_C, then we can assert that the treatment is working. Of course, we would still have to judge whether the size of the improvement is enough to warrant the additional expense and patients' suffering the treatment entails.

If we could take an arbitrarily large number of observations from f_T and f_C, then we know, from the results in previous chapters, that we could determine these distributions with a great deal of accuracy. In practice, however, we are restricted to a relatively small number of observations. For example, in the cited study there were 30 patients in the control group (those who did not receive a transplant) and 52 patients in the treatment group (those who did receive a transplant).

P	X	S	P	X	S	P	X	S
1	49	d	11	1400	a	21	2	d
2	5	d	12	5	d	22	148	d
3	17	d	13	34	d	23	1	d
4	2	d	14	15	d	24	68	d
5	39	d	15	11	d	25	31	d
6	84	d	16	2	d	26	1	d
7	7	d	17	1	d	27	20	d
8	0	d	18	39	d	28	118	a
9	35	d	19	8	d	29	91	a
10	36	d	20	101	d	30	427	a

Table 5.1: Survival times (X) in days and status (S) at the end of the study for each patient (P) in the control group.

For each control patient, the value of X, the number of days they were alive after the date they were determined to be a candidate for a heart transplant until the termination date of the study, was recorded. For a variety of reasons, these patients did not receive new hearts, e.g., they died before a new heart could be found for them. These data, together with an indicator for the status of the patient at the termination date of the study, are presented in Table 5.1. The indicator value $S = a$ denotes that the patient was alive at the end of the study and $S = d$ denotes that the patient was dead.

For each treatment patient, the value of Y, the number of days they waited for the transplant after the date they were determined to be a candidate for a heart transplant, and Z, the number of days they were alive after the date they received the heart transplant until the termination date of the study, were both recorded. The survival times for the treatment group are then given by the values of $Y + Z$. These data, together with an indicator for the status of the patient at the termination date of the study, are presented in Table 5.2.

We cannot compare f_T and f_C directly because we do not know these distributions. But we do have some information about them because we have obtained values from each, as presented in Tables 5.1 and 5.2. So now the question is — how do we use these data to compare f_T and f_C to answer the question of central importance, concerning whether or not the treatment is effective? This is the realm of statistics and statistical theory, namely, providing methods for making inferences about unknown probability distributions based upon observations (samples) obtained from them.

We note that we have simplified this example somewhat, although our discussion presents the essence of the problem. The added complexity comes from the fact that typically statisticians will have available additional data on each patient, such as their age, gender, and disease history. As a particular example of this, in Table 5.2 we have the values of both Y and Z for each patient in the treatment group. As it turns out, this additional information, known as covariates, can be used to make our comparisons more accurate. This will be

P	Y	Z	S	P	Y	Z	S	P	Y	Z	S
1	0	15	d	19	50	1140	a	37	77	442	a
2	35	3	d	20	22	1153	a	38	2	65	d
3	50	624	d	21	45	54	d	39	26	419	a
4	11	46	d	22	18	47	d	40	32	362	a
5	25	127	d	23	4	0	d	41	13	64	d
6	16	61	d	24	1	43	d	42	56	228	d
7	36	1350	d	25	40	971	a	43	2	65	d
8	27	312	d	26	57	868	a	44	9	264	a
9	19	24	d	27	0	44	d	45	4	25	d
10	17	10	d	28	1	780	a	46	30	193	a
11	7	1024	d	29	20	51	d	47	3	196	a
12	11	39	d	30	35	710	a	48	26	63	d
13	2	730	d	31	82	663	a	49	4	12	d
14	82	136	d	32	31	253	d	50	45	103	a
15	24	1379	a	33	40	147	d	51	25	60	a
16	70	1	d	34	9	51	d	52	5	43	a
17	15	836	d	35	66	479	a				
18	16	60	d	36	20	322	d				

Table 5.2: The number of days until transplant (Y), survival times in days after transplant (Z), and status (S) at the end of the study for each patient (P) in the treatment group.

discussed in Chapter 10. ∎

The previous example provides some evidence that questions of great practical importance require the use of statistical thinking and methodology. There are many situations in the physical and social sciences where statistics plays a key role, and the reasons are just like those found in Example 5.1.1. The central ingredient in all of these is that we are faced with uncertainty. This uncertainty is caused both by variation, which can be modeled via probability, and by the fact that we cannot collect enough observations to know the correct probability models precisely. The first four chapters have dealt with building, and using, a mathematical model to deal with the first source of uncertainty. In this chapter we begin to discuss methods for dealing with the second source of uncertainty.

Summary of Section 5.1

- Statistics is applied to situations in which we have questions that cannot be answered definitively, typically because of variation in data.

- Probability is used to model the variation observed in the data. Statistical inference is concerned with using the observed data to help identify the true probability distribution (or distributions) producing this variation and thus gain insight into the answers to the questions of interest.

Exercises

5.1.1 Compute the mean survival times for the control group and for the treatment groups in Example 5.1.1. What do you conclude from these numbers? Do you think it is valid to base your conclusions on the effectiveness of the treatment on these numbers? Explain why or why not.

5.1.2 Are there any unusual observations in the data presented in Example 5.1.1? If so, what effect do you think these observations have on the mean survival times computed in Exercise 5.1.1?

5.1.3 In Example 5.1.1, we can use the status variable S as a covariate. What is the practical significance of this variable?

5.1.4 A student is uncertain about the mark that will be received in a statistics course. The course instructor has made available a database of marks in the course for a number of years. Can you identify a probability distribution that may be relevant to quantifying the student's uncertainty? What covariates might be relevant in this situation?

5.1.5 The following data were generated from an $N(\mu, 1)$ distribution by a student. Unfortunately the student forgot which value of μ was used, so we are uncertain about the correct probability distribution to use to describe the variation in the data.

0.2	−0.7	0.0	−1.9	0.7	−0.3	0.3	0.4
0.3	−0.8	1.5	0.1	0.3	−0.7	−1.8	0.2

Can you suggest a plausible value for μ? Explain your reasoning.

Problem

5.1.6 Can you identify any potential problems with the method we have discussed in Example 5.1.1 for determining whether or not the heart transplant program is effective in extending life?

Computer Problem

5.1.7 Suppose that we want to obtain the distribution of the quantity $Y = X^4 + 2X^3 - 3$ when $X \sim N(0, 1)$. Here we are faced with a form of mathematical uncertainty, because it is very difficult to determine the distribution of Y using mathematical methods. Propose a computer method for approximating the distribution function of Y and estimate $P(Y \in (1, 2))$. What is the relevance of statistical methodology to your approach?

Discussion Topics

5.1.8 Sometimes it is claimed that all uncertainties can and should be modeled using probability. Discuss this issue in the context of Example 5.1.1, namely, indicate all the things you are uncertain about in this example and how you might propose probability distributions to quantify these uncertainties.

5.2   Inference Using a Probability Model

In the first four chapters of this book we have been discussing probability theory. A good part of this discussion has been concerned with the mathematics of probability theory. This tells us how to carry out various calculations associated with the application of the theory. It is important to keep in mind, however, our reasons for introducing probability in the first place. As we discussed in Section 1.1, probability is concerned with measuring or quantifying uncertainty.

Of course, we are uncertain about many things, and we cannot claim that probability is applicable to all of these situations. Let us assume, however, that we are in a situation in which we feel probability is applicable and that we have a probability measure P defined on a collection of subsets of a sample space S for a response s.

In an application of probability, we presume that we know P and are uncertain about a future, or concealed, response value $s \in S$. In such a context we may be required, or may wish, to make an *inference* about the unknown value of s. This can take the form of a *prediction* or *estimate* of a plausible value for s, e.g., under suitable conditions we might take the expected value of s as our prediction. In other contexts, we may be asked to construct a subset that has a high probability of containing s and is in some sense small, e.g., find the region that contains at least 95% of the probability and has the smallest size amongst all such regions. Alternatively, we might be asked to assess whether or not a stated value s_0 is an implausible value from the known P, e.g., assess whether or not s_0 lies in a region assigned low probability by P and so is implausible. These are examples of inferences that are relevant to applications of probability theory.

Example 5.2.1

As a specific application, consider the lifelength X in years of a machine where it is known that $X \sim \text{Exponential}(1)$ (see Figure 5.2.1).

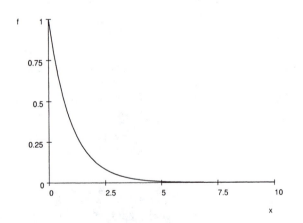

Figure 5.2.1: Plot of the Exponential(1) density f.

Then for a new machine, we might predict its lifelength by $E(X) = 1$ year. Further, from the graph of the Exponential(1) density, it is clear that the smallest interval containing 95% of the probability for X is $(0, c)$ where c satisfies

$$0.95 = \int_0^c e^{-x} \, dx = 1 - e^{-c}$$

or $c = -\ln(0.05) = 2.9957$. This interval gives us a reasonable range of probable lifelengths for the new machine. Finally if we wanted to assess whether or not $x_0 = 5$ is a plausible lifelength for a newly purchased machine we might compute the tail probability as

$$P(X > 5) = \int_5^\infty e^{-x} \, dx = e^{-5} = 0.0067$$

which, in this case, is very small and therefore indicates that $x_0 = 5$ is fairly far out in the tail. The right tail of this density is a region of low probability for this distribution, so $x_0 = 5$ can be considered implausible. It is thus unlikely that a machine will last 5 years and a purchaser would have to plan to replace the machine before the end of such a period. ∎

In some applications, we receive some partial information about the unknown s taking the form $s \in C \subset S$. In such a case, we replace P by the conditional probability measure $P(\cdot \mid C)$ when deriving our inferences. Our reasons for doing this are many, and, in general, we can say that most statisticians agree that it is the right thing to do. It is important to recognize, however, that this step does not proceed from a mathematical theorem but rather can be regarded as a basic axiom or principle of inference. We will refer to this as the *principle of conditional probability*. This principle will play a key role in some later developments.

Example 5.2.2
Suppose that we have a machine whose lifelength is distributed as in Example 5.2.1, and the machine has already been running for 1 year. Then inferences about the lifelength of the machine are based on the conditional distribution given that $X > 1$. The density of this conditional distribution is given by $e^{-(x-1)}$ for $x > 1$. The predicted lifelength is now

$$E(X \mid X > 1) = \int_1^\infty x e^{-(x-1)} \, dx = -x e^{-(x-1)} \Big|_1^\infty + \int_1^\infty e^{-(x-1)} \, dx = 2.$$

The fact that the additional lifelength is the same as the predicted lifelength before the machine starts working is a special characteristic of the Exponential distribution. This will not be true in general (see Exercise 5.2.4).

The tail probability measuring the plausibility of the value $x_0 = 5$ is given by

$$P(X > 5 \mid X > 1) = \int_5^\infty e^{-(x-1)} \, dx = e^{-4} = 0.0183,$$

which indicates that $x_0 = 5$ is a little more plausible in light of the fact that the machine has already survived one year. The shortest interval containing 0.95 of the conditional probability is now of the form $(1, c)$, where c is the solution to

$$0.95 = \int_1^c e^{-(x-1)} \, dx = e \left(e^{-1} - e^{-c} \right),$$

which implies that $c = -\ln \left(e^{-1} - 0.95e^{-1} \right) = 3.9957.$ ∎

Our main point in this section is simply that we are already somewhat familiar with inferential concepts. Further, via the principle of conditional probability, we have a basic rule or axiom governing how we go about making inferences in the context where the probability measure P is known and s is not known.

Summary of Section 5.2

- Probability models are used to model uncertainty about future responses.

- We can use the probability distribution to predict a future response or assess whether or not a given value makes sense as a possible future value from the distribution.

Exercises

5.2.1 Sometimes the *mode of a density* (the point where the density takes its maximum value) is chosen as a predictor for a future value of a response. Determine this predictor in Examples 5.2.1 and 5.2.2 and comment on its suitability as a predictor.

5.2.2 Suppose that it has been decided to use the mean of a distribution to predict a future response. In Example 5.2.1, compute the mean-squared error (expected value of the square of the error between a future value and its predictor) of this predictor, prior to observing the value. What characteristic of the distribution of the lifelength does this correspond to?

5.2.3 Graph the density of the distribution obtained as a mixture of a normal distribution with mean 4 and variance 1 and a normal distribution with mean -4 and variance 1, where the mixture probability is 0.5. Explain why neither the mean nor the mode is a suitable predictor in this case. (Hint: Section 2.5.4.)

5.2.4 Repeat the calculations of Examples 5.2.1 and 5.2.2 when the lifelength of a machine is known to be distributed as $Y = 10X$, where $X \sim \text{Uniform}(0, 1)$.

Problems

5.2.5 Suppose that a fair coin is tossed ten times and the response X measured is the number of times we observe a head.
(a) If you use the expected value of the response as a predictor, then what is the prediction of a future response X?

(b) Using Table D.6 (or a statistical package), compute a shortest interval containing at least 0.95 of the probability for X. Note that it might help to plot the probability function of X first.

(c) What region would you use to assess whether or not a value s_0 is a possible future value? (Hint: What are the regions of low probability for the distribution?) Assess whether or not $x = 8$ is plausible.

5.2.6 In Example 5.2.1, explain (intuitively) why the interval $(0, 2.9957)$ is the shortest interval containing 0.95 of the probability for the lifelength.

5.2.7 (Problem 5.2.5 continued) Suppose we are told that the number of heads observed is an even number. Repeat parts (a), (b), and (c).

5.2.8 Suppose that a response X is distributed Beta(a, b) with $a, b > 1$ fixed (see Problem 2.4.16). Determine the mean and the mode (point where density takes its maximum) of this distribution and assess which is the most accurate predictor of a future X when using mean-squared error, i.e., the expected squared distance between X and the prediction.

5.2.9 Suppose that a response X is distributed $N(0, 1)$ and that we have decided to predict a future value using the mean of the distribution.
(a) Determine the prediction for a future X.
(b) Determine the prediction for a future $Y = X^2$.
(c) Comment on the relationship (or the lack of a relationship) between the answers in parts (a) and (b).

Discussion Topic

5.2.10 Do you think it is realistic for a practitioner to proceed as if they know the true probability distribution for a response in a problem?

5.3   Statistical Models

In a statistical problem, we are faced with uncertainty of a different character than that arising in Section 5.2. In a statistical context, we observe the *data* s, but we are uncertain about P. In such a situation, we want to construct inferences about P based on s. This is the inverse of the situation discussed in Section 5.2.

How we should go about making these *statistical inferences* is probably not at all obvious to you. In fact, there are several possible approaches that we will discuss in subsequent chapters. In this chapter we will develop the basic ingredients of all the approaches.

Common to virtually all approaches to statistical inference is the concept of the *statistical model* for the data s. This takes the form of a set $\{P_\theta : \theta \in \Omega\}$ of probability measures, one of which corresponds to the true unknown probability measure P that produced the data s. In other words, we are asserting that there *is* a random mechanism generating s, and we *know* that the corresponding probability measure P is one of the probability measures in $\{P_\theta : \theta \in \Omega\}$.

The statistical model $\{P_\theta : \theta \in \Omega\}$ corresponds to the information a statistician brings to the application about what the true probability measure is, or at least what he or she is willing to assume about it. The variable θ is called the *parameter* of the model, and the set Ω is called the *parameter space*. Typically, we use models where $\theta \in \Omega$ indexes the probability measures in the model, i.e., $P_{\theta_1} = P_{\theta_2}$ if and only if $\theta_1 = \theta_2$. If the probability measures P_θ can all be presented via probability functions or density functions f_θ (for convenience we will not distinguish between the discrete and continuous case in the notation), then it is common to write the statistical model as $\{f_\theta : \theta \in \Omega\}$.

From the definition of a statistical model, we see that there is a unique value $\theta \in \Omega$, such that P_θ is the true probability measure. We refer to this value as the *true parameter value*. It is obviously equivalent to talk about making inferences about the true parameter value rather than the true probability measure, i.e., an inference about the true value of θ is at once an inference about the true probability distribution. So, for example, we may wish to estimate the true value of θ, construct small regions in Ω that are likely to contain the true value, or assess whether or not the data are in agreement with some particular value θ_0, suggested as being the true value. These are types of inferences, and note they are just like those we discussed in Section 5.2, but the situation here is quite different.

Example 5.3.1
Suppose we have an urn containing 100 chips, each colored either black or white. Suppose further that we are told that there are either 50 or 60 black chips in the urn. The chips are thoroughly mixed, and then two chips are withdrawn without replacement. The goal is to make an inference about the true number of black chips in the urn having observed the data $s = (s_1, s_2)$, where s_i is the color of the ith chip drawn.

In this case, we can take the statistical model to be $\{P_\theta : \theta \in \Omega\}$, where θ is the number of black chips in the urn, so that $\Omega = \{50, 60\}$, and P_θ is the probability measure on

$$S = \{(B, B), (B, W), (W, B), (W, W)\}$$

corresponding to θ. Therefore, P_{50} assigns the probability $50 \cdot 49/(100 \cdot 99)$ to each of the sequences (black, black) and (white, white) and the probability $50 \cdot 50/(100 \cdot 99)$ to each of the sequences (B, W) and (W, B), and P_{60} assigns the probability $60 \cdot 59/(100 \cdot 99)$ to the sequence (B, B), the probability $40 \cdot 39/(100 \cdot 99)$ to the sequence (W, W), and the probability $60 \cdot 40/(100 \cdot 99)$ to each of the sequences (B, W) and (W, B).

The choice of the parameter is somewhat arbitrary, as we could just have easily labelled the possible probability measures as P_1 and P_2, respectively. The parameter is in essence only a label that allows us to distinguish amongst the possible candidates for the true probability measure. It is typical, however, to choose this label conveniently so that it means something in the problem under discussion. ∎

We note some additional terminology in common usage. If a single observed value for a response X has the statistical model $\{f_\theta : \theta \in \Omega\}$, then a sample $(X_1, \ldots, X_n)$ (recall that sample here means that the X_i are independent and identically distributed — see Definition 2.8.6) has its distribution given by the joint density $f_\theta(x_1) f_\theta(x_2) \cdots f_\theta(x_n)$ for some $\theta \in \Omega$. This specifies the statistical model for the response $(X_1, \ldots, X_n)$. We refer to this as the *statistical model for a sample*. Of course, the true value of θ for the statistical model for a sample is the same as that for a single observation. Sometimes, rather than referring to the statistical model for a sample, we speak of a sample from the statistical model $\{f_\theta : \theta \in \Omega\}$.

Note that, wherever possible we will use uppercase letters to denote an unobserved value of a random variable X and lowercase letters to denote the observed value. So an observed sample $(X_1, \ldots, X_n)$ will be denoted $(x_1, \ldots, x_n)$.

Example 5.3.2

Suppose that there are two manufacturing plants for machines. It is known that machines built by the first plant have lifelengths distributed Exponential(1), while machines manufactured by the second plant have lifelengths distributed Exponential(1.5). The densities of these distributions are depicted in Figure 5.3.1.

You have purchased *five* of these machines and you know that all five came from the same plant but do not know which plant. Subsequently, you observe the lifelengths of these machines, obtaining the sample $(x_1, \ldots, x_5)$, and want to make inferences about the true P.

In this case, the statistical model for a single observation comprises two probability measures $\{P_1, P_2\}$, where P_1 is the Exponential(1) probability measure and P_2 is the Exponential(1.5) probability measure. Here we take the parameter to be $\theta \in \Omega = \{1, 2\}$.

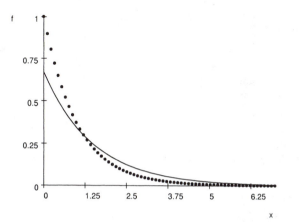

Figure 5.3.1: Plot of the Exponential(1) ($\bullet \bullet \bullet$) and Exponential(1.5) ($-$) densities.

Clearly, longer observed lifelengths favor $\theta = 2$. For example, if

$$(x_1, \ldots, x_5) = (5.0, 3.5, 3.3, 4.1, 2.8),$$

then intuitively we are more certain that $\theta = 2$ than if

$$(x_1, \ldots, x_5) = (2.0, 2.5, 3.0, 3.1, 1.8).$$

The subject of statistical inference is concerned with making statements like this more precise and quantifying our uncertainty concerning the validity of such assertions.

We note again that the quantity θ serves only as a label for the distributions in the model. The value of θ has no interpretation other than as a label and we could just as easily have used different values for the labels. In many applications, however, the parameter θ is taken to be some characteristic of the distribution that takes a unique value for each distribution in the model. Here we could have taken θ to be the mean and then the parameter space would be $\Omega = \{1, 1.5\}$. Notice that we could just as well have used the first quartile, or for that matter any other quantile, to have labelled the distributions, provided that each distribution in the family yields a unique value for the characteristic chosen. Generally, any one-to-one transformation of a parameter is acceptable as a parameterization of a statistical model. When we relabel, we refer to this as a *reparameterization* of the statistical model. ∎

We now consider two important examples of statistical models. These are important because they commonly arise in applications.

Example 5.3.3 *Bernoulli Model*
Suppose that $(x_1, \ldots, x_n)$ is a sample from a Bernoulli(θ) distribution with $\theta \in [0, 1]$ unknown. We could be observing the results of tossing a coin and recording X_i equal to 1 whenever a head is observed on the ith toss and equal to 0 otherwise. Alternatively we could be observing items produced in an industrial process and recording X_i equal to 1 whenever the ith item is defective and 0 otherwise. In a biomedical application, the response $X_i = 1$ might indicate that a treatment on a patient has been successful, while $X_i = 0$ indicates a failure. In all these cases, we want to know the true value of θ, as this tells us something important about the coin we are tossing, the industrial process, or the medical treatment, respectively.

Now suppose that we have no information whatsoever about the true probability θ. Accordingly, we take the parameter space to be $\Omega = [0, 1]$, the set of all possible values for θ. The probability function for the ith sample item is given by

$$f_\theta(x_i) = \theta^{x_i} (1 - \theta)^{1 - x_i},$$

and the probability function for the sample is given by

$$\prod_{i=1}^{n} f_\theta(x_i) = \prod_{i=1}^{n} \theta^{x_i} (1 - \theta)^{1 - x_i} = \theta^{n\bar{x}} (1 - \theta)^{n(1 - \bar{x})}.$$

This specifies the model for a sample.

Note that we could parameterize this model by any 1–1 function of θ. For example, $\alpha = \theta^2$ would work (as it is 1–1 on Ω), as would $\psi = \ln\{\theta/(1-\theta)\}$. ∎

Example 5.3.4 *Location-Scale Normal Model*
Suppose that $(x_1, \ldots, x_n)$ is a sample from an $N(\mu, \sigma^2)$ distribution with $\theta = (\mu, \sigma^2) \in R^1 \times R^+$ unknown, where $R^+ = (0, \infty)$. For example, we may have observations of heights in centimeters of individuals in a population and feel that it is reasonable to assume that the distribution of heights in the population is normal with some unknown mean and standard deviation.

The density for the sample is then given by

$$\prod_{i=1}^n f_{(\mu,\sigma^2)}(x_i) = \left(2\pi\sigma^2\right)^{-n/2} \exp\left\{-\frac{1}{2\sigma^2}\sum_{i=1}^n (x_i - \mu)^2\right\}$$

$$= \left(2\pi\sigma^2\right)^{-n/2} \exp\left\{-\frac{n}{2\sigma^2}(\bar{x} - \mu)^2 - \frac{n-1}{2\sigma^2}s^2\right\},$$

because (Problem 5.3.9)

$$\sum_{i=1}^n (x_i - \mu)^2 = n(\bar{x} - \mu)^2 + \sum_{i=1}^n (x_i - \bar{x})^2 \qquad (5.3.1)$$

where

$$\bar{x} = \frac{1}{n}\sum_{i=1}^n x_i$$

is the *sample mean*, and

$$s^2 = \frac{1}{n-1}\sum_{i=1}^n (x_i - \bar{x})^2$$

is the *sample variance*.

Alternative parameterizations for this model are commonly used. For example, rather than using (μ, σ^2), sometimes (μ, σ^{-2}) or (μ, σ) or $(\mu, \ln\sigma)$ are convenient choices. Note that $\ln\sigma$ ranges in R^1 as σ varies in R^+. ∎

Actually, we might wonder how appropriate the model of Example 5.3.4 is for the distribution of heights in a population, for in any finite population the true distribution is discrete (there are only finitely many students). Of course, a normal distribution may provide a good approximation to a discrete distribution, as in Example 4.4.9. So, in Example 5.3.4, we are also assuming that a continuous probability distribution can provide a close approximation to the true discrete distribution. As it turns out, such approximations can lead to great simplifications in the derivation of inferences, so we use them whenever it is feasible. Such an approximation is, of course, not applicable in Example 5.3.3.

Also note that heights will always be expressed in some specific unit, e.g., centimeters; based on this, we know that the population mean must be in a certain range of values, e.g., $\mu \in (0, 300)$, but the statistical model allows for any value for μ. So we often do have additional information about the true value of the parameter for a model, but it is somewhat imprecise, e.g., we also probably have $\mu \in (100, 300)$. In Chapter 7 we will discuss ways of incorporating such information into our analysis.

Where does the model information $\{P_\theta : \theta \in \Omega\}$ come from in an application? For example, how could we know that heights are approximately normally distributed in Example 5.3.4? Sometimes there is such information based upon previous experience with related applications, but often it is an *assumption* that requires checking before inference procedures can be used. Procedures designed to check such assumptions are referred to as *model-checking* procedures which will be discussed in Chapter 9. In practice, model-checking procedures are required, or else inferences drawn from the data and statistical model can be erroneous if the model is wrong.

Summary of Section 5.3

- In a statistical application, we do not know the distribution of a response but know (or are willing to assume) that the true probability distribution is one of a set of possible distributions $\{f_\theta : \theta \in \Omega\}$, where f_θ is the density or probability function (whichever is relevant) for the response. The set of possible distributions is called the *statistical model*.

- The set Ω is called the *parameter space,* and the variable θ is called the *parameter* of the model. Because each value of θ corresponds to a distinct probability distribution in the model, we can talk about the *true value* of θ, as this gives the true distribution via f_θ.

Exercises

5.3.1 Suppose that there are three coins — one of which is known to be fair, one has probability 1/3 of yielding a head on a single toss, and the other one has probability 2/3 for head on a single toss. A coin is selected (not randomly) and then tossed five times. The goal is to make inference about which of the coins is being tossed based on the sample. Fully describe a statistical model for a single response and for the sample.

5.3.2 Suppose that one face of a symmetrical six-sided die is duplicated but we do not know which one. We do know that if 1 is duplicated, then 2 does not appear; otherwise, 1 does not appear. Describe the statistical model for a single roll.

5.3.3 Suppose we have two populations (I and II), and variable X is known to be distributed $N(10, 2)$ on population I and distributed $N(8, 3)$ on population II. A sample $(X_1, \ldots, X_n)$ is generated from one of the populations; you are not told which population the sample came from, but you are required to draw inferences about the true distribution based on the sample. Describe the statistical model for this problem. Could you parameterize this model by the population mean, by the population variance? Sometimes problems like this are called *classification problems* because making inferences about the true distribution is equivalent to classifying the sample as belonging to one of the populations.

5.3.4 Suppose the situation is as described in Exercise 5.3.3, but now the distribution for population I is $N(10, 2)$ and the distribution for population II is $N(10, 3)$. Could you parameterize the model by the population mean, by the population variance? Justify your answer.

5.3.5 Suppose that a manufacturing process produces batteries whose lifelengths are known to be exponentially distributed but with the mean of the distribution completely unknown. Describe the statistical model for a single observation. Is it possible to parameterize this model by the mean? Is it possible to parameterize this model by the variance? Is it possible to parameterize this model by the *coefficient of variation* (the coefficient of variation of a distribution equals the standard deviation divided by the mean)?

5.3.6 Suppose it is known that a response X is distributed Uniform$[0, \beta]$, where $\beta > 0$ is unknown. Is it possible to parameterize this model by the first quartile of the distribution? (The first quartile of the distribution of a random variable X is the point c satisfying $P(X \leq c) = 0.25$.) Explain why or why not.

Problems

5.3.7 Suppose in Example 5.3.3 we parameterize the model by $\psi = \ln \{\theta/(1-\theta)\}$. Record the statistical model using this parameterization, i.e., record the probability function using ψ as the parameter and record the relevant parameter space.

5.3.8 Suppose in Example 5.3.4 we parameterize the model by $(\mu, \ln \sigma) = (\mu, \psi)$. Record the statistical model using this parameterization, i.e., record the density function using (μ, ψ) as the parameter and record the relevant parameter space.

5.3.9 Establish the identity (5.3.1).

5.3.10 A sample $(X_1, \ldots, X_n)$ is generated from a Bernoulli(θ) distribution with $\theta \in [0, 1]$ unknown, but only $T = \sum_{i=1}^{n} X_i$ is observed by the statistician. Describe the statistical model for the observed data.

5.3.11 Suppose it is known that a response X is distributed $N(\mu, \sigma^2)$, where $\theta = (\mu, \sigma^2) \in R^1 \times R^+$ and θ is completely unknown. Show how to calculate the first quartile of each distribution in this model from the values (μ, σ^2). Is it possible to parameterize the model by the first quartile? Explain your answer.

5.3.12 Suppose that response X is known to be distributed $N(Y, \sigma^2)$, where $Y \sim N(0, \delta^2)$ and $\sigma^2, \delta^2 > 0$ are completely unknown. Describe the statistical model for an observation (X, Y). If Y is not observed, describe the statistical model for X.

Discussion Topics

5.3.13 Explain why you think it is important that statisticians state very clearly what they are assuming any time they carry out a statistical analysis.

5.3.14 Consider the statistical model given by the collection of $N(\mu, \sigma_0^2)$ distributions where $\mu \in R^1$ is considered completely unknown, but σ_0^2 is assumed known. Do you think this is a reasonable model to use in an application? Give your reasons why or why not.

5.4   Data Collection

The developments of Sections 5.2 and 5.3 are based on the observed response s being a realization from a probability measure P. In fact, in many applications, this is an assumption. We are often presented with data that could have been produced in this way, but we cannot always be sure.

When we cannot be sure that the data were produced by a random mechanism, then the statistical analysis of the data is known as an *observational study*. In an observational study, the statistician merely observes the data rather than intervening directly in their generation, to ensure that the randomness assumption holds. For example, suppose that a professor collects data from his students for a study concerned with examining the relationship between grades and part-time employment. Is it reasonable to regard the data collected as having come from a probability distribution? If so, how would we justify this?

It is important that a statistician distinguish carefully between situations that are observational studies and those that are not. As the following discussion illustrates, there are qualifications that must be applied to the analysis of an observational study. While statistical analyses of observational studies are valid and indeed important, we must be aware of their limitations when interpreting their results.

5.4.1   Finite Populations

Suppose that we have a finite set Π of objects, called the *population*, and a real-valued function X (sometimes called a *measurement*) defined on Π. So for each $\pi \in \Pi$ we have a real-valued quantity $X(\pi)$ that measures some aspect or feature of π.

For example, Π could be the set of all students currently enrolled full-time at a particular university, with $X(\pi)$ the height of student π in centimeters. Or, for the same Π, we could take $X(\pi)$ to be the gender of student π, where $X(\pi) = 1$ denotes female and $X(\pi) = 2$ denotes male. Here height is a *quantitative variable*, because its values mean something on a numerical scale, and we can perform arithmetic on these values, e.g., calculate a mean. On the other hand, gender is an example of a *categorical variable* because its values serve only to classify, and any other choice of unique real numbers would have served as well as the ones we have chosen. The first step in a statistical analysis is to determine the types of variables we are working with, because the relevant statistical analysis techniques depend on this.

The population and the measurement together produce a *population distribution* over the population. This is specified by the *population cumulative distribution function* $F_X : R^1 \rightarrow [0, 1]$ where

$$F_X(x) = \frac{|\{\pi : X(\pi) \leq x\}|}{N},$$

with $|A|$ being the number of elements in the set A, and $N = |\Pi|$. Therefore, $F_X(x)$ is the proportion of elements in Π with their measurement less than or equal to x.

Consider the following simple example where we can calculate F_X exactly.

Example 5.4.1
Suppose that Π is a population of $N = 20$ plots of land of the same size. Further suppose that $X(\pi)$ is a measure of the fertility of plot π on a ten-point scale and the following measurements were obtained.

4	8	6	7	8	3	7	5	4	6
9	5	7	5	8	3	4	7	8	3

Then we have

$$F_X(x) = \begin{cases} 0 & x < 3 \\ 3/20 & 3 \leq x < 4 \\ 6/20 & 4 \leq x < 5 \\ 9/20 & 5 \leq x < 6 \\ 11/20 & 6 \leq x < 7 \\ 15/20 & 7 \leq x < 8 \\ 19/20 & 8 \leq x < 9 \\ 1 & 9 \leq x, \end{cases}$$

because, for example, 6 out of the 20 plots have fertility measurements less than or equal to 4. ∎

The goal of a statistician in this context is to know the function F_X as precisely as possible. If we know F_X exactly, then we have identified the distribution of X over Π. One way of knowing the distribution exactly is to conduct a *census*, namely, the statistician goes out and observes $X(\pi)$ for every $\pi \in \Pi$ and then calculates F_X. Sometimes this is feasible, but often it is not possible or even desirable, due to the costs involved in the accurate accumulation of all the measurements — think of how difficult it might be to collect the heights of all the students at your school.

While sometimes a census is necessary, even mandated by law, often a very accurate approximation to F_X can be obtained by selecting a subset

$$\{\pi_1, \ldots, \pi_n\} \subset \Pi$$

for some $n < N$. We then approximate $F_X(x)$ by the *empirical distribution function* defined by

$$
\hat{F}_X(x) = \frac{|\{\pi_i : X(\pi_i) \le x, i = 1, \ldots, n\}|}{n}
$$

$$
= \frac{1}{n} \sum_{i=1}^{n} I_{(-\infty, x]}(X(\pi_i)).
$$

We could also measure more than one aspect of π to produce a *multivariate measurement* $X : \Pi \to R^k$ for some k. For example, if Π is again the population of students, we might have $X(\pi) = (X_1(\pi), X_2(\pi))$, where $X_1(\pi)$ is the height in centimeters of student π and $X_2(\pi)$ is the weight of student π in kilograms. We will discuss multivariate measurements in Chapter 10, where our concern is with relationships amongst variables, but we focus on univariate measurements here.

There are two questions we need to answer now — namely, how should we select the subset $\{\pi_1, \ldots, \pi_n\}$ and how large should n be?

5.4.2   Simple Random Sampling

We will first address the issue of selecting $\{\pi_1, \ldots, \pi_n\}$. Suppose we select this subset according to some given rule based on the unique label that each $\pi \in \Pi$ possesses. For example, if the label is a number, we might order the numbers and then take the n elements with the smallest labels. Or we could order the numbers and take every other element until we have a subset of n, etc.

There are many such rules we could apply, and there is a basic problem with all of them. If we want $\hat{F}_X$ to approximate F_X for the full population then, when we employ a rule, we run the risk of only selecting $\{\pi_1, \ldots, \pi_n\}$ from a subpopulation. For example, if we use student numbers to identify each element of a population of students, and more senior students have lower student numbers, then, when n is much smaller than N and we select the students with smallest student numbers, $\hat{F}_X$ is really only approximating the distribution of X in the population of senior students at best. This distribution could be very different than F_X. Similarly, for any other rule we employ, even if we cannot imagine what the subpopulation could be, there may be such a *selection effect,* or *bias,* induced that renders the estimate invalid.

This is the qualification we need to apply when analyzing the results of observational studies. In an observational study, the data are generated by some rule, typically unknown to the statistician, and this means that any conclusions drawn based on the data $X(\pi_1), ,\ldots, X(\pi_n)$ may not be valid for the full population.

There seems to be only one way to guarantee that selection effects are avoided, namely, the set $\{\pi_1, \ldots, \pi_n\}$ must be selected using randomness. For *simple random sampling,* this means that a random mechanism is used to select the π_i in such a way that each subset of n has probability $1/\binom{N}{n}$ of being chosen. For example, we might place N chips in a bowl, each with a unique label on it corresponding to a population element, and then randomly draw n chips from the bowl without replacement. The labels on the drawn chips identify the individuals that have been selected from Π. Alternatively, for the randomization, we might use a table of random numbers, such as Table D.1 in Appendix D (see Table D.1 for a description of how it is used) or generate random values using a computer algorithm (see Section 2.10).

Note that with simple random sampling $(X(\pi_1),,\ldots,X(\pi_n))$ is random. In particular, when $n = 1$, we then have

$$P(X(\pi_1) \leq x) = F_X(x),$$

namely, the probability distribution of the random variable $X(\pi_1)$ is the same as the population distribution.

Example 5.4.2
Consider the context of Example 5.4.1. When we randomly select the first plot from Π, it is clear that each plot has probability $1/20$ of being selected. Then we have

$$P(X(\pi_1) \leq x) = \frac{|\{\pi : X(\pi) \leq x\}|}{20} = F_X(x)$$

for every $x \in R^1$. ∎

Prior to observing the sample, we also have $P(X(\pi_2) \leq x) = F_X(x)$. Consider, however, the distribution of $X(\pi_2)$ given that $X(\pi_1) = x_1$. Because we have removed one population member, with measurement value x_1, then $NF_X(x) - 1$ is the number of individuals left in Π with $X(\pi) \leq x_1$. Therefore,

$$P(X(\pi_2) \leq x \mid X(\pi_1) = x_1) = \begin{cases} \frac{NF_X(x)-1}{N-1} & x \leq x_1 \\ \frac{NF_X(x)}{N-1} & x > x_1. \end{cases}$$

Note that this is not equal to $F_X(x)$.

So with simple random sampling, $X(\pi_1)$ and $X(\pi_2)$ are not independent. Observe, however, that when N is large, then

$$P(X(\pi_2) \leq x \mid X(\pi_1) = x_1) \approx F_X(x),$$

so that $X(\pi_1)$ and $X(\pi_2)$ are approximately independent and identically distributed (i.i.d.). Similar calculations lead to the conclusion that, when N is large and n is small relative to N, then with simple random sampling from the population, the random variables

$$X(\pi_1), \ldots, X(\pi_n)$$

are approximately i.i.d. and with distribution given by F_X. So we will treat the observed values $(x_1, \ldots, x_n)$ of $(X(\pi_1), \ldots, X(\pi_n))$ as a sample (in the sense of Definition 2.8.6) from F_X. In this text, unless we indicate otherwise, we will always assume that n is small relative to N so that this approximation makes sense.

Under the i.i.d. assumption, the weak law of large numbers (Theorem 4.2.1) implies that the empirical distribution function $\hat{F}_X$ satisfies

$$\hat{F}_X(x) = \frac{1}{n} \sum_{i=1}^n I_{(-\infty, x]}\left(X\left(\pi_i\right)\right) \xrightarrow{P} F_X(x)$$

as $n \to \infty$. So we see that $\hat{F}_X$ can be considered as an estimate of the population cdf F_X.

Whenever the data have been collected using simple random sampling, we will refer to the statistical investigation as a *sampling study*. It is a basic principle of good statistical practice that sampling studies are always to be preferred to observational studies, whenever they are feasible. This is because we can be sure that, with a sampling study, any conclusions we draw based on the sample $\pi_1, , \ldots, \pi_n$ will apply to the population Π of interest. With observational studies, we can never be sure that the sample data have not actually been selected from some proper subset of Π. For example, if you were asked to make inferences about the distribution of student heights at your school but selected some of your friends as your sample, then it is clear that the estimated cdf may be very unlike the true cdf (possibly more of your friends are of one gender than the other).

Often, however, we have no choice but to use observational data for a statistical analysis. Sampling directly from the population of interest may be extremely difficult or even impossible. We can still treat the results of such analyses as a form of evidence, but we must be wary about possible selection effects and acknowledge this possibility. Sampling studies constitute a higher level of statistical evidence than observational studies do, as they avoid the possibility of selection effects.

In Chapter 10 we will discuss *experiments* that constitute the highest level of statistical evidence. Experiments are appropriate when we are investigating the possibility of cause–effect relationships existing amongst variables defined on populations.

The second question we need to address concerns the choice of the sample size n. It seems natural that we would like to choose this as large as possible. On the other hand, there are always costs associated with sampling, and sometimes each sample value is very expensive to obtain. Further, it is often the case that the more data we collect, the more difficulty we have in making sure that the data are not corrupted by various errors that can arise in the collection process. So our answer, concerning how large n need be, is that we want it chosen large enough so that we obtain the accuracy necessary but no larger. Accordingly, the statistician must specify what accuracy is required and then n is determined.

We will see in the subsequent chapters that there are various methods for specifying the required accuracy in a problem and then determining an appropriate value for n. Determining n is a key component in the implementation of a sampling study and is often referred to as a *sample-size calculation*.

If we define

$$f_X(x) = \frac{|\{\pi : X(\pi) = x\}|}{N} = \frac{1}{N} \sum_{\pi \in \Pi} I_{\{x\}}(X(\pi)),$$

namely, $f_X(x)$ is the proportion of population members satisfying $X(\pi) = x$, then we see that f_X plays the role of the probability function because

$$F_X(x) = \sum_{z \leq x} f_X(z).$$

We refer to f_X as the *population relative frequency function*. Now, $f_X(x)$ may be estimated, based on the sample $\{\pi_1, \ldots, \pi_n\}$, by

$$\hat{f}_X(x) = \frac{|\{\pi_i : X(\pi_i) = x, i = 1, \ldots, n\}|}{n} = \frac{1}{n} \sum_{i=1}^{n} I_{\{x\}}(X(\pi_i)),$$

namely, the proportion of sample members π satisfying $X(\pi) = x$.

With categorical variables, $\hat{f}_X(x)$ estimates the population proportion $f_X(x)$ in the category specified by x. With some quantitative variables, however, f_X is not an appropriate quantity to estimate, and an alternative function must be considered.

5.4.3 Histograms

Quantitative variables can be further classified as either discrete or continuous variables. Continuous variables are those that we can measure to an arbitrary precision as we increase the accuracy of a measuring instrument. For example, the height of an individual could be considered to be a continuous variable, whereas the number of years of education an individual possesses would be considered a discrete quantitative variable. For discrete quantitative variables, f_X is an appropriate quantity to describe a population distribution, but we proceed differently with continuous quantitative variables.

Suppose that X is a continuous quantitative variable. In this case it makes more sense to *group* values into intervals, given by

$$(h_1, h_2], (h_2, h_3], \ldots, (h_{m-1}, h_m],$$

where the h_i are chosen to satisfy $h_1 < h_2 < \cdots < h_m$ with (h_1, h_m) effectively covering the range of possible values for X. Then we define

$$h_X(x) = \begin{cases} \frac{|\{\pi : X(\pi) \in (h_i, h_{i+1}]\}|}{N(h_{i+1} - h_i)} & x \in (h_i, h_{i+1}] \\ 0 & \text{otherwise} \end{cases}$$

and refer to h_X as a *density histogram function*. Here $h_X(x)$ is the proportion of population elements π that have their measurement $X(\pi)$ in the interval $(h_i, h_{i+1}]$ containing x, divided by the length of the interval.

In Figure 5.4.1, we have plotted a density histogram based on a sample of 10,000 from an $N(0,1)$ distribution (here we are treating this sample as the full population) and using the values $h_1 = -5, h_2 = -4, \ldots, h_{11} = 5$. Note that the vertical lines are only artifacts of the plotting software and do not represent values of h_X, as these are given by the horizontal lines.

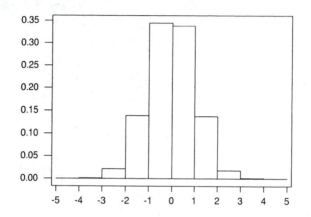

Figure 5.4.1: Density histogram function for a sample of 10,000 from a $N(0,1)$ distribution using the values $h_1 = -5, h_2 = -4, \ldots, h_{11} = 5$.

If $x \in (h_i, h_{i+1}]$, then $h_X(x)(h_{i+1} - h_i)$ gives the proportion of individuals in the population that have their measurement $X(\pi)$ in $(h_i, h_{i+1}]$. Further, we have

$$F_X(h_j) = \int_{-\infty}^{h_j} h_X(x)\, dx$$

for each interval endpoint and

$$F_X(h_j) - F_X(h_i) = \int_{h_i}^{h_j} h_X(x)\, dx$$

when $h_i \leq h_j$. If the intervals $(h_i, h_{i+1}]$ are small, then we expect that

$$F_X(b) - F_X(a) \approx \int_a^b h_X(x)\, dx$$

for any choice of $a < b$.

Now suppose that the lengths $h_{i+1} - h_i$ are small and N is very large. Then it makes sense to imagine a smooth, continuous function f_X, e.g., perhaps a normal or gamma density function, that approximates h_X in the sense that

$$\int_a^b f_X(x)\, dx \approx \int_a^b h_X(x)\, dx$$

for every $a < b$. Then we will also have

$$\int_a^b f_X(x)\, dx \approx F_X(b) - F_X(a)$$

for every $a < b$. We will refer to such an f_X as a density function for the population distribution.

In essence, this is how many continuous distributions arise in practice. In Figure 5.4.2, we have plotted a density histogram for the same values as used in Figure 5.4.1, but this time we used the interval endpoints $h_1 = -5, h_2 = -4.75, \ldots, h_{41} = 5$. We note that Figure 5.4.2 looks much more like a continuous function than does Figure 5.4.1.

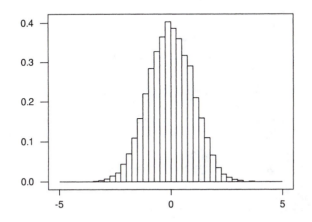

Figure 5.4.2: Density histogram function for a sample of 10,000 from a $N(0, 1)$ distribution using the values $h_1 = -5, h_2 = -4.75, \ldots, h_{41} = 5$.

5.4.4 Survey Sampling

Finite population sampling provides the formulation for a very important application of statistics, namely, *survey sampling* or *polling*. Typically, a survey consists of a set of questions that are asked of a sample $\{\pi_1, \ldots, \pi_n\}$ from a population Π. Each question corresponds to a measurement, so if there are m questions, the response from a respondent π is the m-dimensional vector $(X_1(\pi), X_2(\pi), \ldots, X_m(\pi))$. A very important example of survey sampling is the pre-election polling that is undertaken to predict the outcome of a vote. Also, many consumer product companies engage in extensive market surveys to try to learn what consumers want and so gain information that can lead to improved sales.

Typically, the analysis of the results will be concerned not only with the population distributions of the individual X_i over the population but also the joint population distributions. For example, the joint cumulative distribution function of (X_1, X_2) is given by

$$F_{(X_1,X_2)}(x_1, x_2) = \frac{|\{\pi : X_1(\pi) \le x_1, X_2(\pi) \le x_2\}|}{N},$$

namely, $F_{(X_1,X_2)}(x_1, x_2)$ is the proportion of the individuals in the population whose X_1 measurement is no greater than x_1 and whose X_2 measurement is no greater than x_2. Of course, we can also define the joint distributions of three or more measurements. These joint distributions are what we use to answer questions like, is there a relationship between X_1 and X_2, and if so, what form does it take? This topic will be extensively discussed in Chapter 10. We can also define $f_{(X_1, X_2)}$ for the joint distribution, and joint density histograms are again useful when X_1 and X_2 are both continuous quantitative variables.

Example 5.4.3
Suppose there are four candidates running for the position of mayor in a particular city. A random sample of 1000 voters is selected; they are each asked if they will vote and, if so, which of the four candidates they will vote for. Additionally, the respondents are asked their age. We denote the answer to the question of whether or not they will vote by X_1, with $X_1(\pi) = 1$ meaning yes and $X_1(\pi) = 0$ meaning no. For those voting, we denote by X_2 the response concerning which candidate they will vote for, with $X_2(\pi) = i$ indicating candidate i. Finally, the age in years of the respondent is denoted by X_3. In addition to the distributions of X_1 and X_2, the pollster is also interested in the joint distributions of (X_1, X_3) and (X_2, X_3), as these tell us about the relationship between voter participation and age in the first case and candidate choice and age in the second case. ∎

There are many interesting and important aspects to survey sampling that go well beyond this book. For example, it is often the case with human populations that a randomly selected person will not respond to a survey. This is called *nonresponse error*, and it is a serious selection effect. The sampler must design the study carefully to try and mitigate the effects of nonresponse error. Further, there are variants of simple random sampling (see Problem 5.4.10) that can be preferable in certain contexts, as these increase the accuracy of the results. The design of the actual questionnaire used is also very important, as we must ensure that responses address the issues intended without biasing the results.

Summary of Section 5.4

- Simple random sampling from a population Π means that we randomly select a subset of size n from Π in such way that each subset of n has the same probability, namely, $1/\binom{|\Pi|}{n}$ of being selected.

- Data that arise from a sampling study are generated from the distribution of the measurement of interest X over the whole population Π rather than some subpopulation. This is why sampling studies are preferred to observational studies.

- When the sample size n is small relative to $|\Pi|$, we can treat the observed values of X as a sample from the distribution of X over the population.

Exercises

5.4.1 Suppose that we have a population $\Pi = \{\pi_1, \ldots, \pi_{10}\}$ and quantitative measurement X given by:

i	1	2	3	4	5	6	7	8	9	10
$X(\pi_i)$	1	1	2	1	2	3	3	1	2	4

Calculate F_X, f_X, μ_X, and σ_X^2.

5.4.2 Suppose you take a sample of $n = 3$ (without replacement) from the population in Exercise 5.4.1.
(a) Can you consider this as an approximate i.i.d. sample from the population distribution? Why or why not?
(b) Explain how you would actually physically carry out the sampling from the population in this case. (Hint: Table D.1.)
(c) Using the method you outlined in part (b), generate three samples of size $n = 3$ and calculate $\bar{X}$ for each sample.

5.4.3 Suppose you take a sample of $n = 4$ (with replacement) from the population in Exercise 5.4.1.
(a) Can you consider this as an approximate i.i.d. sample from the population distribution? Why or why not?
(b) Explain how you would actually physically carry out the sampling in this case.
(c) Using the method you outlined in part (b), generate three samples of size $n = 3$ and calculate $\bar{X}$ for each sample.

5.4.4 Suppose that we have a finite population Π and a measurement $X : \Pi \rightarrow \{0, 1\}$ where $|\Pi| = N$ and $|\{\pi : X(\pi) = 0\}| = a$.
(a) Determine $f_X(0)$ and $f_X(1)$. Can you identify this population distribution?
(b) For a simple random sample of size n, determine the probability that $\hat{f}_X(0) = f_X(0)$. (Hint: Hypergeometric distribution.)
(c) Under the assumption of i.i.d. sampling, determine the probability that $\hat{f}_X(0) = f_X(0)$. (Hint: Binomial distribution.)

5.4.5 Suppose the following sample of size of $n = 20$ is obtained from an industrial process.

3.9	7.2	6.9	4.5	5.8	3.7	4.4	4.5	5.6	2.5
4.8	8.5	4.3	1.2	2.3	3.1	3.4	4.8	1.8	3.7

(a) Construct a density histogram for this data set using the intervals $(1, 4.5], (4.5, 5.5], (5.5, 6.5], (6.5, 10]$.
(b) Construct a density histogram for this data set using the intervals $(1, 3.5], (3.5, 4.5], (4.5, 6.5], (6.5, 10]$.
(c) Based on the results of parts (b) and (c), what do you conclude about histograms?

5.4.6 Suppose it is known that in a population of 1000 students, 350 students will vote for party A, 550 students will vote for party B, and the remaining students will vote for party C.
(a) Explain how such information can be obtained.
(b) If we let $X : \Pi \to \{A, B, C\}$ be such that $X(\pi)$ is the party that π will vote for, then explain why we can't represent the population distribution of X by F_X.
(c) Compute f_X.
(d) Explain how one might go about estimating f_X prior to the election.
(e) What is unrealistic about the population distribution specified via f_X? (Hint: Does it seem realistic based on what you know about voting behavior?)

Computer Exercises

5.4.7 Generate a sample of 1000 from an $N(3, 2)$ distribution.
(a) Calculate $\hat{F}_X$ for this sample.
(b) Plot a density histogram based on these data using the intervals of length 1 over the range $(-5, 10)$.
(c) Plot a density histogram based on these data using the intervals of length 0.1 over the range $(-5, 10)$.
(d) Comment on the difference in the look of the histograms in parts (b) and (c). What do you attribute this to?
(e) What limits the size of the intervals we use to group observations when we are plotting histograms?

5.4.8 Suppose we have a population of 10,000 elements, each with a unique label from the set $\{1, \ldots, 10,000\}$.
(a) Generate a sample of 500 labels from this population using simple random sampling.
(b) Generate a sample of 500 labels from this population using i.i.d. sampling.

Problems

5.4.9 Suppose that we have a finite population Π and a measurement $X : \Pi \to \{0, 1, 2\}$ where $|\Pi| = N$ and $|\{\pi : X(\pi) = 0\}| = a$ and $|\{\pi : X(\pi) = 1\}| = b$. This problem generalizes Exercise 5.4.4.
(a) Determine $f_X(0), f_X(1),$ and $f_X(2)$.
(b) For a simple random sample of size n, determine the probability that $\hat{f}_X(0) = f_0, \hat{f}_X(1) = f_1,$ and $\hat{f}_X(2) = f_2$.
(c) Under the assumption of i.i.d. sampling, determine the probability that $\hat{f}_X(0) = f_0, \hat{f}_X(1) = f_1,$ and $\hat{f}_X(2) = f_2$.

5.4.10 Suppose X is a quantitative measurement defined on a finite population.
(a) Prove that the population mean equals $\mu_X = \sum_x x f_X(x)$, i.e., the average of $X(\pi)$ over all population elements π equals μ_X.
(b) Prove that the population variance is given by $\sigma_X^2 = \sum_x (x - \mu_X)^2 f_X(x)$, i.e., the average of $(X(\pi) - \mu_X)^2$ over all population elements π equals σ_X^2.

5.4.11 Suppose we have the situation described in Exercise 5.4.4, and we take a simple random sample of size n from Π where $|\Pi| = N$.

(a) Prove that the mean of $\hat{f}_X(0)$ is given by $f_X(0)$. (Hint: Note that we can write $\hat{f}_X(0) = n^{-1} \sum_{i=1}^n I_{\{0\}}(X(\pi_i))$ and $I_{\{0\}}(X(\pi_i)) \sim \text{Bernoulli}(f_X(0))$.)

(b) Prove that the variance of $\hat{f}_X(0)$ is given by

$$\frac{f_X(0)(1 - f_X(0))}{n} \frac{N - n}{N - 1}. \tag{5.4.1}$$

(Hint: Use the hint in part (a), but note that the $I_{\{0\}}(X(\pi_i))$ are not independent, use Theorem 3.3.4(b) and evaluate $\text{Cov}(I_{\{0\}}(X(\pi_i)), I_{\{0\}}(X(\pi_i)))$ in terms of $f_X(0)$.)

(c) Repeat the calculations in parts (a) and (b), but this time assume that you take a sample of n with replacement. (Hint: Use the hint in Exercise 5.4.4(c).)

(d) Explain why the factor $(N - n)/(N - 1)$ in (5.4.1) is called the *finite sample correction factor*.

5.4.12 Suppose we have a finite population Π and we do not know $|\Pi| = N$. In addition, suppose we have a measurement variable $X : \Pi \to \{0, 1\}$ and we know that $N f_X(0) = a$ where a is known. Based on a simple random sample of n from Π, determine an estimator of N. (Hint: Use a function of $\hat{f}_X(0)$.)

5.4.13 Suppose that X is a quantitative variable defined on a population Π, and we take a simple random sample of size n from Π.

(a) If we estimate the population mean μ_X by the sample mean

$$\bar{X} = \frac{1}{n} \sum_{i=1}^n X(\pi_i),$$

prove that $E(\bar{X}) = \mu_X$ where μ_X is defined in Problem 5.4.10(a). (Hint: What is the distribution of each $X(\pi_i)$?)

(b) Under the assumption that i.i.d. sampling makes sense, show that the variance of $\bar{X}$ equals σ_X^2/n where σ_X^2 is defined in Problem 5.4.10(b).

5.4.14 Suppose we have a finite population Π and we do not know $|\Pi| = N$. In addition, suppose we have a measurement variable $X : \Pi \to R^1$ and we know $T = \sum_\pi X(\pi)$. Based on a simple random sample of n from Π, determine an estimator of N. (Hint: Use a function of $\bar{X}$.)

5.4.15 Under i.i.d. sampling, prove that $\hat{f}_X(x) \overset{D}{\to} f_X(x)$ as $n \to \infty$. (Hint: $\hat{f}_X(x) = n^{-1} \sum_{i=1}^n I_{\{x\}}(X(\pi_i))$.)

Challenge

5.4.16 (*Stratified sampling*) Suppose that X is a quantitative variable defined on a population Π and that we can partition Π into two subpopulations Π_1 and Π_2 such that a proportion p of the full population is in Π_1. Let f_{iX} denote the conditional population distribution of X on Π_i.

(a) Prove that $f_X(x) = p f_{1X}(x) + (1-p) f_{2X}(x)$.

(b) Establish that $\mu_X = p\mu_{1X} + (1-p)\mu_{2X}$ where μ_{iX} is the mean of X on Π_i.

(c) Establish that $\sigma_X^2 = p\sigma_{1X}^2 + (1-p)\sigma_{2X}^2 + p(1-p)(\mu_{1X} - \mu_{2X})^2$.

(d) Suppose that it makes sense to assume i.i.d. sampling whenever we take a sample from either the full population or either of the subpopulations, i.e., the sample sizes we are considering are small relative to the sizes of these populations. We implement stratified sampling by taking a simple random sample of size n_i from subpopulation Π_i. We then estimate μ_X by $p\bar{X}_1 + (1-p)\bar{X}_2$ where $\bar{X}_i$ is the sample mean based on the sample from Π_i. Prove that $E\left(p\bar{X}_1 + (1-p)\bar{X}_2\right) = \mu_X$ and

$$\text{Var}\left(p\bar{X}_1 + (1-p)\bar{X}_2\right) = p^2 \frac{\sigma_{1X}^2}{n_1} + (1-p)^2 \frac{\sigma_{2X}^2}{n_2}.$$

(e) Under the assumptions of part (d), prove that

$$\text{Var}\left(p\bar{X}_1 + (1-p)\bar{X}_2\right) \leq \text{Var}\left(\bar{X}\right)$$

when $\bar{X}$ is based on a simple random sample of size n from the full population and $n_1 = pn, n_2 = (1-p)n$. This is called *proportional stratified sampling*.

(f) Under what conditions is there no benefit to proportional stratified sampling? What do you conclude about situations in which stratified sampling will be most beneficial?

Discussion Topics

5.4.17 Sometimes it is argued that it is possible for a skilled practitioner to pick a more accurate representative sample of a population deterministically rather than by employing simple random sampling. This argument is based in part on the argument that it is always possible with simple random sampling that we could get a very unrepresentative sample through pure chance and that this can be avoided by an expert. Comment on this assertion.

5.4.18 Suppose that it is claimed that a quantitative measurement X defined on a finite population Π is approximately distributed according to a normal distribution with unknown mean and unknown variance. Explain fully what this claim means.

5.5   Some Basic Inferences

Now suppose that we are in a situation involving a measurement X, whose distribution is unknown, and we have obtained the data $(x_1, x_2, \ldots, x_n)$, i.e., observed n values of X. Hopefully, these data were the result of simple random sampling, but perhaps they were collected as part of an observational study. Denote the associated unknown population relative frequency function, or an approximating density, by f_X and the population distribution function by F_X.

What we do now with the data depends on two things. First, we have to determine what we want to know about the underlying population distribution.

Typically, our interest is in only a few characteristics, of this distribution like the mean and variance. Second, we have to use statistical theory to combine the data with the statistical model to make inferences about the characteristics of interest.

We now discuss some typical characteristics of interest. Also, we present some informal estimation methods for these characteristics, known as *descriptive statistics*. These are often used as a preliminary step before more formal inferences are drawn and are justified on simple intuitive grounds. They are called descriptive because they are estimating quantities that *describe* features of the underlying distribution.

5.5.1 Descriptive Statistics

There are a variety of characteristics of distributions on which statisticians often focus. We present some of these in the following examples.

Example 5.5.1 *Estimating Proportions and Cumulative Proportions*
Often we want to make inferences about the value $f_X(x)$ or the value $F_X(x)$ for a specific x. Recall that $f_X(x)$ is the proportion of population members whose X measurement equals x. In general, $F_X(x)$ is the proportion of population members whose X measurement is less than or equal to x.

Now suppose we have a sample $(x_1, x_2, \ldots, x_n)$ from f_X. A natural estimate of $f_X(x)$ is given by $\hat{f}_X(x)$, the proportion of sample values equal to x. A natural estimate of $F_X(x)$ is given by $\hat{F}_X(x) = n^{-1}\sum_{i=1}^{n} I_{(-\infty,x]}(x_i)$, the proportion of sample values less than or equal to x, otherwise known as the empirical distribution function evaluated at x.

Suppose we obtained the following sample of $n = 10$ data values.

1.2	2.1	0.4	3.3	−2.1	4.0	−0.3	2.2	1.5	5.0

In this case $\hat{f}_X(x) = 0.1$ whenever x is a data value and is 0 otherwise. To compute $\hat{F}_X(x)$, we simply count how many sample values are less than or equal to x and divide by $n = 10$. For example, $\hat{F}_X(-3) = 0/10 = 0, \hat{F}_X(0) = 2/10 = 0.2$, and $\hat{F}_X(4) = 9/10 = 0.9$. ∎

An important class of characteristics of the distribution of a continuous variable X is given by the following definition.

Definition 5.5.1 For $p \in [0, 1]$, the *pth quantile* x_p, for the distribution with cdf F_X, is defined to be the smallest number x_p satisfying

$$p \le F_X(x_p).$$

Note that by the definition of the inverse cumulative distribution function (Definition 2.10.1), we can write $x_p = F_X^{-1}(p) = \min\{x : p \le F_X(x)\}$.

When F_X is strictly increasing and continuous, then $F_X^{-1}(p)$ is the unique value x_p satisfying

$$F_X(x_p) = p. \tag{5.5.1}$$

Figure 5.5.1 illustrates the situation when there is a unique solution to (5.5.1). When F_X is not strictly increasing or continuous (as when X is discrete), then there may be more than one, or no, solutions to (5.5.1). Figure 5.5.2 illustrates the situation when there is no solution to (5.5.1).

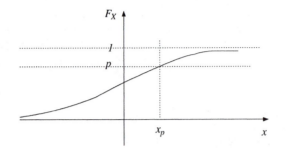

Figure 5.5.1: The pth quantile x_p when there is a unique solution to (5.5.1).

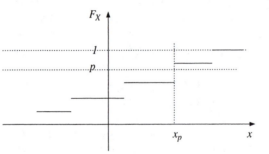

Figure 5.5.2: The pth quantile x_p determined by a cdf F_X when there is no solution to (5.5.1).

So, when X is a continuous measurement, a proportion p of the population have their X measurement less than or equal to x_p. As particular cases, $x_{0.5} = F_X^{-1}(0.5)$ is the *median* while $x_{0.25} = F_X^{-1}(0.25)$ and $x_{0.75} = F_X^{-1}(0.75)$ are the first and third *quartiles*, respectively, of the distribution. You will also see x_p referred to as a *100pth percentile*. For example, if your mark on a test placed you at the 90th percentile, then your mark equals $x_{0.9}$ and 90% of your fellow test takers achieved your mark or lower.

Example 5.5.2 *Estimating Quantiles*
A natural estimate of a population quantile $x_p = F_X^{-1}(p)$ is to use $\hat{x}_p = \hat{F}_X^{-1}(p)$. Note, however, that $\hat{F}_X$ is not continuous, so there may not be a solution to (5.5.1) using $\hat{F}_X$.

Applying Definition 5.5.1, however, leads to the following estimate. First, order the observed sample values $(x_1, \ldots, x_n)$ to obtain the *order statistics* $x_{(1)} < \cdots < x_{(n)}$ (see Section 2.8.4). Then, note that $x_{(i)}$ is the (i/n)th quantile of the empirical distribution, because

$$\hat{F}_X\left(x_{(i)}\right) = \frac{i}{n}$$

and $\hat{F}_X(x) < i/n$ whenever $x < x_{(i)}$. In general, we have that the *sample pth quantile* is $\hat{x}_p = x_{(i)}$ whenever

$$\frac{i-1}{n} < p \le \frac{i}{n}. \tag{5.5.2}$$

There are a number of modifications to this estimate that are sometimes used. For example, if we find i such that (5.5.2) is satisfied and put

$$\tilde{x}_p = x_{(i-1)} + n\left(x_{(i)} - x_{(i-1)}\right)\left(p - \frac{i-1}{n}\right), \tag{5.5.3}$$

then $\tilde{x}_p$ is the linear interpolant between $x_{(i-1)}$ and $x_{(i)}$. When n is even, this definition gives the *sample median* as $\tilde{x}_{0.5} = x_{(n/2)}$, and a similar formula holds when n is odd (Problem 5.5.18). Also see Problem 5.5.19 for more discussion of (5.5.3).

Quite often the sample median is defined to be

$$\check{x}_{0.5} = \begin{cases} x_{((n+1)/2)} & n \text{ odd} \\ \frac{1}{2}\left(x_{(n/2)} + x_{(n/2+1)}\right) & n \text{ even,} \end{cases} \tag{5.5.4}$$

namely, the middle value when n is odd and the average of the two middle values when n is even. For n large enough, all these definitions will yield similar answers. The use of any of these is permissible in an application.

Consider the data in Example 5.5.1. Sorting the data from smallest to largest, the order statistics are given by the following table.

$x_{(1)} = -2.1$	$x_{(2)} = -0.3$	$x_{(3)} = 0.4$	$x_{(4)} = 1.2$	$x_{(5)} = 1.5$
$x_{(6)} = 2.1$	$x_{(7)} = 2.2$	$x_{(8)} = 3.3$	$x_{(9)} = 4.0$	$x_{(10)} = 5.0$

Then, using (5.5.3), the sample median is given by $\tilde{x}_{.5} = x_{(5)} = 1.5$, while the sample quartiles are given by

$$\begin{aligned} \tilde{x}_{.25} &= x_{(2)} + 10\left(x_{(3)} - x_{(2)}\right)(0.25 - 0.2) \\ &= -0.3 + 10\left(0.4 - (-0.3)\right)(0.25 - 0.2) = 0.05 \end{aligned}$$

and

$$\begin{aligned} \tilde{x}_{.75} &= x_{(7)} + 10\left(x_{(8)} - x_{(7)}\right)(0.75 - 0.7) \\ &= 2.2 + 10\left(3.3 - 2.2\right)(0.75 - 0.7) = 2.75. \end{aligned}$$

So in this case, we estimate that 25% of the population under study has an X measurement less than 0.05, etc. ∎

Example 5.5.3 *Measuring Location and Scale of a Population Distribution*
Often we are asked to make inferences about the value of the *population mean*

$$\mu_X = \frac{1}{|\Pi|} \sum_{\pi \in \Pi} X(\pi)$$

and the *population variance*

$$\sigma_X^2 = \frac{1}{|\Pi|} \sum_{\pi \in \Pi} (X(\pi) - \mu_X)^2$$

where Π is a finite population and X is a real-valued measurement defined on it. These are measures of the location and spread of the population distribution about the mean, respectively. Note that calculating a mean or variance makes sense only when X is a quantitative variable.

When X is discrete, we can also write

$$\mu_X = \sum_x x f_X(x)$$

because $|\Pi| f_X(x)$ equals the number of elements $\pi \in \Pi$ with $X(\pi) = x$. In the continuous case, using an approximating density f_X, we can write

$$\mu_X \approx \int_{-\infty}^{\infty} x f_X(x) \, dx.$$

Similar formulas exist for the population variance of X (see Problem 5.4.10).

It will probably occur to you that a natural estimate of the population mean μ_X is given by the sample mean

$$\bar{x} = \frac{1}{n} \sum_{i=1}^{n} x_i.$$

Also a natural estimate of the population standard deviation σ_X^2 is given by the *sample variance*

$$s^2 = \frac{1}{n-1} \sum_{i=1}^{n} (x_i - \bar{x})^2. \tag{5.5.5}$$

Later we will explain why we divided by $n-1$ in (5.5.5) rather than n. Actually, it makes little difference which we use, for even modest values of n. The *sample standard deviation* is given by s, the positive square root of s^2. For the data in Example 5.1.1 we obtain $\bar{x} = 1.73$ and $s = 2.097$.

The population mean μ_X and population standard deviation σ_X serve as a pair, in which μ_X measures where the distribution is located on the real line and σ_X measures how much spread there is in the distribution about μ_X. Clearly, the greater the value of σ_X, the more variability there is in the distribution.

Alternatively, we could use the population median $x_{0.5}$ as a measure of location of the distribution and the *population interquartile range* $x_{0.75} - x_{0.25}$ as a measure of the amount of variability in the distribution around the median. The median and interquartile range are the preferred choice to measure these aspects of the distribution whenever the distribution is *skewed*, i.e., not symmetrical. This is because the median is insensitive to very extreme values while the mean is not. For example, house prices in an area are well known to exhibit a right-skewed distribution. A few houses selling for very high prices will not change the median price but could result in a big change in the mean price.

When we have a symmetric distribution, the mean and median will agree (provided the mean exists). The greater the skewness in a distribution, however, the greater will be the discrepancy between its mean and median. For example, in Figure 5.5.3 we have plotted the density of a $\chi^2(4)$ distribution. This distribution is skewed to the right, and the mean is 4 while the median is 3.3567.

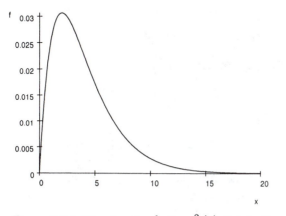

Figure 5.5.3: The density f of a $\chi^2(4)$ distribution.

We estimate the population interquartile range by the *sample interquartile range)* given by $IQR = \tilde{x}_{0.75} - \tilde{x}_{0.25}$. For the data in Example 5.1.1, we obtain the sample median to be $\tilde{x}_{0.5} = 1.5$ while $IQR = 2.75 - .05 = 2.70$.

If we changed the largest value in the sample from $x_{(10)} = 5.0$ to $x_{(10)} = 500.0$ the sample median remains $\tilde{x}_{.5} = 1.5$ but note that the sample mean goes from 1.73 to 51.23! ∎

5.5.2 Plotting Data

It is always a good idea to plot the data. For discrete quantitative variables we can plot $\hat{f}_X$, i.e., plot the sample proportions (relative frequencies). For continuous quantitative variables we introduced the density histogram in section 5.4.3. These plots gives us some idea of the shape of the distribution we are sampling from. For example, we can see if there is any evidence that the distribution is strongly skewed.

We now consider another very useful plot for quantitative variables.

Example 5.5.4 *Boxplots and Outliers*
Another useful plot for quantitative variables is known as a *boxplot*. For example, Figure 5.5.4 gives a boxplot for the data in Example 5.5.1. The line in the center of the box is the median. The line below the median is the first quartile, and the line above the median is the third quartile.

The vertical lines from the quartiles are called *whiskers*, and these run from the quartiles to the *adjacent values*. The adjacent values are given by the greatest value less than or equal to the *upper limit* (the third quartile plus 1.5 times the IQR) and by the least value greater than or equal to the *lower limit* (the first quartile minus 1.5 times the IQR). Values beyond the adjacent values, when these exist, are plotted with a *, in this case there are none. If we changed $x_{(10)} = 5.0$ to $x_{(10)} = 15.0$, however, we see this extreme value plotted as a *, as shown in Figure 5.5.5.

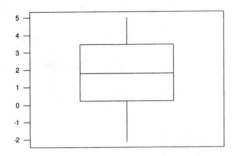

Figure 5.5.4: A boxplot of the data in Example 5.5.1.

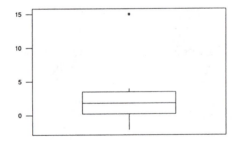

Figure 5.5.5: A boxplot of the data in Example 5.5.1, changing $x_{(10)} = 5.0$ to $x_{(10)} = 15.0$.

Points outside the upper and lower limits, and thus plotted by *, are commonly referred to as *outliers*. An outlier is a value that is extreme with respect to the rest of the observations. Sometimes outliers occur because a mistake has been made in collecting or recording the data, but they also occur simply because we are sampling from a long-tailed distribution. It is often difficult to ascertain which is the case in a particular application, but each such observation should be noted. We have seen in Example 5.5.3 that outliers can have a big impact on statistical analyses. Their effects should be recorded when reporting the results of a statistical analysis. ∎

For categorical variables, it is typical to plot the data in a bar chart, as described in the next example.

Example 5.5.5 *Bar Charts*

For categorical variables, we code the values of the variable as equispaced numbers and then plot constant-width rectangles (the bars) over these values so that the height of the rectangle over a value equals the proportion of times that value is assumed. Such a plot is called a *bar chart*. Note that the values along the x-axis are only labels and not to be treated as numbers that we can do arithmetic on, etc.

For example, suppose we take a simple random sample of 100 students and record their favorite flavor of ice cream (from amongst four possibilities), obtaining the results given in the following table.

Flavor	Count	Proportion
Chocolate	42	0.42
Vanilla	28	0.28
Butterscotch	22	0.22
Strawberry	8	0.08

Coding Chocolate as 1, Vanilla as 2, Butterscotch as 3, and Strawberry as 4, Figure 5.5.6 presents a bar chart of these data. It is typical in bar charts for the bars not to touch. ∎

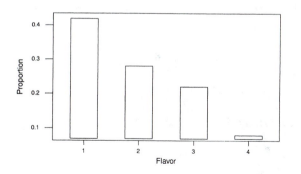

Figure 5.5.6: A bar chart for the data of Example 5.5.5.

5.5.3 Types of Inference

Certainly quoting descriptive statistics and plotting the data are methods used by a statistician to try to learn something about the underlying population distribution. There are difficulties with this approach, however, as we have just chosen these methods based on intuition. Often it is not clear which descriptive statistics we should use. Further, these data summaries make no use of the information that we have about the true population distribution as expressed by the statistical model, namely, $f_X \in \{f_\theta : \theta \in \Omega\}$. Taking account of this information leads us to develop a theory of statistical inference, i.e., specify how we should combine the model information together with the data to make inferences about population quantities. We will do this in Chapters 6, 7, and 8, but first we discuss the types of inferences that are commonly used in applications.

In Section 5.2 we discussed three types of inference in the context of a known probability model as specified by some density or probability function f. We noted that we might want to do any of the following concerning an unobserved response value s.

(i) Estimate (or predict) an unknown response value s via an estimate t.

(ii) Construct a subset C of the sample space S that has a high probability of containing an unknown response value s and was as small as possible in some sense.

(iii) Assess whether or not $s_0 \in S$ is a plausible value from the probability distribution specified by f.

We will refer to (i) as the problem of *estimation* or *prediction*, refer to (ii) as the problem of *credible region* or *confidence region* construction, and refer to (iii) as the problem of *hypothesis assessment*. So estimates, credible or confidence regions, and hypothesis assessment are examples of types of inference. The examples of Section 5.2 show, at least after studying probability theory, that these are intuitively reasonable concepts.

In a statistical application, we do not know f; we know only that $f \in \{f_\theta : \theta \in \Omega\}$, and we observe the data s. We are uncertain about which candidate f_θ is correct, or, equivalently, which of the possible values of θ is correct.

As mentioned in Section 5.5.1, our primary goal may not be knowing the true f_θ, but some characteristic of the true distribution such as its mean, median, or the value of the true distribution function F at a specified value. We will denote this characteristic of interest by $\psi(\theta)$. For example, when the characteristic of interest is the mean of the true distribution of a continuous random variable then

$$\psi(\theta) = \int_{-\infty}^{\infty} x f_\theta(x) \, dx.$$

Alternatively, we might be interested in $\psi(\theta) = F_\theta^{-1}(0.5)$, the median of the distribution of a random variable with distribution function given by F_θ.

Different values of θ lead to possibly different values for the characteristic $\psi(\theta)$. After observing the data s, we want to make inferences about what the correct value is. We will consider the three types of inference we have just identified. In particular, we want to construct estimates $T(s)$ of $\psi(\theta)$, construct credible or confidence regions $C(s)$ for $\psi(\theta)$, and assess the plausibility of a hypothesized value ψ_0 for $\psi(\theta)$.

The *problem of statistical inference* is concerned with determining how we should combine the information in the model $\{f_\theta : \theta \in \Omega\}$ and the data s to carry out these inferences about $\psi(\theta)$.

A very important statistical model for applications is the normal location-scale model introduced in Example 5.3.4. We illustrate some of the ideas discussed in this section via that model.

Example 5.5.6 *Application of the Location-Scale Normal Model*
Suppose that the following simple random sample of the heights (in inches) of
30 students has been collected.

64.9	61.4	66.3	64.3	65.1	64.4	59.8	63.6	66.5	65.0
64.9	64.3	62.5	63.1	65.0	65.8	63.4	61.9	66.6	60.9
61.6	64.0	61.5	64.2	66.8	66.4	65.8	71.4	67.8	66.3

The statistician believes that the distribution of heights in the population can
be well approximated by a normal distribution with some unknown mean and
variance, and she is unwilling to make any further assumptions about the true
distribution. Accordingly the statistical model is given by the family of $N(\mu, \sigma^2)$
distributions where $\theta = (\mu, \sigma^2) \in \Omega = R^1 \times R^+$ is unknown.

Does this statistical model make sense, i.e., is the assumption of normality
appropriate for this situation? The density histogram (based on 12 equal-length
intervals from 59.5 to 71.5) in Figure 5.5.7 looks very roughly normal but the
extreme observation in the right tail might be some grounds for concern. In any
case, we will proceed as if this assumption is reasonable. In Chapter 9 we will
discuss more refined methods for assessing this assumption.

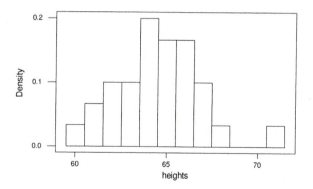

Figure 5.5.7: Density histogram of heights in Example 5.5.6.

Suppose that we are interested in making inferences about the population
mean height, namely, the characteristic of interest is $\psi(\mu, \sigma^2) = \mu$. Alternatively, we might want to make inferences about the 90th percentile of this dis-
tribution, i.e., $\psi(\mu, \sigma^2) = x_{0.90} = \mu + \sigma z_{0.90}$ where $z_{0.90}$ is the 90th percentile
of the $N(0, 1)$ distribution (when $X \sim N(\mu, \sigma^2)$ then $P(X \leq \mu + \sigma z_{0.90}) =
P((X - \mu)/\sigma \leq z_{0.90}) = \Phi(z_{0.90}) = 0.90.)$ So 90% of the population under
study have their height less than $x_{0.90}$ and this value is unknown to us because we do not know the value of (μ, σ^2). Obviously, there are many other
characteristics of the true distribution that we might want to make inferences
about.

Just using our intuition, $T(x_1, \ldots, x_n) = \bar{x}$ seems like a sensible estimate of
μ and $T(x_1, \ldots, x_n) = \bar{x} + s z_{0.90}$ seems like a sensible estimate of $\mu + \sigma z_{0.90}$.
To justify the choice of these estimates we will need the theories developed in

later chapters. In this case, we obtain $\bar{x} = 64.517$, and from (5.5.5) we compute $s = 2.379$. From Table D.2 we obtain $z_{.90} = 1.2816$, so that

$$\bar{x} + sz_{0.90} = 64.517 + 2.379\,(1.2816) = 67.566.$$

How accurate is the estimate $\bar{x}$ of μ? A natural approach to answering this question is to construct a credible interval, based on the estimate, that we believe has a high probability of containing the true value of μ and is as short as possible. For example, the theory in Chapter 6 leads to using confidence intervals for μ, of the form

$$[\bar{x} - sc, \bar{x} + sc]$$

for some choice of the constant c. Notice that $\bar{x}$ is at the center of the interval. The theory in Chapter 6 will show that, in this case, choosing $c = 0.3734$ leads to what is known as a 0.95-confidence interval for μ. We then take the half-length of this interval, namely

$$sc = 2.379\,(0.373\,4) = 0.888,$$

as a measure of the accuracy of the estimate $\bar{x} = 64.517$ of μ. In this case, we have enough information to say that we know the true value of μ to within one inch, at least with "confidence" equal to .95.

Finally, suppose that we have a hypothesized value μ_0 for the population mean height. For example, we may believe that the mean height of the population of individuals under study is the same as the mean height of another population for which this quantity is known to equal $\mu_0 = 65$. Then, based on the observed sample of heights, we want to assess whether or not the value $\mu_0 = 65$ makes sense. If the sample mean height $\bar{x}$ is far from μ_0, this would seem to be evidence against the hypothesized value. In Chapter 6 we will show that we can base our assessment on the value of

$$t = \frac{\bar{x} - \mu_0}{s/\sqrt{n}} = \frac{64.517 - 65}{2.379/\sqrt{30}} = -1.112.$$

If the value of $|t|$ is very large, then we will conclude that we have evidence against the hypothesized value $\mu_0 = 65$. We have to prescribe what we mean by large here, and we will do this in Chapter 6. It turns out that $t = -1.112$ is a plausible value for t, when the true value of μ equals 65, and so we have no evidence against the hypothesis. ∎

Summary of Section 5.5

- Descriptive statistics represent informal statistical methods that a statistician uses to make inference about the distribution of a variable X of interest based on an observed sample from this distribution. These quantities summarize characteristics of the observed sample and can be thought of as estimates of the corresponding unknown population quantities. More formal methods are required to assess the error in these estimates or even to replace them with estimates having greater accuracy.

- It is important to plot the data using relevant plots. These give us some idea of the shape of the population distribution we are sampling from.

- There are three main types of inference: estimates, credible or confidence intervals, and hypothesis assessment.

Exercises

5.5.1 Suppose that the following data were obtained by recording X, the number of customers that arrive at an automatic banking machine during 15 successive one-minute time intervals.

2	1	3	2	0	1	4	2
0	2	3	1	0	0	4	

(a) Record estimates of $f_X(0)$, $f_X(1)$, $f_X(2)$, $f_X(3)$, and $f_X(4)$.
(b) Record estimates of $F_X(0)$, $F_X(1)$, $F_X(2)$, $F_X(3)$, and $F_X(4)$.
(c) Plot $\hat{f}_X$.
(d) Record the mean and variance.
(e) Record the median and IQR and provide a boxplot. Using the rule prescribed in Example 5.5.4, decide if there are any outliers.

5.5.2 Suppose that the following sample of waiting times in minutes for customers in a queue at an automatic banking machine was obtained.

15	10	2	3	1	0	4	5
5	3	3	4	2	1	4	5

(a) Record the empirical distribution function.
(b) Plot $\hat{f}_X$.
(c) Record the mean and variance.
(d) Record the median and IQR and provide a boxplot. Using the rule given in Example 5.5.4, decide if there are any outliers.

5.5.3 Suppose that an experiment was conducted to see if mosquitoes are attracted differentially to different colors. Three different colors of fabric were used and the number of mosquitoes landing on each piece was recorded over a 15-minute interval. The following data were obtained.

	Number of landings
Color 1	25
Color 2	35
Color 3	22

(a) Record estimates of $f_X(1)$, $f_X(2)$, and $f_X(3)$ where we use i for color i.
(b) Does it make sense to estimate $F_X(i)$? Explain why or why not.
(c) Plot a bar chart of these data.

5.5.4 A student is told that his score on a test was at the 90th percentile in the population of all students who took the test. Explain exactly what this means.

5.5.5 Determine the empirical distribution function based on the sample given below.

1.0	−1.2	0.4	1.3	−0.3
−1.4	0.4	−0.5	−0.2	−1.3
0.0	−1.0	−1.3	2.0	1.0
0.9	0.4	2.1	0.0	−1.3

Plot this function. Determine the sample median, the first and third quartiles, and the interquartile range. What is your estimate of $F(1)$?

5.5.6 Consider the density histogram in Figure 5.5.8. If you were asked to record measures of location and spread for the data corresponding to this plot, what would you choose? Justify your answer.

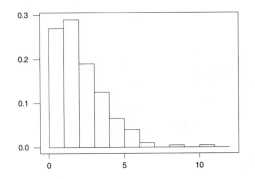

Figure 5.5.8: Density histogram for Exercise 5.5.6.

5.5.7 Suppose that a statistical model is given by the family of $N(\mu, \sigma_0^2)$ distributions where $\theta = \mu \in R^1$ is unknown, while σ_0^2 is known. If our interest is in making inferences about the first quartile of the true distribution, then determine $\psi(\mu)$.

5.5.8 Suppose that a statistical model is given by the family of $N(\mu, \sigma_0^2)$ distributions where $\theta = \mu \in R^1$ is unknown, while σ_0^2 is known. If our interest is in making inferences about the third moment of the distribution, then determine $\psi(\mu)$.

5.5.9 Suppose that a statistical model is given by the family of $N(\mu, \sigma_0^2)$ distributions where $\theta = \mu \in R^1$ is unknown, while σ_0^2 is known. If our interest is in making inferences about the distribution function evaluated at 3, then determine $\psi(\mu)$.

5.5.10 Suppose that a statistical model is given by the family of $N(\mu, \sigma^2)$ distributions where $\theta = (\mu, \sigma^2) \in R^1 \times R^+$ is unknown. If our interest is in making inferences about the first quartile of the true distribution, then determine $\psi(\mu, \sigma^2)$.

5.5.11 Suppose that a statistical model is given by the family of $N(\mu, \sigma^2)$ distributions where $\theta = (\mu, \sigma^2) \in R^1 \times R^+$ is unknown. If our interest is in making inferences about the distribution function evaluated at 3, then determine $\psi(\mu, \sigma^2)$.

5.5.12 Suppose that a statistical model is given by the family of Bernoulli(θ) distributions where $\theta \in \Omega = [0, 1]$. If our interest is in making inferences about the probability that two independent observations from this model are the same, then determine $\psi(\theta)$.

5.5.13 Suppose that a statistical model is given by the family of Bernoulli(θ) distributions where $\theta \in \Omega = [0, 1]$. If our interest is in making inferences about the probability that in two independent observations from this model we obtain a 0 and a 1, then determine $\psi(\theta)$.

5.5.14 Suppose that a statistical model is given by the family of Uniform$[0, \theta]$ distributions where $\theta \in \Omega = (0, \infty)$. If our interest is in making inferences about the coefficient of variation (see Exercise 5.3.5) of the true distribution, then determine $\psi(\theta)$. What do you notice about this characteristic?

5.5.15 Suppose that a statistical model is given by the family of Gamma(α_0, β) distributions where $\theta = \beta \in \Omega = (0, \infty)$. If our interest is in making inferences about the variance of the true distribution, then determine $\psi(\theta)$.

Computer Exercises

5.5.16 Do the following based on the data in Exercise 5.4.5.
(a) Compute the order statistics for these data.
(b) Calculate the empirical distribution function at the data points.
(c) Calculate the sample mean and the sample standard deviation.
(d) Obtain the sample median and the sample interquartile range.
(e) Based on the histograms obtained in Exercise 5.4.5, which set of descriptive statistics do you feel are appropriate for measuring location and spread?
(f) Suppose the first data value was recorded incorrectly as 13.9 rather than as 3.9. Repeat parts (c) and (d) using this data set and compare your answers with those previously obtained. Can you draw any general conclusions about these measures? Justify your reasoning.

5.5.17 Do the following based on the data in Example 5.5.6.
(a) Compute the order statistics for these data.
(b) Plot the empirical distribution function (only at the sample points).
(c) Calculate the sample median and the sample interquartile range and obtain a boxplot. Are there any outliers?
(d) Based on the boxplot, which set of descriptive statistics do you feel are appropriate for measuring location and spread?
(e) Suppose the first data value was recorded incorrectly as 84.9 rather than as 64.9. Repeat parts (c) and (d) using this data set and see if any observations are determined to be outliers.

Problems

5.5.18 Determine a formula for the sample median, based on interpolation (i.e., using (5.5.3)) when n is odd. (Hint: Use the *least integer function* or *ceiling* $\lceil x \rceil$ = smallest integer greater than or equal to x.)

5.4.19 An alternative to the empirical distribution function is to define a distribution function $\tilde{F}$ by $\tilde{F}(x) = 0$ if $x < x_{(1)}$, $\tilde{F}(x) = 1$ if $x \geq x_{(n)}$, $\tilde{F}(x) = \hat{F}(x_{(i)})$ if $x = x_{(i)}$, and

$$\tilde{F}(x) = \hat{F}(x_{(i)}) + \frac{\hat{F}(x_{(i+1)}) - \hat{F}(x_{(i)})}{x_{(i+1)} - x_{(i)}} \left(x - x_{(i)} \right)$$

if $x_{(i)} \leq x \leq x_{(i+1)}$ for $i = 1, \ldots, n$.
(a) Show that $\tilde{F}(x_{(i)}) = \hat{F}(x_{(i)})$ for $i = 1, \ldots, n$ and is increasing from 0 to 1.
(b) Prove that $\tilde{F}$ is continuous on $(x_{(1)}, \infty)$ and right continuous everywhere.
(c) Show that, for $p \in [1/n, 1)$, the value $\tilde{x}_p$ defined in (5.5.3) is the solution to $\tilde{F}(\tilde{x}_p) = p$.

Discussion Topic

5.5.20 Sometimes it is argued that statistics does not need a formal theory to prescribe inferences. Rather statistical practice is better left to the skilled practitioner to decide what is a sensible approach in each problem. Comment on these statements.

Chapter 6

Likelihood Inference

In this chapter we discuss some of the most basic approaches to inference. In essence, we want our inferences to depend only on the model $\{P_\theta : \theta \in \Omega\}$ and the data s. These methods are very minimal in the sense that they require few assumptions. While successful for certain problems, it seems that the additional structure of Chapter 7 or Chapter 8 is necessary in more involved situations.

The likelihood function is one of the most basic concepts in statistical inference. Entire theories of inference have been constructed based on it. We discuss likelihood methods in Sections 6.1, 6.2, 6.3, and 6.5. In Section 6.4 we introduce some distribution-free methods of inference. These are not really examples of likelihood methods, but they follow the same basic idea of having the inferences depend on as few assumptions as possible.

6.1   The Likelihood Function

Likelihood inferences are based only on the data s and the model $\{P_\theta : \theta \in \Omega\}$ — the set of possible probability measures for the system under investigation. From these ingredients we obtain the basic entity of likelihood inference, namely, the likelihood function.

To motivate the definition of the likelihood function, suppose that we have a statistical model in which each P_θ is discrete, given by probability function f_θ. Having observed s, consider the function $L(\cdot \mid s)$ defined on the parameter space Ω and taking values in R^1, given by $L(\theta \mid s) = f_\theta(s)$. We refer to $L(\cdot \mid s)$ as the *likelihood function* determined by the model and the data. The value $L(\theta \mid s)$ is called the *likelihood* of θ. Note that for the likelihood function we are fixing the data and varying the value of the parameter.

We see that $f_\theta(s)$ is just the probability of obtaining the data s when the true value of the parameter is θ. This imposes a belief ordering on Ω, namely, we believe in θ_1 as the true value of θ over θ_2 whenever $f_{\theta_1}(s) > f_{\theta_2}(s)$. This is because the inequality says that the data are more likely under θ_1 than θ_2. We are indifferent between θ_1 and θ_2 whenever $f_{\theta_1}(s) = f_{\theta_2}(s)$. Likelihood inference about θ is based on this ordering.

It is important to remember the correct interpretation of $L(\theta \mid s)$. The value $L(\theta \mid s)$ is the probability of s given that θ is the true value — it is *not* the probability of θ given that we have observed s. Also, it is possible that the value of $L(\theta \mid s)$ is very small for every value of θ. So it is not the actual value of the likelihood that is telling us how much support to give to a particular θ, but rather its value relative to the likelihoods of other possible parameter values.

Example 6.1.1

Suppose that $S = \{1, 2, \ldots\}$ and that the statistical model is $\{P_\theta : \theta \in \{1, 2\}\}$, where P_1 is the uniform distribution on the integers $\{1, \ldots, 10^3\}$ and P_2 is the uniform distribution on $\{1, \ldots, 10^6\}$. Further suppose that we observe $s = 10$. Then $L(1 \mid 10) = 1/10^3$ and $L(2 \mid 10) = 1/10^6$. Both values are quite small, but note that the likelihood supports $\theta = 1$ a thousand times more than it supports $\theta = 2$. ∎

Accordingly, we are only interested in *likelihood ratios* $L(\theta_1 \mid s) / L(\theta_2 \mid s)$ for $\theta_1, \theta_2 \in \Omega$ when it comes to determining inferences for θ based on the likelihood function. This implies that any function that is a positive multiple of $L(\cdot \mid s)$, i.e., $L^*(\cdot \mid s) = cL(\cdot \mid s)$ for some fixed $c > 0$, can serve equally well as a likelihood function. We call two likelihoods equivalent if they are proportional in this way. In general, we refer to any positive multiple of $L(\cdot \mid s)$ as a likelihood function.

Example 6.1.2

Suppose that a coin is tossed $n = 10$ times and that $s = 4$ heads are observed. With no knowledge whatsoever concerning the probability of getting a head on a single toss, the appropriate statistical model for the data is the Binomial$(10, \theta)$ model with $\theta \in \Omega = [0, 1]$. The likelihood function is given by

$$L(\theta \mid 4) = \binom{10}{4} \theta^4 (1 - \theta)^6, \tag{6.1.1}$$

which is plotted in Figure 6.1.1.

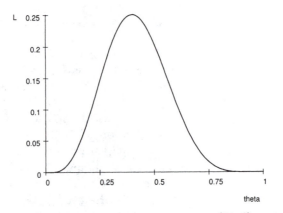

Figure 6.1.1: Likelihood function from the Binomial$(10, \theta)$ model when $s = 4$ is observed.

This likelihood peaks at $\theta = 0.4$ and takes the value 0.2508 there. We will subsequently examine uses of the likelihood to estimate the unknown θ and assess the accuracy of the estimate. Roughly speaking, however, this is based on where the likelihood takes its maximum and how much spread there is in the likelihood about its peak. ∎

There is a range of approaches to obtaining inferences via the likelihood function. At one extreme is the likelihood principle.

> *Likelihood Principle*: If two model, data combinations yield equivalent likelihood functions, then inferences about the unknown parameter must be the same.

This principle dictates that anything we want to say about the unknown value of θ must be based only on $L(\cdot \mid s)$. For many statisticians this is viewed as a very severe proscription. Consider the following example.

Example 6.1.3
Suppose that a coin is tossed in independent tosses until four heads are obtained and the number of tails observed until the fourth head is $s = 6$. Then s is distributed Negative-Binomial$(4, \theta)$, and the likelihood specified by the observed data is

$$L(\theta \mid 4) = \binom{9}{3}\theta^4(1 - \theta)^6.$$

Note that this likelihood function is a positive multiple of (6.1.1).

So the likelihood principle asserts that these two model, data combinations must yield the same inferences about the unknown θ. In effect, the likelihood principle says that we must ignore the fact that the data were obtained in entirely different ways. If, however, we take into account additional model features beyond the likelihood function, then it turns out that we can derive different inferences for the two situations. In particular, assessing a hypothesized value $\theta = \theta_0$ can be carried out in different ways when the sampling method is taken into account. Many statisticians believe that this additional information should be used when deriving inferences. ∎

As an example of an inference derived from a likelihood function, consider a set of the form

$$C(s) = \{\theta : L(\theta \mid s) \geq c\},$$

for some $c \geq 0$. The set $C(s)$ is referred to as a *likelihood region*. It contains all those θ values for which their likelihood is at least c. A likelihood region, for some c, seems like a sensible set to quote as possibly containing the true value of θ. For, if $\theta^* \notin C(s)$, then $L(\theta^* \mid s) < L(\theta \mid s)$ for every $\theta \in C(s)$ and so is not as well-supported by the observed data as any value in $C(s)$. The size of $C(s)$ can then be taken as a measure of how uncertain we are about the true value of θ.

We are left with the problem, however, of choosing a suitable value for c and, as Example 6.1.1 seems to indicate, the likelihood itself doesn't suggest a natural way to do this. In Section 6.3.2 we will discuss a method for choosing c that is based upon additional model properties beyond the likelihood function.

So far in this section we have assumed that our statistical models are comprised of discrete distributions. The definition of the likelihood is quite natural, as $L(\theta \mid s)$ is simply the probability of s occurring when θ is the true value. This interpretation is clearly not directly available, however, when we have a continuous model, because every data point has probability 0 of occurring. Imagine, however, that $f_{\theta_1}(s) > f_{\theta_2}(s)$ and that $s \in R^1$. Then, assuming the continuity of every f_θ at s, we have

$$P_{\theta_1}(V) = \int_a^b f_{\theta_1}(x)\,dx > P_{\theta_2}(V) = \int_a^b f_{\theta_2}(x)\,dx$$

for every interval $V = (a, b)$ containing s that is small enough. We interpret this to mean that the probability of s occurring when θ_1 is true is greater than the probability of s occurring when θ_2 is true. So the data s support θ_1 more than θ_2. A similar interpretation applies when $s \in R^n$ for $n > 1$, and V is a region containing s.

Therefore, in the continuous case we again define the likelihood function by $L(\theta \mid s) = f_\theta(s)$ and interpret the ordering this imposes on the values of θ exactly as we do in the discrete case.[1] Again, two likelihoods will be considered equivalent if one is a positive multiple of the other.

Now consider a very important example.

Example 6.1.4 *Location Normal Model*
Suppose that $(x_1, \ldots, x_n)$ is an observed identically and independently distributed (i.i.d.) sample from an $N(\theta, \sigma_0^2)$ distribution where $\theta \in \Omega = R^1$ is unknown and $\sigma_0^2 > 0$ is known. The likelihood function is given by

$$
\begin{aligned}
L(\theta \mid x_1, \ldots, x_n) &= \prod_{i=1}^n f_\theta(x_i) = \prod_{i=1}^n (2\pi\sigma_0^2)^{-1/2} \exp\left(-\frac{1}{2\sigma_0^2}(x_i - \theta)^2\right) \\
&= (2\pi\sigma_0^2)^{-n/2} \exp\left(-\frac{1}{2\sigma_0^2}\sum_{i=1}^n (x_i - \theta)^2\right) \\
&= (2\pi\sigma_0^2)^{-n/2} \exp\left(-\frac{n}{2\sigma_0^2}(\bar{x} - \theta)^2\right) \exp\left(-\frac{n-1}{2\sigma_0^2}s^2\right).
\end{aligned}
$$

An equivalent, simpler version of the likelihood function is then given by

$$L(\theta \mid x_1, \ldots, x_n) = \exp\left(-\frac{n}{2\sigma_0^2}(\bar{x} - \theta)^2\right),$$

---

[1] Note, however, that whenever we have a situation in which $f_{\theta_1}(s) = f_{\theta_2}(s)$, we could still have $P_{\theta_1}(V) > P_{\theta_2}(V)$ for every V containing s, and small enough. This implies that θ_1 is supported more than θ_2 rather than these two values having equal support as implied by the likelihood. This phenomenon does not occur in the examples we discuss, so we will ignore it here.

and we will use this version.

For example, suppose $n = 25, \sigma_0^2 = 1$, and we observe $\bar{x} = 3.3$. This function is plotted in Figure 6.1.2.

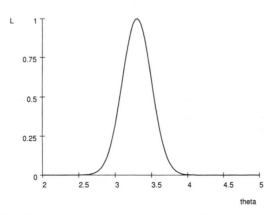

Figure 6.1.2: Likelihood from a location normal model based on a sample of 25 with $\bar{x} = 3.3$.

The likelihood peaks at $\theta = \bar{x} = 3.3$, and the plotted function takes the value 1 there. The likelihood interval

$$C(x) = \{\theta : L(\theta \,|\, x_1, \ldots, x_n) \geq 0.5\} = (3.0645, 3.53548)$$

contains all those θ values whose likelihood is at least 0.5 of the value of the likelihood at its peak.

The location normal model is impractical for many applications, as it assumes that the variance is known while the mean is unknown. For example, if we are interested in the distribution of heights in a population, it seems unlikely that we will know the population variance but not know the population mean. Still, it is an important statistical model, as it is a context where inference methods can be developed fairly easily. The methodology developed for this situation is often used as a paradigm for inference methods in much more complicated models. ∎

The parameter θ need not be one-dimensional. The interpretation of the likelihood is still the same, but it is not possible to plot it, at least when the dimension of θ is greater than two.

Example 6.1.5 *Multinomial Models*
In Example 2.8.5 we introduced multinomial distributions. These arise in applications when we have a categorical response variable s that can take a finite number k of values, say $\{1, \ldots, k\}$, and $P(s = i) = \theta_i$.

Suppose, then, that $k = 3$ and we do not know the value of $(\theta_1, \theta_2, \theta_3)$. In this case, the parameter space is given by

$$\Omega = \{(\theta_1, \theta_2, \theta_3) : \theta_i \geq 0, \text{ for } i = 1, 2, 3, \text{ and } \theta_1 + \theta_2 + \theta_3 = 1\}.$$

Notice that it is really only two-dimensional, because as soon as we know the value of any two of the θ_i, say θ_1 and θ_2, we immediately know the value of the remaining parameter, as $\theta_3 = 1 - \theta_1 - \theta_2$. This fact should always be remembered when we are dealing with multinomial models.

Now suppose we observe a sample of n from this distribution, say $(s_1, \ldots, s_n)$. The likelihood function for this sample is given by

$$L(\theta_1, \theta_2, \theta_3 \mid s_1, \ldots, s_n) = \theta_1^{x_1} \theta_2^{x_2} \theta_3^{x_3} \qquad (6.1.2)$$

where x_i is the number of i's in the sample.

Using the fact that we can treat positive multiples of the likelihood as being equivalent, we see that the likelihood based on the observed counts (x_1, x_2, x_3) (since they arise from a Multinomial$(n, \theta_1, \theta_2, \theta_3)$ distribution) is given by

$$L(\theta_1, \theta_2, \theta_3 \mid x_1, x_2, x_3) = \theta_1^{x_1} \theta_2^{x_2} \theta_3^{x_3}.$$

This is identical to the likelihood (as functions of θ_1, θ_2, and θ_3) for the original sample. It is certainly simpler to deal with the counts rather than the original sample. This is a very important phenomenon in statistics and is characterized by the concept of sufficiency, discussed in the next section. ∎

6.1.1 Sufficient Statistics

The equivalence for inference of positive multiples of the likelihood function leads to a useful equivalence amongst possible data values coming from the same model. For suppose data values s_1 and s_2 are such that $L(\cdot \mid s_1) = cL(\cdot \mid s_2)$ for some $c > 0$. Then, from the point of view of likelihood, we are indifferent whether we obtained the data s_1 or the data s_2, as they lead to the same likelihood ratios.

Consider a function T defined on the sample space S, such that whenever $T(s_1) = T(s_2)$, then

$$L(\cdot \mid s_1) = c(s_1, s_2) L(\cdot \mid s_2)$$

for some constant $c(s_1, s_2) > 0$. We call such a T a *sufficient statistic* for the model. The terminology is motivated by the fact that we need only observe the value t for the function T, as we can pick any value

$$s \in T^{-1}\{t\} = \{s : T(s) = t\}$$

and use the likelihood based on s. All of these choices give the same likelihood ratios. Typically, $T(s)$ will be of lower dimension than s, so we can consider replacing s by $T(s)$ as a *data reduction* and this simplifies the analysis somewhat.

We illustrate the computation of a sufficient statistic in a simple context.

Example 6.1.6
Suppose that $S = \{1, 2, 3, 4\}, \Omega = \{a, b\}$, and the two probability distributions are given by the following table.

	$s = 1$	$s = 2$	$s = 3$	$s = 4$
$\theta = a$	1/2	1/6	1/6	1/6
$\theta = b$	1/4	1/4	1/4	1/4

Then $L(\cdot\,|\,2) = L(\cdot\,|\,3) = L(\cdot\,|\,4)$ (e.g., $L(a\,|\,2) = 1/6$ and $L(b\,|\,2) = 1/4$), and so the data values in $\{2, 3, 4\}$ all give the same likelihood ratios. Therefore, $T : S \longrightarrow \{0, 1\}$ given by $T(1) = 0$ and $T(2) = T(3) = T(4) = 1$ is a sufficient statistic. The model has simplified a bit, as now the sample space for T has only two elements as opposed to four for the original model. ∎

The following result helps identify sufficient statistics.

Theorem 6.1.1 (*Factorization theorem*) If the density (or probability function) for a model factors as $f_\theta(s) = h(s)\, g_\theta(T(s))$, where g_θ and h are nonnegative, then T is a sufficient statistic.

Proof: By hypothesis, it is clear that, when $T(s_1) = T(s_1)$, we have

$$
\begin{aligned}
L(\cdot\,|\,s_1) &= h(s_1)\, g_\theta(T(s_1)) = \frac{h(s_1)\, g_\theta(T(s_1))}{h(s_2)\, g_\theta(T(s_2))} h(s_2)\, g_\theta(T(s_2)) \\
&= \frac{h(s_1)}{h(s_2)} h(s_2)\, g_\theta(T(s_2)) = c(s_1, s_2)\, L(\cdot\,|\,s_2)
\end{aligned}
$$

because $g_\theta(T(s_1)) = g_\theta(T(s_2))$. ∎

Note that the name of this result is motivated by the fact that we have factored f_θ as a product of two functions. The important point about a sufficient statistic T is that we are indifferent between observing the full data s or the value of $T(s)$.

If we can find a sufficient statistic T for a model, such that the value of T can be calculated once we know the likelihood function, then we call T a *minimal sufficient statistic*. So a relevant likelihood function can always be obtained from the value of any sufficient statistic T, but if T is minimal sufficient as well, then we can also obtain the value of T from any likelihood function. It can be shown that a minimal sufficient statistic gives the maximal reduction of the data. Note that the definitions of sufficient statistic and minimal sufficient statistic depend on the model, i.e., different models can give rise to different sufficient and minimal sufficient statistics.

Example 6.1.7 *Location Normal Model*
By the factorization theorem we see immediately, from the discussion in Example 6.1.4, that $\bar{x}$ is a sufficient statistic. Now any likelihood function for this model is a positive multiple of

$$
\exp\left(-\frac{n}{2\sigma_0^2}(\bar{x} - \theta)^2\right).
$$

Notice that any such function of θ is completely specified by the point where it takes its maximum, namely, at $\theta = \bar{x}$. So we have that $\bar{x}$ can be obtained from any likelihood function for this model and is therefore a minimal sufficient statistic. ∎

Example 6.1.8 *Location-Scale Normal Model*
Suppose that $(x_1, \ldots, x_n)$ is a sample from an $N(\mu, \sigma^2)$ distribution in which $\mu \in R^1$ and $\sigma \geq 0$ are unknown. Recall the discussion and application of this model in Examples 5.3.4 and 5.5.6.

The parameter in this model is two-dimensional and is given by $\theta = (\mu, \sigma^2) \in \Omega = R^1 \times (0, \infty)$. Therefore, the likelihood function is given by

$$
\begin{aligned}
L(\theta \mid x_1, \ldots, x_n) &= (2\pi\sigma^2)^{-n/2} \exp\left(-\frac{1}{2\sigma^2} \sum_{i=1}^{n} (x_i - \mu)^2\right) \\
&= (2\pi\sigma^2)^{-n/2} \exp\left(-\frac{n}{2\sigma^2}(\bar{x} - \mu)^2\right) \exp\left(-\frac{n-1}{2\sigma^2}s^2\right).
\end{aligned}
$$

We see immediately, from the factorization theorem, that $(\bar{x}, s^2)$ is a sufficient statistic.

Now, fixing σ^2, any positive multiple of $L(\cdot \mid x_1, \ldots, x_n)$ is maximized, as a function of μ, at $\mu = \bar{x}$. This is independent of σ^2. Fixing μ at $\bar{x}$, we have

$$
L\left((\bar{x}, \sigma^2) \mid x_1, \ldots, x_n\right) = (2\pi\sigma^2)^{-n/2} \exp\left(-\frac{n-1}{2\sigma^2}s^2\right)
$$

is maximized, as a function of σ^2, at the same point as $\ln L\left((\bar{x}, \sigma^2) \mid x_1, \ldots, x_n\right)$, because $\ln$ is a strictly increasing function. Now

$$
\begin{aligned}
\frac{\partial \ln L\left((\bar{x}, \sigma^2) \mid x\right)}{\partial \sigma^2} &= \frac{\partial}{\partial \sigma^2}\left(-\frac{n}{2}\ln\sigma^2 - \frac{n-1}{2\sigma^2}s^2\right) \\
&= -\frac{n}{2\sigma^2} + \frac{n-1}{2\sigma^4}s^2.
\end{aligned}
$$

Setting this equal to 0 yields the solution

$$
\hat{\sigma}^2 = \frac{n-1}{n}s^2,
$$

which is a 1–1 function of s^2. So, given any likelihood function for this model, we can compute $(\bar{x}, s^2)$, which establishes that $(\bar{x}, s^2)$ is a minimal sufficient statistic for the model. In fact, the likelihood is maximized at $(\bar{x}, \hat{\sigma}^2)$ (Problem 6.1.15). ∎

Example 6.1.9 *Multinomial Models*
We saw in Example 6.1.5 that the likelihood function for a sample is given by (6.1.2). It is clear from this that if two different samples have the same counts, then they have the same likelihood, and so the counts (x_1, x_2, x_3) comprise a sufficient statistic.

Now it turns out that this likelihood function is maximized by taking

$$
(\theta_1, \theta_2, \theta_3) = \left(\frac{x_1}{n}, \frac{x_2}{n}, \frac{x_3}{n}\right).
$$

So, given the likelihood, we can compute the counts (the sample size n is assumed known). Therefore, (x_1, x_2, x_3) is a minimal sufficient statistic. ∎

Summary of Section 6.1

- The likelihood function for a model and data shows how the data supports the various possible values of the parameter. It is not the actual value of the likelihood that is important but the ratios of the likelihood at different values of the parameter.

- A sufficient statistic T for a model is any function of the data s such that once we know the value of $T(s)$, then we can determine the likelihood function $L(\cdot \mid s)$ (up to a positive constant multiple).

- A minimal sufficient statistic T for a model is any sufficient statistic such that once we know a likelihood function $L(\cdot \mid s)$ for the model and data, then we can determine $T(s)$.

Exercises

6.1.1 Suppose that a sample of n individuals is being tested for the presence of an antibody in their blood and that the number with the antibody present is recorded. Record an appropriate statistical model for this situation when we assume that the responses from individuals are independent. If we have a sample of 10 and record 3 positives, graph a representative likelihood function.

6.1.2 Suppose that suicides occur in a population at a rate p per person year and that p is assumed completely unknown. If we model the number of suicides observed in a population with a total of N person years as Poisson(Np), then record a representative likelihood function for p when we observe 22 suicides with $N = 30,345$.

6.1.3 Suppose that the lifelengths (in thousands of hours) of light bulbs are distributed Exponential(θ) where $\theta > 0$ is unknown. If we observe $\bar{x} = 5.2$ for a sample of 20 light bulbs, record a representative likelihood function. Why is it that we only need to observe the sample average to obtain a representative likelihood?

6.1.4 Suppose that we take a sample of $n = 100$ students from a university with over 50,000 students enrolled. We classify these students as either living on campus, living off campus with their parents, or living off campus independently. Suppose we observe the counts $(x_1, x_2, x_3) = (34, 44, 22)$. Determine the form of the likelihood function for the unknown proportions of students in the population that are in these categories.

6.1.5 Determine the constant that makes the likelihood functions in Examples 6.1.2 and 6.1.3 equal.

6.1.6 Suppose that $(x_1, \ldots, x_n)$ is a sample from the Bernoulli(θ) distribution where $\theta \in [0, 1]$ is unknown. Determine the likelihood function and a minimal sufficient statistic for this model. (Hint: Maximize the logarithm of the likelihood function.)

6.1.7 Suppose that $(x_1, \ldots, x_n)$ is a sample from the Poisson(θ) distribution where $\theta > 0$ is unknown. Determine the likelihood function and a minimal sufficient statistic for this model. (Hint: Maximize the logarithm of the likelihood function.)

Problems

6.1.8 Show that T defined in Example 6.1.6 is a minimal sufficient statistic. (Hint: Show that once you know the likelihood function, you can determine which of the two possible values for T has occurred.)

6.1.9 Suppose that $S = \{1, 2, 3, 4\}, \Omega = \{a, b, c\}$ where the three probability distributions are given by the following table.

	$s = 1$	$s = 2$	$s = 3$	$s = 4$
$\theta = a$	1/2	1/6	1/6	1/6
$\theta = b$	1/4	1/4	1/4	1/4
$\theta = c$	1/2	1/4	1/4	0

Determine a minimal sufficient statistic for this model. Is the minimal sufficient statistic in Example 6.1.6 sufficient for this model?

6.1.10 Suppose that $(x_1, \ldots, x_n)$ is a sample from the $N(\mu, \sigma_0^2)$ distribution where $\mu \in R^1$ is unknown. Determine the form of likelihood intervals for this model.

6.1.11 Suppose that $(x_1, \ldots, x_n) \in R^n$ is a sample from f_θ where $\theta \in \Omega$ is unknown. Show that the order statistics $(x_{(1)}, \ldots, x_{(n)})$ comprise a sufficient statistic for the model.

6.1.12 Determine a minimal sufficient statistic for a sample of n from the scale gamma model, i.e.,

$$f_\theta(x) = \frac{\theta^{\alpha_0}}{\Gamma(\alpha_0)} x^{\alpha_0 - 1} \exp\{-\theta x\}$$

for $x > 0$, $\theta > 0$ and where $\alpha_0 > 0$ is fixed.

6.1.13 Determine a minimal sufficient statistic for a sample of n from the Uniform$(0, \theta)$ model where $\theta > 0$.

6.1.14 Determine a minimal sufficient statistic for a sample of n from the Uniform(θ_1, θ_2) model where $\theta_1 < \theta_2$.

6.1.15 For the location-scale normal model, establish that the point where the likelihood is maximized is given by $(\bar{x}, \hat{\sigma}^2)$ as defined in Example 6.1.8. (Hint: Show that the second derivative of $\ln L\left((\bar{x}, \sigma^2) \mid x\right)$ with respect to σ^2 is negative at $\hat{\sigma}^2$ and then argue that $(\bar{x}, \hat{\sigma}^2)$ is the maximum.)

6.1.16 Suppose that we have a sample of n from a Bernoulli(θ) distribution where $\theta \in [0, 0.5]$. Determine a minimal sufficient statistic for this model. (Hint: It is easy to establish the sufficiency of $\bar{x}$, but this point will not maximize the likelihood when $\bar{x} > 0.5$, so $\bar{x}$ cannot be obtained from the likelihood by maximization as in Exercise 6.1.6. In general, consider the second derivative of the log of the likelihood at any point $\theta \in (0, 0.5)$ and note that knowing the likelihood means that we can compute any of its derivatives at any values where these exist.)

6.1.17 Suppose that we have a sample of n from the Multinomial$(1, \theta, 2\theta, 3\theta)$ distribution where $\theta \in [0, 1/6]$ is unknown. Determine the form of the likelihood function and show that $x_1 + x_2$ is a minimal sufficient statistic where x_i is the number of sample values corresponding to an observation in the ith category. (Hint: Problem 6.1.16.)

6.1.18 Suppose that we observe s from a statistical model with two densities f_1 and f_2. Show that the likelihood ratio $T(s) = f_1(s)/f_2(s)$ is a minimal sufficient statistic. (Hint: Use the definition of sufficiency directly.)

Challenge

6.1.19 (MV) Consider the location-scale gamma model, i.e.,

$$f_{(\mu,\sigma)}(x) = \frac{1}{\Gamma(\alpha_0)} \left(\frac{x - \mu}{\sigma} \right)^{\alpha_0 - 1} \exp\left\{ -\frac{x - \mu}{\sigma} \right\} \frac{1}{\sigma}$$

for $x > \mu \in R^1, \sigma > 0$ and where $\alpha_0 > 0$ is fixed.
(a) Determine the minimal sufficient statistic for a sample of n when $\alpha_0 = 1$. (Hint: Determine where the likelihood is positive and calculate the partial derivative of the log of the likelihood with respect to μ.)
(b) Determine the minimal sufficient statistic for a sample of n when $\alpha_0 \neq 1$. (Hint: Use Problem 6.1.11, the partial derivative of the log of the likelihood with respect to μ and determine where it is infinite.)

Discussion Topic

6.1.20 How important do you think it is that a statistician try to quantify how much error there is in any inference drawn? For example, if an estimate is being quoted for some unknown quantity, is it important that the statistician give some indication about how accurate (or inaccurate) this inference is?

6.2 Maximum Likelihood Estimation

In Section 6.1 we introduced the likelihood function $L(\cdot \mid s)$ as a basis for making inferences about the unknown true value $\theta \in \Omega$. We now begin to consider the specific types of inferences discussed in Section 5.5.3 and start with estimation.

When we are interested in a point estimate of θ then a value $\hat{\theta}(s)$ that maximizes $L(\theta \mid s)$ is a sensible choice, as this value is the best supported by the data, i.e.,

$$L\left(\hat{\theta}(s) \mid s\right) \geq L(\theta \mid s) \tag{6.2.1}$$

for every $\theta \in \Omega$. We call $\hat{\theta} : S \rightarrow \Omega$ satisfying (6.2.1) a *maximum likelihood estimator*, and the value $\hat{\theta}(s)$ is called a *maximum likelihood estimate*, or *MLE* for short. Notice that, if we use $cL(\cdot \mid s)$ as the likelihood function, for fixed $c > 0$, then $\hat{\theta}(s)$ is also an MLE using this version of the likelihood. So we can use any version of the likelihood to calculate an MLE.

Example 6.2.1
Suppose that the sample space is $S = \{1, 2, 3\}$, the parameter space is $\Omega = \{1, 2\}$, and the model is given by the following table.

	$s = 1$	$s = 2$	$s = 3$
$f_1(s)$	0.3	0.4	0.3
$f_2(s)$	0.1	0.7	0.2

Further suppose we observe $s = 1$. So, for example, we could be presented with one of two bowls of chips containing these proportions of chips labeled 1, 2, and 3. We draw a chip and observe that the chip is labelled 1 and now want to make inferences about which bowl we have been presented with.

In this case, the MLE is given by $\hat{\theta}(1) = 1$, since $0.3 = L(1 \mid 1) > 0.1 = L(2 \mid 1)$. If we had instead observed $s = 2$, then $\hat{\theta}(2) = 2$; if we had observed $s = 3$, then $\hat{\theta}(3) = 1$. ∎

Note that an MLE need not be unique. For example, in Example 6.2.1, if f_2 was defined by $f_2(1) = 0$, $f_2(2) = 0.7$ and $f_2(3) = 0.3$, then an MLE is as given there, but putting $\hat{\theta}(3) = 2$ also gives an MLE.

The MLE has a very important invariance property. Suppose that we *reparameterize* a model via a 1–1 function Ψ defined on Ω. By this we mean that, instead of labelling the individual distributions in the model using $\theta \in \Omega$, we use $\psi \in \Upsilon = \{\Psi(\theta) : \theta \in \Omega\}$. For example, in Example 6.2.1, we could take $\Psi(1) = a$ and $\Psi(2) = b$ so that $\Upsilon = \{a, b\}$. So the model is now given by $\{g_\psi : \psi \in \Upsilon\}$ where $g_\psi = f_\theta$ for the unique value θ such that $\Psi(\theta) = \psi$. We have a new parameter ψ and a new parameter space Υ. Nothing has changed about the probability distributions in the statistical model, only the way they are labelled. We then have the following result.

Theorem 6.2.1 If $\hat{\theta}(s)$ is an MLE for the original parameterization and, if Ψ is a 1–1 function defined on Ω, then $\hat{\psi}(s) = \Psi(\hat{\theta}(s))$ is an MLE in the new parameterization.

Proof: If we select the likelihood function for the new parameterization to be $L^*(\psi \mid s) = g_\psi(s)$, and the likelihood for the original parameterization to be $L(\theta \mid s) = f_\theta(s)$, then we have

$$L^*\left(\hat{\psi}(s) \mid s\right) = g_{\Psi(\hat{\theta}(s))}(s) = f_{\hat{\theta}(s)}(s) = L\left(\hat{\theta}(s) \mid s\right) \geq L(\theta \mid s) = L^*(\Psi(\theta) \mid s)$$

for every $\theta \in \Omega$. This implies that $L^*\left(\hat{\psi}(s) \mid s\right) \geq L^*(\psi \mid s)$ for every $\psi \in \Upsilon$ and establishes the result. ∎

Theorem 6.2.1 shows that no matter how we parameterize the model, the MLE behaves in a consistent way under the reparameterization. This is an important property, and not all estimation procedures satisfy this.

An important issue is the computation of MLE's. In Example 6.2.1 we were able to do this by simply examining the table giving the distributions. With more complicated models, this approach is not possible. In many situations, however, we can use the methods of calculus to compute $\hat{\theta}(s)$. For this we require that $f_\theta(s)$ be a continuously differentiable function of θ so that we can use optimization methods from calculus.

Rather than using the likelihood function, it is often convenient to use the *log-likelihood function* $l(\cdot \mid s)$ defined on Ω by

$$l(\theta \mid s) = \ln f_\theta(s).$$

Note that $\ln(x)$ is a 1–1 increasing function of $x > 0$ and this implies that $L(\hat{\theta}(s) \mid s) \geq L(\theta \mid s)$ for every $\theta \in \Omega$ if and only if $l(\hat{\theta}(s) \mid s) \geq l(\theta \mid s)$ for every $\theta \in \Omega$. So we can maximize $l(\cdot \mid s)$ instead when computing an MLE. The convenience of the log-likelihood arises from the fact that, for a sample $(s_1, \ldots, s_n)$ from $\{f_\theta : \theta \in \Omega\}$, the likelihood function is given by

$$L(\theta \mid s_1, \ldots, s_n) = \prod_{i=1}^{n} f_\theta(s_i)$$

whereas the log-likelihood is given by

$$l(\theta \mid s_1, \ldots, s_n) = \sum_{i=1}^{n} \ln f_\theta(s_i).$$

It is typically much easier to differentiate a sum than a product.

Because we are going to be differentiating the log-likelihood, it is convenient to give a name to this derivative. We define the *score function* $S(\theta \mid s)$ of a model to be the derivative of its log-likelihood function whenever this exists. So when θ is a one-dimensional real-valued parameter, then

$$S(\theta \mid s) = \frac{\partial l(\theta \mid s)}{\partial \theta},$$

provided this partial derivative exists (see Appendix A.5 for a definition of partial derivative). We restrict our attention now to the situation in which θ is one-dimensional.

To obtain the MLE, we must then solve the *score equation*

$$S(\theta \mid s) = 0 \tag{6.2.2}$$

for θ. Of course, a solution to (6.2.2) is not necessarily an MLE, because such a point may be a local minimum or only a local maximum rather than a global maximum. To guarantee that a solution $\hat{\theta}(s)$ is at least a local maximum, we must also check that

$$\left. \frac{\partial S(\theta \,|\, s)}{\partial \theta} \right|_{\theta = \hat{\theta}(s)} = \left. \frac{\partial^2 l(\theta \,|\, s)}{\partial \theta^2} \right|_{\theta = \hat{\theta}(s)} < 0. \qquad (6.2.3)$$

Then we must evaluate $l(\cdot \,|\, s)$ at each local maximum to determine the global maximum.

Let us compute some MLE's using calculus.

Example 6.2.2 *Location Normal Model*
Consider the likelihood function

$$L(\theta \,|\, x_1, \ldots, x_n) = \exp\left(-\frac{n}{2\sigma_0^2} (\bar{x} - \theta)^2 \right),$$

obtained in Example 6.1.4 for a sample $(x_1, \ldots, x_n)$ from the $N(\theta, \sigma_0^2)$ model where $\theta \in R^1$ is unknown and σ_0^2 is known. The log-likelihood function is then

$$l(\theta \,|\, x_1, \ldots, x_n) = -\frac{n}{2\sigma_0^2} (\bar{x} - \theta)^2,$$

and the score function is

$$S(\theta \,|\, x_1, \ldots, x_n) = \frac{n}{\sigma_0^2} (\bar{x} - \theta).$$

The score equation is given by

$$\frac{n}{\sigma_0^2} (\bar{x} - \theta) = 0.$$

Solving this for θ gives the unique solution $\hat{\theta}(x_1, \ldots, x_n) = \bar{x}$. To check that this is a local maximum, we calculate

$$\left. \frac{\partial S(\theta \,|\, x_1, \ldots, x_n)}{\partial \theta} \right|_{\theta = \bar{x}} = -\frac{n}{\sigma_0^2},$$

which is negative, and so indicates that $\bar{x}$ is a local maximum. Because we have only one local maximum, it is also the global maximum and we have indeed obtained the MLE. ∎

Example 6.2.3 *Exponential Model*
Suppose that a lifetime is known to be distributed Exponential(θ) where $\theta > 0$ is unknown. Then based on a sample $(x_1, \ldots, x_n)$, the likelihood is given by

$$L(\theta \,|\, x_1, \ldots, x_n) = \frac{1}{\theta^n} \exp\left(-\frac{n\bar{x}}{\theta} \right),$$

the log-likelihood is given by

$$l(\theta \,|\, x_1, \ldots, x_n) = -n \ln \theta - \frac{n\bar{x}}{\theta},$$

and the score function is given by

$$S\left(\theta \,|\, x_1, \ldots, x_n\right) = -\frac{n}{\theta} + \frac{n\bar{x}}{\theta^2}.$$

Solving the score equation gives $\hat{\theta}\left(x_1, \ldots, x_n\right) = \bar{x}$, and because $\bar{x} > 0$,

$$\left.\frac{\partial S\left(\theta \,|\, x_1, \ldots, x_n\right)}{\partial \theta}\right|_{\theta=\bar{x}} = \left.\frac{n}{\theta^2} - 2\frac{n\bar{x}}{\theta^3}\right|_{\theta=\bar{x}} = -\frac{n}{\bar{x}^2} < 0,$$

so $\bar{x}$ is indeed the MLE. ∎

In both examples just considered, we were able to derive simple formulas for the MLE. This is not always possible. Consider the following example.

Example 6.2.4
Consider a population in which individuals are classified according to one of three types labelled 1, 2, and 3, respectively. Further suppose that the proportions of individuals falling in these categories are known to follow the law $p_1 = \theta, p_2 = \theta^2, p_3 = 1 - \theta - \theta^2$ where

$$\theta \in \left[0, \left(\sqrt{5} - 1\right)/2\right] = [0, 0.618\,03]$$

is unknown. Here p_i denotes the proportion of individuals in the ith class. Note that the requirement that $0 \le \theta + \theta^2 \le 1$ imposes the upper bound on θ and the precise bound is obtained by solving $\theta + \theta^2 - 1 = 0$ for θ using the formula for the roots of a quadratic. Relationships like this, amongst the proportions of the distribution of a categorical variable, often arise in genetics. For example, the categorical variable might serve to classify individuals into different genotypes.

For a sample of n (where n is small relative to the size of the population so that we can assume observations are i.i.d.), the likelihood function is given by

$$L\left(\theta \,|\, s_1, \ldots, s_n\right) = \theta^{x_1} \theta^{2x_2} \left(1 - \theta - \theta^2\right)^{x_3}$$

where x_i denotes the sample count in the ith class. The log-likelihood function is then

$$l\left(\theta \,|\, s_1, \ldots, s_n\right) = \left(x_1 + 2x_2\right)\ln\theta + x_3\ln\left(1 - \theta - \theta^2\right),$$

and the score function is

$$S\left(\theta \,|\, s_1, \ldots, s_n\right) = \frac{\left(x_1 + 2x_2\right)}{\theta} - \frac{x_3\left(1 + 2\theta\right)}{1 - \theta - \theta^2}.$$

The score equation then leads to a solution $\hat{\theta}$ being a root of the quadratic

$$\left(x_1 + 2x_2\right)\left(1 - \theta - \theta^2\right) - x_3\left(\theta + 2\theta^2\right)$$
$$= -\left(x_1 + 2x_2 + 2x_3\right)\theta^2 - \left(x_1 + 2x_2 + x_3\right)\theta + \left(x_1 + 2x_2\right).$$

Using the formula for the roots of a quadratic, we obtain

$$\hat{\theta} = \frac{1}{2(x_1 + 2x_2 + 2x_3)}$$
$$\times \left(-x_1 - 2x_2 - x_3 \pm \sqrt{5x_1^2 + 20x_1x_2 + 10x_1x_3 + 20x_2^2 + 20x_2x_3 + x_3^2} \right).$$

Notice that the formula for the roots does not determine the MLE in a clear way. In fact we cannot even tell if either of the roots lies in $[0, 1]$! So there are four possible values for the MLE at this point — either of the roots or the boundary points 0 and 0.61803.

We can resolve this easily in an application by simply numerically evaluating the likelihood at the four points. For example, if $x_1 = 70, x_2 = 5$, and $x_3 = 25$, then the roots are -1.28616 and 0.47847 so it is immediate that the MLE is $\hat{\theta}(x_1, \ldots, x_n) = 0.47847$. We can see this graphically in the plot of the log-likelihood provided in Figure 6.2.1. ∎

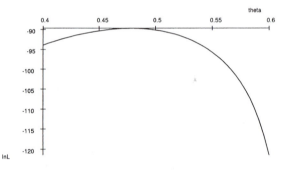

Figure 6.2.1: The log-likelihood function in Example 6.2.4 when $x_1 = 70, x_2 = 5$, and $x_3 = 25$.

In general, the score equation (6.2.2) must be solved numerically, using an iterative routine like Newton-Raphson. Example 6.2.4 demonstrates that we must be very careful not to just accept a solution from such a procedure as the MLE, but check that the fundamental defining property (6.2.1) is satisfied. We also have to be careful that the necessary smoothness conditions are satisfied so that calculus can be used. Consider the following example.

Example 6.2.5 *Uniform$(0, \theta)$ Model*
Suppose $(x_1, \ldots, x_n)$ is a sample from the Uniform$(0, \theta)$ model where $\theta > 0$ is unknown. Then the likelihood function is given by

$$L(\theta \,|\, x_1, \ldots, x_n) = \begin{cases} \theta^{-n} & x_i \leq \theta \text{ for } i = 1, \ldots, n \\ 0 & x_i > \theta \text{ for some } i \end{cases}$$
$$= \theta^{-n} I_{[x_{(n)}, \infty)}(\theta)$$

where $x_{(n)}$ is the largest order statistic from the sample. In Figure 6.2.2, we have graphed this function when $n = 10$ and $x_{(n)} = 1.3$. Notice that the maximum clearly occurs at $x_{(n)}$, and we cannot obtain this value via differentiation, as $L(\cdot \mid x_1, \ldots, x_n)$ is not differentiable there. ∎

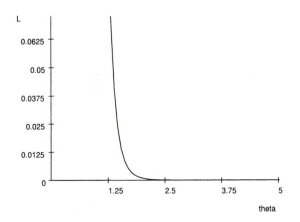

Figure 6.2.2: Plot of the likelihood function in Example 6.2.5 when $n = 10$ and $x_{(10)} = 3.3$.

The lesson of Examples 6.2.4 and 6.2.5 is that we have to be careful when computing MLE's. We now look at an example of a two-dimensional problem where the MLE can be obtained using one-dimensional methods.

Example 6.2.6 *Location-Scale Normal Model*
Suppose that $(x_1, \ldots, x_n)$ is a sample from an $N(\mu, \sigma^2)$ distribution where $\mu \in R^1$ and $\sigma > 0$ are unknown. The parameter in this model is two-dimensional and is given by $\theta = (\mu, \sigma^2) \in \Omega = R^1 \times (0, \infty)$. The likelihood function is then given by

$$L\left(\mu, \sigma^2 \mid x_1, \ldots, x_n\right) = \left(2\pi\sigma^2\right)^{-n/2} \exp\left(-\frac{n}{2\sigma^2}(\bar{x} - \mu)^2\right) \exp\left(-\frac{n-1}{2\sigma^2}s^2\right),$$

as shown in Example 6.1.8. The log-likelihood function is given by

$$l\left(\mu, \sigma^2 \mid x_1, \ldots, x_n\right) = -\frac{n}{2}\ln 2\pi - \frac{n}{2}\ln \sigma^2 - \frac{n}{2\sigma^2}(\bar{x} - \mu)^2 - \frac{n-1}{2\sigma^2}s^2. \quad (6.2.4)$$

As discussed in Example 6.1.8, it is clear that, for fixed σ^2, (6.2.4) is maximized, as a function of μ, by $\hat{\mu} = \bar{x}$. Note that this does not involve σ^2, so this must be the first coordinate of the MLE.

Substituting $\mu = \bar{x}$ into (6.2.4), we obtain

$$-\frac{n}{2}\ln 2\pi - \frac{n}{2}\ln \sigma^2 - \frac{n-1}{2\sigma^2}s^2, \quad (6.2.5)$$

and the second coordinate of the MLE must be the value of σ^2 that maximizes (6.2.5). Differentiating (6.2.5) with respect to σ^2 and setting this equal to 0 gives

$$-\frac{n}{2\sigma^2} + \frac{n-1}{2\left(\sigma^2\right)^2}s^2 = 0. \tag{6.2.6}$$

Solving (6.2.6) for σ^2 leads to the solution

$$\hat{\sigma}^2 = \frac{n-1}{n}s^2 = \frac{1}{n}\sum_{i=1}^{n}\left(x_i - \bar{x}\right)^2.$$

Differentiating (6.2.6) with respect to σ^2, and substituting in $\hat{\sigma}^2$, we see that the second derivative is negative and so $\hat{\sigma}^2$ is a point where the maximum is attained.

Therefore, we have shown that the MLE of $\left(\mu, \sigma^2\right)$ is given by

$$\left(\bar{x}, \frac{1}{n}\sum_{i=1}^{n}\left(x_i - \bar{x}\right)^2\right).$$

In the following section we will show that this result can also be obtained using multidimensional calculus. ∎

So far we have talked about estimating only the full parameter θ for a model. What about estimating a general characteristic of interest $\psi\left(\theta\right)$ for some function ψ defined on the parameter space Ω? Perhaps the obvious answer here is to use the estimate $\hat{\psi}\left(s\right) = \psi(\hat{\theta}\left(s\right))$ where $\hat{\theta}\left(s\right)$ is an MLE of θ. This is sometimes referred to as the *plug-in MLE* of ψ. Notice, however, that the plug-in MLE is not necessarily a true MLE, in the sense that we have a likelihood function for a model indexed by ψ and that takes its maximum value at $\hat{\psi}\left(s\right)$. If ψ is a 1–1 function defined on Ω, then Theorem 6.2.1 establishes that $\hat{\psi}\left(s\right)$ is a true MLE but not otherwise.

If ψ is not 1–1, then we can often find a *complementing function* λ defined on Ω so that (ψ, λ) is a 1–1 function of θ. Then, by Theorem 6.2.1,

$$\left(\hat{\psi}\left(s\right), \hat{\lambda}\left(s\right)\right) = \left(\psi\left(\hat{\theta}\left(s\right)\right), \lambda\left(\hat{\theta}\left(s\right)\right)\right)$$

is the joint MLE, but $\hat{\psi}\left(s\right)$ is still not formally an MLE. Sometimes a plug-in MLE can perform badly, as it ignores the information in $\lambda(\hat{\theta}\left(s\right))$ about the true value of ψ. An example illustrates this phenomenon.

Example 6.2.7 *Sum of Squared Means*
Suppose that $X_i \sim N(\mu_i, 1)$ for $i = 1, \ldots, n$ and that these are independent with the μ_i completely unknown. So here $\theta = (\mu_1, \ldots, \mu_n)$ and $\Omega = R^n$. Suppose that we want to estimate $\psi\left(\theta\right) = \mu_1^2 + \cdots + \mu_n^2$.

The log-likelihood function is given by

$$l\left(\theta \,|\, x_1, \ldots, x_n\right) = -\frac{1}{2}\sum_{i=1}^{n}\left(x_i - \mu_i\right)^2.$$

Clearly this is maximized by $\hat{\theta}(x_1, \ldots, x_n) = (x_1, \ldots, x_n)$. So the plug-in MLE of ψ is given by $\hat{\psi} = \sum_{i=1}^n x_i^2$.

Now observe that

$$E_\theta \left(\sum_{i=1}^n X_i^2 \right) = \sum_{i=1}^n E_\theta \left(X_i^2 \right) = \sum_{i=1}^n \left(\text{Var}_\theta \left(X_i^2 \right) + \mu_i^2 \right) = n + \psi(\theta),$$

where $E_\theta(g)$ refers to the expectation of $g(s)$ when $s \sim f_\theta$. So when n is large, it is likely that $\hat{\psi}$ is far from the true value. An immediate improvement in this estimator is to use $\sum_{i=1}^n x_i^2 - n$ instead. ∎

There have been various attempts to correct problems such as the one illustrated in Example 6.2.7. Typically, these involve modifying the likelihood in some way. We do not pursue this issue further in this text but just advise caution when using plug-in MLE's. Sometimes, as in Example 6.2.6 where we estimate μ by $\bar{x}$ and σ^2 by s^2, they seem appropriate, whereas other times, as in Example 6.2.7, they do not.

6.2.1 The Multidimensional Case (Advanced)

We now consider the situation in which $\theta = (\theta_1, \ldots, \theta_k) \in R^k$ is multidimensional, i.e., $k > 1$. The likelihood and log-likelihood are then defined just as before but the score function is now given by

$$S(\theta \,|\, s) = \begin{pmatrix} \frac{\partial l(\theta \,|\, s)}{\partial \theta_1} \\ \frac{\partial l(\theta \,|\, s)}{\partial \theta_2} \\ \vdots \\ \frac{\partial l(\theta \,|\, s)}{\partial \theta_k} \end{pmatrix},$$

provided all these partial derivatives exist. For the score equation we get

$$\begin{pmatrix} \frac{\partial l(\theta \,|\, s)}{\partial \theta_1} \\ \frac{\partial l(\theta \,|\, s)}{\partial \theta_2} \\ \vdots \\ \frac{\partial l(\theta \,|\, s)}{\partial \theta_k} \end{pmatrix} = \begin{pmatrix} 0 \\ 0 \\ \vdots \\ 0 \end{pmatrix},$$

and we must solve this k-dimensional equation for $(\theta_1, \ldots, \theta_k)$. This is often much more difficult than in the one-dimensional case, and we typically have to resort to numerical methods.

A necessary and sufficient condition for $(\hat{\theta}_1, \ldots, \hat{\theta}_k)$ to be a local maximum, when the log-likelihood has continuous second partial derivatives, is that the matrix of second partial derivatives of the log-likelihood, evaluated at $(\hat{\theta}_1, \ldots, \hat{\theta}_k)$, must be negative definite (equivalently, all of its eigenvalues must be negative). We then must evaluate the likelihood at each of the local maxima obtained to determine the global maximum or MLE.

We will not pursue the numerical computation of MLE's in the multidimensional case any further here, but restrict our attention to a situation in which we carry out the calculations in closed form.

Example 6.2.8 *Location-Scale Normal Model*

We determined the log-likelihood function for this model in (6.2.4). The score function is then

$$S\left(\mu, \sigma^2 \mid x_1, \ldots, x_n\right) = \begin{pmatrix} \frac{\partial S(\theta \mid x_1, \ldots, x_n)}{\partial \mu} \\ \frac{\partial S(\theta \mid x_1, \ldots, x_n)}{\partial \sigma^2} \end{pmatrix}$$

$$= \begin{pmatrix} \frac{n}{\sigma^2}\left(\bar{x} - \mu\right) \\ -\frac{n}{2\sigma^2} + \frac{n}{2\sigma^4}\left(\bar{x} - \mu\right)^2 + -\frac{n-1}{2\sigma^4}s^2 \end{pmatrix}.$$

The score equation is

$$\begin{pmatrix} \frac{n}{\sigma^2}\left(\bar{x} - \mu\right) \\ -\frac{n}{2\sigma^2} + \frac{n}{2\sigma^4}\left(\bar{x} - \mu\right)^2 + \frac{n-1}{2\sigma^4}s^2 \end{pmatrix} = \begin{pmatrix} 0 \\ 0 \end{pmatrix},$$

and the first of these equations immediately implies that $\hat{\mu} = \bar{x}$. Substituting this value for μ into the second equation and solving for σ^2 leads to the solution

$$\hat{\sigma}^2 = \frac{n-1}{n}s^2 = \frac{1}{n}\sum_{i=1}^{n}\left(x_i - \bar{x}\right)^2.$$

From Example 6.2.6, we know that this solution does indeed give the MLE. ∎

Summary of Section 6.2

- An MLE (maximum likelihood estimator) is a value of the parameter θ that maximizes the likelihood function. It is the value of θ that is best supported by the model and data.

- We can often compute an MLE by using the methods of calculus. When applicable, this leads to solving the score equation for θ either explicitly or using numerical algorithms. Always be careful to check that these methods are applicable to the specific problem you are dealing with. Further, always check that any solution to the score equation is a maximum and indeed an absolute maximum.

Exercises

6.2.1 Suppose that $S = \{1, 2, 3, 4\}, \Omega = \{a, b\}$ where the two probability distributions are given by the following table.

	$s = 1$	$s = 2$	$s = 3$	$s = 4$
$\theta = a$	1/2	1/6	1/6	1/6
$\theta = b$	1/3	1/3	1/3	0

Determine the MLE of θ for each possible data value.

6.2.2 If $(x_1, \ldots, x_n)$ is a sample from a Bernoulli(θ) distribution where $\theta \in [0, 1]$ is unknown, then determine the MLE of θ.

6.2.3 If $(x_1, \ldots, x_n)$ is a sample from a Bernoulli(θ) distribution where $\theta \in [0, 1]$ is unknown, then determine the MLE of θ^2.

6.2.4 If $(x_1, \ldots, x_n)$ is a sample from a Poisson(θ) distribution where $\theta \in (0, \infty)$ is unknown, then determine the MLE of θ.

6.2.5 If $(x_1, \ldots, x_n)$ is a sample from a Gamma(α_0, θ) distribution where $\alpha_0 > 0$ and $\theta \in (0, \infty)$ is unknown, then determine the MLE of θ.

6.2.6 Suppose that $(x_1, \ldots, x_n)$ is the result of independent tosses of a coin where we toss until the first head occurs and where the probability of a head on a single toss is $\theta \in [0, 1]$. Determine the MLE of θ.

6.2.7 If $(x_1, \ldots, x_n)$ is a sample from a Beta$(\alpha, 1)$ distribution (see Problem 2.4.16) where $\alpha > 0$ is unknown, then determine the MLE of α. (Hint: Use Problem 2.4.7(a) and assume $\Gamma(\alpha)$ is a differentiable function of α.)

6.2.8 If $(x_1, \ldots, x_n)$ is a sample from a Weibull(β) distribution (see Problem 2.4.11), where $\beta > 0$ is unknown, then determine the score equation for the MLE of β.

6.2.9 If $(x_1, \ldots, x_n)$ is a sample from a Pareto(α) distribution (see Problem 2.4.12), where $\alpha > 0$ is unknown, then determine the MLE of α.

6.2.10 If $(x_1, \ldots, x_n)$ is a sample from a Log-normal(τ) distribution (see Problem 2.6.12), where $\tau > 0$ is unknown, then determine the MLE of τ.

Problems

6.2.11 (*Hardy-Weinberg law*) The Hardy-Weinberg law in genetics says that the proportions of genotypes AA, Aa, and aa are θ^2, $2\theta(1-\theta)$, and $(1-\theta)^2$, respectively where $\theta \in [0, 1]$. Suppose that in a sample of n from the population (small relative to the size of the population), we observe x_1 individuals of type AA, x_2 individuals of type Aa, and x_3 individuals of type aa.
(a) What distribution do the counts (X_1, X_2, X_3) follow?
(b) Record the likelihood function, the log-likelihood function, and the score function for θ.
(c) Record the form of the MLE for θ.

6.2.12 If $(x_1, \ldots, x_n)$ is a sample from an $N(\mu, 1)$ distribution where $\mu \in R^1$ is unknown, determine the MLE of the probability content of the interval $(-\infty, 1)$. Justify your answer.

6.2.13 If $(x_1, \ldots, x_n)$ is a sample from an $N(\mu, 1)$ distribution where $\mu \geq 0$ is unknown, determine the MLE of μ.

6.2.14 Prove that, if $\hat{\theta}(s)$ is the MLE for a model for response s and if T is a sufficient statistic for the model, then $\hat{\theta}(s)$ is also the MLE for the model for $T(s)$.

6.2.15 (MV) Suppose that $(X_1, X_2, X_3) \sim$ Multinomial$(n, \theta_1, \theta_2, \theta_3)$ (Example 6.1.5) where $\Omega = \{(\theta_1, \theta_2, \theta_3) : 0 \leq \theta_i \leq 1, \theta_1 + \theta_2 + \theta_3 = 1\}$ and we observe $(X_1, X_2, X_3) = (x_1, x_2, x_3)$.
(a) Determine the MLE of $(\theta_1, \theta_2, \theta_3)$.
(b) What is the plug-in MLE of $\theta_1 + \theta_2^2 - \theta_3^2$?

6.2.16 If $(x_1, \ldots, x_n)$ is a sample from a $U(\theta_1, \theta_2)$ distribution with $\Omega = \{(\theta_1, \theta_2) \in R^2 : \theta_1 < \theta_2\}$, determine the MLE of (θ_1, θ_2). (Hint: You can't use calculus. Instead, directly determine the maximum over θ_1 when θ_2 is fixed, and then vary θ_2.)

Computer Problem

6.2.17 Suppose that the proportion of left-handed individuals in a population is θ. Based on a simple random sample of 20, you observe four left-handed individuals.
(a) Assuming the sample size is small relative to the population size, plot the log-likelihood function and determine the MLE.
(b) If instead the population size is only 50, then plot the log-likelihood function and determine the MLE. (Hint: Remember that the number of left-handed individuals follows a hypergeometric distribution. This forces θ to be of the form $i/50$ for some integer i between 4 and 36. From a tabulation of the log-likelihood you can obtain the MLE.)

Challenge

6.2.18 If $(x_1, \ldots, x_n)$ is a sample from a distribution with density $f_\theta(x) = (1/2) \exp(-|x - \theta|)$ for $x \in R^1$ and where $\theta \in R^1$ is unknown, then determine the MLE of θ. (Hint: You can't use calculus. Instead, maximize the log-likelihood in each of the intervals $(-\infty, x_{(1)})$, $[x_{(1)} \leq \theta < x_{(2)})$, etc.)

6.3   Inferences Based on the MLE

In Table 6.3.1 we have recorded $n = 66$ measurements of the speed of light (passage time recorded as deviations from $24,800$ nanoseconds between two mirrors 7400 meters apart) made by A. A. Michelson and S. Newcomb in 1882.

28	26	33	24	34	−44	27	16	40	−2	29
22	24	21	25	30	23	29	31	19	24	20
36	32	36	28	25	21	28	29	37	25	28
26	30	32	36	26	30	22	36	23	27	27
28	27	31	27	26	33	26	32	32	24	39
28	24	25	32	25	29	27	28	29	16	23

Table 6.3.1: Speed of light measurements.

Figure 6.3.1 is a boxplot of these data with the variable labeled as x. Notice that there are two outliers at $x = -2$ and $x = -44$. We'll presume there is something very special about these observations and discard them for the remainder of our discussion.

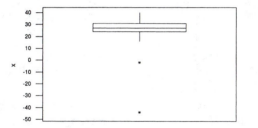

Figure 6.3.1: Boxplot of the data values in Table 6.3.1.

Figure 6.3.2 presents a histogram of these data minus the two data values identified as outliers. Notice that the histogram looks reasonably symmetrical, so it seems plausible to assume that these data are from an $N(\mu, \sigma^2)$ distribution for some values of μ and σ^2. Accordingly, a reasonable statistical model for this data would appear to be the location-scale normal model. In Chapter 9 we will discuss further how to assess the validity of the normality assumption.

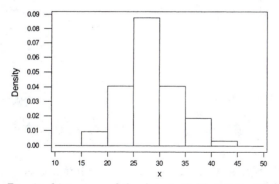

Figure 6.3.2: Density histogram of the data in Table 6.3.1 with the outliers removed.

If we accept that the location-scale normal model makes sense, the question arises concerning how to make inferences about the unknown parameters μ and σ^2. The purpose of this section is to develop methods for handling problems like this. The methods developed in this section depend on special features of the MLE in a given context. In Section 6.5 we develop a more general approach based upon the MLE.

6.3.1 Standard Errors and Bias

Based on the justification for the likelihood, the MLE $\hat{\theta}(s)$ seems like a natural estimate of the true value of θ. Let us suppose that we will then use the plug-in MLE estimate $\hat{\psi}(s) = \psi(\hat{\theta}(s))$ for a characteristic of interest $\psi(\theta)$ (e.g., $\psi(\theta)$ might be the first quartile or the variance).

In an application, we want to know how reliable the estimate $\hat{\psi}\,(s)$ is. In other words, can we expect $\hat{\psi}\,(s)$ to be close to the true value of $\psi\,(\theta)$, or is there a reasonable chance that $\hat{\psi}\,(s)$ is far from the true value? This leads us to consider the sampling distribution of $\hat{\psi}\,(s)$, as this tells us how much variability there will be in $\hat{\psi}\,(s)$ under repeated sampling from the true distribution f_θ. Because we do not know what the true value of θ is, we have to look at the sampling distribution of $\hat{\psi}\,(s)$ for every $\theta \in \Omega$.

To simplify this, we substitute a numerical measure of how concentrated these sampling distributions are about $\psi\,(\theta)$. Perhaps the most commonly used measure, when $\psi\,(\theta) \in R^1$, is the *mean-squared error (MSE)*, given by

$$\text{MSE}_\theta\left(\hat{\psi}\right) = E_\theta\left(\left(\hat{\psi} - \psi\,(\theta)\right)^2\right)$$

for each $\theta \in \Omega$. Clearly, the smaller $\text{MSE}_\theta(\hat{\psi})$ is, the more concentrated the sampling distribution of $\hat{\psi}\,(s)$ is about the value $\psi\,(\theta)$.

Looking at $\text{MSE}_\theta(\hat{\psi})$, as a function of θ, gives us some idea of how reliable $\hat{\theta}\,(s)$ is as an estimate of the true value of θ. Because we do not know the true value of θ, and thus the true value of $\text{MSE}_\theta(\hat{\psi})$, statisticians record an estimate of the mean-squared error at the true value. Often

$$\text{MSE}_{\hat{\theta}(s)}\left(\hat{\psi}\right)$$

is used for this. In other words, we evaluate $\text{MSE}_\theta(\hat{\psi})$ at $\theta = \hat{\theta}\,(s)$ as a measure of the accuracy of the estimate $\hat{\psi}\,(s)$.

The following result gives an important identity for the MSE for a general estimator T.

Theorem 6.3.1 If $\psi\,(\theta) \in R^1$ and T is a real-valued function defined on S such that $E_\theta\,(T)$ exists, then

$$\text{MSE}_\theta\,(T) = \text{Var}_\theta\,(T) + (E_\theta\,(T) - \psi\,(\theta))^2. \qquad (6.3.1)$$

Proof: We have

$$
\begin{aligned}
E_\theta\left((T - \psi\,(\theta))^2\right) &= E_\theta\left((T - E_\theta\,(T) + E_\theta\,(T) - \psi\,(\theta))\right)^2 \\
&= E_\theta\left((T - E_\theta\,(T))^2\right) \\
&\quad + 2E_\theta\left((T - E_\theta\,(T))\,(E_\theta\,(T) - \psi\,(\theta))\right) + (E_\theta\,(T) - \psi\,(\theta))^2 \\
&= \text{Var}_\theta\,(T) + (E_\theta\,(T) - \psi\,(\theta))^2
\end{aligned}
$$

because

$$
\begin{aligned}
E_\theta\left((T - E_\theta\,(T))\,(E_\theta\,(T) - \psi\,(\theta))\right) &= (E_\theta\,(T - E_\theta\,(T)))\,(E_\theta\,(T) - \psi\,(\theta)) \\
&= 0. \quad \blacksquare
\end{aligned}
$$

The second term in (6.3.1) is the square of the *bias* $E_\theta(T) - \psi(\theta)$ in the estimator T. Note that when the bias in an estimator is 0, then the MSE is just the variance. When the bias in an estimator T is 0 for every θ, we call T an *unbiased estimator* of ψ.

Unbiasedness tells us that, in a sense, the sampling distribution of the estimator is centered on the true value. For unbiased estimators,

$$\text{MSE}_{\hat{\theta}(s)}(T) = \text{Var}_{\hat{\theta}(s)}(T)$$

and

$$\text{Sd}_{\hat{\theta}(s)}(T) = \sqrt{\text{Var}_{\hat{\theta}(s)}(T)}$$

is an estimate of the standard deviation of T and is referred to as the *standard error of the estimate* $T(s)$. As a principle of good statistical practice, whenever we quote an estimate of a quantity, we should also provide its standard error, at least when we have an unbiased estimator, as this tells us something about the accuracy of the estimate.

We consider some examples.

Example 6.3.1 *Location Normal Model*
Consider the likelihood function

$$L(\mu \,|\, x_1, \ldots, x_n) = \exp\left(-\frac{n}{2\sigma_0^2}(\bar{x} - \mu)^2\right),$$

obtained in Example 6.1.4 for a sample $(x_1, \ldots, x_n)$ from the $N(\mu, \sigma_0^2)$ model where $\mu \in R^1$ is unknown and $\sigma_0^2 > 0$ is known. Suppose that we want to estimate μ. The MLE of μ was computed in Example 6.2.2 to be $\bar{x}$.

In this case, we can determine the sampling distribution of the MLE exactly from the results in Section 4.6. We have that $\bar{X} \sim N(\mu, \sigma_0^2/n)$ and so $\bar{X}$ is unbiased, and

$$\text{MSE}_\mu(\bar{X}) = \text{Var}_\mu(\bar{X}) = \frac{\sigma_0^2}{n},$$

which is independent of μ. So we do not need to estimate the MSE in this case. The standard error of the estimate is given by

$$\text{Sd}_\mu(\bar{X}) = \frac{\sigma_0}{\sqrt{n}}.$$

Note that the standard error decreases as the population variance σ_0^2 decreases and as the sample size n increases. ∎

Example 6.3.2 *Bernoulli Model*
Suppose that $(x_1, \ldots, x_n)$ is a sample from a Bernoulli(θ) distribution where $\theta \in [0, 1]$ is unknown. Suppose that we wish to estimate θ. The likelihood function is given by

$$L(\theta \,|\, x_1, \ldots, x_n) = \theta^{n\bar{x}}(1 - \theta)^{n(1-\bar{x})},$$

and the MLE of θ is $\bar{x}$ (Exercise 6.2.2), the proportion of successes in the n performances. We have $E_\theta\left(\bar{X}\right) = \theta$ for every $\theta \in [0,1]$, so the MLE is an unbiased estimator of θ.

Therefore,

$$\text{MSE}_\theta\left(\bar{X}\right) = \text{Var}_\theta\left(\bar{X}\right) = \frac{\theta\left(1-\theta\right)}{n},$$

and the estimated MSE is

$$\text{MSE}_{\hat{\theta}}\left(\bar{X}\right) = \frac{\bar{x}\left(1-\bar{x}\right)}{n}.$$

The standard error of the estimate $\bar{x}$ is then given by

$$\text{Sd}_{\hat{\theta}}\left(\bar{X}\right) = \sqrt{\frac{\bar{x}\left(1-\bar{x}\right)}{n}}.$$

Note how this standard error is quite different from the standard error of $\bar{x}$ in Example 6.3.1. ∎

Example 6.3.3 *Application of the Bernoulli Model*
A polling organization is asked to estimate the proportion of households in the population in a specific district who will participate in a proposed recycling program by separating their garbage into various components. The pollsters decided to take a sample of $n = 1000$ from the population of approximately 1.5 million households (we'll say more on how to choose this number later).

Each respondent will indicate either yes or no to a question concerning their participation. Given that the sample size is small relative to the population size, we can assume that we are sampling from a Bernoulli(θ) model where $\theta \in [0,1]$ is the proportion of individuals in the population who will respond yes.

After conducting the sample, there were 790 respondents who replied yes and 210 who responded no. Therefore, the MLE of θ is

$$\hat{\theta} = \bar{x} = \frac{790}{1000} = 0.79$$

and the standard error of the estimate is

$$\sqrt{\frac{\bar{x}\left(1-\bar{x}\right)}{1000}} = \sqrt{\frac{0.79\left(1-0.79\right)}{1000}} = 0.01288.$$

Notice that it is not entirely clear how we should interpret the value 0.01288. Does it mean our estimate 0.79 is highly accurate, modestly accurate, or not accurate at all? We will discuss this further in Section 6.3.2. ∎

Example 6.3.4 *Location-Scale Normal Model*
Suppose that $(x_1, \ldots, x_n)$ is a sample from an $N(\mu, \sigma^2)$ distribution where $\mu \in R^1$ and $\sigma^2 > 0$ are unknown. The parameter in this model is given by $\theta = \left(\mu, \sigma^2\right) \in \Omega = R^1 \times (0, \infty)$. Suppose that we want to estimate $\mu = \psi\left(\mu, \sigma^2\right)$, i.e., just the first coordinate of the full model parameter.

In Example 6.1.8 we determined that the likelihood function is given by

$$L\left(\mu, \sigma^2 \mid x_1, \ldots, x_n\right) = \left(2\pi\sigma^2\right)^{-n/2} \exp\left(-\frac{n}{2\sigma^2}\left(\bar{x} - \mu\right)^2\right) \exp\left(-\frac{n-1}{2\sigma^2}s^2\right).$$

In Example 6.2.6 we showed that the MLE of θ is

$$\left(\bar{x}, \frac{n-1}{n}s^2\right).$$

Further, from Theorem 4.6.6, the sampling distribution of the MLE is given by $\bar{X} \sim N(\mu, \sigma^2/n)$ independent of $(n-1)\,S^2/\sigma^2 \sim \chi^2(n-1)$.

The plug-in MLE of μ is $\bar{x}$. This estimator is unbiased and has

$$\mathrm{MSE}_\theta\left(\bar{X}\right) = \mathrm{Var}_\theta\left(\bar{X}\right) = \frac{\sigma^2}{n}.$$

Since σ^2 is unknown we estimate $\mathrm{MSE}_\theta\left(\bar{X}\right)$ by

$$\mathrm{MSE}_{\hat{\theta}}\left(\bar{X}\right) = \frac{\frac{n-1}{n}s^2}{n} = \frac{n-1}{n^2}s^2 \approx \frac{s^2}{n}.$$

The value s^2/n is commonly used instead of $\mathrm{MSE}_{\hat{\theta}}\left(\bar{X}\right)$, because (Corollary 4.6.2)

$$E_\theta\left(S^2\right) = \sigma^2,$$

i.e., S^2 is an unbiased estimator of σ^2. The quantity $s/\sqrt{n}$ is referred to as the *standard error* of the estimate $\bar{x}$. ∎

Example 6.3.5 *Application of the Location-Scale Normal Model*
In Example 5.5.6 we have a sample of $n = 30$ heights (in centimeters) of students. We calculated $\bar{x} = 64.517$ as our estimate of the mean population height μ. In addition, we obtained the estimate $s = 2.379$ of σ. Therefore, the standard error of the estimate $\bar{x} = 64.517$ is

$$\frac{s}{\sqrt{30}} = \frac{2.379}{\sqrt{30}} = 0.43434.$$

As in Example 6.3.3, we are faced with interpreting exactly what this number means in terms of the accuracy of the estimate. ∎

6.3.2 Confidence Intervals

While the standard error seems like a reasonable quantity for measuring the accuracy of an estimate of $\psi\left(\theta\right)$, its interpretation is not entirely clear at this point. It turns out that this is intrinsically tied up with the idea of a *confidence interval*.

Consider the construction of an interval

$$C\left(s\right) = \left(l\left(s\right), u\left(s\right)\right),$$

based on the data s, that we believe is likely to contain the true value of $\psi(\theta)$. To do this, we have to specify the lower endpoint $l(s)$ and upper endpoint $u(s)$ for each data value s. How should we do this?

One approach is to specify a probability $\gamma \in [0,1]$ and then require that random interval C have the *confidence property* as specified in the following definition.

Definition 6.3.1 An interval $C(s) = (l(s), u(s))$ is a γ-*confidence interval* for $\psi(\theta)$ if

$$P_\theta(\psi(\theta) \in C(s)) = P_\theta(l(s) \leq \psi(\theta) \leq u(s)) \geq \gamma$$

for every $\theta \in \Omega$. We refer to γ as the *confidence level* of the interval.

So C is a γ-confidence interval for $\psi(\theta)$ if, whenever we are sampling from P_θ, the probability that $\psi(\theta)$ is in the interval is at least equal to γ. For a given data set, such an interval either covers $\psi(\theta)$ or it does not. So note that it is not correct to say that a particular instance of a γ-confidence region has probability γ of containing the true value of μ.

If we choose γ to be a value close to 1, then we are highly confident that the true value of $\psi(\theta)$ is in $C(s)$. Of course, we can always take $C(s) = R^1$ (a very big interval!), and we are then 100% confident that the interval contains the true value. But this tells us nothing we didn't already know. So the idea is to try to make use of the information in the data to construct an interval such that we have a high confidence, say $\gamma = 0.95$ or $\gamma = 0.99$, that it contains the true value and is not any longer than necessary. We then interpret the length of the interval as a measure of how accurately the data allow us to know the true value of $\psi(\theta)$.

Consider the following example, which provides one approach to the construction of confidence intervals.

Example 6.3.6 *Location Normal Model*
Suppose we have a sample $(x_1, \ldots, x_n)$ from the $N(\mu, \sigma_0^2)$ model where $\mu \in R^1$ is unknown and $\sigma_0^2 > 0$ is known. The likelihood function is as specified in Example 6.3.1. Suppose that we want a confidence interval for μ.

The reasoning that underlies the likelihood function leads naturally to the following restriction for such a region: if $\mu_1 \in C(x_1, \ldots, x_n)$ and

$$L(\mu_2 \mid x_1, \ldots, x_n) \geq L(\mu_1 \mid x_1, \ldots, x_n),$$

then we should also have $\mu_2 \in C(x_1, \ldots, x_n)$. This restriction is implied by the likelihood because the model and the data support μ_2 at least as well as μ_1. Thus, if we conclude that μ_1 is a plausible value, so is μ_2.

Therefore, $C(x_1, \ldots, x_n)$ is of the form

$$C(x_1, \ldots, x_n) = \{\mu : L(\mu \mid x_1, \ldots, x_n) \geq k\} \qquad (6.3.2)$$

for some constant k, i.e., $C(x_1, \ldots, x_n)$ is a likelihood interval for μ. Then

$$
\begin{aligned}
C(x_1, &\ldots, x_n) \\
&= \left\{ \mu : \exp\left(-\frac{n}{2\sigma_0^2}(\bar{x} - \mu)^2 \right) \geq k \right\} = \left\{ \mu : -\frac{n}{2\sigma_0^2}(\bar{x} - \mu)^2 \geq \ln k \right\} \\
&= \left\{ \mu : (\bar{x} - \mu)^2 \leq -\frac{2\sigma_0^2}{n} \ln k \right\} = \left[\bar{x} - c\frac{\sigma_0}{\sqrt{n}}, \bar{x} + c\frac{\sigma_0}{\sqrt{n}} \right]
\end{aligned}
$$

where $c = \sqrt{-2 \ln k}$.

We are now left to choose the constant k, or equivalently c, so that the interval C is a γ-confidence interval for μ. We choose the smallest c, to make our interval as short as possible. Because

$$
Z = \frac{\bar{X} - \mu}{\sigma_0/\sqrt{n}} \sim N(0, 1), \tag{6.3.3}
$$

we have

$$
\begin{aligned}
\gamma &\leq P_\mu(\mu \in C(x_1, \ldots, x_n)) = P_\mu\left(\bar{X} - c\frac{\sigma_0}{\sqrt{n}} \leq \mu \leq \bar{X} + c\frac{\sigma_0}{\sqrt{n}} \right) \\
&= P_\mu\left(-c \leq \frac{\bar{X} - \mu}{\sigma_0/\sqrt{n}} \leq c \right) = P_\mu\left(\left| \frac{\bar{X} - \mu}{\sigma_0/\sqrt{n}} \right| \leq c \right) \\
&= 1 - 2(1 - \Phi(c)) \tag{6.3.4}
\end{aligned}
$$

for every $\mu \in R^1$, where Φ is the $N(0, 1)$ cumulative distribution function. We have equality in (6.3.4) whenever

$$
\Phi(c) = \frac{1 + \gamma}{2},
$$

and so $c = z_{(1+\gamma)/2}$, where z_α denotes the αth quantile of the $N(0, 1)$ distribution, and this is the smallest c satisfying (6.3.4).

We have shown that the likelihood interval given by

$$
\left[\bar{x} - z_{(1+\gamma)/2}\frac{\sigma_0}{\sqrt{n}}, \bar{x} + z_{(1+\gamma)/2}\frac{\sigma_0}{\sqrt{n}} \right] \tag{6.3.5}
$$

is an exact γ-confidence interval for μ. As these intervals are based on the *z-statistic*, given by (6.3.3), they are called *z-confidence intervals*. For example, if we take $\gamma = 0.95$, then $(1 + \gamma)/2 = 0.975$, and, from a statistical package (or Table D.2 in Appendix D), we obtain $z_{0.975} = 1.96$. Therefore in repeated sampling 95% of the intervals of the form

$$
\left[\bar{x} - 1.96\frac{\sigma_0}{\sqrt{n}}, \bar{x} + 1.96\frac{\sigma_0}{\sqrt{n}} \right]
$$

will contain the true value of μ.

This is illustrated in Figure 6.3.3. Here we have plotted the upper and lower endpoints of the .95-confidence intervals for μ for each of $N = 25$ samples of size $n = 10$ generated from an $N(0, 1)$ distribution. The theory says that when N is large, approximately 95% of these intervals will contain the true value $\mu = 0$. In the plot, coverage means that the lower endpoint (denoted by $\bullet$) must be below the horizontal line at 0 and that the upper endpoint (denoted by $\circ$) must be above this horizontal line. We see that only the fourth and twenty-third confidence intervals do not contain 0, so $23/25 = 92\%$ of the intervals contain 0. As $N \to \infty$, this proportion will converge to 0.95.

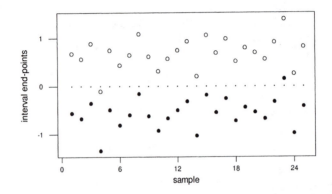

Figure 6.3.3: Plot of 0.95-confidence intervals for $\mu = 0$, where the lower endpoint is denoted by $\bullet$ and the upper endpoint is denoted by $\circ$, for $N = 25$ samples of size $n = 10$ from an $N(0, 1)$ distribution.

Notice that interval (6.3.5) is symmetrical about $\bar{x}$. Accordingly, the half-length of this interval,

$$z_{\frac{1+\gamma}{2}} \frac{\sigma_0}{\sqrt{n}},$$

is a measure of the accuracy estimate $\bar{x}$. The half-length is often referred to as the *margin of error*.

From the margin of error we now see how to interpret the standard error. For the standard error controls the lengths of the confidence intervals for the unknown μ. For example, we know that with probability approximately equal to 1 (actually $\gamma = 0.9974$), the interval $[\bar{x} \pm 3\sigma_0/\sqrt{n}]$ contains the true value of μ. ∎

Example 6.3.6 serves as a standard example for how confidence intervals are often constructed in statistics. Basically, the idea is that we take an estimate and then look at the intervals formed by taking symmetrical intervals around the estimate via multiples of its standard error. We illustrate this via some further examples.

Example 6.3.7 *Bernoulli Model*
Suppose that $(x_1, \ldots, x_n)$ is a sample from a Bernoulli(θ) distribution where $\theta \in [0, 1]$ is unknown and we want a γ-confidence interval for θ. Following

Example 6.3.2, we have that the MLE is $\bar{x}$ (Exercise 6.2.2) and the standard error of this estimate is

$$\sqrt{\frac{\bar{x}\,(1-\bar{x})}{n}}.$$

For this model, likelihood intervals take the form

$$C\left(x_1,\ldots,x_n\right) = \left\{\theta : \theta^{n\bar{x}}\,(1-\theta)^{n(1-\bar{x})} \geq c\right\}.$$

To determine these intervals we have to find the roots of equations of the form

$$\theta^{n\bar{x}}\,(1-\theta)^{n(1-\bar{x})} = c.$$

While numerical root-finding methods can handle this quite easily, this approach is not very tractable when we want to find the appropriate value of c to give a γ-confidence interval.

To avoid these computational complexities, it is common to use an approximate likelihood and confidence intervals based on the central limit theorem. The central limit theorem (see Example 4.4.9) implies that

$$\frac{\sqrt{n}\,(\bar{X}-\theta)}{\sqrt{\theta\,(1-\theta)}} \xrightarrow{D} N(0,1)$$

as $n \to \infty$. Further, a generalization of the central limit theorem (see Section 4.4.2), shows that

$$Z = \frac{\sqrt{n}\,(\bar{X}-\theta)}{\sqrt{\bar{X}\,(1-\bar{X})}} \xrightarrow{D} N(0,1).$$

Therefore, we have

$$
\begin{aligned}
\gamma &= \lim_{n\to\infty} P_\theta \left(-z_{(1+\gamma)/2} \leq \frac{\sqrt{n}\,(\bar{X}-\theta)}{\sqrt{\bar{X}\,(1-\bar{X})}} \leq z_{(1+\gamma)/2} \right) \\
&= \lim_{n\to\infty} P_\theta \left(\bar{X} - z_{(1+\gamma)/2}\sqrt{\frac{\bar{X}\,(1-\bar{X})}{n}} \leq \theta \leq \bar{X} + z_{(1+\gamma)/2}\sqrt{\frac{\bar{X}\,(1-\bar{X})}{n}} \right),
\end{aligned}
$$

and

$$\left[\bar{x} - z_{(1+\gamma)/2}\sqrt{\frac{\bar{x}\,(1-\bar{x})}{n}},\ \bar{x} + z_{(1+\gamma)/2}\sqrt{\frac{\bar{x}\,(1-\bar{x})}{n}} \right] \tag{6.3.6}$$

is an approximate γ-confidence interval for θ. Notice that this takes the same form as the interval in Example 6.3.6, except that the standard error has changed.

For example, if we want an approximate 0.95-confidence interval for θ in Example 6.3.3, then based on the observed $\bar{x} = 0.79$, we obtain

$$\left[0.79 \pm 1.96\sqrt{\frac{0.79\,(1-0.79)}{1000}} \right] = [0.76475, 0.81525].$$

The margin of error in this case equals 0.025245, so we can conclude that we know the true proportion with reasonable accuracy based on our sample. Actually, it may be that this accuracy is not good enough or even too good. We will discuss methods for ensuring that we achieve appropriate accuracy in Section 6.3.4.

The γ-confidence interval derived here for θ is one of many that you will see recommended in the literature. Recall that (6.3.6) is only an approximate γ-confidence interval for θ and, n may need to be large for the approximation to be accurate. In other words, the true confidence level for (6.3.6) will not equal γ and could be far from that value if n is too small. In particular, if the true θ is near 0 or 1, then n may need to be very large. In an actual application, we usually have some idea of a small range of possible values a population proportion θ can take. Accordingly, it is advisable to carry out some simulation studies to assess whether or not (6.3.6) is going to provide an acceptable approximations for θ in that range (see Computer Exercise 6.3.15). ∎

Example 6.3.8 *Location-Scale Normal Model*
Suppose that $(x_1, \ldots, x_n)$ is a sample from an $N(\mu, \sigma^2)$ distribution where $\mu \in R^1$ and $\sigma > 0$ are unknown. The parameter in this model is given by $\theta = (\mu, \sigma^2) \in \Omega = R^1 \times (0, \infty)$. Suppose that we want to form confidence intervals for $\mu = \psi(\mu, \sigma^2)$.

The likelihood function in this case is a function of two variables, μ and σ^2, and so the reasoning we employed in Example 6.3.6 to determine the form of the confidence intervals is not directly applicable. In Example 6.3.4, we developed $s/\sqrt{n}$ as the standard error of the estimate $\bar{x}$ of μ. Accordingly we restrict our attention to confidence intervals of the form

$$C(x_1, \ldots, x_n) = \left[\bar{x} - c\frac{s}{\sqrt{n}}, \bar{x} + c\frac{s}{\sqrt{n}}\right].$$

We then have

$$P_{(\mu, \sigma^2)}\left(\bar{X} - c\frac{S}{\sqrt{n}} \leq \mu \leq \bar{X} + c\frac{S}{\sqrt{n}}\right) = P_{(\mu, \sigma^2)}\left(-c \leq \frac{\bar{X} - \mu}{S/\sqrt{n}} \leq c\right)$$

$$= P_{(\mu, \sigma^2)}\left(\left|\frac{\bar{X} - \mu}{S/\sqrt{n}}\right| \leq c\right) = 1 - 2(1 - G(c; n-1))$$

where $G(\cdot; n-1)$ is the distribution function of

$$T = \frac{\bar{X} - \mu}{S/\sqrt{n}}. \tag{6.3.7}$$

Now, by Theorem 4.6.6,

$$\frac{\bar{X} - \mu}{\sigma/\sqrt{n}} \sim N(0, 1)$$

independent of $(n-1)S^2/\sigma^2 \sim \chi^2(n-1)$. Therefore, by Definition 4.6.2,

$$T = \left(\frac{\bar{X} - \mu}{\sigma/\sqrt{n}}\right) \bigg/ \sqrt{\frac{(n-1)S^2}{\sigma^2}} \bigg/ = \frac{\bar{X} - \mu}{S/\sqrt{n}} \sim t(n-1).$$

So if we take

$$c = t_{(1+\gamma)/2}(n-1),$$

where $t_\alpha(\lambda)$ is the αth quantile of the $t(\lambda)$ distribution,

$$\left[\bar{x} - t_{(1+\gamma)/2}(n-1)\frac{s}{\sqrt{n}}, \bar{x} + t_{(1+\gamma)/2}(n-1)\frac{s}{\sqrt{n}}\right]$$

is an exact γ-confidence interval for μ. The quantiles of the t distributions are available from a statistical package (or Table D.4 in Appendix D). As these intervals are based on the t-*statistic*, given by (6.3.7), they are called t-*confidence intervals*.

These confidence intervals for μ tend to be longer than those obtained in Example 6.3.6, and this reflects the greater uncertainty due to σ being unknown. When $n = 5$, then it can be shown that $\bar{x} \pm 3s/\sqrt{n}$ is a 0.97-confidence interval. When we replace s by the true value of σ, then $\bar{x} \pm 3\sigma/\sqrt{n}$ is a 0.9974-confidence interval.

As already noted, the intervals $\bar{x} \pm cs/\sqrt{n}$ are not likelihood intervals for μ. So the justification for using these must be a little different from that given in Example 6.3.6. In fact, the likelihood is defined for the full parameter $\theta = (\mu, \sigma^2)$ and it is not entirely clear how to extract inferences from it when our interest is in a marginal parameter like μ. There are a number of different attempts at resolving this issue. Here, however, we rely on the intuitive reasonableness of these intervals. In Chapter 7 we will see that these intervals also arise from another approach to inference, and this reinforces our belief that the use of these intervals is appropriate.

In Example 6.3.5 we have a sample of $n = 30$ heights (in centimeters) of students. We calculated $\bar{x} = 64.517$ as our estimate of μ with standard error $s/\sqrt{30} = 0.43434$. Using software (or Table D.4), we obtain $t_{.975}(29) = 2.0452$. So a 0.95-confidence interval for μ is given by

$$[64.517 \pm 2.0452\,(0.43434)] = [63.629, 65.405].$$

The margin of error is 0.888, so we are very confident that the estimate $\bar{x} = 64.517$ is within a centimeter of the true mean height. ∎

6.3.3 Testing Hypotheses and P-Values

As discussed in Section 5.5.3, another class of inference procedures is concerned with what we call *hypothesis assessment*. Suppose there is a theory, conjecture, or hypothesis, that specifies a value for a characteristic of interest $\psi(\theta)$, say $\psi(\theta) = \psi_0$. Often this hypothesis is written $H_0 : \psi(\theta) = \psi_0$ and is referred to as the *null hypothesis*.

The word *null* is used because, as we will see in Chapter 10, the value specified in H_0 is often associated with a treatment having no effect. For example, if we want to assess whether or not a proposed new drug does a better job of treating a particular condition than a standard treatment does, the null hypothesis will often be equivalent to the new drug providing no improvement. Of course we have to show how this can be expressed in terms of some characteristic $\psi(\theta)$ of an unknown distribution, and we will do so in Chapter 10.

The statistician is then charged with assessing whether or not the observed s is in accord with this hypothesis. So we wish to assess the evidence in s for $\psi(\theta) = \psi_0$ being true. A statistical procedure that does this can be referred to as a hypothesis assessment, a *test of significance*, or a *test of hypothesis*. Such a procedure involves measuring how surprising the observed s is when we assume H_0 to be true. It is clear that s is surprising whenever s lies in a region of low probability for each of the distributions specified by the null hypothesis, i.e., for each of the distributions in the model for which $\psi(\theta) = \psi_0$ is true. If we decide that the data is surprising under H_0, then this is evidence against H_0. This assessment is carried out by calculating a probability, called a *P-value*, so that small values of the P-value indicate that s is surprising.

We illustrate this via the following examples.

Example 6.3.9 *Location Normal Model*
Suppose we have a sample $(x_1, \ldots, x_n)$ from the $N(\mu, \sigma_0^2)$ model where $\mu \in R^1$ is unknown and $\sigma_0^2 > 0$ is known, and we have a theory that specifies a value for the unknown mean, say $H_0 : \mu = \mu_0$. Note that, by Corollary 4.6.1, when H_0 is true, the sampling distribution of the MLE is given by $\bar{X} \sim N(\mu_0, \sigma_0^2/n)$.

So one method of assessing whether or not the hypothesis H_0 makes sense is to compare the observed value $\bar{x}$ with this distribution. If $\bar{x}$ is in a region of low probability for the $N(\mu_0, \sigma_0^2/n)$ distribution, then this is evidence that H_0 is false. Because the density of the $N(\mu_0, \sigma_0^2/n)$ distribution is unimodal, the regions of low probability for this distribution occur in its tails. The farther out in the tails $\bar{x}$ lies, the more surprising this will be when H_0 is true, and so the more evidence we will have against H_0.

In Figure 6.3.4 we have plotted a density of the MLE together with an observed value $\bar{x}$ that lies far in the right tail of the distribution. This would clearly be a surprising value from this distribution.

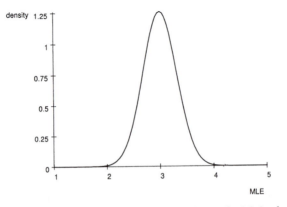

Figure 6.3.4: Plot of the density of the MLE in Example 6.3.9 when $\mu_0 = 3, \sigma_0^2 = 1$, and $n = 10$ together with the observed value $\bar{x} = 4.2$ (○).

So we want to measure how far out in the tails of the $N(\mu_0, \sigma_0^2/n)$ distribution the value $\bar{x}$ is. We can do this by computing the probability of observing a value of $\bar{x}$ as far, or farther, away from the center of the distribution under H_0 as $\bar{x}$. The center of this distribution is given by μ_0. Because

$$Z = \frac{\bar{X} - \mu_0}{\sigma_0/\sqrt{n}} \sim N(0, 1) \tag{6.3.8}$$

under H_0, the P-value is then given by

$$
\begin{aligned}
P_{\mu_0}\left(\left|\bar{X} - \mu_0\right| \geq \left|\bar{x} - \mu_0\right|\right) &= P_{\mu_0}\left(\left|\frac{\bar{X} - \mu_0}{\sigma_0/\sqrt{n}}\right| \geq \left|\frac{\bar{x} - \mu_0}{\sigma_0/\sqrt{n}}\right|\right) \\
&= 2\left[1 - \Phi\left(\left|\frac{\bar{x} - \mu_0}{\sigma_0/\sqrt{n}}\right|\right)\right],
\end{aligned}
$$

where Φ denotes the $N(0, 1)$ distribution function. If the P-value is small, then we have evidence that $\bar{x}$ is a surprising value, because this tells us that $\bar{x}$ is out in a tail of the $N(\mu_0, \sigma_0^2/n)$ distribution. Because this P-value is based on the statistic Z defined in (6.3.8), this is referred to as the *z-test* procedure. ∎

Example 6.3.10 *Application of the z-Test*
We generated the following sample of $n = 10$ from an $N(26, 4)$ distribution.

29.0651	27.3980	23.4346	26.3665	23.4994
28.6592	25.5546	29.4477	28.0979	25.2850

Even though we know the true value of μ, let us suppose we do not and test the hypothesis $H_0 : \theta = 25$. To assess this, we compute (using a statistical package to evaluate Φ) the P-value

$$
\begin{aligned}
2\left[1 - \Phi\left(\frac{|\bar{x} - \theta_0|}{\sigma_0/\sqrt{n}}\right)\right] &= 2\left[1 - \Phi\left(\frac{|26.6808 - 25|}{2/\sqrt{10}}\right)\right] \\
&= 2\left(1 - \Phi(2.6576)\right) = 0.0078
\end{aligned}
$$

which is quite small. For example, if the hypothesis H_0 is correct, then, in repeated sampling, we would see data giving a value at least as surprising as what we have observed only 0.78% of the time. So we conclude that we have evidence against H_0 being true, which, of course, is appropriate in this case.

If you do not use a statistical package for the evaluation of $\Phi(2.6576)$, then you will have to use Table D.2 of Appendix D to get an approximation. For example, rounding 2.6576 to 2.66, Table D.2 gives $\Phi(2.66) = 0.9961$ and the approximate P-value is $2(1 - 0.9961) = 0.0078$. In this case, the approximation is exact to four decimal places. ∎

Example 6.3.11 *Bernoulli Model*
Suppose that $(x_1, \ldots, x_n)$ is a sample from a Bernoulli(θ) distribution where $\theta \in [0, 1]$ is unknown, and we want to test $H_0 : \theta = \theta_0$. As in Example 6.3.7,

when H_0 is true, we have

$$Z = \frac{\sqrt{n}\,(\bar{X} - \theta_0)}{\sqrt{\theta_0\,(1 - \theta_0)}} \xrightarrow{D} N(0,1)$$

as $n \to \infty$. So we can test this hypothesis by computing the approximate P-value

$$P\left(|Z| \geq \left|\frac{\sqrt{n}\,(\bar{x} - \theta_0)}{\sqrt{\theta_0\,(1 - \theta_0)}}\right|\right) \approx 2\left[1 - \Phi\left(\left|\frac{\sqrt{n}\,(\bar{x} - \theta_0)}{\sqrt{\theta_0\,(1 - \theta_0)}}\right|\right)\right]$$

when n is large.

As a specific example, suppose that a psychic claims the ability to predict the value of a randomly tossed fair coin. To test this, a coin was tossed 100 times and the psychic's guesses were recorded as successes or failures. A total of 54 successes were observed.

If the psychic has no predictive ability, then we would expect the successes to occur randomly, just like heads occur when we toss the coin. Therefore, we want to test the null hypothesis that the probability θ of a success occurring is equal to $1/2$. This is equivalent to saying that the psychic has no predictive ability. The MLE is 0.54 and the approximate P-value is given by

$$2\left[1 - \Phi\left(\left|\frac{\sqrt{100}\,(0.54 - 0.5)}{\sqrt{0.5\,(1 - 0.5)}}\right|\right)\right] = 2\,(1 - \Phi\,(0.8)) = 2\,(1 - 0.7881) = 0.4238,$$

and we would appear to have no evidence that H_0 is false, i.e., no reason to doubt that the psychic has no predictive ability. ∎

Often cut-off values like 0.05 or 0.01 are used to determine whether the results of a test are significant or not. For example, if the P-value is less than 0.05 then the results are said to be *statistically significant* at the 5% level. There is nothing sacrosanct about the 0.05 level, however, and different values can be used depending on the application. For example, if the result of concluding that we have evidence against H_0 is that something very expensive or important will take place, then naturally we might demand that the cut-off value be much smaller than 0.05.

It is also important to point out here the difference between statistical significance and *practical significance*. Consider the situation in Example 6.3.9, when the true value of μ is $\mu_1 \neq \mu_0$, but μ_1 is so close to μ_0 that, practically speaking, they are indistinguishable. By the strong law of large numbers we have that $\bar{X} \xrightarrow{a.s} \mu_1$ as $n \to \infty$ and therefore

$$\left|\frac{\bar{X} - \mu_0}{\sigma_0/\sqrt{n}}\right| \xrightarrow{a.s.} \infty.$$

This implies that

$$2\left[1 - \Phi\left(\left|\frac{\bar{X} - \mu_0}{\sigma_0/\sqrt{n}}\right|\right)\right] \xrightarrow{a.s.} 0.$$

We conclude that, if we take a large enough sample size n, we will inevitably conclude that $\mu \neq \mu_0$ because the P-value of the z-test goes to 0. Of course, this is correct because the hypothesis is false. In spite of this, we do not want to conclude that, just because we have statistical significance, the difference between the true value and μ_0 is of any practical importance. If we examine the observed difference $|\bar{x} - \mu_0|$ as well, however, we will not make this mistake. The issue of practical significance is something we should always be aware of when conducting a test of significance.

Another approach to testing hypotheses is via confidence intervals. For example, if we have a γ-confidence interval $C(s)$ for $\psi(\theta)$ and $\psi_0 \notin C(s)$ then this seems like clear evidence against $H_0 : \psi(\theta) = \psi_0$, at least when γ is close to 1. It turns out that in many problems, the approach to testing via confidence intervals is equivalent to using P-values with a specific cut-off for the P-value to determine statistical significance. We illustrate this equivalence using the z-test and z-confidence intervals.

Example 6.3.12 *An Equivalence Between z-Tests and z-Confidence Intervals*
Observe that

$$1 - \gamma \leq 2 \left[1 - \Phi \left(\left| \frac{\bar{x} - \mu_0}{\sigma_0/\sqrt{n}} \right| \right) \right]$$

if and only if

$$\Phi \left(\left| \frac{\bar{x} - \mu_0}{\sigma_0/\sqrt{n}} \right| \right) \leq \frac{1 + \gamma}{2}.$$

This is true, if and only if

$$\frac{|\bar{x} - \mu_0|}{\sigma_0/\sqrt{n}} \leq z_{(1+\gamma)/2},$$

which holds if and only if

$$\mu_0 \in \left[\bar{x} - z_{(1+\gamma)/2} \frac{\sigma_0}{\sqrt{n}}, \bar{x} + z_{(1+\gamma)/2} \frac{\sigma_0}{\sqrt{n}} \right].$$

This implies that the γ-confidence interval for μ comprises those values μ_0 for which the P-value for the hypothesis $H_0 : \mu = \mu_0$ is greater than $1 - \gamma$.

Therefore, the P-value, based on the z-statistic, for the null hypothesis $H_0 : \mu = \mu_0$, will be smaller than $1 - \gamma$ if and only if μ_0 is not in the γ-confidence interval for μ derived in Example 6.3.6. For example, if we decide that for any P-values less than $1 - \gamma = 0.05$ we will declare the results statistically significant, then we know the results will be significant whenever the 0.95-confidence interval for μ does not contain μ_0. For the data of Example 6.3.10, a 0.95-confidence interval is given by $[25.441, 27.920]$. As this interval does not contain $\mu_0 = 25$, we reject the null hypothesis at the .05 level.

We can apply the same reasoning for tests about θ when sampling from a Bernoulli(θ) model. For the data in Example 6.3.11, we obtain the 0.95-confidence interval

$$\bar{x} \pm z_{0.975} \sqrt{\frac{\bar{x}(1 - \bar{x})}{n}} = 0.54 \pm 1.96 \sqrt{\frac{0.54(1 - 0.54)}{100}} = [0.44231, 0.63769],$$

which includes the value $\theta_0 = 0.5$, so we have no evidence against the null hypothesis of no predictive ability for the psychic at the 0.05 level. ∎

We now consider several examples pertaining to the important location-scale normal model.

Example 6.3.13 *Location-Scale Normal Model*
Suppose that $(x_1, \ldots, x_n)$ is a sample from an $N(\mu, \sigma^2)$ distribution where $\mu \in R^1$ and $\sigma > 0$ are unknown, and suppose we want to test the null hypothesis $H_0 : \mu = \mu_0$. In Example 6.3.8 we obtained a γ-confidence interval for μ. This was based on the t-statistic given by (6.3.7). So we base our test on this statistic also. In fact it can be shown that the test we derive here is equivalent to using the confidence intervals to assess the hypothesis as described in Example 6.3.12.

As in Example 6.3.8, we can prove that when the null hypothesis is true then,

$$T = \frac{\bar{X} - \mu_0}{S/\sqrt{n}} \tag{6.3.9}$$

is distributed $t(n-1)$. The t distributions are unimodal, with the mode at 0, and the regions of low probability are given by the tails. So we test, or assess this hypothesis by computing the probability of observing a value as far, or farther away, from 0 as (6.3.9). Therefore, the P-value is given by

$$P_{(\mu_0, \sigma^2)}\left(|T| \geq \left|\frac{\bar{x} - \mu_0}{s/\sqrt{n}}\right|\right) = 2\left[1 - G\left(\left|\frac{\bar{x} - \mu_0}{s/\sqrt{n}}\right| ; n-1\right)\right]$$

where $G(\cdot ; n-1)$ is the distribution function of the $t(n-1)$ distribution. We then have evidence against H_0 whenever this probability is small. This procedure is called the *t-test*. Again, it is a good idea to look at the difference $|\bar{x} - \mu_0|$, when we conclude that H_0 is false, to determine whether or not the detected difference is of practical importance.

Consider now the data in Example 6.3.10, and let us pretend that we do not know μ or σ^2. Then we have $\bar{x} = 26.6808$ and $s = \sqrt{4.8620} = 2.2050$, and so to test $H_0 : \mu = 25$, the value of the t-statistic is

$$t = \frac{\bar{x} - \mu_0}{s/\sqrt{n}} = \frac{26.6808 - 25}{2.2050/\sqrt{10}} = 2.4105.$$

From a statistics package (or Table D.4) we obtain $t_{0.975}(9) = 2.2622$, so we have obtained a statistically significant result at the 5% level and conclude that we have evidence against $H_0 : \mu = 25$. Using a statistical package we can determine the precise value of the P-value to be 0.039 in this case. ∎

Example 6.3.14 *Inferences for the Variance*
Suppose that $(x_1, \ldots, x_n)$ is a sample from an $N(\mu, \sigma^2)$ distribution where $\mu \in R^1$ and $\sigma > 0$ are unknown, and we want to make inferences about the population variance σ^2. The plug-in MLE is given by $(n-1)s^2/n$, which is the average of the squared deviations of the data values from $\bar{x}$. Often s^2 is recommended as the estimate because it has the unbiasedness property, and we will

use this here. An expression can be determined for the standard error of this estimate, but, as it is somewhat complicated, we will not pursue this further here.

We can form a γ-confidence interval for σ^2 using $(n-1)S^2/\sigma^2 \sim \chi^2(n-1)$ (Theorem 4.6.6). There are a number of possibilities for this interval, but one is to note that, letting $\chi^2_\alpha(\lambda)$ denote the αth quantile for the $\chi^2(\lambda)$ distribution, then

$$
\begin{aligned}
\gamma &= P_{(\mu,\sigma^2)}\left(\chi^2_{(1-\gamma)/2}(n-1) \le \frac{(n-1)S^2}{\sigma^2} \le \chi^2_{(1+\gamma)/2}(n-1)\right) \\
&= P_{(\mu,\sigma^2)}\left(\frac{(n-1)S^2}{\chi^2_{(1+\gamma)/2}(n-1)} \le \sigma^2 \le \frac{(n-1)S^2}{\chi^2_{(1-\gamma)/2}(n-1)}\right)
\end{aligned}
$$

for every $(\mu,\sigma^2) \in R^1 \times (0,\infty)$. So

$$
\left[\frac{(n-1)s^2}{\chi^2_{(1+\gamma)/2}(n-1)}, \frac{(n-1)s^2}{\chi^2_{(1-\gamma)/2}(n-1)}\right]
$$

is an exact γ-confidence interval for σ^2. To test a hypothesis such as $H_0 : \sigma = \sigma_0$ at the $1-\gamma$ level, we need only see whether or not σ_0^2 is in the interval. The smallest value of γ such that σ_0^2 is in the interval is the P-value for this hypothesis assessment procedure.

For the data in Example 6.3.10, let us pretend that we do not know that $\sigma^2 = 4$. Here $n = 10$ and $s^2 = 4.8620$. From a statistics package (or Table D.3 in Appendix D) we obtain $\chi^2_{0.025}(9) = 2.700, \chi^2_{0.975}(9) = 19.023$. So a 0.95-confidence interval for σ^2 is given by

$$
\left[\frac{(n-1)s^2}{\chi^2_{(1+\gamma)/2}(n-1)}, \frac{(n-1)s^2}{\chi^2_{(1-\gamma)/2}(n-1)}\right] = \left[\frac{9(4.8620)}{19.023}, \frac{9(4.8620)}{2.700}\right]
$$

$$
= [2.3003, 16.207].
$$

The length of the interval indicates that there is a reasonable degree of uncertainty concerning the true value of σ^2. We see, however, that a test of $H_0 : \sigma = 2$ would not reject this hypothesis at the 5% level because the value 4 is in the 0.95-confidence interval. ∎

All of the tests that we have discussed so far in this section for a characteristic of interest $\psi(\theta)$ have been *two-sided tests*. This means that the null hypothesis specified the value of $\psi(\theta)$ to be a single value ψ_0. Sometimes, however, we want to test a null hypothesis of the form $H_0 : \psi(\theta) \le \psi_0$ or $H_0 : \psi(\theta) \ge \psi_0$. To carry out such tests, we use the same test statistics as we have developed in the various examples here, but compute the P-value in a way that reflects the one-sided nature of the null. These are known as *one-sided tests*. We illustrate a one-sided test using the location normal model.

Example 6.3.15 *One-Sided Tests*
Suppose we have a sample $(x_1, \ldots, x_n)$ from the $N(\mu, \sigma_0^2)$ model where $\mu \in R^1$ is unknown and $\sigma_0^2 > 0$ is known. Suppose further that it is hypothesized that $H_0 : \mu \le \mu_0$ is true, and we wish to assess this after observing the data.

We will base our test on the z-statistic

$$Z = \frac{\bar{X} - \mu_0}{\sigma_0/\sqrt{n}} = \frac{\bar{X} - \mu + \mu - \mu_0}{\sigma_0/\sqrt{n}} = \frac{\bar{X} - \mu}{\sigma_0/\sqrt{n}} + \frac{\mu - \mu_0}{\sigma_0/\sqrt{n}}.$$

So Z is the sum of a random variable having an $N(0, 1)$ distribution and the constant $\sqrt{n}\,(\mu - \mu_0)\,/\sigma_0$, which implies that

$$Z \sim N\left(\frac{\mu - \mu_0}{\sigma_0/\sqrt{n}}, 1\right).$$

Note that

$$\frac{\mu - \mu_0}{\sigma_0/\sqrt{n}} \le 0$$

if and only if H_0 is true.

This implies that, when the null hypothesis is false, we will tend to see values of Z in the right tail of the $N(0, 1)$ distribution; when the null hypothesis is true, we will tend to see values of Z that are reasonable for the $N(0, 1)$ distribution, or in the left tail of this distribution. Accordingly, to test H_0, we compute the P-value

$$P\left(Z \ge \frac{\bar{x} - \mu_0}{\sigma_0/\sqrt{n}}\right) = 1 - \Phi\left(\frac{\bar{x} - \mu_0}{\sigma_0/\sqrt{n}}\right),$$

with $Z \sim N(0, 1)$ and conclude that we have evidence against H_0 when this is small. Using the same reasoning, the P-value for the null hypothesis $H_0 : \mu \ge \mu_0$, equals

$$P\left(Z \le \frac{\bar{x} - \mu_0}{\sigma_0/\sqrt{n}}\right) = \Phi\left(\frac{\bar{x} - \mu_0}{\sigma_0/\sqrt{n}}\right).$$

For more discussion of one-sided tests and confidence intervals, see Problems 6.3.17 through 6.3.24. ∎

6.3.4 Sample Size Calculations: Confidence Intervals

Quite often a statistician is asked to determine the sample size n to ensure that with a very high probability the results of a statistical analysis will yield definitive results. For example, suppose we are going to take a sample of size n from a population Π and want to estimate the population mean μ so that the estimate is within 0.5 of the true mean with probability at least 0.95. This means that we want the half-length, or margin of error, of the 0.95-confidence interval for the mean to be guaranteed to be less than 0.5.

We consider such problems in the following examples. Note that in general, *sample size calculations* are the domain of experimental design, which we will discuss more extensively in Chapter 10.

First we consider the problem of selecting the sample size to ensure that a confidence interval is shorter than some prescribed value.

Example 6.3.16 *The Length of a Confidence Interval for a Mean*
Suppose we are in the situation described in Example 6.3.6, in which we have a
sample $(x_1, \ldots, x_n)$ from the $N(\mu, \sigma_0^2)$ model with $\mu \in R^1$ unknown and $\sigma_0^2 > 0$
known. Further suppose that the statistician is asked to determine n so that
the margin of error for a γ-confidence interval for the population mean μ is no
greater than a prescribed value $\delta > 0$. This entails that n be chosen so that

$$z_{(1+\gamma)/2} \frac{\sigma_0}{\sqrt{n}} \leq \delta$$

or equivalently, so that

$$n \geq \sigma_0^2 \left(\frac{z_{(1+\gamma)/2}}{\delta} \right)^2.$$

For example, if $\sigma_0^2 = 10, \gamma = 0.95$, and $\delta = 0.5$, then the smallest possible value
for n is 154.

Now consider the situation described in Example 6.3.8, in which we have
a sample $(x_1, \ldots, x_n)$ from the $N(\mu, \sigma^2)$ model with $\mu \in R^1$ and $\sigma^2 > 0$ both
unknown. In this case, we want n so that

$$t_{(1+\gamma)/2}(n-1) \frac{s}{\sqrt{n}} \leq \delta$$

which entails

$$n \geq s^2 \left(\frac{t_{(1+\gamma)/2}(n-1)}{\delta} \right)^2.$$

But note this also depends on the unobserved value of s, so we cannot determine
a value of n.

Often, however, we can determine an upper bound on the population standard deviation, say $\sigma \leq b$. For example, suppose we are measuring human
heights in centimeters. Then we have a pretty good idea of upper and lower
bounds on the possible heights we will actually obtain. Therefore, with the
normality assumption, the interval given by the population mean plus or minus
three standard deviations, must be contained within the interval given by the
upper and lower bounds. So dividing the length of this interval by 6 gives a
plausible upper bound b for the value of σ. In any case, when we have such an
upper bound we can expect that $s \leq b$, at least if we choose b conservatively.
Therefore we take n to satisfy

$$n \geq b^2 \left(\frac{t_{(1+\gamma)/2}(n-1)}{\delta} \right)^2.$$

Note that we need to evaluate $t_{(1+\gamma)/2}(n-1)$ for each n as well. It is wise to be
fairly conservative in our choice of n in this case, i.e., do not choose the smallest
possible value. ∎

Example 6.3.17 *The Length of a Confidence Interval for a Proportion*
Suppose we are in the situation described in Example 6.3.2, in which we have a
sample $(x_1, \ldots, x_n)$ from the Bernoulli(θ) model and $\theta \in [0, 1]$ is unknown. The

statistician is required to specify the sample size n so that the margin of error of a γ-confidence interval for θ is no greater than a prescribed value δ. So, from Example 6.3.7, we want n to satisfy

$$z_{(1+\gamma)/2} \sqrt{\frac{\bar{x}(1-\bar{x})}{n}} \leq \delta, \qquad (6.3.10)$$

and this entails

$$n \geq \bar{x}(1-\bar{x}) \left(\frac{z_{(1+\gamma)/2}}{\delta} \right)^2.$$

Because this also depends on the unobserved $\bar{x}$, we cannot determine n. Note, however, that $0 \leq \bar{x}(1-\bar{x}) \leq 1/4$ for every $\bar{x}$ (plot this function) and this upper bound is achieved when $\bar{x} = 1/2$. Therefore, if we determine n so that

$$n \geq \frac{1}{4} \left(\frac{z_{(1+\gamma)/2}}{\delta} \right)^2,$$

then we know that (6.3.10) is satisfied. For example, if $\gamma = 0.95, \delta = 0.1$, the smallest possible value of n is 97; if $\gamma = 0.95, \delta = 0.01$, the smallest possible value of n is 9604. ∎

6.3.5   Sample Size Calculations: Power

Suppose the purpose of a study is to assess a specific hypothesis $H_0 : \psi(\theta) = \psi_0$ and it is has been decided that the results will be declared statistically significant whenever the P-value is less than α. Suppose that the statistician is asked to choose n, so that the P-value obtained is smaller than α, with probability at least β_0, at some specific θ_1 such that $\psi(\theta_1) \neq \psi_0$. The probability that the P-value is less than α for a specific value of θ is called the *power* of the test at θ. We will denote this by $\beta(\theta)$ and call β the *power function* of the test. The notation β is not really complete, as it suppresses the dependence of β on ψ, α, n, and the test procedure, but we will assume that these are clear in a particular context. The problem the statistician is presented with can then be stated as: find n so that $\beta(\theta_1) \geq \beta_0$.

The power function of a test is a measure of the sensitivity of the test to detect departures from the null hypothesis. We choose α small ($\alpha = 0.05, 0.01$, etc.) so that we do not erroneously declare that we have evidence against the null hypothesis when the null hypothesis is in fact true. When $\psi(\theta) \neq \psi_0$, then $\beta(\theta)$ is the probability that the test does the right thing and detects that H_0 is false.

For any test procedure it is a good idea to examine its power function, perhaps for several choices of α, to see how good the test is at detecting departures. For it can happen that we do not find any evidence against a null hypothesis when it is false, because the sample size is too small. In such a case, the power will be small at θ values that represent practically significant departures from H_0. To avoid this problem, we should always choose a value ψ_1 that represents a practically significant departure from ψ_0 and then determine n so that we reject H_0 with high probability when $\psi(\theta) = \psi_1$.

We consider now the computation and use of the power function in several examples.

Example 6.3.18 *The Power Function in the Location Normal Model*
For the two-sided z-test in Example 6.3.9, we have

$$\beta(\mu) = P_\mu \left(2 \left[1 - \Phi \left(\left| \frac{\bar{X} - \mu_0}{\sigma_0/\sqrt{n}} \right| \right) \right] < \alpha \right)$$

$$= P_\mu \left(\Phi \left(\left| \frac{\bar{X} - \mu_0}{\sigma_0/\sqrt{n}} \right| \right) > 1 - \frac{\alpha}{2} \right) = P_\mu \left(\left| \frac{\bar{X} - \mu_0}{\sigma_0/\sqrt{n}} \right| > z_{1-\alpha/2} \right)$$

$$= P_\mu \left(\frac{\bar{X} - \mu_0}{\sigma_0/\sqrt{n}} > z_{1-\alpha/2} \right) + P_\mu \left(\frac{\bar{X} - \mu_0}{\sigma_0/\sqrt{n}} < -z_{1-\alpha/2} \right)$$

$$= P_\mu \left(\frac{\bar{X} - \mu}{\sigma_0/\sqrt{n}} > \frac{\mu_0 - \mu}{\sigma_0/\sqrt{n}} + z_{1-\alpha/2} \right) + P_\mu \left(\frac{\bar{X} - \mu}{\sigma_0/\sqrt{n}} < \frac{\mu_0 - \mu}{\sigma_0/\sqrt{n}} - z_{1-\alpha/2} \right)$$

$$= 1 - \Phi \left(\frac{\mu_0 - \mu}{\sigma_0/\sqrt{n}} + z_{1-\alpha/2} \right) + \Phi \left(\frac{\mu_0 - \mu}{\sigma_0/\sqrt{n}} - z_{1-\alpha/2} \right). \tag{6.3.11}$$

Notice that
$$\beta(\mu) = \beta(\mu_0 + (\mu - \mu_0)) = \beta(\mu_0 - (\mu - \mu_0)),$$

so β is symmetric about μ_0 (put $\delta = \mu - \mu_0$ and $\mu = \mu_0 + \delta$ in the expression for $\beta(\mu)$ and we get the same value).

Differentiating (6.3.11) with respect to $\sqrt{n}$, we obtain

$$\left[\varphi \left(\frac{\mu_0 - \mu}{\sigma_0/\sqrt{n}} - z_{1-\alpha/2} \right) - \varphi \left(\frac{\mu_0 - \mu}{\sigma_0/\sqrt{n}} + z_{1-\alpha/2} \right) \right] \frac{\mu_0 - \mu}{\sigma_0} \tag{6.3.12}$$

where φ is the density of the $N(0,1)$ distribution. We can establish that (6.3.12) is always nonnegative (Challenge 6.3.26). This implies that $\beta(\mu)$ is increasing in n, so we need only solve $\beta(\mu_1) = \beta_0$ for n (the solution may not be an integer) to determine a suitable sample size (all larger values of n will give a larger power).

For example, when $\sigma_0 = 1, \alpha = 0.05, \beta_0 = 0.99$ and $\mu_1 = \mu_0 \pm 0.1$ we must find n satisfying

$$1 - \Phi \left(\sqrt{n}(0.1) + 1.96 \right) + \Phi \left(\sqrt{n}(0.1) - 1.96 \right) = 0.80. \tag{6.3.13}$$

(Note the symmetry of β about μ_0 means we will get the same answer if we used $\mu_0 - 0.1$ here instead of $\mu_0 + 0.1$.) Tabulating (6.3.13) as a function of n using a statistical package, determines that $n = 785$ is the smallest value achieving the required bound.

Also observe that the derivative of (6.3.11) with respect to μ is given by

$$\left[\varphi \left(\frac{\mu_0 - \mu}{\sigma_0/\sqrt{n}} + z_{1-\alpha/2} \right) - \varphi \left(\frac{\mu_0 - \mu}{\sigma_0/\sqrt{n}} - z_{1-\alpha/2} \right) \right] \frac{\sqrt{n}}{\sigma_0}. \tag{6.3.14}$$

This is positive when $\mu > \mu_0$, negative when $\mu < \mu_0$, and takes the value 0 when $\mu = \mu_0$ (Challenge 6.3.27). From (6.3.11) we have that $\beta(\mu) \to 1$ as

$\mu \to \pm\infty$. These facts establish that β takes its minimum value at μ_0 and that it is increasing as we move away from μ_0. Therefore, once we have determined n so that the power is at least β_0 at some μ_1, we know that the power is at least β_0 for all values of μ satisfying $|\mu_0 - \mu| \geq |\mu_0 - \mu_1|$.

As an example of this, consider Figure 6.3.5, where we have plotted the power function when $n = 10, \mu_0 = 0, \sigma_0 = 1$, and $\alpha = 0.05$ so that

$$\beta(\mu) = 1 - \Phi\left(\sqrt{10}\mu + 1.96\right) + \Phi\left(\sqrt{10}\mu - 1.96\right).$$

Notice the symmetry about $\mu_0 = 0$ and the fact that $\beta(\mu)$ increases as μ moves away from 0. We obtain $\beta(1.2) = 0.967$ so that when $\mu = \pm 1.2$, the probability that the P-value for testing $H_0 : \mu = 0$ will be less than .05 is .967. Of course, as we increase n, this graph will rise even more steeply to 1 as we move away from 0.

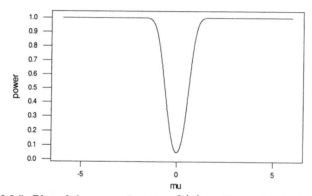

Figure 6.3.5: Plot of the power function $\beta(\mu)$ for Example 6.3.18 when $\alpha = 0.05$, $\mu_0 = 0$, and $\sigma_0 = 1$ is assumed known.

Many statistical packages contain the power function as a built-in function for various tests. This is very convenient for examining the sensitivity of the test and determining sample sizes. ∎

Example 6.3.19 *The Power Function for θ in the Bernoulli Model*
For the two-sided test in Example 6.3.11, we have that the power function is given by

$$\beta(\theta) = P_\theta\left(2\left[1 - \Phi\left(\left|\frac{\sqrt{n}(\bar{X} - \theta_0)}{\sqrt{\theta_0(1 - \theta_0)}}\right|\right)\right] < \alpha\right).$$

Under the assumption that we choose n large enough so that $\bar{X}$ is approximately distributed $N(\theta, \theta(1 - \theta)/n)$, the approximate calculation of this power function can be approached as in Example 6.3.18, when we put $\sigma_0 = \theta(1 - \theta)$. We do not pursue this calculation further here but note that many statistical packages will evaluate β as a built-in function. ∎

Example 6.3.20 *The Power Function in the Location-Scale Normal Model*
For the two-sided t-test in Example 6.3.13. we have

$$\beta_n\left(\mu, \sigma^2\right) \;=\; P_{\left(\mu, \sigma^2\right)}\left(2\left[1 - G\left(\left|\frac{\bar{X} - \mu_0}{S/\sqrt{n}}\right|; n-1\right)\right] < \alpha\right)$$

$$= \; P_{\left(\mu, \sigma^2\right)}\left(\left|\frac{\bar{X} - \mu_0}{S/\sqrt{n}}\right| > t_{1-\frac{\alpha}{2}}\left(n-1\right)\right)$$

where $G\left(\cdot\,; n-1\right)$ is the cdf of the $t\left(n-1\right)$ distribution. Notice that it is a
function of both μ and σ^2. In particular, we have to specify both μ and σ^2
and then determine n so that $\beta_n\left(\mu, \sigma^2\right) \geq \beta_0$. Many statistical packages will
have the calculation of this power function as a built-in function so that an
appropriate n can be determined using this. Alternatively, we can use Monte
Carlo methods to approximate the distribution function of

$$\left|\frac{\bar{X} - \mu_0}{S/\sqrt{n}}\right|$$

when sampling from the $N(\mu, \sigma^2)$, for a variety of values of n, to determine an
appropriate value. ∎

Summary of Section 6.3

- The MLE $\hat{\theta}$ is the best supported value of the parameter θ by the model
  and data. As such, it makes sense to base the derivation of inferences
  about some characteristic $\psi\left(\theta\right)$ on the MLE. These inferences include estimates and their standard errors, confidence intervals, and the assessment
  of hypotheses via P-values.

- An important aspect of the design of a sampling study is to decide on
  the size n of the sample to ensure that the results of the study produce
  sufficiently accurate results. Prescribing the half-lengths of confidence
  intervals (margins of error) or the power of a test are two techniques for
  doing this.

Exercises

6.3.1 Suppose measurements (in centimeters) are taken using an instrument.
There is error in the measuring process and a measurement is assumed to be
distributed $N(\mu, \sigma_0^2)$, where μ is the exact measurement and $\sigma_0^2 = 0.5$. If the
following $(n = 10)$ measurements 4.7, 5.5, 4.4, 3.3, 4.6, 5.3, 5.2, 4.8, 5.7, 5.3 were
obtained, assess the hypothesis $H_0 : \mu = 5$ by computing the relevant P-value.
Also compute a 0.95-confidence interval for the unknown μ.

6.3.2 Suppose in Exercise 6.3.1 we drop the assumption that $\sigma_0^2 = 0.5$. Then
assess the hypothesis $H_0 : \mu = 5$ and compute a 0.95-confidence interval for μ.

6.3.3 Marks on an exam in a statistics course are assumed to be normally distributed with unknown mean but with variance equal to 5. A sample of four students is selected, and their marks are 52, 63, 64, 84. Assess the hypothesis $H_0 : \mu = 60$ by computing the relevant P-value and compute a 0.95-confidence interval for the unknown μ.

6.3.4 Suppose in Exercise 6.3.3 we drop the assumption that the population variance is 5. Assess the hypothesis $H_0 : \mu = 60$ by computing the relevant P-value and compute a 0.95-confidence interval for the unknown μ.

6.3.5 Suppose that in Exercise 6.3.3 we had only observed one mark and that this was 52. Assess the hypothesis $H_0 : \mu = 60$ by computing the relevant P-value and compute a 0.95-confidence interval for the unknown μ. Is it possible to compute a P-value and construct a 0.95-confidence interval for μ without the assumption that we know the population variance? Explain your answer and, if your answer is no, determine the minimum sample size n for which inference is possible without the assumption that the population variance is known.

6.3.6 Assume that the speed of light data in Table 6.3.1 is a sample from an $N(\mu, \sigma^2)$ distribution for some unknown values of μ and σ^2. Determine a 0.99-confidence interval for μ. Assess the null hypothesis $H_0 : \mu = 24$.

6.3.7 A manufacturer wants to assess whether or not rods are being constructed appropriately, where the diameter of the rods is supposed to be 1.0 cm, and the variation in the diameters is known to be distributed $N(\mu, 0.1)$. The manufacturer is willing to tolerate a deviation of the population mean from this value of no more than 0.1 cm, i.e., if the population mean is within the interval 1.0 ± 0.1 cm, then the manufacturing process is performing correctly. A sample of $n = 500$ rods is taken, and the average diameter of these rods is found to be $\bar{x} = 1.05$ cm, with $s^2 = 0.083$ cm^2. Are these results statistically significant? Are the results practically significant? Justify your answers.

6.3.8 A polling firm conducts a poll to determine what proportion θ of voters in a given population will vote in an upcoming election. A random sample of $n = 250$ was taken from the population, and the proportion answering yes was 0.62. Assess the hypothesis $H_0 : \theta = 0.65$ and construct an approximate 0.90-confidence interval for θ.

6.3.9 A coin was tossed $n = 1000$ times, and the proportion of heads observed was 0.51. Do we have evidence to conclude that the coin is unfair?

6.3.10 How many times must we toss a coin to ensure that a 0.95-confidence interval, for the probability of heads on a single toss, has length less than 0.1, 0.05, and 0 .01, respectively?

6.3.11 Suppose that a possibly biased die is rolled 30 times and that the face containing 2 pips comes up 10 times. Do we have evidence to conclude that the die is biased?

6.3.12 Suppose that a measurement on a population can be assumed to be distributed $N(\mu, 2)$ where $\mu \in R^1$ is unknown and that the size of the population is very large. A researcher wants to determine a 0.95-confidence interval for μ that is no longer than 1. What is the minimum sample size that will guarantee this?

Computer Exercises

6.3.13 Suppose that a measurement on a population can be assumed to follow the $N(\mu, \sigma^2)$ distribution where $(\mu, \sigma^2) \in R^1 \times (0, \infty)$ is unknown and the size of the population is very large. A very conservative upper bound on σ is given by 5. A researcher wants to determine a .95-confidence interval for μ that is no longer than 1. Determine a sample size that will guarantee this? (Hint: Start with a large sample approximation.)

6.3.14 Suppose that a measurement on a population can be assumed to be distributed $N(\mu, 2)$ where $\mu \in R^1$ is unknown and that the size of the population is very large. A researcher wants to assess a null hypothesis $H_0 : \mu = \mu_0$ and wants to make sure that the probability is at least 0.80 that the P-value is less than 0.05 when $\mu = \mu_0 \pm 0.5$. What is the minimum sample size that will guarantee this? (Hint: Tabulate the power as a function of the sample size n.)

6.3.15 Generate 10^3 samples of size $n = 5$ from the Bernoulli(0.5) distribution. For each of these samples, calculate (6.3.6) with $\gamma = 0.95$ and record the proportion of intervals that contain the true value. What do you notice? Repeat this simulation with $n = 20$. What do you notice?

6.3.16 Generate 10^4 samples of size $n = 5$ from the $N(0, 1)$ distribution. For each of these samples, calculate the interval $(\bar{x} - s/\sqrt{5}, \bar{x} + s/\sqrt{5})$, where s is the sample standard deviation, and compute the proportion of times this interval contains μ. Repeat this simulation with $n = 10$ and 100 and compare your results.

Problems

6.3.17 (*One-sided confidence intervals for means*) Suppose that $(x_1, \ldots, x_n)$ is a sample from a $N(\mu, \sigma_0^2)$ distribution where $\mu \in R^1$ is unknown and σ_0^2 is known. Suppose we want to make inferences about the interval $\psi(\mu) = (-\infty, \mu)$. Consider the problem of finding an interval $C(x_1, \ldots, x_n) = (-\infty, u(x_1, \ldots, x_n))$ that covers the interval $(-\infty, \mu)$ with probability at least γ. So we want u such that

$$P_\mu(\mu \leq u(X_1, \ldots, X_n)) \geq \gamma$$

for every μ. Note that $(-\infty, \mu) \subset (-\infty, u(x_1, \ldots, x_n))$ if and only if $\mu \leq u(x_1, \ldots, x_n)$ and so $C(x_1, \ldots, x_n)$ is called a left-sided γ-confidence interval for μ. Obtain an exact left-sided γ-confidence interval for μ using $u(x_1, \ldots, x_n) = \bar{x} + k(\sigma_0/\sqrt{n})$, i.e., find the k which gives this property.

6.3.18 (*One-sided hypotheses for means*) Suppose that $(x_1, \ldots, x_n)$ is a sample from a $N(\mu, \sigma_0^2)$ distribution where μ is unknown and σ_0^2 is known. Suppose we want to assess the hypothesis $H_0 : \mu \leq \mu_0$. Under these circumstances we say that the observed value $\bar{x}$ is surprising if $\bar{x}$ occurs in a region of low probability for every distribution in H_0. Therefore, a sensible P-value for this problem is $\max_{\mu \in H_0} P_\mu \left(\bar{X} > \bar{x} \right)$. Show that this leads to the P-value

$$1 - \Phi \left(\frac{\bar{x} - \mu_0}{\sigma_0/\sqrt{n}} \right).$$

6.3.19 Determine the form of the power function associated with the hypothesis assessment procedure of Problem 6.3.18, when we declare a test result as being statistically significant whenever the P-value is less than α.

6.3.20 Repeat Problems 6.3.17 and 6.3.18, but this time obtain a right-sided γ-confidence interval for μ and assess the hypothesis $H_0 : \mu \geq \mu_0$.

6.3.21 Repeat Problems 6.3.17 and 6.3.18, but this time do not assume the population variance is known. In particular determine k so that $u\left(x_1, \ldots, x_n\right) = \bar{x} + k\left(s/\sqrt{n}\right)$ gives an exact left-sided γ-confidence interval for μ and show that the P-value for testing $H_0 : \mu \leq \mu_0$ is given by

$$1 - G \left(\frac{\bar{x} - \mu_0}{\sigma_0/\sqrt{n}} \, ; n - 1 \right).$$

6.3.22 (*One-sided confidence intervals for variances*) Suppose that $(x_1, \ldots, x_n)$ is a sample from the $N(\mu, \sigma^2)$ distribution where $\left(\mu, \sigma^2\right) \in R^1 \times (0, \infty)$ is unknown, and we want a γ-confidence interval of the form

$$C\left(x_1, \ldots, x_n\right) = \left(0, u\left(x_1, \ldots, x_n\right)\right)$$

for σ^2. If $u\left(x_1, \ldots, x_n\right) = ks^2$, then determine k so that this interval is an exact γ-confidence interval.

6.3.23 (*One-sided hypotheses for variances*) Suppose that $(x_1, \ldots, x_n)$ is a sample from the $N(\mu, \sigma^2)$ distribution where $\left(\mu, \sigma^2\right) \in R^1 \times (0, \infty)$ is unknown, and we want to assess the hypothesis $H_0 : \sigma^2 \leq \sigma_0^2$. Argue that the sample variance s^2 is surprising if s^2 is large and therefore, a sensible P-value for this problem is to compute $\max_{(\mu, \sigma^2) \in H_0} P_\mu \left(S^2 > s^2 \right)$. Show that this leads to the P-value

$$1 - H \left(\frac{(n-1)s^2}{\sigma_0^2} \, ; n - 1 \right)$$

where $H\left(\cdot \, ; n - 1 \right)$ is the distribution function of the $\chi^2 \left(n - 1 \right)$ distribution.

6.3.24 Determine the form of the power function associated with the hypothesis assessment procedure of Problem 6.3.23, for computing the probability that the P-value is less than α.

6.3.25 Repeat Exercise 6.3.7, but this time do not assume that the population variance is known. In this case, the manufacturer deems the process to be under control if the population standard deviation is less than or equal to 0.1 and the population mean is in the interval 1.0 $\pm$0.1 cm. Use Problem 6.3.23 for the test concerning the population variance.

Challenges

6.3.26 Prove that (6.3.12) is always nonnegative. (Hint: Use the facts that φ is symmetric about 0, increases to the left of 0, and decreases to the right of 0.)

6.3.27 Establish that (6.3.14) is positive when $\mu > \mu_0$, negative when $\mu < \mu_0$, and takes the value 0 when $\mu = \mu_0$.

Discussion Topic

6.3.28 Discuss the following statement: The accuracy of the results of a statistical analysis are so important that we should always take the largest possible sample size.

6.4 Distribution-Free Methods

The likelihood methods that we have been discussing all depend on the assumption that the true distribution lies in $\{P_\theta : \theta \in \Omega\}$. There is typically nothing that guarantees that the assumption $\{P_\theta : \theta \in \Omega\}$ is correct. If the distribution we are sampling from is far different from any of the distributions in $\{P_\theta : \theta \in \Omega\}$, then methods of inference that depend on this assumption, like likelihood methods, can be very misleading. So it is important in any application that we check that our assumptions make sense. We will discuss the topic of model checking in Chapter 9.

Another approach to this problem is to take the model $\{P_\theta : \theta \in \Omega\}$ as large as possible, reflecting the fact that we may have very little information about what the true distribution is like. For example, inferences based on the Bernoulli(θ) model with $\theta \in \Omega = [0, 1]$ really specify no information about the true distribution, because this model includes all the possible distributions on the sample space $S = \{0, 1\}$. Inference methods that are suitable when $\{P_\theta : \theta \in \Omega\}$ is very large are sometimes called *distribution-free*, to reflect the fact that very little information is specified in the model about the true distribution.

For finite sample spaces, it is straight-forward to adopt the distribution-free approach, as with the just cited Bernoulli model, but when the sample space is infinite, things are more complicated. In fact, sometimes it is very difficult to determine inferences about characteristics of interest when the model is very big. Further, if we have

$$\{P_\theta : \theta \in \Omega_1\} \subset \{P_\theta : \theta \in \Omega\},$$

then, when the smaller model contains the true distribution, methods based on the smaller model will make better use of the information in the data about the true value in Ω_1 than will methods using the bigger model $\{P_\theta : \theta \in \Omega\}$. So there is a trade-off between taking too big a model and taking too precise a model. This is an issue that a statistician must always be concerned with.

We now consider some examples of distribution-free inferences. In some cases, the inferences have approximate sampling properties, while in other cases the inferences have exact sampling properties for very large models.

6.4.1 Method of Moments

Suppose we take $\{P_\theta : \theta \in \Omega\}$ to be the set of all distributions on R^1 that have their first l moments, and we want to make inferences about the moments

$$\mu_i = E_\theta\left(X^i\right),$$

for $i = 1, \ldots, l$ based on a sample $(x_1, \ldots, x_n)$. The natural sample analog of the population moment μ_i is the ith *sample moment*

$$m_i = \frac{1}{n}\sum_{j=1}^{n} x_j^i,$$

which would seem to be a sensible estimator.

In particular, we have that $E_\theta\left(M_i\right) = \mu_i$ for every $\theta \in \Omega$, so m_i is unbiased, and the weak and strong laws of large numbers establish that m_i converges to μ_i as n increases. Further, the central limit theorem establishes that

$$\frac{M_i - \mu_i}{\sqrt{\text{Var}_\theta\left(M_i\right)}} \xrightarrow{D} N(0,1)$$

as $n \to \infty$ provided that $\text{Var}_\theta\left(M_i\right) < \infty$. Now, because $X_1, \ldots, X_n$ are i.i.d., we have that

$$\text{Var}_\theta\left(M_i\right) = \frac{1}{n^2}\sum_{j=1}^{n}\text{Var}_\theta\left(X_j^i\right) = \frac{1}{n}\text{Var}_\theta\left(X_1^i\right) = \frac{1}{n}E_\theta\left(\left(X_1^i - \mu_i\right)^2\right)$$

$$= \frac{1}{n}E_\theta\left(X_1^{2i} - 2\mu_i^i X_1 + \mu_i^2\right) = \frac{1}{n}\left(\mu_{2i} - \mu_i^2\right),$$

so we have that $\text{Var}_\theta\left(M_i\right) < \infty$, provided that $i \le l/2$. In this case, we can estimate $\mu_{2i} - \mu_i^2$ by

$$s_i^2 = \frac{1}{n-1}\sum_{j=1}^{n}\left(x_j^i - m_i\right)^2,$$

as we can simply treat $\left(x_1^i, \ldots, x_n^i\right)$ as a sample from a distribution with mean μ_i and variance $\mu_{2i} - \mu_i^2$. So, as with inferences for the population mean based on the z-statistic, we have that

$$m_i \pm z_{(1+\gamma)/2}\frac{s_i}{\sqrt{n}}$$

is an approximate γ-confidence interval for μ_i whenever $i \le l/2$ and n is large. Also, we can test hypotheses $H_0 : \mu_i = \mu_{i0}$ in exactly the same fashion, as we did this for the population mean using the z-statistic.

Notice that the model $\{P_\theta : \theta \in \Omega\}$ is very large and these approximate inferences are appropriate for every distribution in the model. A cautionary note is that estimation of moments becomes more difficult as the order of the moments rises. Very large sample sizes are required for the accurate estimation of high-order moments.

The general *method of moments principle* allows us to make inference about characteristics that are functions of moments. This takes the following form.

Method of moments principle: A function $\psi(\mu_1, \ldots, \mu_k)$ of the first $k \leq l$ moments is estimated by $\psi(m_1, \ldots, m_k)$.

When ψ is continuously differentiable and nonzero at $(\mu_1, \ldots, \mu_k)$, and $k \leq l/2$, then it can be proved that $\psi(M_1, \ldots, M_k)$ is asymptotically normal with mean given by $\psi(\mu_1, \ldots, \mu_k)$ and variance given by an expression involving the variances and covariances of $M_1, \ldots, M_k$ and the partial derivatives of ψ. We do not pursue this topic further here but note that, in the case $k = 1$ and $l = 2$, these conditions lead to the so-called *delta theorem*, which says that

$$\frac{\sqrt{n}\left(\psi\left(\bar{X}\right) - \psi(\mu_1)\right)}{\left|\psi'\left(\bar{X}\right)\right| s} \xrightarrow{D} N(0,1) \tag{6.4.1}$$

as $n \to \infty$, provided that ψ is continuously differentiable at μ_1 and $\psi'(\mu_1) \neq 0$; see *Approximation Theorems of Mathematical Statistics* by R. J. Serfling (John Wiley & Sons, New York, 1980) for a proof of this result. This result provides approximate confidence intervals and tests for $\psi(\mu_1)$.

Example 6.4.1 *Inference about a Characteristic Using the Method of Moments* Suppose $(x_1, \ldots, x_n)$ is a sample from a distribution with unknown mean μ and variance σ^2, and we want to construct a γ-confidence interval for $\psi(\mu) = 1/\mu^2$. Then $\psi'(\mu) = -2/\mu^3$, so the delta theorem says that

$$\frac{\sqrt{n}\left(1/\bar{X}^2 - 1/\mu^2\right)}{2s/\bar{X}^3} \xrightarrow{D} N(0,1)$$

as $n \to \infty$. Therefore,

$$\left(\frac{1}{\bar{x}}\right)^2 \pm 2\frac{s}{\sqrt{n}\bar{x}^3} z_{(1+\gamma)/2}$$

is an approximate γ-confidence interval for $\psi(\mu) = 1/\mu$.

Notice that if $\mu = 0$, then this confidence interval is not valid, because ψ is not continuously differentiable at 0. So if you think the population mean could be 0, or even close to 0, this would not be an appropriate choice of confidence interval for ψ. ∎

6.4.2 Bootstrapping

Suppose that $\{P_\theta : \theta \in \Omega\}$ is the set of all distributions on R^1 and $(x_1, \ldots, x_n)$ is a sample from some unknown distribution with cdf F. Then the empirical distribution function

$$\hat{F}(x) = \frac{1}{n}\sum_{i=1}^{n} I_{(-\infty, x]}(x_i),$$

introduced in Section 5.4.2, is a natural estimator of the cdf $F(x)$.

We have

$$E_\theta\left(\hat{F}(x)\right) = \frac{1}{n}\sum_{i=1}^{n} E_\theta\left(I_{(-\infty, x]}(X_i)\right) = \frac{1}{n}\sum_{i=1}^{n} F(x) = F(x)$$

for every $\theta \in \Omega$ so that $\hat{F}$ is unbiased for F. The weak and strong laws of large numbers then establish the consistency of $\hat{F}(x)$ for $F(x)$ as $n \to \infty$. Observing that the $I_{(-\infty, x]}(x_i)$ constitute a sample from the Bernoulli($F(x)$) distribution, we have that the standard error of $\hat{F}(x)$ is given by

$$\sqrt{\frac{\hat{F}(x)\left(1 - \hat{F}(x)\right)}{n}}.$$

These facts can be used to form approximate confidence intervals and test hypotheses for $F(x)$, just as in Examples 6.3.7 and 6.3.11.

Observe that $\hat{F}(x)$ prescribes a distribution on the set $\{x_1, \ldots, x_n\}$, e.g., if the sample values are distinct, this probability distribution puts mass $1/n$ on each x_i. Suppose we are interested in estimating $\psi(\theta) = T(F_\theta)$ where T is a function of the distribution F_θ. We use this notation to emphasize that $\psi(\theta)$ corresponds to some characteristic of the distribution rather than just being an arbitrary mathematical function of θ. For example, $T(F_\theta)$ could be a moment of F_θ, a quantile of F_θ, etc.

Now suppose we have an estimator $\hat{\psi}(x_1, \ldots, x_n)$ that is being proposed for inferences about $\psi(\theta)$. Naturally, we are interested in the accuracy of $\hat{\psi}$, and we could choose to measure this by

$$\text{MSE}_\theta\left(\hat{\psi}\right) = \left(E_\theta\left(\hat{\psi}\right) - \psi(\theta)\right)^2 + \text{Var}_\theta\left(\hat{\psi}\right). \tag{6.4.2}$$

Then, to assess the accuracy of our estimate $\hat{\psi}(x_1, \ldots, x_n)$, we need to estimate (6.4.2).

When n is large, we expect $\hat{F}$ to be close to F, so a natural estimate of $\psi(\theta)$ is $T(\hat{F})$, i.e., simply compute the same characteristic of the empirical distribution. Then we estimate the square of the bias in $\hat{\psi}$ by

$$\left(\hat{\psi} - T(\hat{F})\right)^2. \tag{6.4.3}$$

To estimate the variance of $\hat{\psi}$, we use

$$\text{Var}_{\hat{F}}\left(\hat{\psi}\right) = E_{\hat{F}}\left(\hat{\psi}^2\right) - E_{\hat{F}}^2\left(\hat{\psi}\right)$$

$$= \frac{1}{n^n} \sum_{i_1=1}^{n} \cdots \sum_{i_n=1}^{n} \hat{\psi}^2(x_{i_1}, \ldots, x_{i_n})$$

$$- \left(\frac{1}{n^n} \sum_{i_1=1}^{n} \cdots \sum_{i_n=1}^{n} \hat{\psi}(x_{i_1}, \ldots, x_{i_n})\right)^2, \tag{6.4.4}$$

i.e., we treat $x_1, \ldots, x_n$ as i.i.d. random values with cdf given by $\hat{F}$. So to calculate an estimate of (6.4.2), we simply have to calculate $\text{Var}_{\hat{F}}(\hat{\psi})$. This is rarely feasible, however, because the sums in (6.4.4) involve n^n terms. For even very modest sample sizes, like $n = 10$, this cannot be carried out, even on a computer.

The solution to this problem is to approximate (6.4.4) by drawing m independent samples of size n from $\hat{F}$, evaluating $\hat{\psi}$ for each of these samples to obtain $\hat{\psi}_1, \ldots, \hat{\psi}_m$, and then use the sample variance

$$\widehat{\mathrm{Var}}_{\hat{F}}\left(\hat{\psi}\right) = \frac{1}{m-1}\left\{\sum_{i=1}^{m}\hat{\psi}_i^2 - \left(\frac{1}{m}\sum_{i=1}^{m}\hat{\psi}_i\right)^2\right\} \tag{6.4.5}$$

as the estimate. The m samples from $\hat{F}$ are referred to as *bootstrap samples*, and this technique is referred to as *bootstrapping*. Combining (6.4.3) and (6.4.5) gives an estimate of $\mathrm{MSE}_\theta(\hat{\psi})$.

Example 6.4.2 *The Sample Median as an Estimator of the Population Mean* Suppose we want to estimate the location of a unimodal, symmetric distribution. While the sample mean might seem like the obvious choice for this, it turns out that for some distributions there are better estimators. This is because the distribution we are sampling may have long tails, i.e., produce extreme values that are far from the center of the distribution. This implies that the sample average itself could be highly influenced by a few extreme observations and so be a poor estimate of the true mean.

Not all estimators suffer from this defect. For example, if we are sampling from a symmetric distribution, then either the sample mean or the sample median could serve as an estimator of the population mean. But, as we have previously discussed, the sample median is not influenced by extreme values, i.e., it does not change as we move the smallest (or largest) values away from the rest of the data, and this is not the case for the sample mean.

A problem with working with the sample median $\hat{x}_{0.5}$, rather than the sample mean $\bar{x}$, is that the sampling distribution for $\hat{x}_{0.5}$ is typically more difficult to study than that of $\bar{x}$. In this situation, bootstrapping becomes useful. If we are estimating the population mean $T(F_\theta)$ by using the sample median (which is appropriate when we know the distribution we were sampling from is symmetric), then the estimate of the squared bias in the sample median is given by

$$\left(\hat{\psi} - T(\hat{F})\right)^2 = (\hat{x}_{0.5} - \bar{x})^2,$$

because $\hat{\psi} = \hat{x}_{0.5}$ and $T(\hat{F}) = \bar{x}$ (the mean of the empirical distribution is $\bar{x}$). To calculate (6.4.5) we have to generate m samples of size n from $\{x_1, \ldots, x_n\}$ (with replacement) and calculate $\hat{x}_{0.5}$ for each sample.

To illustrate, suppose that we have a sample of size $n = 15$ given by the following table.

-2.0	-0.2	-5.2	-3.5	-3.9
-0.6	-4.3	-1.7	-9.5	1.6
-2.9	0.9	-1.0	-2.0	3.0

Then, using the definition of $\hat{x}_{0.5}$ given by (5.5.4), $\hat{\psi} = -2.000$ and $\bar{x} = -2.087$. Therefore, the estimate of the squared bias (6.4.3) equals $(-2.000 + 2.087)^2 = 7.569 \times 10^{-3}$. Using a statistical package, we generated $m = 10^3$ samples of size $n = 15$ from the distribution that has probability $1/15$ at each of the sample points and obtained

$$\widehat{\text{Var}}_{\hat{F}}\left(\hat{\psi}\right) = 0.770866.$$

Based on $m = 10^4$ samples, we obtained

$$\widehat{\text{Var}}_{\hat{F}}\left(\hat{\psi}\right) = 0.718612,$$

and based on $m = 10^5$ samples we obtained

$$\widehat{\text{Var}}_{\hat{F}}\left(\hat{\psi}\right) = 0.704928.$$

Because these estimates appear to be stabilizing, we take this as our estimate. So in this case, the bootstrap estimate of the MSE of the sample median at the true value of θ is given by

$$\widehat{\text{MSE}}_{\theta}\left(\hat{\psi}\right) = 0.007569 + 0.704928 = 0.71250.$$

Note that the estimated MSE of the sample average is given by $s^2 = 0.704928$, so the sample mean and sample median appear to be providing equivalent accuracy in this problem. In Figure 6.4.1 we have plotted a density histogram of the sample medians obtained from the $m = 10^5$ bootstrap samples. Note that the histogram is very skewed. See Appendix B for more details on how these computations were carried out.

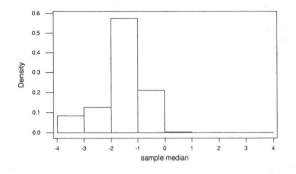

Figure 6.4.1: A density histogram of $m = 10^5$ sample medians, each obtained from a bootstrap sample of size $n = 15$ from the data in Example 6.4.2.

Even with the very small sample size here, it has been necessary to use the computer to carry out our calculations. To evaluate (6.4.4) exactly would have required computing the median of 15^{15} (roughly 4.4×10^{17}) samples, which is clearly impossible even using the computer. So the bootstrap is a very useful device. ∎

The validity of the bootstrapping technique depends upon $\hat{\psi}$ having its first two moments. So the family $\{P_\theta : \theta \in \Omega\}$ must be appropriately restricted, but we can see that the technique is very general.

In general, it is not clear how to choose m. Perhaps the most direct method is to implement bootstrapping for successively higher values of m and stop when we see that the results stabilize for several values. This is what we did in Example 6.4.2, but it must be acknowledged that this approach is not foolproof as we could have a sample $(x_1, \ldots, x_n)$ such the estimate (6.4.5) is very slowly convergent. More details about the bootstrap can be found in *An Introduction to the Bootstrap*, by B. Efron, B. and R. J. Tibshirani, Chapman and Hall, New York, 1993.

6.4.3 The Sign Statistic and Inferences about Quantiles

Suppose that $\{P_\theta : \theta \in \Omega\}$ is the set of all distributions on R^1 such that the associated distribution functions are continuous. Suppose we want to make inferences about a pth quantile of P_θ. We denote this quantile by $x_p(\theta)$ so that, when the distribution function associated with P_θ is denoted by F_θ, we have $p = F_\theta(x_p(\theta))$. Note that continuity implies there is always a solution in x to $p = F_\theta(x)$, and $x_p(\theta)$ is the smallest solution.

Recall the definitions and discussion of estimation of these quantities in Example 5.5.2 based on a sample $(x_1, \ldots, x_n)$. For simplicity, let us restrict attention to the cases where $p = i/n$ for some $i \in \{1, \ldots, n\}$. In this case, we have that $\hat{x}_p = x_{(i)}$ is the natural estimate of x_p.

Now consider assessing the evidence in the data concerning the hypothesis $H_0 : x_p(\theta) = x_0$. For testing this hypothesis we can use the *sign test statistic* given by $S = \sum_{i=1}^{n} I_{(-\infty, x_0]}(x_i)$. So S is the number of sample values less than or equal to x_0.

Notice that when H_0 is true, $I_{(-\infty, x_0]}(x_1), \ldots, I_{(-\infty, x_0]}(x_n)$ is a sample from the Bernoulli(p) distribution. This implies that, when H_0 is true, $S \sim$ Binomial(n, p).

So we can test H_0 by computing the observed value of S, denoted S_o, and seeing if this value lies in a region of low probability for the Binomial(n, p) distribution. Because the binomial distribution is unimodal, the regions of low probability correspond to the left and right tails of this distribution. See, for example, Figure 6.4.2, where we have plotted the probability function of a Binomial$(20, 0.7)$ distribution.

The P-value is therefore obtained by computing the probability of the set

$$\left\{ i : \binom{n}{i} p^i (1-p)^{n-i} \leq \binom{n}{S_o} p^{S_o} (1-p)^{n-S_o} \right\} \qquad (6.4.6)$$

using the Binomial(n, p) probability distribution. This is a measure of how far out in the tails the observed value S_o is (see Figure 6.4.2). Notice that this

P-value is completely independent of θ and is thus valid for the entire model. Tables of binomial probabilities (Table D.6 in Appendix D), or built-in functions available in most statistical packages, can be used to calculate this P-value.

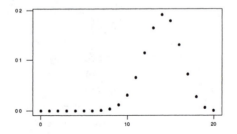

Figure 6.4.2: Plot of the Binomial$(20, 0.7)$ probability function.

When n is large, we have that, under H_0,

$$Z = \frac{S - np}{\sqrt{np(1-p)}} \xrightarrow{D} N(0,1)$$

as $n \to \infty$. Therefore, an approximate P-value is given by (just as in Example 6.3.11)

$$2\left\{1 - \Phi\left(\left|\frac{S_o - 0.5 - np}{\sqrt{np(1-p)}}\right|\right)\right\}$$

where we have replaced S_o by $S_o - 0.5$ as a correction for continuity (see Example 4.4.9 for discussion of the correction for continuity).

A special case arises when $p = 1/2$, i.e., when we are making inferences about an unknown population median $x_{0.5}(\theta)$. In this case, the distribution of S under H_0 is Binomial$(n, 1/2)$. Because the Binomial$(n, 1/2)$ is unimodal and symmetrical about $n/2$, (6.4.6) becomes

$$\{i : |S_o - n/2| \le |i - n/2|\}.$$

If we want a γ-confidence interval for $x_{0.5}(\theta)$, then we can use the duality between tests, which always reject when the P-value is less than or equal to $1 - \gamma$, and γ-confidence intervals (see Example 6.3.12). For this, let j be the smallest integer greater than $n/2$ satisfying

$$P\left(\{i : |i - n/2| \ge j - n/2\}\right) \le 1 - \gamma, \tag{6.4.7}$$

where P is the Binomial$(n, 1/2)$ distribution. If $S \in \{i : |i - n/2| \ge j - n/2\}$, we will reject $H_0 : x_{0.5}(\theta) = x_0$ at the $1 - \gamma$ level and will not otherwise. This leads to the γ-confidence interval, namely, the set of all those values $x_{0.5}$ such that the null hypothesis $H_0 : x_{0.5}(\theta) = x_{0.5}$ is not rejected at the $1 - \gamma$ level, equaling

$$C(x_1, \ldots, x_n) = \left\{ x_0 : \left| \sum_{i=1}^{n} I_{(-\infty, x_0]}(x_i) - n/2 \right| < j - n/2 \right\}$$

$$= \left\{ x_0 : n - j < \sum_{i=1}^{n} I_{(-\infty, x_0]}(x_i) < j \right\} = [x_{(n-j+1)}, x_{(j)}) \qquad (6.4.8)$$

because, for example, $n - j < \sum_{i=1}^{n} I_{(-\infty, x_0]}(x_i)$ if and only if $x_0 \geq x_{(n-j+1)}$.

Example 6.4.3 *Application of the Sign Test*

Suppose we have the following sample of size $n = 10$ from a continuous random variable X, and we wish to test the hypothesis $H_0 : x_{0.5}(\theta) = 0$.

0.44	−0.06	0.43	−0.16	−2.13
1.15	1.08	5.67	−4.97	0.11

The boxplot in Figure 6.4.3 indicates that it is very unlikely that this sample came from a normal distribution, as there are two extreme observations. So it is appropriate to measure the location of the distribution of X by the median.

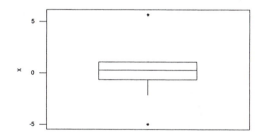

Figure 6.4.3: Boxplot of the data in Example 6.4.3.

In this case the sample median (using (5.5.4)) is given by $(0.11 + 0.43)/2 = 0.27$. The sign statistic for the null is given by

$$S = \sum_{i=1}^{10} I_{(-\infty, 0]}(x_i) = 4.$$

The P-value is given by

$$P(\{i : |4 - 5| \leq |i - 5|\}) = P(\{i : |i - 5| \geq 1\}) = 1 - P(\{i : |i - 5| < 1\})$$

$$= 1 - P(\{5\}) = 1 - \binom{10}{5}\left(\frac{1}{2}\right)^{10} = 1 - 0.24609 = 0.75391,$$

and we have no reason to reject the null hypothesis.

Now suppose that we want a 0.95-confidence interval for the median. Using software (or Table D.6), we calculate

$$\binom{10}{5}\left(\tfrac{1}{2}\right)^{10} = 0.24609 \qquad \binom{10}{4}\left(\tfrac{1}{2}\right)^{10} = 0.20508$$
$$\binom{10}{3}\left(\tfrac{1}{2}\right)^{10} = 0.11719 \qquad \binom{10}{2}\left(\tfrac{1}{2}\right)^{10} = 4.3945 \times 10^{-2}$$
$$\binom{10}{1}\left(\tfrac{1}{2}\right)^{10} = 9.7656 \times 10^{-3} \qquad \binom{10}{0}\left(\tfrac{1}{2}\right)^{10} = 9.7656 \times 10^{-4}.$$

We will use these values to compute the value of j in (6.4.7).

Using the symmetry of the Binomial$(10, 1/2)$ distribution, the values of $P\left(\{i : |i - n/2| \ge j - n/2\}\right)$ are computed as follows. For $j = 10$ we have that (6.4.7) equals

$$P\left(\{i : |i - 5| \ge 5\}\right) = P\left(\{0, 10\}\right) = 2\binom{10}{0}\left(\frac{1}{2}\right)^{10} = 1.9531 \times 10^{-3},$$

and note that $1.9531 \times 10^{-3} < 1 - 0.95 = 0.05$. For $j = 9$, we have that (6.4.7) equals

$$P\left(\{i : |i - 5| \ge 4\}\right) = P\left(\{0, 1, 9, 10\}\right)$$
$$= 2\binom{10}{0}\left(\frac{1}{2}\right)^{10} + 2\binom{10}{1}\left(\frac{1}{2}\right)^{10} = 2.1484 \times 10^{-2},$$

which is also less than 0.05. For $j = 8$, we have that (6.4.7) equals

$$P\left(\{i : |i - 5| \ge 3\}\right) = P\left(\{0, 1, 2, 8, 9, 10\}\right)$$
$$= 2\binom{10}{0}\left(\frac{1}{2}\right)^{10} + 2\binom{10}{1}\left(\frac{1}{2}\right)^{10} + 2\binom{10}{2}\left(\frac{1}{2}\right)^{10} = 0.10938,$$

and this is greater than 0.05. Therefore, the appropriate value is $j = 9$, and a 0.95-confidence interval for the median is given by $\left[x_{(2)}, x_{(9)}\right) = [-0.16, 1.15)$. ■

There are many other distribution-free methods for a variety of statistical situations. While some of these are discussed in the problems, we leave a thorough study of such methods to further courses in statistics.

Summary of Section 6.4

- Distribution-free methods of statistical inference are methods that are appropriate when we feel we can make only very minimal assumptions about the distribution we are sampling from.

- The method of moments, bootstrapping, and methods of inference based on the sign statistic are three distribution-free methods that are applicable in different circumstances.

Exercises

6.4.1 Suppose that we obtained the following sample from a distribution that we know has its first six moments. Determine an approximate 0.95-confidence interval for μ_3.

3.27	−1.24	3.97	2.25	3.47	−0.09	7.45	6.20	3.74	4.12
1.42	2.75	−1.48	4.97	8.00	3.26	0.15	−3.64	4.88	4.55

6.4.2 Determine the method of moments estimator of the population variance. Is this estimator unbiased for the population variance? Justify your answer.

6.4.3 (*Coefficient of variation*) The coefficient of variation for a population measurement with nonzero mean is given by σ/μ, where μ is the population mean and σ is the population standard deviation. What is the method of moments estimate of the coefficient of variation? Prove that the coefficient of variation is invariant under rescalings of the distribution, i.e., under transformations of the form $T(x) = cx$ for constant $c > 0$. It is this invariance that leads to the coefficient of variation being an appropriate measure of sampling variability in certain problems, as it is independent of the units we use for the measurement.

6.4.4 For the context described in Exercise 6.4.1, determine an approximate 0.95 confidence interval for $\exp(\mu_1)$.

6.4.5 Verify that the third moment of an $N(\mu, \sigma^2)$ distribution is given by $\mu_3 = \mu^3 + 3\mu\sigma^2$. Because the normal distribution is specified by its first two moments, any characteristic of the normal distribution can be estimated by simply plugging in the MLE estimates of μ and σ^2. Compare the method of moments estimator of μ_3 with this plug-in MLE estimator, i.e., determine whether they are the same or not.

6.4.6 Suppose we have the sample data 1.48, 4.10, 2.02, 56.59, 2.98, 1.51, 76.49, 50.25, 43.52, 2.96. Consider this as a sample from a normal distribution with unknown mean and variance, and assess the hypothesis that the population median (which is the same as the mean in this case) is 3. Also carry out a sign test that the population median is 3 and compare the results. Plot a boxplot for these data. Does this support the assumption that we are sampling from a normal distribution? Which test do you think is more appropriate? Justify your answer.

6.4.7 Determine the empirical distribution function based on the sample given below.

1.06	−1.28	0.40	1.36	−0.35
−1.42	0.44	−0.58	−0.24	−1.34
0.00	−1.02	−1.35	2.05	1.06
0.98	0.38	2.13	−0.03	−1.29

Using the empirical cdf, determine the sample median, the first and third quartiles, and the interquartile range. What is your estimate of $F(2)$?

Computer Exercises

6.4.8 For the data of Exercise 6.4.7, assess the hypothesis that the population median is 0. State a 0.95-confidence interval for the population median. What is the exact coverage probability of this interval?

6.4.9 For the data of Exercise 6.4.7, assess the hypothesis that the first quartile of the distribution we are sampling from is -1.0.

6.4.10 With a bootstrap sample size of $m = 1000$, use bootstrapping to estimate the MSE of the plug-in MLE estimator of μ_3 for the normal distribution, using the sample data in Exercise 6.4.1. Determine whether $m = 1000$ is a large enough sample for accurate results.

6.4.11 For the data of Exercise 6.4.1, use the plug-in MLE to estimate the first quartile of an $N(\mu, \sigma^2)$ distribution. Use bootstrapping to estimate the MSE of this estimate for $m = 10^3$ and $m = 10^4$ (use (5.5.3) to compute the first quartile of the empirical distribution).

6.4.12 For the data of Exercise 6.4.1, use the plug-in MLE to estimate $F(3)$ for an $N(\mu, \sigma^2)$ distribution. Use bootstrapping to estimate the MSE of this estimate for $m = 10^3$ and $m = 10^4$.

Problems

6.4.13 Prove that when $(x_1, \ldots, x_n)$ is a sample of distinct values from a distribution on R^1, then the ith moment of the empirical distribution on R^1 (i.e., the distribution with cdf given by $\hat{F}$) is m_i.

6.4.14 Suppose that $(x_1, \ldots, x_n)$ is a sample from a distribution on R^1. Determine the general form of the ith moment of $\hat{F}$, i.e., in contrast to Problem 6.4.13, we are now allowing for several of the data values to be equal.

6.4.15 (*Variance stabilizing transformations*) From the delta theorem, we have that $\psi(M_1)$ is asymptotically normal with mean $\psi(\mu_1)$ and variance

$$\left(\psi'(\mu_1)\right)^2 \frac{\sigma^2}{n}$$

when ψ is continuously differentiable, $\psi'(\mu_1) \neq 0$, and M_1 is asymptotically normal with mean μ_1 and variance σ^2/n. In some applications, it is important to choose the transformation ψ so that the asymptotic variance does not depend upon the mean μ_1, i.e., $\left(\psi'(\mu_1)\right)^2 \sigma^2$ is constant as μ_1 varies (note that σ^2 may change as μ_1 changes). Such transformations are known as variance stabilizing transformations.
(a) If we are sampling from a Poisson(λ) distribution, then show that $\psi(x) = \sqrt{x}$ is variance stabilizing.
(b) If we are sampling from a Bernoulli(θ) distribution, then show that $\psi(x) = \arcsin \sqrt{x}$ is variance stabilizing.
(c) If we are sampling from a distribution on $(0, \infty)$ whose variance is proportional to the square of its mean, then show that $\psi(x) = \ln(x)$ is variance stabilizing.

Challenges

6.4.16 Suppose that X has an absolutely continuous distribution on R^1 with density f that is symmetrical about its median. Assuming that the median is 0, prove that $|X|$ and

$$sgn\,(X) = \begin{cases} -1 & x < 0 \\ 0 & x = 0 \\ 1 & x > 0 \end{cases}$$

are independent, with $|X|$ having density $2f$ and $sgn\,(X)$ uniformly distributed on $\{-1,1\}$.

6.4.17 (*Fisher signed deviation statistic*) Suppose that $(x_1, \ldots, x_n)$ is a sample from an absolutely continuous distribution on R^1 with density that is symmetrical about its median. Suppose we want to assess the hypothesis $H_0 : x_{0.5}\,(\theta) = x_0$.

One possibility for this is to use the Fisher signed deviation test based on the statistic S^+. The observed value of S^+ is given by

$$S_o^+ = \sum_{i=1}^{n} |x_i - x_0|\, sgn\,(x_i - x_0).$$

We then assess H_0 by comparing S_o^+ with the conditional distribution of S^+ given the absolute deviations $|x_1 - x_0|, \ldots, |x_n - x_0|$. If a value S_o^+ occurs near the smallest or largest possible value for S^+, under this conditional distribution, then we assert that we have evidence against H_0. We measure this by computing the P-value given by the conditional probability of obtaining a value as far, or farther, from the center of the conditional distribution of S^+ using the conditional mean as the center. This is an example of a *randomization test*, as the distribution for the test statistic is determined by randomly modifying the observed data (in this case by randomly changing the signs of the deviations of the x_i from x_0).

(a) Prove that $S_o^+ = n\,(\bar{x} - x_0)$.

(b) Prove that the P-value described above does not depend upon which distribution we are sampling from in the model. Prove that the conditional mean of S^+ is 0 and the conditional distribution of S^+ is symmetric about this value.

(c) Use the Fisher signed deviation test statistic to assess the hypothesis $H_0 : x_{0.5}\,(\theta) = 2$ when the data are 2.2, 1.5, 3.4, 0.4, 5.3, 4.3, 2.1, and you declare the results to be statistically significant if the P-value is less than or equal to 0.05. (Hint: Based on the results obtained in part (b), you need only compute probabilities for the extreme values of S^+.)

(d) Show that using the Fisher signed deviation test statistic to assess the hypothesis $H_0 : x_{0.5}\,(\theta) = x_0$ is equivalent to the following randomized t-test statistic hypothesis assessment procedure. For this, we compute the conditional distribution of

$$T = \frac{(\bar{X} - x_0)}{S/\sqrt{n}}$$

when the $|X_i - x_0| = |x_i - x_0|$ are fixed and the $sgn\,(X_i - x_0)$ are i.i.d. uniform on $\{-1, 1\}$. Compare the observed value of the t-statistic with this distribution as we did for the Fisher signed deviation test statistic. (Hint: Show that $\sum_{i=1}^{n}(x_i - \bar{x})^2 = \sum_{i=1}^{n}(x_i - x_0)^2 - n\,(\bar{x} - x_0)^2$ and that large absolute values of T correspond to large absolute values of S^{+}.)

6.5 Large Sample Behavior of the MLE (Advanced)

As we saw in Examples 6.3.7 and 6.3.11, implementing exact sampling procedures based on the MLE can be difficult. In those examples, because the MLE was the sample average and we could use the central limit theorem, large sample theory allowed us to work out approximate procedures. In fact, there is some general large sample theory available for the MLE that allows us to obtain approximate sampling inferences. This is the content of this section. The results we develop are all for the case when θ is one-dimensional. Similar results exist for the higher-dimensional problems, but we leave those to a later course.

In Section 6.3 the basic issue was the need to measure the accuracy of the MLE. One approach is to plot the likelihood and look at how concentrated the likelihood is about its peak, with a more highly concentrated likelihood implying greater accuracy for the MLE. There are several problems with this. In particular, the appearance of the likelihood will depend greatly on how we choose the scales for the axes. With appropriate choices, we can make a likelihood look as concentrated or as diffuse as we want. Also, when θ is more than two-dimensional, we can't even plot the likelihood. One solution, when the likelihood is a smooth function of θ, is to compute a numerical measure of how concentrated the log-likelihood is at its peak. The quantity typically used for this is called the *observed Fisher information* and is given by

$$\hat{I}(s) = -\left.\frac{\partial^2 l\,(\theta \mid s)}{\partial \theta^2}\right|_{\theta = \hat{\theta}(s)} \tag{6.5.1}$$

where $\hat{\theta}(s)$ is the MLE. The larger the observed Fisher information is, the more peaked the likelihood function is at its maximum value. We will show that the observed Fisher information is estimating a quantity of considerable importance in statistical inference.

Suppose that the response X is real-valued, θ is real-valued, and the model $\{f_\theta : \theta \in \Omega\}$ satisfies the following regularity conditions:

$$\frac{\partial^2 \ln f_\theta\,(x)}{\partial \theta^2} \text{ exists for each } x, \tag{6.5.2}$$

$$E_\theta\,(S\,(\theta \mid X)) = \int_{-\infty}^{\infty} \frac{\partial \ln f_\theta\,(x)}{\partial \theta} f_\theta\,(x)\,dx = 0, \tag{6.5.3}$$

$$\int_{-\infty}^{\infty} \frac{\partial}{\partial\theta} \left\{ \frac{\partial \ln f_\theta(x)}{\partial\theta} f_\theta(x) \right\} dx = 0, \tag{6.5.4}$$

and

$$\int_{-\infty}^{\infty} \left| \frac{\partial^2 \ln f_\theta(x)}{\partial\theta^2} \right| f_\theta(x) \, dx < \infty. \tag{6.5.5}$$

Note that we have

$$\frac{\partial f_\theta(x)}{\partial\theta} = \frac{\partial \ln f_\theta(x)}{\partial\theta} f_\theta(x),$$

so we can write (6.5.3) equivalently as

$$\int_{-\infty}^{\infty} \frac{\partial f_\theta(x)}{\partial\theta} \, dx = 0.$$

Also note that (6.5.4) can be written as

$$\begin{aligned}
0 &= \int_{-\infty}^{\infty} \frac{\partial}{\partial\theta} \left\{ \frac{\partial l(\theta \mid x)}{\partial\theta} f_\theta(x) \right\} dx \\
&= \int_{-\infty}^{\infty} \left\{ \frac{\partial^2 l(\theta \mid x)}{\partial\theta^2} + \left(\frac{\partial l(\theta \mid x)}{\partial\theta} \right)^2 \right\} f_\theta(x) \, dx \\
&= \int_{-\infty}^{\infty} \left\{ \frac{\partial^2 l(\theta \mid x)}{\partial\theta^2} + S^2(\theta \mid x) \right\} f_\theta(x) \, dx = E_\theta \left(\frac{\partial^2 l(\theta \mid X)}{\partial\theta^2} + S^2(\theta \mid X) \right).
\end{aligned}$$

This together with (6.5.3) and (6.5.5), implies that we can write (6.5.4) equivalently as

$$\mathrm{Var}_\theta \left(S(\theta \mid X) \right) = E_\theta \left(S^2(\theta \mid X) \right) = E_\theta \left(-\frac{\partial^2}{\partial\theta^2} l(\theta \mid X) \right).$$

We call the function

$$I(\theta) = \mathrm{Var}_\theta \left(S(\theta \mid X) \right),$$

when it exists, the *Fisher information* of the model.

Our arguments above have proven the following result.

Theorem 6.5.1 If (6.5.2) and (6.5.3) are satisfied, then $E_\theta \left(S(\theta \mid X) \right) = 0$. If, in addition, (6.5.4) and (6.5.5) are satisfied, then

$$I(\theta) = \mathrm{Var}_\theta \left(S(\theta \mid X) \right) = E_\theta \left(-\frac{\partial^2 l(\theta \mid X)}{\partial\theta^2} \right).$$

Now we see why $\hat{I}$ is called the observed Fisher information, as it is a natural estimate of the Fisher information at the true value θ. We note that there is another natural estimate of the Fisher information at the true value, given by $I(\hat{\theta})$. We call this the *plug-in Fisher estimation*.

When we have a sample $(x_1, \ldots, x_n)$ from f_θ, then

$$S\left(\theta \mid x_1, \ldots, x_n\right) = \frac{\partial}{\partial \theta} \ln \prod_{i=1}^{n} f_\theta\left(x_i\right) = \sum_{i=1}^{n} \frac{\partial \ln f_\theta\left(x_i\right)}{\partial \theta} = \sum_{i=1}^{n} S\left(\theta \mid x_i\right).$$

So, if (6.5.3) holds for the basic model, then $E_\theta\left(S\left(\theta \mid X_1, \ldots, X_n\right)\right) = 0$ and (6.5.3) also holds for the sampling model. Further, if (6.5.4) holds for the basic model, then

$$\begin{aligned}
0 &= \sum_{i=1}^{n} E_\theta\left(\frac{\partial^2}{\partial \theta^2} \ln f_\theta\left(X_i\right)\right) + \sum_{i=1}^{n} E_\theta\left(S^2\left(\theta \mid X_i\right)\right) \\
&= E_\theta\left(\frac{\partial^2}{\partial \theta^2} l\left(\theta \mid X_1, \ldots, X_n\right)\right) + \operatorname{Var}_\theta\left(S\left(\theta \mid X_1, \ldots, X_n\right)\right),
\end{aligned}$$

which implies

$$\operatorname{Var}_\theta\left(S\left(\theta \mid X_1, \ldots, X_n\right)\right) = -E_\theta\left(\frac{\partial^2}{\partial \theta^2} l\left(\theta \mid X_1, \ldots, X_n\right)\right) = nI(\theta),$$

because $l\left(\theta \mid x_1, \ldots, x_n\right) = \sum_{i=1}^{n} \ln f_\theta\left(x_i\right)$. Therefore, (6.5.4) holds for the sampling model as well, and the Fisher information for the sampling model is given by the sample size times the Fisher information for the basic model. We have established the following result.

Corollary 6.5.1 Under i.i.d. sampling from a model with Fisher information $I(\theta)$, the Fisher information for a sample of size n is given by $nI(\theta)$.

The conditions necessary for Theorem 6.5.1 to apply do not hold in general and have to be checked in each example. There are, however, many models where these conditions do hold.

Example 6.5.1 *Nonexistence of the Fisher Information*
If $X \sim U(0, \theta)$, then $f_\theta\left(x\right) = \theta^{-1} I_{(0,\theta)}(x)$, and this is not differentiable at $\theta = x$ for any x. Indeed, if we ignored the lack of differentiability at $\theta = x$ and wrote

$$\frac{\partial f_\theta\left(x\right)}{\partial \theta} = -\frac{1}{\theta^2} I_{(0,\theta)}(x),$$

then

$$\int_{-\infty}^{\infty} \frac{\partial f_\theta\left(x\right)}{\partial \theta} \, dx = -\int_{-\infty}^{\infty} \frac{1}{\theta^2} I_{(0,\theta)}(x) \, dx = -\frac{1}{\theta} \neq 0.$$

So we cannot define the Fisher information for this model. ∎

Example 6.5.2 *Location Normal*
Suppose we have a sample $(x_1, \ldots, x_n)$ from an $N(\theta, \sigma_0^2)$ distribution where $\theta \in R^1$ is unknown and σ_0^2 is known. We saw in Example 6.2.2 that

$$S\left(\theta \mid x_1, \ldots, x_n\right) = \frac{n}{\sigma_0^2}\left(\bar{x} - \theta\right)$$

and therefore

$$\frac{\partial^2}{\partial\theta^2} l\left(\theta \mid x_1, \ldots, x_n\right) = -\frac{n}{\sigma_0^2},$$

$$nI(\theta) = E_\theta \left(-\frac{\partial^2}{\partial\theta^2} l\left(\theta \mid X_1, \ldots, X_n\right) \right) = \frac{n}{\sigma_0^2}.$$

We also determined in Example 6.2.2 that the MLE is given by $\hat{\theta}\left(x_1, \ldots, x_n\right) = \bar{x}$. Then the plug-in Fisher information is

$$nI(\bar{x}) = \frac{n}{\sigma_0^2},$$

while the observed Fisher information is

$$\hat{I}\left(x_1, \ldots, x_n\right) = -\left.\frac{\partial^2 l\left(\theta \mid x_1, \ldots, x_n\right)}{\partial\theta^2}\right|_{\theta=\bar{x}} = \frac{n}{\sigma_0^2}.$$

In this case, there is no need to estimate the Fisher information, but it is comforting that both of our estimates give the exact value. ∎

We now state, without proof, some theorems about the large sample behavior of the MLE under repeated sampling from the model. First we state the consistency of the MLE as an estimator of the true value of θ.

Theorem 6.5.2 Under regularity conditions (like those specified above) for the model $\{f_\theta : \theta \in \Omega\}$, the MLE $\hat{\theta}$ exists a.s. and $\hat{\theta} \overset{a.s.}{\to} \theta$ as $n \to \infty$.

Proof: See *Approximation Theorems of Mathematical Statistics* by R. J. Serfling (John Wiley & Sons, New York, 1980) for the proof of this result. ∎

We see that Theorem 6.5.2 serves as a kind of strong law for the MLE. It also turns out that when the sample size is large, then the sampling distribution of the MLE is approximately normal.

Theorem 6.5.3 Under regularity conditions (like those specified above) for the model $\{f_\theta : \theta \in \Omega\}$, then $(nI(\theta))^{1/2}(\hat{\theta} - \theta) \overset{D}{\to} N(0, 1)$ as $n \to \infty$.

Proof: See *Approximation Theorems of Mathematical Statistics* by R. J. Serfling (John Wiley & Sons, New York, 1980) for the proof of this result. ∎

We see that Theorem 6.5.3 serves as a kind of central limit theorem for the MLE. To make this result fully useful to us for inference, we need the following corollary to this theorem.

Corollary 6.5.2 When I is a continuous function of θ, then

$$\left(nI(\hat{\theta})\right)^{1/2}(\hat{\theta} - \theta) \overset{D}{\to} N(0, 1).$$

In Corollary 6.5.2 we have estimated the Fisher information $I(\theta)$ by the plug-in Fisher estimation $I(\hat{\theta})$. Often it is very difficult to evaluate the function

I. In such a case, we instead estimate $nI(\theta)$ by the observed Fisher information $\hat{I}(x_1, \ldots, x_n)$. A result such as Corollary 6.5.2 again holds in this case.

From Corollary 6.5.2 we can devise large sample approximate inference methods based on the MLE. For example, the approximate standard error of the MLE is

$$\left(nI(\hat{\theta})\right)^{-1/2}.$$

An approximate γ-confidence interval is given by

$$\hat{\theta} \pm \left(nI(\hat{\theta})\right)^{-1/2} z_{(1+\gamma)/2}.$$

Finally, if we want to assess the hypothesis $H_0 : \theta = \theta_0$, we can do this by computing the approximate P-value

$$2\left\{1 - \Phi\left((nI(\theta_0))^{1/2}\left|\hat{\theta} - \theta_0\right|\right)\right\}.$$

Notice that we are using Theorem 6.5.3 for the P-value, rather than Corollary 6.5.2, as, when H_0 is true, we know the asymptotic variance of the MLE is $(nI(\theta_0))^{-1}$, so we do not have to estimate this quantity.

When evaluating I is difficult, we can replace $nI(\hat{\theta})$ by $\hat{I}(x_1, \ldots, x_n)$ in the above expressions for the confidence interval and P-value. We now see very clearly the significance of the observed information. Of course, as we move from using $nI(\theta)$ to $nI(\hat{\theta})$ to $\hat{I}(x_1, \ldots, x_n)$, we expect that larger sample sizes n are needed to make the normality approximation accurate.

We consider some examples.

Example 6.5.3 *Location Normal Model*
Using the Fisher information derived in Example 6.5.2, the approximate γ-confidence interval based on the MLE is

$$\hat{\theta} \pm \left(nI(\hat{\theta})\right)^{-1/2} z_{(1+\gamma)/2} = \bar{x} \pm \frac{\sigma_0}{\sqrt{n}}.$$

This is just the z-confidence interval derived in Example 6.3.6. Rather than being an approximate γ-confidence interval, the coverage is exact in this case. Similarly, the approximate P-value corresponds to the z-test and the P-value is exact. ∎

Example 6.5.4 *Bernoulli Model*
Suppose that $(x_1, \ldots, x_n)$ is a sample from a Bernoulli(θ) distribution where $\theta \in [0, 1]$ is unknown. The likelihood function is given by

$$L(\theta \,|\, x_1, \ldots, x_n) = \theta^{n\bar{x}} (1 - \theta)^{n(1-\bar{x})},$$

and the MLE of θ is $\bar{x}$. The log-likelihood is

$$l(\theta \,|\, x_1, \ldots, x_n) = n\bar{x} \ln \theta + n(1 - \bar{x}) \ln(1 - \theta),$$

the score function is given by

$$S\left(\theta \mid x_1, \ldots, x_n\right) = \frac{n\bar{x}}{\theta} - \frac{n\left(1 - \bar{x}\right)}{1 - \theta},$$

and

$$\frac{\partial}{\partial\theta}S\left(\theta \mid x_1, \ldots, x_n\right) = -\frac{n\bar{x}}{\theta^2} - \frac{n\left(1 - \bar{x}\right)}{\left(1 - \theta\right)^2}.$$

Therefore, the Fisher information for the sample is

$$nI\left(\theta\right) = E_\theta\left(-\frac{\partial}{\partial\theta}S\left(\theta \mid X_1, \ldots, X_n\right)\right) = E_\theta\left(\frac{n\bar{X}}{\theta^2} + \frac{n\left(1 - \bar{X}\right)}{\left(1 - \theta\right)^2}\right) = \frac{n}{\theta\left(1 - \theta\right)}$$

and the plug-in Fisher information is

$$nI\left(\bar{x}\right) = \frac{n}{\bar{x}\left(1 - \bar{x}\right)}.$$

Note that the plug-in Fisher information is the same as the observed Fisher information in this case.

So an approximate γ-confidence interval is given by

$$\hat{\theta} \pm \left(nI(\hat{\theta})\right)^{-1/2} z_{(1+\gamma)/2} = \bar{x} \pm z_{(1+\gamma)/2}\sqrt{\frac{\bar{x}\left(1 - \bar{x}\right)}{n}},$$

which is precisely the interval obtained in Example 6.3.7 using large sample considerations based on the central limit theorem. Similarly, we obtain the same P-value as in Example 6.3.11 when testing $H_0 : \theta = \theta_0$. ∎

Example 6.5.5 *Poisson Model*
Suppose that $(x_1, \ldots, x_n)$ is a sample from a Poisson(λ) distribution where $\lambda > 0$ is unknown. The likelihood function is given by

$$L\left(\lambda \mid x_1, \ldots, x_n\right) = \lambda^{n\bar{x}}e^{-n\lambda}.$$

The log-likelihood is

$$l\left(\lambda \mid x_1, \ldots, x_n\right) = n\bar{x}\ln\lambda - n\lambda,$$

the score function is given by

$$S\left(\lambda \mid x_1, \ldots, x_n\right) = \frac{n\bar{x}}{\lambda} - n,$$

and

$$\frac{\partial}{\partial\lambda}S\left(\lambda \mid x_1, \ldots, x_n\right) = -\frac{n\bar{x}}{\lambda^2}.$$

From this we deduce that the MLE of λ is $\hat{\lambda} = \bar{x}$.

Therefore, the Fisher information for the sample is

$$nI(\lambda) = E_\lambda\left(-\frac{\partial}{\partial\lambda}S(\lambda\,|\,X_1,\ldots,X_n)\right) = E_\lambda\left(\frac{n\bar{X}}{\lambda^2}\right) = \frac{n}{\lambda},$$

and the plug-in Fisher information is

$$nI(\bar{x}) = \frac{n}{\bar{x}}.$$

Note that the plug-in Fisher information is the same as the observed Fisher information in this case.

So an approximate γ-confidence interval is given by

$$\hat{\lambda} \pm \left(nI(\hat{\lambda})\right)^{-1/2} z_{(1+\gamma)/2} = \bar{x} \pm z_{(1+\gamma)/2}\sqrt{\frac{\bar{x}}{n}}.$$

Similarly, the approximate P-value for testing $H_0 : \lambda = \lambda_0$ is given by

$$2\left\{1 - \Phi\left((nI(\lambda_0))^{1/2}\left|\hat{\lambda} - \lambda_0\right|\right)\right\} = 2\left\{1 - \Phi\left(\left(\frac{n}{\lambda_0}\right)^{1/2}|\bar{x} - \lambda_0|\right)\right\}.$$

Note that we have used the Fisher information evaluated at λ_0 for this test. ∎

Summary of Section 6.5

- Under regularity conditions on the statistical model with parameter θ, we can define the Fisher information $I(\theta)$ for the model.

- Under regularity conditions on the statistical model it can be proved that, when θ is the true value of the parameter, then the MLE is consistent for θ and that the MLE is approximately normally distributed with mean given by θ and with variance given by $(nI(\theta))^{-1}$.

- The Fisher information $I(\theta)$ can be estimated by plugging in the MLE or by using the observed Fisher information. These estimates lead to practically useful inferences for θ in many problems.

Exercises

6.5.1 If $(x_1,\ldots,x_n)$ is a sample from an $N(\mu_0,\sigma^2)$ distribution where μ_0 is known and $\sigma^2 \in (0,\infty)$ is unknown, determine the Fisher information.

6.5.2 If $(x_1,\ldots,x_n)$ is a sample from a Gamma(α_0,θ) distribution where α_0 is known and $\theta \in (0,\infty)$ is unknown, determine the Fisher information.

6.5.3 If $(x_1,\ldots,x_n)$ is a sample from a Pareto(α) distribution (see Exercise 6.2.9) where $\alpha > 0$ is unknown, determine the Fisher information.

6.5.4 Suppose that the number of calls arriving at an answering service during a given hour of the day is Poisson(λ) where $\lambda \in (0,\infty)$ is unknown. The number of calls actually received during this hour was recorded for 20 days and the following data were obtained.

9	10	8	12	11	12	5	13	9	9
7	5	16	13	9	5	13	8	9	10

Construct an approximate 0.95-confidence interval for λ. Assess the hypothesis that this is a sample from a Poisson(11) distribution. If you are going to decide that the hypothesis is false when the P-value is less than 0.05, then compute an approximate power for this procedure when $\lambda = 10$.

6.5.5 Suppose that the lifelengths in hours of light-bulbs from a manufacturing process are known to be distributed Gamma$(2, \theta)$ distribution where $\theta \in (0, \infty)$ is unknown. A random sample of 27 bulbs was taken and their lifelengths measured with the following data obtained.

336.87	2750.71	2199.44	292.99	1835.55	1385.36	2690.52
710.64	2162.01	1856.47	2225.68	3524.23	2618.51	361.68
979.54	2159.18	1908.94	1397.96	914.41	1548.48	1801.84
1016.16	1666.71	1196.42	1225.68	2422.53	753.24	

Determine an approximate 0.90-confidence interval for θ.

6.5.6 Repeat the analysis of Exercise 6.5.5, but this time assume that the lifelengths are distributed Gamma$(1, \theta)$. Comment on the differences in the two analyses.

6.5.7 Suppose that incomes (measured in thousands of dollars) above \$20K, can be assumed to be Pareto(α) where $\alpha > 0$ is unknown, for a particular population. A sample of 20 is taken from the population and the following data were obtained.

21.265	20.857	21.090	20.047	20.019	32.509	21.622	20.693
20.109	23.182	21.199	20.035	20.084	20.038	22.054	20.190
20.488	20.456	20.066	20.302				

Construct an approximate 0.95-confidence interval for α. Assess the hypothesis that the mean income in this population is \$25K.

6.5.8 Suppose that $(x_1, \ldots, x_n)$ is a sample from an Exponential(θ) distribution. Construct an approximate left-sided γ-confidence interval for θ. (See Exercise 6.3.17.)

6.5.9 Suppose that $(x_1, \ldots, x_n)$ is a sample from a Geometric(θ) distribution. Construct an approximate left-sided γ-confidence interval for θ. (See Exercise 6.3.17.)

6.5.10 Suppose that $(x_1, \ldots, x_n)$ is a sample from a Negative-Binomial(r, θ) distribution. Construct an approximate left-sided γ-confidence interval for θ. (See Exercise 6.3.17.)

Problems

6.5.11 In Exercise 6.5.1, verify that (6.5.2), (6.5.3), (6.5.4), and (6.5.5) are satisfied.

6.5.12 In Exercise 6.5.2, verify that (6.5.2), (6.5.3), (6.5.4), and (6.5.5) are satisfied.

6.5.13 In Exercise 6.5.3, verify that (6.5.2), (6.5.3), (6.5.4), and (6.5.5) are satisfied.

6.5.14 Suppose that sampling from the model $\{f_\theta : \theta \in \Omega\}$ satisfies (6.5.2), (6.5.3), (6.5.4), and (6.5.5). Prove that $n^{-1}\hat{I} \overset{a.s.}{\to} I(\theta)$ as $n \to \infty$.

6.5.15 (MV) When $\theta = (\theta_1, \theta_2)$, then, under appropriate regularity conditions for the model $\{f_\theta : \theta \in \Omega\}$, the *Fisher information matrix* is defined by

$$I(\theta) = \begin{pmatrix} E_\theta\left(-\frac{\partial^2}{\partial\theta_1^2}l(\theta \mid X)\right) & E_\theta\left(-\frac{\partial^2}{\partial\theta_1\partial\theta_2}l(\theta \mid X)\right) \\ E_\theta\left(-\frac{\partial^2}{\partial\theta_1\partial\theta_2}l(\theta \mid X)\right) & E_\theta\left(-\frac{\partial^2}{\partial\theta_2^2}l(\theta \mid X)\right) \end{pmatrix}.$$

If $(X_1, X_2, X_3) \sim \text{Multinomial}(1, \theta_1, \theta_2, \theta_3)$ (Example 6.1.5), then determine the Fisher information for this model. Recall that $\theta_3 = 1 - \theta_1 - \theta_2$ and so is determined from (θ_1, θ_2).

6.5.16 (MV) Generalize Problem 6.5.15 to the case where

$$(X_1, \ldots, X_k) \sim \text{Multinomial}(1, \theta_1, \ldots, \theta_k).$$

6.5.17 (MV) Using the definition of the Fisher information matrix in Exercise 6.5.15, determine the Fisher information for the Bivariate Normal$(\mu_1, \mu_2, 1, 1, 0)$ model where $\mu_1, \mu_2 \in R^1$ are unknown.

6.5.18 (MV) Extending the definition in Exercise 6.5.15 to the three-dimensional case, determine the Fisher information for the Bivariate Normal$(\mu_1, \mu_2, \sigma^2, \sigma^2, 0)$ model where $\mu_1, \mu_2 \in R^1$, and $\sigma^2 > 0$ are unknown.

Challenge

6.5.19 Suppose that model $\{f_\theta : \theta \in \Omega\}$ satisfies (6.5.2), (6.5.3), (6.5.4), (6.5.5), and has Fisher information $I(\theta)$. If $\Psi : \Omega \to R^1$ is 1–1, and Ψ and Ψ^{-1} are continuously differentiable, then, putting $\Upsilon = \{\Psi(\theta) : \theta \in \Omega\}$, prove that the model given by $\{g_\psi : \psi \in \Upsilon\}$, satisfies the regularity conditions and its Fisher information at ψ is given by $I\left(\Psi^{-1}(\psi)\right)\left(\left(\Psi^{-1}\right)'(\psi)\right)^2$.

Discussion Topic

6.5.20 The method of moments inference methods discussed in Section 6.4.1 are essentially large sample methods based upon the central limit theorem. The large sample methods in Section 6.5 are based upon the form of the likelihood function. Which methods do you think will be more likely to be correct when we know very little about the form of the distribution we are sampling from? In what sense will your choice be "more correct"?

Chapter 7

Bayesian Inference

In Chapter 5 we introduced the basic concepts of inference. At the heart of the theory of inference is the concept of the statistical model $\{f_\theta : \theta \in \Omega\}$ that describes the statistician's uncertainty about how the observed data were produced. Chapter 6 dealt with the analysis of this uncertainty based on the model and the data alone. In some cases this seemed quite successful, but we note that we only dealt with some of the simpler contexts there.

If we accept the principle that, to be amenable to analysis, all uncertainties need to be described by probabilities, then the prescription of a model alone is incomplete, as this does not tell us how to make probability statements about the unknown true value of θ. In this chapter we complete the description so that all uncertainties are described by probabilities. This leads to a probability distribution for θ, and, in essence, we are in the situation of Section 5.2 with the parameter now playing the role of the unobserved response. This is the Bayesian approach to inference.

Many statisticians prefer to develop statistical theory without the additional ingredients necessary for a full probability description of the unknowns. In part, this is motivated by the desire to avoid the prescription of the additional model ingredients necessary for the Bayesian formulation. Of course, we would prefer to have our statistical analysis proceed based on the fewest and weakest model assumptions as possible. For example, in Section 6.4 we introduced distribution-free methods. A price is paid for this weakening, however, and this typically manifests itself in ambiguities about how inference should proceed. The Bayesian formulation in essence removes the ambiguity, but at the price of a more involved model.

The Bayesian approach to inference is sometimes presented as antagonistic to methods that are based on repeated sampling properties (often referred to as *frequentist methods*), as discussed, for example, in Chapter 6. The approach taken in this text, however, is that the Bayesian model arises naturally from the statistician assuming more ingredients for the model. It is up to the statistician to decide what ingredients can be justified and then use appropriate methods. We must be wary of *all* model assumptions, as when they are inappropriate, this may invalidate our inferences. Model checking will be taken up in Chapter 9.

7.1   The Prior and Posterior Distributions

The *Bayesian model* for inference contains the statistical model $\{f_\theta : \theta \in \Omega\}$ for the data $s \in S$ and adds to this the *prior probability measure* Π for θ. The prior describes the statistician's beliefs about the true value of the parameter θ *a priori*, i.e., before observing the data. For example, if $\Omega = [0, 1]$ and θ equals the probability of getting a head on the toss of a coin, then the prior density π plotted in Figure 7.1.1 indicates that the statistician has some belief that the true value of θ is around 0.5 but this information is not very precise.

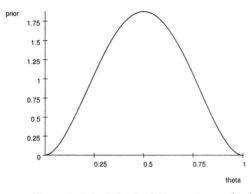

Figure 7.1.1: A fairly diffuse prior on [0,1].

On the other hand, the prior density π plotted in Figure 7.1.2 is indicating that the statistician has very precise information about the true value of θ. If the statistician knows nothing about the true value of θ, then using the uniform distribution on $[0, 1]$ might seem appropriate.

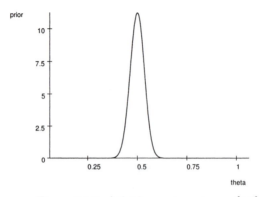

Figure 7.1.2: A fairly precise prior on [0,1].

It is important to remember that the probabilities prescribed by the prior represent beliefs. They do not in general correspond to long-run frequencies, although they could in certain circumstances. A natural question to ask is: Where do these beliefs come from in an application? An easy answer is to say that they come from previous experience with the random system under investigation or perhaps with related systems. To be honest, however, this is rarely the case, and one has to admit that the prior, as well as the statistical model, is often a somewhat arbitrary construction used to drive the statistician's investigations. This raises the issue as to whether or not the inferences derived have any relevance to the practical context, if the model ingredients suffer from this arbitrariness. This is where the concept of model checking comes into play. We will discuss this topic in Chapter 9. At this point, we will assume that all the ingredients make sense, but remember that in an application these must be checked, if the inferences taken are to be practically meaningful.

We note that the ingredients of the Bayesian formulation for inference prescribe a marginal distribution for θ, namely, the prior Π, and a set of conditional distributions for the data s given θ, namely, $\{f_\theta : \theta \in \Omega\}$. By the law of total probability (Theorems 2.3.1 and 2.8.1), these ingredients specify a joint distribution for (s, θ), namely,

$$\pi(\theta) f_\theta(s)$$

where π denotes the probability or density function associated with Π. When the prior distribution is absolutely continuous, the marginal distribution for s is given by

$$m(s) = \int_\Omega \pi(\theta) f_\theta(s) \, d\theta$$

and is referred to as the *prior predictive distribution* of the data. When the prior distribution of θ is discrete we replace, as usual, the integral by a sum.

If we did not observe any data, the prior predictive distribution is the relevant distribution for making probability statements about the unknown value of s. Similarly, the prior π is the relevant distribution to use in making probability statements about θ, before we observe s. Inference about these unobserved quantities then proceeds as described in Section 5.2.

Recall now the principle of conditional probability, namely, $P(A)$ is replaced by $P(A \mid C)$ after we are told that C is true. Therefore, after observing the data, the relevant distribution to use in making probability statements about θ is the conditional distribution of θ given s. We denote this conditional probability measure by $\Pi(\cdot \mid s)$. This probability measure has density, or probability function (whichever is relevant), given by

$$\pi(\theta \mid s) = \frac{\pi(\theta) f_\theta(s)}{m(s)}, \qquad (7.1.1)$$

i.e., the joint density of (s, θ) divided by the marginal density of s. This conditional distribution is called the *posterior distribution* of θ.

Sometimes this use of conditional probability is referred to as an application of Bayes' Theorem (Theorem 1.5.2). This is because we can think of a value of θ being selected first according to π, and then s is generated from f_θ. We then want to make probability statements about the first stage, having observed the outcome of the second stage. It is important to remember, however, that choosing to use the posterior distribution for probability statements about θ is an axiom, or principle, and not a theorem.

We note that in (7.1.1) the prior predictive of the data s plays the role of the *inverse normalizing constant* for the posterior density. By this we mean that the posterior density of θ is proportional to $\pi(\theta) f_\theta(s)$, as a function of θ, and to convert this into a proper density function we need only divide by $m(s)$. In many examples we do not need to compute the inverse normalizing constant. This is because we recognize the functional form, as a function of θ, of the posterior from the expression $\pi(\theta) f_\theta(s)$ and so immediately deduce the posterior probability distribution of θ. Also, there are Monte Carlo methods, such as those discussed in Chapter 4, that allow us to sample from $\pi(\theta \,|\, x)$ without knowing $m(x)$ (also see Section 7.3).

We consider some applications of Bayesian inference.

Example 7.1.1 *Bernoulli Model*
Suppose that we observe a sample $(x_1, \ldots, x_n)$ from the Bernoulli(θ) distribution with $\theta \in [0,1]$ unknown. For the prior we take π to be equal to a Beta(α, β) density (see Problem 2.4.16). Then the posterior of θ is proportional to the likelihood

$$\prod_{i=1}^{n} \theta^{x_i} (1-\theta)^{1-x_i} = \theta^{n\bar{x}} (1-\theta)^{n(1-\bar{x})}$$

times the prior

$$B^{-1}(\alpha, \beta) \theta^{\alpha-1} (1-\theta)^{\beta-1}.$$

This product is proportional to

$$\theta^{n\bar{x}+\alpha-1} (1-\theta)^{n(1-\bar{x})+\beta-1}.$$

We immediately recognize this as the unnormalized density of a

$$\text{Beta}(n\bar{x} + \alpha, n(1-\bar{x}) + \beta)$$

distribution. So in this example we did not need to compute $m(x_1, \ldots, x_n)$ to obtain the posterior.

As a specific case, suppose that we observe $n\bar{x} = 10$ in a sample of $n = 40$ and $\alpha = \beta = 1$, i.e., we have a uniform prior on θ. Then the posterior of θ is given by the Beta$(11, 31)$ distribution. We plot the posterior density in Figure 7.1.3 as well as the prior.

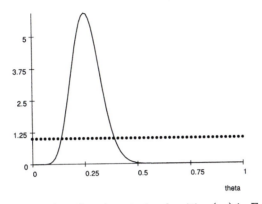

Figure 7.1.3: Prior ($\bullet \, \bullet \, \bullet$) and posterior densities ($-$) in Example 7.1.1.

The spread of the posterior distribution gives us some idea of how precise any probability statements we make about θ will be. Note how much information the data have added, as reflected in the graphs of the prior and posterior densities. ∎

Example 7.1.2 *Location Normal Model*
Suppose that $(x_1, \ldots, x_n)$ is a sample from an $N(\mu, \sigma_0^2)$ distribution where $\mu \in R^1$ is unknown and σ_0^2 is known. The likelihood function is then given by

$$L\left(\mu \,|\, x_1, \ldots, x_n\right) = \exp\left(-\frac{n}{2\sigma_0^2}\left(\bar{x} - \mu\right)^2\right).$$

Suppose we take the prior distribution of μ to be an $N(\mu_0, \tau_0^2)$ for some specified choice of μ_0 and τ_0^2. The posterior density of μ is then proportional to

$$\exp\left\{-\frac{1}{2\tau_0^2}\left(\mu - \mu_0\right)^2\right\} \exp\left\{-\frac{n}{2\sigma_0^2}\left(\bar{x} - \mu\right)^2\right\}$$

$$= \exp\left\{-\frac{1}{2\tau_0^2}\left(\mu^2 - 2\mu\mu_0 + \mu_0^2\right) - \frac{n}{2\sigma_0^2}\left(\bar{x}^2 - 2\mu\bar{x} + \mu^2\right)\right\}$$

$$= \exp\left\{-\frac{1}{2}\left(\frac{1}{\tau_0^2} + \frac{n}{\sigma_0^2}\right)\left[\mu^2 - 2\left(\frac{1}{\tau_0^2} + \frac{n}{\sigma_0^2}\right)^{-1}\left(\frac{\mu_0}{\tau_0^2} + \frac{n}{\sigma_0^2}\bar{x}\right)\mu\right]\right\}$$

$$\times \exp\left\{-\frac{\mu_0^2}{2\tau_0^2} - \frac{n\bar{x}^2}{2\sigma_0^2}\right\}$$

$$= \exp\left\{-\frac{1}{2}\left(\frac{1}{\tau_0^2} + \frac{n}{\sigma_0^2}\right)\left(\mu - \left(\frac{1}{\tau_0^2} + \frac{n}{\sigma_0^2}\right)^{-1}\left(\frac{\mu_0}{\tau_0^2} + \frac{n}{\sigma_0^2}\bar{x}\right)\right)^2\right\}$$

$$\times \exp\left\{\frac{1}{2}\left(\frac{1}{\tau_0^2} + \frac{n}{\sigma_0^2}\right)^{-1}\left(\frac{\mu_0}{\tau_0^2} + \frac{n}{\sigma_0^2}\bar{x}\right)^2\right\}$$

$$\times \exp\left\{-\frac{1}{2}\left(\frac{\mu_0^2}{\tau_0^2} + \frac{n\bar{x}^2}{\sigma_0^2}\right)\right\}. \tag{7.1.2}$$

We immediately recognize this, as a function of μ, as being proportional to the density of a

$$N\left(\left(\frac{1}{\tau_0^2}+\frac{n}{\sigma_0^2}\right)^{-1}\left(\frac{\mu_0}{\tau_0^2}+\frac{n}{\sigma_0^2}\bar{x}\right),\left(\frac{1}{\tau_0^2}+\frac{n}{\sigma_0^2}\right)^{-1}\right)$$

distribution.

Notice that the posterior mean is a weighted average of the prior mean μ_0 and the sample mean $\bar{x}$, with weights

$$\left(\frac{1}{\tau_0^2}+\frac{n}{\sigma_0^2}\right)^{-1}\frac{1}{\tau_0^2}\quad\text{and}\quad\left(\frac{1}{\tau_0^2}+\frac{n}{\sigma_0^2}\right)^{-1}\frac{n}{\sigma_0^2},$$

respectively. This implies that the posterior mean lies between the prior mean and the sample mean.

Further, the posterior variance is smaller than the variance of the sample mean. So if the information expressed by the prior is accurate, inferences about μ based on the posterior will be more accurate than those based on the sample mean alone. Note that the more diffuse the prior is, namely, the larger τ_0^2 is, the less influence the prior has. For example, when $n=20$ and $\sigma_0^2=1,\tau_0^2=1$, then the ratio of the posterior variance to the sample mean variance is $20/21\approx0.95$. So there has been a 5% improvement due to the use of prior information.

For example, suppose that $\sigma_0^2=1,\mu_0=0,\tau_0^2=2$, and, for $n=10$, we observe $\bar{x}=1.2$. Then the prior is an $N(0,2)$ distribution, while the posterior is an

$$N\left(\left(\frac{1}{2}+\frac{10}{1}\right)^{-1}\left(\frac{0}{2}+\frac{10}{1}1.2\right),\left(\frac{1}{2}+\frac{10}{1}\right)^{-1}\right)=N(1.1429,9.523\,8\times10^{-2})$$

distribution. These densities are plotted in Figure 7.1.4. Notice that the posterior is quite concentrated when compared to the prior, so we have learned a lot from the data. ∎

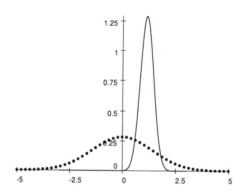

Figure 7.1.4: Plot of the $N(0,2)$ prior ($\bullet\,\bullet\,\bullet$) and the $N\left(1.1429,9.523\,8\times10^{-2}\right)$ posterior (—) in Example 7.1.2.

Example 7.1.3 *Multinomial Model*

Suppose we have a categorical response s that takes k possible values, say, $s \in S = \{1, \ldots, k\}$. For example, suppose we have a bowl containing chips labelled one of $1, \ldots, k$, a proportion θ_i of the chips are labelled i and we randomly draw a chip observing its label.

When the θ_i are unknown, the statistical model is given by

$$\left\{ p_{(\theta_1, \ldots, \theta_k)} : (\theta_1, \ldots, \theta_k) \in \Omega \right\}$$

where $p_{(\theta_1, \ldots, \theta_k)}(i) = P(s = i) = \theta_i$ and

$$\Omega = \{(\theta_1, \ldots, \theta_k) : 0 \le \theta_i \le 1, i = 1, \ldots, k \text{ and } \theta_1 + \cdots + \theta_k = 1\}.$$

Note that the parameter space is really only $(k - 1)$-dimensional because, for example, $\theta_k = 1 - \theta_1 - \cdots - \theta_{k-1}$, namely, once we have determined $k - 1$ of the θ_i, the remaining value is specified.

Now suppose we observe a sample $(s_1, \ldots, s_n)$ from this model. Let the frequency (count) of the ith category in the sample be denoted by x_i. Then, from Example 2.8.5, we see that the likelihood is given by

$$L(\theta_1, \ldots, \theta_k \mid (s_1, \ldots, s_n)) = \theta_1^{x_1} \theta_2^{x_2} \cdots \theta_k^{x_k}.$$

For the prior we assume that $(\theta_1, \ldots, \theta_{k-1}) \sim Dirichlet(\alpha_1, \alpha_2, \ldots, \alpha_k)$ with density (see Problem 2.7.13) given by

$$\frac{\Gamma(\alpha_1 + \cdots + \alpha_k)}{\Gamma(\alpha_1) \cdots \Gamma(\alpha_k)} \theta_1^{\alpha_1 - 1} \theta_2^{\alpha_2 - 1} \cdots \theta_k^{\alpha_k - 1} \tag{7.1.3}$$

for $(\theta_1, \ldots, \theta_k) \in \Omega$ (recall that $\theta_k = 1 - \theta_1 - \cdots - \theta_k$). The α_i are nonnegative constants chosen by the statistician to reflect their beliefs about the unknown value of $(\theta_1, \ldots, \theta_k)$. The choice $\alpha_1 = \alpha_2 = \cdots = \alpha_k = 1$ corresponds to a uniform distribution, as then (7.1.3) is constant on Ω.

The posterior density of $(\theta_1, \ldots, \theta_{k-1})$ is then proportional to

$$\theta_1^{x_1 + \alpha_1 - 1} \theta_2^{x_2 + \alpha_2 - 1} \cdots \theta_k^{x_k + \alpha_k - 1}$$

for $(\theta_1, \ldots, \theta_k) \in \Omega$. From (7.1.3) we immediately deduce that the posterior distribution of $(\theta_1, \ldots, \theta_{k-1})$ is Dirichlet$(x_1 + \alpha_1, x_2 + \alpha_2, \ldots, x_k + \alpha_k)$. ∎

Example 7.1.4 *Location-Scale Normal Model*

Suppose that $(x_1, \ldots, x_n)$ is a sample from an $N(\mu, \sigma^2)$ distribution where $\mu \in R^1$ and $\sigma \ge 0$ are unknown. The likelihood function is then given by

$$L(\mu, \sigma^2 \mid x_1, \ldots, x_n) = (2\pi\sigma^2)^{-n/2} \exp\left(-\frac{n}{2\sigma^2}(\bar{x} - \mu)^2\right) \exp\left(-\frac{n-1}{2\sigma^2} s^2\right).$$

Suppose we put the following prior on (μ, σ^2). First we specify that

$$\mu \mid \sigma^2 \sim N(\mu_0, \tau_0^2 \sigma^2),$$

i.e., the conditional prior distribution of μ given σ^2 is normal with mean μ_0 and variance $\tau_0^2 \sigma^2$, and then specify the marginal prior distribution of σ^2 as

$$\frac{1}{\sigma^2} \sim \text{Gamma}(\alpha_0, \beta_0). \tag{7.1.4}$$

Sometimes (7.1.4) is referred to by saying that σ^2 is distributed *inverse Gamma*. The values $\mu_0, \tau_0^2, \alpha_0,$ and β_0 are selected by the statistician to reflect their prior beliefs.

From this we can deduce (see Section 7.5 for the full derivation) that the posterior distribution of (μ, σ^2) is given by

$$\mu \mid \sigma^2, x_1, \ldots, x_n \sim N\left(\mu_x, \left(n + \frac{1}{\tau_0^2}\right)^{-1} \sigma^2\right) \tag{7.1.5}$$

and

$$\frac{1}{\sigma^2} \mid x_1, \ldots, x_n \sim \text{Gamma}(\alpha_0 + n/2, \beta_x) \tag{7.1.6}$$

where

$$\mu_x = \left(n + \frac{1}{\tau_0^2}\right)^{-1}\left(\frac{\mu_0}{\tau_0^2} + n\bar{x}\right) \tag{7.1.7}$$

and

$$\beta_x = \beta_0 + \frac{n}{2}\bar{x}^2 + \frac{\mu_0^2}{2\tau_0^2} + \frac{n-1}{2}s^2 - \frac{1}{2}\left(n + \frac{1}{\tau_0^2}\right)^{-1}\left(\frac{\mu_0}{\tau_0^2} + n\bar{x}\right)^2. \tag{7.1.8}$$

To generate a value (μ, σ^2) from the posterior, we can make use of the method of composition (Problem 2.10.13) by first generating σ^2 using (7.1.6) and then using (7.1.5) to generate μ. We will discuss this further in Section 7.3.

Notice that as $\tau_0 \to \infty$, that is, as the prior on μ becomes increasingly diffuse, the conditional posterior distribution of μ given σ^2 converges in distribution to an $N(\bar{x}, \sigma^2/n)$ distribution, because

$$\mu_x \to \bar{x} \tag{7.1.9}$$

and

$$\left(n + \frac{1}{\tau_0^2}\right)^{-1} \to \frac{1}{n}. \tag{7.1.10}$$

Further, as $\tau_0 \to \infty$ and $\beta_0 \to 0$, the marginal posterior of $1/\sigma^2$ converges in distribution to a $\text{Gamma}(\alpha_0 + n/2, (n-1)s^2/2)$ distribution, because

$$\beta_x \to (n-1)s^2/2. \tag{7.1.11}$$

Actually, it doesn't really seem to make sense to let $\tau_0 \to \infty$ and $\beta_0 \to 0$ in the prior distribution of (μ, σ^2), as the prior does not converge to a proper probability distribution. The idea here, however, is that we think of taking τ_0 large and β_0 small, so that the posterior inferences are approximately those obtained from the limiting posterior. There is still a need to choose α_0, however, even in the diffuse case, as the limiting inferences are dependent on this quantity. ∎

Summary of Section 7.1

- Bayesian inference adds the prior probability distribution to the sampling model for the data as an additional ingredient to be used in determining inferences about the unknown value of the parameter.

- Having observed the data, the principle of conditional probability leads to the posterior distribution of the parameter as the basis for inference.

- Inference about marginal parameters is handled by marginalizing the full posterior.

Exercises

7.1.1 Suppose that $S = \{1, 2\}, \Omega = \{1, 2, 3\}$, and the class of probability distributions for the response s is given by the following table.

	$s = 1$	$s = 2$
$f_1(s)$	1/2	1/2
$f_2(s)$	1/3	2/3
$f_3(s)$	3/4	1/4

If we use the prior $\pi(\theta)$ given by the table

	$\theta = 1$	$\theta = 2$	$\theta = 3$
$\pi(\theta)$	1/5	2/5	2/5

then determine the posterior distribution of θ for each possible sample of size 2.

7.1.2 In Example 7.1.1, determine the posterior mean and variance of θ.

7.1.3 In Example 7.1.2, what is the posterior probability that θ is positive, given that $n = 10, \bar{x} = 1$ when $\sigma_0^2 = 1, \theta_0 = 0$, and $\tau_0^2 = 10$? Compare this with the prior probability of this event.

7.1.4 Suppose that $(x_1, \ldots, x_n)$ is a sample from a Poisson(λ) distribution with $\lambda \geq 0$ unknown. If we use the prior distribution for λ given by the Gamma(α, β) distribution, then determine the posterior distribution of λ.

7.1.5 Suppose that $(x_1, \ldots, x_n)$ is a sample from a $U(0, \theta)$ distribution with $\theta > 0$ unknown. If the prior distribution of θ is Gamma(α, β), then obtain the form of the posterior density of θ.

7.1.6 Find the posterior mean and variance of θ_i in Example 7.1.3 when $k = 3$. (Hint: See Problems 3.2.16 and 3.3.20.)

7.1.7 Suppose we have a sample

6.56	6.39	3.30	3.03	5.31	5.62	5.10	2.45	8.24	3.71
4.14	2.80	7.43	6.82	4.75	4.09	7.95	5.84	8.44	9.36

from an $N(\mu, \sigma^2)$ distribution and we determine that a prior specified by $\mu \mid \sigma^2 \sim N(3, 4\sigma^2), \sigma^{-2} \sim \text{Gamma}(1, 1)$ is appropriate. Determine the posterior distribution of $(\mu, 1/\sigma^2)$.

Computer Exercises

7.1.8 In Example 7.1.2, when $\mu_0 = 2, \tau_0^2 = 1, \sigma_0^2 = 1, n = 20$, and $\bar{x} = 8.2$, generate a sample of 10^4 (or as large as possible) from the posterior distribution of μ and estimate the posterior probability that the coefficient of variation is greater than 1, i.e., the posterior probability that $\sigma_0/\mu > 0.125$. Estimate the error in your approximation.

7.1.9 In Example 7.1.2, when $\mu_0 = 2, \tau_0^2 = 1, \sigma_0^2 = 1, n = 20$, and $\bar{x} = 8.2$, generate a sample of 10^4 (or as large as possible) from the posterior distribution of μ and estimate the posterior expectation of the coefficient of variation σ_0/μ. Estimate the error in your approximation.

7.1.10 In Example 7.1.1, plot the prior and posterior densities on the same graph and compare them, when $n = 30, \bar{x} = 0.73, \alpha = 3$, and $\beta = 3$. (Hint: Calculate the logarithm of the posterior density and then exponentiate this. You will need the *log-gamma function* defined by $\ln \Gamma(\alpha)$ for $\alpha > 0$.)

Problems

7.1.11 Suppose that the prior of a real-valued parameter θ is given by the $N(\theta_0, \tau^2)$ distribution. Show that this distribution does not converge to a probability distribution as $\tau \to \infty$. (Hint: Consider the limits of the distribution functions.)

7.1.12 Suppose that $(x_1, \ldots, x_n)$ is a sample from $\{f_\theta : \theta \in \Omega\}$ and that we have a prior π. Show that if we observe a further sample $(x_{n+1}, \ldots, x_{n+m})$, then the posterior you obtain from using the posterior $\pi(\cdot \mid x_1, \ldots, x_n)$ as a prior, and then conditioning on $(x_{n+1}, \ldots, x_{n+m})$, is the same as the posterior obtained using the prior π and conditioning on $(x_1, \ldots, x_n, x_{n+1}, \ldots, x_{n+m})$. This is the *Bayesian updating* property.

7.1.13 In Example 7.1.1, determine $m(x)$. If you were asked to generate a value from this distribution, how would you do it? (Hint: For the generation part use the theorem of total probability.)

7.1.14 Prove that the posterior distribution depends on the data only through the value of a sufficient statistic.

Computer Problems

7.1.15 For the data of Exercise 7.1.7, plot the prior and posterior densities of σ^2 over $(0, 10)$ on the same graph and compare them. (Hint: Evaluate the logarithms of the densities first and then plot the exponential of these values.)

7.1.16 In Example 7.1.4, when $\mu_0 = 0, \tau_0^2 = 1, \alpha_0 = 2, \beta_0 = 1, n = 20, \bar{x} = 8.2$, and $s^2 = 2.1$, generate a sample of 10^4 (or as large as is feasible) from the posterior distribution of σ^2 and estimate the posterior probability that $\sigma >$ 0.125. Estimate the error in your approximation.

7.1.17 In Example 7.1.4, when $\mu_0 = 0, \tau_0^2 = 1, \alpha_0 = 2, \beta_0 = 1, n = 20, \bar{x} = 8.2$, and $s^2 = 2.1$, generate a sample of 10^4 (or as large as is feasible) from the posterior distribution of (μ, σ^2) and estimate the posterior expectation of σ. Estimate the error in your approximation.

Discussion Topic

7.1.18 One of the objections raised concerning Bayesian inference methodology is that it is subjective in nature. Comment on this and the role of subjectivity in scientific investigations.

7.2 Inferences Based on the Posterior

In Section 7.1 we determined the posterior distribution of θ as the fundamental object of Bayesian inference. In essence, the principle of conditional probability asserts that the posterior distribution $\pi(\theta \mid s)$ contains all the relevant information in the sampling model $\{f_\theta : \theta \in \Omega\}$, the prior π and the data x, about the unknown true value of θ. While this is a major step forward, it does not completely tell us how to make the types of inferences we discussed in Section 5.5.3. In particular, we must specify how to compute estimates, credible regions, and carry out hypothesis assessment. That is what we will do in this section. It turns out that there are often several plausible ways of proceeding, but they all have the common characteristic that they are based on the posterior.

In general, we are interested in specifying inferences about a real-valued characteristic of interest $\psi(\theta)$. One of the great advantages of the Bayesian approach is that inferences about ψ are determined in the same way as inferences about the full parameter θ, but with the marginal posterior distribution for ψ replacing the full posterior. This situation can be compared with the likelihood methods of Chapter 6, where it is not always entirely clear how we should proceed to determine inferences about ψ based upon the likelihood. Still, we have paid a price for this in requiring the addition of another model ingredient, namely, the prior.

So we need to determine the posterior distribution of ψ. This can be a difficult task in general, even if we have a closed-form expression for $\pi(\theta \mid s)$. When the posterior distribution of θ is discrete, the posterior probability function of ψ is given by

$$\omega(\psi_0 \mid s) = \sum_{\{\theta : \psi(\theta) = \psi_0\}} \pi(\theta \mid s).$$

When the posterior distribution of θ is absolutely continuous, we can often find a complementing function $\lambda(\theta)$ so that $h(\theta) = (\psi(\theta), \lambda(\theta))$ is 1–1 and such that the methods of Section 2.9.2 can be applied. Then, denoting the inverse of this transformation by $\theta = h^{-1}(\psi, \lambda)$, the methods of Section 2.9.2 show that the marginal posterior distribution of ψ has density given by

$$\omega\left(\psi_0 \mid s\right) = \int \pi\left(h^{-1}\left(\psi_0, \lambda\right) \mid s\right)\left|J\left(h^{-1}\left(\psi_0, \lambda\right)\right)\right|^{-1} d\lambda \qquad (7.2.1)$$

where J denotes the Jacobian derivative of this transformation (see Problem 7.2.29). Evaluating (7.2.1) can be difficult, and we will generally avoid doing so here. An example illustrates how we can sometimes avoid directly implementing (7.2.1) and still obtain the marginal posterior distribution of ψ.

Example 7.2.1 *Location-Scale Normal Model*
Suppose that $(x_1, \ldots, x_n)$ is a sample from an $N(\mu, \sigma^2)$ distribution where $\mu \in R^1$ and $\sigma \geq 0$ are unknown, and we use the prior given in Example 7.1.4. The posterior distribution for (μ, σ^2) is then given by (7.1.5) and (7.1.6).

Suppose we are primarily interested in $\psi\left(\mu, \sigma^2\right) = \sigma^2$. We see immediately that the marginal posterior of σ^2 is prescribed by (7.1.6) and we have no further work to do, unless we wanted a form for the marginal posterior density of σ^2. We can use the methods of Section 2.6 for this (see Exercise 7.2.4).

If we want the marginal posterior distribution of $\psi\left(\mu, \sigma^2\right) = \mu$, then things are not quite so simple because (7.1.5) only prescribes the conditional posterior distribution of μ given σ^2. We can, however, avoid the necessity to implement (7.2.1). Note that (7.1.5) implies that

$$Z = \frac{\mu - \mu_x}{\left(n + 1/\tau_0^2\right)^{-1/2} \sigma} \mid \sigma^2, x_1, \ldots, x_n \sim N(0, 1)$$

where μ_x is given in (7.1.7). Because this distribution does not involve σ^2, the posterior distribution of Z is independent of the posterior distribution of σ. Now if $X \sim \text{Gamma}(\alpha, \beta)$, then $Y = 2\beta X \sim \text{Gamma}(\alpha, 1/2) = \chi^2(2\alpha)$ (see Problem 4.6.13 for the definition of the general chi-squared distribution) and so, from (7.1.6),

$$2\frac{\beta_x}{\sigma^2} \mid x_1, \ldots, x_n \sim \chi^2\left(2\alpha_0 + n\right),$$

where β_x is given in (7.1.8). Therefore (using Problem 4.6.14) we conclude, as we are dividing an $N(0, 1)$ variable by the square root of an independent $\chi^2\left(2\alpha_0 + n\right)$ random variable divided by its degrees of freedom, that the posterior distribution of

$$T = \frac{Z}{\sqrt{\left(2\frac{\beta_x}{\sigma^2}\right)/\left(2\alpha_0 + n\right)}} = \frac{\mu - \mu_x}{\sqrt{\frac{2\beta_x}{\left(2\alpha_0 + n\right)\left(n + 1/\tau_0^2\right)}}}$$

is $t\left(2\alpha_0 + n\right)$. Equivalently, we can say that the posterior distribution of μ is the same as

$$\mu_x + \sqrt{\frac{1}{2\alpha_0 + n}}\sqrt{\frac{2\beta_x}{n + 1/\tau_0^2}}\, T.$$

where $T \sim t\left(n + 2\alpha_0\right)$. By (7.1.9), (7.1.10), and (7.1.11), we have that the posterior distribution of μ converges to the distribution of

$$\bar{x} + \sqrt{\frac{n-1}{(2\alpha_0 + n)} \frac{s}{\sqrt{n}}} \, T,$$

as $\tau_0 \to \infty$ and $\beta_0 \to 0$.

In other cases, we cannot avoid the use of (7.2.1) if we want the marginal posterior density of ψ. For example, suppose we are interested in the posterior distribution of the coefficient of variation (we exclude the line given by $\mu = 0$ from the parameter space)

$$\psi = \psi\left(\mu, \sigma^{-2}\right) = \frac{\sigma}{\mu} = \frac{1}{\mu}\left(\frac{1}{\sigma^2}\right)^{-1/2}.$$

Then a complementing function to ψ is given by

$$\lambda = \lambda\left(\mu, \sigma^{-2}\right) = \frac{1}{\sigma^2},$$

and it can be shown (see Section 7.5) that

$$J\left(\theta\left(\psi, \lambda\right)\right) = \psi^{-2}\lambda^{-1/2}.$$

If we let $\pi(\cdot \,|\, \lambda^{-1}, x_1, \ldots, x_n)$ denote the posterior density of μ given λ and $\rho(\cdot \,|\, x_1, \ldots, x_n)$ the posterior density of λ then, from (7.2.1), the marginal density of ψ is given by

$$\psi^{-2} \int_0^\infty \pi\left(\psi^{-1}\lambda^{-1/2} \,|\, \lambda^{-1}, x_1, \ldots, x_n\right) \rho(\lambda \,|\, x_1, \ldots, x_n) \, \lambda^{-1/2} \, d\lambda. \quad (7.2.2)$$

Without writing this out (Problem 7.2.16), we note that we are left with a rather messy integral to evaluate. ∎

In some cases, integrals such as (7.2.2) can be evaluated in closed form; in other cases they cannot. While it is convenient to have a closed form for a density, often this is not necessary, as we can use Monte Carlo methods to approximate posterior probabilities and expectations of interest. We will return to this in Section 7.3. We should always remember that our goal, in implementing Bayesian inference methods, is not to find the marginal posterior densities of quantities of interest, but rather to have a computational algorithm that allows us to implement our inferences.

Under fairly weak conditions, it can be shown that the posterior distribution of θ converges, as the sample size increases, to a distribution degenerate at the true value. This is very satisfying, as it indicates that Bayesian inference methods are consistent.

7.2.1   Estimation

Suppose now that we want to calculate an estimate of a characteristic of interest $\psi(\theta)$. We base this on the posterior distribution of this quantity. There are several different approaches to this problem.

Perhaps the most natural estimate is to obtain the posterior density (or probability function when relevant) of ψ and use the *posterior mode* $\hat{\psi}$, i.e., the point where the posterior probability or density function of ψ takes its maximum. In the discrete case, this is the value of ψ with the greatest posterior probability; in the continuous case, it is the value that has the greatest amount of posterior probability in short intervals containing it.

To calculate the posterior mode we need to maximize $\omega(\psi\,|\,s)$ as a function of ψ. Note that it is equivalent to maximize $m(s)\omega(\psi\,|\,s)$ so that we do not need to compute the inverse normalizing constant to implement this. In fact, we can conveniently choose to maximize any function that is a 1–1 increasing function of $\omega(\cdot\,|\,s)$, and we will get the same answer. In general, $\omega(\cdot\,|\,s)$ may not have a unique mode, but typically there is only one.

An alternative estimate is commonly used and has a natural interpretation. This is given by the posterior mean

$$E(\psi(\theta)\,|\,s),$$

whenever this exists. When the posterior distribution of ψ is symmetrical about its mode, and the expectation exists, then the posterior expectation is the same as the posterior mode, but otherwise these estimates will be different. If we want the estimate to reflect where the central mass of probability lies then in cases where $\omega(\cdot\,|\,s)$ is highly skewed, perhaps the mode is a better choice than the mean. We will see in Chapter 8, however, that there are other ways of justifying the posterior mean as an estimate.

We now consider some examples.

Example 7.2.2 *Bernoulli Model*
Suppose we observe a sample $(x_1, \ldots, x_n)$ from the Bernoulli(θ) distribution with $\theta \in [0, 1]$ unknown and we place a Beta(α, β) prior on θ. In Example 7.1.1 we determined the posterior distribution of θ to be Beta($n\bar{x} + \alpha, n(1 - \bar{x}) + \beta$). Let us suppose that the characteristic of interest is $\psi(\theta) = \theta$.

The posterior expectation of θ is given by

$$E(\theta\,|\,x_1, \ldots, x_n)$$

$$= \int_0^1 \theta\, \frac{\Gamma(n + \alpha + \beta)}{\Gamma(n\bar{x} + \alpha)\,\Gamma(n(1 - \bar{x}) + \beta)}\, \theta^{n\bar{x} + \alpha - 1} (1 - \theta)^{n(1 - \bar{x}) + \beta - 1}\, d\theta$$

$$= \frac{\Gamma(n + \alpha + \beta)}{\Gamma(n\bar{x} + \alpha)\,\Gamma(n(1 - \bar{x}) + \beta)} \int_0^1 \theta^{n\bar{x} + \alpha} (1 - \theta)^{n(1 - \bar{x}) + \beta - 1}\, d\theta$$

$$= \frac{\Gamma(n + \alpha + \beta)}{\Gamma(n\bar{x} + \alpha)\,\Gamma(n(1 - \bar{x}) + \beta)}\, \frac{\Gamma(n\bar{x} + \alpha + 1)\,\Gamma(n(1 - \bar{x}) + \beta)}{\Gamma(n + \alpha + \beta + 1)}$$

$$= \frac{n\bar{x} + \alpha}{n + \alpha + \beta}.$$

When we have a uniform prior, the posterior expectation is given by

$$E\left(\theta \,|\, x\right) = \frac{n\bar{x} + 1}{n + 2}.$$

To determine the posterior mode, we need to maximize

$$\ln \theta^{n\bar{x}+\alpha-1} \left(1 - \theta\right)^{n(1-\bar{x})+\beta-1}$$
$$= \left(n\bar{x} + \alpha - 1\right) \ln \theta + \left(n\left(1 - \bar{x}\right) + \beta - 1\right) \ln\left(1 - \theta\right).$$

This function has first derivative

$$\frac{n\bar{x} + \alpha - 1}{\theta} - \frac{n\left(1 - \bar{x}\right) + \beta - 1}{1 - \theta}$$

and second derivative

$$-\frac{n\bar{x} + \alpha - 1}{\theta^2} - \frac{n\left(1 - \bar{x}\right) + \beta - 1}{\left(1 - \theta\right)^2}.$$

Setting the first derivative equal to 0 and solving gives the solution

$$\hat{\theta} = \frac{n\bar{x} + \alpha - 1}{n + \alpha + \beta - 2}.$$

Now, if $\alpha \geq 1, \beta \geq 1$, we see that the second derivative is always negative and so $\hat{\theta}$ is the unique posterior mode. The restriction on the choice of $\alpha \geq 1, \beta \geq 1$ implies that the prior has a mode in $(0, 1)$ rather than at 0 or 1. Note that when $\alpha = 1, \beta = 1$, namely, we put a uniform prior on θ, the posterior mode is $\hat{\theta} = \bar{x}$ and this is the same as the maximum likelihood estimate (MLE).

The posterior is highly skewed whenever $n\bar{x} + \alpha$ and $n\left(1 - \bar{x}\right) + \beta$ are far apart (plot Beta densities to see this). So, in such a case, we might consider the posterior mode as a more sensible estimate of θ. Note that when n is large, the mode and the mean will be very close together and in fact very close to the MLE $\bar{x}$. ∎

Example 7.2.3 *Location Normal Model*
Suppose that $(x_1, \ldots, x_n)$ is a sample from an $N(\mu, \sigma_0^2)$ distribution where $\mu \in R^1$ is unknown and σ_0^2 is known, and we take the prior distribution on μ to be $N(\mu, \sigma_0^2)$. Let us suppose that the characteristic of interest is $\psi\left(\mu\right) = \mu$.

In Example 7.1.2 we showed that the posterior distribution of μ is given by the

$$N\left(\left(\frac{1}{\tau_0^2} + \frac{n}{\sigma_0^2}\right)^{-1}\left(\frac{\mu_0}{\tau_0^2} + \frac{n}{\sigma_0^2}\bar{x}\right), \left(\frac{1}{\tau_0^2} + \frac{n}{\sigma_0^2}\right)^{-1}\right)$$

distribution. Because this distribution is symmetric about its mode, and the mean exists, the posterior mode and mean agree and equal

$$\left(\frac{1}{\tau_0^2} + \frac{n}{\sigma_0^2}\right)^{-1}\left(\frac{\mu_0}{\tau_0^2} + \frac{n}{\sigma_0^2}\bar{x}\right).$$

This is a weighted average of the prior mean and the sample mean and lies between these two values.

When n is large, we see that this estimator is approximately equal to the sample mean $\bar{x}$, which we also know to be the MLE for this situation. Further, when we take the prior to be very diffuse, namely, when τ_0^2 is very large, then again this estimator is close to the sample mean.

Also observe that the ratio of the sampling variance of $\bar{x}$ to the posterior variance of μ is given by

$$\frac{\sigma_0^2}{n}\left(\frac{1}{\tau_0^2} + \frac{n}{\sigma_0^2}\right) = 1 + \frac{\sigma_0^2}{n\tau_0^2},$$

and this is always greater than 1. The closer τ_0^2 is to 0, the larger this ratio is. Further, as $\tau_0^2 \to 0$, the Bayesian estimate converges to μ_0.

If we are pretty confident that the population mean μ is close to the prior mean μ_0, we will take τ_0^2 small so that the bias in the Bayesian estimate will be small and its variance will be much smaller than the sampling variance of $\bar{x}$. In such a situation, the Bayesian estimator improves on accuracy over the sample mean. Of course, if we are not very confident that μ is close to the prior mean μ_0, then we choose a large value for τ_0^2, and the Bayesian estimator is basically the MLE. ■

Example 7.2.4 *Multinomial Model*
Suppose we have a sample $(s_1, \ldots, s_n)$ from the model discussed in Example 7.1.3 and we place a Dirichlet$(\alpha_1, \alpha_2, \ldots, \alpha_k)$ distribution on $(\theta_1, \ldots, \theta_{k-1})$. The posterior distribution of $(\theta_1, \ldots, \theta_{k-1})$ is then

$$\text{Dirichlet}\left(x_1 + \alpha_1, x_2 + \alpha_2, \ldots, x_k + \alpha_k\right)$$

where x_i is the number of responses in the ith category.

Now suppose that we are interested in estimating $\psi(\theta) = \theta_1$, the probability that a response is in the first category. It can be shown (Problem 7.2.19) that, if $(\theta_1, \ldots, \theta_{k-1})$ is distributed Dirichlet$(\alpha_1, \alpha_2, \ldots, \alpha_k)$, then θ_i is distributed Dirichlet$(\alpha_i, \alpha_{-i}) = \text{Beta}(\alpha_i, \alpha_{-i})$ where $\alpha_{-i} = \alpha_1 + \alpha_2 + \cdots + \alpha_k - \alpha_i$. This result implies that the marginal posterior distribution of θ_1 is

$$\text{Beta}\left(x_1 + \alpha_1, x_2 + \cdots + x_k + \alpha_2 + \cdots + \alpha_k\right).$$

Then, assuming that each $\alpha_i \geq 1$, and using the argument in Example 7.2.2 and $x_1 + \cdots + x_k = n$, the marginal posterior mode of θ_1 is

$$\hat{\theta}_1 = \frac{x_1 + \alpha_1 - 1}{n - 2 + \alpha_1 + \cdots + \alpha_k}.$$

When the prior is the uniform, or $\alpha_1 = \cdots = \alpha_k = 1$, then

$$\hat{\theta}_1 = \frac{x_1}{n + k - 2}.$$

As in Example 7.2.2, we compute the posterior expectation to be

$$E\left(\theta_1 \mid x\right) = \frac{x_1 + \alpha_1}{n + \alpha_1 + \cdots + \alpha_k}.$$

The posterior distribution is highly skewed whenever $x_1 + \alpha_1$ and $x_2 + \cdots + x_k + \alpha_2 + \cdots + \alpha_k$ are far apart.

From Problem 7.2.20, we have that the plug-in MLE of θ_1 is x_1/n. When n is large, the Bayesian estimates are close to this value, so there is no conflict between the estimates. Notice, however, that when the prior is uniform, then $\alpha_1 + \cdots + \alpha_k = k$, so the plug-in MLE and the Bayesian estimates will be quite different when k is large relative to n. In fact, the posterior mode will always be smaller than the plug-in MLE when $k > 2$ and $x_1 > 0$. This is a situation in which the Bayesian and frequentist approaches to inference differ.

At this point, the decision about which estimate to use is left with the practitioner, as theory does not seem to provide a clear answer. We can be comforted by the fact that the estimates will not differ by much in many contexts of practical importance. ∎

Example 7.2.5 *Location-Scale Normal Model*
Suppose that $(x_1, \ldots, x_n)$ is a sample from an $N(\mu, \sigma^2)$ distribution where $\mu \in R^1$ and $\sigma > 0$ are unknown, and we use the prior given in Example 7.1.4. Let us suppose that the characteristic of interest is $\psi\left(\mu, \sigma^2\right) = \mu$.

In Example 7.2.1, we derived the marginal posterior distribution of μ to be the same as the distribution of

$$\mu_x + \sqrt{\frac{1}{2\alpha_0 + n}} \sqrt{\frac{2\beta_x}{n + 1/\tau_0^2}}\, T$$

where $T \sim t\left(n + 2\alpha_0\right)$. This is a $t\left(n + 2\alpha_0\right)$ distribution relocated to have its mode at μ_x and rescaled by the factor

$$\sqrt{\frac{1}{2\alpha_0 + n}} \sqrt{\frac{2\beta_x}{n + 1/\tau_0^2}}.$$

So the marginal posterior mode of μ is

$$\mu_x = \left(n + \frac{1}{\tau_0^2}\right)^{-1} \left(\frac{\mu_0}{\tau_0^2} + n\bar{x}\right).$$

Because a t distribution is symmetric about its mode, this is also the posterior mean of μ, provided that $n + 2\alpha_0 > 1$, as a $t\left(\lambda\right)$ distribution has a mean only when $\lambda > 1$ (Problem 4.6.16). This will always be the case as the sample size $n \geq 1$. Again, μ_x is a weighted average of the prior mean μ_0 and the sample average $\bar{x}$.

The marginal posterior mode and expectation can also be obtained for $\psi\left(\mu, \sigma^2\right) = \sigma^2$. These computations are left to the reader (see Exercise 7.2.4). ∎

One issue that we have not yet addressed is how we will assess the accuracy of Bayesian estimates. Naturally, this is based upon the posterior distribution and how concentrated the posterior distribution is about the quantity in question. In the case of the posterior mean, this means that we compute the posterior variance as a measure of spread for the posterior distribution of ψ about its mean. For the posterior mode, we will discuss this issue further in Section 7.2.3.

Example 7.2.6 *Posterior Variances*
In Example 7.2.2 the posterior variance of θ is given by (Exercise 7.2.6)

$$\frac{(n\bar{x} + \alpha)(n(1 - \bar{x}) + \beta)}{(n + \alpha + \beta)^2 (n + \alpha + \beta + 1)}.$$

Notice that the posterior variance converges to 0 as $n \to \infty$.

In Example 7.2.3 the posterior variance is given by $\left(1/\tau_0^2 + n/\sigma_0^2\right)^{-1}$. Notice that the posterior variance converges to 0 as $\tau_0^2 \to 0$ and converges to σ_0^2/n, the sampling variance of $\bar{x}$, as $\tau_0^2 \to \infty$.

In Example 7.2.4 the posterior variance of θ_1 is given by (Exercise 7.2.7)

$$\frac{(x_1 + \alpha_1)(x_2 + \cdots + x_k + \alpha_2 + \cdots + \alpha_k)}{(n + \alpha_1 + \cdots + \alpha_k)^2 (n + \alpha_1 + \cdots + \alpha_k + 1)}.$$

Notice that the posterior variance converges to 0 as $n \to \infty$.

In Example 7.2.5 the posterior variance of μ is given by (Problem 7.2.22)

$$\left(\frac{1}{n + 2\alpha_0}\right)\left(\frac{2\beta_x}{n + 1/\tau_0^2}\right)\left(\frac{n + 2\alpha_0}{n + 2\alpha_0 - 2}\right) = \left(\frac{2\beta_x}{n + 1/\tau_0^2}\right)\left(\frac{1}{n + 2\alpha_0 - 2}\right),$$

provided $n + 2\alpha_0 > 2$, because the variance of a $t(\lambda)$ distribution is $\lambda/(\lambda - 2)$ when $\lambda > 2$ (Problem 4.6.16). Notice that the posterior variance goes to 0 as $n \to \infty$. ∎

7.2.2   Credible Intervals

A *credible interval* for a real-valued parameter $\psi(\theta)$ is an interval $C(s) = [l(s), u(s)]$ that we believe will contain the true value of ψ. As with the sampling theory approach, we specify a probability γ, and then find an interval $C(s)$ satisfying

$$\Pi(\psi(\theta) \in C(s) \,|\, s) = \Pi(\{\theta : l(s) \leq \psi(\theta) \leq u(s)\} \,|\, s) \geq \gamma. \qquad (7.2.3)$$

We then refer to $C(s)$ as a γ-credible interval for ψ.

Naturally, we try to find a γ-credible interval $C(s)$ so that $\Pi(\psi(\theta) \in C(s) \,|\, s)$ is as close to γ as possible, and such that $C(s)$ is as short as possible. This leads to the consideration of *HPD intervals*, or *highest posterior density intervals*, which are of the form

$$C(s) = \{\psi : \omega(\psi \,|\, s) \geq c\}$$

where $\omega\left(\cdot\mid s\right)$ is the marginal posterior density of ψ and where c is chosen as large as possible so that (7.2.3) is satisfied. In Figure 7.2.1 we have plotted an example of an HPD interval for a given value of c.

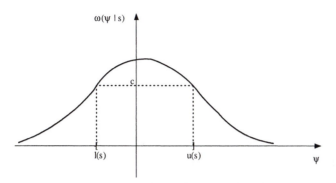

Figure 7.2.1: An HPD interval $C(s) = [l(s), u(s)] = \{\psi : \omega\left(\psi\mid s\right) \geq c\}$.

Clearly, $C(s)$ contains the mode whenever $c \leq \max_{\psi} \omega\left(\psi\mid s\right)$. We can take the length of an HPD interval as a measure of the accuracy of the mode of $\omega\left(\cdot\mid s\right)$ as an estimator of $\psi\left(\theta\right)$. The length of a 0.95-credible interval for ψ will serve the same purpose as the margin of error does with confidence intervals.

Consider now some applications of the concept of credible interval.

Example 7.2.7 *Location Normal Model*

Suppose that $(x_1, \ldots, x_n)$ is a sample from an $N(\mu, \sigma_0^2)$ distribution where $\mu \in R^1$ is unknown and σ_0^2 is known, and we take the prior distribution on μ to be $N(\mu_0, \tau_0^2)$. In Example 7.1.2 we showed that the posterior distribution of μ is given by the

$$N\left(\left(\frac{1}{\tau_0^2} + \frac{n}{\sigma_0^2}\right)^{-1} \left(\frac{\mu_0}{\tau_0^2} + \frac{n}{\sigma_0^2}\bar{x}\right), \left(\frac{1}{\tau_0^2} + \frac{n}{\sigma_0^2}\right)^{-1}\right)$$

distribution. Since this distribution is symmetric about its mode (also mean) $\hat{\mu}$, a shortest γ-HPD interval is of the form

$$\hat{\mu} \pm \left(\frac{1}{\tau_0^2} + \frac{n}{\sigma_0^2}\right)^{-1/2} c$$

where c is such that

$$
\begin{aligned}
\gamma &= \Pi\left(\mu \in \left[\hat{\mu} \pm \left(\frac{1}{\tau_0^2} + \frac{n}{\sigma_0^2}\right)^{-1/2} c\right] \mid x_1, \ldots, x_n\right) \\
&= \Pi\left(-c \leq \left(\frac{1}{\tau_0^2} + \frac{n}{\sigma_0^2}\right)^{1/2} (\mu - \hat{\mu}) \leq c \mid x_1, \ldots, x_n\right).
\end{aligned}
$$

Since

$$\left(\frac{1}{\tau_0^2} + \frac{n}{\sigma_0^2}\right)^{1/2} (\mu - \hat{\mu}) \mid x_1, \ldots, x_n \sim N(0, 1),$$

we have $\gamma = \Phi(c) - \Phi(-c)$ where Φ is the standard normal cumulative distribution function (cdf). This immediately implies that $c = z_{(1+\gamma)/2}$ and the γ-HPD interval is given by

$$\left(\frac{1}{\tau_0^2} + \frac{n}{\sigma_0^2}\right)^{-1} \left(\frac{\mu_0}{\tau_0^2} + \frac{n}{\sigma_0^2}\bar{x}\right) \pm \left(\frac{1}{\tau_0^2} + \frac{n}{\sigma_0^2}\right)^{-1/2} z_{(1+\gamma)/2}.$$

Note that as $\tau_0^2 \to \infty$, namely, as the prior becomes increasingly diffuse, this interval converges to the interval

$$\bar{x} \pm \frac{\sigma_0}{\sqrt{n}} z_{(1+\gamma)/2},$$

which is also the γ-confidence interval derived in Chapter 6 for this problem. So under a diffuse normal prior, the Bayesian and frequentist approaches agree. ∎

Example 7.2.8 *Location-Scale Normal Model*
Suppose that $(x_1, \ldots, x_n)$ is a sample from an $N(\mu, \sigma^2)$ distribution where $\mu \in R^1$ and $\sigma \geq 0$ are unknown, and we use the prior given in Example 7.1.4. In Example 7.2.1, we derived the marginal posterior distribution of μ to be the same as

$$\mu_x + \sqrt{\frac{1}{2\alpha_0 + n}} \sqrt{\frac{2\beta_x}{n + 1/\tau_0^2}} \, T,$$

where $T \sim t(2\alpha_0 + n)$. Because this distribution is symmetric about its mode μ_x, a γ-HPD interval is of the form

$$\mu_x \pm \sqrt{\frac{1}{2\alpha_0 + n}} \sqrt{\frac{2\beta_x}{n + 1/\tau_0^2}} \, c$$

where c satisfies

$$\gamma = \Pi\left(\mu \in \left[\mu_x \pm \sqrt{\frac{1}{2\alpha_0 + n}} \sqrt{\frac{2\beta_x}{n + 1/\tau_0^2}} \, c\right] \,\Big|\, x_1, \ldots, x_n\right)$$

$$= \Pi\left(-c \leq \left(\frac{2\beta_x}{(2\alpha_0 + n)(n + 1/\tau_0^2)}\right)^{-1/2} (\mu - \hat{\mu}) \leq c \,\Big|\, x_1, \ldots, x_n\right)$$

$$= G_{2\alpha_0 + n}(c) - G_{2\alpha_0 + n}(-c).$$

Here, $G_{2\alpha_0 + n}$ is the $t(2\alpha_0 + n)$ cdf, and therefore $c = t_{(1+\gamma)/2}(2\alpha_0 + n)$.

Using (7.1.9), (7.1.10), and (7.1.11) we have that this interval converges to the interval

$$\bar{x} \pm \sqrt{\frac{n-1}{(2\alpha_0 + n)}} \frac{s}{\sqrt{n}} t(n + 2\alpha_0)$$

as $\tau_0 \to \infty$ and $\beta_0 \to 0$. Note that this is a little different from the γ-confidence interval we obtained for μ in Example 6.3.8, but when α_0/n is small, they are virtually identical. ∎

In the examples we have considered so far, we could obtain closed-form expressions for the HPD intervals. In general, this is not the case. In such situations, we have to resort to numerical methods to obtain the HPD intervals. We do not pursue this topic further here.

There are other methods of deriving credible intervals. For example, a common method of obtaining a γ-credible interval for ψ is to take the interval $[\psi_l, \psi_r]$ where ψ_l is a $(1 - \gamma)/2$ quantile for the posterior distribution of ψ, and ψ_r is a $1 - (1 - \gamma)/2$ quantile for this distribution. Alternatively, we could form one-sided intervals. These credible intervals avoid the more extensive computations that may be needed for HPD intervals.

7.2.3 Hypothesis Testing and Bayes Factors

Suppose now that we want to assess the evidence in the observed data concerning the hypothesis $H_0 : \psi(\theta) = \psi_0$. It seems clear how we should assess this, namely, compute the posterior probability

$$\Pi(\psi(\theta) = \psi_0 \mid s) \tag{7.2.4}$$

and, if this is small, then conclude that we have evidence against H_0. We will see further justification for this approach in Chapter 8.

Example 7.2.9
Suppose that we want to assess the evidence concerning whether or not $\theta \in A$. If we let $\psi = I_A$, then we are assessing the hypothesis $H_0 : \psi(\theta) = 1$ and

$$\Pi(\psi(\theta) = 1 \mid s) = \Pi(A \mid s).$$

So in this case, we simply compute the posterior probability that $\theta \in A$. ∎

There can be a problem, however, with using (7.2.4) to assess a hypothesis. For when the prior distribution of ψ is absolutely continuous, then we have that $\Pi(\psi(\theta) = \psi_0 \mid s) = 0$ for all data s. Therefore, we would always find evidence against H_0 no matter what is observed, and this doesn't seem sensible. In general, if the value ψ_0 is assigned small prior probability then it can happen that this value also has a small posterior probability no matter what data are observed.

To avoid this problem, there is an alternative approach to hypothesis assessment that is sometimes used. Recall that, if ψ_0 is a surprising value for the posterior distribution of ψ then this is evidence that H_0 is false. The value ψ_0 is surprising whenever it occurs in a region of low probability for the posterior distribution of ψ. A region of low probability will correspond to a region where the posterior density $\omega(\cdot \mid s)$ is relatively low. So, one possible method for assessing this is by computing the *(Bayesian) P-value*

$$\Pi(\{\theta : \omega(\psi(\theta) \mid s) \leq \omega(\psi_0 \mid s)\} \mid s). \tag{7.2.5}$$

Note that, when $\omega(\cdot \mid s)$ is unimodal, then (7.2.5) corresponds to computing a tail probability. If the probability (7.2.5) is small, then ψ_0 is surprising, at least with respect to our posterior beliefs. When we decide to reject H_0 whenever the P-value is less than $1 - \gamma$, then this approach is equivalent to computing a γ-HPD region for ψ and rejecting H_0 whenever ψ_0 is not in the region.

Example 7.2.10 *(Example 7.2.9 continued)*.
Applying the P-value approach to this problem, we see that $\psi(\theta) = I_A(\theta)$ has posterior given by the Bernoulli($\Pi(A \mid s)$) distribution. Therefore, $\omega(\cdot \mid s)$ is defined by $\omega(0 \mid s) = 1 - \Pi(A \mid s) = \Pi(A^c \mid s)$ and $\omega(1 \mid s) = \Pi(A \mid s)$.

Now $\psi_0 = 1$, so

$$\{\theta : \omega(\psi(\theta) \mid s) \leq \omega(1 \mid s)\} = \{\theta : \omega(I_A(\theta) \mid s) \leq \Pi(A \mid s)\}$$
$$= \begin{cases} \Omega & \Pi(A \mid s) \geq \Pi(A^c \mid s) \\ A & \Pi(A \mid s) < \Pi(A^c \mid s). \end{cases}$$

Therefore, (7.2.5) becomes

$$\Pi(\{\theta : \omega(\psi(\theta) \mid s) \leq \omega(1 \mid s)\} \mid s) = \begin{cases} 1 & \Pi(A \mid s) \geq \Pi(A^c \mid s) \\ \Pi(A \mid s) & \Pi(A \mid s) < \Pi(A^c \mid s), \end{cases}$$

and so again we have evidence against H_0 whenever $\Pi(A \mid s)$ is small. ∎

We see from Examples 7.2.9 and 7.2.10 that computing the P-value (7.2.5) is essentially equivalent to using (7.2.4), whenever the marginal parameter ψ takes only two values. This is not the case whenever ψ takes more than two values, and the statistician has to decide which method is more appropriate in such a context.

As previously noted, when the prior distribution of ψ is absolutely continuous, then (7.2.4) is always 0, no matter what data are observed. As the following example illustrates, there is also a difficulty with using (7.2.5) in such a situation.

Example 7.2.11
Suppose that the posterior distribution of θ is Beta(2, 1), i.e., $\omega(\theta \mid s) = 2\theta$ when $0 \leq \theta \leq 1$, and we want to assess $H_0 : \theta = 3/4$. Then $\omega(\theta \mid s) \leq \omega(3/4 \mid s)$ if and only if $\theta \leq 3/4$, and (7.2.5) is given by

$$\int_0^{3/4} 2\theta \, d\theta = 9/16.$$

On the other hand, suppose we make a 1–1 transformation to $\rho = \theta^2$ so that the hypothesis is now $H_0 : \rho = 9/16$. The posterior distribution of ρ is Beta(1, 1). Since the posterior density of ρ is constant, this implies that the posterior density at every possible value is less than or equal to the posterior density evaluated at $9/16$. Therefore (7.2.5) equals 1 and we would never find evidence against H_0 using this parameterization.

This example shows that our assessment of H_0 via (7.2.5) depends on the parameterization used, and this doesn't seem appropriate. ∎

The difficulty in using (7.2.5), as demonstrated in Example 7.2.11, only occurs with continuous posterior distributions. So, to avoid this problem, it is often recommended that the hypothesis to be tested always be assigned a positive prior probability. As demonstrated in Example 7.2.10 the approach via (7.2.5) is then essentially equivalent to using (7.2.4) to assess H_0.

In problems where it seems natural to use continuous priors this is accomplished by taking the prior Π to be a mixture of probability distributions, as discussed in Section 2.5.4, namely, the prior distribution equals

$$\Pi = p\Pi_1 + (1-p)\,\Pi_2$$

where $\Pi_1\,(\psi\,(\theta) = \psi_0) = 1$ and $\Pi_2\,(\psi\,(\theta) = \psi_0) = 0$, i.e., Π_1 is degenerate at ψ_0 and Π_2 is continuous at ψ_0. Then

$$\Pi\,(\psi\,(\theta) = \psi_0) = p\Pi_1\,(\psi\,(\theta) = \psi_0) + (1-p)\,\Pi_2\,(\psi\,(\theta) = \psi_0) = p > 0$$

is the prior probability that H_0 is true.

The prior predictive for the data s is then given by

$$m(s) = pm_1(s) + (1-p)m_2(s)$$

where m_i is the prior predictive obtained via prior Π_i (Problem 7.2.28). This implies (Problem 7.2.28) that the posterior probability measure for θ, when using the prior Π, is

$$
\begin{aligned}
&\Pi\,(A\,|\,s)\\
&= \frac{pm_1(s)}{pm_1(s) + (1-p)m_2(s)}\Pi_1\,(A\,|\,s) + \frac{(1-p)\,m_2(s)}{pm_1(s) + (1-p)m_2(s)}\Pi_2\,(A\,|\,s)
\end{aligned}
$$

$$(7.2.6)$$

where $\Pi_i\,(\cdot\,|\,s)$ is the posterior measure obtained via the prior Π_i. Note that this a mixture of the posterior probability measures $\Pi_1\,(\cdot\,|\,s)$ and $\Pi_2\,(\cdot\,|\,s)$ with mixture probabilities

$$\frac{pm_1(s)}{pm_1(s) + (1-p)m_2(s)} \quad \text{and} \quad \frac{(1-p)\,m_2(s)}{pm_1(s) + (1-p)m_2(s)}.$$

Now $\Pi_1\,(\cdot\,|\,s)$ is degenerate at ψ_0 (if the prior is degenerate at a point then the posterior must be degenerate at that point too) and $\Pi_2\,(\cdot\,|\,s)$ is continuous at ψ_0. Therefore,

$$\Pi\,(\psi\,(\theta) = \psi_0\,|\,s) = \frac{pm_1(s)}{pm_1(s) + (1-p)m_2(s)}$$

$$(7.2.7)$$

and we use this probability to assess H_0.

The following example illustrates this approach.

Example 7.2.12 *Location Normal Model*
Suppose that $(x_1, \ldots, x_n)$ is a sample from a $N(\mu, \sigma_0^2)$ distribution where $\mu \in R^1$ is unknown and σ_0^2 is known, and we want to assess the hypothesis H_0 : $\mu = \mu_0$. As in Example 7.1.2, we will take the prior for μ to be an $N(\mu_0, \tau_0^2)$ distribution. Given that we are assessing whether or not $\mu = \mu_0$, it seems reasonable to place the mode of the prior at the hypothesized value. The choice of the hyperparameter τ_0^2 then reflects the degree of our prior belief that H_0 is true. We let Π_2 denote this prior probability measure, i.e., Π_2 is the $N(\mu_0, \sigma_0^2)$ probability measure.

If we use Π_2 as our prior, then, as shown in Example 7.1.2, the posterior distribution of μ is absolutely continuous. This implies that (7.2.4) is 0. So, following the preceding discussion, we consider instead the prior $\Pi = p\Pi_1 + (1 - p)\Pi_2$ obtained by mixing Π_2 with a probability measure Π_1 degenerate at μ_0. Then $\Pi_1(\{\mu_0\}) = 1$ and so $\Pi(\{\mu_0\}) = p$. As shown in Example 7.1.2, under Π_2 the posterior distribution of μ is

$$N\left(\left(\frac{1}{\tau_0^2} + \frac{n}{\sigma_0^2}\right)^{-1}\left(\frac{\mu_0}{\tau_0^2} + \frac{n}{\sigma_0^2}\bar{x}\right), \left(\frac{1}{\tau_0^2} + \frac{n}{\sigma_0^2}\right)^{-1}\right),$$

while the posterior under Π_1 is the distribution degenerate at μ_0. We now need to evaluate (7.2.7), and we will do this in Example 7.2.14. ∎

Bayes Factors

Bayes factors comprise another method of hypothesis assessment. In a probability model with sample space S and probability measure P, the *odds in favor* of event $A \subset S$ is defined to be

$$\frac{P(A)}{P(A^c)},$$

namely, the ratio of the probability of A to the probability of A^c. Obviously, large values of the odds in favor of A indicate that there is a strong belief that A is true.

The *Bayes factor* BF_{H_0} *in favor of the hypothesis* H_0 is defined, whenever the prior probability of H_0 is not 0 or 1, to be the ratio of the *posterior odds* in favor of H_0 to the *prior odds* in favor of H_0, or

$$BF_{H_0} = \left\{\frac{\Pi(\psi(\theta) = \psi_0 \mid s)}{1 - \Pi(\psi(\theta) = \psi_0 \mid s)}\right\} / \left\{\frac{\Pi(\psi(\theta) = \psi_0)}{1 - \Pi(\psi(\theta) = \psi_0)}\right\}. \qquad (7.2.8)$$

So the Bayes factor in favor of H_0 is measuring the degree to which the data have changed the odds in favor of the hypothesis. If BF_{H_0} is small, then the data are providing evidence against H_0 and evidence in favor of H_0 when BF_{H_0} is large.

There is a relationship between the posterior probability of H_0 being true and BF_{H_0}. From (7.2.8), we obtain

$$\Pi(\psi(\theta) = \psi_0 \mid s) = \frac{rBF_{H_0}}{1 + rBF_{H_0}} \qquad (7.2.9)$$

where

$$r = \frac{\Pi\left(\psi\left(\theta\right) = \psi_0\right)}{1 - \Pi\left(\psi\left(\theta\right) = \psi_0\right)}$$

is the prior odds in favor of H_0. So, when BF_{H_0} is small, then $\Pi\left(\psi\left(\theta\right) = \psi_0 \mid s\right)$ is small and conversely.

One reason for using Bayes factors to assess hypotheses is the following result. This establishes a connection with likelihood ratios.

Theorem 7.2.1 If the prior Π is a mixture $\Pi = p\Pi_1 + (1-p)\Pi_2$ where $\Pi_1\left(A\right) = 1, \Pi_2\left(A^C\right) = 1$ and we want to assess the hypothesis $H_0 : \theta \in A$, then

$$BF_{H_0} = \frac{m_1(s)}{m_2(s)}$$

where m_i is the prior predictive of the data under Π_i.

Proof: Recall that, if a prior concentrates all of its probability on a set, then the posterior concentrates all of its probability on this set, too. Then using (7.2.6) we have

$$BF_{H_0} = \frac{\Pi\left(A \mid s\right)}{1 - \Pi\left(A \mid s\right)} \Big/ \frac{\Pi\left(A\right)}{1 - \Pi\left(A\right)} = \frac{pm_1(s)}{(1-p)m_2(s)} \Big/ \frac{p}{1-p} = \frac{m_1(s)}{m_2(s)}. \blacksquare$$

Interestingly, Theorem 7.2.1 indicates that the Bayes factor is independent of p. We note, however, that it is not immediately clear how to interpret the value of BF_{H_0}. In particular, how large does BF_{H_0} have to be to provide strong evidence in favor of H_0? One approach to this problem is to use (7.2.9) as this gives the posterior probability of H_0 which is directly interpretable and so we can calibrate the Bayes factor. Note, however, that this requires the specification of p.

Example 7.2.13 *Location Normal Model (Example 7.2.12 continued)*
We now compute the prior predictive under Π_2. We have that the joint density of $(x_1, \ldots, x_n)$ given μ equals

$$\left(2\pi\sigma_0^2\right)^{-n/2} \exp\left(-\frac{n-1}{2\sigma_0^2}s^2\right) \exp\left(-\frac{n}{2\sigma_0^2}\left(\bar{x} - \mu\right)^2\right)$$

and so

$$m_2(x_1, \ldots, x_n)$$
$$= \int_{-\infty}^{\infty} \left\{ \begin{array}{c} \left(2\pi\sigma_0^2\right)^{-n/2} \exp\left(-\frac{n-1}{2\sigma_0^2}s^2\right) \exp\left(-\frac{n}{2\sigma_0^2}\left(\bar{x} - \mu\right)^2\right) \\ \times \left(2\pi\tau_0^2\right)^{-1/2} \exp\left(-\frac{1}{2\tau_0^2}\left(\mu - \mu_0\right)^2\right) \end{array} \right\} d\mu$$
$$= \left(2\pi\sigma_0^2\right)^{-n/2} \exp\left(-\frac{n-1}{2\sigma_0^2}s^2\right)$$
$$\times \tau_0^{-1}\left(2\pi\right)^{-1/2} \int_{-\infty}^{\infty} \exp\left(-\frac{n}{2\sigma_0^2}\left(\bar{x} - \mu\right)^2\right) \exp\left(-\frac{1}{2\tau_0^2}\left(\mu - \mu_0\right)^2\right) d\mu.$$

Then using (7.1.2), we have

$$\tau_0^{-1} (2\pi)^{-1/2} \int_{-\infty}^{\infty} \exp\left(-\frac{n}{2\sigma_0^2}(\bar{x} - \mu)^2\right) \exp\left(-\frac{1}{2\tau_0^2}(\mu - \mu_0)^2\right) d\mu$$

$$= \tau_0^{-1} \exp\left(\frac{1}{2}\left(\frac{1}{\tau_0^2} + \frac{n}{\sigma_0^2}\right)^{-1}\left(\frac{\mu_0}{\tau_0^2} + \frac{n}{\sigma_0^2}\bar{x}\right)^2\right)$$

$$\times \exp\left(-\frac{1}{2}\left(\frac{\mu_0^2}{\tau_0^2} + \frac{n\bar{x}^2}{\sigma_0^2}\right)\right)\left(\frac{n}{\sigma_0^2} + \frac{1}{\tau_0^2}\right)^{-1/2}. \tag{7.2.10}$$

Therefore

$$m_2(x_1, \ldots, x_n)$$

$$= \left(2\pi\sigma_0^2\right)^{-n/2} \exp\left(-\frac{n-1}{2\sigma_0^2}s^2\right) \tau_0^{-1} \exp\left(\frac{1}{2}\left(\frac{1}{\tau_0^2} + \frac{n}{\sigma_0^2}\right)^{-1}\left(\frac{\mu_0}{\tau_0^2} + \frac{n}{\sigma_0^2}\bar{x}\right)^2\right)$$

$$\times \exp\left(-\frac{1}{2}\left(\frac{\mu_0^2}{\tau_0^2} + \frac{n\bar{x}^2}{\sigma_0^2}\right)\right)\left(\frac{n}{\sigma_0^2} + \frac{1}{\tau_0^2}\right)^{-1/2}.$$

Because Π_1 is degenerate at μ_0, it is immediate that the prior predictive under Π_1 is given by

$$m_1(x_1, \ldots, x_n) = \left(2\pi\sigma_0^2\right)^{-n/2} \exp\left(-\frac{n-1}{2\sigma_0^2}s^2\right) \exp\left(-\frac{n}{2\sigma_0^2}(\bar{x} - \mu_0)^2\right).$$

Therefore, BF_{H_0} equals

$$\exp\left(-\frac{n}{2\sigma_0^2}(\bar{x} - \mu_0)^2\right)$$

divided by (7.2.10).

For example, suppose that $\mu_0 = 0, \tau_0^2 = 2, \sigma_0^2 = 1, n = 10$, and $\bar{x} = 0.2$. Then

$$\exp\left(-\frac{n}{2\sigma_0^2}(\bar{x} - \mu_0)^2\right) = \exp\left(-\frac{10}{2}(0.2)^2\right) = 0.81873,$$

while (7.2.10) equals

$$\frac{1}{\sqrt{2}} \exp\left(\frac{1}{2}\left(\frac{1}{2} + 10\right)^{-1}(10(0.2))^2\right) \exp\left(-\frac{10(0.2)^2}{2}\right)\left(10 + \frac{1}{2}\right)^{-1/2}$$

$$= 0.21615.$$

So

$$BF_{H_0} = \frac{0.81873}{0.21615} = 3.787\,8,$$

which gives some evidence in favor of $H_0 : \mu = \mu_0$. If we suppose that $p = 1/2$, so we are completely indifferent between H_0 being true and not being true, then $r = 1$, and (7.2.9) gives

$$\Pi\left(\mu = 0 \mid x_1, \ldots, x_n\right) = \frac{3.787\,8}{1 + 3.787\,8} = 0.79114,$$

indicating a large degree of support for H_0. ∎

7.2.4 Prediction

Prediction problems arise when we have an unobserved response value t in a sample space T and observed response $s \in S$. Further, we have the statistical model $\{P_\theta : \theta \in \Omega\}$ for s and the conditional statistical model $\{Q_\theta(\cdot \,|\, s) : \theta \in \Omega\}$ for t given s. We assume that both models have the same true value of $\theta \in \Omega$. The objective is to construct a *prediction* $\tilde{t}(s) \in T$, of the unobserved value t, based on the observed data s.

If we denote the conditional density or probability function (whichever is relevant) of t by $q_\theta(\cdot \,|\, s)$, the joint distribution of (θ, s, t) is given by

$$q_\theta(t \,|\, s)\, f_\theta(s)\, \pi(\theta).$$

Then, once we have observed s (we'll assume here that the distributions of θ and t are absolutely continuous, and if not, we replace integrals by sums), the conditional density of (t, θ), given s, is

$$\frac{q_\theta(t \,|\, s)\, f_\theta(s)\, \pi(\theta)}{\int_\Omega \int_T q_\theta(t \,|\, s)\, f_\theta(s)\, \pi(\theta)\, dt\, d\theta} = \frac{q_\theta(t \,|\, s)\, f_\theta(s)\, \pi(\theta)}{\int_\Omega f_\theta(s)\, \pi(\theta)\, d\theta} = \frac{q_\theta(t \,|\, s)\, f_\theta(s)\, \pi(\theta)}{m(s)}.$$

Then the marginal posterior distribution of t, known as the *posterior predictive* of t, is given by

$$q(t \,|\, s) = \int_\Omega \frac{q_\theta(t \,|\, s)\, f_\theta(s)\, \pi(\theta)}{m(s)}\, d\theta = \int_\Omega q_\theta(t \,|\, s)\, \pi(\theta \,|\, s)\, d\theta.$$

Notice that the posterior predictive of t is obtained by averaging the conditional density of t, given (θ, s), with respect to the posterior distribution of θ.

Now that we have obtained the posterior predictive distribution of t, we can use it to select an estimate of the unobserved value. Again, we could choose the posterior mode $\hat{t}$ or the posterior expectation $E(t \,|\, x) = \int_T t\, q(t \,|\, s)\, dt$ as our prediction, whichever is deemed most relevant.

Example 7.2.14 *Bernoulli Model*
Suppose we want to predict the next independent outcome X_{n+1}, having observed a sample $(x_1, \ldots, x_n)$ from the Bernoulli(θ) and $\theta \sim$ Beta(α, β). Here the future observation is independent of the observed data. The posterior predictive probability function of X_{n+1} at t is then given by

$q(t \,|\, x_1, \ldots, x_n)$

$$= \int_0^1 \theta^t (1-\theta)^{1-t}\, \frac{\Gamma(n+\alpha+\beta)}{\Gamma(n\bar{x}+\alpha)\, \Gamma(n(1-\bar{x})+\beta)}\, \theta^{n\bar{x}+\alpha-1} (1-\theta)^{n(1-\bar{x})+\beta-1}\, d\theta$$

$$= \frac{\Gamma(n+\alpha+\beta)}{\Gamma(n\bar{x}+\alpha)\, \Gamma(n(1-\bar{x})+\beta)} \int_0^1 \theta^{n\bar{x}+\alpha+t-1} (1-\theta)^{n(1-\bar{x})+\beta+(1-t)-1}\, d\theta$$

$$= \frac{\Gamma(n+\alpha+\beta)}{\Gamma(n\bar{x}+\alpha)\, \Gamma(n(1-\bar{x})+\beta)}\, \frac{\Gamma(n\bar{x}+\alpha+t)\, \Gamma(n(1-\bar{x})+\beta+1-t)}{\Gamma(n+\alpha+\beta+1)}$$

$$= \begin{cases} \frac{n\bar{x}+\alpha}{n+\alpha+\beta} & t=1 \\ \frac{n(1-\bar{x})+\beta}{n+\alpha+\beta} & t=0, \end{cases}$$

which is the probability function of a Bernoulli$((n\bar{x} + \alpha) / (n + \alpha + \beta))$ distribution.

Using the posterior mode as the predictor, i.e., maximizing $q(t \mid x_1, \ldots, x_n)$ for t, leads to the prediction

$$\hat{t} = \begin{cases} 1 & \max\left\{\frac{n\bar{x}+\alpha}{n+\alpha+\beta}, \frac{n(1-\bar{x})+\beta}{n+\alpha+\beta}\right\} = \frac{n\bar{x}+\alpha}{n+\alpha+\beta} \\ 0 & \max\left\{\frac{n\bar{x}+\alpha}{n+\alpha+\beta}, \frac{n(1-\bar{x})+\beta}{n+\alpha+\beta}\right\} = \frac{n(1-\bar{x})+\beta}{n+\alpha+\beta}. \end{cases}$$

The posterior expectation predictor is given by

$$E(t \mid x_1, \ldots, x_n) = \frac{n\bar{x} + \alpha}{n + \alpha + \beta}.$$

Note that the posterior mode takes a value in $\{0, 1\}$, and the future X_{n+1} will be in this set, too. The posterior mean can be any value in $[0, 1]$. ∎

Example 7.2.15 *Location Normal Model*
Suppose that $(x_1, \ldots, x_n)$ is a sample from an $N(\mu, \sigma_0^2)$ distribution where $\mu \in R^1$ is unknown and σ_0^2 is known, and we use the prior given in Example 7.1.2. Suppose we want to predict a future observation X_{n+1}, but this time X_{n+1} is from the

$$N\left(\bar{x}, \left(\frac{1}{\tau_0^2} + \frac{n}{\sigma_0^2}\right)^{-1} \sigma_0^2\right) \tag{7.2.11}$$

distribution. So the future observation is not independent of the observed data, in this case, but it is independent of the parameter. A simple calculation (Exercise 7.2.9) shows that (7.2.11) is the posterior predictive distribution of t and so we would predict t by $\bar{x}$, as this is both the posterior mode and mean. ∎

We can also construct a γ-*prediction region* $C(s)$ for a future value t from the model $\{q_\theta(\cdot \mid s) : \theta \in \Omega\}$. A γ-prediction region for t satisfies $Q(C(s) \mid s) \geq \gamma$, where $Q(\cdot \mid s)$ is the posterior predictive measure for t. One approach to constructing $C(s)$ is to apply the HPD concept to $q(t \mid s)$. We illustrate this via several examples.

Example 7.2.16 *Bernoulli Model (Example 7.2.14 continued)*
Suppose that we want a γ-prediction region for a future value X_{n+1}. In Example 7.2.14 we derived the posterior predictive of X_{n+1} to be a

$$\text{Bernoulli}\left(\frac{n\bar{x} + \alpha}{n + \alpha + \beta}\right)$$

distribution. Accordingly, a γ-prediction region for t, derived via the HPD concept, is given by

$$C(x_1, \ldots, x_n) = \begin{cases} \{0, 1\} & \max\left\{\frac{n\bar{x}+\alpha}{n+\alpha+\beta}, \frac{n(1-\bar{x})+\beta}{n+\alpha+\beta}\right\} \leq \gamma \\ \{1\} & \gamma \leq \max\left\{\frac{n\bar{x}+\alpha}{n+\alpha+\beta}, \frac{n(1-\bar{x})+\beta}{n+\alpha+\beta}\right\} = \frac{n\bar{x}+\alpha}{n+\alpha+\beta} \\ \{0\} & \gamma \leq \max\left\{\frac{n\bar{x}+\alpha}{n+\alpha+\beta}, \frac{n(1-\bar{x})+\beta}{n+\alpha+\beta}\right\} = \frac{n(1-\bar{x})+\beta}{n+\alpha+\beta}. \end{cases}$$

We see that this predictive region contains just the mode or encompasses all possible values for X_{n+1}. In the latter case, this is not an informative inference. ∎

Example 7.2.17 *Location Normal (Example 7.2.15 continued)*
Suppose we want a γ-prediction interval for a future observation X_{n+1} from a

$$N\left(\bar{x}, \left(\frac{1}{\tau_0^2} + \frac{n}{\sigma_0^2}\right)^{-1}\sigma_0^2\right)$$

distribution. As this is also the posterior predictive distribution of X_{n+1} and is symmetric about $\bar{x}$, a γ-prediction interval for X_{n+1}, derived via the HPD concept, is given by

$$\bar{x} \pm \left(\frac{1}{\tau_0^2} + \frac{n}{\sigma_0^2}\right)^{-1/2}\sigma_0\, z_{(1+\gamma)/2}. \quad ∎$$

Summary of Section 7.2

- Based on the posterior distribution of a parameter, we can obtain estimates of the parameter (posterior modes or means), construct credible intervals for the parameter (HPD intervals), and assess hypotheses about the parameter (posterior probability of the hypothesis, Bayesian P-values, Bayes factors).

- A new type of inference was discussed in this section, namely, prediction problems where we are concerned with predicting a value from a model.

Exercises

7.2.1 For the model discussed in Example 7.1.1, derive the posterior mean of $\psi = \theta^m$ where $m > 0$.

7.2.2 For the model discussed in Example 7.1.2, determine the posterior distribution of the third quartile $\psi = \mu + \sigma_0 z_{0.75}$. Determine the posterior mode and the posterior expectation of ψ.

7.2.3 In Example 7.2.1, determine the posterior expectation and mode of $1/\sigma^2$.

7.2.4 In Example 7.2.1, determine the posterior expectation and mode of σ^2. (Hint: You will need the posterior density of σ^2 to determine the mode.)

7.2.5 Carry out the calculations to verify the posterior mode and posterior expectation of θ_1 in Example 7.2.4.

7.2.6 Establish that the variance of the θ in Example 7.2.2 is as given in Example 7.2.6. Prove that this goes to 0 as $n \to \infty$.

7.2.7 Establish that the variance of θ_1 in Example 7.2.4 is as given in Example 7.2.6. Prove that this goes to 0 as $n \to \infty$.

7.2.8 In Example 7.2.14, which of the two predictors derived there do you find more sensible? Why?

7.2.9 In Example 7.2.15, prove that the posterior predictive distribution for X_{n+1} is as stated. (Hint: Write the posterior predictive distribution density as an expectation.)

7.2.10 Suppose that $(x_1, \ldots, x_n)$ is a sample from the Exponential(λ) distribution where $\lambda > 0$ is unknown and $\lambda \sim$ Gamma(α_0, β_0). Determine the mode of posterior distribution of λ. Also determine the posterior expectation and posterior variance of λ.

7.2.11 Suppose that $(x_1, \ldots, x_n)$ is a sample from the Exponential(λ) distribution where $\lambda > 0$ is unknown and $\lambda \sim$ Gamma(α_0, β_0). Determine the mode of posterior distribution of a future independent observation X_{n+1}. Also determine the posterior expectation of X_{n+1} and posterior variance of X_{n+1}. (Hint: Problems 3.2.12 and 3.3.16.)

7.2.12 Suppose that in a population of students in a course with a large enrollment, the mark, out of 100, on a final exam is approximately distributed $N(\mu, 9)$. The instructor places the prior $\mu \sim N(65, 1)$ on the unknown parameter. A sample of 10 marks is obtained as given below.

46	68	34	86	75	56	77	73	53	64

(a) Determine the posterior mode and a 0.95-credible interval for μ. What does this interval tell you about the accuracy of the estimate?
(b) Use the 0.95-credible interval for μ to test the hypothesis $H_0 : \mu = 65$.
(c) Suppose we assign prior probability 0.5 to $\mu = 65$. Using the mixture prior $\Pi = 0.5\Pi_1 + 0.5\Pi_2$, where Π_1 is degenerate at $\mu = 65$ and Π_2 is the $N(65, 1)$ distribution, compute the posterior probability of the null hypothesis.
(d) Compute the Bayes factor in favor of $H_0 : \mu = 65$ when using the mixture prior.

7.2.13 A manufacturer believes that a machine produces rods with lengths in centimeters distributed $N(\mu_0, \sigma^2)$ where μ_0 is known and $\sigma^2 > 0$ is unknown and that the prior distribution $1/\sigma^2 \sim$ Gamma(α_0, β_0) is appropriate.
(a) Determine the posterior distribution of σ^2 based on a sample $(x_1, \ldots, x_n)$.
(b) Determine the posterior mean of σ^2.
(c) Indicate how you would assess the hypothesis $H_0 : \sigma^2 \leq \sigma_0^2$.

Problems

7.2.14 Suppose that $(x_1, \ldots, x_n)$ is a sample from the Uniform$(0, \theta)$ distribution where $\theta > 0$ is unknown, and we have $\theta \sim$ Gamma(α_0, β_0). Determine the mode of the posterior distribution of θ. (Hint: The posterior is not differentiable at $\theta = x_{(n)}$.)

7.2.15 Suppose that $(x_1, \ldots, x_n)$ is a sample from the Uniform$(0, \theta)$ distribution where $\theta \in (0, 1)$ is unknown, and we have $\theta \sim$ Uniform$(0, 1)$. Determine the form of the γ-credible interval for θ based upon the HPD concept.

7.2.16 In Example 7.2.1, write out in full the integral given in (7.2.2).

7.2.17 (MV) In Example 7.2.1, write out in full the integral that you would need to evaluate if you wanted to compute the posterior density of the third quartile of the population distribution, i.e., $\psi = \mu + \sigma z_{0.75}$.

7.2.18 Consider the location normal model discussed in Example 7.1.2 and the population coefficient of variation $\psi = \sigma_0/\mu$.
(a) Show that the posterior expectation of ψ does not exist. (Hint: Show that we can write the posterior expectation as

$$\int_{-\infty}^{\infty} \frac{\sigma_0}{a + bz} \frac{1}{\sqrt{2\pi}} e^{-z^2/2} \, dz$$

where $b > 0$, and show that this integral does not exist by considering the behavior of the integrand at $z = -a/b$.)
(a) Determine the posterior density of ψ.
(b) Show that you can determine the posterior mode of ψ by evaluating the posterior density at two specific points. (Hint: Proceed by maximizing the logarithm of the posterior density using the methods of calculus.)

7.2.19 (MV) Suppose that $(\theta_1, \ldots, \theta_{k-1}) \sim \text{Dirichlet}(\alpha_1, \alpha_2, \ldots, \alpha_k)$.
(a) Prove that $(\theta_1, \ldots, \theta_{k-2}) \sim \text{Dirichlet}(\alpha_1, \alpha_2, \ldots, \alpha_{k-1} + \alpha_k)$. (Hint: In the integral to integrate out θ_{k-1}, make the transformation $\theta_{k-1} \to \theta_{k-1}/(1 - \theta_1 - \cdots - \theta_{k-2})$.)
(b) Prove that $\theta_1 \sim \text{Beta}(\alpha_1, \alpha_2 + \cdots + \alpha_k)$. (Hint: Use part (a).)
(c) Suppose that $(i_1, \ldots, i_k)$ is a permutation of $(1, \ldots, k)$. Prove that $(\theta_{i_1}, \ldots, \theta_{i_{k-1}}) \sim \text{Dirichlet}(\alpha_{i_1}, \alpha_{i_2}, \ldots, \alpha_{i_k})$. (Hint: What is the Jacobian of this transformation?)
(d) Prove that $\theta_i \sim \text{Beta}(\alpha_i, \alpha_{-i})$. (Hint: Use parts (b) and (c).)

7.2.20 (MV) In Example 7.2.4, show that the plug-in MLE of θ_1 is given by x_1/n, i.e., find the the MLE of $(\theta_1, \ldots, \theta_k)$ and determine the first coordinate. (Hint: Show there is a unique solution to the score equations and then use the facts that the log-likelihood is bounded above and goes to $-\infty$ whenever $\theta_i \to 0$.)

7.2.21 Compare the results obtained in Exercises 7.2.3 and 7.2.4. What do you conclude about the invariance properties of these estimation procedures? (Hint: Consider Theorem 6.2.1.)

7.2.22 In Example 7.2.5, establish that the posterior variance of μ is as stated in Example 7.2.6. (Hint: Problem 4.6.16.)

7.2.23 In a prediction problem, as described in Section 7.2.4, derive the form of the prior predictive density for t when the joint density of (θ, s, t) is

$$q_\theta(t \mid s) f_\theta(s) \pi(\theta)$$

(assume s and θ are real-valued).

7.2.24 In Example 7.2.15, derive the posterior predictive probability function of (X_{n+1}, X_{n+2}), having observed $x_1, \ldots, x_n$ when $X_1, \ldots, X_n, X_{n+1}, X_{n+2}$ are independently and identically distributed (i.i.d.) Bernoulli(θ).

7.2.25 In Example 7.2.16, derive the posterior predictive distribution for X_{n+1}, having observed $x_1, \ldots, x_n$ when $X_1, \ldots, X_n, X_{n+1}$ are i.i.d. $N(\mu, \sigma_0^2)$. (Hint: We can write $X_{n+1} = \mu + \sigma_0 Z$ where $Z \sim N(0,1)$ is independent of the posterior distribution of μ.)

7.2.26 For the context of Example 7.2.1, prove that the posterior predictive distribution of an additional future observation X_{n+1} from the population distribution has the same distribution as

$$\mu_x + \sqrt{\frac{2\beta_x \left((n + 1/\tau_0^2)^{-1} + 1 \right)}{(2\alpha_0 + n)}} \, T$$

where $T \sim t(2\alpha_0 + n)$. (Hint: Note that we can write $X_{n+1} = \mu + \sigma U$ where $U \sim N(0,1)$ independent of $X_1, \ldots, X_n, \mu, \sigma$ and then reason as in Example 7.2.1.)

7.2.27 In Example 7.2.1, determine the form of an exact γ-prediction interval for an additional future observation X_{n+1} from the population distribution, based on the HPD concept. (Hint: Use Problem 7.2.26.)

7.2.28 Suppose that Π_1 and Π_2 are discrete probability distributions on the parameter space Ω. Prove that when the prior Π is a mixture $\Pi = p\Pi_1 + (1-p)\Pi_2$, then the prior predictive for the data s is given by $m(s) = pm_1(s) + (1-p)m_2(s)$, and the posterior probability measure is given by (7.2.6).

7.2.29 (MV) Suppose that $\theta = (\theta_1, \theta_2) \in R^2$ and $h(\theta_1, \theta_2) = (\psi(\theta), \lambda(\theta)) \in R^2$. Assume that h satisfies the necessary conditions and establish (7.2.1). (Hint: Theorem 2.9.2.)

Challenge

7.2.30 Another way to assess the null hypothesis $H_0 : \psi(\theta) = \psi_0$ is to compute the P-value

$$\Pi \left(\frac{\omega(\psi(\theta) \mid s)}{\omega(\psi(\theta))} \le \frac{\omega(\psi_0 \mid s)}{\omega(\psi_0)} \,\Big|\, s \right) \tag{7.2.12}$$

where ω is the marginal prior density or probability function of ψ. The quantity $\omega(\psi_0 \mid s)/\omega(\psi_0)$ is a measure of how the data s have changed our *a priori* belief that ψ_0 is the true value of ψ. When (7.2.12) is small, ψ_0 is a surprising value for ψ, as this indicates that the data have increased our belief more for other values of ψ. Prove that (7.2.12) is invariant under 1–1 continuously differentiable transformations of ψ.

7.3   Bayesian Computations

In virtually all the examples in this chapter so far, we have been able to work out the exact form of the posterior distributions and carry out a number of important computations using these. It often occurs, however, that we cannot derive any convenient form for the posterior distribution. Further, even when we can derive the posterior distribution, there can arise computations that cannot be carried out exactly — e.g., recall the discussion in Example 7.2.1 that led to the integral (7.2.2). These calculations involve evaluating complicated sums or integrals. Therefore, when we go to apply Bayesian inference in a practical example, we need to have available methods for approximating these quantities.

The subject of approximating integrals is an extensive topic that we cannot deal with fully here.[1] We will, however, introduce several approximation methods that arise very naturally in Bayesian inference problems.

7.3.1   Asymptotic Normality of the Posterior

In many circumstances, it turns out that the posterior distribution of $\theta \in R^1$ is approximately normally distributed. We can then use this to compute approximate credible regions for the true value of θ, carry out hypothesis assessment, etc. One such result says that, under conditions that we will not describe here, when $(x_1, \ldots, x_n)$ is a sample from f_θ then

$$\Pi\left(\frac{\theta - \hat{\theta}(x_1, \ldots, x_n)}{\hat{\sigma}(x_1, \ldots, x_n)/\sqrt{n}} \leq z \,|\, x_1, \ldots, x_n \right) \to \Phi(z)$$

as $n \to \infty$ where $\hat{\theta}(x_1, \ldots, x_n)$ is the posterior mode, and

$$\hat{\sigma}^2(x_1, \ldots, x_n) = \left(-\left. \frac{\partial^2 \ln L(\theta \,|\, x_1, \ldots, x_n)\,\pi(\theta)}{\partial \theta^2} \right|_{\theta=\hat{\theta}} \right)^{-1}.$$

Note that this result is similar to Theorem 6.5.3 for the MLE. When θ is k-dimensional, there is a similar but more complicated result.

7.3.2   Sampling from the Posterior

Typically, there are many things we want to compute as part of implementing a Bayesian analysis. Many of these can be written as expectations with respect to the posterior distribution of θ. For example, we might want to compute the posterior probability content of a subset $A \subset \Omega$, namely,

$$\Pi(A \,|\, s) = E(I_A(\theta) \,|\, s).$$

---

[1] See, for example, *Approximating Integrals via Monte Carlo and Deterministic Methods* by M. Evans and T. Swartz (Oxford University Press, Oxford, 2000).

More generally, we want to be able to compute the posterior expectation of some arbitrary function $w(\theta)$, namely

$$E(w(\theta) \,|\, s). \tag{7.3.1}$$

It would certainly be convenient if we could compute all of these quantities exactly, but quite often we cannot. In fact, it is not really necessary that we evaluate (7.3.1) exactly. This is because we naturally expect any inference we make about the true value of the parameter to be subject to sampling error (different data sets of the same size lead to different inferences). Carrying out our computations to a much higher degree of precision than what the sampling error contributes, is not necessary. For example, if the sampling error allows us to know the value of a parameter to within only ± 0.1 units, then there is no point in computing an estimate to many more digits of accuracy.

In light of this, many of the computational problems associated with implementing Bayesian inference are effectively solved if we can sample from the posterior for θ. For when this is possible, we simply generate an i.i.d. sequence $\theta_1, \theta_2, \ldots, \theta_N$ from the posterior distribution of θ and estimate (7.3.1) by

$$\bar{w} = \frac{1}{N} \sum_{i=1}^{N} w(\theta_i).$$

We know then, from the strong law of large numbers (Theorem 4.3.2), that $\bar{w} \overset{a.s.}{\to} E(w(\theta) \,|\, x)$ as $N \to \infty$.

Of course, for any given N, the value of $\bar{w}$ only approximates (7.3.1); we would like to know that we have chosen N large enough so that the approximation is appropriately accurate. When $E(w^2(\theta) \,|\, s) < \infty$, then the central limit theorem (Theorem 4.4.3) tells us that

$$\frac{\bar{w} - E(w(\theta) \,|\, s)}{\sigma_w / \sqrt{N}} \overset{D}{\to} N(0,1)$$

as $N \to \infty$, where $\sigma_w^2 = \text{Var}(w(\theta) \,|\, s)$. In general, we do not know the value of σ_w^2, but we can estimate it by

$$s_w^2 = \frac{1}{N-1} \sum_{i=1}^{N} (w(\theta_i) - \bar{w})^2$$

when $w(\theta)$ is a quantitative variable, and by $s_w^2 = \bar{w}(1 - \bar{w})$ when $w = I_A$ for $A \subset \Omega$. As shown in Section 4.4.2, in either case, s_w^2 is a consistent estimate of σ_w^2. Then, by Corollary 4.4.4, we have that

$$\frac{\bar{w} - E(w(\theta) \,|\, s)}{s_w / \sqrt{N}} \overset{D}{\to} N(0,1)$$

as $N \to \infty$.

From this result we know that

$$\bar{w} \pm 3 \frac{s_w}{\sqrt{N}}$$

is an approximate 100% confidence interval for $E(w(\theta) \,|\, s)$, so we can look at $3s_w/\sqrt{N}$ to determine whether or not N is large enough for the accuracy required.

One caution concerning this approach to assessing error is that $3s_w/\sqrt{N}$ is itself subject to error, as s_w is an estimate of σ_w, and so this could be misleading. A common recommendation then is to monitor the value of $3s_w/\sqrt{N}$ for successively larger values of N and only stop the sampling when it is clear that the value of $3s_w/\sqrt{N}$ is small enough for the accuracy desired and appears to be declining appropriately. Even this approach, however, will not give a guaranteed bound on the accuracy of the computations, so it is necessary to be cautious.

It is also important to remember that application of these results requires that $\sigma_w^2 < \infty$. For a bounded w this is always true, as any bounded random variable always has a finite variance. For an unbounded w, however, this must be checked and sometimes this is very difficult to do.

We consider now an example where it is possible to exactly sample from the posterior.

Example 7.3.1 *Location-Scale Normal*
Suppose that $(x_1, \ldots, x_n)$ is a sample from an $N(\mu, \sigma^2)$ distribution where $\mu \in R^1$ and $\sigma \geq 0$ are unknown, and we use the prior given in Example 7.1.4. The posterior distribution for (μ, σ^2) developed there is

$$\mu \,|\, \sigma^2, x_1, \ldots, x_n \sim N\left(\mu_x, \left(n + \frac{1}{\tau_0^2}\right)^{-1} \sigma^2\right) \tag{7.3.2}$$

and

$$\frac{1}{\sigma^2} \,|\, x_1, \ldots, x_n \sim \text{Gamma}\left(\alpha_0 + n/2, \beta_x\right) \tag{7.3.3}$$

where μ_x is given by (7.1.7) and β_x is given by (7.1.8).

Most statistical packages have built-in generators for gamma distributions and for the normal distribution. Accordingly, it is very easy to generate a sample $\left(\mu_1, \sigma_1^2\right), \ldots, \left(\mu_N, \sigma_N^2\right)$ from this posterior. We simply generate a value for $1/\sigma_i^2$ from the specified gamma distribution; then, given this value, we generate the value of μ_i from the specified normal distribution.

Suppose, then, that we want to derive the posterior distribution of the coefficient of variation $\psi = \sigma/\mu$. To do this we generate N values from the joint posterior of (μ, σ^2), using (7.3.2) and (7.3.3), and compute ψ for each of these. We then know immediately that $\psi_1, \ldots, \psi_N$ is a sample from the posterior distribution of ψ.

As a specific numerical example, suppose that we observed the following sample $(x_1, \ldots, x_{15})$.

11.6714	1.8957	2.1228	2.1286	1.0751
8.1631	1.8236	4.0362	6.8513	7.6461
1.9020	7.4899	4.9233	8.3223	7.9486

Here $\bar{x} = 5.2$ and $s = 3.3$. Suppose further that the prior is specified by $\mu_0 = 4, \tau_0^2 = 2, \alpha_0 = 2$, and $\beta_0 = 1$.

From (7.1.7) we have

$$\mu_x = \left(15 + \frac{1}{2}\right)^{-1} \left(\frac{4}{2} + 15 \cdot 5.2\right) = 5.161,$$

and from (7.1.8),

$$\beta_x = 1 + \frac{15}{2}(5.2)^2 + \frac{4^2}{2 \cdot 2} + \frac{14}{2}(3.3)^2 - \frac{1}{2}\left(15 + \frac{1}{2}\right)^{-1}\left(\frac{4}{2} + 15 \cdot 5.2\right)^2$$
$$= 77.578.$$

Therefore, we generate

$$\frac{1}{\sigma^2} \mid x_1, \ldots, x_n \sim \text{Gamma}(9.5, 77.578),$$

followed by

$$\mu \mid \sigma^2, x_1, \ldots, x_n \sim N(5.161, (15.5)^{-1}\sigma^2).$$

See Appendix B for some code that can be used to generate from this joint distribution.

In Figure 7.3.1 we have plotted a sample of $N = 200$ values of (μ, σ^2) from this joint posterior. In Figure 7.3.2 we have plotted a density histogram of the 200 values of ψ that arise from this sample.

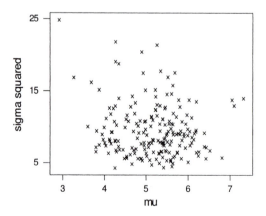

Figure 7.3.1: A sample of 200 values of (μ, σ^2) from the joint posterior in Example 7.3.1 when $n = 15, \bar{x} = 5.2, s = 3.3, \mu_0 = 4, \tau_0^2 = 2, \alpha_0 = 2$, and $\beta_0 = 1$.

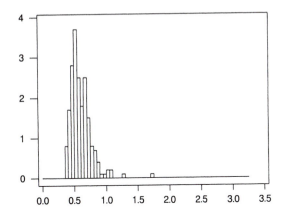

Figure 7.3.2: A density histogram of 200 values from the posterior distribution of ψ in Example 7.3.1.

A sample of 200 is not very large, so we next generated a sample of $N = 10^3$ values from the posterior distribution of ψ. A density histogram of these values is provided in Figure 7.3.3. In Figure 7.3.4 we have provided a density histogram based on a sample of $N = 10^4$ values. We can see from this that at $N = 10^3$ the basic shape of the distribution has been obtained, although the right tail is not being very accurately estimated. Things look better in the right tail for $N = 10^4$, but note there are still some extreme values quite disconnected from the main mass of values. As is characteristic of most distributions, we will need very large values of N to accurately estimate the tails. In any case, we have learned that this distribution is skewed to the right with a long right tail.

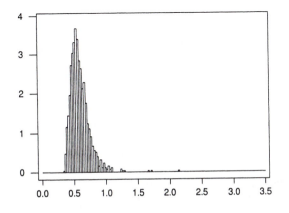

Figure 7.3.3: A density histogram of 1000 values from the posterior distribution of ψ in Example 7.3.1.

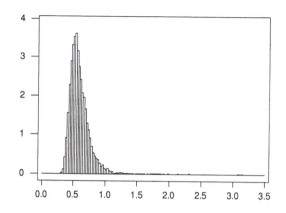

Figure 7.3.4: A density histogram of $N = 10^4$ values from the posterior distribution of ψ in Example 7.3.1.

Suppose that we want to estimate

$$\Pi\left(\psi \leq 0.5 \,|\, x_1, \ldots, x_n\right) = E\left(I_{(-\infty, 0.5)}\left(\psi\right) \,|\, x_1, \ldots, x_n\right).$$

Now $w = I_{(-\infty, 0.5)}$ is bounded so its posterior variance exists. In the following table, we have recorded the estimates for each N together with the standard error based on each of the generated samples. We have included some code for computing these estimates and their standard errors in Appendix B. Based on the results from $N = 10^4$, it would appear that this posterior probability is in the interval $0.289 \pm 3\,(0.0045) = [0.2755, 0.3025]$.

N	Estimate of $\Pi\left(\psi \leq .5 \,|\, x_1, \ldots, x_n\right)$	Standard Error
200	0.265	0.0312
10^3	0.271	0.0141
10^4	0.289	0.0045

This example also demonstrates an important point. It would be very easy for us to calculate the sample mean of the values of ψ generated from its posterior distribution and then consider this as an estimate of the posterior mean of ψ. But Problem 7.2.18 suggests (see Problem 7.3.10) that this mean will not exist. Accordingly, a Monte Carlo estimate of this quantity doesn't make any sense! So we must always check first that any expectation we want to estimate exists, before we proceed with some estimation procedure. ∎

When we cannot sample directly from the posterior, then the methods of the following section are needed.

7.3.3 Sampling from the Posterior Using Gibbs Sampling (Advanced)

Sampling from the posterior, as described in Section 7.3.2, is very effective, when it can be implemented. Unfortunately, it is often difficult or even impossible to do this directly, as we did in Example 7.3.1. There are, however, a number of algorithms that allow us to approximately sample from the posterior. One of these, known as *Gibbs sampling*, is applicable in many statistical contexts.

To describe this algorithm, suppose that we want to generate samples from the joint distribution of $(Y_1, \ldots, Y_k) \in R^k$. Further suppose that we can generate from each of the full conditional distributions $Y_i \mid Y_{-i} = y_{-i}$ where

$$Y_{-i} = (Y_1, \ldots, Y_{i-1}, Y_{i+1}, \ldots, Y_k),$$

namely, we can generate from the conditional distribution of Y_i given the values of all the other coordinates. The Gibbs sampler then proceeds iteratively as follows.

1. Specify an initial value $(y_{1(0)}, \ldots, y_{k(0)})$ for $(Y_1, \ldots, Y_k)$.

2. For $N > 0$, generate $Y_{i(N)}$ from its conditional distribution given $(y_{1(N)}, \ldots, y_{i-1(N)}, y_{i+1(N-1)}, \ldots, y_{k(N-1)})$ for each $i = 1, \ldots, k$.

For example, if $k = 3$, we first specify $(y_{1(0)}, y_{2(0)}, y_{3(0)})$. Then we generate

$$
\begin{aligned}
Y_{1(1)} \mid Y_{2(0)} &= y_{2(0)}, & Y_{3(0)} &= y_{3(0)} \\
Y_{2(1)} \mid Y_{1(1)} &= y_{1(1)}, & Y_{3(0)} &= y_{3(0)} \\
Y_{3(1)} \mid Y_{1(1)} &= y_{1(1)}, & Y_{2(1)} &= y_{2(1)}
\end{aligned}
$$

to obtain $(Y_{1(1)}, Y_{2(1)}, Y_{3(1)})$. Next we generate

$$
\begin{aligned}
Y_{1(2)} \mid Y_{2(1)} &= y_{2(1)}, & Y_{3(1)} &= y_{3(1)} \\
Y_{2(2)} \mid Y_{1(2)} &= y_{1(2)}, & Y_{3(1)} &= y_{3(1)} \\
Y_{3(2)} \mid Y_{1(2)} &= y_{1(2)}, & Y_{2(2)} &= y_{2(2)}
\end{aligned}
$$

to obtain $(Y_{1(2)}, Y_{2(2)}, Y_{3(2)})$, etc. Note that we actually did not need to specify $Y_{1(0)}$, as it is never used.

It can then be shown (see Section 11.3) that, in fairly general circumstances, $(Y_{1(N)}, \ldots, Y_{k(N)})$ converges in distribution to the joint distribution of $(Y_1, \ldots, Y_k)$ as $N \to \infty$. So for large N, we have that the distribution of $(Y_{1(N)}, \ldots, Y_{k(N)})$ is approximately the same as the joint distribution of $(Y_1, \ldots, Y_k)$ that we want to sample from. So Gibbs sampling provides an approximate method for sampling from a distribution of interest.

Further, and this is the result that is most relevant for simulations, it can be shown that, under conditions,

$$\bar{w} = \frac{1}{N} \sum_{i=1}^{N} w\left(Y_{1(i)}, \ldots, Y_{k(i)}\right)$$

converges almost surely to $E(w(Y_1, \ldots, Y_k))$. Estimation of the variance of $\bar{w}$ is different than in the i.i.d. case, where we used the sample variance, because now the $w(Y_{1(i)}, \ldots, Y_{k(i)})$ terms are not independent.

There are several approaches to estimating the variance of $\bar{w}$, but perhaps the most commonly used is the technique of *batching*. For this we divide the sequence

$$w(Y_{1(0)}, \ldots, Y_{k(0)}), \ldots, w(Y_{1(N)}, \ldots, Y_{k(N)})$$

into N/m nonoverlapping sequential batches of size m (assuming here that N is divisible by m), calculate the mean in each batch obtaining $\bar{w}_1, \ldots, \bar{w}_{N/m}$, and then estimate the variance of $\bar{w}$ by

$$\frac{s_b^2}{N/m} \tag{7.3.4}$$

where s_b^2 is the sample variance obtained from the batch means, i.e.,

$$s_b^2 = \frac{1}{N/m - 1} \sum_{i=1}^{N/m} (\bar{w}_i - \bar{w})^2.$$

It can be shown that $(Y_{1(i)}, \ldots, Y_{k(i)})$ and $(Y_{1(i+m)}, \ldots, Y_{k(i+m)})$ are approximately independent for m large enough. Accordingly, we choose the batch size m large enough so that the batch means are approximately independent, but not so large as to leave very few degrees of freedom for the estimation of the variance. Under ideal conditions, $\bar{w}_1, \ldots, \bar{w}_{N/m}$ is an i.i.d. sequence with sample mean

$$\bar{w} = \frac{1}{N/m} \sum_{i=1}^{N/m} \bar{w}_i$$

and, as usual, we estimate the variance of $\bar{w}$ by (7.3.4).

Sometimes even Gibbs sampling cannot be directly implemented because we cannot obtain algorithms to generate from all the full conditionals. There are a variety of techniques for dealing with this, but in many statistical applications the technique of *latent variables* often works. For this we search for some random variables, say $(V_1, \ldots, V_l)$, where each Y_i is a function of $(V_1, \ldots, V_l)$ and such that we can apply Gibbs sampling to the joint distribution of $(V_1, \ldots, V_l)$. We illustrate Gibbs sampling via latent variables in the following example.

Example 7.3.2 *Location-Scale Student*

Suppose now that $(x_1, \ldots, x_n)$ is a sample from a distribution that is of the form $X = \mu + \sigma Z$ where $Z \sim t(\lambda)$ (see Section 4.6.2 and Problem 4.6.14). If $\lambda > 2$, then μ is the mean and $\sigma (\lambda/\lambda - 2)^{1/2}$ is the standard deviation of the distribution (Problem 4.6.16). Note that $\lambda = \infty$ corresponds to normal variation, while $\lambda = 1$ corresponds to Cauchy variation.

We will fix λ at some specified value to reflect the fact that we are interested in modeling situations in which the variable under consideration has a distribution with longer tails than the normal distribution. Typically, this manifests itself in a histogram of the data with a roughly symmetric shape but exhibiting a few extreme values out in the tails, so a $t(\lambda)$ distribution might be appropriate.

Suppose we place the prior on (μ, σ^2) given by $\mu \,|\, \sigma^2 \sim N(\mu_0, \tau_0^2 \sigma^2)$ and $1/\sigma^2 \sim \text{Gamma}(\alpha_0, \beta_0)$. The likelihood function is given by

$$\left(\frac{1}{\sigma^2}\right)^{n/2} \prod_{i=1}^{n} \left[1 + \frac{1}{\lambda}\left(\frac{x_i - \mu}{\sigma}\right)^2\right]^{-(\lambda+1)/2}, \tag{7.3.5}$$

and so the posterior density of $(\mu, 1/\sigma^2)$ is proportional to

$$\left(\frac{1}{\sigma^2}\right)^{n/2} \prod_{i=1}^{n} \left[1 + \frac{1}{\lambda}\left(\frac{x_i - \mu}{\sigma}\right)^2\right]^{-(\lambda+1)/2}$$
$$\times \left(\frac{1}{\sigma^2}\right)^{1/2} \exp\left(-\frac{1}{2\tau_0^2 \sigma^2}(\mu - \mu_0)^2\right) \left(\frac{1}{\sigma^2}\right)^{\alpha_0 - 1} \exp\left(-\frac{\beta_0}{\sigma^2}\right).$$

This distribution is not immediately recognizable, and it is not at all clear how to generate from it.

It is natural, then, to see if we can implement Gibbs sampling. To do this directly, we need an algorithm to generate from the posterior of μ given the value of σ^2, and an algorithm to generate from the posterior of σ^2 given μ. Unfortunately, neither of these conditional distributions is amenable to the techniques discussed in Section 2.10, so we cannot implement Gibbs sampling directly.

Recall, however, that when $V \sim \chi^2(\lambda) = \text{Gamma}(\lambda/2, 1/2)$ (Problem 4.6.13) independent of $Y \sim N(\mu, \sigma^2)$, then (Problem 4.6.14)

$$Z = \frac{Y - \mu}{\sigma\sqrt{V/\lambda}} \sim t(\lambda).$$

Therefore, writing

$$X = \mu + \sigma Z = \mu + \sigma\frac{Y - \mu}{\sigma\sqrt{V/\lambda}} = \mu + \frac{Y - \mu}{\sqrt{V/\lambda}},$$

we have that $X \,|\, V = v \sim N(\mu, \sigma^2\lambda/v)$.

We now introduce the n *latent* or *hidden variables* $(V_1, \ldots, V_n)$, which are i.i.d. $\chi^2(\lambda)$ and suppose $X_i \,|\, V_i = v_i \sim N(\mu, \sigma^2\lambda/v_i)$. The V_i are considered latent because they are not really part of the problem formulation but have been added here for convenience (as we shall see). Then, noting that there is a factor $v_i^{1/2}$ associated with the density of $X_i \,|\, V_i = v_i$, the joint density of the values $(X_1, V_1), \ldots, (X_n, V_n)$ is proportional to

$$\left(\frac{1}{\sigma^2}\right)^{n/2} \prod_{i=1}^{n} \exp\left(-\frac{v_i}{2\sigma^2\lambda}(x_i - \mu)^2\right) v_i^{(\lambda/2)-(1/2)} \exp\left(-\frac{v_i}{2}\right).$$

From the above argument, the marginal joint density of $(X_1, \ldots, X_n)$ (after integrating out the v_i's) is proportional to (7.3.3), namely a sample of n from the distribution specified by $X = \mu + \sigma Z$ where $Z \sim t(\lambda)$. With the same prior structure as before, we have that the joint density of

$$\left(X_1, V_1\right), \ldots, \left(X_n, V_n\right), \mu, 1/\sigma^2$$

is proportional to

$$\left(\frac{1}{\sigma^2}\right)^{n/2} \prod_{i=1}^{n} \exp\left(-\frac{v_i}{2\sigma^2 \lambda}\left(x_i - \mu\right)^2\right) v_i^{(\lambda/2)-(1/2)} \exp\left(-\frac{v_i}{2}\right)$$

$$\times \left(\frac{1}{\sigma^2}\right)^{1/2} \exp\left(-\frac{1}{2\tau_0^2 \sigma^2}\left(\mu - \mu_0\right)^2\right) \left(\frac{1}{\sigma^2}\right)^{\alpha_0 - 1} \exp\left(-\frac{\beta_0}{\sigma^2}\right). \qquad (7.3.6)$$

In (7.3.6) treat $x_1, \ldots, x_n$ as constants (we observed these values) and consider the conditional distributions of each of the variables $V_1, \ldots, V_n, \mu, 1/\sigma^2$ given all the other variables. From (7.3.6) we have that the full conditional density of μ is proportional to

$$\exp\left\{-\frac{1}{2\sigma^2}\left(\sum_{i=1}^{n} \frac{v_i}{\lambda}\left(x_i - \mu\right)^2 + \frac{1}{\tau_0^2}\left(\mu - \mu_0\right)^2\right)\right\},$$

which is proportional to

$$\exp\left\{-\frac{1}{2\sigma^2}\left[\left(\sum_{i=1}^{n} \frac{v_i}{\lambda}\right) + \frac{1}{\tau_0^2}\right]\mu^2 + \frac{2}{2\sigma^2}\left[\left(\sum_{i=1}^{n} \frac{v_i}{\lambda} x_i\right) + \frac{\mu_0}{\tau_0^2}\right]\mu\right\}.$$

From this we immediately deduce that

$$\mu \,|\, x_1, \ldots, x_n, v_1, \ldots, v_n, \sigma^2$$

$$\sim N\left(r(v_1, \ldots, v_n)\left[\left(\sum_{i=1}^{n} \frac{v_i}{\lambda} x_i\right) + \frac{\mu_0}{\tau_0^2}\right], r(v_1, \ldots, v_n)\sigma^2\right).$$

where

$$r(v_1, \ldots, v_n) = \left[\left(\sum_{i=1}^{n} \frac{v_i}{\lambda}\right) + \frac{1}{\tau_0^2}\right]^{-1}.$$

From (7.3.6) we have that the conditional density of $1/\sigma^2$ is proportional to

$$\left(\frac{1}{\sigma^2}\right)^{(n/2)+\alpha_0-(1/2)} \exp\left\{-\left(\begin{array}{c}\sum_{i=1}^{n} \frac{v_i}{\lambda}\left(x_i - \mu\right)^2 \\ +\frac{1}{\tau_0^2}\left(\mu - \mu_0\right)^2 + 2\beta_0\end{array}\right)\frac{1}{2\sigma^2}\right\},$$

and we immediately deduce that

$$\frac{1}{\sigma^2} \,|\, x_1, \ldots, x_n, v_1, \ldots, v_n, \mu$$

$$\sim \text{Gamma}\left(\frac{n}{2} + \alpha_0 + \frac{1}{2}, \frac{1}{2}\left(\sum_{i=1}^{n} \frac{v_i}{\lambda}\left(x_i - \mu\right)^2 + \frac{1}{\tau_0^2}\left(\mu - \mu_0\right)^2 + 2\beta_0\right)\right).$$

Finally the conditional density of V_i is proportional to

$$v_i^{(\lambda/2)-(1/2)} \exp\left\{-\left[\frac{(x_i-\mu)^2}{2\sigma^2\lambda}+\frac{1}{2}\right]v_i\right\},$$

and it is immediate that

$$V_i \,|\, x_1,\dots,x_n,\, v_1,\dots,v_{i-1},v_{i+1},\dots v_n, \mu, \sigma^2$$

$$\sim \text{Gamma}\left(\frac{\lambda}{2}+\frac{1}{2},\frac{1}{2}\left(\frac{(x_i-\mu)^2}{\sigma^2\lambda}+1\right)\right).$$

We can now easily generate from all these distributions and implement a Gibbs sampling algorithm. As we are not interested in the values of $V_1,\dots,V_n$, we simply discard these as we iterate.

Let us now consider a specific computation using the same data and prior as in Example 7.3.1. The analysis of Example 7.3.1 assumed that the data were coming from a normal distribution, but now we are going to assume that the data are a sample from a $\mu + \sigma t(3)$ distribution, i.e., $\lambda = 3$. We again consider approximating the posterior distribution of the coefficient of variation $\psi = \sigma/\mu$.

We carry out the Gibbs sampling iteration in the order $v_1,\dots,v_n,\mu,1/\sigma^2$. This implies that we only need starting values for μ and σ^2 (the full conditionals of the v_i do not depend on the other v_j). We take the starting value of μ to be $\bar{x} = 5.2$ and the starting value of σ to be $s = 3.3$. For each generated value of (μ,σ^2), we calculate ψ to obtain the sequence $\psi_1,\psi_2,\dots,\psi_N$.

The values $\psi_1,\psi_2,\dots,\psi_N$ are not i.i.d. from the posterior of ψ. The best we can say is that

$$\psi_m \xrightarrow{D} \psi \sim \omega\left(\cdot \,|\, x_1,\dots,x_n\right)$$

as $m \to \infty$, where $\omega\left(\cdot \,|\, x_1,\dots,x_n\right)$ is the posterior density of ψ. Also values sufficiently far apart in the sequence, will be like i.i.d. values from $\omega\left(\cdot \,|\, x_1,\dots,x_n\right)$. Therefore, one approach is to determine an appropriate value m and then extract $\psi_m,\psi_{2m},\psi_{3m},\dots$ as an approximate i.i.d. sequence from the posterior. Often it is difficult to determine an appropriate value for m, however.

In any case, it is known that, under fairly weak conditions,

$$\bar{w} = \frac{1}{N}\sum_{i=1}^{N} w\left(\psi_i\right) \xrightarrow{a.s.} E\left(w\left(\psi\right) \,|\, x_1,\dots,x_n\right)$$

as $N \to \infty$. So we can use the whole sequence $\psi_1,\psi_2,\dots,\psi_N$ and record a density histogram for ψ, just as we did in Example 7.3.1. The value of the density histogram between two cutpoints will converge almost surely to the correct value as $N \to \infty$. We will have to take N larger when using the Gibbs sampling algorithm, however, than with i.i.d. sampling, to achieve the same accuracy. For many examples, the effect of the deviation of the sequence from being i.i.d. is very small, so N will not have to be much larger. We always need to be cautious, however, and the general recommendation is to compute estimates for successively higher values of N, only stopping when the results seem to have stabilized.

In Figure 7.3.5 we have plotted the density histogram of the ψ values that resulted from 10^4 iterations of the Gibbs sampler. In this case, plotting the density histogram of ψ based upon $N = 5 \times 10^4$ and $N = 8 \times 10^4$ resulted in only minor deviations from this plot. Note that this density looks very similar to that plotted in Example 7.3.1, but it is not quite so peaked and it has a shorter right tail.

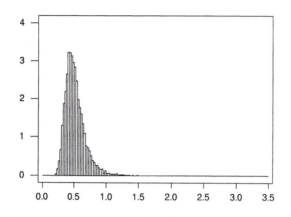

Figure 7.3.5: A density histogram of $N = 10^4$ values of ψ generated sequentially via Gibbs sampling in Example 7.3.2.

We can also estimate $\Pi\left(\psi \leq 0.5 \mid x_1, \ldots, x_n\right)$, just as we did in Example 7.3.1, by recording the proportion of values in the sequence that are smaller than 0.5, i.e., $w\left(\psi\right) = I_A\left(\psi\right)$ where $A = \{\theta : \psi \leq 0.5\}$. In this case we obtained the estimate 0.5441, which is quite different from the value obtained in Example 7.3.1. So using a $t\left(3\right)$ distribution to describe the variation in the response has made a big difference in the results.

Of course, we must also quantify how accurate we believe our estimate is. Using a batch size of $m = 10$, we obtained the standard error of the estimate 0.5441 to be 0.00639. When we took the batch size to be $m = 20$ the standard error of the mean is 0.00659; with a batch size of $m = 40$, the standard error of the mean is 0.00668. So we feel quite confident that we are assessing the error in the estimate appropriately. Again, under conditions, we have that $\bar{w}$ is asymptotically normal so that in this case we can assert that the interval $0.5441 \pm 3(0.0066) = [0.5243, 0.5639]$ contains the true value of $\Pi\left(\psi \leq 0.5 \mid x_1, \ldots, x_n\right)$ with virtual certainty.

See Appendix B for some code that was used to implement the Gibbs sampling algorithm described here. ∎

It is fair to say that the introduction of Gibbs sampling has resulted in a revolution in statistical applications due to the wide variety of previously intractable problems that it successfully handles. There are a number of modifications and closely related algorithms. We refer the interested reader to Chapter 11, where the general theory of what is called Markov chain Monte Carlo (MCMC) is discussed.

Summary of Section 7.3

- Implementation of Bayesian inference often requires the evaluation of complicated integrals or sums.

- If, however, we can sample from the posterior of the parameter, this will often lead to sufficiently accurate approximations to these integrals or sums via Monte Carlo.

- It is often difficult to sample exactly from a posterior distribution of interest. In such circumstances, Gibbs sampling can prove to be an effective method for generating an approximate sample from this distribution.

Exercise

7.3.1 Suppose that we have the following sample from an $N(\mu, 2)$ distribution where μ is unknown.

2.6	4.2	3.1	5.2	3.7	3.8	5.6	1.8	5.3	4.0
3.0	4.0	4.1	3.2	2.2	3.4	4.5	2.9	4.7	5.2

If the prior on μ is Uniform$(2, 6)$, determine an approximate 0.95-credible interval for μ based upon the large sample results described in Section 7.3.1.

Computer Exercises

7.3.2 In the context of Example 7.3.1, construct a density histogram of the posterior distribution of $\psi = \mu + \sigma z_{0.25}$, i.e., the population first quartile, using $N = 5 \times 10^3$ and $N = 10^4$, and compare the results. Estimate the posterior mean of this distribution and assess the error in your approximation. (Hint: Modify the program in Appendix B.)

7.3.3 Suppose that a manufacturer takes a random sample of manufactured items and tests each item as to whether it is defective or not. The responses are felt to be i.i.d. Bernoulli(θ) where θ is the probability that the item is defective. The manufacturer places a Beta$(0.5, 10)$ distribution on θ. If a sample of $n = 100$ items is taken and 5 defectives are observed, then, using a Monte Carlo sample with $N = 1000$, estimate the posterior probability that $\theta < 0.1$ and assess the error in your estimate.

7.3.4 Suppose that lifelengths (in years) of a manufactured item are known to follow an Exponential(λ) distribution where $\lambda > 0$ is unknown and for the prior we take $\lambda \sim$ Gamma$(10, 2)$. Suppose that the lifelengths 4.3, 6.2, 8.4, 3.1, 6.0, 5.5, and 7.8 were observed.

(a) Using a Monte Carlo sample of size $N = 10^3$, approximate the posterior probability that $\lambda \in [3, 6]$ and assess the error of your estimate.

(b) Using a Monte Carlo sample of size $N = 10^3$, approximate the posterior probability function of $\lfloor 1/\lambda \rfloor$ ($\lfloor x \rfloor$ equals the greatest integer less than or equal to x).

(c) Using a Monte Carlo sample of size $N = 10^3$, approximate the posterior expectation of $\lfloor 1/\lambda \rfloor$ and assess the error in your approximation.

7.3.5 Generate a sample of $n = 10$ from a Pareto(2) distribution. Now pretend you only know that you have a sample from a Pareto(α) distribution where $\alpha > 0$ is unknown and place a Gamma(2, 1) prior on α. Using a Monte Carlo sample of size $N = 10^4$, approximate the posterior expectation of $1/(\alpha + 1)$ based on the observed sample, and assess the accuracy of your approximation by quoting an interval that contains the exact value with virtual certainty. (Hint: Problem 2.10.8.)

Problems

7.3.6 Suppose $X_1, \ldots, X_n$ is a sample from the model $\{f_\theta : \theta \in \Omega\}$ and all the regularity conditions of Section 6.5 apply. Assume that the prior $\pi(\theta)$ is a continuous function of θ and that the posterior mode $\hat{\theta}(X_1, \ldots, X_n) \overset{a.s.}{\to} \theta$ when $X_1, \ldots, X_n$ is a sample from f_θ (the latter assumption holds under very general conditions).

(a) Using the fact that, if $Y_n \overset{a.s.}{\to} Y$ and g is a continuous function then $g(Y_n) \overset{a.s.}{\to} g(Y)$, prove that

$$-\frac{1}{n} \left.\frac{\partial^2 \ln L(\theta \,|\, x_1, \ldots, x_n) \pi(\theta)}{\partial \theta^2}\right|_{\theta=\hat{\theta}} \overset{a.s.}{\to} I(\theta)$$

when $X_1, \ldots, X_n$ is a sample from f_θ.

(b) Explain to what extent the large sample approximate methods of Section 7.3.1 depend upon the prior if the assumptions just described apply.

7.3.7 In Exercise 7.3.5, explain why the interval you constructed to contain the posterior mean of $1/(\alpha + 1)$ with virtual certainty may or may not contain the true value of $1/(\alpha + 1)$.

7.3.8 Suppose that (X, Y) is distributed Bivariate Normal$(\mu_1, \mu_2, \sigma_1, \sigma_2, \rho)$. Determine a Gibbs sampling algorithm to generate from this distribution. Assume that you have an algorithm for generating from univariate normal distributions. Is this the best way to sample from this distribution? (Hint: Problem 2.8.23.)

7.3.9 Suppose that the joint density of (X, Y) is given by $f_{X,Y}(x, y) = 8xy$ for $0 < x < y < 1$. Fully describe a Gibbs sampling algorithm for this distribution. In particular, indicate how you would generate all random variables. Can you design an algorithm to generate exactly from this distribution?

7.3.10 In Example 7.3.1, prove that the posterior mean of $\psi = \sigma/\mu$ does not exist. (Hint: Use Problem 7.2.18 and the theorem of total expectation to split the integral into two parts, where one part has value ∞ and the other part has value $-\infty$.)

Computer Problem

7.3.11 In the context of Example 7.3.2, construct a density histogram of the posterior distribution of $\psi = \mu + \sigma z_{0.25}$, i.e., the population first quartile, using $N = 10^4$. Estimate the posterior mean of this distribution and assess the error in your approximation.

7.4 Choosing Priors

The issue of selecting a prior for a problem is an important one. Of course, the idea is that we choose a prior to reflect our *a priori* beliefs about the true value of θ. Because this will typically vary from statistician to statistician, this is often criticized as being too subjective for scientific studies. It should be remembered, however, that the sampling model $\{f_\theta : \theta \in \Omega\}$ is also a subjective choice by the statistician. These choices are guided by the statistician's judgement. What then justifies one choice of a statistical model or prior over another?

In effect, when statisticians choose a prior and a model, they are prescribing a joint distribution for (θ, s). The only way to assess whether or not a good choice was made is to check whether the observed s is reasonable given this choice. If s is surprising, when compared to the distribution prescribed by the model and prior, then we have evidence against the statistician's choices. Methods designed to assess this are called model-checking procedures and will be discussed in Chapter 9. At this point, however, we should recognize the subjectivity that enters into statistical analyses, but take some comfort that we have a methodology for checking whether or not the choices made by the statistician make sense.

One approach to choosing the prior is based on mathematical convenience. We have the following definition.

Definition 7.4.1 The family of priors $\{\pi_\tau : \tau \in \Upsilon\}$ for the parameter θ of the model $\{f_\theta : \theta \in \Omega\}$ is *conjugate*, if for all data $s \in S$ and all $\tau \in \Upsilon$ the posterior $\pi_\tau(\cdot \mid s) \in \{\pi_\tau : \tau \in \Upsilon\}$.

Conjugacy is usually a great convenience as we start with some choice $\tau_0 \in T$ for the prior, and then we find the relevant $\tau \in T$ for the posterior. Often this does not require much computation. While it can be criticized as a mere mathematical convenience, it has to be acknowledged that many conjugate families $\{\pi_\tau : \tau \in T\}$ offer sufficient variety to allow the expression of a wide spectrum of prior beliefs.

Example 7.4.1 *Conjugate Families*
In Example 7.1.1 we have effectively shown that the family of all Beta distributions is conjugate for sampling from the Bernoulli model. In Example 7.1.2 it is shown that the family of normal priors is conjugate for sampling from the

location normal model. In Example 7.1.3 it is shown that the family of Dirichlet distributions is conjugate for Multinomial models. In Example 7.1.4 it is shown that the family of priors specified there is conjugate for sampling from the location-scale normal model. ∎

Often a statistician will consider a particular family $\{\pi_\tau : \tau \in T\}$ of priors for a problem, not necessarily conjugate, and try to select a suitable prior $\pi_{\tau_0} \in \{\pi_\tau : \tau \in T\}$. In such a context the parameter τ is called a *hyperparameter*.

There are various approaches to selecting τ_0. When the choice of τ_0 is based on the data s, these methods are referred to as *empirical Bayesian methods*. For example, one empirical Bayesian method is to compute the prior predictive $m_\tau (s)$ for the data s, and then base the choice of τ on these values. Note that the prior predictive is a likelihood function for τ (as it is the density or probability function for the observed s), and so the methods of Chapter 6 apply for inference about τ. For example, we could select the value of τ, that maximizes $m_\tau (s)$.

Another approach consists of putting yet another prior distribution ω, called a *hyperprior*, on τ. This approach is commonly called *hierarchical Bayes*. So in this situation, the posterior density of θ is equal to

$$\pi (\theta \mid s) = \int \frac{f_\theta (s) \pi_\tau (\theta) \omega (\tau)}{m(s)} \, d\tau = \int \frac{f_\theta (s) \pi_\tau (\theta)}{m_\tau (s)} \frac{m_\tau (s) \omega (\tau)}{m(s)} \, d\tau$$

where $m(s) = \int \int f_\theta (s) \pi_\tau (\theta) \omega (\tau) \, d\theta \, d\tau$ and $m_\tau (s) = \int f_\theta (s) \pi_\tau (\theta) \, d\theta$ (assuming τ is continuous with prior given by ω). Note that $m_\tau (s) \omega (\tau) / m(s)$ is the posterior density of τ and $f_\theta (s) \pi_\tau (\theta) / m_\tau (s)$ is the posterior density of θ given τ. Of course, this merely postpones the inevitable, as we are left with the question of how to choose the hyperprior ω and so on.

One attempt to stop this endless chain is to prescribe a *default prior* based on ignorance. The idea here is to give a rule such that, if a statistician has no prior beliefs about the value of a parameter, or hyperparameter, then a prior is prescribed that reflects this. In the hierarchical Bayes approach, one continues up the chain until the statistician declares their ignorance, and a default prior completes the specification.

Unfortunately, just how ignorance is to be expressed turns out to be a rather subtle issue. In many cases, the default priors turn out to be *improper*, i.e., the integral or sum of the prior over the whole parameter space equals ∞, so it is not a probability distribution. The interpretation of an improper prior is not at all clear, and the use of these is somewhat controversial.

There have been numerous difficulties associated with the use of improper priors, and perhaps this is not surprising. In particular, it is important to note that there is no reason in general for the posterior of θ to exist as a proper probability distribution when π is improper. If an improper prior is being used, then we should always check to make sure the posterior is proper, as inferences will not make sense if we are using an improper posterior.

When using an improper prior π, it is completely equivalent to instead use the prior $c\pi$ for any $c > 0$, for the posterior under π is proper if and only if the posterior under $c\pi$ is proper and then the posteriors are identical (Exercise 7.4.6).

The following example illustrates the use of an improper prior.

Example 7.4.2 *Location Normal Model with an Improper Prior*
Suppose that $(x_1, \ldots, x_n)$ is a sample from an $N(\mu, \sigma_0^2)$ distribution where $\mu \in \Omega = R^1$ is unknown and σ_0^2 is known. Many arguments for default priors in this context lead to the choice $\pi(\mu) = 1$ which is clearly improper.

Proceeding as in Example 7.1.2, namely, pretending that this π is a proper probability density, we get that the posterior density of μ is proportional to

$$\exp\left(-\frac{n}{2\sigma_0^2}(\bar{x} - \mu)^2\right).$$

This immediately implies that the posterior distribution of μ is $N(\bar{x}, \sigma_0^2/n)$. Note that this is the same as the limiting posterior obtained in Example 7.1.2 as $\tau_0 \to \infty$, although the point of view is quite different. ∎

One commonly used method of selecting a default prior is to use, when it is available, the prior given by $I^{1/2}(\theta)$ when $\theta \in R^1$ (and by $(\det I(\theta))^{1/2}$ in the multidimensional case), where I is the Fisher information for the statistical model as defined in Section 6.5. This is referred to as *Jeffreys' prior*. Note that Jeffreys' prior is dependent on the model.

Jeffreys' prior has an important invariance property. From Challenge 6.5.19, we have that, under some regularity conditions, if we make a 1–1 transformation of the real-valued parameter θ via $\psi = \Psi(\theta)$, then the Fisher information of ψ is given by

$$I\left(\Psi^{-1}(\psi)\right)\left(\left(\Psi^{-1}\right)'(\psi)\right)^2.$$

Therefore, the default Jeffreys' prior for ψ is

$$I^{1/2}\left(\Psi^{-1}(\psi)\right)\left|\left(\Psi^{-1}\right)'(\psi)\right|. \tag{7.4.1}$$

Now we see that, if we had started with the default prior $I^{1/2}(\theta)$ for θ and made the change of variable to ψ, then this prior transforms to (7.4.1) by Theorems 2.6.2 and 2.6.3. A similar result can be obtained when θ is multidimensional.

Jeffreys' prior often turns out to be improper, as the next example illustrates.

Example 7.4.3 *Location Normal (Example 7.4.2 continued)*
In this case, Jeffreys' prior is given by $\pi(\theta) = \sqrt{n}/\sigma_0$, which gives the same posterior as in Example 7.4.2. Note that Jeffreys' prior is effectively a constant and so the prior of Example 7.4.2 is equivalent to Jeffreys' prior. ∎

The methods for selecting a prior discussed so far have been methods for making choices that do not involve the statistician's subjective beliefs about the true value of θ. Other methods try to make explicit use of expert opinion when this is available. These are called methods of *prior elicitation*, and they involve the statistician asking questions of the expert in such a way that these specify a prior from some family of priors. For example, suppose we are sampling from an $N(\mu, \sigma_0^2)$ distribution with μ unknown and σ_0^2 known, and we restricted attention to priors for μ in the family $\{N(\mu_0, \tau_0^2) : \mu_0 \in R^1, \tau_0^2 > 0\}$. Then asking an expert to specify two quantiles of his or her prior distribution for μ (Exercise 7.4.10) specifies a prior in this family. So we might ask an expert to specify a number μ_0 such that the true value of μ was as likely to be greater than as less than μ_0, so that μ_0 is the median of the prior. We might also ask the expert to specify a value v_0 such that there is 99% certainty that the true value of μ is less than v_0. This of course is the .99-quantile of their prior.

Summary of Section 7.4

- To implement Bayesian inference, the statistician must choose a prior as well as the sampling model for the data.

- These choices must be checked if the inferences obtained are supposed to have practical validity. This topic is discussed in Chapter 9.

- Various techniques have been devised to allow for automatic selection of a prior. These include empirical Bayes methods, hierarchical Bayes, and the use of default priors to express ignorance.

- Default priors are often improper. We must always check that an improper prior leads to a proper posterior.

Exercises

7.4.1 Prove that the family $\{\text{Gamma}(\alpha, \beta) : \alpha > 0, \beta > 0\}$ is a conjugate family of priors with respect to sampling from the model given by Pareto(λ) distributions with $\lambda > 0$.

7.4.2 Prove that the family $\{\pi_{\alpha,\beta}(\theta) : \alpha > 1, \beta > 0\}$ of priors given by

$$\pi_{\alpha,\beta}(\theta) = \frac{\theta^{-\alpha} I_{[\beta,\infty)}(\theta)}{(\alpha - 1)\beta^{\alpha-1}}$$

is a conjugate family of priors with respect to sampling from the model given by the Uniform$(0, \theta)$ distributions with $\theta > 0$.

7.4.3 Suppose that the statistical model is given by

	$p_\theta(1)$	$p_\theta(2)$	$p_\theta(3)$	$p_\theta(3)$
$\theta = a$	1/3	1/6	1/3	1/6
$\theta = b$	1/2	1/4	1/8	1/8

and that we consider the family of priors given by

	$\pi_\tau(a)$	$\pi_\tau(b)$
$\tau = 1$	1/2	1/2
$\tau = 2$	1/3	2/3

and we observe the sample $x_1 = 1, x_2 = 1, x_2 = 3$.
(a) If we use the maximum value of the prior predictive for the data to determine the value of τ, and hence the prior, which prior is selected here?
(b) Determine the posterior of θ based on the selected prior.

7.4.4 For the situation described in Exercise 7.4.3, put a uniform prior on the hyperparameter τ and determine the posterior of θ. (Hint: Theorem of total probability.)

7.4.5 For the model for proportions described in Example 7.1.1, determine the prior predictive density. If $n = 10$ and $n\bar{x} = 7$, which of the priors given by $(\alpha, \beta) = (1, 1)$ or $(\alpha, \beta) = (5, 5)$ would the prior predictive criterion select for further inferences about θ?

7.4.6 Prove that when using an improper prior π, the posterior under π is proper if and only if the posterior under $c\pi$ is proper for $c > 0$, and then the posteriors are identical.

7.4.7 Determine Jeffreys' prior for the Bernoulli(θ) model and determine the posterior distribution of θ based upon this prior.

7.4.8 Suppose that we are sampling from a Uniform$(0, \theta), \theta > 0$ model and we want to use the improper prior $\pi(\theta) \equiv 1$.
(a) Does the posterior exist in this context?
(b) Does Jeffreys' prior exist in this context?

Computer Exercise

7.4.9 Consider the situation discussed in Exercise 7.4.5.
(a) If we observe $n = 10, n\bar{x} = 7$, and we are using a symmetric prior, i.e., $\alpha = \beta$, plot the prior predictive as a function of α in the range $(0, 20)$ (you will need a statistical package that provides evaluations of the gamma function for this). Does this graph clearly select a value for α?
(b) If we observe $n = 10, n\bar{x} = 9$, plot the prior predictive as a function of α in the range $(0, 20)$. Compare this plot with that of part (a).

Problems

7.4.10 Show that a distribution in the family $\{N(\mu_0, \tau_0^2) : \mu_0 \in R^1, \tau_0^2 > 0\}$ is completely determined once we specify two quantiles of the distribution.

7.4.11 (*Scale normal model*) Consider the family of $N(\mu_0, \sigma^2)$ distributions, where μ_0 is known and $\sigma^2 > 0$ is unknown. Determine Jeffreys' prior for this model.

7.4.12 Suppose that for the location-scale normal model described in Example 7.1.4, we use the prior formed by the Jeffreys' prior for the location model (just a constant) times the Jeffreys' prior for the scale normal model. Determine the posterior distribution of (μ, σ^2).

7.4.13 Consider the location normal model described in Example 7.1.2.

(a) Determine the prior predictive density m. (Hint: Write down the joint density of the sample and μ. Use (7.1.2) to integrate out μ and don't worry about getting m into a recognizable form.)

(b) How would you generate a value $(X_1, \ldots, X_n)$ from this distribution?

(c) Are $X_1, \ldots, X_n$ mutually independent? Justify your answer. (Hint: Write $X_i = \mu + \sigma_0 Z_i, \mu = \mu_0 + \tau_0 Z$ where $Z, Z_1, \ldots, Z_n$ are i.i.d. $N(0, 1)$.)

7.4.14 Consider Example 7.3.2, but this time use the prior $\pi(\mu, \sigma^2) = 1/\sigma^2$. Develop the Gibbs sampling algorithm for this situation. (Hint: Simply adjust each full conditional in Example 7.3.2 appropriately.)

Computer Problem

7.4.15 Using the formulation described in Problem 7.4.12 and the data in the following table

| 2.6 | 4.2 | 3.1 | 5.2 | 3.7 | 3.8 | 5.6 | 1.8 | 5.3 | 4.0 |
| 3.0 | 4.0 | 4.1 | 3.2 | 2.2 | 3.4 | 4.5 | 2.9 | 4.7 | 5.2 |

generate a sample of size $N = 10^4$ from the posterior. Plot a density histogram estimate of the posterior density of $\psi = \sigma/\mu$ based upon this sample.

Challenge

7.4.16 When $\theta = (\theta_1, \theta_2)$, the Fisher information matrix $I(\theta_1, \theta_2)$ is defined in Problem 6.5.15. The Jeffreys' prior is then defined as $(\det I(\theta_1, \theta_2))^{1/2}$. Determine Jeffreys' prior for the location-scale normal model and compare this with the prior used in Problem 7.4.12.

Discussion Topic

7.4.17 Using empirical Bayes methods to determine a prior violates the Bayesian principle that all unknowns should be assigned probability distributions. Comment on this. Is the hierarchical Bayesian approach a solution to this problem?

7.5 Further Proofs (Advanced)

Derivation of the Posterior Distribution in Example 7.1.4

The likelihood function is given by

$$L(\theta \mid x_1, \ldots, x_n) = (2\pi\sigma^2)^{-n/2} \exp\left(-\frac{n}{2\sigma^2}(\bar{x} - \mu)^2\right) \exp\left(-\frac{n-1}{2\sigma^2}s^2\right).$$

The prior on (μ, σ^2) is given by $\mu \mid \sigma^2 \sim N(\mu_0, \tau_0^2 \sigma^2)$ and $1/\sigma^2 \sim \text{Gamma}(\alpha_0, \beta_0)$ where $\mu_0, \tau_0^2, \alpha_0$ and β_0 are fixed and known.

The posterior density of (μ, σ^{-2}) is then proportional to the likelihood times the joint prior. Therefore, retaining only those parts of the likelihood and the prior that depend on μ and σ^2, the joint posterior density is proportional to

$$\left\{ \left(\frac{1}{\sigma^2}\right)^{n/2} \exp\left(-\frac{n}{2\sigma^2}(\bar{x}-\mu)^2\right) \exp\left(-\frac{n-1}{2\sigma^2}s^2\right) \right\}$$

$$\times \left\{ \left(\frac{1}{\sigma^2}\right)^{1/2} \exp\left(-\frac{1}{2\tau_0^2\sigma^2}(\mu-\mu_0)^2\right) \right\} \left\{ \left(\frac{1}{\sigma^2}\right)^{\alpha_0-1} \exp\left(-\frac{\beta_0}{\sigma^2}\right) \right\}$$

$$= \exp\left(-\frac{n}{2\sigma^2}(\bar{x}-\mu)^2 - \frac{1}{2\tau_0^2\sigma^2}(\mu-\mu_0)^2\right) \left(\frac{1}{\sigma^2}\right)^{\alpha_0+n/2-1/2}$$

$$\times \exp\left(-\left[\beta_0 + \frac{n-1}{2}s^2\right]\frac{1}{\sigma^2}\right)$$

$$= \exp\left(-\frac{1}{2\sigma^2}\left[\left(n+\frac{1}{\tau_0^2}\right)\mu^2 - 2\left(n\bar{x}+\frac{\mu_0}{\tau_0^2}\right)\mu\right]\right)$$

$$\times \left(\frac{1}{\sigma^2}\right)^{\alpha_0+n/2-1/2} \exp\left(-\left[\beta_0 + \frac{n}{2}\bar{x}^2 + \frac{1}{2\tau_0^2}\mu_0^2 + \frac{n-1}{2}s^2\right]\frac{1}{\sigma^2}\right)$$

$$= \left(\frac{1}{\sigma^2}\right)^{1/2} \exp\left(-\frac{1}{2\sigma^2}\left(n+\frac{1}{\tau_0^2}\right)\left[\mu - \left(n+\frac{1}{\tau_0^2}\right)^{-1}\left(\frac{\mu_0}{\tau_0^2}+n\bar{x}\right)\right]^2\right)$$

$$\times \left(\frac{1}{\sigma^2}\right)^{\alpha_0+n/2-1} \exp\left(-\left[\begin{array}{c}\beta_0 + \frac{n}{2}\bar{x}^2 + \frac{1}{2\tau_0^2}\mu_0^2 + \frac{n-1}{2}s^2 \\ -\frac{1}{2}\left(n+\frac{1}{\tau_0^2}\right)^{-1}\left(\frac{\mu_0}{\tau_0^2}+n\bar{x}\right)^2\end{array}\right]\frac{1}{\sigma^2}\right).$$

From this we deduce that the posterior distribution of (μ,σ^2) is given by

$$\mu\,|\,\sigma^2, x \sim N\left(\mu_x, \left(n+\frac{1}{\tau_0^2}\right)^{-1}\sigma^2\right)$$

and

$$\frac{1}{\sigma^2}\,|\,x \sim \text{Gamma}\left(\alpha_0+n/2, \beta_x\right)$$

where

$$\mu_x = \left(n+\frac{1}{\tau_0^2}\right)^{-1}\left(\frac{\mu_0}{\tau_0^2}+n\bar{x}\right)$$

and

$$\beta_x = \beta_0 + \frac{n}{2}\bar{x}^2 + \frac{\mu_0^2}{2\tau_0^2} + \frac{n-1}{2}s^2 - \frac{1}{2}\left(n+\frac{1}{\tau_0^2}\right)^{-1}\left(\frac{\mu_0}{\tau_0^2}+n\bar{x}\right)^2. \quad \blacksquare$$

Example 7.2.1 *Derivation of* $J\left(\theta\left(\psi_0, \lambda\right)\right)$
Here we have that

$$\psi = \psi\left(\mu,\sigma^{-2}\right) = \frac{\sigma}{\mu} = \frac{1}{\mu}\left(\frac{1}{\sigma^2}\right)^{-1/2}.$$

and
$$\lambda = \lambda\left(\mu, \sigma^{-2}\right) = \frac{1}{\sigma^2}.$$

We have that

$$\left|\det\begin{pmatrix} \frac{\partial \psi}{\partial \mu} & \frac{\partial \psi}{\partial \left(\frac{1}{\sigma^2}\right)} \\ \frac{\partial \lambda}{\partial \mu} & \frac{\partial \lambda}{\partial \left(\frac{1}{\sigma^2}\right)} \end{pmatrix}\right| = \left|\det\begin{pmatrix} -\mu^{-2}\left(\frac{1}{\sigma^2}\right)^{-1/2} & -\frac{1}{2}\mu^{-1}\left(\frac{1}{\sigma^2}\right)^{-3/2} \\ 0 & 1 \end{pmatrix}\right|$$

$$= \left|\det\begin{pmatrix} -\psi^2\lambda^{1/2} & -\frac{1}{2}\psi\lambda^{-1} \\ 0 & 1 \end{pmatrix}\right| = \psi^2\lambda^{1/2}.$$

and so
$$J\left(\theta\left(\psi_0, \lambda\right)\right) = \left(\psi^2\lambda^{1/2}\right)^{-1}. \quad \blacksquare$$

Chapter 8

Optimal Inferences

In Chapter 5 we introduced the basic ingredient of statistical inference — the statistical model. In Chapter 6 inference methods were developed based on the model alone via the likelihood function. In Chapter 7 we added the prior distribution on the parameter, and this led to the posterior distribution as the basis for deriving inference methods.

With both the likelihood and the posterior, however, the inferences were derived largely based on intuition. For example, when we had a characteristic of interest $\psi(\theta)$, there was nothing in the theory in Chapters 6 and 7 that forced us to choose a particular estimator, confidence or credible interval, or testing procedure. A complete theory of statistical inference, however, would totally prescribe our inferences.

One attempt to resolve this issue is to introduce a performance measure on inferences and then choose an inference that does best with respect to this measure. For example, we might choose to measure the performance of estimators by their MSE and then try to obtain an estimator that had the smallest possible MSE. This is the optimality approach to inference, and it has been quite successful in a number of problems. In this chapter we will consider several successes for the optimality approach to deriving inferences.

Sometimes the performance measure that we use can be considered to be based on what is called a *loss function*. Loss functions form the basis for yet another approach to statistical inference called *decision theory*. While it is not always the case that a performance measure is based on a loss function, this holds in some of the most important problems in statistical inference. Decision theory provides a general framework in which to discuss these problems. A brief introduction to decision theory is provided in Section 8.4 as an advanced topic.

8.1 Optimal Unbiased Estimation

Suppose we want to estimate the real-valued characteristic $\psi(\theta)$ for the model $\{f_\theta : \theta \in \Omega\}$. If we have observed the data s, an estimate is a value $T(s)$ that the statistician hopes will be close to the true value of $\psi(\theta)$. The error in the estimate is given by $|T(s) - \psi(\theta)|$. For a variety of reasons (mostly to do with mathematics) it is more convenient to consider the squared error $(T(s) - \psi(\theta))^2$.

Of course, we would like this squared error to be as small as possible. Because we do not know the true value of θ, this leads us to consider the distributions of the squared error, when s has distribution given by f_θ, for each $\theta \in \Omega$. We would then like to choose the estimator T so that these distributions are as concentrated as possible about 0. A convenient measure of the concentration of these distributions about 0 is given by their means, or

$$\mathrm{MSE}_\theta(T) = E_\theta\left((T - \psi(\theta))^2\right), \qquad (8.1.1)$$

called the *mean-squared error*.

An optimal estimator of $\psi(\theta)$ is then a T that minimizes (8.1.1) for every $\theta \in \Omega$. In other words, T would be optimal if, for any other estimator T^* defined on S, we have that

$$\mathrm{MSE}_\theta(T) \le \mathrm{MSE}_\theta(T^*)$$

for each θ. Unfortunately, it can be shown that, except in very artificial circumstances, there is no such T, so we need to modify our optimization problem.

This modification takes the form of restricting the estimators T that we will entertain as possible choices for the inference. Consider an estimator T such that $E_\theta(T)$ does not exist or is infinite. It can then be shown that (8.1.1) is infinite (Challenge 8.1.21). So we will first restrict our search to those T for which $E_\theta(T)$ is finite for every θ.

Further restrictions on the types of estimators that we consider make use of the following result.

Theorem 8.1.1 If T is such that $E\left(T^2\right)$ is finite, then

$$E\left((T - c)^2\right) = \mathrm{Var}(T) + (E(T) - c)^2,$$

and this is minimized by taking $c = E(T)$.

Proof: We have that

$$E\left((T - c)^2\right)$$
$$= E\left((T - E(T) + E(T) - c)^2\right)$$
$$= E\left((T - E(T))^2\right) + 2E(T - E(T))(E(T) - c) + (E(T) - c)^2$$
$$= \mathrm{Var}(T) + (E(T) - c)^2, \qquad (8.1.2)$$

because $E(T - E(T)) = E(T) - E(T) = 0$. As $(E(T) - c)^2 \ge 0$, and $\mathrm{Var}(T)$ does not depend on c, the value of (8.1.2) is minimized by taking $c = E(T)$. ∎

We will prove that, when we are looking for T to minimize (8.1.1), we can further restrict our attention to estimators T that depend on the data only through the value of a sufficient statistic. This simplifies our search, as sufficiency often results in a reduction of the dimension of the data (recall the discussion and examples in Section 6.1.1). First, however, we need the following property of sufficiency.

Theorem 8.1.2 A statistic U is sufficient for a model if and only if the conditional distribution of the data s given $U = u$ is the same for every $\theta \in \Omega$.

Proof: See Section 8.5 for the proof of this result. ∎

The implication of this result is that information in the data s beyond the value of $U(s) = u$ can tell us nothing about the true value of θ, because this information comes from a distribution that does not depend on the parameter. Notice that Theorem 8.1.2 is a characterization of sufficiency alternative to that provided in Section 6.1.1.

Consider a simple example that illustrates the content of Theorem 8.1.2.

Example 8.1.1
Suppose that $S = \{1, 2, 3, 4\}, \Omega = \{a, b\}$ where the two probability distributions are given by the following table.

	$s = 1$	$s = 2$	$s = 3$	$s = 4$
$\theta = a$	1/2	1/6	1/6	1/6
$\theta = b$	1/4	1/4	1/4	1/4

Then $L(\cdot \mid 2) = L(\cdot \mid 3) = L(\cdot \mid 4)$, and so $U : S \longrightarrow \{0, 1\}$ given by $U(1) = 0$ and $U(2) = U(3) = U(4) = 1$ is a sufficient statistic.

As we must have $s = 1$ when we observe $U(s) = 0$, the conditional distribution of the response s, given $U(s) = 0$, is degenerate at 1 (i.e., all the probability mass is at the point 1) for both $\theta = a$ and $\theta = b$. When $\theta = a$, the conditional distribution of the response s, given $U(s) = 1$, places 1/3 of its mass at each of the points in $\{2, 3, 4\}$ and similarly when $\theta = b$. So given $U(s) = 1$, the conditional distributions are as in the following table.

	$s = 1$	$s = 2$	$s = 3$	$s = 4$
$\theta = a$	0	1/3	1/3	1/3
$\theta = b$	0	1/3	1/3	1/3

Thus we see that indeed the conditional distributions are independent of θ. ∎

We now combine Theorems 8.1.1 and 8.1.2 to show that we can restrict our attention to estimators T that depend on the data only through the value of a sufficient statistic U. By Theorem 8.1.2 we can denote the conditional probability measure for s, given $U(s) = u$, by $P(\cdot \mid U = u)$, i.e., this probability measure does not depend on θ.

For estimator T of $\psi(\theta)$, such that $E_\theta(T)$ is finite for every θ, put $T_U(s)$ equal to the conditional expectation of T given the value of $U(s)$, namely,

$$T_U(s) = E_{P(\cdot \mid U = U(s))}(T),$$

i.e., T_U is the average value of T when we average using $P\left(\cdot\,|\,U=U(s)\right)$. Notice that $T_U\left(s_1\right) = T_U\left(s_2\right)$ whenever $U(s_1) = U(s_2)$ (because $P\left(\cdot\,|\,U=U(s_1)\right) = P\left(\cdot\,|\,U=U(s_2)\right)$), and so T_U depends on the data s only through the value of $U(s)$.

Theorem 8.1.3 (*Rao-Blackwell*) Suppose that U is a sufficient statistic and $E_\theta\left(T^2\right)$ is finite for every θ. Then $\mathrm{MSE}_\theta\left(T_U\right) \le \mathrm{MSE}_\theta\left(T\right)$ for every $\theta \in \Omega$.

Proof: Let $P_{\theta,U}$ denote the marginal probability measure of U induced by P_θ. By the theorem of total expectation (Theorem 3.5.2), we have that

$$\mathrm{MSE}_\theta\left(T\right) = E_{P_{\theta,U}}\left(E_{P\left(\cdot\,|\,U=u\right)}\left(\left(T - \psi\left(\theta\right)\right)^2\right)\right)$$

where $E_{P\left(\cdot\,|\,U=u\right)}\left(\left(T - \psi\left(\theta\right)\right)\right)^2$ denotes the conditional MSE of T given $U = u$. Now by Theorem 8.1.1,

$$E_{P\left(\cdot\,|\,U=u\right)}\left(\left(T - \psi\left(\theta\right)\right)^2\right) = \mathrm{Var}_{P\left(\cdot\,|\,U=u\right)}\left(T\right) + \left(E_{P\left(\cdot\,|\,U=u\right)}\left(T\right) - \psi\left(\theta\right)\right)^2$$
$$(8.1.3)$$

As both terms in (8.1.3) are nonnegative, and recalling the definition of T_U, we have

$$\mathrm{MSE}_\theta\left(T\right) = E_{P_{\theta,U}}\left(\mathrm{Var}_{P\left(\cdot\,|\,U=u\right)}\left(T\right)\right) + E_{P_{\theta,U}}\left(\left(T_U\left(s\right) - \psi\left(\theta\right)\right)^2\right)$$
$$\ge E_{P_{\theta,U}}\left(\left(T_U\left(s\right) - \psi\left(\theta\right)\right)^2\right).$$

Now $\left(T_U\left(s\right) - \psi\left(\theta\right)\right)^2 = E_{P\left(\cdot\,|\,U=u\right)}\left(\left(T_U\left(s\right) - \psi\left(\theta\right)\right)^2\right)$ (Theorem 3.5.4) and so, by the theorem of total expectation,

$$E_{P_{\theta,U}}\left(\left(T_U\left(s\right) - \psi\left(\theta\right)\right)^2\right) = E_{P_{\theta,U}}\left(E_{P\left(\cdot\,|\,U=u\right)}\left(\left(T_U\left(s\right) - \psi\left(\theta\right)\right)^2\right)\right)$$
$$= E_{P_\theta}\left(\left(T_U\left(s\right) - \psi\left(\theta\right)\right)^2\right) = \mathrm{MSE}_\theta\left(T_U\right)$$

and the theorem is proved. ∎

Theorem 8.1.3 shows that we can always improve on (or at least make no worse) any estimator T that possesses a finite second moment, by replacing $T(s)$ by the estimate $T_U\left(s\right)$. This process is sometimes referred to as the *Rao-Blackwellization* of an estimator.

Notice that putting $E = E_\theta$ and $c = \psi\left(\theta\right)$ in Theorem 8.1.1 implies that

$$\mathrm{MSE}_\theta\left(T\right) = \mathrm{Var}_\theta\left(T\right) + \left(E_\theta\left(T\right) - \psi\left(\theta\right)\right)^2. \qquad (8.1.4)$$

So the MSE of T can be decomposed as the sum of the variance of T plus the squared bias of T (this was also proved in Theorem 6.3.1).

Theorem 8.1.1 has another important implication, for (8.1.4) is minimized by taking $\psi\left(\theta\right) = E_\theta\left(T\right)$. This indicates that, on average, the estimator T comes closer (in terms of squared error) to $E_\theta\left(T\right)$ than to any other value. So, if we are sampling from the distribution specified by θ, $T\left(s\right)$ is a natural estimate of

$E_\theta(T)$. Therefore, for a general characteristic $\psi(\theta)$, it makes sense to restrict attention to estimators that have bias equal to 0. This leads to the following definition.

Definition 8.1.1 An estimator T of $\psi(\theta)$ is *unbiased* if $E_\theta(T) = \psi(\theta)$ for every $\theta \in \Omega$.

Notice that, for unbiased estimators with finite second moment, (8.1.4) becomes

$$\mathrm{MSE}_\theta(T) = \mathrm{Var}_\theta(T).$$

Therefore, our search for an optimal estimator has become the search for an unbiased estimator with smallest variance. If such an estimator exists, we refer to it as a *uniformly minimum variance unbiased (UMVU)* estimator for $\psi(\theta)$.

It is important to note that the Rao-Blackwell theorem (Theorem 8.1.3) also applies to unbiased estimators. This is because the Rao-Blackwellization of an unbiased estimator yields an unbiased estimator, as the following result demonstrates.

Theorem 8.1.4 (*Rao-Blackwell for unbiased estimators*) If T has finite second moment, is unbiased for $\psi(\theta)$, and U is a sufficient statistic, then $E_\theta(T_U) = \psi(\theta)$ for every $\theta \in \Omega$ (so T_U is also unbiased for $\psi(\theta)$) and $\mathrm{Var}_\theta(T_U) \leq \mathrm{Var}_\theta(T)$.

Proof: Using the theorem of total expectation (Theorem 3.5.2), we have

$$E_\theta(T_U) = E_{P_{\theta,U}}(T_U) = E_{P_{\theta,U}}\left(E_{P(\cdot\,|\,U=u))}(T)\right) = E_\theta(T) = \psi(\theta).$$

So T_U is unbiased for $\psi(\theta)$ and $\mathrm{MSE}_\theta(T) = \mathrm{Var}_\theta(T)$, $\mathrm{MSE}_\theta(T_U) = \mathrm{Var}_\theta(T_U)$. Applying Theorem 8.1.3 gives $\mathrm{Var}_\theta(T_U) \leq \mathrm{Var}_\theta(T)$. ∎

There are many situations in which the theory of unbiased estimation leads to good estimators. The following example illustrates, however, that in some problems there are no unbiased estimators and so the theory has some limitations.

Example 8.1.2 *The Nonexistence of an Unbiased Estimator*
Suppose that $(x_1, \ldots, x_n)$ is a sample from the Bernoulli(θ) and we wish to find a UMVU estimator of $\psi(\theta) = \theta/(1-\theta)$, the odds in favor of a success occurring. From Theorem 8.1.4 we can restrict our search to unbiased estimators T that are functions of the sufficient statistic $n\bar{x}$.

Such a T satisfies $E_\theta\left(T\left(n\bar{X}\right)\right) = \theta/(1-\theta)$ for every $\theta \in [0,1]$. Recalling that $n\bar{X} \sim \mathrm{Binomial}(n, \theta)$, this implies that

$$\frac{\theta}{1-\theta} = \sum_{k=0}^n T(k) \binom{n}{k} \theta^k (1-\theta)^{n-k}$$

for every $\theta \in [0,1]$. By the binomial theorem, we have

$$(1-\theta)^{n-k} = \sum_{l=0}^{n-k} \binom{n-k}{l} (-1)^l \theta^l.$$

Substituting this into the preceding expression for $\theta/(1-\theta)$ and writing this in terms of powers of θ leads to

$$\frac{\theta}{1-\theta} = \sum_{m=0}^{n} \left(\sum_{k=0}^{m} T(k) \binom{n}{k} (-1)^{m-k} \right) \theta^m. \qquad (8.1.5)$$

Now the left-hand side of (8.1.5) goes to ∞ as $\theta \to 1$, but the right-hand side is a polynomial in θ and this is bounded in $[0, 1]$. Therefore, an unbiased estimator of ψ cannot exist. ∎

If a characteristic $\psi(\theta)$ has an unbiased estimator, then it is said to be *estimable*. It should be kept in mind, however, that just because a parameter is not estimable does not mean that we can't estimate it! For example, ψ in Example 8.1.2 is a 1–1 function of θ, so the MLE of ψ is given by $\bar{x}/(1-\bar{x})$ (Theorem 6.2.1), and this seems like a perfectly satisfactory estimator, even if it is biased.

In certain circumstances, if an unbiased estimator exists, and is a function of a sufficient statistic U, then there is only one such estimator — and so must be UMVU. We need the concept of completeness to establish this.

Definition 8.1.2 A statistic U is *complete* if any function h of U, which satisfies $E_\theta(h(U)) = 0$ for every $\theta \in \Omega$, also satisfies $h(U(s)) = 0$ with probability 1 for each $\theta \in \Omega$ (i.e., $P_\theta(\{s : h(U(s)) = 0\}) = 1$ for every $\theta \in \Omega$).

In probability theory we treat two functions as equivalent if they differ only on a set having probability content 0, as the probability of the functions taking different values at an observed response value is 0. So in Definition 8.1.2 we need not distinguish between h and the constant 0. Therefore, a statistic U is complete if the only unbiased estimator of 0, based on U, is given by 0 itself.

We can now derive the following result.

Theorem 8.1.5 (*Lehmann-Scheffé*) If U is a complete sufficient statistic, and if T depends on the data only through the value of U, has finite second moment for every θ, and is unbiased for $\psi(\theta)$, then T is UMVU.

Proof: Suppose that T^* is also an unbiased estimator of $\psi(\theta)$. By Theorem 8.1.4 we can assume that T^* depends on the data only through the value of U. Then there exist functions h and h^* such that $T(s) = h(U(s))$ and $T^*(s) = h^*(U(s))$ and

$$0 = E_\theta(T) - E_\theta(T^*) = E_\theta(h(U)) - E_\theta(h^*(U)) = E_\theta(h(U) - h^*(U)).$$

By the completeness of U we have that $h(U) = h^*(U)$ with probability 1 for each $\theta \in \Omega$, which implies that $T = T^*$ with probability 1 for each $\theta \in \Omega$. This says there is essentially only one unbiased estimator for $\psi(\theta)$ based on U, and so it must be UMVU. ∎

The Rao-Blackwell theorem for unbiased estimators (Theorem 8.1.4) and the Lehmann-Scheffé theorem together provide a method for obtaining a UMVU estimator of $\psi(\theta)$. Suppose we can find an unbiased estimator T that has finite second moment. If we also have a complete sufficient statistic U, then by Theorem 8.1.4 $T_U(s) = E_{P(\cdot \,|\, U=U(s))}(T)$ is unbiased for $\psi(\theta)$ and it depends on the data only through the value of U, because $T_U(s_1) = T_U(s_2)$ whenever $U(s_1) = U(s_2)$. Therefore, by Theorem 8.1.5, T_U is UMVU for $\psi(\theta)$.

It is not necessary in a given problem that a complete sufficient statistic exist. In fact, it can be proved that the only candidate for this is a minimal sufficient statistic (recall the definition in Section 6.1.1). So in a given problem, we must obtain a minimal sufficient statistic and then determine whether or not it is complete. We illustrate this via an example.

Example 8.1.3 *Location Normal*
Suppose that $(x_1, \ldots, x_n)$ is a sample from an $N(\mu, \sigma_0^2)$ distribution where $\mu \in R^1$ is unknown and $\sigma_0^2 > 0$ is known. In Example 6.1.7 we showed that $\bar{x}$ is a minimal sufficient statistic for this model.

In fact, $\bar{x}$ is also complete for this model. The proof of this is a bit involved and is presented in Section 8.5.

Given that $\bar{x}$ is a complete, minimal sufficient statistic, this implies that $T(\bar{x})$ is a UMVU estimator of its mean $E_\mu(T(\bar{X}))$ whenever T has a finite second moment for every $\mu \in R^1$. In particular, $\bar{x}$ is the UMVU estimator of μ because $E_\mu(\bar{X}) = \mu$ and $E_\mu(\bar{X}^2) = (\sigma_0^2/n) + \mu^2 < \infty$. Further, $\bar{x} + \sigma_0 z_p$ is the UMVU estimator of $E_\mu(\bar{X} + \sigma_0 z_p) = \mu + \sigma_0 z_p$ (the pth quantile of the true distribution). ∎

The arguments needed to show the completeness of a minimal sufficient statistic in a problem are often similar to the one required in Example 8.1.3 (see Challenge 8.1.22). Rather than pursue such technicalities here, we quote some important examples in which the minimal sufficient statistic is complete.

Example 8.1.4 *Location-Scale Normal*
Suppose that $(x_1, \ldots, x_n)$ is a sample from an $N(\mu, \sigma^2)$ distribution where $\mu \in R^1$ and $\sigma > 0$ are unknown. The parameter in this model is two-dimensional and is given by $(\mu, \sigma^2) \in R^1 \times (0, \infty)$.

We showed, in Example 6.1.8, that $(\bar{x}, s^2)$ is a minimal sufficient statistic for this model. In fact, it can be shown that $(\bar{x}, s^2)$ is a complete minimal sufficient statistic. Therefore, $T(\bar{x}, s^2)$ is a UMVU estimator of $E_\theta(T(\bar{X}, S^2))$ whenever the second moment of $T(\bar{x}, s^2)$ is finite for every (μ, σ^2). In particular, $\bar{x}$ is the UMVU estimator of μ and s^2 is UMVU for σ^2. ∎

Example 8.1.5 *Distribution-Free Models*
Suppose that $(x_1, \ldots, x_n)$ is a sample from some continuous distribution on R^1. The statistical model comprises all continuous distributions on R^1.

It can be shown that the order statistics $(x_{(1)}, \ldots, x_{(n)})$ make up a complete minimal sufficient statistic for this model. Therefore, $T(x_{(1)}, \ldots, x_{(n)})$ is UMVU for

$$E_\theta(T(X_{(1)}, \ldots, X_{(n)}))$$

whenever

$$E_\theta(T^2(X_{(1)}, \ldots, X_{(n)})) < \infty \qquad (8.1.6)$$

for every continuous distribution. In particular, if $T : R^n \to R^1$ is bounded, then this is the case. For example, if

$$T\left(x_{(1)}, \ldots, x_{(n)}\right) = \frac{1}{n} \sum_{i=1}^{n} I_A\left(x_{(i)}\right),$$

the relative frequency of the event A in the sample, then $T\left(x_{(1)}, \ldots, x_{(n)}\right)$ is UMVU for $E_\theta(T\left(X_{(1)}, \ldots, X_{(n)}\right)) = P_\theta\left(A\right)$.

Now change the model assumption so that $(x_1, \ldots, x_n)$ is a sample from some continuous distribution on R^1 that possesses its first m moments. Again, it can be shown that the order statistics make up a complete minimal sufficient statistic. Therefore, $T(x_{(1)}, \ldots, x_{(n)})$ is UMVU for $E_\theta(T\left(X_{(1)}, \ldots, X_{(n)}\right))$ whenever (8.1.6) holds for every continuous distribution possessing its first m moments. For example, if $m = 2$, then this implies that $T\left(x_{(1)}, \ldots, x_{(n)}\right) = \bar{x}$ is UMVU for $E_\theta(\bar{X})$. When $m = 4$, we have that s^2 is UMVU for the population variance (Exercise 8.1.2). ∎

8.1.1 The Cramer-Rao Inequality (Advanced)

There is a fundamental inequality that holds for the variance of an estimator T. This is given by the *Cramer-Rao inequality* (sometimes called the *information inequality*). It is a corollary to the following inequality.

Theorem 8.1.6 (*Covariance inequality*) Suppose that $T, U_\theta : S \to R^1$ and $E_\theta\left(T^2\right) < \infty, 0 < E_\theta\left(U_\theta^2\right) < \infty$ for every $\theta \in \Omega$. Then

$$\text{Var}_\theta\left(T\right) \geq \frac{\left(\text{Cov}_\theta\left(T, U_\theta\right)\right)^2}{\text{Var}_\theta\left(U_\theta\right)}$$

for every $\theta \in \Omega$. Equality holds if and only if

$$T\left(s\right) = E_\theta\left(T\right) + \frac{\text{Cov}_\theta\left(T, U_\theta\right)}{\text{Var}_\theta\left(U_\theta\right)}\left(U_\theta\left(s\right) - E_\theta\left(U_\theta\left(s\right)\right)\right)$$

with probability 1 for every $\theta \in \Omega$ (i.e., if and only if $T\left(s\right)$ and $U_\theta\left(s\right)$ are linearly related).

Proof: This result follows immediately from the Cauchy–Schwartz inequality (Theorem 3.6.3). ∎

Now suppose that Ω is an open subinterval of R^1 and we take

$$U_\theta(s) = S\left(\theta \mid s\right) = \frac{\partial \ln f_\theta\left(s\right)}{\partial \theta}, \tag{8.1.7}$$

i.e., U_θ is the score function. Assume that the conditions discussed in Section 6.5 hold, so that $E_\theta\left(S\left(\theta \mid s\right)\right) = 0$ for all θ, and, Fisher's information $I\left(\theta\right) = \text{Var}_\theta\left(S\left(\theta \mid s\right)\right)$ is finite. Then using

$$\frac{\partial \ln f_\theta\left(s\right)}{\partial \theta} = \frac{\partial f_\theta\left(s\right)}{\partial \theta}\frac{1}{f_\theta\left(s\right)},$$

we have

$$
\begin{aligned}
&\text{Cov}_\theta (T, U_\theta) \\
&= E_\theta \left(T(s) \frac{\partial \ln f_\theta (s)}{\partial \theta} \right) = E_\theta \left(T(s) \frac{\partial f_\theta (s)}{\partial \theta} \frac{1}{f_\theta (s)} \right) \\
&= \sum_s \left(T(s) \frac{\partial f_\theta (s)}{\partial \theta} \frac{1}{f_\theta (s)} \right) f_\theta (s) = \frac{\partial}{\partial \theta} \sum_s T(s) f_\theta (s) = \frac{\partial E_\theta (T)}{\partial \theta}, \quad (8.1.8)
\end{aligned}
$$

in the discrete case, assuming conditions like those discussed in section 6.5 so we can pull the partial derivative through the sum. A similar argument gives the equality (8.1.8) in the continuous case as well.

The covariance inequality, applied with U_θ specified as in (8.1.7) and using (8.1.8), gives the following result.

Corollary 8.1.1 (*Cramer-Rao or information inequality*) Under conditions,

$$
\text{Var}_\theta (T) \geq \left(\frac{\partial E_\theta (T)}{\partial \theta} \right)^2 (I(\theta))^{-1}
$$

for every $\theta \in \Omega$. Equality holds if and only if

$$
T(s) = E_\theta (T) + \frac{\partial E_\theta (T)}{\partial \theta} (I(\theta))^{-1} S(\theta \mid s)
$$

with probability 1 for every $\theta \in \Omega$.

The Cramer-Rao inequality provides a fundamental lower bound on the variance of an estimator T. From (8.1.4) we know that the variance is a relevant measure of the accuracy of an estimator only when the estimator is unbiased, and so we restate Corollary 8.1.1 for this case.

Corollary 8.1.2 Under the conditions of Corollary 8.1.1, when T is an unbiased estimator of $\psi(\theta)$,

$$
\text{Var}_\theta (T) \geq \left(\psi'(\theta) \right)^2 (I(\theta))^{-1}
$$

for every $\theta \in \Omega$. Equality holds if and only if

$$
T(s) = \psi(\theta) + \psi'(\theta) (I(\theta))^{-1} S(\theta \mid s) \tag{8.1.9}
$$

with probability 1 for every $\theta \in \Omega$.

Notice that when $\psi(\theta) = \theta$, then Corollary 8.1.2 says that the variance of the unbiased estimator T is bounded below by the reciprocal of the Fisher information. More generally, when ψ is a 1–1, smooth transformation, (using Challenge 6.5.20) we have that the variance of an unbiased T is again bounded below by the reciprocal of the Fisher information, but this time the model uses the parameterization in terms of $\psi(\theta)$.

Corollary 8.1.2 has several interesting implications. First, if we obtain an unbiased estimator T with variance at the lower bound, then we know immediately that it is UMVU. Second, we know that any unbiased estimator that achieves the lower bound is of the form given in (8.1.9). Note that, the right-hand side of (8.1.9) must be independent of θ, for this to be an estimator. If this is not the case then there are no UMVU estimators whose variance achieves the lower bound. The following example demonstrates that there are cases where UMVU estimators exist, but their variance does not achieve the lower bound.

Example 8.1.6 *Poisson(λ) Model*
Suppose that $(x_1, \ldots, x_n)$ is a sample from the Poisson(λ) distribution where $\lambda > 0$ is unknown. The log-likelihood is given by $l(\lambda \,|\, x_1, \ldots, x_n) = n\bar{x} \ln \lambda - n\lambda$ and so the score function is given by $S(\lambda \,|\, x_1, \ldots, x_n) = n\bar{x}/\lambda - n$. Now

$$\frac{\partial S(\lambda \,|\, x_1, \ldots, x_n)}{\partial \lambda} = -\frac{n\bar{x}}{\lambda^2},$$

and so

$$I(\lambda) = E_\lambda\left(\frac{n\bar{x}}{\lambda^2}\right) = \frac{n}{\lambda}.$$

Suppose we are estimating λ. Then the Cramer-Rao lower bound is given by $I^{-1}(\lambda) = \lambda/n$. Noting that $\bar{x}$ is unbiased for λ and that $\text{Var}_\lambda(\bar{X}) = \lambda/n$, we see immediately that $\bar{x}$ is UMVU and achieves the lower bound.

Now suppose that we are estimating $\psi(\lambda) = e^{-\lambda} = P_\lambda(\{0\})$. The Cramer-Rao lower bound equals $\lambda e^{-2\lambda}/n$ and

$$
\begin{aligned}
\psi(\lambda) + \psi'(\lambda) I^{-1}(\lambda) S(\lambda \,|\, x_1, \ldots, x_n) &= e^{-\lambda} - e^{-\lambda}\left(\frac{\lambda}{n}\right)\left(\frac{n\bar{x}}{\lambda} - n\right) \\
&= e^{-\lambda}(1 - \bar{x} + \lambda),
\end{aligned}
$$

and this is clearly not independent of λ. So there does not exist a UMVU estimator for ψ that attains the lower bound.

Does there exist a UMVU estimator for ψ? Observe that when $n = 1$, then $I_{\{0\}}(x_1)$ is an unbiased estimator of ψ. As it turns out, $\bar{x}$ is (for every n) a complete minimal sufficient statistic for this model, so by the Lehmann-Scheffé theorem $I_{\{0\}}(x_1)$ is UMVU for ψ. Further $I_{\{0\}}(X_1)$ has variance

$$P_\lambda(X_1 = 0)(1 - P_\lambda(X_1 = 0)) = e^{-\lambda}\left(1 - e^{-\lambda}\right)$$

since $I_{\{0\}}(X_1) \sim \text{Bernoulli}(e^{-\lambda})$. This implies that $e^{-\lambda}(1 - e^{-\lambda}) > \lambda e^{-2\lambda}$.

In general we have that

$$\frac{1}{n}\sum_{i=1}^{n} I_{\{0\}}(x_i)$$

is an unbiased estimator of ψ but it is not a function of $\bar{x}$, so we cannot apply the Lehmann-Scheffé theorem. But we can Rao-Blackwellize this estimator. Therefore, the UMVU estimator of ψ is given by

$$\frac{1}{n}\sum_{i=1}^{n} E\left(I_{\{0\}}(X_i) \,|\, \bar{X} = \bar{x}\right).$$

To determine this estimator in closed form, we reason as follows. The conditional probability function of $(X_1, \ldots, X_n)$ given $\bar{X} = \bar{x}$, because $n\bar{X}$ is distributed Poisson$(n\lambda)$, is

$$
\left\{ \frac{\lambda^{x_1}}{x_1!} \cdots \frac{\lambda^{x_n}}{x_n!} e^{-n\lambda} \right\} \left\{ \frac{(n\lambda)^{n\bar{x}}}{(n\bar{x})!} e^{-n\lambda} \right\}^{-1} = \binom{n\bar{x}}{x_1 \ldots x_n} \left(\frac{1}{n} \right)^{x_1} \cdots \left(\frac{1}{n} \right)^{x_n},
$$

i.e., $(X_1, \ldots, X_n)$ given $\bar{X} = \bar{x}$ is distributed Multinomial$(n\bar{x}, 1/n, \ldots, 1/n)$. Accordingly, the UMVU estimator is given by

$$
E\left(I_{\{0\}}(X_1) \mid \bar{X} = \bar{x} \right) = P\left(X_1 = 0 \mid \bar{X} = \bar{x} \right) = \left(1 - \frac{1}{n} \right)^{n\bar{x}}
$$

because $X_i \mid \bar{X} = \bar{x} \sim$ Binomial$(n\bar{x}, 1/n)$ for each $i = 1, \ldots, n$.

Certainly it is not at all obvious from the functional form that this estimator is unbiased, let alone UMVU. So this result can be viewed as a somewhat remarkable application of the theory. ∎

Recall now Theorems 6.5.2 and 6.5.3. The implications of these results, with some additional conditions, are that the MLE of θ is asymptotically unbiased for θ and that the asymptotic variance of the MLE is at the information lower bound. This is often interpreted to mean that, with large samples, the MLE makes full use of the information about θ contained in the data.

Summary of Section 8.1

- An estimator comes closest (using squared distance) on average to its mean (Theorem 8.1.1), so we can restrict attention to unbiased estimators for quantities of interest.

- The Rao-Blackwell theorem says that we can restrict attention to functions of a sufficient statistic when looking for an estimator minimizing MSE.

- When a sufficient statistic is complete, then any function of that sufficient statistic is UMVU for its mean.

- The Cramer-Rao lower bound gives a lower bound on the variance of an unbiased estimator and a method for obtaining an estimator that has variance at this lower bound when such an estimator exists.

Exercises

8.1.1 Suppose that a statistical model is given by the two distributions in the following table.

	$s = 1$	$s = 2$	$s = 3$	$s = 4$
$f_a(s)$	1/3	1/6	1/12	5/12
$f_b(s)$	1/2	1/4	1/6	1/12

If $T : \{1, 2, 3, 4\} \to \{1, 2, 3, 4\}$ is defined by $T(1) = T(2) = 1$ and $T(s) = s$ otherwise, then prove that T is a sufficient statistic. Derive the conditional distributions of s given $T(s)$ and show that these are independent of θ.

8.1.2 Suppose $(x_1, \ldots, x_n)$ is a sample from a distribution with mean μ and variance σ^2. Prove that $s^2 = (n-1)^{-1} \sum_{i=1}^{n} (x_i - \bar{x})^2$ is unbiased for σ^2.

8.1.3 Suppose that $(x_1, \ldots, x_n)$ is a sample from an $N(\mu, \sigma_0^2)$ distribution where $\mu \in R^1$ is unknown and σ_0^2 is known. Determine a UMVU estimator of the second moment $\mu^2 + \sigma_0^2$.

8.1.4 Suppose that $(x_1, \ldots, x_n)$ is a sample from an $N(\mu, \sigma_0^2)$ distribution where $\mu \in R^1$ is unknown and σ_0^2 is known. Determine a UMVU estimator of the first quartile $\mu + \sigma_0 z_{0.25}$.

8.1.5 Suppose that $(x_1, \ldots, x_n)$ is a sample from an $N(\mu, \sigma_0^2)$ distribution where $\mu \in R^1$ is unknown and σ_0^2 is known. Is $2\bar{x} + 3$ a UMVU estimator of anything? If so, what is it UMVU for? Justify your answer.

8.1.6 Suppose that $(x_1, \ldots, x_n)$ is a sample from a Bernoulli(θ) distribution where $\theta \in [0, 1]$ is unknown. Determine a UMVU estimator of θ (use the fact that a minimal sufficient statistic for this model is complete).

8.1.7 Suppose that $(x_1, \ldots, x_n)$ is a sample from a Gamma(α_0, β) distribution where α_0 is known and $\beta > 0$ is unknown. Using the fact that $\bar{x}$ is a complete sufficient statistic (see Challenge 8.1.22), determine a UMVU estimator of β^{-1}.

8.1.8 Suppose that $(x_1, \ldots, x_n)$ is a sample from an $N(\mu_0, \sigma^2)$ distribution where μ_0 is known and $\sigma^2 > 0$ is unknown. Show that $\sum_{i=1}^{n} (x_i - \mu_0)^2$ is a sufficient statistic for this problem. Using the fact that it is complete, determine a UMVU estimator for σ^2.

Problems

8.1.9 Suppose that $(x_1, \ldots, x_n)$ is a sample from a Uniform$(0, \theta)$ distribution where $\theta > 0$ is unknown. Show that $x_{(n)}$ is a sufficient statistic and determine its distribution. Using the fact that $x_{(n)}$ is complete, determine a UMVU estimator of θ.

8.1.10 Suppose that $(x_1, \ldots, x_n)$ is a sample from a Bernoulli(θ) distribution where $\theta \in [0, 1]$ is unknown. Then determine the conditional distribution of $(x_1, \ldots, x_n)$ given the value of the sufficient statistic $\bar{x}$.

8.1.11 Prove that $L(\theta, a) = (\theta - a)^2$ satisfies

$$L(\theta, \alpha a_1 + (1 - \alpha) a_2) \leq \alpha L(\theta, a_1) + (1 - \alpha) L(\theta, a_2)$$

when a ranges in a subinterval of R^1. Use this result together with Jensen's inequality (Theorem 3.6.4) to prove the Rao-Blackwell Theorem.

8.1.12 Prove that $L(\theta, a) = |\theta - a|$ satisfies

$$L(\theta, \alpha a_1 + (1 - \alpha) a_2) \leq \alpha L(\theta, a_1) + (1 - \alpha) L(\theta, a_2)$$

when a ranges in a subinterval of R^1. Use this result together with Jensen's inequality (Theorem 3.6.4) to prove the Rao-Blackwell Theorem for absolute error. (Hint: First show that $|x + y| \le |x| + |y|$ for any x and y.)

8.1.13 Suppose that $(x_1, \ldots, x_n)$ is a sample from an $N(\mu, \sigma^2)$ distribution where $(\mu, \sigma^2) \in R^1 \times (0, \infty)$ is unknown. Show that the optimal estimator (in the sense of minimizing the MSE), of the form cs^2 for σ^2, is given by $c = (n-1)/(n+1)$. Determine the bias of this estimator and show that it goes to 0 as $n \to \infty$.

8.1.14 Prove that if a statistic T is complete for a model and $U = h(T)$ for a 1–1 function h, then U is also complete.

8.1.15 Suppose that $(x_1, \ldots, x_n)$ is a sample from an $N(\mu, \sigma^2)$ distribution where $(\mu, \sigma^2) \in R^1 \times (0, \infty)$ is unknown. Derive a UMVU estimator of the standard deviation σ. (Hint: Calculate the expected value of the sample standard deviation s.)

8.1.16 Suppose that $(x_1, \ldots, x_n)$ is a sample from an $N(\mu, \sigma^2)$ distribution where $(\mu, \sigma^2) \in R^1 \times (0, \infty)$ is unknown. Derive a UMVU estimator of the first quartile $\mu + \sigma z_{0.25}$. (Hint: Problem 8.1.15.)

8.1.17 Suppose that $(x_1, \ldots, x_n)$ is a sample from an $N(\mu, \sigma^2)$ distribution where $\theta \in \Omega = \{\mu_1, \mu_2\}$ is unknown and $\sigma_0^2 > 0$ is known. Establish that $\bar{x}$ is a minimal sufficient statistic for this model but is not complete.

8.1.18 Suppose that $(x_1, \ldots, x_n)$ is a sample from an $N(\mu, \sigma^2)$ distribution where $\mu \in R^1$ is unknown and σ_0^2 is known. Determine the information lower bound, for an unbiased estimator, when we are considering estimating the second moment $\mu^2 + \sigma_0^2$. Does the UMVU estimator in Exercise 8.1.3 attain the information lower bound?

8.1.19 Suppose that $(x_1, \ldots, x_n)$ is a sample from a Gamma(α_0, β) distribution where α_0 is known and $\beta > 0$ is unknown. Determine the information lower bound for the estimation of β^{-1} using unbiased estimators, and determine if the UMVU estimator obtained in Exercise 8.1.7 attains this.

8.1.20 Suppose that $(x_1, \ldots, x_n)$ is a sample from the distribution with density $f_\theta(x) = \theta x^{\theta - 1}$ for $x \in [0, 1]$ and $\theta > 0$ is unknown. Determine the information lower bound for estimating θ using unbiased estimators. Does a UMVU estimator with variance at the lower bound exist for this problem?

Challenges

8.1.21 If X is a random variable such that $E(X)$ either does not exist or is infinite, then show that $E((X - c)^2) = \infty$ for any constant c.

8.1.22 Suppose that $(x_1, \ldots, x_n)$ is a sample from a Gamma(α_0, β) distribution where α_0 is known and $\beta > 0$ is unknown. Show that $\bar{x}$ is a complete minimal sufficient statistic.

8.2 Optimal Hypothesis Testing

Suppose we want to assess a hypothesis about the real-valued characteristic $\psi(\theta)$ for the model $\{f_\theta : \theta \in \Omega\}$. Typically, this will take the form $H_0 : \psi(\theta) = \psi_0$ where we have specified a value for ψ. After observing data s, we want to assess whether or not we have evidence against H_0.

In Section 6.3.3 we discussed methods for assessing such a hypothesis based on the plug-in MLE for $\psi(\theta)$. These involved computing a P-value, as a measure of how surprising the data s are when the null hypothesis is assumed to be true. If s is surprising for each of the distributions f_θ for which $\psi(\theta) = \psi_0$, then we have evidence against H_0. The development of such procedures was largely based on the intuitive justification for the likelihood function.

Closely associated with a specific procedure for computing a P-value is the concept of a *power function* $\beta(\theta)$, as defined in Section 6.3.5. For this we specified a *critical value* α, such that we declare the results of the test statistically significant whenever the P-value is less than or equal to α. The power $\beta(\theta)$ is then the probability of the P-value being less than or equal to α when we are sampling from f_θ. The greater the value of $\beta(\theta)$, when $\psi(\theta) \neq \psi_0$, then the better the procedure is at detecting departures from H_0. The power function is thus a measure of the sensitivity of the testing procedure to detecting departures from H_0.

Recall the following fundamental example.

Example 8.2.1 *Location Normal Model*
Suppose we have a sample $(x_1, \ldots, x_n)$ from the $N(\mu, \sigma_0^2)$ model where $\mu \in R^1$ is unknown and $\sigma_0^2 > 0$ is known, and we want to assess the null hypothesis $H_0 : \mu = \mu_0$. In Example 6.3.9 we showed that a sensible test for this problem is based on the z-statistic

$$z = \frac{\bar{x} - \mu_0}{\sigma_0/\sqrt{n}},$$

with $Z \sim N(0,1)$ under H_0. The P-value is then given by

$$P_{\mu_0}\left(|Z| \geq \left|\frac{\bar{x} - \mu_0}{\sigma_0/\sqrt{n}}\right|\right) = 2\left[1 - \Phi\left(\left|\frac{\bar{x} - \mu_0}{\sigma_0/\sqrt{n}}\right|\right)\right],$$

where Φ denotes the $N(0,1)$ distribution function.

In Example 6.3.18 we showed that, for critical value α, the power function of the z-test is given by

$$
\begin{aligned}
\beta(\mu) &= P_\mu\left(2\left[1 - \Phi\left(\left|\frac{\bar{X} - \mu_0}{\sigma_0/\sqrt{n}}\right|\right)\right] < \alpha\right) = P_\mu\left(\Phi\left(\left|\frac{\bar{X} - \mu_0}{\sigma_0/\sqrt{n}}\right|\right) > 1 - \frac{\alpha}{2}\right) \\
&= 1 - \Phi\left(\frac{\mu_0 - \mu}{\sigma_0/\sqrt{n}} + z_{1-(\alpha/2)}\right) + \Phi\left(\frac{\mu_0 - \mu}{\sigma_0/\sqrt{n}} - z_{1-(\alpha/2)}\right),
\end{aligned}
$$

because $\bar{X} \sim N(\mu, \sigma_0^2/n)$.

We see that specifying a value for α specifies a set of data values

$$R = \left\{(x_1, \ldots, x_n) : \Phi\left(\left|\frac{\bar{x} - \mu_0}{\sigma_0/\sqrt{n}}\right|\right) > 1 - \frac{\alpha}{2}\right\}$$

such that the results of the test are determined to be statistically significant whenever $(x_1, \ldots, x_n) \in R$. Using the fact that Φ is 1–1 increasing, we can also write R as

$$
\begin{aligned}
R &= \left\{ (x_1, \ldots, x_n) : \left| \frac{\bar{x} - \mu_0}{\sigma_0/\sqrt{n}} \right| > \Phi^{-1}\left(1 - \frac{\alpha}{2}\right) \right\} \\
&= \left\{ (x_1, \ldots, x_n) : \left| \frac{\bar{x} - \mu_0}{\sigma_0/\sqrt{n}} \right| > z_{1-(\alpha/2)} \right\}.
\end{aligned}
$$

Further, the power function is given by $\beta(\mu) = P_\mu(R)$ and $\beta(\mu_0) = P_{\mu_0}(R) = \alpha$. ∎

We now adopt a different point of view. We are going to look for tests that are optimal for testing the null hypothesis $H_0 : \psi(\theta) = \psi_0$. First we will assume that, having observed the data s, we will decide to either accept or reject H_0. Our performance measure for assessing testing procedures will then be the probability that the testing procedure makes an error.

Sometimes hypothesis testing problems for real-valued parameters are distinguished as being one-sided or two-sided. For example, if θ is real-valued, then $H_0 : \theta = \theta_0$ versus $H_a : \theta \neq \theta_0$ is a two-sided testing problem, while $H_0 : \theta \leq \theta_0$ versus $H_a : \theta > \theta_0$ or $H_0 : \theta \geq \theta_0$ versus $H_a : \theta < \theta_0$ are examples of one-sided problems. Notice, however, that if we define

$$
\psi(\theta) = I_{(\theta_0, \infty)}(\theta),
$$

then $H_0 : \theta \leq \theta_0$ versus $H_a : \theta > \theta_0$ is equivalent to the problem $H_0 : \psi(\theta) = 0$ versus $H_a : \psi(\theta) \neq 0$. Similarly, if we define

$$
\psi(\theta) = I_{(-\infty, \theta_0)}(\theta),
$$

then $H_0 : \theta \geq \theta_0$ versus $H_a : \theta < \theta_0$ is equivalent to the problem $H_0 : \psi(\theta) = 0$ versus $H_a : \psi(\theta) \neq 0$. So the formulation we have adopted for testing problems about a general ψ includes the one-sided problems as special cases.

There are two types of error. We can make a *type I error* — rejecting H_0 when it is true — or make a *type II error* — accepting H_0 when H_0 is false. Note that, if we reject H_0, then this implies that we are accepting the *alternative hypothesis* $H_a : \psi(\theta) \neq \psi_0$.

It turns out that, except in very artificial circumstances, there are no testing procedures that simultaneously minimize the probabilities of making the two kinds of errors. Accordingly, we will place an upper bound α, called the *critical value*, on the probability of making a type I error. We then search among those tests whose probability of making a type I error is less than or equal to α, for a testing procedure that minimizes the probability of making a type II error.

One approach to specifying a testing procedure is to select a subset $R \subset S$ before we observe s. We then *reject* H_0 whenever $s \in R$ and accept H_0 whenever $s \notin R$. The set R is referred to as a *rejection region*.

A rejection region R satisfying

$$
P_\theta(R) \leq \alpha \tag{8.2.1}
$$

whenever $\psi(\theta) = \psi_0$ is called a *size α rejection region*. So (8.2.1) expresses the bound on the probability of making a type I error.

Among all size α rejection regions R, we want to find the one (if it exists) that will minimize the probability of making a type II error. This is equivalent to finding the size α rejection region R that maximizes the probability of rejecting the null hypothesis when it is false. This probability is the power of R and is given $\beta(\theta) = P_\theta(R)$ whenever $\psi(\theta) \neq \psi_0$.

To fully specify the optimality approach to testing hypotheses, we need one additional ingredient. Observe that our search for an optimal size α rejection region R is equivalent to finding the indicator function I_R that satisfies $E_\theta(I_R) = P_\theta(R) \leq \alpha$, when $\psi(\theta) = \psi_0$, and maximizes $\beta(\theta) = E_\theta(I_R) = P_\theta(R)$, when $\psi(\theta) \neq \psi_0$. It turns out that, in a number of problems, there is no such rejection region.

On the other hand, there is often a solution to the more general problem of finding a function $\varphi : S \to [0, 1]$ satisfying

$$E_\theta(\varphi) \leq \alpha, \tag{8.2.2}$$

when $\psi(\theta) = \psi_0$, and maximizing

$$\beta(\theta) = E_\theta(\varphi),$$

when $\psi(\theta) \neq \psi_0$. We call $\varphi : S \to [0, 1]$ a *test function* and β the power function associated with the test function φ. Note that $\varphi = I_R$ is a test function. If φ satisfies (8.2.2) when $\psi(\theta) = \psi_0$, it is called a *size α test function*. If φ satisfies $E_\theta(\varphi) = \alpha$ when $\psi(\theta) = \psi_0$, it is called an *exact size α test function*. A size α test function φ that maximizes $\beta(\theta) = E_\theta(\varphi)$ when $\psi(\theta) \neq \psi_0$ is called a *uniformly most powerful (UMP) size α test function*.

For observed data s, we interpret $\varphi(s) = 0$ to mean that we accept H_0 and interpret $\varphi(s) = 1$ to mean that we reject H_0. In general, we interpret $\varphi(s)$ to be the conditional probability that we reject H_0 given the data s. Operationally this means that, after we observe s, we generate a Bernoulli($\varphi(s)$) random variable. If we get a 1, we reject H_0; if we get a 0, we accept H_0. Therefore, by the theorem of total expectation, $E_\theta(\varphi)$ is the unconditional probability of rejecting H_0. The randomization that occurs when $0 < \varphi(s) < 1$ may seem somewhat counterintuitive, but it is forced on us by our search for a UMP size α test, as we can increase power by doing this in certain problems.

Therefore, for a testing problem specified by a null hypothesis $H_0 : \psi(\theta) = \psi_0$ and a critical value α, we want to find a UMP size α test function φ. Note that a UMP size α test function φ_0 for $H_0 : \psi(\theta) = \psi_0$ is characterized by (letting β_φ denote the power function of φ)

$$\beta_{\varphi_0}(\theta) \leq \alpha,$$

when $\psi(\theta) = \psi_0$, and

$$\beta_{\varphi_0}(\theta) \geq \beta_\varphi(\theta),$$

when $\psi(\theta) \neq \psi_0$, for any other size α test function φ.

Still, this optimization problem does not have a solution in general. In certain problems, however, an optimal solution can be found. The following result gives one of these. It is fundamental to the entire theory of optimal hypothesis testing.

Theorem 8.2.1 (*Neyman–Pearson*) Suppose that $\Omega = \{\theta_0, \theta_1\}$ and that we want to test $H_0 : \theta = \theta_0$. Then an exact size α test function φ_0 of the form

$$\varphi_0(s) = \begin{cases} 1 & f_{\theta_1}(s) / f_{\theta_0}(s) > c_0 \\ \gamma & f_{\theta_1}(s) / f_{\theta_0}(s) = c_0 \\ 0 & f_{\theta_1}(s) / f_{\theta_0}(s) < c_0, \end{cases} \tag{8.2.3}$$

exists for some $\gamma \in [0, 1]$ and $c_0 \geq 0$. This test is UMP size α.

Proof: See Section 8.5 for the proof of this result. ∎

The following result can be established by a simple extension of the proof of the Neyman–Pearson theorem.

Corollary 8.2.1 If φ is a UMP size α test, then $\varphi(s) = \varphi_0(s)$ everywhere except possibly on the *boundary* $B = \{s : f_{\theta_1}(s) / f_{\theta_0}(s) = c_0\}$. Further φ has exact size α unless the power of a UMP size α test equals 1.

Proof: See Challenge 8.2.15. ∎

Notice the intuitive nature of the test given by the Neyman–Pearson theorem. For (8.2.3) indicates that we categorically reject H_0 as being true when the likelihood ratio of θ_1 versus θ_0 is greater than the constant c_0, and we accept H_0 when it is smaller. When the likelihood ratio equals c_0, we randomly decide to reject H_0 with probability γ. Also, Corollary 8.2.1 says that a UMP size α test is basically unique, although there are possibly different randomization strategies on the boundary.

The proof of the Neyman–Pearson theorem reveals that c_0 is the smallest real number such that

$$P_{\theta_0}\left(\frac{f_{\theta_1}(s)}{f_{\theta_0}(s)} > c_0\right) \leq \alpha \tag{8.2.4}$$

and

$$\gamma = \begin{cases} \dfrac{\alpha - P_{\theta_0}\left(\frac{f_{\theta_1}(s)}{f_{\theta_0}(s)} > c_0\right)}{P_{\theta_0}\left(\frac{f_{\theta_1}(s)}{f_{\theta_0}(s)} = c_0\right)} & P_{\theta_0}\left(\frac{f_{\theta_1}(s)}{f_{\theta_0}(s)} = c_0\right) \neq 0 \\ 0 & \text{otherwise.} \end{cases} \tag{8.2.5}$$

We use (8.2.4) and (8.2.5) to calculate c_0 and γ, and so determine the UMP size α test, in a particular problem.

Note that the test is nonrandomized whenever $P_{\theta_0}(f_{\theta_1}(s) / f_{\theta_0}(s) > c_0) = \alpha$, as then $\gamma = 0$, i.e., we categorically accept or reject H_0 after seeing the data. This always occurs whenever the distribution of $f_{\theta_1}(s) / f_{\theta_0}(s)$ is continuous when $s \sim P_{\theta_0}$. Interestingly, it can happen that the distribution of the ratio is not continuous even when the distribution of s is continuous (see Problem 8.2.9).

Before considering some applications of the Neyman–Pearson theorem, we establish the analog of the Rao-Blackwell theorem for hypothesis testing problems. Given the value of the sufficient statistic $U(s) = u$ we denote the conditional probability measure for the response s by $P(\cdot \,|\, U = u)$ (by Theorem 8.1.2 this probability measure does not depend on θ). For test function φ put $\varphi_U(s)$ equal to the conditional expectation of φ given the value of $U(s)$, namely,

$$\varphi_U(s) = E_{P(\cdot \,|\, U=U(s))}(\varphi).$$

Theorem 8.2.2 Suppose that U is a sufficient statistic and φ is a size α test function for $H_0 : \psi(\theta) = \psi_0$. Then φ_U is a size α test function for $H_0 : \psi(\theta) = \psi_0$ that depends on the data only through the value of U. Further, φ and φ_U have the same power function.

Proof: It is clear that $\varphi_U(s_1) = \varphi_U(s_2)$ whenever $U(s_1) = U(s_2)$, and so φ_U depends on the data only through the value of U. Now let $P_{\theta,U}$ denote the marginal probability measure of U induced by P_θ. Then by the theorem of total expectation we have

$$E_\theta(\varphi) = E_{P_{\theta,U}}\left(E_{P(\cdot \,|\, U=u)}(\varphi)\right) = E_{P_{\theta,U}}(\varphi_U) = E_\theta(\varphi_U).$$

Now $E_\theta(\varphi) \leq \alpha$ when $\psi(\theta) = \psi_0$, which implies that $E_\theta(\varphi_U) \leq \alpha$ when $\psi(\theta) = \psi_0$, and $\beta(\theta) = E_\theta(\varphi) = E_\theta(\varphi_U)$ when $\psi(\theta) \neq \psi_0$. ∎

This result allows us to restrict our search for a UMP size α test to those test functions that depend on the data only through the value of a sufficient statistic.

We now consider some applications of the Neyman–Pearson theorem. The following example shows that this result can lead to solutions to much more general problems than the simple case it addresses.

Example 8.2.2 *Optimal Hypothesis Testing in the Location Normal Model*
Suppose that $(x_1, \ldots, x_n)$ is a sample from an $N(\mu, \sigma_0^2)$ distribution where $\mu \in \Omega = \{\mu_0, \mu_1\}$ and $\sigma_0^2 > 0$ is known, and we want to test $H_0 : \mu = \mu_0$ versus $H_a : \mu = \mu_1$. The likelihood function is given by

$$L(\mu \,|\, x_1, \ldots, x_n) = \exp\left(-\frac{n}{2\sigma_0^2}(\bar{x} - \mu)^2\right),$$

and $\bar{x}$ is a sufficient statistic for this restricted model.

By Theorem 8.2.2, we can restrict our attention to test functions that depend on the data through $\bar{x}$. Now $\bar{X} \sim N(\mu, \sigma_0^2)$ so that

$$
\begin{aligned}
\frac{f_{\mu_1}(\bar{x})}{f_{\mu_0}(\bar{x})} &= \frac{\exp\left(-\frac{n}{2\sigma_0^2}(\bar{x} - \mu_1)^2\right)}{\exp\left(-\frac{n}{2\sigma_0^2}(\bar{x} - \mu_0)^2\right)} \\
&= \exp\left(-\frac{n}{2\sigma_0^2}\left(\bar{x}^2 - 2\bar{x}\mu_1 + \mu_1^2 - \bar{x}^2 + 2\bar{x}\mu_0 - \mu_0^2\right)\right) \\
&= \exp\left(\frac{n}{\sigma_0^2}(\mu_1 - \mu_0)\bar{x}\right)\exp\left(-\frac{n}{2\sigma_0^2}(\mu_1^2 - \mu_0^2)\right).
\end{aligned}
$$

Therefore,

$$P_{\mu_0} \left(\frac{f_{\mu_1}\left(\bar{X}\right)}{f_{\mu_0}\left(\bar{X}\right)} > c_0 \right)$$

$$= P_{\mu_0} \left(\exp\left(\frac{n}{\sigma_0^2}(\mu_1 - \mu_0)\bar{X} \right) \exp\left(-\frac{n}{2\sigma_0^2}(\mu_1^2 - \mu_0^2) \right) > c_0 \right)$$

$$= P_{\mu_0} \left(\exp\left(\frac{n}{\sigma_0^2}(\mu_1 - \mu_0)\bar{X} \right) > c_0 \exp\left(\frac{n}{2\sigma_0^2}(\mu_1^2 - \mu_0^2) \right) \right)$$

$$= P_{\mu_0} \left((\mu_1 - \mu_0)\bar{X} > \frac{\sigma_0^2}{n} \ln\left\{ c_0 \exp\left(\frac{n}{2\sigma_0^2}(\mu_1^2 - \mu_0^2) \right) \right\} \right)$$

$$= \begin{cases} P_{\mu_0}\left(\frac{\bar{X}-\mu_0}{\sigma_0/\sqrt{n}} > c_0' \right) & \mu_1 > \mu_0 \\ P_{\mu_0}\left(\frac{\bar{X}-\mu_0}{\sigma_0/\sqrt{n}} < c_0' \right) & \mu_1 < \mu_0 \end{cases}$$

where

$$c_0' = \frac{\sqrt{n}}{\sigma_0}\left\{ \frac{\sigma_0^2}{n(\mu_1 - \mu_0)} \ln\left\{ c_0 \exp\left(\frac{n}{2\sigma_0^2}(\mu_1^2 - \mu_0^2) \right) \right\} - \mu_0 \right\}.$$

Using (8.2.4), when $\mu_1 > \mu_0$, we select c_0 so that $c_0' = z_{1-\alpha}$; when $\mu_1 < \mu_0$, we select c_0 so that $c_0' = z_\alpha$. These choices imply that

$$P_{\mu_0} \left(\frac{f_{\mu_1}\left(\bar{X}\right)}{f_{\mu_0}\left(\bar{X}\right)} > c_0 \right) = \alpha$$

and, by (8.2.5), $\gamma = 0$.

So the UMP size α test is nonrandomized. When $\mu_1 > \mu_0$, the test is given by

$$\varphi_0\left(\bar{x}\right) = \begin{cases} 1 & \bar{x} \geq \mu_0 + \frac{\sigma_0}{\sqrt{n}} z_{1-\alpha} \\ 0 & \bar{x} < \mu_0 + \frac{\sigma_0}{\sqrt{n}} z_{1-\alpha}. \end{cases} \tag{8.2.6}$$

When $\mu_1 < \mu_0$, the test is given by

$$\varphi_0^*\left(\bar{x}\right) = \begin{cases} 1 & \bar{x} \leq \mu_0 + \frac{\sigma_0}{\sqrt{n}} z_\alpha \\ 0 & \bar{x} > \mu_0 + \frac{\sigma_0}{\sqrt{n}} z_\alpha. \end{cases} \tag{8.2.7}$$

Notice that the test function in (8.2.6) does not depend in any way upon μ_1. The subsequent implication is that this test function is UMP size α for $H_0 : \mu = \mu_0$ versus $H_a : \mu = \mu_1$ for any $\mu_1 > \mu_0$. This implies that φ_0 is UMP size α for $H_0 : \mu = \mu_0$ versus the alternative $H_a : \mu > \mu_0$.

Further, we have

$$\beta_{\varphi_0}(\mu) = P_\mu\left(\bar{X} \geq \mu_0 + \frac{\sigma_0}{\sqrt{n}} z_{1-\alpha} \right) = P_\mu\left(\frac{\bar{X}-\mu}{\sigma_0/\sqrt{n}} \geq \frac{\mu_0 - \mu}{\sigma_0/\sqrt{n}} + z_{1-\alpha} \right)$$

$$= 1 - \Phi\left(\frac{\mu_0 - \mu}{\sigma_0/\sqrt{n}} + z_{1-\alpha} \right).$$

Note that this is increasing in μ which implies that φ_0 is a size α test function for $H_0 : \mu \leq \mu_0$ versus $H_a : \mu > \mu_0$. Observe that, if φ is a size α test function for $H_0 : \mu \leq \mu_0$ versus $H_a : \mu > \mu_0$, then it is also a size α test for $H_0 : \mu = \mu_0$ versus $H_a : \mu > \mu_0$. From this we conclude that φ_0 is UMP size α for $H_0 : \mu \leq \mu_0$ versus $H_a : \mu > \mu_0$. Similarly (Problem 8.2.4), it can be shown that φ_0^* in (8.2.7) is UMP size α for $H_0 : \mu \geq \mu_0$ versus $H_a : \mu < \mu_0$.

We might wonder if a UMP size α test exists for the two-sided problem $H_0 : \mu = \mu_0$ versus $H_a : \mu \neq \mu_0$. Suppose that φ is a size α UMP test for this problem. Then φ is also size α for $H_0 : \mu = \mu_0$ versus $H_a : \mu = \mu_1$ when $\mu_1 > \mu_0$. Using Corollary 8.2.1 and the preceding developments (which also shows that there does not exist a test of the form (8.2.3) having power equal to 1 for this problem), this implies that $\varphi = \varphi_0$ (the boundary B has probability 0 here). But φ is also UMP size α for $H_0 : \mu = \mu_0$ versus $H_a : \mu = \mu_1$ when $\mu_1 < \mu_0$ and so, by the same reasoning, $\varphi = \varphi_0^*$. But clearly $\varphi_0 \neq \varphi_0^*$, so there is no UMP size α test for the two-sided problem.

Intuitively we would expect that the size α test given by

$$
\varphi(\bar{x}) =
\begin{cases}
1 & \left| \frac{\bar{x} - \mu_0}{\sigma_0/\sqrt{n}} \right| \geq z_{1-\alpha/2} \\[2ex]
0 & \left| \frac{\bar{x} - \mu_0}{\sigma_0/\sqrt{n}} \right| < z_{1-\alpha/2}
\end{cases}
\tag{8.2.8}
$$

would be a good test to use, but it is not UMP size α. It turns out, however, that the test in (8.2.8) is UMP size α among all tests satisfying $\beta_\varphi(\mu_0) \leq \alpha$ and $\beta_\varphi(\mu) \geq \alpha$ when $\mu \neq \mu_0$. ∎

A test φ that satisfies $\beta_\varphi(\theta) \leq \alpha$, when $\psi(\theta) = \psi_0$, and $\beta_\varphi(\theta) \geq \alpha$, when $\psi(\theta) \neq \psi_0$, is said to be an *unbiased size α test* for the hypothesis testing problem $H_0 : \psi(\theta) = \psi_0$. So (8.2.8) is a UMP unbiased size α test. An unbiased test has the property that the probability of rejecting the null hypothesis, when the null hypothesis is false, is always greater than the probability of rejecting the null hypothesis, when the null hypothesis is true. This seems like a very reasonable property. In particular, it can be proved that any UMP size α is always an unbiased size α test (Problem 8.2.6). We do not pursue the theory of unbiased tests further in this text.

We now consider an example which shows that we cannot dispense with the use of randomized tests.

Example 8.2.3 *Optimal Hypothesis Testing in the Bernoulli(θ) Model*
Suppose that $(x_1, \ldots, x_n)$ is a sample from a Bernoulli(θ) distribution where $\theta \in \Omega = \{\theta_0, \theta_1\}$, and we want to test $H_0 : \theta = \theta_0$ versus $H_a : \theta = \theta_1$, where $\theta_1 > \theta_0$. Then $n\bar{x}$ is a minimal sufficient statistic and, by Theorem 8.2.2, we can restrict our attention to test functions that depend on the data only through $n\bar{x}$.

Now $n\bar{X} \sim \text{Binomial}(n, \theta)$, so

$$
\frac{f_{\theta_1}(n\bar{x})}{f_{\theta_0}(n\bar{x})} = \frac{\theta_1^{n\bar{x}}(1-\theta_1)^{n-n\bar{x}}}{\theta_0^{n\bar{x}}(1-\theta_0)^{n-n\bar{x}}} = \left(\frac{\theta_1}{\theta_0}\right)^{n\bar{x}}\left(\frac{1-\theta_1}{1-\theta_0}\right)^{n-n\bar{x}}.
$$

Therefore,

$$P_{\theta_0} \left(\frac{f_{\theta_1}(n\bar{X})}{f_{\theta_0}(n\bar{X})} > c_0 \right)$$

$$= P_{\theta_0} \left(\left(\frac{\theta_1}{\theta_0} \right)^{n\bar{X}} \left(\frac{1-\theta_1}{1-\theta_0} \right)^{n-n\bar{X}} > c_0 \right)$$

$$= P_{\theta_0} \left(\left(\frac{\theta_1}{1-\theta_1} \frac{1-\theta_0}{\theta_0} \right)^{n\bar{X}} > c_0 \left(\frac{1-\theta_1}{1-\theta_0} \right)^{-n} \right)$$

$$= P_{\theta_0} \left(n\bar{X} \left[\ln \left(\frac{\theta_1}{1-\theta_1} \frac{1-\theta_0}{\theta_0} \right) \right] > \ln c_0 \left(\frac{1-\theta_1}{1-\theta_0} \right)^{-n} \right)$$

$$= P_{\theta_0} \left(n\bar{X} > \frac{\ln c_0 \left(\frac{1-\theta_1}{1-\theta_0} \right)^{-n}}{\ln \left(\frac{\theta_1}{1-\theta_1} \frac{1-\theta_0}{\theta_0} \right)} \right) = P_{\theta_0} \left(n\bar{X} > c_0' \right)$$

because

$$\ln \left(\frac{\theta_1}{1-\theta_1} \frac{1-\theta_0}{\theta_0} \right) > 0$$

as $\theta/(1-\theta)$ is increasing in θ, which implies $\theta_1/(1-\theta_1) > \theta_0/(1-\theta_0)$.
Now, using (8.2.4), we choose c_0 so that c_0' is an integer satisfying

$$P_{\theta_0} \left(n\bar{X} > c_0' \right) \le \alpha \text{ and } P_{\theta_0} \left(n\bar{X} > c_0' - 1 \right) > \alpha.$$

Because $n\bar{X} \sim \text{Binomial}(n, \theta_0)$ is a discrete distribution, we see that, in general, we will not be able to achieve $P_{\theta_0} \left(n\bar{X} > c_0' \right) = \alpha$ exactly. So, using (8.2.5),

$$\gamma = \frac{\alpha - P_{\theta_0} \left(n\bar{X} > c_0' \right)}{P_{\theta_0} \left(n\bar{X} = c_0' \right)}$$

will not be equal to 0. Then

$$\varphi_0 (n\bar{x}) = \begin{cases} 1 & n\bar{x} > c_0' \\ \gamma & n\bar{x} = c_0' \\ 0 & n\bar{x} < c_0' \end{cases}$$

is UMP size α for $H_0 : \theta = \theta_0$ versus $H_a : \theta = \theta_1$. Note that we can use statistical software (or Table D.6) for the binomial distribution to obtain c_0'.

For example, suppose $n = 6$ and $\theta_0 = 0.25$. The following table gives the values of the Binomial(6, .25) distribution function to 3 decimal places.

x	0	1	2	3	4	5	6
$F(x)$	0.178	0.534	0.831	0.962	0.995	1.000	1.000

Therefore, if $\alpha = 0.05$, we have that $c_0' = 2$ because $P_{0.25}\left(n\bar{X} > 2\right) = 1 - 0.962 = 0.038$ and $P_{0.25}\left(n\bar{X} > 1\right) = 1 - 0.534 = 0.466$. This implies that

$$\gamma = \frac{0.05 - (1 - 0.962)}{0.962} = 0.012.$$

So with this test, we reject $H_0 : \theta = \theta_0$ categorically if the number of successes is greater than 2, accept $H_0 : \theta = \theta_0$ categorically when the number of successes is less than 2, and when the number of 1's equals 2, we randomly reject $H_0 : \theta = \theta_0$ with probability 0.012 (e.g., generate $U \sim \text{Uniform}(0,1)$ and reject whenever $U \leq 0.012$).

Notice that the test φ_0 does not involve θ_1, so indeed it is UMP size α for $H_0 : \theta = \theta_0$ versus $H_a : \theta > \theta_0$. Further, using Problem 8.2.10, we have

$$P_\theta\left(n\bar{X} > c_0'\right) = \sum_{k=c_0'+1}^{n} \binom{n}{k} \theta^k (1-\theta)^{n-k}$$

$$= 1 - \frac{\Gamma(n+1)}{\Gamma(c_0'+1)\Gamma(n-c_0')} \int_\theta^1 u^{c_0'} (1-u)^{n-c_0'-1} \, du.$$

Because

$$\int_\theta^1 u^{c_0'} (1-u)^{n-c_0'-1} \, du$$

is decreasing in θ, we must have that $P_\theta\left(n\bar{X} > c_0'\right)$ is increasing in θ. Arguing as in Example 8.2.2, we conclude that φ_0 is UMP size α for $H_0 : \theta \leq \theta_0$ versus $H_a : \theta > \theta_0$.

Similarly, we obtain a UMP size α test for $H_0 : \theta \leq \theta_0$ versus $H_a : \theta > \theta_0$. As in Example 8.2.2, there is no UMP size α test for $H_0 : \theta = \theta_0$ versus $H_a : \theta \neq \theta_0$, but there is a UMP unbiased size α test for this problem. ∎

8.2.1 Likelihood Ratio Tests (Advanced)

In the examples considered so far, the Neyman–Pearson theorem has led to solutions to problems in which H_0 or H_a are not just single values of the parameter, even though the theorem was only stated for the single-value case. We also noted, however, that this is not true in general (for example, the two-sided problems discussed in Examples 8.2.2 and 8.2.3).

The method of *generalized likelihood ratio tests* for $H_0 : \psi(\theta) = \psi_0$ has been developed to deal with the general case. This is motivated by the Neyman–Pearson theorem, for observe that in (8.2.3),

$$\frac{f_{\theta_1}(s)}{f_{\theta_0}(s)} = \frac{L(\theta_1 \mid s)}{L(\theta_0 \mid s)}.$$

Therefore, (8.2.3) can be thought of as being based on the ratio of the likelihood at θ_1 to the likelihood at θ_0, and we reject when the likelihood gives much more support to θ_1 than to θ_0. The amount of the additional support required for rejection is determined by c_0. The larger c_0 is, the larger the likelihood $L(\theta_1 \mid s)$ has to be relative to $L(\theta_0 \mid s)$ before we reject $H_0 : \theta = \theta_0$.

Denote the overall MLE of θ by $\hat{\theta}(s)$, and the MLE, when $\theta \in H_0$, by $\hat{\theta}_{H_0}(s)$. So we have

$$L(\theta \mid s) \leq L\left(\hat{\theta}_{H_0}(s) \mid s\right)$$

for all θ such that $\psi(\theta) = \psi_0$. The generalized likelihood ratio test then rejects H_0 when

$$\frac{L\left(\hat{\theta}(s) \mid s\right)}{L\left(\hat{\theta}_{H_0}(s) \mid s\right)} \tag{8.2.9}$$

is large, as this indicates evidence against H_0 being true.

How do we determine when (8.2.9) is large enough to reject? Denoting the observed data by s_0, we do this by computing the P-values

$$P_\theta\left(\frac{L\left(\hat{\theta}(s) \mid s\right)}{L\left(\hat{\theta}_{H_0}(s) \mid s\right)} > \frac{L\left(\hat{\theta}(s_0) \mid s_0\right)}{L\left(\hat{\theta}_{H_0}(s_0) \mid s_0\right)}\right) \tag{8.2.10}$$

when $\theta \in H_0$. Small values of (8.2.10) are evidence against H_0. Of course, when $\psi(\theta) = \psi_0$ for more than one value of θ, then it is not clear which value of (8.2.10) to use. It can be shown, however, that under conditions such as those discussed in Section 6.5, if s corresponds to a sample of n values from a distribution, then

$$2\ln\frac{L\left(\hat{\theta}(s) \mid s\right)}{L\left(\hat{\theta}_{H_0}(s) \mid s\right)} \xrightarrow{D} \chi^2\left(\dim\Omega - \dim H_0\right)$$

as $n \to \infty$, whenever the true value of θ is in H_0. Here $\dim\Omega$ and $\dim H_0$ are the dimensions of these sets. This leads us to a test that rejects H_0 whenever

$$2\ln\frac{L\left(\hat{\theta}(s) \mid s_0\right)}{L\left(\hat{\theta}_{H_0}(s) \mid s_0\right)} \tag{8.2.11}$$

is greater than a particular quantile of the $\chi^2\left(\dim\Omega - \dim H_0\right)$ distribution.

For example, suppose that in a location-scale normal model we are testing $H_0 : \mu = \mu_0$. Then $\Omega = R^1 \times [0, \infty)$, $H_0 = \{\mu_0\} \times [0, \infty)$, $\dim\Omega = 2$, $\dim H_0 = 1$, and, for a size 0.05 test, we reject whenever (8.2.11) is greater than $\chi^2_{0.95}(1)$. Note that, strictly speaking, likelihood ratio tests are not derived via optimality considerations. We will not discuss likelihood ratio tests further in this text.

Summary of Section 8.2

- In searching for an optimal hypothesis testing procedure, we place an upper bound on the probability of making a type I error (rejecting H_0 when it is true) and search for a test that minimizes the probability of making a type II error (accepting H_0 when it is false).

- The Neyman–Pearson theorem prescribes an optimal size α test when H_0 and H_a each specify a single value for the full parameter θ.

- Sometimes the Neyman–Pearson theorem leads to solutions to hypothesis testing problems when the null or alternative hypotheses allow for more than one possible value for θ, but in general we must resort to likelihood ratio tests for such problems.

Exercises

8.2.1 Suppose that a statistical model is given by the two distributions in the following table.

	$s = 1$	$s = 2$	$s = 3$	$s = 4$
$f_a(s)$	1/3	1/6	1/12	5/12
$f_b(s)$	1/2	1/4	1/6	1/12

Determine the MP size 0.10 test for testing $H_0 : \theta = a$ versus $H_a : \theta = b$. What is the power of this test? Repeat this with the size equal to 0.05.

8.2.2 Suppose for the hypothesis testing problem of Exercise 8.2.1, a statistician decides to generate $U \sim \text{Uniform}(0, 1)$ and reject H_0 whenever $U \leq 0.05$. Show that this test has size 0.05. Explain why this is not a good choice of test and why the test derived in Exercise 8.2.1 is better. Provide numerical evidence for this.

8.2.3 Suppose that an investigator knows that an industrial process yields a response variable that follows an $N(1, 2)$ distribution. Some changes have been made in the industrial process, and the investigator believes that these have possibly made a change in the mean of the response (not the variance), increasing its value. The investigator wants the probability of a type I error occurring to be less than 1%. Determine an appropriate testing procedure for this problem based on a sample of size 10.

Problems

8.2.4 Prove that φ_0^* in (8.2.5) is UMP size α for $H_0 : \mu \geq \mu_0$ versus $H_a : \mu < \mu_0$.

8.2.5 Prove that the test function $\varphi(s) = \alpha$ for every $s \in S$ is an exact size α test function. What is the interpretation of this test function?

8.2.6 Using the test function in Problem 8.2.5, show that a UMP size α test is also a UMP unbiased size α test.

8.2.7 Suppose that $(x_1, \ldots, x_n)$ is a sample from a Gamma(α_0, β) distribution where α_0 is known and $\beta > 0$ is unknown. Determine the UMP size α test for testing $H_0 : \beta = \beta_0$ versus $H_a : \beta = \beta_1$ where $\beta_1 > \beta_0$. Is this test UMP size α for $H_0 : \beta \leq \beta_0$ versus $H_a : \beta > \beta_0$?

8.2.8 Suppose that $(x_1, \ldots, x_n)$ is a sample from an $N(\mu_0, \sigma^2)$ distribution where μ_0 is known and $\sigma^2 > 0$ is unknown. Determine the UMP size α test for testing $H_0 : \sigma^2 = \sigma_0^2$ versus $H_a : \sigma^2 = \sigma_1^2$ where $\sigma_0^2 > \sigma_1^2$. Is this test UMP size α for $H_0 : \sigma^2 \leq \sigma_0^2$ versus $H_a : \sigma^2 > \sigma_0^2$?

8.2.9 Suppose that $(x_1, \ldots, x_n)$ is a sample from a Uniform$(0, \theta)$ distribution where $\theta > 0$ is unknown. Determine the UMP size α test for testing $H_0 : \theta = \theta_0$ versus $H_a : \theta = \theta_1$ where $\theta_0 < \theta_1$. Is this test function UMP size α for $H_0 : \theta \leq \theta_0$ versus $H_a : \theta > \theta_0$?

8.2.10 Suppose that F is the distribution function for the Binomial(n, θ) distribution. Then prove that

$$F(x) = \frac{\Gamma(n+1)}{\Gamma(x+1)\Gamma(n-x)} \int_\theta^1 y^x (1-y)^{n-x-1} \, dy$$

for $x = 0, 1, \ldots, n-1$. This establishes a relationship between the binomial probability distribution and the beta function. (Hint: Integration by parts.)

8.2.11 Suppose that F is the distribution function for the Poisson(λ) distribution. Then prove that

$$F(x) = \frac{1}{x!} \int_\lambda^\infty y^x e^{-y} \, dy$$

for $x = 0, 1, \ldots$. This establishes a relationship between the Poisson probability distribution and the gamma function. (Hint: Integration by parts.)

8.2.12 Suppose that $(x_1, \ldots, x_n)$ is a sample from a Poisson(λ) distribution where $\lambda > 0$ is unknown. Determine the UMP size α test for $H_0 : \lambda = \lambda_0$ versus $H_a : \lambda = \lambda_1$ where $\lambda_0 < \lambda_1$. Is this test function UMP size α for $H_0 : \lambda \leq \lambda_0$ versus $H_a : \lambda > \lambda_0$? (Hint: You will need the result of Problem 8.2.11.)

8.2.13 Suppose that $(x_1, \ldots, x_n)$ is a sample from an $N(\mu, \sigma^2)$ distribution where $(\mu, \sigma^2) \in R^1 \times (0, \infty)$ is unknown. Derive the form of the exact size α likelihood ratio test for testing $H_0 : \mu = \mu_0$ versus $H_0 : \mu \neq \mu_0$.

8.2.14 (*Optimal confidence intervals*) Suppose that for model $\{f_\theta : \theta \in \Omega\}$ we have a UMP size α test function φ_{ψ_0} for $H_0 : \psi(\theta) = \psi_0$, for each possible value of ψ_0. Suppose further that each φ_{ψ_0} only takes values in $\{0, 1\}$, i.e., each φ_{ψ_0} is a nonrandomized size α test function.
(a) Prove that

$$C(s) = \left\{ \psi_0 : \varphi_{\psi_0}(s) = 0 \right\}$$

satisfies

$$P_\theta(\psi(\theta) \in C(s)) \geq 1 - \alpha$$

for every $\theta \in \Omega$. Conclude that $C(s)$ is a $(1-\alpha)$-confidence set for $\psi(\theta)$.

(b) If C^* is a $(1-\alpha)$-confidence set for $\psi(\theta)$, then prove that the test function defined by

$$\varphi_{\psi_0}^*(s) = \begin{cases} 1 & \psi_0 \notin C(s) \\ 0 & \psi_0 \in C(s) \end{cases}$$

is size α for $H_0 : \psi(\theta) = \psi_0$.

(c) Suppose that for each value ψ_0, the test function φ_{ψ_0} is UMP size α for testing $H_0 : \psi(\theta) = \psi_0$ versus $H_0 : \psi(\theta) \neq \psi_0$. Then prove that

$$P_\theta (\psi(\theta^*) \in C(s)) \tag{8.2.10}$$

is minimized, when $\psi(\theta) \neq \psi_0$, among all $(1-\alpha)$-confidence sets for $\psi(\theta)$. The probability (8.2.10) is the probability of C containing the false value $\psi(\theta^*)$, and a $(1-\alpha)$-confidence region that minimizes this probability when $\psi(\theta) \neq \psi_0$ is called a *uniformly most accurate (UMA)* $(1-\alpha)$-confidence region for $\psi(\theta)$.

Challenge

8.2.15 Prove Corollary 8.2.1 in the discrete case.

8.3 Optimal Bayesian Inferences

We now add the prior probability measure Π with density π. As we will see, this completes the specification of an optimality problem, as now there is always a solution. Solutions to Bayesian optimization problems are known as *Bayes rules*.

In Section 8.1 the unrestricted optimization problem was to find the estimator T of $\psi(\theta)$ that minimizes $\mathrm{MSE}_\theta(T) = E_\theta((T - \psi(\theta))^2)$, for each $\theta \in \Omega$. The Bayesian version of this problem is to minimize

$$E_\Pi (\mathrm{MSE}_\theta(T)) = E_\Pi \left(E_\theta \left((T - \psi(\theta))^2 \right) \right). \tag{8.3.1}$$

By the theorem of total expectation (Theorem 3.5.2), (8.3.1) is the expected value of the squared error $(T(s) - \psi(\theta))^2$ under the joint distribution on (θ, s) induced by the conditional distribution for s, given θ (the sampling model), and by the marginal distribution for θ (the prior distribution of θ). Again by the theorem of total expectation, we can write this as

$$E_\Pi (\mathrm{MSE}_\theta(T)) = E_M \left(E_{\Pi(\cdot\,|\,s)} \left((T - \psi(\theta))^2 \right) \right) \tag{8.3.2}$$

where $\Pi(\cdot\,|\,s)$ denotes the posterior probability measure for θ given the data s (the conditional distribution of θ given s), and M denotes the prior predictive for s (the marginal distribution of s).

We have the following result.

Theorem 8.3.1 When (8.3.1) is finite, a Bayes rule is given by

$$T(s) = E_{\Pi(\cdot \,|\, s)}\left(\psi\left(\theta\right)\right),$$

namely, the posterior expectation of $\psi\left(\theta\right)$.

Proof: First consider the expected posterior squared error

$$E_{\Pi(\cdot \,|\, s)}\left(\left(T'\left(s\right) - \psi\left(\theta\right)\right)^2\right)$$

of an estimate $T'\left(s\right)$. By Theorem 8.1.1 this is minimized by taking $T'\left(s\right)$ equal to $T(s) = E_{\Pi(\cdot \,|\, s)}\left(\psi\left(\theta\right)\right)$ (note that the "random" quantity here is θ).

Now suppose that T' is any estimator of $\psi\left(\theta\right)$. Then we have just shown that

$$0 \le E_{\Pi(\cdot \,|\, s)}\left(\left(T\left(s\right) - \psi\left(\theta\right)\right)^2\right) \le E_{\Pi(\cdot \,|\, s)}\left(\left(T'\left(s\right) - \psi\left(\theta\right)\right)^2\right)$$

and thus,

$$E_{\Pi}\left(\mathrm{MSE}_{\theta}\left(T\right)\right) = E_M\left(E_{\Pi(\cdot \,|\, s)}\left(\left(T\left(s\right) - \psi\left(\theta\right)\right)^2\right)\right)$$
$$\le E_M\left(E_{\Pi(\cdot \,|\, s)}\left(\left(T'\left(s\right) - \psi\left(\theta\right)\right)^2\right)\right) = E_{\Pi}\left(\mathrm{MSE}_{\theta}\left(T'\right)\right).$$

Therefore, T minimizes (8.3.1) and is a Bayes rule. ∎

So we see that, under mild conditions, the optimal Bayesian estimation problem always has a solution and there is no need to restrict ourselves to unbiased estimators, etc.

For the hypothesis testing problem $H_0 : \psi\left(\theta\right) = \psi_0$, we want to find the test function φ that minimizes the prior probability of making an error (type I or type II). Such a φ is a Bayes rule. We have the following result.

Theorem 8.3.2 A Bayes rule for the hypothesis testing problem $H_0 : \psi\left(\theta\right) = \psi_0$ is given by

$$\varphi_0\left(s\right) = \begin{cases} 1 & \Pi\left(\{\psi\left(\theta\right) = \psi_0\} \,|\, s\right) \le \Pi\left(\{\psi\left(\theta\right) \ne \psi_0\} \,|\, s\right) \\ 0 & \text{otherwise.} \end{cases}$$

Proof: Consider test function φ and let $I_{\{\psi(\theta)=\psi_0\}}\left(\theta\right)$ denote the indicator function of the set $\{\theta : \psi\left(\theta\right) = \psi_0\}$ (so $I_{\{\psi(\theta)=\psi_0\}}(\theta) = 1$ when $\psi\left(\theta\right) = \psi_0$ and equals 0 otherwise). Observe that $\varphi\left(s\right)$ is the probability of rejecting H_0, having observed s, and this is an error when $I_{\{\psi(\theta)=\psi_0\}}\left(\theta\right) = 1$; $1 - \varphi\left(s\right)$ is the probability of accepting H_0, having observed s, and this is an error when $I_{\{\psi(\theta)=\psi_0\}}\left(\theta\right) = 0$. Therefore, given s and θ, the probability of making an error is

$$e\left(\theta, s\right) = \varphi\left(s\right) I_{\{\psi(\theta)=\psi_0\}}\left(\theta\right) + \left(1 - \varphi\left(s\right)\right)\left(1 - I_{\{\psi(\theta)=\psi_0\}}\left(\theta\right)\right).$$

By the theorem of total expectation, the prior probability of making an error (taking the expectation of $e\left(\theta, s\right)$ under the joint distribution of (θ, s)) is

$$E_M\left(E_{\Pi(\cdot \,|\, s)}\left(e\left(\theta, s\right)\right)\right). \tag{8.3.3}$$

As in the proof of Theorem 8.3.1, if we can find φ that minimizes $E_{\Pi(\cdot \mid s)}\left(e\left(\theta, s\right)\right)$ for each s, then φ also minimizes (8.3.3) and is a Bayes rule.

Using Theorem 3.5.4 to pull $\varphi(s)$ through the conditional expectation, and the fact that $E_{\Pi(\cdot \mid s)}\left(I_A\left(\theta\right)\right) = \Pi\left(A \mid s\right)$ for any event A, then

$$E_{\Pi(\cdot \mid s)}\left(e\left(\theta, s\right)\right)$$
$$= \varphi(s)\,\Pi\left(\{\psi\left(\theta\right) = \psi_0\} \mid s\right) + \left(1 - \varphi(s)\right)\left(1 - \Pi\left(\{\psi\left(\theta\right) = \psi_0\} \mid s\right)\right).$$

Because $\varphi(s) \in [0, 1]$, we have

$$\min\left\{\Pi\left(\{\psi\left(\theta\right) = \psi_0\} \mid s\right), 1 - \Pi\left(\{\psi\left(\theta\right) = \psi_0\} \mid s\right)\right\}$$
$$\leq \varphi(s)\,\Pi\left(\{\psi\left(\theta\right) = \psi_0\} \mid s\right) + \left(1 - \varphi(s)\right)\left(1 - \Pi\left(\{\psi\left(\theta\right) = \psi_0\} \mid s\right)\right).$$

Therefore, the minimum value of $E_{\Pi(\cdot \mid s)}\left(e\left(\theta, s\right)\right)$ is attained by $\varphi(s) = \varphi_0(s)$. ∎

Observe that Theorem 8.3.2 says that the Bayes rule rejects H_0 whenever the posterior probability of the null hypothesis is less than or equal to the posterior probability of the alternative. This is an intuitively satisfying result.

The following problem does arise with this approach, however. We have

$$\Pi\left(\{\psi\left(\theta\right) = \psi_0\} \mid s\right) = \frac{E_\Pi\left(I_{\{\theta:\psi(\theta)=\psi_0\}}\,f_\theta\left(s\right)\right)}{m(s)}$$
$$\leq \frac{\max_{\{\theta:\psi(\theta)=\psi_0\}}\,f_\theta\left(s\right)\Pi\left(\{\psi\left(\theta\right) = \psi_0\}\right)}{m(s)}. \qquad (8.3.4)$$

When $\Pi\left(\{\psi\left(\theta\right) = \psi_0\}\right) = 0$, (8.3.4) implies that $\Pi\left(\{\psi\left(\theta\right) = \psi_0\} \mid s\right) = 0$ no matter what data the s is. Therefore, using the Bayes rule, we would always reject H_0 no matter what data s are obtained which doesn't seem sensible. As discussed in Section 7.2.3, we have to be careful to make sure we use a prior Π that assigns positive mass to H_0 if we are going to use the optimal Bayes approach to a hypothesis testing problem.

Summary of Section 8.3

- Optimal Bayesian procedures are obtained by minimizing the expected performance measure using the posterior distribution.

- In estimation problems, when using squared error as the performance measure, the posterior mean is optimal.

- In hypothesis testing problems, when minimizing the probability of making an error as the performance measure, then computing the posterior probability of the null hypothesis and accepting H_0 when this is greater than $1/2$, is optimal.

Exercises

8.3.1 Suppose that $S = \{1, 2, 3\}, \Omega = \{1, 2\}$, with data distributions given by the following table. We place a uniform prior on θ and want to estimate θ.

	$s = 1$	$s = 2$	$s = 3$
$f_1(s)$	1/6	1/6	2/3
$f_2(s)$	1/4	1/4	1/2

Using a Bayes rule, test the hypothesis $H_0 : \theta = 2$ when $s = 2$ is observed.

8.3.2 For the situation described in Exercise 8.3.1, determine the Bayes rule estimator of θ when using expected squared error as our performance measure for estimators.

8.3.3 Suppose that we have a sample $(x_1, \ldots, x_n)$ from an $N(\mu, \sigma_0^2)$ where μ is unknown and σ_0^2 is known, and we want to estimate μ using expected squared error as our performance measure for estimators. If we use the prior distribution $\mu \sim N(\mu, \tau_0^2)$, then determine the Bayes rule for this problem. Determine the limiting Bayes rule as $\tau_0 \to \infty$.

8.3.4 Suppose that we observe a sample $(x_1, \ldots, x_n)$ from a Bernoulli(θ) distribution where θ is completely unknown, and we want to estimate θ using expected squared error as our performance measure for estimators. If we use the prior distribution $\theta \sim \text{Beta}(\alpha, \beta)$, then determine a Bayes rule for this problem.

8.3.5 Suppose that $(x_1, \ldots, x_n)$ is a sample from a Gamma(α_0, β) distribution where α_0 is known and $\beta \sim \text{Gamma}(\tau_0, \upsilon_0)$ where τ_0 and υ_0 are known. If we want to estimate β using expected squared error as our performance measure for estimators, then determine the Bayes rule. Use the weak (or strong) law of large numbers to determine what this estimator converges to as $n \to \infty$.

8.3.6 For the situation described in Exercise 8.3.5, determine the Bayes rule for estimating β^{-1} when using expected squared error as our performance measure for estimators.

Problems

8.3.7 Suppose that $\Omega = \{\theta_1, \theta_2\}$, that we put a prior π on Ω, and that we want to estimate θ. Suppose our performance measure for estimators is the probability of making an incorrect choice of θ. If the model is denoted $\{f_\theta : \theta \in \Omega\}$, then obtain the form of the Bayes rule when data s are observed.

8.3.8 For the situation described in Exercise 8.3.1, use the Bayes rule obtained via the method of Problem 8.3.7 to estimate θ when $s = 2$. What advantage does this estimate have over that obtained in Exercise 8.3.2?

8.3.9 Suppose that $(x_1, \ldots, x_n)$ is a sample from an $N(\mu, \sigma^2)$ distribution where $(\mu, \sigma^2) \in R^1 \times (0, \infty)$ is unknown, and and want to estimate μ using expected squared error as our performance measure for estimators. If we use the prior distribution given by

$$\mu \,|\, \sigma^2 \sim N(\mu_0, \tau_0^2 \sigma^2),$$

and

$$\frac{1}{\sigma^2} \sim \text{Gamma}\left(\alpha_0, \beta_0\right)$$

where $\mu_0, \tau_0^2, \alpha_0$, and β_0 are fixed and known, then determine the Bayes rule for μ.

8.3.10 (*Model selection*) Generalize Problem 8.3.7 to the case $\Omega = \{\theta_1, \ldots, \theta_k\}$.

Challenge

8.3.11 In Section 7.2.4 we described the Bayesian prediction problem. Using the notation found there, suppose we wish to predict $t \in R^1$ using a predictor $\tilde{T}(s)$. If we assess the accuracy of a predictor by

$$E\left(\left(\tilde{T}(s) - t\right)^2\right) = E_\Pi\left(E_{P_\theta}\left(E_{Q_\theta(\cdot \,|\, s)}\left(\left(\tilde{T}(s) - t\right)^2\right)\right)\right)$$

then determine the prior predictor that minimizes this quantity (assume all relevant expectations are finite). If we observe s_0, then determine the best predictor. (Hint: Assume all the probability measures are discrete.)

8.4 Decision Theory (Advanced)

To determine an optimal inference we chose a performance measure and then attempted to find an inference, of a given type, that has optimal performance with respect to this measure. For example, when considering estimates of a real-valued characteristic of interest $\psi(\theta)$, we took the performance measure to be MSE and then searched for the estimator that minimizes this for each value of θ.

Decision theory is closely related to the optimal approach to deriving inferences, but it is a little more specialized. In the decision framework we take the point of view that, in any statistical problem, the statistician is faced with making a decision, e.g., deciding on a particular value for $\psi(\theta)$. Further, associated with a decision is the notion of a loss incurred whenever the decision is incorrect. A decision rule is a procedure, based on the observed data s, that the statistician uses to select a decision. The decision problem is then to find a decision rule that minimizes the average loss incurred.

There are a number of real-world contexts in which losses are an obvious part of the problem, e.g., the monetary losses associated with various insurance plans that an insurance company may consider offering. So the decision theory approach has many applications. It is clear in many practical problems, however, that losses (as well as performance measures) are somewhat arbitrary components of a statistical problem, often chosen simply for convenience. In such circumstances the approaches to deriving inferences described in Chapters 6 and 7 are preferred by many statisticians.

So the *decision theory model* for inference adds another ingredient to the sampling model (or to the sampling model and prior) to derive inferences —

the loss function. To formalize this, we conceive of a set of possible actions or decisions that the statistician could take after observing the data s. This set of possible actions is denoted by $\mathcal{A}$ and is called the *action space*. To connect these actions with the statistical model, there is a *correct action function* $A : \Omega \rightarrow \mathcal{A}$ such that $A(\theta)$ is the correct action to take when θ is the true value of the parameter. Of course, because we do not know θ, we do not know the correct action $A(\theta)$, so there is uncertainty involved in our decision. Consider a simple example.

Example 8.4.1
Suppose you are told that an urn containing 100 balls has either 50 white and 50 black balls or 60 white and 40 black balls. Five balls are drawn from the urn without replacement and their colors are observed. The statistician's job is to make a decision about the true proportion of white balls in the urn based on these data.

The statistical model then comprises two distributions $\{P_1, P_2\}$ where, using parameter space $\Omega = \{1, 2\}$, P_1 is the Hypergeometric$(100, 50, 5)$ distribution (see Example 2.3.7) and P_2 is the Hypergeometric$(100, 60, 5)$ distribution. The action space is $\mathcal{A} = \{0.5, 0.6\}$, and $A : \Omega \rightarrow \mathcal{A}$ is given by $A(1) = 0.5$ and $A(2) = 0.6$. The data is given by the colors of the five balls drawn. ∎

We suppose now that there is also a loss or penalty $L(\theta, a)$ incurred when we select action $a \in \mathcal{A}$ and θ is true. If we select the correct action, then the loss is 0; it is greater than 0 otherwise. Thus, a *loss function* is a function L defined on $\Omega \times \mathcal{A}$ and taking values in $[0, \infty)$ such that $L(\theta, a) = 0$ if and only if $a = A(\theta)$. Sometimes the loss can be an actual monetary loss. Actually, decision theory is a little more general than what we have just described, as we can allow for negative losses (gains or profits), but the restriction to nonnegative losses is suitable for purely statistical applications.

In a specific problem, the statistician chooses a loss function that is believed to lead to reasonable statistical procedures. This choice is dependent on the particular application. Consider some examples.

Example 8.4.2 (*Example 8.4.1 continued*)
Perhaps a sensible choice in this problem would be

$$L(\theta, a) = \begin{cases} 1 & \theta = 1, a = 0.6 \\ 2 & \theta = 2, a = 0.5 \\ 0 & \text{otherwise.} \end{cases}$$

Here we have decided that selecting $a = 0.5$ when it is not correct, is a more serious error than selecting $a = 0.6$ when it is not correct. If we want to treat errors symmetrically, then we could take

$$L(\theta, a) = I_{\{(1, 0.6), (2, 0.5)\}}(\theta, a),$$

i.e., the losses are 1 or 0. ∎

Example 8.4.3 *Estimation as a Decision Problem*

Suppose we have a marginal parameter $\psi(\theta)$ of interest, and we want to specify an estimate $T(s)$ after observing $s \in S$. Here the action space is $\mathcal{A} = \{\psi(\theta) : \theta \in \Omega\}$ and $A(\theta) = \psi(\theta)$. Naturally, we want $T(s) \in \mathcal{A}$.

For example, suppose $(x_1, \ldots, x_n)$ is a sample from an $N(\mu, \sigma^2)$ distribution where $(\mu, \sigma^2) \in \Omega = R^1 \times R^+$ is unknown, and we want to estimate $\psi(\mu, \sigma^2) = \mu$. In this case, $\mathcal{A} = R^1$ and a possible estimator is the sample average $T(x_1, \ldots, x_n) = \bar{x}$.

There are many possible choices for the loss function. Perhaps a natural choice is to use

$$L(\theta, a) = |\psi(\theta) - a|, \tag{8.4.1}$$

the absolute deviation between $\psi(\theta)$ and a. Alternatively, it is common to use

$$L(\theta, a) = (\psi(\theta) - a)^2, \tag{8.4.2}$$

the squared deviations between $\psi(\theta)$ and a.

We refer to (8.4.2) as *squared error loss*. Notice that (8.4.2) is just the square of the Euclidean distance between $\psi(\theta)$ and a. It might seem more natural to actually use the distance (8.4.1) as the loss function. It turns out, however, that there are a number of mathematical conveniences that arise from using squared distance. ∎

Example 8.4.4 *Hypothesis Testing as a Decision Problem*

In this problem we have a characteristic of interest $\psi(\theta)$ and want to assess the plausibility of the value ψ_0 after viewing the data s. In a hypothesis testing problem, this is written as $H_0 : \psi(\theta) = \psi_0$ versus $H_a : \psi(\theta) \neq \psi_0$. As in Section 8.2, we refer to H_0 as the null hypothesis and to H_a as the alternative hypothesis.

The purpose of a hypothesis testing procedure is to decide which of H_0 or H_a is true based on the observed data s. So in this problem, the action space is $\mathcal{A} = \{H_0, H_a\}$ and the correct action function is

$$A(\theta) = \begin{cases} H_0 & \psi(\theta) = \psi_0 \\ H_a & \psi(\theta) \neq \psi_0. \end{cases}$$

An alternative, and useful, way of thinking of the two hypotheses is as subsets of Ω. We write $H_0 = \psi^{-1}\{\psi_0\}$ as the subset of all θ values that make the null hypothesis true, and $H_a = H_0^c$ is the subset of all θ values that make the null hypothesis false. Then, based on the data s, we want to decide if the true value of θ is in H_0 or if θ is in H_a. If H_0 (or H_a) is composed of a single point then it is called a *simple hypothesis* or a *point hypothesis*; otherwise, it is referred to as a *composite hypothesis*.

For example, suppose that $(x_1, \ldots, x_n)$ is a sample from an $N(\mu, \sigma^2)$ distribution where $\theta = (\mu, \sigma^2) \in \Omega = R^1 \times R^+$, $\psi(\theta) = \mu$, and we want to test the null hypothesis $H_0 : \mu = \mu_0$ versus the alternative $H_a : \mu \neq \mu_0$. Then $H_0 = \{\mu_0\} \times R^+$ and $H_a = \{\mu_0\}^c \times R^+$. For the same model, let

$$\psi(\theta) = I_{(-\infty, \mu_0] \times R^+}(\mu, \sigma^2),$$

i.e., ψ is the indicator function for the subset $(-\infty, \mu_0] \times R^+$. Then testing $H_0 : \psi = 1$ versus the alternative $H_a : \psi = 0$ is equivalent to testing that the mean is less than or equal to μ_0 versus the alternative that it is greater than μ_0. This one-sided hypothesis testing problem is often denoted as $H_0 : \mu \le \mu_0$ versus $H_a : \mu > \mu_0$.

There are a number of possible choices for the loss function, but the most commonly used is of the form

$$L(\theta, a) = \begin{cases} 0 & \theta \in H_0, a = H_0 \text{ or } \theta \in H_a, a = H_a \\ b & \theta \notin H_0, a = H_0 \\ c & \theta \notin H_a, a = H_a. \end{cases}$$

If we reject H_0 when H_0 is true (a type I error), we incur a loss of c; if we accept H_0 when H_0 is false (a type II error), we incur a loss of b. When $b = c$, we can take $b = c = 1$ and produce the commonly used *0–1 loss function*. ∎

A statistician faced with a decision problem — i.e., a model, action space, correct action function, and loss function — must now select a rule for choosing an element of the action space $\mathcal{A}$ when the data s are observed. A *decision function* is a procedure that specifies how an action is to be selected in the action space $\mathcal{A}$.

Definition 8.4.1 A *nonrandomized decision function d* is a function $d : S \to \mathcal{A}$.

So after observing s, we decide that the appropriate action is $d(s)$. Actually, we will allow our decision procedures to be a little more general than this, as we permit a random choice of an action after observing s.

Definition 8.4.2 A *decision function* δ is such that $\delta(s, \cdot)$ is a probability measure on the action space $\mathcal{A}$ for each $s \in S$ (so $\delta(s, A)$ is the probability that the action taken is in $A \subset \mathcal{A}$).

Operationally, after observing s, a random mechanism, with distribution specified by $\delta(s, \cdot)$, is used to select the action from the set of possible actions. Notice that if $\delta(s, \cdot)$ is a probability measure degenerate at the point $d(s)$ (so $\delta(s, \{d(s)\}) = 1$) for each s, then δ is equivalent to the nonrandomized decision function d and conversely (Problem 8.4.8).

The use of randomized decision procedures may seem rather unnatural, but, as we will see, sometimes they are an essential ingredient of decision theory. In many estimation problems, the use of randomized procedures provides no advantage, but this is not the case in hypothesis testing problems. We let D denote the set of all decision functions δ for the specific problem of interest.

The decision problem is to choose a decision function $\delta \in D$. The selected δ will then be used to generate decisions in applications. We base this choice on how the various decision functions δ perform with respect to the loss function. Intuitively, we want to choose δ to make the loss as small as possible. For a particular δ, because $s \sim f_\theta$ and $a \sim \delta(s, \cdot)$, the loss $L(\theta, a)$ is a random quantity. Therefore, rather than minimizing specific losses, we speak instead about minimizing some aspect of the distribution of the losses for each $\theta \in \Omega$. Perhaps a reasonable choice is to minimize the average loss. Accordingly, we define the risk function associated with $\delta \in D$ as the average loss incurred by δ. The risk function plays a central role in determining an appropriate decision function for a problem.

Definition 8.4.3 The *risk function* associated with decision function δ is given by

$$R_\delta (\theta) = E_\theta \left(E_{\delta(s,\cdot)} (L (\theta, a)) \right). \tag{8.4.3}$$

Notice that to calculate the risk function we first calculate the average of $L (\theta, a)$, based on s fixed and $a \sim \delta (s, \cdot)$, and then average this conditional average with respect to $s \sim f_\theta$. By the theorem of total expectation, this is the average loss. When $\delta (s, \cdot)$ is degenerate at $d(s)$ for each s, then (8.4.3) simplifies (Problem 8.4.8) to

$$R_\delta (\theta) = E_\theta (L (\theta, d (s))).$$

Consider the following examples.

Example 8.4.5
Suppose that $S = \{1, 2, 3\}, \Omega = \{1, 2\}$, and the distributions are given by the following table.

	$s = 1$	$s = 2$	$s = 3$
$f_1 (s)$	1/3	1/3	1/3
$f_2 (s)$	1/2	1/2	0

Further suppose that $\mathcal{A} = \Omega, A(\theta) = \theta$, and the loss function is given by $L (\theta, a) = 1$ when $\theta \neq a$ but is 0 otherwise.

Now consider the decision function δ specified by the following table.

	$a = 1$	$a = 2$
$\delta (1, \{a\})$	1/4	3/4
$\delta (2, \{a\})$	1/4	3/4
$\delta (3, \{a\})$	1	0

So when we observe $s = 1$, we randomly choose the action $a = 1$ with probability 1/4 and choose the action $a = 2$ with probability 3/4, etc. Notice that this decision function does the sensible thing and selects the decision $a = 1$ when we observe $s = 3$, as we know unequivocally that $\theta = 1$ in this case.
We have

$$E_{\delta(1,\cdot)} (L (\theta, a)) = \frac{1}{4}L(\theta, 1) + \frac{3}{4}L(\theta, 2)$$

$$E_{\delta(2,\cdot)} (L (\theta, a)) = \frac{1}{4}L(\theta, 1) + \frac{3}{4}L(\theta, 2)$$

$$E_{\delta(3,\cdot)} (L (\theta, a)) = L(\theta, 1),$$

so the risk function of δ is then given by

$$
\begin{aligned}
R_\delta(1) &= E_1\left(E_{\delta(s,\cdot)}\left(L(1,a)\right)\right) \\
&= \frac{1}{3}\left(\frac{1}{4}L(1,1) + \frac{3}{4}L(1,2)\right) + \frac{1}{3}\left(\frac{1}{4}L(1,1) + \frac{3}{4}L(1,2)\right) + \frac{1}{3}L(1,1) \\
&= \frac{3}{12} + \frac{3}{12} + 0 = \frac{1}{2}
\end{aligned}
$$

and

$$
\begin{aligned}
R_\delta(2) &= E_2\left(E_{\delta(s,\cdot)}\left(L(2,a)\right)\right) \\
&= \frac{1}{2}\left(\frac{1}{4}L(2,1) + \frac{3}{4}L(2,2)\right) + \frac{1}{2}\left(\frac{1}{4}L(2,1) + \frac{3}{4}L(2,2)\right) + 0L(2,1) \\
&= \frac{1}{8} + \frac{1}{8} + 0 = \frac{1}{4}. \quad \blacksquare
\end{aligned}
$$

Example 8.4.6 *Estimation*
We will restrict our attention to nonrandomized decision functions and note that these are also called estimators. The risk function associated with estimator T and loss function (8.4.1) is given by

$$
R_T(\theta) = E_\theta\left(|\psi(\theta) - T|\right)
$$

and is called the *mean absolute deviation* (MAD). The risk function associated with the estimator T and loss function (8.4.2) is given by

$$
R_T(\theta) = E_\theta\left((\psi(\theta) - T)^2\right)
$$

and is called the MSE.

We want to choose the estimator T to minimize $R_T(\theta)$ for every $\theta \in \Omega$. Note that, when using (8.4.2), this decision problem is exactly the same as the optimal estimation problem discussed in Section 8.1. $\blacksquare$

Example 8.4.7 *Hypothesis Testing*
We note that for a given decision function δ for this problem, and a data value s, the distribution $\delta(s,\cdot)$ is characterized by $\varphi(s) = \delta(s, H_a)$, which is the probability of rejecting H_0 when s has been observed. This is because the probability measure $\delta(s,\cdot)$ is concentrated on two points, so we need only give its value at one of these to completely specify it. We call φ the *test function* associated with δ and observe that a decision function for this problem is also specified by a test function φ.

We have immediately that

$$
E_{\delta(s,\cdot)}\left(L(\theta,a)\right) = (1 - \varphi(s))\,L(\theta, H_0) + \varphi(s)\,L(\theta, H_a) \tag{8.4.4}
$$

and therefore, when using the $0 - 1$ loss function,

$$
\begin{aligned}
R_\delta(\theta) &= E_\theta\left((1 - \varphi(s))\,L(\theta, H_0) + \varphi(s)\,L(\theta, H_a)\right) \\
&= L(\theta, H_0) + E_\theta\left(\varphi(s)\right)\left(L(\theta, H_a) - L(\theta, H_0)\right) \\
&= \begin{cases} E_\theta\left(\varphi(s)\right) & \theta \in H_0 \\ 1 - E_\theta\left(\varphi(s)\right) & \theta \in H_a. \end{cases}
\end{aligned}
$$

Recall that in Section 6.3.5 we introduced the power function associated with a hypothesis assessment procedure that rejected H_0 whenever the P-value was smaller than some prescribed value. The power function, evaluated at θ, is the probability that such a procedure rejects H_0 when θ is the true value. Because $\varphi(s)$ is the conditional probability, given s, that H_0 is rejected, the theorem of total expectation implies that $E_\theta(\varphi(s))$ equals the unconditional probability that we reject H_0 when θ is the true value. So in general we refer to the function

$$\beta_\varphi(\theta) = E_\theta(\varphi(s))$$

as the *power function* of the decision procedure δ or, equivalently, as the power function of the test function φ.

Therefore, minimizing the risk function in this case is equivalent to choosing φ to minimize $\beta(\theta)$ for every $\theta \in H_0$ and to maximize $\beta_\varphi(\theta)$ for every $\theta \in H_a$. Accordingly, this decision problem is exactly the same as the optimal inference problem discussed in Section 8.2. ∎

Once we have written down all the ingredients for a decision problem, it is then clear what form a solution to the problem will take. In particular, any decision function δ_0 that satisfies

$$R_{\delta_0}(\theta) \le R_\delta(\theta)$$

for every $\theta \in \Omega$ and $\delta \in D$ is an *optimal decision function* and is a solution. If two decision functions have the same risk functions, then, from the point of view of decision theory, they are equivalent. So it is conceivable that there might be more than one solution to a decision problem.

Actually, it turns out that an optimal decision function exists only in extremely unrealistic cases, namely, the data always tell us categorically what the correct decision is (Problem 8.4.9). We do not really need statistical inference for such situations. For example, suppose we have two coins — coin A has two heads and coin B has two tails. As soon as we observe an outcome from a coin toss, we know exactly which coin was tossed and there is no need for statistical inference.

Still, we can identify some decision rules that we do not want to use. For example, if $\delta \in D$ is such that there exists $\delta_0 \in D$ satisfying $R_{\delta_0}(\theta) \le R_\delta(\theta)$ for every θ, and if there is at least one θ for which $R_{\delta_0}(\theta) < R_\delta(\theta)$, then naturally we strictly prefer δ_0 to δ. A decision function δ is said to be *admissible* if there is no δ_0 that is strictly preferred to it. A consequence of decision theory is that we use only admissible decision functions. Still, there are many admissible decision functions and typically none is optimal. Further, a procedure that is only admissible may be a very poor choice (Challenge 8.4.11).

There are several routes out of this impasse for decision theory. One approach is to use *reduction principles*. By this we mean that we look for an optimal decision function in some subclass $D_0 \subset D$ that is considered appropriate. So we look for a $\delta_0 \in D_0$ such that $R_{\delta_0}(\theta) \le R_\delta(\theta)$ for every $\theta \in \Omega$ and $\delta \in D_0$. Consider the following example.

Example 8.4.8 *Size α Tests for Hypothesis Testing*

Consider a hypothesis testing problem H_0 versus H_a. Recall that in Section 8.2 we restricted attention to those test functions φ that satisfy $E_\theta (\varphi) \leq \alpha$ for every $\theta \in H_0$. Such a φ is called a size α test function for this problem. So in this case we are restricting to the class D_0 of all decision functions δ for this problem, which correspond to size α test functions.

In Section 8.2 we showed that sometimes there is an optimal $\delta \in D_0$. For example, when H_0 and H_a are simple, the Neyman–Pearson theorem (Theorem 8.2.1) provides an optimal φ, and thus δ, defined by $\delta (s, H_a) = \varphi (s)$ is optimal. We also showed in Section 8.2, however, that in general there is no optimal size α test function φ and so there is no optimal $\delta \in D_0$. In this case, further reduction principles are necessary. ∎

Another approach to selecting a $\delta \in D$ is based on choosing one particular real-valued characteristic of the risk function of δ and ordering the decision functions based on that. There are several possibilities.

One way to do this is to introduce a prior π into the problem and then look for the decision procedure $\delta \in D$ that has smallest average risk

$$r_\delta = E_\pi (R_\delta (\theta)) .$$

This approach is called *Bayesian decision theory*. The quantity r_δ is called the *Bayes risk*, and a rule with smallest Bayes risk is called a *Bayes rule*. We derived Bayes rules for several problems in Section 8.3. Interestingly, Bayesian decision theory always effectively produces an answer to a decision problem. This is a very desirable property for any theory of statistics.

Another way to order decision functions uses the maximum (or supremum) risk. So for a decision function δ, we calculate

$$\max_{\theta \in \Omega} R_\delta (\theta)$$

(or $\sup_{\theta \in \Omega} R_\delta (\theta)$) and then select the $\delta \in D$ that minimizes this quantity. Such a δ is called a *minimax decision function* (δ has the smallest, largest risk or the smallest, worst behavior). Again, this approach will always effectively produce an answer to a decision problem (see Problem 8.4.10).

Much more can be said about decision theory than this brief introduction to the basic concepts. Many interesting, general results have been established for the decision theoretic approach to statistical inference.

Summary of Section 8.4

- The decision theoretic approach to statistical inference introduces an action space $\mathcal{A}$ and a loss function L.

- A decision function δ prescribes a probability distribution $\delta (s, \cdot)$ on $\mathcal{A}$. The statistician generates a decision in $\mathcal{A}$ using this distribution after observing s.

- The problem in decision theory is to select δ; for this, the risk function $R_\delta(\theta)$ is used. The value $R_\delta(\theta)$ is the average loss incurred when using the decision function δ, and the goal is to minimize risk.

- Typically, no optimal decision function δ exists. So, to select a δ, various reduction criteria are used to reduce the class of possible decision functions, or the decision functions are ordered using some real-valued characteristic of their risk functions, e.g., maximum risk or average risk with respect to some prior.

Exercises

8.4.1 Suppose that we observe a sample $(x_1, \ldots, x_n)$ from a Bernoulli(θ) distribution where θ is completely unknown, and we want to estimate θ using squared error loss. Write out all the ingredients of this decision problem. Consider the estimator $T(x_1, \ldots, x_n) = \bar{x}$ and calculate its risk function. Graph the risk function when $n = 10$.

8.4.2 Suppose that we have a sample $(x_1, \ldots, x_n)$ from a Poisson(λ) distribution where λ is completely unknown, and we want to estimate λ using squared error loss. Write out all the ingredients of this decision problem. Consider the estimator $T(x_1, \ldots, x_n) = \bar{x}$ and calculate its risk function. Graph the risk function when $n = 25$.

8.4.3 Suppose that we have a sample $(x_1, \ldots, x_n)$ from an $N(\mu, \sigma_0^2)$ where μ is unknown and σ_0^2 is known, and we want to estimate μ using squared error loss. Write out all the ingredients of this decision problem. Consider the estimator $T(x_1, \ldots, x_n) = \bar{x}$ and calculate its risk function. Graph the risk function when $n = 25, \sigma_0^2 = 2$.

8.4.4 Suppose that we observe a sample $(x_1, \ldots, x_n)$ from a Bernoulli(θ) distribution where θ is completely unknown, and we want to test the null hypothesis that $\theta = 1/2$ versus the alternative that it is not equal to this quantity, and we use $0-1$ loss. Write out all the ingredients of this decision problem. Suppose that we reject the null hypothesis whenever we observe $n\bar{x} \in \{0, 1, n-1, n\}$. Determine the form of the test function and its associated power function. Graph the power function when $n = 10$.

8.4.5 Consider the decision problem with sample space $S = \{1, 2, 3, 4\}$, parameter space $\Omega = \{a, b\}$, with the parameter indexing the distributions given in the following table.

	$s = 1$	$s = 2$	$s = 3$	$s = 4$
$f_a(s)$	$1/4$	$1/4$	0	$1/2$
$f_b(s)$	$1/2$	0	$1/4$	$1/4$

Suppose that the action space $\mathcal{A} = \Omega$, with $A(\theta) = \theta$, and the loss function is given by $L(\theta, a) = 1$ when $a \neq A(\theta)$ and is equal to 0 otherwise.

(a) Calculate the risk function of the deterministic decision function given by $d(1) = d(2) = d(3) = a$ and $d(4) = b$.
(b) Is d in part (a) optimal?

Computer Exercises

8.4.6 Suppose that we have a sample $(x_1, \ldots, x_n)$ from a Poisson(λ) distribution where λ is completely unknown, and we want to test the hypothesis that $\lambda \leq \lambda_0$ versus the alternative that $\lambda > \lambda_0$, using the $0 - 1$ loss function. Write out all the ingredients of this decision problem. Suppose that we decide to reject the null hypothesis whenever $n\bar{x} > \lfloor n\lambda_0 + 2\sqrt{n\lambda_0} \rfloor$ and randomly reject the null hypothesis with probability $1/2$ when $n\bar{x} = \lfloor n\lambda_0 + 2\sqrt{n\lambda_0} \rfloor$. Determine the form of the test function and its associated power function. Graph the power function when $\lambda_0 = 1$ and $n = 5$.

8.4.7 Suppose that we have a sample $(x_1, \ldots, x_n)$ from an $N(\mu, \sigma_0^2)$ distribution where μ is unknown and σ_0^2 is known, and we want to test the null hypothesis that the mean response is μ_0 versus the alternative that the mean response is not equal to μ_0, using the $0 - 1$ loss function. Write out all the ingredients of this decision problem. Suppose that we decide to reject whenever $\bar{x} \notin [\mu_0 - 2\sigma_0/\sqrt{n}, \mu_0 + 2\sigma_0/\sqrt{n}]$. Determine the form of the test function and its associated power function. Graph the power function when $\mu_0 = 0, \sigma_0 = 3$, and $n = 10$.

Problems

8.4.8 Prove that a decision function δ that gives a probability measure $\delta(s, \cdot)$ degenerate at $d(s)$ for each $s \in S$ is equivalent to specifying a function $d : S \to \mathcal{A}$ and conversely. For such a δ, prove that $R_\delta(\theta) = E_\theta(L(\theta, d(s)))$.

8.4.9 Suppose we have a decision problem and that each probability distribution in the model is discrete.
(a) Prove that δ is optimal in D if and only if $\delta(s, \cdot)$ is degenerate at $A(\theta)$ for each s for which $P_\theta(\{s\}) > 0$.
(b) Prove that if there exist $\theta_1, \theta_2 \in \Omega$ such that $A(\theta_1) \neq A(\theta_2)$, and $P_{\theta_1}, P_{\theta_2}$ are not concentrated on disjoint sets, then there is no optimal $\delta \in D$.

8.4.10 If decision function δ has constant risk and is admissible, then prove that δ is minimax.

Challenge

8.4.11 Suppose we have a decision problem in which $\theta_0 \in \Omega$ is such that $P_{\theta_0}(C) = 0$ implies that $P_\theta(C) = 0$ for every $\theta \in \Omega$. Further assume that there is no optimal decision function (see Problem 8.4.9). Then prove that the nonrandomized decision function d given by $d(s) \equiv A(\theta_0)$ is admissible. What does this result tell you about the concept of admissibility?

Discussion Topics

8.4.12 Comment on the following statement: A natural requirement for any theory of inference is that it produce an answer for every inference problem posed. Have we discussed any theories so far that you believe will satisfy this?

8.4.13 Decision theory produces a decision in a given problem. It says nothing about how likely it is that the decision is in error. Some statisticians argue that a valid approach to inference must include some quantification of our uncertainty concerning any statement we make about an unknown, as only then can a recipient judge the reliability of the inference. Comment on this.

8.5   Further Proofs (Advanced)

Theorem 8.1.2 A statistic U is sufficient for a model if and only if the conditional distribution of the data s given $U = u$ is the same for every $\theta \in \Omega$.

Proof: We prove this in the discrete case so that $f_\theta(s) = P_\theta(\{s\})$. The general case requires more mathematics, and we leave that to a further course.

Let u be such that $P_\theta(U^{-1}\{u\}) > 0$ where $U^{-1}\{u\} = \{s : U(s) = u\}$, so $U^{-1}\{u\}$ is the set of values of s such that $U(s) = u$. We have

$$P_\theta(s = s_1 \,|\, U = u) = \frac{P_\theta(s = s_1, U = u)}{P_\theta(U = u)}. \qquad (8.5.1)$$

Whenever $s_1 \notin U^{-1}\{u\}$,

$$P_\theta(s = s_1, U = u) = P_\theta(\{s_1\} \cap \{s : U(s) = u\}) = P_\theta(\phi) = 0$$

independently of θ. Therefore, $P_\theta(s = s_1 \,|\, U = u) = 0$ independently of θ.

So let us suppose that $s_1 \in U^{-1}\{u\}$. Then

$$P_\theta(s = s_1, U = u) = P_\theta(\{s_1\} \cap \{s : U(s) = u\}) = P_\theta(\{s_1\}) = f_\theta(s_1).$$

If U is a sufficient statistic, the factorization theorem (Theorem 6.1.1) implies $f_\theta(s) = h(s) g_\theta(U(s))$ for some h and g. Therefore, since

$$P_\theta(U = u) = \sum_{s \in U^{-1}\{u\}} f_\theta(s),$$

(8.5.1) equals

$$\frac{f_\theta(s_1)}{\sum_{s \in U^{-1}\{u\}} f_\theta(s)} = \frac{f_\theta(s_1)}{\sum_{s \in U^{-1}\{u\}} c(s, s_1) f_\theta(s_1)} = \frac{1}{\sum_{s \in U^{-1}\{u\}} c(s, s_1)}$$

where

$$\frac{f_\theta(s)}{f_\theta(s_1)} = \frac{h(s)}{h(s_1)} = c(s, s_1).$$

We conclude that (8.5.1) is independent of θ.

Conversely, if (8.5.1) is independent of θ, then for $s_1, s_2 \in U^{-1}\{u\}$ we have

$$P_\theta(U = u) = \frac{P_\theta(s = s_2)}{P_\theta(s = s_2 \,|\, U = u)},$$

and so

$$
\begin{aligned}
f_\theta(s_1) &= P_\theta(s = s_1) = P_\theta(s = s_1 \,|\, U = u)\, P_\theta(U = u) \\
&= P_\theta(s = s_1 \,|\, U = u)\, \frac{P_\theta(s = s_2)}{P_\theta(s = s_2 \,|\, U = u)} \\
&= \frac{P_\theta(s = s_1 \,|\, U = u)}{P_\theta(s = s_2 \,|\, U = u)}\, f_\theta(s_2) = c(s_1, s_2) f_\theta(s_2)
\end{aligned}
$$

where

$$
c(s_1, s_2) = \frac{P_\theta(s = s_1 \,|\, U = u)}{P_\theta^{-1}(s = s_2 \,|\, U = u)}.
$$

By the definition of sufficiency in Section 6.1.1, this establishes the sufficiency of U. $\blacksquare$

Example 8.1.3 *Completeness of $\bar{x}$ in the Location Normal Model*
Suppose that $(x_1, \ldots, x_n)$ is a sample from an $N(\mu, \sigma_0^2)$ distribution where $\mu \in R^1$ is unknown and $\sigma_0^2 > 0$ is known. In Example 6.1.6 we showed that $\bar{x}$ is a minimal sufficient statistic.

Suppose that the function h is such that $E_\mu(h(\bar{x})) = 0$ for every $\mu \in R^1$. Then defining

$$
h^+(\bar{x}) = \max(0, h(\bar{x})) \ \text{ and } \ h^-(\bar{x}) = \max(0, -h(\bar{x})),
$$

we have $h(\bar{x}) = h^+(\bar{x}) - h^-(\bar{x})$. Therefore, putting

$$
c^+(\mu) = E_\mu(h^+(\bar{X})) \ \text{ and } \ c^-(\mu) = E_\mu(h^-(\bar{X})),
$$

we must have

$$
E_\mu(h(\bar{X})) = E_\mu(h^+(\bar{X})) - E_\mu(h^-(\bar{X})) = c^+(\mu) - c^-(\mu) = 0,
$$

and so $c^+(\mu) = c^-(\mu)$. Because h^+ and h^- are nonnegative functions, we have that $c^+(\mu) \geq 0$ and $c^-(\mu) \geq 0$.

If $c^+(\mu) = 0$, then we have that $h^+(\bar{x}) = 0$ with probability 1, because a nonnegative function has mean 0 if and only if it is 0 with probability 1 (Challenge 3.3.22). Then $h^-(\bar{x}) = 0$ with probability 1 also, and we conclude that $h(\bar{x}) = 0$ with probability 1.

If $c^+(\mu_0) > 0$, then $h^+(\bar{x}) > 0$ for all $\bar{x}$ in a set A having positive probability with respect to the $N(\mu_0, \sigma_0^2/n)$ distribution (otherwise $h^+(\bar{x}) = 0$ with probability 1, which implies, as above, that $c^+(\mu_0) = 0$). This implies that $c^+(\mu) > 0$ for every μ, because every $N(\mu, \sigma_0^2/n)$ distribution assigns positive probability to A as well (you can think of A as a subinterval of R^1).

Now note that

$$
g^+(\bar{x}) = h^+(\bar{x})\, \frac{1}{\sqrt{2\pi}\sigma_0} \exp\left(-\frac{n\bar{x}^2}{2\sigma_0^2}\right)
$$

is nonnegative and is strictly positive on A. We can write

$$
c^+(\mu) = E_\mu\left(h^+\left(\bar{X}\right)\right) = \int_{-\infty}^{\infty} h^+(\bar{x}) \frac{1}{\sqrt{2\pi}\sigma_0} \exp\left(-\frac{n(\bar{x}-\mu)^2}{2\sigma_0^2}\right) d\bar{x}
$$

$$
= \exp\left(-\frac{n\mu^2}{2\sigma_0^2}\right) \int_{-\infty}^{\infty} \exp\left(\frac{n\mu}{\sigma_0^2}\bar{x}\right) g^+(\bar{x})\, d\bar{x}. \tag{8.5.2}
$$

Putting $\mu = 0$ establishes that $0 < \int_{-\infty}^{\infty} g^+(\bar{x})\, d\bar{x} < \infty$, because $0 < c^+(\mu) < \infty$ for every μ. Therefore,

$$
\frac{g^+(\bar{x})}{\int_{-\infty}^{\infty} g^+(\bar{x})\, d\bar{x}}
$$

is a probability density of a distribution concentrated on $A^+ = \{\bar{x} : h(\bar{x}) > 0\}$. Further, using (8.5.2) and the definition of moment-generating function in Section 3.4,

$$
\frac{c^+(\mu) \exp\left(\frac{n\mu^2}{2\sigma_0^2}\right)}{\int_{-\infty}^{\infty} g^+(\bar{x})\, d\bar{x}} \tag{8.5.3}
$$

is the moment-generating function of this distribution evaluated at $n\mu/\sigma_0^2$.

Similarly we define

$$
g^-(\bar{x}) = h^-(\bar{x}) \frac{1}{\sqrt{2\pi}\sigma_0} \exp\left(-\frac{n\bar{x}^2}{2\sigma_0^2}\right)
$$

so that

$$
\frac{g^-(\bar{x})}{\int_{-\infty}^{\infty} g^-(\bar{x})\, d\bar{x}}
$$

is a probability density of a distribution concentrated on $A^- = \{\bar{x} : h(\bar{x}) < 0\}$. Also,

$$
\frac{c^-(\mu) \exp\left(\frac{n\mu^2}{2\sigma_0^2}\right)}{\int_{-\infty}^{\infty} g^-(\bar{x})\, d\bar{x}} \tag{8.5.4}
$$

is the moment-generating function of this distribution evaluated at $n\mu/\sigma_0^2$.

Because $c^+(\mu) = c^-(\mu)$ we have that (setting $\mu = 0$)

$$
\int_{-\infty}^{\infty} g^+(\bar{x})\, d\bar{x} = \int_{-\infty}^{\infty} g^-(\bar{x})\, d\bar{x}.
$$

This implies that (8.5.3) equals (8.5.4) for every μ, and so the moment-generating functions of these two distributions are the same everywhere. By Theorem 3.4.6 these distributions must be the same. But this is impossible, as the distribution given by g^+ is concentrated on A^+ while the distribution given by g^- is concentrated on A^- and $A^+ \cap A^- = \phi$. Accordingly, we conclude that we cannot have $c^+(\mu) > 0$ and we are done. ∎

Theorem 8.2.1 (*Neyman–Pearson*) Suppose that $\Omega = \{\theta_0, \theta_1\}$ and that we want to test $H_0 : \theta = \theta_0$. Then an exact size α test function φ_0 of the form

$$\varphi_0(s) = \begin{cases} 1 & f_{\theta_1}(s)/f_{\theta_0}(s) > c_0 \\ \gamma & f_{\theta_1}(s)/f_{\theta_0}(s) = c_0 \\ 0 & f_{\theta_1}(s)/f_{\theta_0}(s) < c_0 \end{cases} \qquad (8.5.5)$$

exists for some $\gamma \in [0, 1]$ and $c_0 \geq 0$. This test is UMP size α.

Proof: We develop the proof of this result in the discrete case. The proof in the more general context is similar.

First we note that $\{s : f_{\theta_0}(s) = f_{\theta_1}(s) = 0\}$ has P_θ measure equal to 0 for both $\theta = \theta_0$ and $\theta = \theta_1$. Accordingly, without loss we can remove this set from the sample space and assume hereafter that $f_{\theta_0}(s)$ and $f_{\theta_1}(s)$ cannot be simultaneously 0. Therefore, the ratio $f_{\theta_1}(s)/f_{\theta_0}(s)$ is always defined.

Suppose that $\alpha = 1$. Then putting $c = 0$ and $\gamma = 1$ in (8.5.5), we see that $\varphi_0(s) \equiv 1$, and so $E_{\theta_1}(\varphi_0) = 1$. Therefore, φ_0 is UMP size α, because no test can have power greater than 1.

Suppose that $\alpha = 0$. Putting $c_0 = \infty$ and $\gamma = 1$ in (8.5.5), we see that $\varphi_0(s) = 0$ if and only if $f_{\theta_0}(s) > 0$ (if $f_{\theta_0}(s) = 0$, then $f_{\theta_1}(s)/f_{\theta_0}(s) = \infty$ and conversely). So φ_0 is the indicator function for the set $A = \{s : f_{\theta_0}(s) = 0\}$, and therefore $E_{\theta_0}(\varphi_0) = 0$. Further, any size 0 test function φ must be 0 on A^c to have $E_{\theta_0}(\varphi) = 0$. On A we have that $0 \leq \varphi(s) \leq 1 = \varphi_0(s)$ and so $E_{\theta_1}(\varphi) \leq E_{\theta_1}(\varphi_0)$. Therefore, φ_0 is UMP size α.

Now assume that $0 < \alpha < 1$. Consider the distribution function of the likelihood ratio when $\theta = \theta_0$, namely,

$$1 - \alpha^*(c) = P_{\theta_0}\left(\frac{f_{\theta_1}(s)}{f_{\theta_0}(s)} \leq c\right).$$

So $1 - \alpha^*(c)$ is a nondecreasing function of c with $1 - \alpha^*(-\infty) = 0$ and $1 - \alpha^*(\infty) = 1$.

Let c_0 be the smallest value of c such that $1 - \alpha \leq 1 - \alpha^*(c)$ (recall that $1 - \alpha^*(c)$ is right continuous because it is a distribution function). Then we have that $1 - \alpha^*(c_0 - 0) = 1 - \lim_{\varepsilon \searrow 0} \alpha^*(c_0 - \varepsilon) \leq 1 - \alpha \leq 1 - \alpha^*(c_0)$ and (using the fact that the jump in a distribution function at a point equals the probability of the point)

$$\begin{aligned} P_{\theta_0}\left(\frac{f_{\theta_1}(s)}{f_{\theta_0}(s)} = c_0\right) &= (1 - \alpha^*(c_0)) - (1 - \alpha^*(c_0 - 0)) \\ &= \alpha^*(c_0 - 0) - \alpha^*(c_0). \end{aligned}$$

Using this value of c_0 in (8.5.5), put

$$\gamma = \begin{cases} \frac{\alpha - \alpha^*(c_0)}{\alpha^*(c_0 - 0) - \alpha^*(c_0)} & \alpha^*(c_0 - 0) \neq \alpha^*(c_0) \\ 0 & \text{otherwise,} \end{cases}$$

and note that $\gamma \in [0, 1]$. Then we have

$$
\begin{aligned}
E_{\theta_0}(\varphi_0) &= \gamma P_{\theta_0}\left(\frac{f_{\theta_1}(s)}{f_{\theta_0}(s)} = c_0\right) + P_{\theta_0}\left(\frac{f_{\theta_1}(s)}{f_{\theta_0}(s)} > c_0\right) \\
&= \alpha - \alpha^*(c_0) + \alpha^*(c_0) = \alpha,
\end{aligned}
$$

so φ_0 has exact size α.

Now suppose that φ is another size α test and $E_{\theta_1}(\varphi) \geq E_{\theta_1}(\varphi_0)$. We partition the sample space as $S = S_0 \cup S_1 \cup S_2$ where

$$
\begin{aligned}
S_0 &= \{s : \varphi_0(s) - \varphi(s) = 0\}, \\
S_1 &= \{s : \varphi_0(s) - \varphi(s) < 0\}, \\
S_2 &= \{s : \varphi_0(s) - \varphi(s) > 0\}.
\end{aligned}
$$

Note that

$$
S_1 = \{s : \varphi_0(s) - \varphi(s) < 0, f_{\theta_1}(s)/f_{\theta_0}(s) \leq c_0\}
$$

because $f_{\theta_1}(s)/f_{\theta_0}(s) > c_0$ implies $\varphi_0(s) = 1$, which implies $\varphi_0(s) - \varphi(s) = 1 - \varphi(s) \geq 0$ as $0 \leq \varphi(s) \leq 1$. Also

$$
S_2 = \{s : \varphi_0(s) - \varphi(s) > 0, f_{\theta_1}(s)/f_{\theta_0}(s) \geq c_0\}
$$

because $f_{\theta_1}(s)/f_{\theta_0}(s) < c_0$ implies $\varphi_0(s) = 0$, which implies $\varphi_0(s) - \varphi(s) = -\varphi(s) \leq 0$ as $0 \leq \varphi(s) \leq 1$.

Therefore,

$$
\begin{aligned}
0 &\geq E_{\theta_1}(\varphi_0) - E_{\theta_1}(\varphi) = E_{\theta_1}(\varphi_0 - \varphi) \\
&= E_{\theta_1}(I_{S_1}(s)(\varphi_0(s) - \varphi(s))) + E_{\theta_1}(I_{S_2}(s)(\varphi_0(s) - \varphi(s))).
\end{aligned}
$$

Now note that

$$
E_{\theta_1}(I_{S_1}(s)(\varphi_0(s) - \varphi(s))) = \sum_{s \in S_1}(\varphi_0(s) - \varphi(s))f_{\theta_1}(s)
$$

$$
\geq c_0 \sum_{s \in S_1}(\varphi_0(s) - \varphi(s))f_{\theta_0}(s) = c_0 E_{\theta_0}(I_{S_1}(s)(\varphi_0(s) - \varphi(s)))
$$

because $\varphi_0(s) - \varphi(s) < 0$ and $f_{\theta_1}(s)/f_{\theta_0}(s) \leq c_0$ when $s \in S_1$. Similarly, we have that

$$
E_{\theta_1}(I_{S_2}(s)(\varphi_0(s) - \varphi(s))) = \sum_{s \in S_2}(\varphi_0(s) - \varphi(s))f_{\theta_1}(s)
$$

$$
\geq c_0 \sum_{s \in S_2}(\varphi_0(s) - \varphi(s))f_{\theta_0}(s) = c_0 E_{\theta_0}(I_{S_2}(s)(\varphi_0(s) - \varphi(s)))
$$

because $\varphi_0(s) - \varphi(s) > 0$ and $f_{\theta_1}(s)/f_{\theta_0}(s) \geq c_0$ when $s \in S_2$.

Combining these inequalities we obtain

$$
\begin{aligned}
0 \geq E_{\theta_1}(\varphi_0) - E_{\theta_1}(\varphi) &\geq c_0 E_{\theta_0}(\varphi_0 - \varphi) \\
&= c_0(E_{\theta_0}(\varphi_0) - E_{\theta_0}(\varphi)) = c_0(\alpha - E_{\theta_0}(\varphi)) \geq 0
\end{aligned}
$$

because $E_{\theta_0}(\varphi) \leq 0$. Therefore, $E_{\theta_1}(\varphi_0) = E_{\theta_1}(\varphi)$, which proves that φ_0 is MP among all size α tests. ∎

Chapter 9

Model Checking

The statistical inference methods developed in Chapters 6 through 8 all depend on various assumptions. For example, in Chapter 6 we assumed that the data s were generated from a distribution in the statistical model $\{P_\theta : \theta \in \Omega\}$. In Chapter 7 we also assumed that our uncertainty concerning θ could be described by a prior probability distribution Π. As such, any inferences drawn are of questionable validity if these assumptions do not make sense in a particular application.

In fact, all statistical methodology is based on assumptions and these must be checked if we want to feel confident that our inferences are relevant. We refer to the process of checking these assumptions as *model checking*, and it is the topic of this chapter. Obviously, this is of enormous importance in applications of statistics, and good statistical practice demands that effective model checking be carried out. Methods range from fairly informal graphical methods to more elaborate hypothesis assessment, and we will discuss a number of these.

9.1 Checking the Sampling Model

Frequency-based inference methods start with a statistical model $\{f_\theta : \theta \in \Omega\}$, for the true distribution that generated the data s. This means that we are assuming the true distribution for the observed data is in this set. If this assumption is not true, then it seems reasonable to question the relevance of any subsequent inferences we make about θ.

Except in relatively rare circumstances, we can never know categorically that a model is correct. The most we can hope for is that we can assess whether or not the observed data s could plausibly have arisen from the model.

If the observed data are surprising for each distribution in the model, then we have evidence that the model is incorrect. This leads us to think in terms of computing a P-value to check the correctness of the model. Of course, in this situation the null hypothesis is that the model is correct; the alternative is that the model could be any of the other possible models for the type of data we are dealing with.

We recall now our discussion of P-values in Chapter 6, where we distinguished between practical significance and statistical significance. It was noted there that, while a P-value may indicate that a null hypothesis is false, in practical terms the deviation from the null hypothesis may be so small as to be immaterial for the application. When the sample size gets large, it is inevitable that any reasonable approach via P-values will detect such a deviation and indicate that the null hypothesis is false. This is also true when we are carrying out model checking using P-values. The resolution of this is to estimate, in some fashion, the size of the deviation of the model from correctness, and so determine whether or not the model will be adequate for the application. Even if we ultimately accept the use of the model, it is still valuable to know, however, that we have detected evidence of model incorrectness when this is the case.

One P-value approach to model checking entails specifying a *discrepancy statistic* $D : S \to R^1$ that measures deviations from the model under consideration. Typically, large values of D are meant to indicate that a deviation has occurred. The actual value $D(s)$ is, of course, not necessarily an indication of this. The relevant issue is whether or not the observed value $D(s)$ is surprising under the assumption that the model is correct. Therefore, we must assess whether or not $D(s)$ lies in a region of low probability for the distribution of this quantity when the model is correct. For example, consider the density of a potential D statistic plotted in Figure 9.1.1. Here a value $D(s)$ in the left tail (near 0), right tail (out past 15), or between the two modes (in the interval from about 7 to 9) all would indicate that the model is incorrect, because such values have a low probability of occurrence when the model is correct.

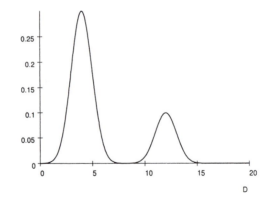

Figure 9.1.1: Plot of a density for a discrepancy statistic D.

The above discussion places the restriction that, when the model is correct, D must have a single distribution, i.e., the distribution cannot depend on θ. For many commonly used discrepancy statistics, this distribution is unimodal. A value in the right tail then indicates a lack of fit, or *underfitting* by the model (the discrepancies are unnaturally large); a value in the left tail then indicates *overfitting* by the model (the discrepancies are unnaturally small).

There are two general methods available for obtaining a single distribution for the computation of P-values. One method requires that D be ancillary. A statistic whose distribution under the model does not depend upon θ is called *ancillary*. If D is ancillary, then it has a single distribution specified by the model. If $D(s)$ is a surprising value for this distribution, then we have evidence against the model being true.

It is not the case that any ancillary D will serve as a useful discrepancy statistic. For example, if D is a constant, then it is ancillary but it is obviously not useful for model checking. So we have to be careful in choosing D.

Quite often we can find useful ancillary statistics for a model by looking at *residuals*. Loosely speaking, residuals are based on the information in the data that is left over after we have fit the model. If we have used all the relevant information in the data for fitting, then the residuals should contain no useful information for inference about the parameter θ. Example 9.1.1 will illustrate more clearly what we mean by residuals. Residuals play a major role in model checking.

The second method works with any discrepancy statistic D. For this we use the conditional distribution of D, given the value of a sufficient statistic T. By Theorem 8.1.2 this conditional distribution is the same for every value of θ. If $D(s)$ is a surprising value for this distribution, then we have evidence against the model being true.

Sometimes the two approaches we have just described agree, but not always. Consider some examples.

Example 9.1.1 *Location Normal*
Suppose we assume that $(x_1, \ldots, x_n)$ is a sample from an $N(\mu, \sigma_0^2)$ distribution where $\mu \in R^1$ is unknown and σ_0^2 is known. We know that $\bar{x}$ is a minimal sufficient statistic for this problem (Example 6.1.7). Also, $\bar{x}$ represents the fitting of the model to the data, as it is the estimate of the unknown parameter value μ.

Now consider

$$r = r(x_1, \ldots, x_n) = (r_1, \ldots, r_n) = (x_1 - \bar{x}, \ldots, x_n - \bar{x})$$

as one possible definition of the residual. Note that we can reconstruct the original data from the values of $\bar{x}$ and r.

It turns out that $R = (X_1 - \bar{X}, \ldots, X_n - \bar{X})$ has a distribution that is independent of μ with $E(R_i) = 0$ and $\text{Cov}(R_i, R_j) = \sigma_0^2 (\delta_{ij} - 1/n)$ for every i, j ($\delta_{ij} = 1$ when $i = j$ and 0 otherwise). Moreover, R is independent of $\bar{X}$ and $R_i \sim N(0, \sigma_0^2 (1 - 1/n))$ (Problems 9.1.12 and 9.1.13).

Accordingly, we have that r is ancillary and so is any discrepancy statistic D that depends on the data only through r. Furthermore, the conditional distribution of $D(R)$ given $\bar{X} = \bar{x}$ is the same as the marginal distribution of $D(R)$ because they are independent. Therefore, the two approaches to obtaining a P-value agree here, whenever the discrepancy statistic depends on the data only through r.

By Theorem 4.6.6 we have that

$$D\left(R\right) = \frac{1}{\sigma_0^2} \sum_{i=1}^n R_i^2 = \frac{1}{\sigma_0^2} \sum_{i=1}^n \left(X_i - \bar{X}\right)^2$$

is distributed $\chi^2\left(n-1\right)$, and so this is a possible discrepancy statistic. Therefore, the P-value

$$P\left(D > D\left(r\right)\right), \tag{9.1.1}$$

where $D \sim \chi^2\left(n-1\right)$, provides an assessment of whether or not the model is correct.

Note that values of (9.1.1) near 0 or near 1 are both evidence against the model, as both indicate that $D\left(r\right)$ is in a region of low probability when assuming the model is correct. A value near 0 indicates that $D\left(r\right)$ is in the right tail, while a value near 1 indicates that $D(r)$ is in the left tail.

The necessity of examining the left tail of the distribution of $D(r)$, as well as the right, is seen as follows. Consider the situation where we are in fact sampling from an $N(\mu, \sigma^2)$ distribution where σ^2 is much smaller than σ_0^2. In this case we expect $D\left(r\right)$ to be a value in the left tail, because $E\left(D\left(R\right)\right) = (n-1)\sigma^2/\sigma_0^2$.

There are obviously many other choices that could be made for the D statistic. At present, there is not a theory that prescribes one choice over another. One caution should be noted, however. The choice of a statistic D cannot be based upon looking at the data first. Doing so invalidates the computation of the P-value as described above, as then we must condition on the data feature that led us to choose that particular D. ∎

Example 9.1.2 *Location-Scale Normal*
Suppose we assume that $(x_1, \ldots, x_n)$ is a sample from an $N(\mu, \sigma^2)$ distribution where $(\mu, \sigma^2) \in R^1 \times (0, \infty)$ is unknown. We know that $(\bar{x}, s^2)$ is a minimal sufficient statistic for this model (Example 6.1.8). Consider

$$r = r\left(x_1, \ldots, x_n\right) = (r_1, \ldots, r_n) = \left(\frac{x_1 - \bar{x}}{s}, \ldots, \frac{x_n - \bar{x}}{s}\right)$$

as one possible definition of the residual. Note that we can reconstruct the data from the values of $(\bar{x}, s^2)$ and r.

It turns out that R has a distribution that is independent of (μ, σ^2) (and so is ancillary — see Challenge 9.1.20) and is also independent of $(\bar{X}, S^2)$. So again, the two approaches to obtaining a P-value agree here, as long as the discrepancy statistic depends on the data only through r.

One possible discrepancy statistic is given by

$$D\left(r\right) = -\frac{1}{n} \sum_{i=1}^n \ln\left(\frac{r_i^2}{n-1}\right).$$

To use this statistic for model checking, we need to obtain its distribution when the model is correct. Then we compare the observed value $D\left(r\right)$ with this distribution, to see if it is surprising.

We can do this via simulation. Because the distribution of $D(R)$ is independent of (μ, σ^2), we can generate N samples of size n from the $N(0,1)$ distribution (or any other normal distribution) and calculate $D(R)$ for each sample. Then we look at histograms of the simulated values to see if $D(r)$, from the original sample, is a surprising value, i.e., see if it lies in a region of low probability like a left or right tail.

For example, suppose we observed the sample

$$\boxed{-2.08 \quad -0.28 \quad 2.01 \quad -1.37 \quad 40.08}$$

obtaining the value $D(r) = 4.93$. Then simulating 10^4 values from the distribution of D, under the assumption of model correctness, we obtained the density histogram given in Figure 9.1.2. See Appendix B for some code used to carry out this simulation. The value $D(r) = 4.93$ is out in the right tail and so indicates that the sample is not from a normal distribution. In fact, only 0.0057 of the simulated values are larger, so this is definite evidence against the model being correct.

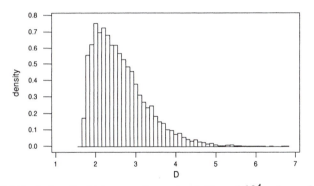

Figure 9.1.2: A density histogram for a simulation of 10^4 values of D in Example 9.1.2.

Obviously there are other possible functions of r that we could use for model checking. ∎

The following examples present contexts where the two approaches to computing a P-value for model checking are not the same.

Example 9.1.3 *Location-Scale Cauchy*
Suppose we assume that $(x_1, \ldots, x_n)$ is a sample from the distribution given by $\mu + \sigma Z$, where $Z \sim t(1)$ and $(\mu, \sigma^2) \in R^1 \times (0, \infty)$ is unknown. This time $(\bar{x}, s^2)$ is not a minimal sufficient statistic, but the statistic r defined in Example 9.1.2 is still ancillary (Challenge 9.1.20). We can again simulate values from the distribution of R (just generate samples from the $t(1)$ distribution and compute r for each sample) to estimate P-values for any discrepancy statistic such as the $D(r)$ discussed in Example 9.1.2. ∎

Example 9.1.4 *Fisher's Exact Test*
Suppose that we take a sample of n from a population of students and observe the values $(a_1, b_1), \ldots, (a_n, b_n)$ where a_i is gender ($A = 1$ indicating male, $A = 2$ indicating female) and b_i is a categorical variable for part-time employment status ($B = 1$ indicating employed, $B = 2$ indicating unemployed). So each individual is being categorized into one of four categories, namely,

$$\text{Category 1, when } A = 1, \ B = 1,$$
$$\text{Category 2, when } A = 1, \ B = 2,$$
$$\text{Category 3, when } A = 2, \ B = 1,$$
$$\text{Category 4, when } A = 2, \ B = 2.$$

Suppose our model for this situation is that A and B are independent with $P(A = 1) = \alpha_1, P(B = 1) = \beta_1$ where $\alpha_1 \in [0, 1]$ and $\beta_1 \in [0, 1]$ are completely unknown. Then letting X_{ij} denote the count for the category where $A = i, B = j$, Example 2.8.5 gives that

$$(X_{11}, X_{12}, X_{21}, X_{22}) \sim \text{Multinomial} (n, \alpha_1\beta_1, \alpha_1\beta_2, \alpha_2\beta_1, \alpha_2\beta_2).$$

As we will see in Chapter 10, this model is equivalent to saying that there is no relationship between gender and employment status.

Denoting the observed cell counts by $(x_{11}, x_{12}, x_{21}, x_{22})$, the likelihood function is given by

$$(\alpha_1\beta_1)^{x_{11}} (\alpha_1\beta_2)^{x_{12}} (\alpha_2\beta_1)^{x_{21}} (\alpha_2\beta_2)^{x_{22}}$$
$$= \alpha_1^{x_{11}+x_{12}} (1 - \alpha_1)^{n-x_{11}-x_{12}} \beta_1^{x_{11}+x_{21}} (1 - \beta_1)^{n-x_{11}-x_{21}}$$
$$= \alpha_1^{x_{1\cdot}} (1 - \alpha_1)^{n-x_{1\cdot}} \beta_1^{x_{\cdot 1}} (1 - \beta_1)^{n-x_{\cdot 1}}$$

where $(x_{1\cdot}, x_{\cdot 1}) = (x_{11} + x_{12}, x_{11} + x_{21})$. Therefore, the MLE (Problem 9.1.14) is given by

$$\left(\hat{\alpha}_1, \hat{\beta}_1\right) = \left(\frac{x_{1\cdot}}{n}, \frac{x_{\cdot 1}}{n}\right).$$

Note that $\hat{\alpha}_1$ is the proportion of males in the sample and $\hat{\beta}_1$ is the proportion of employed in the sample. Because $(x_{1\cdot}, x_{\cdot 1})$ determines the likelihood function and can be calculated from the likelihood function, we have that $(x_{1\cdot}, x_{\cdot 1})$ is a minimal sufficient statistic.

In this example, a natural definition of residual does not seem readily apparent. So we consider looking at the conditional distribution of the data, given the minimal sufficient statistic. The conditional distribution of the sample $(A_1, B_1), \ldots, (A_n, B_n)$, given the values $(x_{1\cdot}, x_{\cdot 1})$, is the uniform distribution on the set of all samples where the restrictions

$$
\begin{aligned}
x_{11} + x_{12} &= x_{1\cdot}, \\
x_{11} + x_{21} &= x_{\cdot 1}, \\
x_{11} + x_{12} + x_{21} + x_{22} &= n,
\end{aligned}
\tag{9.1.2}
$$

are satisfied. Notice that, given $(x_1., x_{.1})$, all of the other values in (9.1.2) are determined, when we specify a value for x_{11}.

It can be shown that the number of such samples is equal to (Problem 9.1.15)

$$\binom{n}{x_1.}\binom{n}{x_{.1}}.$$

Now the number of samples with prescribed values for $x_1., x_{.1}$, and $x_{11} = i$ is given by

$$\binom{n}{x_1.}\binom{x_1.}{i}\binom{n - x_1.}{x_{.1} - i}.$$

Therefore the conditional probability function of x_{11}, given $(x_1., x_{.1})$, is

$$P(x_{11} = i \mid x_1., x_{.1}) = \frac{\binom{n}{x_1.}\binom{x_1.}{i}\binom{n - x_1.}{x_{.1} - i}}{\binom{n}{x_1.}\binom{n}{x_{.1}}} = \frac{\binom{x_1.}{i}\binom{n - x_1.}{x_{.1} - i}}{\binom{n}{x_{.1}}}.$$

This is the Hypergeometric$(n, x_{.1}, x_1.)$ probability function.

So we have evidence against the model holding whenever x_{11} is out in the tails of this distribution. Assessing this requires a tabulation of this distribution, or the use of a statistical package with the hypergeometric distribution function built in.

As a simple numerical example, suppose that we took a sample of $n = 20$ students, obtaining $x_{.1} = 12$ unemployed, $x_1. = 6$ males and $x_{11} = 2$ employed males. Then the Hypergeometric$(20, 12, 6)$ probability function is given by the following table.

i	0	1	2	3	4	5	6
$p(i)$	0.001	0.017	0.119	0.318	0.358	0.163	0.024

The probability of getting a value as far, or farther, out in the tails than $x_{11} = 2$ is equal to the probability of observing a value of x_{11} with probability of occurrence as small or smaller, than $x_{11} = 2$. This P-value equals

$$(0.119 + 0.017 + 0.001) + 0.024 = 0.161.$$

Therefore, we have no evidence against the model of independence between A and B. Of course, the sample size is quite small here.

There is another approach here to testing the independence of A and B. In particular, we could only assume the independence of the initial unclassified sample and then we always have

$$(X_{11}, X_{12}, X_{21}, X_{22}) \sim \text{Multinomial}(n, \alpha_{11}, \alpha_{12}, \alpha_{21}, \alpha_{22}),$$

where the α_{ij} comprise an unknown probability distribution. Given this model, we could then test for the independence of A and B. We will discuss this in Section 10.2. ∎

Another approach to model checking proceeds as follows. We enlarge the model to include more distributions and then test the null hypothesis that the true model is the submodel we initially started with. If we can apply the methods of Section 8.2 to come up with a uniformly most powerful (UMP) test of this null hypothesis, then we will have a check of departures from the model of interest — at least as expressed by the possible alternatives in the enlarged model. If the model passes such a check, however, we are still required to check the validity of the enlarged model. This can be viewed as a technique for generating relevant discrepancy statistics D.

9.1.1 Residual and Probability Plots

There is another more informal approach to checking model correctness that is often used when we have residuals available. These methods involve various plots of the residuals that should exhibit specific characteristics if the model is correct. While this approach lacks the rigor of the P-value approach, it is good at demonstrating gross deviations from model assumptions. We illustrate this via some examples.

Example 9.1.5 *Location and Location-Scale Normal Models*
Using the residuals for the location normal model discussed in Example 9.1.1, we have that $E(R_i) = 0$ and $\mathrm{Var}(R_i) = \sigma_0^2 (1 - 1/n)$. We standardize these values so that they also have variance 1, and so obtain the standardized residuals $(r_1^*, \ldots, r_n^*)$ given by

$$r_i^* = \sqrt{\frac{n}{\sigma_0^2 (n-1)}} \, (x_i - \bar{x}). \qquad (9.1.3)$$

The standardized residuals are distributed $N(0,1)$ and, assuming that n is reasonably large, it can be shown that they are approximately independent. Accordingly, we can think of $r_1^*, \ldots, r_n^*$ as an approximate sample from the $N(0,1)$ distribution.

Therefore, a plot of the points (i, r_i^*) should not exhibit any discernible pattern. Further, all the values in the y-direction should lie in $(-3, 3)$, unless of course n is very large, in which case we might expect a few values outside this interval. A discernible pattern, or several extreme values, can be taken as some evidence that the model assumption is not correct. It always has to be borne in mind, however, that any observed pattern could have arisen simply from sampling variability when the true model is correct. Simulating a few of these residual plots (just generating several samples of n from the $N(0,1)$ distribution and obtaining a residual plot for each sample) will give us some idea of whether or not the observed pattern is unusual.

Figure 9.1.3 shows a plot of the standardized residuals (9.1.3) for a sample of 100 from the $N(0,1)$ distribution. Figure 9.1.4 shows a plot of the standardized residuals for a sample of 100 from the distribution given by $3^{-1/2}Z$ where $Z \sim t(3)$. Note that a $t(3)$ distribution has mean 0 and variance equal to 3, so $\mathrm{Var}(3^{-1/2}Z) = 1$ (Problem 4.6.16). Figure 9.1.5 shows the standardized residuals for a sample of 100 from an Exponential(1) distribution.

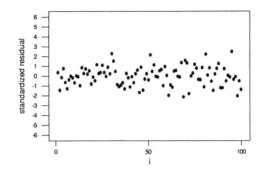

Figure 9.1.3: A plot of the standardized residuals for a sample of 100 from an $N(0, 1)$ distribution.

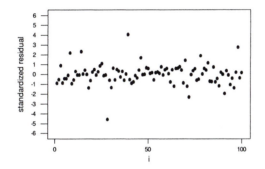

Figure 9.1.4: A plot of the standardized residuals for a sample of 100 from $X = 3^{-1/2}Z$ where $Z \sim t(3)$.

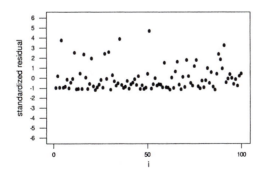

Figure 9.1.5: A plot of the standardized residuals for a sample of 100 from an Exponential(1) distribution.

Note that the distributions of the standardized residuals for all these samples have mean 0 and variance equal to 1. The difference in Figures 9.1.3 and 9.1.4 is due to the fact that the t distribution has much longer tails. This is reflected in the fact that a few of the standardized residuals are outside $(-3, 3)$ in Figure 9.1.4 but not in Figure 9.1.3. Even though the two distributions are quite different, e.g., the $N(0, 1)$ distribution has all of its moments while the $3^{-1/2} t(3)$ distribution has only two moments; the plots of the standardized residuals are otherwise very similar. The difference in Figures 9.1.3 and 9.1.5 is due to the asymmetry in the Exponential(1) distribution, as it is skewed to the right.

Using the residuals for the location-scale normal model discussed in Example 9.1.2, we define the standardized residuals $r_1^*, \ldots, r_n^*$ by

$$r_i^* = \sqrt{\frac{n}{s^2 (n-1)}} (x_i - \bar{x}). \tag{9.1.4}$$

Here the unknown variance is estimated by s^2. Again, it can be shown that when n is large, then $(r_1^*, \ldots, r_n^*)$ is an approximate sample from the $N(0, 1)$ distribution. So we plot the values (i, r_i^*) and interpret the plot just as we described for the location normal model. ∎

It is very common in statistical applications to assume some basic form for the distribution of the data, e.g., we might assume we are sampling from a normal distribution with some mean and variance. To assess such an assumption the use of a *probability plot* has proven to be very useful.

To illustrate, suppose that $(x_1, \ldots, x_n)$ is a sample from an $N(\mu, \sigma^2)$ distribution. Then it can be shown that when n is large, the expectation of the ith order statistic satisfies

$$E\left(X_{(i)}\right) \approx \mu + \sigma \Phi^{-1} \left(i/(n+1)\right). \tag{9.1.5}$$

If the data value x_j corresponds to order statistic $x_{(i)}$ (i.e., $x_{(i)} = x_j$), then we call $\Phi^{-1}\left(i/(n+1)\right)$ the *normal score* of x_j in the sample. Then (9.1.5) indicates that if we plot the points $\left(x_{(i)}, \Phi^{-1}\left(i/(n+1)\right)\right)$, these should lie approximately on a line with intercept μ and slope σ. We call such a plot a *normal probability plot* or *normal quantile plot*. Similar plots can be obtained for other distributions.

Example 9.1.6 *Location-Scale Normal*

For example, suppose that we want to assess whether or not the following data set can be considered a sample of size $n = 10$ from some normal distribution.

| 2.00 | 0.28 | 0.47 | 3.33 | 1.66 | 8.17 | 1.18 | 4.15 | 6.43 | 1.77 |

The order statistics and associated normal scores for this sample are given in the following table.

i	1	2	3	4	5
$x_{(i)}$	0.28	0.47	1.18	1.66	1.77
$\Phi^{-1}\left(i/\left(n+1\right)\right)$	-1.34	-0.91	-0.61	-0.35	-0.12
i	6	7	8	9	10
$x_{(i)}$	2.00	3.33	4.15	6.43	8.17
$\Phi^{-1}\left(i/\left(n+1\right)\right)$	0.11	0.34	0.60	0.90	1.33

The values

$$\left(x_{(i)}, \Phi^{-1}\left(i/\left(n+1\right)\right)\right)$$

are then plotted in Figure 9.1.6. There is some definite deviation from a straight line here, but note that it is difficult to tell whether this is unexpected in a sample of this size from a normal distribution. Again simulating a few samples of the same size (say from a $N(0,1)$) and looking at their normal probability plots is recommended. In this case, we conclude that the plot in Figure 9.1.6 looks reasonable. ∎

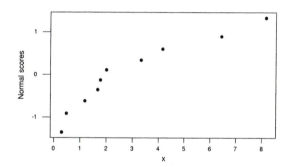

Figure 9.1.6: Normal probability plot of the data in Example 9.1.6.

We will see in Chapter 10 that the use of normal probability plots of standardized residuals is an important part of model checking for more complicated models. So, while they are not really needed here, we consider some of the characteristics of such plots when we are assessing whether or not a sample is from a location normal or location-scale normal model.

Assume that n is large so that we can consider the standardized residuals, given by (9.1.3) or (9.1.4) as an approximate sample from the $N(0,1)$ distribution. Then a normal probability plot of the standardized residuals should be approximately linear with y-intercept approximately equal to 0 and slope approximately equal to 1. If we get a substantial deviation from this, then we have evidence that the assumed model is incorrect.

In Figure 9.1.7 we have plotted a normal probability plot of the standardized residuals for a sample of $n = 25$ from an $N(0,1)$ distribution. In Figure 9.1.8 we have plotted a normal probability plot of the standardized residuals for a sample of $n = 25$ from the distribution given by $X = 3^{-1/2}Z$ where $Z \sim t(3)$.

Both distributions have mean 0 and variance 1, so the difference in the normal probability plots is due to other distributional differences.

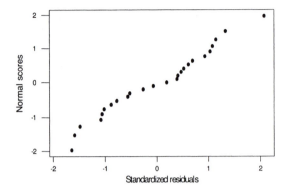

Figure 9.1.7: Normal probability plot of the standardized residuals of a sample of 25 from an $N(0,1)$ distribution.

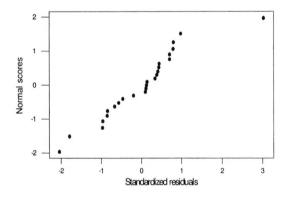

Figure 9.1.8: Normal probability plot of the standardized residuals of a sample of 25 from $X = 3^{-1/2}Z$ where $Z \sim \mathrm{t}(3)$.

9.1.2   The Chi-Squared Goodness of Fit Test

The chi-squared goodness of fit test has an important historical place in any discussion of assessing model correctness. We use this test to assess whether or not a categorical random variable X, which takes its values in the finite sample space $\{1, 2, \ldots, k\}$, has a specified probability measure P, after having observed a sample $(x_1, \ldots, x_n)$. When we have a random variable that is discrete and takes infinitely many values, then we partition the possible values into k categories and let X simply indicate which category has occurred. If we have a random variable that is quantitative, then we partition R^1 into k subintervals and let X indicate in which interval the response occurred. In effect, we want to check whether or not a specific probability model, as given by P, is correct for X based on an observed sample.

Let $(X_1, \ldots, X_k)$ be the observed counts or frequencies of $1, \ldots, k$, respectively. If P is correct then, from Example 2.8.5,

$$(X_1, \ldots, X_k) \sim \text{Multinomial}(n, p_1, \ldots, p_k)$$

where $p_i = P(\{i\})$. This implies that $E(X_i) = np_i$ and $\text{Var}(X_i) = np_i(1 - p_i)$ (recall that $X_i \sim \text{Binomial}(n, p_i)$). From this we deduce that

$$R_i = \frac{X_i - np_i}{\sqrt{np_i(1 - p_i)}} \xrightarrow{D} N(0, 1) \tag{9.1.6}$$

as $n \to \infty$ (see Example 4.4.9).

For finite n, the distribution of R_i, when the model is correct, is dependent on P, but the limiting distribution is not, so we can think of the R_i as standardized residuals when n is large. Therefore, it would seem that a reasonable discrepancy statistic is given by the sum of the squares of the standardized residuals with $\sum_{i=1}^{k} R_i^2$ approximately distributed $\chi^2(k)$. The restriction $f_1 + \cdots + f_k = n$ holds, however, so the R_i are not independent and the limiting distribution is not $\chi^2(k)$. We do, however, have the following result, that provides a similar discrepancy statistic.

Theorem 9.1.1 If $(X_1, \ldots, X_k) \sim \text{Multinomial}(n, p_1, \ldots, p_k)$ then

$$X^2 = \sum_{i=1}^{k} (1 - p_i) R_i^2 = \sum_{i=1}^{k} \frac{(X_i - np_i)^2}{np_i} \xrightarrow{D} \chi^2(k - 1)$$

as $n \to \infty$.

The proof of this result is a little too involved for this text, so we will just make use of it.

We refer to X^2 as the *chi-squared statistic*. The process of assessing the correctness of the model by computing the P-value $P(X^2 \geq X_0^2)$, where $X^2 \sim \chi^2(k - 1)$ and X_0^2 is the observed value of the chi-squared statistic, is referred to as the *chi-squared goodness of fit test*. Small P-values near 0 and large P-values near 1 are both evidence of the incorrectness of the probability model. Small P-values indicate that some of the residuals are too large, whereas large P-values indicate that the residuals are unnaturally small. As an extreme example, imagine a situation in which all the residuals are 0. This would indicate a very poor fit for the model in general (unless one of the $p_i = 1$), and would make us suspect that randomness was not an appropriate ingredient for the situation being modeled.

Note that the ith term of the chi-squared statistic can be written as

$$\frac{(X_i - np_i)^2}{np_i} = \frac{(\text{number in the } i\text{th cell} - \text{expected number in the } i\text{th cell})^2}{\text{expected number in the } i\text{th cell}}.$$

It is recommended, for example, in *Statistical Methods* by G. Snedecor and W. Cochran (Iowa State Press, 6th ed., 1967) that grouping (combining cells) be employed to ensure that $E(X_i) = np_i \geq 1$ for every i, as simulations have shown that this improves the accuracy of the approximation.

We consider an important application.

Example 9.1.7 *Testing the Accuracy of a Random Number Generator*
In effect, every Monte Carlo simulation can be considered to be a set of mathematical operations applied to a stream of numbers $U_1, U_2, \ldots$ in $[0, 1]$ that are supposed to i.i.d. Uniform$(0, 1)$. Of course, they cannot satisfy this requirement exactly, because they are generated according to some deterministic function. Typically, a function $f : [0, 1]^m \to [0, 1]$ is chosen and is applied iteratively to obtain the sequence. So we select $U_1, \ldots, U_m$ as initial *seed values* and then $U_{m+1} = f(U_1, \ldots, U_m), U_{m+2} = f(U_2, \ldots, U_{m+1})$, etc. There are many possibilities for f, and a great deal of research and study have gone into selecting functions that will produce sequences that adequately mimic the properties of an i.i.d. Uniform$(0, 1)$ sequence.

Of course, it is always possible that the underlying f used in a particular statistical package, or other piece of software, is very poor. In such a case the results of the simulations can be grossly in error. How do we assess whether a particular f is good or not? One approach is to run a battery of statistical tests to see whether the sequence is behaving as we know an ideal sequence would.

For example, if the sequence $U_1, U_2, \ldots$ is i.i.d. Uniform$(0, 1)$, then

$$\lceil 10U_1 \rceil, \lceil 10U_2 \rceil, \ldots$$

is i.i.d. Uniform$\{1, 2, \ldots, 10\}$ ($\lceil x \rceil$ denotes the smallest integer greater than x, e.g., $\lceil 3.2 \rceil = 4$). So we can test the adequacy of the underlying function f by generating $U_1, \ldots, U_n$ for large n, putting $x_i = \lceil 10U_i \rceil$, and then carrying out a chi-squared goodness of fit test with the 10 categories $\{1, \ldots, 10\}$ with each cell probability equal to $1/10$.

Doing this using a popular statistical package (with $n = 10^4$) gave the following table of counts x_i and standardized residuals r_i as specified in (9.1.6).

i	x_i	r_i
1	993	-0.23333
2	1044	1.46667
3	1061	2.03333
4	1021	0.70000
5	1017	0.56667
6	973	-0.90000
7	975	-0.83333
8	965	-1.16667
9	996	-0.13333
10	955	-1.50000

All the standardized residuals look reasonable as possible values from an $N(0, 1)$

distribution. Further,

$$X_0^2 = (1 - 0.1) \left\{ \begin{array}{l} (-0.23333)^2 + (1.46667)^2 + (2.03333)^2 \\ + (0.70000)^2 + (0.56667)^2 + (-0.90000)^2 \\ + (-0.83333)^2 + (-1.16667)^2 + (-0.13333)^2 \\ + (-1.50000)^2 \end{array} \right\}$$

$$= 11.0560$$

gives the P-value $P\left(X^2 \geq 11.0560\right) = 0.27190$ when $X^2 \sim \chi^2(9)$. This indicates that we have no evidence that the random number generator is defective.

Of course, the story does not end with a single test like this. Many other features of the sequence should be tested. For example, we might want to investigate the independence properties of the sequence and so test if each possible combination of (i, j) occurs with probability $1/100$, etc. ∎

More generally, we will not have a prescribed probability distribution P for X but rather a statistical model $\{P_\theta : \theta \in \Omega\}$ where each P_θ is a probability measure on the finite set $\{1, 2, \ldots, k\}$. Then based on the sample from the model, we have that

$$(X_1, \ldots, X_k) \sim \text{Multinomial}\left(n, p_1(\theta), \ldots, p_k(\theta)\right)$$

where $p_i(\theta) = P_\theta(\{i\})$.

Perhaps a natural way to assess whether or not this model fits the data is to find the MLE $\hat{\theta}$ from the likelihood function

$$L(\theta \,|\, x_1, \ldots, x_k) = (p_1(\theta))^{x_1} \cdots (p_k(\theta))^{x_k},$$

and then look at the standardized residuals

$$r_i(\hat{\theta}) = \frac{x_i - np_i(\hat{\theta})}{\sqrt{np_i(\hat{\theta})(1 - p_i(\hat{\theta}))}}.$$

We have the following result, which we state without proof.

Theorem 9.1.2 Under conditions (similar to those discussed in Section 6.5) we have that $R_i(\hat{\theta}) \xrightarrow{D} N(0, 1)$ and

$$X^2 = \sum_{i=1}^{k} \left(1 - p_i(\hat{\theta})\right) R_i^2(\hat{\theta}) = \sum_{i=1}^{k} \frac{(X_i - np_i(\hat{\theta}))^2}{np_i(\hat{\theta})} \xrightarrow{D} \chi^2(k - 1 - \dim \Omega)$$

as $n \to \infty$.

By $\dim \Omega$ we mean the dimension of the set Ω. Loosely speaking, this is the minimum number of coordinates required to specify a point in the set, e.g., a line requires 1 coordinate (positive or negative distance from a fixed point), a circle requires 1 coordinate, a plane in R^3 requires 2 coordinates, etc. Of course this result implies that the number of cells must satisfy $k > 1 + \dim \Omega$.

Consider an example.

Example 9.1.8 *Testing for Exponentiality*
Suppose that a sample of lifelengths of light bulbs (measured in thousands of hours) is supposed to be from an Exponential(θ) distribution where $\theta \in \Omega = (0, \infty)$ is unknown. So here dim $\Omega = 1$, and we require at least 2 cells for the chi-squared test. The manufacturer expects that most bulbs will last at least 1000 hours, 50% will last less than 2000 hours, and most will have failed by 3000 hours. So based on this, we partition the sample space as

$$(0, \infty) = (0, 1] \cup (1, 2] \cup (2, 3] \cup (3, \infty).$$

Suppose that a sample of $n = 30$ light bulbs was taken and that the counts $x_1 = 5$, $x_2 = 16$, $x_3 = 8$, and $x_4 = 1$ were obtained for the four intervals, respectively. Then the likelihood function based on these counts is given by

$$L(\theta \,|\, x_1, \ldots, x_4) = \left(1 - e^{-\theta}\right)^5 \left(e^{-\theta} - e^{-2\theta}\right)^{16} \left(e^{-2\theta} - e^{-3\theta}\right)^8 \left(e^{-3\theta}\right)^1,$$

because, for example, the probability of a value falling in $(1, 2]$ is $e^{-\theta} - e^{-2\theta}$, and we have $x_2 = 16$ observations in this interval. Figure 9.1.9 is a plot of the log-likelihood.

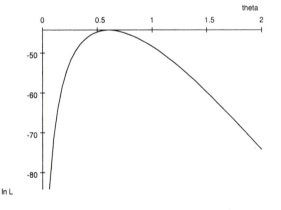

Figure 9.1.9: Plot of the log-likelihood function in Example 9.1.8.

By successively plotting the likelihood on shorter and shorter intervals, the MLE was determined to be $\hat{\theta} = 0.603535$. This value leads to the probabilities

$$
\begin{aligned}
p_1(\hat{\theta}) &= 1 - e^{-0.603535} = 0.453125, \\
p_2(\hat{\theta}) &= e^{-0.603535} - e^{-2(0.603535)} = 0.247803, \\
p_3(\hat{\theta}) &= e^{-2(0.603535)} - e^{-3(0.603535)} = 0.135517, \\
p_4(\hat{\theta}) &= e^{-3(0.603535)} = 0.163555,
\end{aligned}
$$

the fitted values

$$30p_1(\hat{\theta}) = 13.59375,$$
$$30p_2(\hat{\theta}) = 7.43409,$$
$$30p_3(\hat{\theta}) = 4.06551,$$
$$30p_4(\hat{\theta}) = 4.90665,$$

and the standardized residuals

$$r_1(\hat{\theta}) = (5 - 13.59375)/\sqrt{30(0.453125)(1 - 0.453125)} = -3.151875,$$
$$r_2(\hat{\theta}) = (16 - 7.43409)/\sqrt{30(0.247803)(1 - 0.247803)} = 3.622378,$$
$$r_3(\hat{\theta}) = (8 - 4.06551)/\sqrt{30(0.135517)(1 - 0.135517)} = 2.098711,$$
$$r_4(\hat{\theta}) = (1 - 4.90665)/\sqrt{30(0.163555)(1 - 0.163555)} = -1.928382.$$

Note that two of the standardized residuals look large. Finally we compute

$$\begin{aligned}
X_0^2 &= (1 - 0.453125)(-3.151875)^2 + (1 - 0.247803)(3.622378)^2 \\
&\quad + (1 - 0.135517)(2.098711)^2 + (1 - 0.163555)(-1.928382)^2 \\
&= 22.221018
\end{aligned}$$

and

$$P\left(X^2 \geq 22.221018\right) = 0.0000$$

when $X^2 \sim \chi^2(2)$. Therefore we have strong evidence that the Exponential(θ) model is not correct for these data and we would not use this model to make inference about θ.

Note that we used the MLE of θ based on the count data and not the original sample! If we were instead to use the MLE for θ based on the original sample (in this case equal to $\bar{x}$ and so much easier to compute) then Theorem 9.1.2 would no longer be valid. ∎

The chi-squared goodness of fit test is but one of many discrepancy statistics that have been proposed for model checking in the statistical literature. The general approach is to select a discrepancy statistic D, like X^2, such that the exact or asymptotic distribution of D is independent of θ and known. We then compute a P-value based on D. The *Kolmogorov-Smirnov test* and the *Cramer-von Mises test* are further examples of such discrepancy statistics, but we do not discuss these here.

9.1.3 Prediction and Cross-Validation

Perhaps the most rigorous test that a scientific model or theory can be subjected to is assessing how well it predicts new data after it has been fit to an independent data set. In fact, this is a crucial step in the acceptance of any new empirically developed scientific theory — to be accepted, it must predict new results beyond the data that led to its formulation.

If a model does not do a good job at predicting new data, then it is reasonable to say that we have evidence against the model being correct. If the model is too simple, then the fitted model will underfit the observed data and also the future data. If the model is too complicated, then the model will overfit the original data, and this will be detected when we consider the new data in light of this fitted model.

In statistical applications, we typically cannot wait until new data are generated to check the model. So statisticians use a technique called *cross-validation*. For this we split an original data set $x_1, \ldots, x_n$ into two parts: the *training set* T, comprising k of the data values and used to fit the model; and the *validation set* V, which comprises the remaining $n - k$ data values. Based on the training data, we construct predictors of various aspects of the validation data. Using the discrepancies between the predicted and actual values, we then assess whether or not the validation set V is surprising as a possible future sample from the model.

Of course, there are

$$\binom{n}{k}$$

possible such splits of the data and we wouldn't want to make a decision based on just one of these. So a cross-validational analysis will have to take this into account. Further, we will have to decide how to measure the discrepancies between T and V and choose a value for k. We do not pursue this topic any further in this text.

9.1.4   What Do We Do When a Model Fails?

So far we have been concerned with determining whether or not an assumed model is appropriate given observed data. Suppose the result of our model checking is that we decide a particular model is *inappropriate*. What do we do now?

Perhaps the obvious response is to say that we have to come up with a more appropriate model — one that will pass our model checking. It is not obvious how we should go about this, but statisticians have devised some techniques.

One of the simplest techniques is the *method of transformations*. For example, suppose that we observe a sample $y_1, \ldots, y_n$ from the distribution given by $Y = \exp(X)$ with $X \sim N(\mu, \sigma^2)$. A normal probability plot based on the y_i, as in Figure 9.1.10, will detect evidence of the nonnormality of the distribution. Transforming these y_i values to $\ln y_i$ will, however, yield a reasonable looking normal probability plot, as in Figure 9.1.11.

So in this case, a simple transformation of the sample yields a data set that passes this check. In fact, this is something statisticians commonly do. Several transformations from the family of *power transformations* given by Y^p for $p \neq 0$, or the logarithm transformation $\ln Y$, are tried to see if a distributional assumption can be satisfied by a transformed sample. We will see some applications of this in Chapter 10. Surprisingly, this simple technique often works, although there are no guarantees that it always will.

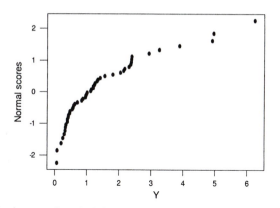

Figure 9.1.10: A normal probability plot of a sample of $n = 50$ from the distribution given by $Y = \exp(X)$ with $X \sim N(0,1)$.

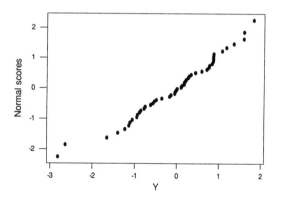

Figure 9.1.11: A normal probability plot of a sample of $n = 50$ from the distribution given by $\ln Y$, where $Y = \exp(X)$ and $X \sim N(0,1)$.

Summary of Section 9.1

- Model checking is a key component of the practical application of statistics.

- One approach to model checking involves choosing a discrepancy statistic D and then assessing whether or not the observed value of D is surprising by computing a P-value.

- Computation of the P-value requires that the distribution of D be known under the assumption that the model is correct. There are two approaches to accomplishing this. One involves choosing D to be ancillary, and the other involves computing the P-value using the conditional distribution of the data given the minimal sufficient statistic.

- The chi-squared goodness of fit statistic is a commonly used discrepancy statistic. For large samples it is approximately ancillary.

- There are also many informal methods of model checking based on various plots of residuals.

- If a model is rejected, then there are several techniques for modifying the model. These typically involve transformations of the data. Also, a model that fails a model-checking procedure may still be useful, if the deviation from correctness is small.

Exercises

9.1.1 Suppose that the following sample is assumed to be from an $N(\theta, 4)$ distribution with $\theta \in R^1$ unknown.

1.8	2.1	−3.8	−1.7	−1.3	1.1	1.0	0.0	3.3	1.0
−0.4	−0.1	2.3	−1.6	1.1	−1.3	3.3	−4.9	−1.1	1.9

Check this model using the discrepancy statistic of Example 9.1.1.

9.1.2 Suppose that the following sample is assumed to be from an $N(\theta, 2)$ distribution with θ unknown.

−0.4	1.9	−0.3	−0.2	0.0	0.0	−0.1	−1.1	2.0	0.4

(a) Plot the standardized residuals.
(b) Construct a normal probability plot of the standardized residuals.
(c) What conclusions do you draw based on the results of parts (a) and (b)?

9.1.3 Suppose that the following sample is assumed to be from an $N(\mu, \sigma^2)$ distribution where $\mu \in R^1$ and $\sigma^2 > 0$ are unknown.

14.0	9.4	12.1	13.4	6.3	8.5	7.1	12.4	13.3	9.1

(a) Plot the standardized residuals.
(b) Construct a normal probability plot of the standardized residuals.
(c) What conclusions do you draw based on the results of parts (a) and (b)?

9.1.4 Suppose that the following table was obtained from classifying members of a sample of $n = 10$ from a student population according to the classification variables A and B where $A = 0, 1$ indicates male, female and $B = 0, 1$ indicates conservative, liberal.

	$B = 0$	$B = 1$
$A = 0$	2	1
$A = 1$	3	4

Check the model that says that gender and political orientation are independent, using Fisher's exact test.

9.1.5 The following sample of $n = 20$ is supposed to be from a Uniform$(0, 1)$ distribution.

0.11	0.56	0.72	0.18	0.26	0.32	0.42	0.22	0.96	0.04
0.45	0.22	0.08	0.65	0.32	0.88	0.76	0.32	0.21	0.80

After grouping the data, using a partition of five equal-length intervals, carry out the chi-squared goodness of fit test to assess whether or not we have evidence against this assumption. Record the standardized residuals.

9.1.6 Suppose that a die is tossed 1000 times, and the following frequencies are obtained for the number of pips up when the die comes to a rest.

x_1	x_2	x_3	x_4	x_5	x_6
163	178	142	150	183	184

Using the chi-squared goodness of fit test, assess whether we have evidence that this is not a symmetrical die. Record the standardized residuals.

Computer Exercises

9.1.7 For the data of Exercise 9.1.1, present a normal probability plot of the standardized residuals and comment on it.

9.1.8 Generate 25 samples from the $N(0, 1)$ distribution with $n = 10$ and look at their normal probability plots. Draw any general conclusions.

9.1.9 Suppose that the following table was obtained from classifying members of a sample on $n = 100$ from a student population according to the classification variables A and B where $A = 0, 1$ indicates male, female and $B = 0, 1$ indicates conservative, liberal.

	$B = 0$	$B = 1$
$A = 0$	20	15
$A = 1$	36	29

Check the model that gender and political orientation are independent using Fisher's exact test.

9.1.10 Using a statistical package, generate a sample of $n = 1000$ from the Binomial$(10, 0.2)$ distribution. Then, using the chi-squared goodness of fit test, check that this sample is indeed from this distribution. Use grouping to ensure $E(X_i) = np_i \geq 1$. What would you conclude if you got a P-value close to 0 or close to 1?

Problems

9.1.11 (*Multivariate normal distribution*) A random vector $Y = (Y_1, \ldots, Y_k)$ is said to have a multivariate normal distribution with mean vector $\mu \in R^k$ and variance matrix $\Sigma = (\sigma_{ij}) \in R^{k \times k}$ if

$$a_1 Y_1 + \cdots + a_k Y_k \sim N\left(\sum_{i=1}^{k} a_i \mu_i, \sum_{i=1}^{k}\sum_{j=1}^{k} a_i a_j \sigma_{ij}\right)$$

for every choice of $a_1, \ldots, a_k \in R^1$. We write $Y \sim N_k(\mu, \Sigma)$. Prove that $E(Y_i) = \mu_i$, $\text{Cov}(Y_i, Y_j) = \sigma_{ij}$ and that $Y_i \sim N(\mu_i, \sigma_{ii})$. (Hint: Theorem 3.3.4.)

9.1.12 In Example 9.1.1, prove that the residual $R = (R_1, \ldots, R_n)$ is distributed multivariate normal (see Problem 9.1.11) with mean vector $\mu = (0, \ldots, 0)$ and variance matrix $\Sigma = (\sigma_{ij}) \in R^{k \times k}$ where $\sigma_{ij} = -\sigma_0^2/n$ when $i \neq j$ and $\sigma_{ii} = \sigma_0^2(1 - 1/n)$. (Hint: Theorem 4.6.1.)

9.1.13 If $Y = (Y_1, \ldots, Y_k)$ is distributed multivariate normal with mean vector $\mu \in R^k$ and variance matrix $\Sigma = (\sigma_{ij}) \in R^{k \times k}$, and if $X = (X_1, \ldots, X_l)$ is distributed multivariate normal with mean vector $\nu \in R^l$ and variance matrix $\Upsilon = (\tau_{ij}) \in R^{l \times l}$, then it can be shown that Y and X are independent whenever $\sum_{i=1}^{k} a_i Y_i$ and $\sum_{i=1}^{l} b_i X_i$ are independent for every choice of $(a_1, \ldots, a_k)$ and $(b_1, \ldots, b_l)$. Use this fact to show that, in Example 9.1.1, $\bar{X}$ and R are independent. (Hint: Theorem 4.6.2 and Problem 9.1.12.)

9.1.14 In Example 9.1.4, prove that $(\hat{\alpha}_1, \hat{\beta}_1) = (x_1./n, x_{\cdot 1}/n)$ is the MLE.

9.1.15 In Example 9.1.4, prove that the number of samples satisfying the constraints (9.1.2) equals

$$\binom{n}{x_1.}\binom{n}{x_{\cdot 1}}.$$

(Hint: Using i for the count x_{11}, show that the number of such samples equals

$$\binom{n}{x_1.}\sum_{i=\max\{0, x_1. + x_{\cdot 1} - n\}}^{\min\{x_1., x_{\cdot 1}\}}\binom{x_1.}{i}\binom{n - x_1.}{x_{\cdot 1} - i}$$

and sum this using the fact that the sum of Hypergeometric$(n, x_{\cdot 1}, x_1.)$ probabilities equals 1.)

Computer Problems

9.1.16 For the data of Exercise 9.1.3, carry out a simulation to estimate the P-value for the discrepancy statistic of Example 9.1.2. Plot a density histogram of the simulated values. (Hint: See Appendix B for appropriate code.)

9.1.17 When $n = 10$, generate 10^4 values of the discrepancy statistic in Example 9.1.2 when we have a sample from an $N(0, 1)$ distribution. Plot these in a density histogram. Repeat this, but this time generate from a Cauchy distribution. Compare the histograms (don't forget to make sure both plots have the same scales).

9.1.18 The following data are supposed to have come from an Exponential(θ) distribution where $\theta > 0$ is unknown.

1.5	1.6	1.4	9.7	12.1	2.7	2.2	1.6	6.8	0.1
0.8	1.7	8.0	0.2	12.3	2.2	0.2	0.6	10.1	4.9

Check this model using a chi-squared goodness of fit test based on the intervals $(-\infty, 2.0], (2.0, 4.0], (4.0, 6.0], (6.0, 8.0], (8.0, 10.0], (10.0, \infty)$. (Hint: Calculate the MLE by plotting the log-likelihood over successively smaller intervals.)

9.1.19 The following table, taken from *Introduction to the Practice of Statistics* by D. Moore and G. McCabe (W. H. Freeman, New York, 1999), gives the measurements in milligrams of daily calcium intake for 38 women between the ages of 18 and 24 years.

808	882	1062	970	909	802	374	416	784	997
651	716	438	1420	1425	948	1050	976	572	403
626	774	1253	549	1325	446	465	1269	671	696
1156	684	1933	748	1203	2433	1255	110		

(a) Suppose that the model specifies a location normal model for these data with $\sigma_0^2 = (500)^2$. Carry out a chi-squared goodness of fit test on these data using the intervals $(-\infty, 600], (600, 1200], (1200, 1800], (1800, \infty)$. (Hint: Plot the log-likelihood over successively smaller intervals to determine the MLE to about one decimal place. To determine the initial range for plotting, use the overall MLE of μ minus three standard errors to the overall MLE plus three standard errors.)

(b) Compare the MLE of μ obtained in part (a) with the ungrouped MLE.

(c) It would be more realistic to assume that the variance σ^2 is unknown as well. Record the log-likelihood for the grouped data. (More sophisticated numerical methods are needed to find the MLE of (μ, σ^2) in this case.)

Challenge

9.1.20 (MV) Prove that when $(x_1, \ldots, x_n)$ is a sample from the distribution given by $\mu + \sigma Z$ where Z has a known distribution and $(\mu, \sigma^2) \in R^1 \times (0, \infty)$ is unknown, then the statistic

$$r(x_1, \ldots, x_n) = \left(\frac{x_1 - \bar{x}}{s}, \ldots, \frac{x_n - \bar{x}}{s} \right)$$

is ancillary. (Hint: Write a sample element as $x_i = \mu + \sigma z_i$ and then show that $r(x_1, \ldots, x_n)$ can be written as a function of the z_i.)

9.2 Checking the Bayesian Model

Bayesian methods add the prior probability measure Π to the statistical model $\{P_\theta : \theta \in \Omega\}$, for the subsequent statistical analysis. The methods of Section

9.1 are designed to check that the observed data can realistically be assumed to have come from a distribution in $\{P_\theta : \theta \in \Omega\}$. When we add the prior, we are in effect saying that our knowledge about the true distribution leads us to assign the prior predictive probability M, given by $M(A) = E_\Pi(P_\theta(A))$ for $A \subset \Omega$, to describe the process generating the data. So it would seem, then, that a sensible Bayesian model-checking approach would be to compare the observed data s with the distribution given by M, to see if it is surprising or not.

Suppose that we were to conclude that the Bayesian model was incorrect after deciding that s is a surprising value from M. This only tells us, however, that the probability measure M is unlikely to have produced the data and not that the model $\{P_\theta : \theta \in \Omega\}$ was wrong. Consider the following example.

Example 9.2.1 *Prior-Data Conflict*
Suppose that we obtain a sample consisting of $n = 20$ values of $s = 1$ from the model with $\Omega = \{1, 2\}$ and probability functions for the basic response given by the following table.

	$s = 0$	$s = 1$
$f_1(s)$	0.9	0.1
$f_2(s)$	0.1	0.9

Then the probability of obtaining this sample from f_2 is given by $(0.9)^{20} = 0.12158$, which is a reasonable value, so we have no evidence against the model $\{f_1, f_2\}$.

Suppose we place a prior on Ω given by $\Pi(\{1\}) = 0.9999$ so that we are virtually certain that $\theta = 1$. Then the probability of getting these data from the prior predictive M is

$$(0.9999)(0.1)^{20} + (0.0001)(0.9)^{20} = 1.2158 \times 10^{-5}.$$

The prior probability of observing a sample of 20, whose prior predictive probability is no greater than 1.2158×10^{-5}, can be calculated (using statistical software to tabulate the prior predictive) to be approximately 0.04. This tells us that the observed data are "in the tails" of the prior predictive and thus are surprising. This leads us to conclude that we have evidence that M is incorrect.

So in this example, checking the model $\{f_\theta : \theta \in \Omega\}$ leads us to conclude that it is plausible for the data observed. On the other hand, checking the model given by M leads us to the conclusion that the Bayesian model is implausible. ∎

The lesson of Example 9.2.1 is that we can have model failure in the Bayesian context in two ways. First, the data s may be surprising in light of the model $\{f_\theta : \theta \in \Omega\}$. Second, even when the data are plausibly from this model, the prior and the data may conflict. This conflict will occur whenever the prior assigns most of its probability to distributions in the model for which the data are surprising. In either situation inferences drawn from the Bayesian model may be flawed.

If, however, the prior assigns positive probability (or density) to every possible value of θ then, the consistency results for Bayesian inference mentioned in Chapter 7 indicate that a large amount of data will overcome a prior-data conflict (see Example 9.2.3). So the existence of a prior-data conflict does not necessarily mean that our inferences are in error. Still, it is useful to know whether or not this conflict exists, as it is often difficult to detect whether or not we have sufficient data to avoid the problem.

Therefore, we should first use the checks discussed in Section 9.1 to ensure that the data s is plausibly from the model $\{f_\theta : \theta \in \Omega\}$. If we accept the model, then we look for any prior-data conflict. We now consider how to go about this.

The prior predictive distribution of any ancillary statistic is the same as its distribution under the sampling model, i.e., its prior predictive distribution is not affected by the choice of the prior. So the observed value of any ancillary statistic cannot tell us anything about the existence of a prior-data conflict. We conclude from this that, if we are going to use some function of the data to assess whether or not there is prior-data conflict, then its marginal distribution has to depend upon θ.

We now show that the prior predictive conditional distribution of the data given a minimal sufficient statistic T is independent of the prior.

Theorem 9.2.1 Suppose T is a sufficient statistic for the model $\{f_\theta : \theta \in \Omega\}$ for data s. Then the conditional prior predictive distribution of the data s given T is independent of the prior π.

Proof: We will prove this in the case that each sample distribution f_θ and the prior π are discrete. A similar argument can be developed for the more general case.

By Theorem 6.1.1 (factorization theorem) we have that

$$f_\theta (s) = h (s) g_\theta (T(s))$$

for some functions g_θ and h. Therefore the prior predictive probability function of s is given by

$$m(s) = h (s) \sum_{\theta \in \Omega} g_\theta (T(s)) \pi (\theta).$$

The prior predictive probability function of T at t is given by

$$m^* (t) = \sum_{\{s : T(s) = t\}} h (s) \sum_{\theta \in \Omega} g_\theta (t) \pi (\theta).$$

Therefore, the conditional prior predictive probability function of the data s given $T(s) = t$ is

$$m(s \,|\, T = t) = \frac{h (s) \sum_{\theta \in \Omega} g_\theta (t) \pi (\theta)}{\sum_{\{s' : T(s') = t\}} h (s') \sum_{\theta \in \Omega} g_\theta (t) \pi (\theta)} = \frac{h (s)}{\sum_{\{s' : t(s') = t\}} h (s')}$$

which is independent of π. ∎

So, from Theorem 9.2.1, we conclude that any aspects of the data, beyond the value of a minimal sufficient statistic, can tell us nothing about the existence of a prior-data conflict. Therefore, if we want to base our check for a prior-data conflict on the prior predictive, then we must use the prior predictive for a minimal sufficient statistic. Consider the following example.

Example 9.2.2 *Checking a Bernoulli Model Using the Prior Predictive*
Suppose that $(x_1, \ldots, x_n)$ is a sample from a Bernoulli(θ) model where $\theta \in [0, 1]$ is unknown, and θ is given a Beta(α, β) prior distribution. Then we have that the sample count $y = \sum_{i=1}^{n} x_i$ is a minimal sufficient statistic, and it is distributed Binomial(n, θ). Therefore, the prior predictive probability function for y is given by

$$
\begin{aligned}
m(y) &= \binom{n}{y} \int_0^1 \theta^y (1-\theta)^{n-y} \frac{\Gamma(\alpha+\beta)}{\Gamma(\alpha)\Gamma(\beta)} \theta^{\alpha-1} (1-\theta)^{\beta-1} \, d\theta \\
&= \frac{\Gamma(n+1)}{\Gamma(y+1)\Gamma(n-y+1)} \frac{\Gamma(\alpha+\beta)}{\Gamma(\alpha)\Gamma(\beta)} \frac{\Gamma(y+\alpha)\Gamma(n-y+\beta)}{\Gamma(n+\alpha+\beta)} \\
&\propto \frac{\Gamma(y+\alpha)\Gamma(n-y+\beta)}{\Gamma(y+1)\Gamma(n-y+1)}.
\end{aligned}
$$

Now observe that when $\alpha = \beta = 1$, then $m(y) = 1/(n+1)$, i.e., the prior predictive of y is Uniform$\{0, 1, \ldots, n\}$, and no values of y are surprising. This is not unexpected, as with the uniform prior on θ, we are implicitly saying that any count y is reasonable.

On the other hand, when $\alpha = \beta = 2$, the prior puts more weight around $1/2$. The prior predictive is then proportional to $(y+1)(n-y+1)$. This prior predictive is plotted in Figure 9.2.1 when $n = 20$. Note that counts near 0 or 20 lead to evidence that there is a conflict between the data and the prior. For example, if we obtain the count $y = 3$, we can assess how surprising this value is by computing the probability of obtaining a value with a lower probability of occurrence. Using the symmetry of the prior predictive, we have that this probability equals (using statistical software for the computation) $m(0) + m(2) + m(19) + m(20) = 0.0688876$. Therefore, the observation $y = 3$ is not surprising at the 5% level.

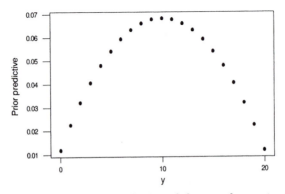

Figure 9.2.1: Plot of the prior predictive of the sample count y in Example 9.2.2 when $\alpha = \beta = 2$ and $n = 20$.

Suppose now that $n = 50$ and $\alpha = 2, \beta = 4$. The mean of this prior is $2/(2+4) = 1/3$ and the prior is right-skewed. The prior predictive is plotted in Figure 9.2.2. Clearly, values of y near 50 give evidence against the model in this case. For example, if we observe $y = 35$, then the probability of getting a count with smaller probability of occurrence is given by (using statistical software for the computation) $m(36) + \cdots + m(50) = 0.0500457$. Only values more extreme than this would provide evidence against the model at the 5% level. ∎

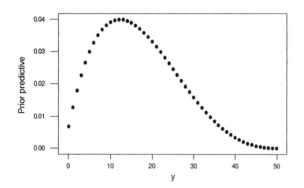

Figure 9.2.2: Plot of the prior predictive of the sample count y in Example 9.2.2 when $\alpha = 2, \beta = 4$ and $n = 50$.

Another possibility for model checking in this context is to look at the posterior predictive distribution of the data. Consider, however, the following example.

Example 9.2.3 (*Example 9.2.1 continued*)
Recall that, in Example 9.2.1 we concluded that a prior-data conflict existed. Note, however, that the posterior probability of $\theta = 2$ is

$$\frac{(0.0001)\,(0.9)^{20}}{(0.9999)\,(0.1)^{20} + (0.0001)\,(0.9)^{20}} \approx 1.$$

Therefore, the posterior predictive probability of the observed sequence of 20 values of 1 is 0.12158 and this does not indicate that there is any prior-data conflict. We note, however, that in this example the amount of data are sufficient to overwhelm the prior and thus we are led to a sensible inference about θ. ∎

The problem with using the posterior predictive to assess whether or not a prior-data conflict exists is that we have an instance of the so-called *double use of the data*. For we have fit the model, i.e., constructed the posterior predictive, using the observed data, and then we tried to use this posterior predictive to assess whether or not a prior-data conflict exists. The double use of the data results in overly optimistic assessments of the validity of the Bayesian model and will often not detect discrepancies. We will not discuss posterior model checking further in this text.

Summary of Section 9.2

- In Bayesian inference there are two potential sources of model incorrect-ness. First, the sampling model for the data may be incorrect. Second, even if the sampling model is incorrect, the prior may conflict with the data in the sense that most of the prior probability is assigned to distrib-utions in the model for which the data are surprising.

- So we first check for the correctness of the sampling model using the methods of Section 9.1. If we do not find evidence against the sampling model, we next check for prior-data conflict by seeing if the observed value of a minimal sufficient statistic is surprising or not, with respect to the prior predictive distribution of this quantity.

- Even if a prior-data conflict exists, posterior inferences may still be valid if we have enough data.

Exercises

9.2.1 Suppose that we observe the value $s = 2$ from the model given by the following table.

	$s = 1$	$s = 2$	$s = 3$
$f_1(s)$	1/3	1/3	1/3
$f_2(s)$	1/3	0	2/3

(a) Do the observed data lead us to doubt the validity of the model? Explain why or why not.
(b) Suppose the prior, given by $\pi(1) = 0.3$, is placed on the parameter $\theta \in \{1, 2\}$. Is there any evidence of a prior-data conflict? (Hint: Compute the prior predictive for each possible data set and assess whether or not the observed data set is surprising.)
(c) Repeat part (b) using the prior given by $\pi(1) = 0.01$.

9.2.2 Suppose that a sample of $n = 6$ is taken from a Bernoulli(θ) distribution where θ has a Beta(3, 3) prior distribution. If the value $n\bar{x} = 2$ is obtained, then determine whether there is any prior-data conflict.

Computer Exercise

9.2.3 Suppose that a sample of $n = 20$ is taken from a Bernoulli(θ) distribution where θ has a Beta(3, 3) prior distribution. If the value $n\bar{x} = 6$ is obtained, then determine whether there is any prior-data conflict.

Problems

9.2.4 Suppose that $(x_1, \ldots, x_n)$ is a sample from an $N(\mu, \sigma_0^2)$ distribution where $\mu \sim N(\mu_0, \tau_0^2)$. Determine the prior predictive distribution of $\bar{x}$.

9.2.5 Suppose that $(x_1, \ldots, x_n)$ is a sample from an Exponential(θ) distribution where $\theta \sim$ Gamma(α_0, β_0). Determine the prior predictive distribution of $\bar{x}$.

9.2.6 Suppose that $(s_1, \ldots, s_n)$ is a sample from a Multinomial$(1, \theta_1, \ldots, \theta_k)$ distribution where $(\theta_1, \ldots, \theta_{k-1}) \sim$ Dirichlet$(\alpha_1, \ldots, \alpha_k)$. Determine the prior predictive distribution of $(x_1, \ldots, x_k)$ where x_i is the count in the ith category.

9.2.7 Suppose that $(x_1, \ldots, x_n)$ is a sample from a Uniform$(0, \theta)$ distribution where θ has prior density given by $\pi_{\alpha, \beta}(\theta) = \theta^{-\alpha} I_{[\beta, \infty)}(\theta) / (\alpha - 1) \beta^{\alpha - 1}$, where $\alpha > 1, \beta > 0$. Determine the prior predictive distribution of $x_{(n)}$.

Challenge

9.2.8 Suppose that $X_1, \ldots, X_n$ is a sample from an $N(\mu, \sigma^2)$ distribution where $\mu \,|\, \sigma^2 \sim N(\mu_0, \tau_0^2 \sigma^2)$ and $1/\sigma^2 \sim$ Gamma(α_0, β_0), then determine a form for the prior predictive density of $(\bar{X}, S^2)$ that you could evaluate without integrating. (Hint: Use the algebraic manipulations found in Section 7.5.)

9.3 The Problem with Multiple Checks

As we have mentioned throughout this text, model checking is a part of good statistical practice. In other words, one should always be wary of the value of statistical work where the investigators have not engaged in, and reported the results of, reasonably rigorous model checking. It is really the job of those who report statistical results to convince us that their models are reasonable for the data collected, bearing in mind the effects of both underfitting and overfitting.

In this chapter we have reported *some* of the possible model-checking approaches available. We have focused on the main categories of procedures and perhaps the most often used methods from within these. There are many others. At this point, we can't say that any one approach is the best possible method. Perhaps greater insight along these lines will come with further research into the topic, and a clearer recommendation can be made.

One recommendation that can be made now, however, is that it is not reasonable to go about model checking by implementing every possible model-checking procedure you can. A simple example illustrates the folly of such an approach.

Example 9.3.1
Suppose that $(x_1, \ldots, x_n)$ is supposed to be a sample from the $N(0, 1)$ distribution. Suppose that we decide to check this model by computing the P-values

$$P_i = P\left(X_i^2 \geq x_i^2\right)$$

for $i = 1, \ldots, n$ where $X_i^2 \sim \chi^2(1)$. Further, we will decide that the model is incorrect if the minimum of these P-values is less than 0.05.

Now consider the repeated sampling behavior of this method when the model is correct. We have that

$$\min\{P_1, \ldots, P_n\} < 0.05$$

if and only if

$$\max\{x_1^2, \ldots, x_n^2\} \geq \chi_{0.95}^2(1),$$

and so

$$P\left(\min\left\{P_1,\dots,P_n\right\}<.05\right)$$
$$=P\left(\max\left\{X_1^2,\dots,X_n^2\right\}\geq\chi_{0.95}^2\left(1\right)\right)=1-P\left(\max\left\{X_1^2,\dots,X_n^2\right\}\leq\chi_{0.05}^2\left(1\right)\right)$$
$$=1-\prod_{i=1}^{n}P\left(X_i^2\leq\chi_{0.95}^2\left(1\right)\right)=1-\left(0.95\right)^n\to1$$

as $n\to\infty$. This tells us that if n is large enough, we will reject the model with virtual certainty even though it is correct! Note that n does not have to be very large for there to be an appreciable probability of making an error. For example, when $n=10$, the probability of making an error is 0.40; when $n=20$ the probability of making an error is 0.64; and when $n=100$, the probability of making an error is 0.99. ∎

We can learn an important lesson from Example 9.3.1. For, if we carry out too many model-checking procedures, we are almost certain to find something wrong, even if the model is correct. The cure for this is that before actually observing the data (so that our choices are not determined by the actual data obtained), we decide on a few relevant model-checking procedures to be carried out and implement only these.

The problem we have been discussing here is sometimes referred to as the problem of *multiple comparisons*, and it comes up in other situations as well, e.g., see Section 10.4.1 where multiple means are compared via pairwise tests for differences in the means. One approach for avoiding the multiple-comparisons problem is to simply lower the cut-off for the P-value so that the probability of making a mistake is appropriately small. For example, if we decided in Example 9.3.1 that evidence against the model is only warranted when an individual P-value is smaller than 0.0001, then the probability of making a mistake is 0.01 when $n=100$. A difficulty with this approach here is that our model-checking procedures will not be independent generally, and it doesn't always seem possible to determine an appropriate cut-off for the individual P-values.

Summary of Section 9.3

- Carrying out too many model checks is not a good idea, as we will invariably find something that leads us to conclude that the model is incorrect. Rather than engaging in a "fishing expedition" where we just keep on checking the model, it is better to choose a few procedures before we see the data, and use these, and only these, for the model checking.

Chapter 10

Relationships Among Variables

In this chapter, we are concerned with perhaps the most important application of statistical inference. This is the problem of analyzing whether or not a relationship exists among variables, and what form the relationship takes. As a particular instance of this, recall the example and discussion in Section 5.1.

Many of the most important problems in science and society are concerned with relationships among variables. For example, what is the relationship between the amount of carbon dioxide placed into the atmosphere and global temperatures? What is the relationship between class size and scholastic achievement by students? What is the relationship between weight and carbohydrate intake in humans? What is the relationship between lifelength and the dosage of a certain drug for cancer patients? These are all examples of questions whose answers involve relationships among variables. We will see that statistics plays a key role in answering such questions.

In Section 10.1 we provide a precise definition of what it means for variables to be related, and we distinguish between two broad categories of relationship, namely, association and cause–effect. Also, we discuss some of the key ideas involved in collecting data when we want to determine if a cause–effect relationship exists. In the remaining sections, we examine the various statistical methodologies that are used to analyze data when we are concerned with relationships.

We emphasize the use of frequentist methodologies in this chapter. We give some examples of the Bayesian approach, but there are some complexities involved with the distributional problems associated with Bayesian methods that are best avoided at this stage. Sampling algorithms for the Bayesian approach have been developed, along the lines of those discussed in Chapter 7 (see also Chapter 11), but their full discussion would take us beyond the scope of this text. It is worth noting, however, that Bayesian analyses with diffuse priors will often yield results very similar to those obtained via the frequentist approach.

As discussed in Chapter 9, model checking is an important feature of any statistical analysis. For the models used in this chapter, a full discussion of the more rigorous P-value approach to model checking requires more development than we can accomplish in this text. As such, we emphasize the informal approach to model checking, via residual and probability plots. This should not be interpreted as a recommendation that these are the preferred methods for such models.

10.1 Related Variables

Consider a population Π with two variables $X, Y : \Pi \to R^1$ defined on it. What does it mean to say that the variables X and Y are related? Perhaps our first inclination is to say that there must be a formula relating the two variables, such as $Y = a + bX^2$ for some choice of constants a and b, or $Y = \exp(X)$, etc. But consider a population Π of humans and suppose $X(\pi)$ is the weight of π in kilograms and $Y(\pi)$ is the height of individual $\pi \in \Pi$ in centimeters. From our experience, we know that taller people tend to be heavier, so we believe that there is some kind of relationship between height and weight. We know, too, that there cannot be an exact formula that describes this relationship, because people with the same weight will often have different heights, and people with the same height will often have different weights.

If we think of all the people with a given weight x, then there will be a distribution of heights for all those individuals π that have weight x. We call this distribution the conditional distribution of Y given that $X = x$.

We can now express what we mean by our intuitive idea that X and Y are related, for, as we change the value of the weight that we condition on, we expect the conditional distribution to change. In particular, as x increases, we expect that the location of the conditional distribution will increase, although other features of the distribution may change as well. For example, in Figure 10.1.1 we provide a possible plot of two approximating densities for the conditional distributions of Y given $X = 70$ kg and the conditional distribution of Y given $X = 90$ kg.

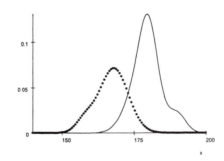

Figure 10.1.1: Plot of two approximating density functions for the conditional distribution of Y given $X = 70$ kg ($\bullet\bullet\bullet$) and the conditional distribution of Y given $X = 90$ kg ($-$).

We see that the conditional distribution has shifted up when X goes from 70 to 90 kg but also that the shape of the distribution has changed somewhat as well. So we can say that a relationship definitely exists between X and Y, at least in this population. Notice that, as defined so far, X and Y are not random variables but become so when we randomly select π from the population. In that case, the conditional distributions referred to become the conditional probability distributions of the random variable Y, given that we observe $X = 70$ and $X = 90$, respectively.

We will adopt the following definition to precisely specify what we mean when we say that variables are related.

Definition 10.1.1 Variables X and Y are *related variables* if there is any change in the conditional distribution of Y, given $X = x$, as x changes.

We could instead define what it means for variables to be *unrelated*. We say that variables X and Y are unrelated if they are independent. This is equivalent to Definition 10.1.1, because two variables are independent if and only if the conditional distribution of one given the other does not depend on the condition (Exercise 10.1.1).

There is an apparent asymmetry in Definition 10.1.1. Because the definition considers only the conditional distribution of Y given X and not the conditional distribution of X given Y. But, if there is a change in the conditional distribution of Y given $X = x$, as we change x, then by the above comment X and Y are not independent, so there must be a change in the conditional distribution of X given $Y = y$, as we change y (also see Problem 10.1.18).

Notice that the definition is applicable no matter what kind of variables we are dealing with. So both could be quantitative variables, or both categorical variables, or one could be a quantitative variable while the other is a categorical variable.

Definition 10.1.1 says that X and Y are related if *any* change is observed in the conditional distribution. In reality, this would mean that there is practically always a relationship between variables X and Y. It seems likely that we will always detect some difference if we carry out a census and calculate all the relevant conditional distributions. This is where the idea of the *strength of a relationship among variables* becomes relevant, for if we see large changes in the conditional distributions, then we can say that a strong relationship exists. If we see only very small changes, then we can say that a very weak relationship exists that is perhaps of no practical importance.

If a relationship exists between two variables, then its form is completely described by the set of conditional distributions of Y given X. Sometimes it may be necessary to describe the relationship using all these conditional distributions. In many problems, however, we look for a simpler presentation. In fact, we often assume a statistical model that prescribes a simple form for how the conditional distributions change as we change X. Consider the following example.

Example 10.1.1 *Simple Normal Linear Model*
In Section 10.3.2 we will discuss the simple normal regression model where
the conditional distribution of quantitative variable Y, given the quantitative
variable $X = x$, is assumed to be distributed

$$N(\beta_1 + \beta_2 x, \sigma^2)$$

where β_1, β_2, and σ^2 are unknown. For example, Y could be the blood pressure
of an individual and X the amount of salt the person consumed each day.

In this case, the conditional distributions have constant shape and change,
as x changes, through the conditional mean. The mean moves along the line
given by $\beta_1 + \beta_2 x$ for some intercept β_1 and slope β_2. If this model is correct,
then the variables are unrelated if and only if $\beta_2 = 0$, as this is the only situation
in which the conditional distributions can remain constant as we change x. ∎

Statistical models, like that described in Example 10.1.1, can be wrong.
There is nothing that requires that two quantitative variables *must* be related
in that way. For example, the conditional variance of Y can vary with x, and the
very shape of the conditional distribution can vary with x, too. The model of
Example 10.1.1 is an example of a simplifying assumption that is appropriate in
many practical contexts. Methods, such as those discussed in Chapter 9, must
be employed to check model assumptions before accepting statistical inferences
based on such a model, however. We will always consider model checking as
part of our discussion of the various models used to examine the relationship
among variables.

Often, we think of Y as a dependent variable (depending on X) and of X as
an independent variable (free to vary). Our goal, then, is to predict the value of
Y given the value of X. In this situation, we call Y the *response variable* and
X the *predictor variable*.

Sometimes, though, there is really nothing to distinguish the roles of X and
Y. For example, suppose that X is the weight of an individual in kilograms and
Y is the height in centimeters. We could then think of predicting weight from
height or conversely. It is then immaterial which we choose to condition on.

In many applications, there is more than one response variable and more
than one predictor variable X. We will not consider the situation in which we
have more than one response variable, but we will consider the case in which
$X = (X_1, \ldots, X_k)$ is k-dimensional. Here the various predictors that comprise
X could be all categorical, all quantitative, or some mixture of categorical and
quantitative variables.

The definition of a relationship existing between response variable Y and the
set of predictors $(X_1, \ldots, X_k)$ is exactly as in Definition 10.1.1. In particular, a
relationship exists between Y and $(X_1, \ldots, X_k)$ if there is any change in the conditional distribution of Y given $(X_1, \ldots, X_k) = (x_1, \ldots, x_k)$ when $(x_1, \ldots, x_k)$
is varied. If such a relationship exists, then the form of the relationship is
specified by the full set of conditional distributions. Again, statistical models
are often used where simplifying assumptions are made about the form of the
relationship. Consider the following example.

Example 10.1.2 *The Normal Linear Model with k Predictors*
In Section 10.3.4 we will discuss the multiple normal regression model. For this, the conditional distribution of quantitative variable Y, given that the quantitative predictors $(X_1, \ldots, X_k) = (x_1, \ldots, x_k)$, is assumed to be the

$$N(\beta_1 + \beta_2 x_1 + \cdots + \beta_{k+1} x_k, \sigma^2)$$

distribution, where $\beta_1, \ldots, \beta_{k+1}$, and σ^2 are unknown. For example, Y could be blood pressure, X_1 the amount of daily salt intake, X_2 the age of the individual, X_3 the weight of the individual, etc.

In this case, the conditional distributions have constant shape and change, as $(x_1, \ldots, x_k)$ changes, through the conditional mean, which changes according to the function $\beta_1 + \beta_2 x_1 + \cdots + \beta_{k+1} x_k$. Notice that, if this model is correct, then the variables are unrelated if and only if $\beta_2 = \cdots = \beta_{k+1} = 0$, as this is the only situation in which the conditional distributions can remain constant as we change $(x_1, \ldots, x_k)$. ∎

When we split a set of variables $Y, X_1, \ldots, X_k$ into response Y and predictors $(X_1, \ldots, X_k)$, we are implicitly saying that we are directly interested only in the conditional distributions of Y given $(X_1, \ldots, X_k)$. There may be relationships among the predictors $X_1, \ldots, X_k$, however, and these can be of interest.

For example, suppose we have two predictors X_1 and X_2, and the conditional distribution of X_1 given X_2 is virtually degenerate at a value $a + cX_2$ for some constants a and c. Then it is not a good idea to include both X_1 and X_2 in a model, such as that discussed in Example 10.1.2, as this can make the analysis very sensitive to small changes in the data. This is known as the problem of *multicollinearity*. The effect of multicollinearity, and how to avoid it, will not be discussed any further in this text. This is, however, a topic of considerable practical importance.

10.1.1 Cause–Effect Relationships and Experiments

Suppose now that we have variables X and Y defined on a population Π and have concluded that a relationship exists according to Definition 10.1.1. This may be based on having conducted a full census of Π, or, more typically, we will have drawn a simple random sample from Π and then used the methods of the remaining sections of this chapter to conclude that such a relationship exists. If Y is playing the role of the response and if X is the predictor, then we often want to be able to assert that changes in X are *causing* the observed changes in the conditional distributions of Y. Of course, if there are no changes in the conditional distributions, then there is no relationship between X and Y, and thus no *cause–effect relationship* either.

For example, suppose that the amount of carbon dioxide gas being released in the atmosphere is increasing, and we observe that mean global temperatures are rising. If we have reason to believe that the amount of carbon dioxide released can have an effect on temperature, then perhaps it is sensible to believe that the increase in carbon dioxide emissions is causing the observed increase in mean global temperatures. As another example, for many years it has been observed that smokers suffer from respiratory diseases much more frequently than do nonsmokers. It seems reasonable, then, to conclude that smoking causes an increased risk for respiratory disease. On the other hand, suppose we consider the relationship between the weight and height. It seems clear that a relationship exists, but it does not make any sense to say that changes in one of the variables is causing the changes in the conditional distributions of the other.

When can we say that an observed relationship between X and Y is a cause–effect relationship? If a relationship exists between X and Y, then we know that there are at least two values x_1 and x_2 such that $f_Y(\cdot \mid X = x_1) \neq f_Y(\cdot \mid X = x_2)$, i.e., these two conditional distributions are not equal. If we wish to say that this difference is caused by the change in X, then we have to know categorically that there is no other variable Z defined on Π that *confounds* with X. The following example illustrates the idea of two variables confounding.

Example 10.1.3

Suppose that Π is a population of students such that most females hold a part-time job and most males do not. A researcher is interested in the distribution of grades, as measured by GPA, and is looking to see if there is a relationship between GPA and gender. On the basis of the data collected, the researcher observes a difference in the conditional distribution of GPA given gender and concludes that a relationship exists between these variables. It seems clear, however, that an assertion of a cause–effect relationship existing between GPA and gender is not warranted, as the difference in the conditional distributions could also be attributed to the difference in part-time work status rather than gender. In this example, part-time work status and gender are confounded. ∎

A more careful analysis might be able to rescue the situation described in Example 10.1.3, for if X and Z denote the confounding variables, then we could collect data on Z as well and examine the conditional distributions $f_Y(\cdot \mid X = x, Z = z)$. In Example 10.1.3, these will be the conditional distributions of GPA, given gender and part-time work status. If these conditional distributions change as we change x, for some fixed value of z, then we could assert that a cause–effect relationship exists between X and Y *provided* there are no further confounding variables. Of course, there are probably still more confounding variables, and we really should be conditioning on all of them. This brings up the point that, in any practical application, we almost certainly will never even know all the potential confounding variables.

Fortunately, there is sometimes a way around the difficulties raised by confounding variables. Suppose that we can *control* the value of the variable X for any $\pi \in \Pi$, i.e., we can *assign* the value x to π so that $X(\pi) = x$, for any of the possible values of x. In Example 10.1.3 this would mean that we could assign a part-time work status to any student in the population. Now consider the following idealized situation. Imagine assigning every element $\pi \in \Pi$ the value $X(\pi) = x_1$ and then carrying out a census to obtain the conditional distribution $f_Y(\cdot \mid X = x_1)$. Now imagine assigning every $\pi \in \Pi$ the value $X(\pi) = x_2$ and then carrying out a census to obtain the conditional distribution $f_Y(\cdot \mid X = x_2)$. If there is any difference in $f_Y(\cdot \mid X = x_1)$ and $f_Y(\cdot \mid X = x_2)$, then the only possible reason is that the value of X differs. Therefore, if $f_Y(\cdot \mid X = x_1) \neq f_Y(\cdot \mid X = x_2)$, we can assert that a cause–effect relationship exists.

A difficulty with the above argument is that typically we can never construct $f_Y(\cdot \mid X = x_1)$ and $f_Y(\cdot \mid X = x_2)$. But in fact, we may be able to sample from them, and then the methods of statistical inference become available to us to infer whether or not there is any difference. Suppose that we take a random sample $\pi_1, \ldots, \pi_{n_1+n_2}$ from Π and randomly assign n_1 of these the value $X = x_1$ with the remaining π's assigned the value x_2. We obtain the Y values $y_{11}, \ldots, y_{1n_1}$ for those π's assigned the value x_1 and obtain the Y values $y_{21}, \ldots, y_{2n_2}$ for those π's assigned the value x_2. Then it is apparent that $y_{11}, \ldots, y_{1n_1}$ is a sample from $f_Y(\cdot \mid X = x_1)$ and $y_{21}, \ldots, y_{2n_2}$ is a sample from $f_Y(\cdot \mid X = x_2)$. In fact, provided that $n_1 + n_2$ is small relative to the population size, then we can consider these as i.i.d. samples from these conditional distributions.

So we see that in certain circumstances, it is possible to collect data in such a way that we can make inferences about whether or not a cause–effect relationship exists. We now specify the characteristics of the relevant data collection technique.

First, if our inferences are to apply to a population Π, then we must have a random sample from that population. This is just the characteristic of what we called a sampling study in Section 5.4, and we must do this to avoid any selection effects. So if the purpose of a study is to examine the relationship between the duration of migraine headaches and the dosage of a certain drug, the investigator must have a random sample from the population of migraine headache sufferers.

Second, we must be able to *assign* any possible value of the predictor variable X to any selected π. If we cannot do this, then there may be hidden confounding variables (sometimes called *lurking variables*) that are influencing the conditional distributions of Y. So in a study of the effects of the dosage of a drug on migraine headaches, the investigator must be able to impose the dosage on each participant in the study.

Third, after deciding what values of X we will use in our study, we must randomly allocate these values to members of the sample. This is done to avoid the possibility of selection effects. So, after deciding what dosages to use in the study of the effects of the dosage of a drug on migraine headaches, and how many participants will receive each dosage, the investigator must randomly select the individuals who will receive each dosage. This will (hopefully) avoid selection effects, such as only the healthiest individuals getting the lowest dosage, etc.

When these requirements are met, we refer to the data collection process as an *experiment*. Statistical inference based on data collected via an experiment has the capability of inferring that cause–effect relationships exist, so this represents an important and powerful scientific tool.

Combining this discussion with Section 5.4, we see a hierarchy of data collection methods. Observational studies reside at the bottom of the hierarchy. Inferences drawn from observational studies must be taken with a degree of caution, for selection effects could mean that the results do not apply to the population intended, and the existence of confounding variables means that we cannot make inferences about cause–effect relationships. For sampling studies, we know that any inferences drawn will be about the appropriate population, but the existence of confounding variables again causes difficulties for any statements about the existence of cause–effect relationships, e.g., just taking random samples of males and females from the population Π of Example 10.1.3 will not avoid the confounding variables. At the top of the hierarchy reside experiments.

It is probably apparent that it is often impossible to conduct an experiment. In Example 10.1.3, we cannot assign the value of gender, and so nothing can be said about the existence of a cause–effect relationship between GPA and gender.

There are many notorious examples in which assertions are made about the existence of cause–effect relationships but no experiment is possible. For example, there have been a number of studies conducted where differences have been noted among the IQ distributions of various racial groups. It is impossible, however, to control the variable racial origin, so it is impossible to assert that the observed differences in the conditional distributions of IQ, given race, are caused by changes in race.

Another example concerns smoking and lung cancer in humans. It has been pointed out that it is impossible to conduct an experiment, as we cannot assign values of the predictor variable (perhaps different amounts of smoking) to humans at birth and then observe the response, namely, whether someone contracts lung cancer or not. This raises an important point. We do not simply reject the results of analyses based on observational studies or sampling studies because the data did not arise from an experiment. Rather we treat these as evidence, potentially flawed evidence, but still evidence. Think of eyewitness evidence in a court of law suggesting that a crime was committed by a certain individual. Eyewitness evidence may be unreliable, but if two or three unconnected eyewitnesses give similar reports, then our confidence grows in the reliability of the evidence. Similarly, if many observational and sampling studies seem to indicate that smoking leads to an increased risk for contracting lung cancer, then our confidence grows that a cause–effect relationship does indeed exist. Further, if we can identify potentially confounding variables, then observational or sampling studies can be conducted taking these into account, increasing our confidence still more. Ultimately, we may not be able to definitively settle the issue via an experiment, but it is still possible to build overwhelming evidence that smoking and lung cancer do have a cause–effect relationship.

10.1.2 Design of Experiments

Suppose we have a response Y and a predictor X (sometimes called a *factor* in experimental contexts) defined on a population Π, and we want to collect data to determine if a cause–effect relationship exists between them. Following the discussion in Section 10.1.1 we will conduct an experiment. There are now a number of decisions to be made, and our choices constitute what we call the *design* of the experiment.

For example, we are going to assign values of X to the sampled elements, now called *experimental units*, $\pi_1, \ldots, \pi_n$ from Π. Which of the possible values of X should we use? When X can take only a small finite number of values, then it is natural to use these values. On the other hand, when the number of possible values of X is very large or even infinite, as with quantitative predictors, then we have to choose values of X to use in the experiment.

Let us suppose that we have chosen the values $x_1, \ldots, x_k$ for X. We refer to $x_1, \ldots, x_k$ as the *levels* of X; any particular assignment x_i to a π_j in the sample, will be called a *treatment*. Typically, we will choose the levels so that they span the possible range of X fairly uniformly. For example, if X is temperature in degrees Celsius, and we want to examine the relationship between Y and X for X in the range $[0, 100]$ then, using $k = 5$ levels, we might take $x_1 = 0, x_2 = 25, x_3 = 50, x_4 = 75$, and $x_5 = 100$.

Having chosen the levels of X, we then have to choose how many treatments of each level we are going to use in the experiment, i.e., decide how many response values n_i we are going to observe at level x_i for $i = 1, \ldots, k$.

In any experiment, we will have a finite amount of resources (money, time, etc.) at our disposal, and this determines the sample size n from Π. The question then is how should we choose the n_i so that $n_1 + \ldots + n_k = n$? If we know nothing about the conditional distributions $f_Y(\cdot \mid X = x_i)$, then it makes sense to use *balance*, namely, choose $n_1 = \cdots = n_k$.

On the other hand, suppose we know that some of the $f_Y(\cdot \mid X = x_i)$ will exhibit greater variability than others. For example, we might measure variability by the variance of $f_Y(\cdot \mid X = x_i)$. Then it makes sense to allocate more treatments to the levels of X where the response is more variable. This is because it will take more observations to make accurate inferences about characteristics of such a $f_Y(\cdot \mid X = x_i)$ than for the less variable conditional distributions.

As discussed in Sections 6.3.4 and 6.3.5, we also want to choose the n_i so that any inferences we make have desired accuracy. Methods for choosing the sample sizes n_i, similar to those discussed in Chapter 7, have been developed for these more complicated designs, but we will not discuss these any further here.

Suppose, then, that we have determined $\{(x_1, n_1), \ldots, (x_k, n_k)\}$. We refer to this set of ordered pairs as the *experimental design*.

Consider some examples.

Example 10.1.4

Suppose that Π is a population of students at a given university. The administration is concerned with determining the value of each student being assigned an academic advisor. The response variable Y will be a rating that a student assigns on a scale of 1 to 10 (completely dissatisfied to completely satisfied with their university experience) at the end of a given semester. We treat Y as a quantitative variable. A random sample of $n = 100$ students is selected from Π, and 50 of these are randomly selected to receive advisers while the remaining 50 are not assigned advisers.

Here the predictor X is a categorical variable that indicates whether or not the student has an advisor. There are only $k = 2$ levels, and both are used in the experiment. If $x_1 = 0$ denotes no advisor and $x_2 = 1$ denotes having an advisor, then $n_1 = n_2 = 50$ and we have a balanced experiment. The experimental design is given by

$$\{(0, 50), (1, 50)\}.$$

At the end of the experiment, we want to use the data to make inferences about the conditional distributions $f_Y(\cdot \mid X = 0)$ and $f_Y(\cdot \mid X = 1)$ to determine if a cause–effect relationship exists. The methods of Section 10.4 will be relevant for this. ■

Example 10.1.5
Suppose that Π is a population of dairy cows. A feed company is concerned with the relationship between weight gain, measured in kilograms, over a specific time period and the amount of a supplement, measured in grams/liter, of an additive put into the cows' feed. Here the response Y is the weight gain and it is a quantitative variable. The predictor X is the concentration of the additive. Suppose X can plausibly range between 0 and 2, so it is also a quantitative variable.

The experimenter decides on using $k = 4$ levels with $x_1 = 0.00, x_2 = 0.66, x_3 = 1.32$, and $x_4 = 2.00$. Further, the sample sizes $n_1 = n_2 = n_3 = n_4 = 10$ were determined to be appropriate. So the balanced experimental design is given by

$$\{(0.00, 10), (0.66, 10), (1.32, 10), (2.00, 10)\}.$$

At the end of the experiment, we want to make inferences about the conditional distributions $f_Y(\cdot \mid X = 0.00)$, $f_Y(\cdot \mid X = 0.66)$, $f_Y(\cdot \mid X = 1.32)$, and $f_Y(\cdot \mid X = 2.00)$. The methods of Section 10.3 are relevant for this. ■

Notice that in Example 10.1.5 we included the level $X = 0$, which corresponds to no application of the additive. This is called a *control treatment*, as it gives a baseline against which we can assess the effect of the predictor. In many experiments, it is important to include a control treatment.

In medical experiments, there is often a *placebo effect* — namely, a disease sufferer given any treatment will often record an improvement in symptoms. The placebo effect is believed to be due to the fact that a sufferer will start to feel better simply because someone is paying attention to the condition. Accordingly, in any experiment to determine the efficacy of a drug in alleviating disease symptoms, it is important that a control treatment be used as well. For example, if we want to investigate whether or not a given drug alleviates migraine headaches, then among the dosages we select for the experiment, we should make sure that we include a pill containing none of the drug (the so-called *sugar pill*), so that we can assess the extent of the placebo effect. Of course, the recipients should not know whether they are receiving the sugar pill or the drug. This is called a *blind* experiment. If we also conceal the identity of the treatment from the experimenters, so as to avoid any biasing of the results on their part, then this is known as a *double-blind* experiment.

In Example 10.1.5 we assumed that it is possible to take a sample from the population of all dairy cows. Strictly speaking, this is necessary if we want to avoid selection effects and make sure that our inferences apply to the population of interest. In practice, however, taking a sample of experimental units from the full population of interest is often not feasible. For example, many medical experiments are conducted on animals, and these are definitely not random samples from the population of the particular animal in question, e.g., rats.

In such cases, however, we simply recognize the possibility that selection effects or lurking variables could render invalid the conclusions drawn from such analyses when they are to be applied to the population of interest. But we still regard the results as evidence concerning the phenomenon under study. It is the job of the experimenter to come as close as possible to the idealized situation specified by a valid experiment, and so, for example, randomization is still employed when assigning treatments to experimental units so that selection effects are avoided as much as possible.

In the experiments we have discussed so far, there has been one predictor. In many practical contexts, there is more than one predictor. Suppose then, that there are two predictors X and W, and that we have decided on the levels $x_1, \ldots, x_k$ for X and the levels $w_1, \ldots, w_l$ for W. One possibility is to look at the conditional distributions $f_Y(\cdot \mid X = x_i)$ for $i = 1, \ldots, k$ and $f_Y(\cdot \mid W = w_j)$ for $j = 1, \ldots, l$ to determine whether X and W individually have a relationship with the response Y. Such an approach, however, ignores the effect of the two predictors together. In particular, the way the conditional distributions $f_Y(\cdot \mid X = x, W = w)$ change as we change x may depend on w; when this is the case, we say that there is an *interaction* between the predictors.

To investigate the possibility of an interaction existing between X and W, we must sample from each of the kl distributions $f_Y(\cdot \mid X = x_i, W = w_j)$ for $i = 1, \ldots, k$ and $j = 1, \ldots, l$. The experimental design then takes the form

$$\{(x_1, w_1, n_{11}), (x_2, w_1, n_{21}), \ldots, (x_k, w_l, n_{kl})\}$$

where n_{ij} gives the number of applications of the treatment (x_i, w_j). We say that the two predictors X and W are *completely crossed* in such a design because each value of X used in the experiment occurs with each value of W used in the experiment. Of course, we can extend all of this discussion to the case where there are more than two predictors.

Example 10.1.6

Suppose we have a population Π of students at a particular university and we are investigating the relationship between the response Y, given by a student's grade in calculus, and the predictors W and X. The predictor W is the number of hours of academic advising given monthly to a student; it can take the values $0, 1$, or 2. The predictor X indicates class size, where $X = 0$ indicates small class size and $X = 1$ indicates large class size. So we have a quantitative response Y, a quantitative predictor W taking three values, and a categorical predictor X taking two values. The crossed values of the predictors (W, X) are given by the set

$$\{(0,0),(1,0),(2,0),(0,1),(1,1),(2,1)\},$$

and so there are six treatments. To conduct the experiment, the university then takes a random sample of $6n$ students and randomly assigns n students to each treatment. ∎

Sometimes we include additional predictors in an experimental design even when we are not primarily interested in their effects on the response Y. We do this because we *know* that such a variable has a relationship with Y. Including such predictors allows us to condition on their values and so investigate more precisely the relationship Y has with the remaining predictors. We refer to such a variable as a *blocking variable*.

Example 10.1.7

Suppose the response variable Y is yield of wheat in bushels per acre and the predictor variable X is an indicator variable for which of three types of wheat is being planted in an agricultural study. Each type of wheat is going to be planted on a plot of land, where all the plots are of the same size, but it is known that the plots used in the experiment will vary considerably with respect to their fertility. Note that such an experiment is another example of a situation in which it is impossible to randomly sample the experimental units (the plots) from the full population of experimental units.

Suppose the experimenter can group the available experimental units into plots of low fertility and of high fertility. We call these two classes of fields *blocks*. Let W indicate the type of plot. So W is a categorical variable taking two values. It then seems clear that the conditional distributions $f_Y(\cdot \mid X = x, W = w)$ will be much less variable than the conditional distributions $f_Y(\cdot \mid X = x)$.

In this case, W is serving as a blocking variable. The experimental units in a particular block, the one of low fertility or the one of high fertility, are more homogeneous than the full set of plots, so variability will be reduced and inferences will be more accurate. ∎

Summary of Section 10.1

- We say two variables are related if the conditional distribution of one given the other changes at all, as we change the value of the conditioning variable.

- To conclude that a relationship between two variables is a cause–effect relationship, we must make sure that we have taken account of (through conditioning) all confounding variables.

- Statistics provides a practical way of avoiding the effects of confounding variables via conducting an experiment. For this we must be able to assign the values of the predictor variable to experimental units sampled from the population of interest.

- The design of experiments is concerned with determining methods of collecting the data so that the analysis of the data will lead to accurate inferences concerning questions of interest.

Exercises

10.1.1 Prove that discrete random variables X and Y are unrelated if and only if X and Y are independent.

10.1.2 Suppose that two variables X and Y defined on a finite population Π are functionally related as $Y = g(X)$ for some unknown nonconstant function g. Explain how this situation is covered by Definition 10.1.1, i.e., the definition will lead us to conclude that X and Y are related. What about the situation in which $g(x) = c$ for some value c for every x? (Hint: Use the relative frequency functions of the variables.)

10.1.3 Suppose that a census is conducted on a population and the joint distribution of (X, Y) is obtained as in the following table.

	$Y = 1$	$Y = 2$	$Y = 3$
$X = 1$	0.15	0.18	0.40
$X = 2$	0.12	0.09	0.06

Determine whether or not a relationship exists between Y and X.

10.1.4 Suppose that a census is conducted on a population and the joint distribution of (X, Y) is obtained as in the following table.

	$Y = 1$	$Y = 2$	$Y = 3$
$X = 1$	1/6	1/6	1/3
$X = 2$	1/12	1/12	1/6

Determine whether or not a relationship exists between Y and X.

10.1.5 Suppose that X is a random variable and $Y = X^2$. Determine whether or not X and Y are related. What happens when X has a degenerate distribution?

10.1.6 Suppose that a researcher wants to investigate the relationship between birth weight and performance on a standardized test administered to children at two years of age. If a relationship is found, can this be claimed to be a cause–effect relationship? Explain why or why not?

10.1.7 Suppose that a large study of all doctors in Canada was undertaken to determine the relationship between various lifestyle choices and lifelength. If the conditional distribution of lifelength given various smoking habits changes, then discuss what can be concluded from this study.

10.1.8 Suppose that a teacher wanted to determine whether an open- or closed-book exam was a more appropriate way to examine students on a particular topic. The response variable is the grade obtained on the exam out of 100. Discuss how the teacher could go about answering this question.

10.1.9 Suppose that a researcher wanted to determine whether or not there is a cause–effect relationship between the type of political ad (negative or positive) seen by a voter from a particular population and the way the voter votes. Discuss your advice to the researcher about how they should conduct their study.

10.1.10 If two random variables have a nonzero correlation are they necessarily related? Explain why or why not.

10.1.11 An experimenter wants to determine the relationship between weight change Y over a specified period and the use of a specially designed diet. The predictor variable X is a categorical variable indicating whether or not a person is on the diet. A total of 200 volunteers signed on for the study, with a random selection of 100 of these being given the diet and the remaining 100 continuing on their usual diet.
(a) Record the experimental design.
(b) If the results of the study are to be applied to the population of all humans, what concerns do you have about how the study was conducted?
(c) It is felt that the amount of weight lost or gained also is dependent on the initial weight W of a participant. How would you propose that the experiment be altered to take this into account?

10.1.12 A study will be conducted, involving the population of people aged 15 to 19 in a particular country, to determine whether a relationship exists between the response Y (amount spent in dollars in a week on music compact disks) and the predictors W (gender) and X (age in years).
(a) If observations are to be taken from every possible conditional distribution of Y given the two factors, then how many such conditional distributions are there?
(b) Identify the types of each variable involved in the study.
(c) Suppose there are enough funds available to monitor 2000 members of the population. How would you recommend that these resources be allocated among the various combinations of factors?
(d) If a relationship is found between the response and the predictors, can this be claimed to be a cause–effect relationship? Explain why or why not.
(e) Suppose that in addition, it was believed that family income would likely have an effect on Y and families can be classified into low and high income. Indicate how you would modify the study to take this into account.

10.1.13 A random sample of 100 households, from the set of all households containing two or more members in a given geographical area, is selected and their television viewing habits are monitored for six months. A random selection of 50 of the households is sent a brochure each week advertising a certain program. The purpose of the study is to determine whether there is any relationship between exposure to the brochure and whether or not this program is watched.

(a) Identify suitable response and predictor variables.

(b) If a relationship is found, can this be claimed to be a cause–effect relationship? Explain why or why not.

10.1.14 Suppose that we have a quantitative response variable Y and two categorical predictor variables W and X, both taking values in $\{0, 1\}$. Suppose that the conditional distributions of Y are given by

$$Y \,|\, W = 0, X = 0 \sim N\,(3, 5)$$
$$Y \,|\, W = 1, X = 0 \sim N\,(3, 5)$$
$$Y \,|\, W = 0, X = 1 \sim N\,(4, 5)$$
$$Y \,|\, W = 1, X = 1 \sim N\,(4, 5).$$

Does W have a relationship with Y? Does X have a relationship with Y? Explain your answers.

10.1.15 Suppose that we have a quantitative response variable Y and two categorical predictor variables W and X, both taking values in $\{0, 1\}$. Suppose that the conditional distributions of Y are given by

$$Y \,|\, W = 0, X = 0 \sim N\,(2, 5)$$
$$Y \,|\, W = 1, X = 0 \sim N\,(3, 5)$$
$$Y \,|\, W = 0, X = 1 \sim N\,(4, 5)$$
$$Y \,|\, W = 1, X = 1 \sim N\,(4, 5).$$

Does W have a relationship with Y? Does X have a relationship with Y? Explain your answers.

10.1.16 Do the predictors interact in Exercise 10.1.14? Do the predictors interact in Exercise 10.1.15? Explain your answers.

Problems

10.1.17 If there is more than one predictor involved in an experiment, do you think it is preferable for the predictors to interact or not? Explain your answer. Can the experimenter control whether or not predictors interact?

10.1.18 Prove directly, using Definition 10.1.1, that when X and Y are related variables defined on a finite population Π, then Y and X are also related.

10.1.19 Suppose that X, Y, Z are independent $N(0, 1)$ random variables and that $U = X + Z, V = Y + Z$. Determine whether or not the variables U and V are related. (Hint: Calculate $\mathrm{Cov}(U, V)$.)

10.1.20 Suppose that $(X, Y, Z) \sim \mathrm{Multinomial}(n, 1/3, 1/3, 1/3)$. Are X and Y related?

10.1.21 Suppose that $(X, Y) \sim$ Bivariate Normal$(\mu_1, \mu_2, \sigma_1, \sigma_2, \rho)$. Show that X and Y are unrelated if and only if $\mathrm{Corr}(X, Y) = 0$.

10.1.22 Suppose that (X, Y, Z) have probability function $p_{X,Y,Z}$. If Y is related to X but not to Z, then prove that $p_{X,Y,Z}\,(x, y, z) = p_{Y|X}\,(y \,|\, x)\,p_{X|Z}(x \,|\, z)p_Z\,(z)$.

10.2 Categorical Response and Predictors

There are two possible situations when we have a single categorical response Y and a single categorical predictor X. The categorical predictor is either random or deterministic and this is determined by how we sample. We examine these two situations separately.

10.2.1 Random Predictor

We consider the situation in which X is categorical, taking values in $\{1, \ldots, a\}$, and Y is categorical, taking values in $\{1, \ldots, b\}$. If we take a sample $\pi_1, \ldots, \pi_n$ from the population, then the values $X(\pi_i) = x_i$ are random, as are the values $Y(\pi_i) = y_j$.

Suppose that the sample size n is very small relative to the population size (so we can assume that i.i.d. sampling is applicable). Then, letting $\theta_{ij} = P(X = i, Y = j)$, we obtain the likelihood function (Problem 10.2.11)

$$L(\theta_{11}, \ldots, \theta_{ab} \mid (x_1, y_1), \ldots, (x_n, y_n)) = \prod_{i=1}^{a} \prod_{j=1}^{b} \theta_{ij}^{f_{ij}}, \qquad (10.2.1)$$

where f_{ij} is the number of sample values with $(X, Y) = (i, j)$. An easy computation (Problem 10.2.12) shows that the MLE of $(\theta_{11}, \ldots, \theta_{kl})$ is given by $\hat{\theta}_{ij} = f_{ij}/n$ and that the standard error of this estimate (because the incidence of a sample member falling in the (i, j)th cell is distributed Bernoulli(θ_{ij}) and using Example 6.3.2) is given by

$$\sqrt{\frac{\hat{\theta}_{ij}\left(1 - \hat{\theta}_{ij}\right)}{n}}.$$

We are interested in whether or not there is a relationship between X and Y. To answer this, we look at the conditional distributions of Y given X. The conditional distributions of Y given X are, using $\theta_{i.} = \theta_{i1} + \cdots + \theta_{ib} = P(X = i)$, given in the following table.

	$Y = 1$	$\cdots$	$Y = b$
$X = 1$	$\theta_{11}/\theta_{1.}$	$\cdots$	$\theta_{b1}/\theta_{1.}$
$\vdots$	$\vdots$		$\vdots$
$X = a$	$\theta_{1a}/\theta_{a.}$	$\cdots$	$\theta_{ba}/\theta_{a.}$

Then estimating $\theta_{ij}/\theta_{i.}$ by $\hat{\theta}_{ij}/\hat{\theta}_{i.} = f_{ij}/f_{i.}$, where $f_{i.} = f_{i1} + \cdots + f_{ib}$, the estimated condition distributions are given in the following table.

	$Y = 1$	$\cdots$	$Y = b$
$X = 1$	$f_{11}/f_{1.}$	$\cdots$	$f_{1b}/f_{1.}$
$\vdots$	$\vdots$		$\vdots$
$X = a$	$f_{a1}/f_{a.}$	$\cdots$	$f_{ab}/f_{a.}$

If we conclude that there is a relationship between X and Y, then we look at the table of estimated conditional distributions to determine the form of the relationship, i.e., how the conditional distributions change as we change the value of X we are conditioning on.

How then do we infer whether or not a relationship exists between X and Y? No relationship exists between Y and X if and only if the conditional distributions of Y given $X = x$ do not change with x. This is the case if and only if X and Y are independent, and this is true if and only if

$$\theta_{ij} = P(X = i, Y = j) = P(X = i)P(Y = j) = \theta_{i\cdot}\theta_{\cdot j},$$

for every i and j where $\theta_{\cdot j} = \theta_{1j} + \cdots + \theta_{aj} = P(Y = j)$. Therefore, to assess whether or not there is a relationship between X and Y, it is equivalent to assess the null hypothesis $H_0 : \theta_{ij} = \theta_{i\cdot}\theta_{\cdot j}$ for every i and j.

How should we assess whether or not the observed data are surprising when H_0 holds? The methods of Section 9.1.2, and in particular Theorem 9.1.2, can be applied here, as we have that

$$(F_{11}, F_{12}, \ldots, F_{ab}) \sim \text{Multinomial}(n, \theta_{1\cdot}\theta_{\cdot 1}, \theta_{1\cdot}\theta_{\cdot 2}, \ldots, \theta_{a\cdot}\theta_{\cdot b})$$

when H_0 holds.

To apply Theorem 9.1.2 we need the MLE of the parameters of the model under H_0. The likelihood, when H_0 holds, is

$$L(\theta_{1\cdot}, \ldots, \theta_{a\cdot}, \theta_{\cdot 1}, \ldots, \theta_{\cdot b} \mid (x_1, y_1), \ldots, (x_n, y_n)) = \prod_{i=1}^{a} \prod_{j=1}^{b} (\theta_{i\cdot}\theta_{\cdot j})^{f_{ij}}. \quad (10.2.2)$$

From this we deduce (Problem 10.2.13) that the MLE's of the $\theta_{i\cdot}$ and $\theta_{\cdot j}$ are given by $\hat{\theta}_{i\cdot} = f_{i\cdot}/n$ and $\hat{\theta}_{\cdot j} = f_{\cdot j}/n$. Therefore, the relevant chi-squared statistic is

$$X^2 = \sum_{i=1}^{a} \sum_{j=1}^{b} \frac{\left(f_{ij} - n\hat{\theta}_{i\cdot}\hat{\theta}_{\cdot j}\right)^2}{n\hat{\theta}_{i\cdot}\hat{\theta}_{\cdot j}}.$$

Under H_0 the parameter space has dimension $(a-1)+(b-1) = a+b-2$, and so we compare the observed value of X^2 with the $\chi^2((a-1)(b-1))$ distribution, because $ab - 1 - a - b + 2 = (a-1)(b-1)$.

Consider an example.

Example 10.2.1 *Piston Ring Data*

The following table gives the counts of piston ring failures where variable Y is the compressor number and variable X is the leg position based on a sample of $n = 166$. These data were taken from *Statistical Methods in Research and Production* by O. L. Davies (Hafner Publishers, New York, 1961).

Here Y takes 4 values and X takes 3 values (N = North, C = Central, and S=South).

	$Y = 1$	$Y = 2$	$Y = 3$	$Y = 4$
$X = $ N	17	11	11	14
$X = $ C	17	9	8	7
$X = $ S	12	13	19	28

The question of interest is whether or not there is any relation between compressor and leg position. Because $f_{1\cdot} = 53, f_{2\cdot} = 41$ and $f_{3\cdot} = 72$ the conditional distributions of Y given X are estimated as in the rows of the following table.

	$Y = 1$	$Y = 2$	$Y = 3$	$Y = 4$
$X = $N	$17/53 = 0.321$	$11/53 = 0.208$	$11/53 = 0.208$	$14/53 = 0.264$
$X = $C	$17/41 = 0.415$	$9/41 = 0.222$	$8/41 = 0.195$	$7/41 = 0.171$
$X = $S	$12/72 = 0.167$	$13/72 = 0.181$	$19/72 = 0.264$	$28/72 = 0.389$

Comparing the rows, it certainly looks like there is a difference in the conditional distributions, but we must assess whether or not the observed differences can be explained as due to sampling error. To see if the observed differences are real, we carry out the chi-squared test.

Under the null hypothesis of independence, the MLE's are given by

$$\hat{\theta}_{\cdot 1} = \tfrac{46}{166}, \quad \hat{\theta}_{\cdot 2} = \tfrac{33}{166}, \quad \hat{\theta}_{\cdot 3} = \tfrac{38}{166}, \quad \hat{\theta}_{\cdot 4} = \tfrac{49}{166}$$

for the Y probabilities, and

$$\hat{\theta}_{1 \cdot} = \tfrac{53}{166}, \quad \hat{\theta}_{2 \cdot} = \tfrac{41}{166}, \quad \hat{\theta}_{3 \cdot} = \tfrac{72}{166}$$

for the X probabilities. Then the estimated expected counts $n\hat{\theta}_{i\cdot}\hat{\theta}_{\cdot j}$ are given by the following table.

	$Y = 1$	$Y = 2$	$Y = 3$	$Y = 4$
$X = $N	14.6867	10.5361	12.1325	15.6446
$X = $C	11.3614	8.1506	9.3855	12.1024
$X = $S	19.9518	14.3133	16.4819	21.2530

The standardized residuals (using (9.1.6))

$$\frac{f_{ij} - n\hat{\theta}_{i\cdot}\hat{\theta}_{\cdot j}}{\sqrt{n\hat{\theta}_{i\cdot}\hat{\theta}_{\cdot j}\left(1 - \hat{\theta}_{i\cdot}\hat{\theta}_{\cdot j}\right)}}$$

are as in the following table.

	$Y = 1$	$Y = 2$	$Y = 3$	$Y = 4$
$X = $N	0.6322	0.1477	-0.3377	-0.4369
$X = $C	1.7332	0.3051	-0.4656	-1.5233
$X = $S	-1.8979	-0.3631	0.6536	1.5673

All of the standardized residuals seem reasonable, and we have that $X^2 = 11.7223$ with $P\left(\chi^2(6) > 11.7223\right) = 0.0685$, which is not unreasonably small.

So, while there may be some indication that the null hypothesis of no relationship is false, this evidence is not overwhelming. Accordingly, in this case we may assume that Y and X are independent and use the estimates of cell probabilities obtained under this assumption. ∎

We must also be concerned with model checking, i.e., is the model that we have assumed for the data $(x_1, y_1), \ldots, (x_n, y_n)$ correct? If these observations are i.i.d., then indeed the model is correct, as that is all that is being effectively assumed. So we need to check that the observations are a plausible i.i.d. sample. Because the minimal sufficient statistic is given by $(f_{11}, \ldots, f_{ab})$, such a test could be based on the conditional distribution of the sample $(x_1, y_1), \ldots, (x_n, y_n)$ given $(f_{11}, \ldots, f_{ab})$. The distribution theory for such tests is computationally difficult to implement, however, and we do not pursue this topic further in this text.

10.2.2 Deterministic Predictor

Consider again the situation in which X is categorical, taking values in $\{1, \ldots, a\}$, and Y is categorical, taking values in $\{1, \ldots, b\}$. But now suppose that we take a sample $\pi_1, \ldots, \pi_n$ from the population, where we have specified that n_i sample members have the value $X = i$, etc. This could be by assignment, when we are trying to determine if a cause–effect relationship exists, or we might have b populations $\Pi_1, \ldots, \Pi_b$ and we want to see if there is any difference in the distribution of Y between populations. Note that $n_1 + \cdots + n_b = n$.

In both cases, we again want to make inferences about the conditional distributions of Y given X as represented by the following table.

	$Y = 1$	$\cdots$	$Y = b$
$X = 1$	$\theta_{1|X=1}$	$\cdots$	$\theta_{b|X=1}$
$\vdots$	$\vdots$		$\vdots$
$X = a$	$\theta_{1|X=a}$	$\cdots$	$\theta_{b|X=a}$

A difference in the conditional distributions means that there is a relationship between Y and X. If we denote the number of observations in the ith sample that have $Y = j$ by f_{ij}, then assuming the sample sizes are small relative to the population sizes, the likelihood function is given by

$$L\left(\theta_{1|X=1}, \ldots, \theta_{b|X=a} \mid (x_1, y_1), \ldots, (x_n, y_n)\right) = \prod_{i=1}^{a} \prod_{j=1}^{b} \left(\theta_{j|X=i}\right)^{f_{ij}}, \quad (10.2.3)$$

and the MLE is given by $\hat{\theta}_{j|X=i} = f_{ij}/n_i$ (Problem 10.2.14).

There is no relationship between Y and X if and only if the conditional distributions do not vary as we vary X, or if and only if

$$H_0 : \theta_{j|X=1} = \ldots = \theta_{j|X=a} = \theta_j$$

for all $j = 1, \ldots, b$ for some probability distribution $\theta_1, \ldots, \theta_b$. Under H_0 the likelihood function is given by

$$L\left(\theta_1, \ldots, \theta_b \mid (x_1, y_1), \ldots, (x_n, y_n)\right) = \prod_{j=1}^{b} \theta_j^{f_{\cdot j}}, \quad (10.2.4)$$

and the MLE of θ_j is given by $\hat{\theta}_j = f_{.j}/n$ (Problem 10.2.15). Then, applying Theorem 9.1.2, we have that the statistic

$$X^2 = \sum_{i=1}^{a} \sum_{j=1}^{b} \frac{\left(f_{ij} - n_i \hat{\theta}_j\right)^2}{n_i \hat{\theta}_j}$$

has an approximate $\chi^2\left((a-1)(b-1)\right)$ distribution under H_0 because there are $a(b-1)$ free parameters in the full model, $(b-1)$ parameters in the independence model, and $a(b-1) - (b-1) = (a-1)(b-1)$.

Consider an example.

Example 10.2.2

This example is taken from a famous applied statistics book, *Statistical Methods*, 6th ed., by G. Snedecor and W. Cochran (Iowa State Press, Ames, 1967). Individuals were classified according to their blood type Y (O, A, B, and AB, although the AB individuals were eliminated, as they were small in number) and also classified according to X, their disease status (peptic ulcer = P, gastric cancer = G, or control = C). So we have three populations; namely, those suffering from a peptic ulcer, those suffering from gastric cancer, and those suffering from neither. We suppose further that the individuals involved in the study can be considered as random samples from the respective populations.

The data are given in the following table.

	$Y = O$	$Y = A$	$Y = B$	Total
$X = P$	983	679	134	1796
$X = G$	383	416	84	883
$X = C$	2892	2625	570	6087

The estimated conditional distributions of Y given X are then as follows.

	$Y = O$	$Y = A$	$Y = B$
$X = P$	$983/1796 = 0.547$	$679/1796 = 0.378$	$134/1796 = 0.075$
$X = G$	$383/883 = 0.434$	$416/883 = 0.471$	$84/883 = 0.095$
$X = C$	$2892/6087 = 0.475$	$2625/6087 = 0.431$	$570/6087 = 0.093$

We now want to assess whether or not there is any evidence for concluding that a difference exists among these conditional distributions. Under the null hypothesis that no difference exists, the MLE's of the probabilities $\theta_1 = P(Y = O)$, $\theta_2 = P(Y = A)$, and $\theta_3 = P(Y = B)$ are given by

$$\hat{\theta}_1 = \frac{983 + 383 + 2892}{1796 + 883 + 6087} = 0.4857,$$

$$\hat{\theta}_2 = \frac{679 + 416 + 2625}{1796 + 883 + 6087} = 0.4244,$$

$$\hat{\theta}_3 = \frac{134 + 84 + 570}{1796 + 883 + 6087} = 0.0899.$$

Then the estimated expected counts $n_i\hat\theta_j$ are given by the following table.

	$Y = O$	$Y = A$	$Y = B$
$X = P$	872.3172	762.2224	161.4604
$X = G$	428.8731	374.7452	79.3817
$X = C$	2956.4559	2583.3228	547.2213

The standardized residuals (using (9.1.6))

$$\frac{f_{ij} - n_i\hat\theta_j}{\sqrt{n_i\hat\theta\left(1 - \hat\theta_j\right)}}$$

are given by the following table.

	$Y = O$	$Y = A$	$Y = B$
$X = P$	5.2219	−3.9705	−2.2643
$X = G$	−3.0910	2.8111	0.5441
$X = C$	−1.659 2	1.0861	1.0227

We have that $X^2 = 40.5434$ and $P\left(\chi^2(4) > 40.5434\right) = 0.0000$, so we have strong evidence against the null hypothesis of no relationship existing between Y and X. Observe the large residuals when $X = P$ and $Y = O$, $Y = A$.

We are left with examining the conditional distributions to ascertain what form the relationship between Y and X takes. A useful tool in this regard is to plot the conditional distributions in bar charts as we have done in Figure 10.2.1. From this we see that the peptic ulcer population has greater proportion of blood type O than the other populations. ∎

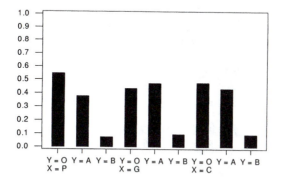

Figure 10.2.1: Plot of the conditional distributions of Y, given X, in Example 10.2.2.

10.2.3 Bayesian Formulation

We now add a prior density π for the unknown values of the parameters of the models discussed in Sections 10.2.1 and 10.2.2. Depending on how we choose

π, and depending on the particular computation we want to carry out, we could be faced with some difficult computational problems. Of course, we have the Monte Carlo methods available in such circumstances, and these can often render a computation fairly straightforward.

The most common choice of prior in these circumstances is to choose a conjugate prior. Because the likelihoods discussed in this section are as in Example 7.1.3, we see immediately that Dirichlet priors will be conjugate for the full model in Section 10.2.1 and that products of independent Dirichlet priors will be conjugate for the full model in Section 10.2.2.

In Section 10.2.1 the general likelihood, i.e., no restrictions on the θ_{ij}, is of the form

$$L\left(\theta_{11},\ldots,\theta_{ab}\mid (x_1,y_1),\ldots,(x_n,y_n)\right) = \prod_{i=1}^{a}\prod_{j=1}^{b}\theta_{ij}^{f_{ij}}.$$

If we place a Dirichlet$(\alpha_{11},\ldots,\alpha_{ab})$ prior on the parameter, then the posterior density is proportional to

$$\prod_{i=1}^{a}\prod_{j=1}^{b}\theta_{ij}^{f_{ij}+\alpha_{ij}-1},$$

so the posterior is a Dirichlet$(f_{11}+\alpha_{11},\ldots,f_{ab}+\alpha_{ab})$ distribution.

In Section 10.2.2 the general likelihood is of the form

$$L\left(\theta_{1|X=1},\ldots,\theta_{b|X=a}\mid (x_1,y_1),\ldots,(x_n,y_n)\right) = \prod_{i=1}^{a}\prod_{j=1}^{b}\left(\theta_{j|X=i}\right)^{f_{ij}}.$$

Because $\sum_{j=1}^{b}\theta_{j|X=i} = 1$ for each $i = 1,\ldots,a$, we must place a prior on each distribution $\left(\theta_{1|X=i},\ldots,\theta_{b|X=i}\right)$. If we choose the prior on the ith distribution to be Dirichlet$\left(\alpha_{1|i},\ldots,\alpha_{a|i}\right)$, then the posterior density is proportional to

$$\prod_{i=1}^{a}\prod_{j=1}^{b}\theta_{j|i}^{f_{ij}+\alpha_{j|i}-1}.$$

We recognize this as the product of independent Dirichlet distributions, with the posterior distribution on $\left(\theta_{1|X=i},\ldots,\theta_{b|X=i}\right)$ equal to a

$$\text{Dirichlet}\left(f_{i1}+\alpha_{1|i},\ldots,f_{ib}+\alpha_{b|i}\right)$$

distribution.

A special and important case of the Dirichlet priors corresponds to the situation in which we feel that we have no information about the parameter. In such a situation, it makes sense to choose all the parameters of the Dirichlet to be 1, so that the priors are all uniform.

There are many characteristics of a Dirichlet distribution that can be evaluated in closed form, e.g., the expectation of any polynomial (Problem 10.2.16), but still there will be many quantities for which exact computations will not be available. It turns out that we can always easily generate samples from Dirichlet distributions, provided we have access to a generator for beta distributions. This is available with most statistical packages. We now discuss how to do this.

Example 10.2.3 *Generating from a Dirichlet*$(\alpha_1, \ldots, \alpha_k)$ *Distribution*
The technique we discuss here is a commonly used method for generating from multivariate distributions. If we want to generate a value of the random vector $(X_1, \ldots, X_k)$, then we can proceed as follows. First, generate a value x_1 from the marginal distribution of X_1. Next, generate a value x_2 from the conditional distribution of X_2 given $X_1 = x_1$. Next, generate a value x_3 from the conditional distribution of X_3 given that $X_1 = x_1$ and $X_2 = x_2$, etc.

If the distribution of X is discrete, then we have that the probability of a particular vector of values $(x_1, x_2, \ldots, x_k)$ arising via this scheme is

$$P(X_1 = x_1) P(X_2 = x_2 \mid X_1 = x_1) P(X_k = x_k \mid X_1 = x_1, \ldots, X_{k-1} = x_{k-1}).$$

Expanding each of these conditional probabilities, we obtain

$$P(X_1 = x_1) \frac{P(X_1=x_1, X_2=x_2)}{P(X_1=x_1)} \cdots \frac{P(X_1=x_1, \ldots, X_{k-1}=x_{k-1}, X_k=x_k)}{P(X_1=x_1, \ldots, X_{k-1}=x_{k-1})},$$

which equals $P(X_1 = x_1, \ldots, X_{k-1} = x_{k-1}, X_k = x_k)$, and so $(x_1, x_2, \ldots, x_k)$ is a value from the joint distribution of $(X_1, \ldots, X_k)$. This approach also works for absolutely continuous distributions, and the proof is the same but uses density functions instead.

In the case of $(X_1, \ldots, X_{k-1}) \sim$ Dirichlet$(\alpha_1, \ldots, \alpha_k)$, we have that (Challenge 10.2.19) $X_1 \sim$ Beta$(\alpha_1, \alpha_2 + \cdots + \alpha_k)$ and X_i given $X_1 = x_1, \ldots, X_{i-1} = x_{i-1}$ has the same distribution as $(1 - x_1 - \cdots - x_{i-1}) U_i$ where

$$U_i \sim \text{Beta}(\alpha_i, \alpha_{i+1} + \cdots + \alpha_k)$$

and $U_2, \ldots, U_{k-1}$ are independent. Note that $X_k = 1 - X_1 - \cdots - X_{k-1}$ for any Dirichlet distribution. So we generate $X_1 \sim$ Beta$(\alpha_1, \alpha_2 + \cdots + \alpha_k)$, generate $U_2 \sim$ Beta$(\alpha_2, \alpha_3 + \cdots + \alpha_k)$ and put $X_2 = (1 - X_1) U_2$, generate $U_3 \sim$ Beta$(\alpha_3, \alpha_4 + \cdots + \alpha_k)$ and put $X_3 = (1 - X_1 - X_2) U_3$, etc.

Below we present a table of a sample of $n = 5$ values from a Dirichlet$(2, 3, 1, 1.5)$ distribution.

	X_1	X_2	X_3	X_4
1	0.116159	0.585788	0.229019	0.069034
2	0.166639	0.566369	0.056627	0.210366
3	0.411488	0.183686	0.326451	0.078375
4	0.483124	0.316647	0.115544	0.084684
5	0.117876	0.147869	0.418013	0.316242

Appendix B contains the code used for this. It can be modified to generate from any Dirichlet distribution. ∎

Summary of Section 10.2

- In this section we have considered the situation in which we have a categorical response variable and a categorical predictor variable.

- We distinguished two situations. The first arises when the value of the predictor variable is not assigned, and the second arises when it is.

- In both cases, the test of the null hypothesis that no relationship exists involved the chi-squared test.

Exercises

10.2.1 The following table gives the counts of accidents for two successive years in a particular city.

	June	July	August
Year 1	60	100	80
Year 2	80	100	60

Is there any evidence of a difference in the distributions of accidents for these months between the two years?

10.2.2 The following data are from a study by Linus Pauling (1971) ("The significance of the evidence about ascorbic acid and the common cold", *Proceedings of the National Academy of Sciences* **68**, p. 2678) concerned with examining the relationship between taking vitamin C and the incidence of colds. Of 279 participants in the study 140 received a placebo (sugar pill) and 139 received vitamin C.

	No cold	Cold
Placebo	31	109
Vitamin C	17	122

Assess the null hypothesis that there is no relationship between taking vitamin C and the incidence of the common cold.

10.2.3 A simulation experiment is carried out to see if there is any relationship between the first and second digits of a random variable generated from a Uniform$(0, 1)$ distribution. A total of 1000 uniforms were generated, and the first and second digits were each recorded as a 0 if they were in $\{0, 1, 2, 3, 4\}$ and as a 1 otherwise. The cross-classified data are given in the following table.

	Second digit 0	Second digit 1
First digit 0	240	250
First digit 1	255	255

Assess the null hypothesis that there is no relationship between the digits.

10.2.4 Grades in a first-year calculus course were obtained for randomly selected students at two universities and classified as pass or fail. The following data were obtained.

	Fail	Pass
University 1	33	143
University 2	22	263

Is there any evidence of a relationship between calculus grades and university?

10.2.5 The following data are recorded in *Statistical Methods for Research Workers* by R. A. Fisher (Hafner Press, New York, 1922) and show the classifications of 3883 Scottish children by gender (X) and hair color (Y).

	$Y = $ fair	$Y = $ red	$Y = $ medium	$Y = $ dark	$Y = $ jet black
$X = $ m	592	119	849	504	36
$X = $ f	544	97	677	451	14

(a) Is there any evidence for a relationship between hair color and gender?
(b) Plot the appropriate bar chart(s).
(c) Record the residuals and relate these to the results in parts (a) and (b). What do you conclude about the size of any deviation from independence?

10.2.6 Suppose we have a controllable predictor X that takes four different values, and we measure a binary-valued predictor Y. A random sample of 100 was taken from the population and the value of X was randomly assigned to each individual in such a way that there are 25 sample members taking each of the possible values of X. Suppose that the following data were obtained.

	$X = 1$	$X = 2$	$X = 3$	$X = 4$
$Y = 0$	12	10	16	14
$Y = 1$	13	15	9	11

(a) Assess whether or not there is evidence for a cause–effect relationship to exist between X and Y.
(b) Explain why it is possible in this example to assert that any evidence found that a relationship exists is evidence that a cause–effect relationship exists.

10.2.7 Write out in full how you would generate a value from a Dirichlet$(1, 1, 1, 1)$ distribution.

Problems

10.2.8 In Example 10.2.1, place a uniform prior on the parameters (a Dirichlet distribution with all parameters equal to 1) and then determine the posterior distribution of the parameters.

10.2.9 In Example 10.2.2, place a uniform prior on the parameters of each population (a Dirichlet distribution with all parameters equal to 1) and such that the three priors are independent and then determine the posterior distribution.

10.2.10 In a 2×2 table with probabilities θ_{ij}, prove that the row and column variables are independent if and only if

$$\frac{\theta_{11}\theta_{22}}{\theta_{12}\theta_{21}} = 1,$$

namely, we have independence if and only if the *cross-ratio* equals 1.

10.2.11 Establish that the likelihood in (10.2.1) is correct when the population size is infinite (or when we are sampling with replacement from the population).

10.2.12 (MV) Prove that the MLE of $(\theta_{11}, \ldots, \theta_{ab})$ in (10.2.1) is given by $\hat{\theta}_{ij} = f_{ij}/n$. Assume that $f_{ij} > 0$ for every i, j. (Hint: Use the facts that a continuous function on this parameter space Ω must achieve its maximum at some point in Ω and, if the function is continuously differentiable at such a point, then all its first-order partial derivatives are zero there. This will allow you to conclude that the unique solution to the score equations must be the point where the log-likelihood is maximized. Try the case where $a = 2, b = 2$ first.)

10.2.13 (MV) Prove that the MLE of $(\theta_1, \ldots, \theta_a, \theta_{\cdot 1}, \ldots, \theta_{\cdot b})$ in (10.2.2) is given by $\hat{\theta}_{i\cdot} = f_{i\cdot}/n$ and $\hat{\theta}_{\cdot j} = f_{\cdot j}/n$. Assume that $f_{i\cdot} > 0, f_{\cdot j} > 0$ for every i, j. (Hint: Use the hint in Problem 10.2.12.)

10.2.14 (MV) Prove that the MLE of $(\theta_{1|X=1}, \ldots, \theta_{b|X=a})$ in (10.2.3) is given by $\hat{\theta}_{j|X=i} = f_{ij}/n_i$. Assume that $f_{ij} > 0$ for every i, j. (Hint: Use the hint in Problem 10.2.12.)

10.2.15 (MV) Prove that the MLE of $(\theta_1, \ldots, \theta_b)$ in (10.2.4) is given by $\hat{\theta}_j = f_{\cdot j}/n$. Assume that $f_{\cdot j} > 0$ for every i, j. (Hint: Use the hint in Problem 10.2.12.)

10.2.16 Suppose that $X = (X_1, \ldots, X_{k-1}) \sim \text{Dirichlet}(\alpha_1, \ldots, \alpha_k)$. Determine $E(X_1^{l_1} \cdots X_k^{l_k})$ in terms of the gamma function, when $l_i > 0$ for $i = 1, \ldots, k$.

Computer Problems

10.2.17 Suppose that $(\theta_1, \theta_2, \theta_3, \theta_4) \sim \text{Dirichlet}(1, 1, 1, 1)$, as in Exercise 10.2.7. Generate a sample of size $N = 10^4$ from this distribution and use this to estimate the expectations of the θ_i. Compare these estimates with their exact values. (Hint: There is some relevant code in Appendix B for the generation; see Appendix C for formulas for the exact values of these expectations.)

10.2.18 For Problem 10.2.8, generate a sample of size $N = 10^4$ from the posterior distribution of the parameters and use this to estimate the posterior expectations of the cell probabilities. Compare these estimates with their exact values. (Hint: There is some relevant code in Appendix B for the generation; see Appendix C for formulas for the exact values of these expectations.)

Challenge

10.2.19 (MV) Establish the validity of the method discussed in Example 10.2.3 for generating from a $\text{Dirichlet}(\alpha_1, \ldots, \alpha_k)$ distribution.

10.3 Quantitative Response and Predictors

When the response and predictor variables are all categorical, it can be difficult to formulate simple models that adequately describe the relationship between the variables. We are left with recording the conditional distributions and plotting these in bar charts. When the response variable is quantitative, however, useful models have been formulated that give a precise mathematical expression for the form of the relationship that may exist. We will study these kinds of models in the next three sections. This section concentrates on the situation in which all the variables are quantitative.

10.3.1 The Method of Least Squares

The *method of least squares* is a general method for obtaining an estimate of a distribution mean. It does not require specific distributional assumptions and so can be thought of as a distribution-free method (see Section 6.4).

Suppose that we have a random variable Y, and we want to estimate $E(Y)$ based on a sample $(y_1, \ldots, y_n)$. The *least-squares principle* says that we select the point $t(y_1, \ldots, y_n)$, in the set of possible values for $E(Y)$, that minimizes the sum of squared deviations (hence, "least squares") given by

$$\sum_{i=1}^{n} (y_i - t(y_1, \ldots, y_n))^2.$$

Such an estimate is called a *least-squares estimate*. Note that a least-squares estimate is defined for every sample size, even $n = 1$.

To implement least squares, we must find the minimizing point $t(y_1, \ldots, y_n)$. Perhaps a first guess at this value is the sample average $\bar{y}$. Because

$$\sum_{i=1}^{n} (y_i - \bar{y})(\bar{y} - t(y_1, \ldots, y_n)) = (\bar{y} - t(y_1, \ldots, y_n)) \left(\sum_{i=1}^{n} y_i - n\bar{y} \right) = 0$$

we have

$$\sum_{i=1}^{n} (y_i - t(y_1, \ldots, y_n))^2$$

$$= \sum_{i=1}^{n} (y_i - \bar{y} + \bar{y} - t(y_1, \ldots, y_n))^2$$

$$= \sum_{i=1}^{n} (y_i - \bar{y})^2 + 2 \sum_{i=1}^{n} (y_i - \bar{y})(\bar{y} - t(y_1, \ldots, y_n)) + \sum_{i=1}^{n} (\bar{y} - t(y_1, \ldots, y_n))^2$$

$$= \sum_{i=1}^{n} (y_i - \bar{y})^2 + n(\bar{y} - t(y_1, \ldots, y_n))^2. \tag{10.3.1}$$

Therefore, the smallest possible value of (10.3.1) is $\sum_{i=1}^{n} (y_i - \bar{y})^2$, and this is assumed by taking $t(y_1, \ldots, y_n) = \bar{y}$. Note, however, that $\bar{y}$ might not be a possible value for $E(Y)$ and that, in such a case, it will not be the least-squares estimate. In general, (10.3.1) says that the least-squares estimate is the value $t(y_1, \ldots, y_n)$ that is closest to $\bar{y}$ and is a possible value for $E(Y)$.

Consider the following example.

Example 10.3.1
Suppose that Y has one of the distributions on $S = \{0, 1\}$ given in the following table.

	$y = 0$	$y = 1$
$p_1(y)$	$1/2$	$1/2$
$p_2(y)$	$1/3$	$2/3$

Then the mean of Y is given by

$$E_1(Y) = 0\left(\frac{1}{2}\right) + 1\left(\frac{1}{2}\right) = \frac{1}{2} \text{ or } E_2(Y) = 0\left(\frac{1}{3}\right) + 1\left(\frac{2}{3}\right) = \frac{2}{3}.$$

Now suppose we observe the sample $(0, 0, 1, 1, 1)$ and so $\bar{y} = 3/5$. Because the possible values for $E(Y)$ are in $\{1/2, 2/3\}$, we see that $t(0, 0, 1, 1, 1) = 2/3$, because $(3/5 - 2/3)^2 = 0.004$ while $(3/5 - 1/2)^2 = 0.01$. ∎

Whenever the set of possible values for $E(Y)$ is an interval (a, b), however, and $P(Y \in (a, b)) = 1$, then $\bar{y} \in (a, b)$. This implies that $\bar{y}$ is the least-squares estimator of $E(Y)$. So we see that in quite general circumstances, $\bar{y}$ is the least-squares estimate.

There is an equivalence between least squares and the maximum likelihood method when we are dealing with normal distributions.

Example 10.3.2 *Least Squares with Normal Distributions*
Suppose that $(y_1, \ldots, y_n)$ is a sample from an $N(\mu, \sigma_0^2)$ distribution where μ is unknown. Then the MLE of μ is obtained by finding the value of μ that maximizes

$$L(\mu \,|\, y_1, \ldots, y_n) = \exp\left\{-\frac{n}{2\sigma_0^2}(\bar{y} - \mu)^2\right\}.$$

Equivalently, the MLE maximizes the log-likelihood

$$l(\mu \,|\, y_1, \ldots, y_n) = -\frac{n}{2\sigma_0^2}(\bar{y} - \mu)^2.$$

So we need to find the value of μ that minimizes $(\bar{y} - \mu)^2$ just as with least squares.

In the case of the normal location model, we see that the least-squares estimate and the MLE of θ agree. This equivalence is true in general for normal models (e.g., the location-scale normal model), at least when we are considering estimates of location parameters. ∎

Some of the most important applications of least squares arise when we have that the response is a random vector $Y = (Y_1, \ldots, Y_n)' \in R^n$ (the prime $'$ indicates that we consider Y as a column), and we observe a single observation

$y = (y_1, \ldots, y_n)' \in R^n$. The expected value of $Y \in R^n$ is defined to be the vector of expectations of its component random variables, namely,

$$E(Y) = \begin{pmatrix} E(Y_1) \\ \vdots \\ E(Y_n) \end{pmatrix} \in R^n.$$

The least-squares principle then says that, based on the single observation $y = (y_1, \ldots, y_n)$, we must find

$$t(y) = t(y_1, \ldots, y_n) = (t_1(y_1, \ldots, y_n), \ldots, t_n(y_1, \ldots, y_n))',$$

in the set of possible values for $E(Y)$ (a subset of R^n), that minimizes

$$\sum_{i=1}^{n} (y_i - t_i(y_1, \ldots, y_n))^2. \tag{10.3.2}$$

So $t(y)$ is the possible value for $E(Y)$ that is closest to y (note that the squared distance between two points $x, y \in R^n$ is given by $\sum_{i=1}^{n} (x_i - y_i)^2$).

As is common in statistical applications, suppose that there are predictor variables that may be related to Y and whose values are observed. In this case we will replace $E(Y)$ by its conditional mean, given the observed values of the predictors. The least-squares estimate of the conditional mean is then the value $t(y_1, \ldots, y_n)$, in the set of possible values for the conditional mean of Y, that minimizes (10.3.2). We will use this definition in the following sections.

Finding the minimizing value of $t(y)$ in (10.3.2) can be a challenging optimization problem when the set of possible values for the mean is complicated. We will now apply least squares to some important problems where the least-squares solution can be found in closed form.

10.3.2 The Simple Linear Regression Model

Suppose we have a single quantitative response variable Y and a single quantitative predictor X, e.g., Y could be blood pressure measured in pounds per square inch and X could be age in years. To study the relationship between these variables, we examine the conditional distributions of Y, given $X = x$, to see how these change as we change x.

We might choose to examine a particular characteristic of these distributions to see how it varied with x. Perhaps the most commonly used characteristic is the conditional mean of Y given $X = x$, or $E(Y \mid X = x)$ (see Section 3.5).

In the *regression model*, we *assume* that the conditional distributions have constant shape and change, as we change x, through the conditional mean. In the *simple linear regression model*, we assume that the only way that the conditional mean can change is via the relationship

$$E(Y \mid X = x) = \beta_1 + \beta_2 x,$$

for some unknown values of $\beta_1 \in R^1$ (the intercept term) and $\beta_2 \in R^1$ (the slope coefficient). We also refer to β_1 and β_2 as the *regression coefficients*.

Suppose that we observe the independent values $(x_1, y_1), \ldots, (x_n, y_n)$ for (X, Y). Then, using the simple linear regression model, we have that

$$E\left(\begin{pmatrix} Y_1 \\ \vdots \\ Y_n \end{pmatrix} \middle| X_1 = x_1, \ldots, X_n = x_n \right) = \begin{pmatrix} \beta_1 + \beta_2 x_1 \\ \vdots \\ \beta_1 + \beta_2 x_n \end{pmatrix}. \tag{10.3.3}$$

Equation (10.3.3) tells us that the conditional expected value of the response $(Y_1, \ldots, Y_n)'$ is in a particular subset of R^n. Further, (10.3.2) becomes

$$\sum_{i=1}^{n} (y_i - t_i(y))^2 = \sum_{i=1}^{n} (y_i - \beta_1 - \beta_2 x_i)^2 \tag{10.3.4}$$

and we must find the values of β_1 and β_2 that minimize (10.3.4). These values are called the *least-squares estimates* of β_1 and β_2.

Before we show how to do this, consider an example.

Example 10.3.3

Suppose we obtained the following $n = 10$ data points (x_i, y_i).

$(3.9, 8.9)$	$(2.6, 7.1)$	$(2.4, 4.6)$	$(4.1, 10.7)$	$(-0.2, 1.0)$
$(5.4, 12.6)$	$(0.6, 3.3)$	$(-5.6, -10.4)$	$(-1.1, -2.3)$	$(-2.1, -1.6)$

In Figure 10.3.1 we have plotted these points together with the line $y = 1 + x$.

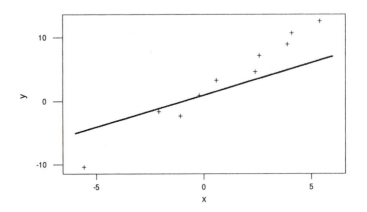

Figure 10.3.1: A plot of the data points (x_i, y_i) (+) and the line $y = 1 + x$ in Example 10.3.3.

Notice that with $\beta_1 = 1$ and $\beta_2 = 1$, then

$$(y_i - \beta_1 - \beta_2 x_i)^2 = (y_i - 1 - x_i)^2$$

is the squared vertical distance between the point (x_i, y_i) and the point on the line with the same x value. So (10.3.4) is the sum of these squared deviations, and in this case equals

$$(8.9 - 1 - 3.9)^2 + (7.1 - 1 - 2.6)^2 + \cdots + (-1.6 - 1 + 2.1)^2 = 141.15.$$

If $\beta_1 = 1$ and $\beta_2 = 1$ were the least-squares estimates, then 141.15 would be equal to the smallest possible value of (10.3.4). In this case, it turns out (see Example 10.3.4) that the least squares estimates are given by the values $\beta_1 = 1.33, \beta_2 = 2.06$, and the minimized value of (10.3.4) is given by 8.46 which is much smaller than 141.15.

So we see that, in finding the least-squares estimates, we are in essence finding the line $\beta_1 + \beta_2 x$ that best fits the data, in the sense that the sum of squared vertical deviations of the observed points to the line is minimized. ∎

Scatter Plots

As part of Example 10.3.3, we plotted the points $(x_1, y_1), \ldots, (x_n, y_n)$ in a graph. This is called a *scatter plot*, and it is a recommended first step, as part of any analysis of the relationship between quantitative variables X and Y. A scatter plot can give us a very general idea of whether or not a relationship exists and what form it might take.

It is important to remember, however, that the appearance of such a plot is highly dependent on the scales we choose for the axes. For example, we can make a scatter plot look virtually flat (and so indicate that no relationship exists) by choosing to place too wide a range of tick marks on the y-axis. So we must always augment a scatter plot with a statistical analysis based on numbers.

Least-Squares Estimates, Predictions, and Standard Errors

For the simple linear regression model, we can work out exact formulas for the least-squares estimates of β_1 and β_2.

Theorem 10.3.1 Suppose that $E(Y \mid X = x) = \beta_1 + \beta_2 x$, and we observe the independent values $(x_1, y_1), \ldots, (x_n, y_n)$ for (X, Y). Then the least-squares estimates of β_1 and β_2 are given by $b_1 = \bar{y} - b_2 \bar{x}$ and

$$b_2 = \frac{\sum_{i=1}^n (x_i - \bar{x})(y_i - \bar{y})}{\sum_{i=1}^n (x_i - \bar{x})^2},$$

respectively, whenever $\sum_{i=1}^n (x_i - \bar{x})^2 \neq 0$.

Proof: The proof of this result can be found in Section 10.6. ∎

We call the line $y = b_1 + b_2 x$ the *least-squares line*, or *best-fitting line*, and $b_1 + b_2 x$ is the least-squares estimate of $E(Y \mid X = x)$. Note that $\sum_{i=1}^n (x_i - \bar{x})^2 = 0$ if and only if $x_1 = \cdots = x_n$. In such a case we cannot use least-squares to estimate β_1 and β_2, although we can still estimate $E(Y \mid X = x)$ (Problem 10.3.12).

Now that we have estimates b_1, b_2 of the regression coefficients, we want to use these for inferences about β_1 and β_2. These estimates have the unbiasedness property.

Theorem 10.3.2 If $E(Y \mid X = x) = \beta_1 + \beta_2 x$, and we observe the independent values $(x_1, y_1), \ldots, (x_n, y_n)$ for (X, Y), then

(i) $E(B_1 \mid X_1 = x_1, \ldots, X_n = x_n) = \beta_1$,

(ii) $E(B_2 \mid X_1 = x_1, \ldots, X_n = x_n) = \beta_2$.

Proof: The proof of this result can be found in Section 10.6. ∎

Note that Theorem 10.3.2 and the theorem of total expectation imply that $E(B_1) = \beta_1$ and $E(B_2) = \beta_2$ unconditionally as well.

Adding the assumption that the conditional variances exist, we have the following theorem.

Theorem 10.3.3 If $E(Y \mid X = x) = \beta_1 + \beta_2 x$, $\mathrm{Var}(Y \mid X = x) = \sigma^2$ for every x, and we observe the independent values $(x_1, y_1), \ldots, (x_n, y_n)$ for (X, Y), then

(i) $\mathrm{Var}(B_1 \mid X_1 = x_1, \ldots, X_n = x_n) = \sigma^2(1/n + \bar{x}^2 / \sum_{i=1}^{n}(x_i - \bar{x})^2)$,

(ii) $\mathrm{Var}(B_2 \mid X_1 = x_1, \ldots, X_n = x_n) = \sigma^2 / \sum_{i=1}^{n}(x_i - \bar{x})^2$,

(iii) $\mathrm{Cov}(B_1, B_2 \mid X_1 = x_1, \ldots, X_n = x_n) = -\sigma^2 \bar{x} / \sum_{i=1}^{n}(x_i - \bar{x})^2$.

Proof: See Section 10.6 for the proof of this result. ∎

For the least-squares estimate $b_1 + b_2 x$ of the mean $E(Y \mid X = x) = \beta_1 + \beta_2 x$, we have the following result.

Corollary 10.3.1

$$\mathrm{Var}(B_1 + B_2 x \mid X_1 = x_1, \ldots, X_n = x_n) = \sigma^2 \left(\frac{1}{n} + \frac{(x - \bar{x})^2}{\sum_{i=1}^{n}(x_i - \bar{x})^2} \right).$$

$$(10.3.5)$$

Proof: See Section 10.6 for the proof of this result. ∎

A natural predictor of a future value of Y, when $X = x$, is given by the conditional mean $E(Y \mid X = x) = \beta_1 + \beta_2 x$. Because we do not know the values of β_1 and β_2, we use the estimated mean $b_1 + b_2 x$ as the predictor.

When we are predicting Y at an x value that lies within the range of the observed values of X, we refer to this as an *interpolation*. When we want to predict at an x value that lies outside this range, we refer to this as an *extrapolation*. Extrapolations are much less reliable than interpolations. The farther away x is from the observed range of X values, then, intuitively, the less reliable we feel such a prediction will be. Such considerations should always be borne in mind. From (10.3.5), we see that the variance of the prediction at the value $X = x$ increases as x moves away from $\bar{x}$. So to a certain extent, the standard error does reflect this increased uncertainty, but note that its form is based upon the assumption that the simple linear regression model is correct. Even if we accept the simple linear regression model based on the observed data (we will discuss model checking later in this section), this model may fail to apply for very different values of x, and so the predictions would be in error.

We want to use the results of Theorem 10.3.3 and Corollary 10.3.1 to calculate standard errors of the least-squares estimates. Because we do not know

σ^2, however, we need an estimate of this quantity as well. The following result shows that

$$s^2 = \frac{1}{n-2} \sum_{i=1}^{n} (y_i - b_1 - b_2 x_i)^2 \qquad (10.3.6)$$

is an unbiased estimate of σ^2

Theorem 10.3.4 If $E(Y \mid X = x) = \beta_1 + \beta_2 x$, $\text{Var}(Y \mid X = x) = \sigma^2$ for every x, and we observe the independent values $(x_1, y_1), \ldots, (x_n, y_n)$ for (X, Y) then

$$E\left(S^2 \mid X_1 = x_1, \ldots, X_n = x_n\right) = \sigma^2.$$

Proof: See Section 10.6 for the proof of this result. ∎

Therefore, the standard error of b_1 is then given by

$$s \left(\frac{1}{n} + \frac{\bar{x}^2}{\sum_{i=1}^{n} (x_i - \bar{x})^2} \right)^{1/2},$$

and the standard error of b_2 is then given by

$$s \left(\sum_{i=1}^{n} (x_i - \bar{x})^2 \right)^{-1/2}.$$

Under further assumptions, these standard errors can be interpreted just as we interpreted standard errors of estimates of the mean in the location and location-scale normal models.

Example 10.3.4 *(Example 10.3.3 continued)*
Using the data in Example 10.3.3, and the formulas of Theorem 10.3.1, we obtain $b_1 = 1.33, b_2 = 2.06$ as the least squares estimates of the intercept and slope, respectively. So the least-squares line is given by $1.33 + 2.06x$. Using (10.3.6), we obtain $s^2 = 1.06$ as the estimate of σ^2.

Using the formulas of Theorem 10.3.3, the standard error of b_1 is 0.3408, while the standard error of b_2 is 0.1023.

The prediction of Y at $X = 2.0$ is given by $1.33 + 2.06(2) = 5.45$. Using Corollary 10.3.1, this estimate has standard error 0.341. This prediction is an interpolation. ∎

The ANOVA Decomposition and the F Statistic

The following result gives a useful decomposition of the *total sum of squares* $\sum_{i=1}^{n} (y_i - \bar{y})^2$.

Lemma 10.3.1 If $(x_1, y_1), \ldots, (x_n, y_n)$ are such that $\sum_{i=1}^{n} (x_i - \bar{x})^2 \neq 0$, then

$$\sum_{i=1}^{n} (y_i - \bar{y})^2 = b_2^2 \sum_{i=1}^{n} (x_i - \bar{x})^2 + \sum_{i=1}^{n} (y_i - b_1 - b_2 x_i)^2.$$

Proof: The proof of this result can be found in Section 10.6. ∎

We refer to

$$b_2^2 \sum_{i=1}^{n} (x_i - \bar{x})^2$$

as the *regression sum of squares (RSS)* and refer to

$$\sum_{i=1}^{n} (y_i - b_1 - b_2 x_i)^2$$

as the *error sum of squares (ESS)*.

If we think of the total sum of squares as measuring the total observed variation in the response values y_i, then Lemma 10.3.1 provides a decomposition of this variation into the RSS, measuring changes in the response due to changes in the predictor, and the ESS, measuring changes in the response due to the contribution of random error.

It is common to write this decomposition in an *analysis of variance table (ANOVA)*.

Source	Df	Sum of Squares	Mean Square
X	1	$b_2^2 \sum_{i=1}^{n} (x_i - \bar{x})^2$	$b_2^2 \sum_{i=1}^{n} (x_i - \bar{x})^2$
Error	$n-2$	$\sum_{i=1}^{n} (y_i - b_1 - b_2 x_i)^2$	s^2
Total	$n-1$	$\sum_{i=1}^{n} (y_i - \bar{y})^2$	

Here Df stands for degrees of freedom (we will discuss how the Df entries are calculated in Section 10.3.4). The entries in the Mean Square column are calculated by dividing the corresponding sum of squares by the Df entry.

To see the significance of the ANOVA table, note that, from Theorem 10.3.3,

$$E\left(B_2^2 \sum_{i=1}^{n} (x_i - \bar{x})^2 \mid X_1 = x_1, \ldots, X_n = x_n \right) = \sigma^2 + \beta_2^2 \sum_{i=1}^{n} (x_i - \bar{x})^2,$$

$$(10.3.7)$$

which is equal to σ^2 if and only if $\beta_2 = 0$ (we are always assuming here that the x_i vary). Given that the simple linear regression model is correct, we have that $\beta_2 = 0$ if and only if there is no relationship between the response and the predictor. Therefore, $b_2^2 \sum_{i=1}^{n} (x_i - \bar{x})^2$ is an unbiased estimator of σ^2 if and only if $\beta_2 = 0$. Because s^2 is always an unbiased estimate of σ^2 (Theorem 10.3.4), a sensible statistic to use in assessing $H_0 : \beta_2 = 0$, is given by

$$F = \frac{RSS}{ESS/(n-2)} = \frac{b_2^2 \sum_{i=1}^{n} (x_i - \bar{x})^2}{s^2}, \qquad (10.3.8)$$

as this is the ratio of two unbiased estimators of σ^2 when H_0 is true. We then conclude that we have evidence against H_0 when F is large, as (10.3.7) also shows that the numerator will tend to be larger than σ^2 when H_0 is false. We refer to (10.3.8) as the *F-statistic*. We will subsequently discuss the sampling distribution of F to see how to determine when the value F is so large as to be evidence against H_0.

Example 10.3.5 *(Example 10.3.3 continued)*
Using the data of Example 10.3.3, we obtain

$$\sum_{i=1}^{n} (y_i - \bar{y})^2 = 437.01,$$

$$b_2^2 \sum_{i=1}^{n} (x_i - \bar{x})^2 = 428.55,$$

$$\sum_{i=1}^{n} (y_i - b_1 - b_2 x_i)^2 = 437.01 - 428.55 = 8.46,$$

and so

$$F = \frac{b_2^2 \sum_{i=1}^{n} (x_i - \bar{x})^2}{s^2} = \frac{428.55}{1.06} = 404.29.$$

Note that F is much bigger than 1, and this seems to indicate that there is a linear effect due to X. ∎

The Coefficient of Determination and Correlation

Lemma 10.3.1 implies that

$$R^2 = \frac{b_2^2 \sum_{i=1}^{n} (x_i - \bar{x})^2}{\sum_{i=1}^{n} (y_i - \bar{y})^2}$$

satisfies $0 \leq R^2 \leq 1$. Therefore, the closer R^2 is to 1, the more of the observed total variation in the response is accounted for by changes in the predictor. In fact, we interpret R^2, called the *coefficient of determination*, as the proportion of the observed variation in the response explained by changes in the predictor via the simple linear regression.

The coefficient of determination is an important descriptive statistic, for, even if we conclude that a relationship does exist, it can happen that most of the observed variation is due to error. If we want to use the model to predict further values of the response, then the coefficient of determination tells us whether we can expect highly accurate predictions or not. A value of R^2 near 1 means highly accurate predictions while a value near 0 means that predictions will not be very accurate.

Example 10.3.6 *(Example 10.3.3 continued)*
Using the data of Example 10.3.1 we obtain $R^2 = 0.981$. Therefore, 98.1% of the observed variation in Y can be explained by the changes in X through the linear relation. This indicates that we can expect fairly accurate predictions when using this model, at least when we are predicting within the range of the observed X values. ∎

Recall that in Section 3.3 we defined the correlation coefficient between random variables X and Y to be

$$\rho_{XY} = \text{Corr}(X, Y) = \frac{\text{Cov}(X, Y)}{\text{Sd}(X)\,\text{Sd}(Y)}.$$

In Corollary 3.6.1 we proved that $-1 \le \rho_{XY} \le 1$ with $\rho_{XY} = \pm 1$ if and only if $Y = a \pm cX$ for some constants $a \in R^1$ and $c > 0$. So ρ_{XY} can be taken as a measure of the extent to which a linear relationship exists between X and Y.

If we don't know the joint distribution of (X, Y), then we will have to estimate ρ_{XY}. Based on the observations $(x_1, y_1), \ldots, (x_n, y_n)$, the natural estimate to use is the *sample correlation coefficient*

$$r_{xy} = \frac{s_{xy}}{s_x s_y},$$

where

$$s_{xy} = \frac{1}{n-1} \sum_{i=1}^{n} (x_i - \bar{x})(y_i - \bar{y})$$

is the *sample covariance* estimating $\text{Cov}(X, Y)$, and s_x, s_y are the sample standard deviations for the X and Y variables, respectively. Then $-1 \le r_{xy} \le 1$ with $r_{xy} = \pm 1$ if and only if $y_i = a \pm c x_i$ for some constants $a \in R^1$ and $c > 0$, for every i (the proof is the same as in Corollary 3.6.1 using the joint distribution that puts probability mass $1/n$ at each point (x_i, y_i) — see Problem 3.6.10).

The following result shows that the coefficient of determination is the square of the correlation between the observed X and Y values.

Theorem 10.3.5 If $(x_1, y_1), \ldots, (x_n, y_n)$ are such that $\sum_{i=1}^{n} (x_i - \bar{x})^2 \ne 0$, $\sum_{i=1}^{n} (y_i - \bar{y})^2 \ne 0$, then $R^2 = r_{xy}^2$.
Proof: We have

$$r_{xy}^2 = \frac{\left(\sum_{i=1}^{n} (x_i - \bar{x})(y_i - \bar{y})\right)^2}{\sum_{i=1}^{n} (x_i - \bar{x})^2 \sum_{i=1}^{n} (y_i - \bar{y})^2} = b_2^2 \frac{\sum_{i=1}^{n} (x_i - \bar{x})^2}{\sum_{i=1}^{n} (y_i - \bar{y})^2} = R^2,$$

where we have used the formula for b_2 given in Theorem 10.3.1. ∎

Confidence Intervals and Testing Hypotheses

We need to make some further assumptions in order to discuss the sampling distributions of the various statistics that we have introduced. We have the following results.

Theorem 10.3.6 If Y, given $X = x$, is distributed $N(\beta_1 + \beta_2 x, \sigma^2)$, and we observe the independent values $(x_1, y_1), \ldots, (x_n, y_n)$ for (X, Y), then the conditional distributions of B_1, B_2, and S^2, given $X_1 = x_1, \ldots, X_n = x_n$, are as follows.

(i) $B_1 \sim N\left(\beta_1, \sigma^2\left(1/n + \bar{x}^2 / \sum_{i=1}^{n} (x_i - \bar{x})^2\right)\right)$

(ii) $B_2 \sim N\left(\beta_2, \sigma^2 / \sum_{i=1}^{n} (x_i - \bar{x})^2\right)$

(iii)
$$B_1 + B_2 x \sim N\left(\beta_1 + \beta_2 x, \sigma^2 \left(\frac{1}{n} + \frac{(x - \bar{x})^2}{\sum_{i=1}^{n} (x_i - \bar{x})^2}\right)\right)$$

(iv) $(n-2) S^2/\sigma^2 \sim \chi^2 (n-2)$ independent of (B_1, B_2)

Proof: The proof of this result can be found in Section 10.6. ∎

Corollary 10.3.2

(i) $(B_1 - \beta_1) / \left(S\left(1/n + \bar{x}^2 / \sum_{i=1}^{n} (x_i - \bar{x})^2\right)^{1/2}\right) \sim t(n-2)$

(ii) $(B_2 - \beta_2) \left(\sum_{i=1}^{n} (x_i - \bar{x})^2\right)^{1/2} / S \sim t(n-2)$

(iii)
$$\frac{B_1 + B_2 x - \beta_1 - \beta_2 x}{S\left(\left(\frac{1}{n} + (x - \bar{x})^2\right) / \sum_{i=1}^{n} (x_i - \bar{x})^2\right)^{1/2}} \sim t(n-2)$$

(iv) If F is defined as in (10.3.8), then $H_0 : \beta_2 = 0$ is true if and only if $F \sim F(1, n-2)$.

Proof: The proof of this result can be found in Section 10.6. ∎

Using Corollary 10.3.2(i), we have that

$$b_1 \pm s \left(1/n + \bar{x}^2 / \sum_{i=1}^{n} (x_i - \bar{x})^2\right)^{1/2} t_{(1+\gamma)/2} (n-2)$$

is an exact γ-confidence interval for β_1. Also from Corollary 10.3.2(ii)

$$b_2 \pm s \left(\sum_{i=1}^{n} (x_i - \bar{x})^2\right)^{-1/2} t_{(1+\gamma)/2} (n-2)$$

is an exact γ-confidence interval for β_2.

From Corollary 10.3.2(iv) we can test $H_0 : \beta_2 = 0$ by computing the P-value

$$P\left(F \geq \frac{b_2^2 \sum_{i=1}^{n} (x_i - \bar{x})^2}{s^2}\right), \tag{10.3.9}$$

where $F \sim F(1, n-2)$, to see whether or not the observed value (10.3.8) is surprising. This is sometimes called the *ANOVA test*. Note that Corollary 10.3.2(ii) implies that we can also test $H_0 : \beta_2 = 0$ by computing the P-value

$$P\left(|T| \geq \frac{b_2 \left(\sum_{i=1}^{n} (x_i - \bar{x})^2\right)^{1/2}}{s}\right) \tag{10.3.10}$$

where $T \sim t(n-2)$. The proof of Corollary 10.3.2(iv) reveals that (10.3.9) and (10.3.10) are equal.

Example 10.3.7 *(Example 10.3.3 continued)*
Using software or Table D.4, we obtain $t_{0.975}(8) = 2.306$. Then, using the data of Example 10.3.1, we obtain a 0.95-confidence interval for β_1 as

$$b_1 \pm s \left(1/n + \bar{x}^2 / \sum_{i=1}^{n} (x_i - \bar{x})^2 \right)^{1/2} t_{(1+\gamma)/2}(n-2)$$

$$= 1.33 \pm (0.3408)(2.306) = [0.544, 2.116]$$

and a 0.95-confidence interval for β_2 as

$$b_2 \pm s \left(\sum_{i=1}^{n} (x_i - \bar{x})^2 \right)^{-1/2} t_{(1+\gamma)/2}(n-2)$$

$$= 2.06 \pm (0.1023)(2.306) = [1.824, 2.296] \, .$$

The 0.95-confidence interval for β_2 does not include 0, so we would reject the null hypothesis $H_0 : \beta_2 = 0$ and conclude that there is evidence of a relationship between X and Y. This is confirmed by the F-test of this null hypothesis, as it gives the P-value $P(F \geq 404.29) = 0.000$ when $F \sim F(1, 8)$.

Analysis of Residuals

In an application of the simple regression model, we must check to make sure that the assumptions make sense in light of the data we have collected. Model checking is based on the residuals $y_i - b_1 - b_2 x_i$ (after standardization) just as discussed in Section 9.1. Note that the ith residual is just the difference between the observed value y_i at x_i and the predicted value $b_1 + b_2 x_i$ at x_i.

From the proof of Theorem 10.3.4 we have the following result.

Corollary 10.3.3
(i) $E(Y_i - B_1 - B_2 x_i \mid X_1 = x_1, \ldots, X_n = x_n) = 0$
(ii) $\text{Var}(Y_i - B_1 - B_2 x_i \mid X_1 = x_1, \ldots, X_n = x_n) = \sigma^2 \left(1 - \frac{1}{n} - \frac{(x_i - \bar{x})^2}{\sum_{i=1}^{n}(x_i - \bar{x})^2} \right)$
This leads to the definition of the ith standardized residual as

$$\frac{y_i - b_1 - b_2 x_i}{s \left(1 - \frac{1}{n} - (x_i - \bar{x})^2 / \sum_{j=1}^{n}(x_j - \bar{x})^2 \right)^{1/2}} . \tag{10.3.11}$$

Corollary 10.3.3 says that (10.3.11), with σ replacing s, is a value from a distribution with conditional mean 0 and conditional variance 1. Further, when the conditional distribution of the response given the predictors is normal, then the conditional distribution of this quantity is $N(0, 1)$ (Problem 10.3.14). These results are approximately true for (10.3.11) for large n. Further, it can be shown (Problem 10.3.13) that

$$\text{Cov}\left(Y_i - B_1 - B_2 x_i, Y_j - B_1 - B_2 x_j \mid X_1 = x_1, \ldots, X_n = x_n\right)$$

$$= -\sigma^2 \left(\frac{1}{n} + \frac{(x_i - \bar{x})(x_j - \bar{x})}{\sum_{k=1}^{n}(x_k - \bar{x})^2}\right)$$

Therefore, under the normality assumption, the residuals are approximately independent when n is large and

$$\frac{x_i - \bar{x}}{\sqrt{\sum_{k=1}^{n}(x_k - \bar{x})^2}} \to 0$$

as $n \to \infty$. This will be the case whenever $\text{Var}(X)$ is finite (Challenge 10.3.20) or, in the design context, when the values of the predictor are chosen accordingly. So one approach to model checking here is to see if the values given by (10.3.11) look at all like a sample from the $N(0, 1)$ distribution. For this we can use the plots discussed in Chapter 9.

Example 10.3.8 *(Example 10.3.3 continued)*
Using the data of Example 10.3.3 we obtain the following standardized residuals.

-0.49643	0.43212	-1.73371	1.00487	0.08358
0.17348	0.75281	-0.28430	-1.43570	1.51027

These are plotted against the predictor x in Figure 10.3.2.

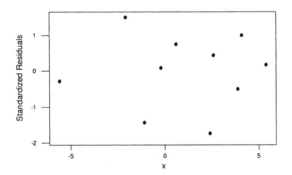

Figure 10.3.2: Plot of the standardized residuals in Example 10.3.8.

It is recommended that we plot the standardized residuals against the predictor, as this may reveal some underlying relationship that has not been captured by the model. This residual plot looks reasonable. In Figure 10.3.3 we have a normal probability plot of the standardized residuals. These points lie close to the line through the origin with slope equal to 1, so we conclude that we have no evidence against the model here. ∎

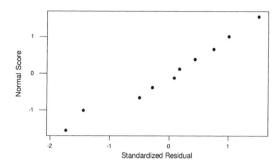

Figure 10.3.3: Normal probability plot of the standardized residuals in Example 10.3.8.

What do we do if model checking leads to a failure of the model? As discussed in Chapter 9, perhaps the most common approach is to consider making various transformations of the data to see if there is a simple modification of the model that will pass. We can make transformations, not only to the response variable Y, but to the predictor variable X as well.

An Application of Simple Linear Regression Analysis

The following data set is taken from *Statistical Methods* by G. Snedecor and W. Cochran (Iowa State University Press, Ames, 1967) and gives the record speed Y in miles per hour at the Indianapolis Memorial Day car races in the years 1911–1941, excepting the years 1917–1918. We have coded the year X starting at 0 in 1911 and incrementing by 1 for each year. There are $n = 29$ data points (x_i, y_i). The goal of the analysis is to obtain the least-squares line and, if warranted, make inferences about the regression coefficients. We take the normal simple linear regression model as our statistical model. Note that this is an observational study.

Year	Speed	Year	Speed	Year	Speed
0	74.6	12	91.0	22	104.2
1	78.7	13	98.2	23	104.9
2	75.9	14	101.1	24	106.2
3	82.5	15	95.9	25	109.1
4	89.8	16	97.5	26	113.6
5	83.3	17	99.5	27	117.2
8	88.1	18	97.6	28	115.0
9	88.6	19	100.4	29	114.3
10	89.6	20	96.6	30	115.1
11	94.5	21	104.1		

Using Theorem 10.3.1, we obtain the least-squares line as $y = 77.5681 + 1.27793x$. This line, together with a *scatter plot* of the values (x_i, y_i), is plotted in Figure 10.3.4. The fit looks quite good, but this is no guarantee of model correctness and we must carry out some form of model checking.

Figure 10.3.5 is a plot of the standardized residuals against the predictor. This plot looks reasonable, with no particularly unusual pattern apparent. Figure 10.3.6 is a normal probability plot of the standardized residuals. The curvature in the center might give rise to some doubt about the normality assumption. We generated a few samples of $n = 29$ from an $N(0, 1)$ distribution, however, and looking at the normal probability plots (always recommended) reveals that this is not much cause for concern. Of course, we should also carry out a P-value based model checking procedure as well, based upon the standardized residuals, but we do not pursue this topic further here.

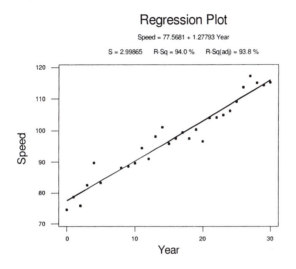

Figure 10.3.4: A scatter plot of the data together with a plot of the least squares line.

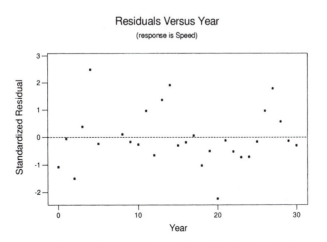

Figure 10.3.5: A plot of the standardized residuals against the predictor.

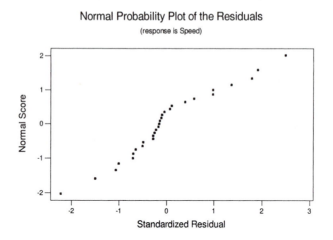

Figure 10.3.6: A normal probability plot of the standardized residuals.

Based on the results of our model checking, we decide to proceed to inferences about the regression coefficients. The estimates and their standard errors are given in the following table, where we have used the estimate of σ^2 given by $s^2 = (2.999)^2$, to compute the standard errors. We have also recorded the t-statistics appropriate for testing each of the hypotheses $H_0 : \beta_1 = 0$ and $H_0 : \beta_2 = 0$.

Coefficient	Estimate	Standard Error	t-statistic
β_1	77.568	1.118	69.39
β_2	1.278	0.062	20.55

From this we see that the P-value for assessing $H_0 : \beta_2 = 0$ is given by

$$P(|T| \geq 20.55) = 0.000,$$

when $T \sim t(27)$, and so we have strong evidence against H_0. It seems clear that there is a strong positive relationship between Y and X. Since the 0.975 point of the $t(27)$ distribution equals 2.0518, a 0.95-confidence interval for β_2 is given by

$$1.278 \pm (0.062)\, 2.0518 = [1.1508, 1.4052].$$

The ANOVA decomposition is given in the following table.

Source	Df	Sum of Squares	Mean Square
Regression	1	3797.0	3797.0
Error	27	242.8	9.0
Total	28	4039.8	

Accordingly, we have that $F = 3797.0/9.0 = 421.888$ and, as $F \sim F(1, 27)$ when $H_0 : \beta_2 = 0$ is true, $P(F > 421.888) = 0.000$, which simply confirms (as it must) what we got from the preceding t-test.

The coefficient of determination is given by $R^2 = 3797.0/4039.8 = 0.94$. Therefore, 94% of the observed variation in the response variable can be explained by the changes in the predictor through the simple linear regression. The value of R^2 indicates that the fitted model will be an excellent predictor of future values, provided that the value of X that we want to predict at, is in the range (or close to it) of the values of X used to fit the model. ∎

10.3.3 Bayesian Simple Linear Model (Advanced)

For the Bayesian formulation of the simple linear regression model with normal error, we need to add a prior distribution for the unknown parameters of the model, namely, β_1, β_2, and σ^2. There are many possible choices for this. A relevant prior is dependent on the application.

To help simplify the calculations, we reparameterize, the model as follows. Let $\alpha_1 = \beta_1 + \beta_2 \bar{x}$ and $\alpha_2 = \beta_2$. It is then easy to show (Problem 10.3.17) that

$$
\begin{aligned}
\sum_{i=1}^{n} (y_i - \beta_1 - \beta_2 x_i)^2 &= \sum_{i=1}^{n} (y_i - \alpha_1 - \alpha_2 (x_i - \bar{x}))^2 \\
&= \sum_{i=1}^{n} ((y_i - \bar{y}) - (\alpha_1 - \bar{y}) - \alpha_2 (x_i - \bar{x}))^2 \\
&= \sum_{i=1}^{n} (y_i - \bar{y})^2 + n(\alpha_1 - \bar{y})^2 + \alpha_2^2 \sum_{i=1}^{n} (x_i - \bar{x})^2 \\
&\quad - 2\alpha_2 \sum_{i=1}^{n} (x_i - \bar{x})(y_i - \bar{y}).
\end{aligned} \tag{10.3.12}
$$

The likelihood function, using this reparameterization, then equals

$$
\left(2\pi\sigma^2\right)^{-n/2} \exp\left(-\frac{1}{2\sigma^2} \sum_{i=1}^{n} (y_i - \alpha_1 - \alpha_2 (x_i - \bar{x}))^2\right).
$$

From (10.3.12), and putting

$$
\begin{aligned}
c_x^2 &= \sum_{i=1}^{n} (x_i - \bar{x})^2, \\
c_y^2 &= \sum_{i=1}^{n} (y_i - \bar{y})^2, \\
c_{xy} &= \sum_{i=1}^{n} (x_i - \bar{x})(y_i - \bar{y}),
\end{aligned}
$$

we can write this as

$$\left(2\pi\sigma^2\right)^{-n/2} \exp\left(-\frac{c_y^2}{2\sigma^2}\right) \exp\left(-\frac{n}{2\sigma^2}(\alpha_1 - \bar{y})^2\right)$$

$$\times \exp\left(-\frac{1}{2\sigma^2}\left\{\alpha_2^2 c_x^2 - 2\alpha_2 c_{xy}\right\}\right)$$

$$= \left(2\pi\sigma^2\right)^{-n/2} \exp\left(-\frac{c_y^2 - c_x^2 a^2}{2\sigma^2}\right) \exp\left(-\frac{n}{2\sigma^2}(\alpha_1 - \bar{y})^2\right)$$

$$\times \exp\left(-\frac{c_x^2}{2\sigma^2}(\alpha_2 - a)^2\right),$$

where the last equality follows from $\alpha_2^2 c_x^2 - 2\alpha_2 c_{xy} = c_x^2(\alpha_2 - a)^2 - c_x^2 a^2$ with $a = c_{xy}/c_x^2$.

This implies that, whenever the prior distribution on (α_1, α_2), is such that α_1 and α_2 are independent given σ^2, then the posterior distributions of α_1 and α_2 are also independent given σ^2. Note also that $\bar{y}$ and a are the least-squares estimates (as well as the MLE's) of α_1 and α_2, respectively (Problem 10.3.17).

Now suppose that we take the prior to be

$$\alpha_1 \,|\, \alpha_2, \sigma^2 \;\sim\; N\left(\mu_1, \tau_1^2 \sigma^2\right),$$
$$\alpha_2 \,|\, \sigma^2 \;\sim\; N\left(\mu_2, \tau_2^2 \sigma^2\right),$$
$$\frac{1}{\sigma^2} \;\sim\; \text{Gamma}\left(\kappa, \nu\right).$$

Note that α_1 and α_2 are independent given σ^2.

As it turns out, this prior is conjugate, so we can easily determine an exact form for the posterior distribution (Problem 10.3.18). The joint posterior of $\left(\alpha_1, \alpha_2, 1/\sigma^2\right)$ is given by

$$\alpha_1 \,|\, \alpha_2, \sigma^2 \;\sim\; N\left(\left(n + \frac{1}{\tau_1^2}\right)^{-1}\left(n\bar{y} + \frac{\mu_1}{\tau_1^2}\right), \left(n + \frac{1}{\tau_1^2}\right)^{-1}\sigma^2\right),$$

$$\alpha_2 \,|\, \sigma^2 \;\sim\; N\left(\left(c_x^2 + \frac{1}{\tau_2^2}\right)^{-1}\left(c_x^2 a + \frac{\mu_2}{\tau_2^2}\right), \left(c_x^2 + \frac{1}{\tau_2^2}\right)^{-1}\sigma^2\right),$$

$$\frac{1}{\sigma^2} \;\sim\; \text{Gamma}\left(\kappa + \frac{n}{2}, \nu_{xy}\right),$$

where

$$\nu_{xy} = \frac{1}{2}\left\{\begin{array}{l} c_y^2 - c_x^2 a^2 + \left[n\bar{y}^2 + \frac{\mu_1^2}{\tau_1^2} - \left(n + \frac{1}{\tau_1^2}\right)^{-1}\left(n\bar{y} + \frac{\mu_1}{\tau_1^2}\right)^2\right] \\ \; + \left[c_x^2 a^2 + \frac{\mu_2^2}{\tau_2^2} - \left(c_x^2 + \frac{1}{\tau_2^2}\right)^{-1}\left(c_x^2 a + \frac{\mu_2}{\tau_2^2}\right)^2\right] \end{array}\right\} + \nu.$$

Of course, we must select the values of the hyperparameters $\mu_1, \tau_1, \mu_2, \tau_2, \kappa$, and ν to fully specify the prior.

Now observe that for a diffuse analysis, i.e., when we have little or no prior information about the parameters, we let $\tau_1 \to \infty, \tau_2 \to \infty$, and $\nu \to 0$, and the posterior converges to

$$\alpha_1 \,|\, \alpha_2, \sigma^2 \;\sim\; N\left(\bar{y}, \frac{\sigma^2}{n}\right),$$

$$\alpha_2 \,|\, \sigma^2 \;\sim\; N\left(a, \frac{\sigma^2}{c_x^2}\right),$$

$$\frac{1}{\sigma^2} \;\sim\; \text{Gamma}\left(\kappa + \frac{n}{2}, \nu_{xy}\right)$$

where $\nu_{xy} = \frac{1}{2}\left\{c_y^2 - c_x^2 a^2\right\}$. But this still leaves us with the necessity of choosing the hyperparameter κ. We will see, however, that this choice has only a small effect on the analysis when n is not too small.

We can easily work out the marginal posterior distribution of the α_i. For example, in the diffuse case, the marginal posterior density of α_2 is proportional to

$$\int_0^\infty \left(\frac{1}{\sigma^2}\right)^{1/2} \exp\left\{-\frac{c_x^2}{2\sigma^2}(\alpha_2 - a)^2\right\} \left(\frac{1}{\sigma^2}\right)^{\kappa+(n/2)-1} \exp\left\{-\frac{\nu_{xy}}{\sigma^2}\right\} d\left(\frac{1}{\sigma^2}\right)$$

$$= \int_0^\infty \left(\frac{1}{\sigma^2}\right)^{\kappa+(n/2)-(1/2)} \exp\left\{-\left(\nu_{xy} + \frac{c_x^2}{2}(\alpha_2 - a)^2\right)\frac{1}{\sigma^2}\right\} d\left(\frac{1}{\sigma^2}\right).$$

Making the change of variable $1/\sigma^2 \to w$, where

$$w = \left(\nu_{xy} + \frac{c_x^2}{2}(\alpha_2 - a)^2\right)\frac{1}{\sigma^2}$$

in the preceding integral, shows that the marginal posterior density of α_2 is proportional to

$$\left(1 + \frac{c_x^2}{2\nu_{xy}}(\alpha_2 - a)^2\right)^{-(\kappa+(n+1)/2)} \int_0^\infty w^{\kappa+(n/2)-(1/2)} \exp\left\{-w\right\} dw,$$

and this is proportional to

$$\left(1 + \frac{c_x^2}{2\nu_{xy}}(\alpha_2 - a)^2\right)^{-(2\kappa+n+1)/2}.$$

This establishes (Problem 4.6.14) that the posterior distribution of α_2 is specified by

$$\sqrt{2\kappa + n}\,\frac{\alpha_2 - a}{\sqrt{2\nu_{xy}/c_x^2}} \;\sim\; t\left(2\kappa + n\right).$$

So a γ-HPD (highest posterior density) interval for α_2 is given by

$$a \pm \frac{1}{\sqrt{2\kappa + n}}\sqrt{\frac{2\nu_{xy}}{c_x^2}}\,t_{(1+\gamma)/2}\left(2\kappa + n\right).$$

Note that these intervals will not change much as we change κ, provided that n is not too small.

We consider an application of a Bayesian analysis for such a model.

Example 10.3.9 *Haavelmo's Data on Income and Investment*
The data for this example was taken from *An Introduction to Bayesian Inference in Econometrics* by A. Zellner (Wiley Classics, New York, 1996). The response variable Y is income in U.S. dollars per capita (deflated), and the predictor variable X is investment in dollars per capita (deflated) for the United States for the years 1922–1941. The data are provided in the following table.

Year	Income	Investment	Year	Income	Investment
1922	433	39	1932	372	22
1923	483	60	1933	381	17
1924	479	42	1934	419	27
1925	486	52	1935	449	33
1926	494	47	1936	511	48
1927	498	51	1937	520	51
1928	511	45	1938	477	33
1929	534	60	1939	517	46
1930	478	39	1940	548	54
1931	440	41	1941	629	100

In Figure 10.3.7 we present a normal probability plot of the standardized residuals, obtained via a least-squares fit. In Figure 10.3.8 we present a plot of the standardized residuals against the predictor. Both plots indicate that the model assumptions are reasonable.

Suppose now that we analyze these data using the limiting diffuse prior with $\kappa = 2$. Here we have that $\bar{y} = 483$, $c_y^2 = 64993$, $c_x^2 = 5710.55$, $c_{xy} = 17408.3$, so that $a = 17408.3/5710.55 = 3.05$ and $\nu_{xy} = (64993 - 17408.3)/2 = 23792.35$. The posterior is then given by

$$\alpha_1 \mid \alpha_2, \sigma^2 \quad \sim \quad N\left(483, \frac{\sigma^2}{20}\right),$$

$$\alpha_2 \mid \sigma^2 \quad \sim \quad N\left(3.05, \frac{\sigma^2}{5710.55}\right),$$

$$\frac{1}{\sigma^2} \quad \sim \quad \text{Gamma}\,(12, 23792.35)\,.$$

The primary interest here is in the investment multiplier α_2. By the above results, a 0.95-HPD interval for α_2, using $t_{0.975}(24) = 2.0639$, is given by

$$a \pm \frac{1}{\sqrt{2\kappa + n}} \sqrt{\frac{2\nu_{xy}}{r_x^2}}\, t_{(1+\gamma)/2}\,(2\kappa + n - 1)$$

$$= 3.05 \pm \frac{1}{\sqrt{24}} \sqrt{\frac{2 \cdot 23792.35}{5710.55}}\, t_{0.975}\,(24) = 3.05 \pm (.589)\, 2.0639$$

$$= (1.834, 4.266)\,. \quad \blacksquare$$

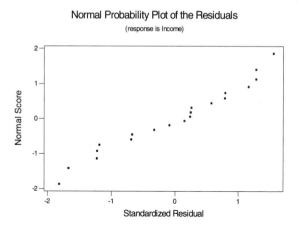

Figure 10.3.7: Normal probability plot of the standardized residuals in Example 10.3.9.

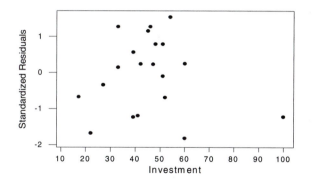

Figure 10.3.8: Plot of the standardized residuals against the predictor variable in Example 10.3.9.

10.3.4 The Multiple Linear Regression Model (Advanced)

We now consider the situation in which we have a quantitative response Y and quantitative predictors $X_1, \ldots, X_k$. For the regression model, we assume that the conditional distributions of Y, given the predictors, have constant shape and change, as the predictors change, through the conditional mean $E(Y \mid X_1 = x_1, \ldots, X_k = x_k)$. For the linear regression model, we assume that this conditional mean is of the form

$$E(Y \mid X_1 = x_1, \ldots, X_k = x_k) = \beta_1 x_1 + \cdots + \beta_k x_k. \qquad (10.3.13)$$

This is linear in the unknown $\beta_i \in R^1$ for $i = 1, \ldots, k$.

We will only develop the broad outline of the analysis of the multiple linear regression model here. All results will be stated without proofs provided. The proofs can be found in more advanced texts. It is important to note, however, that all of these results are just analogs of the results we developed by elementary methods in Section 10.3.2, for the simple linear regression model.

Matrix Formulation of the Least-Squares Problem

For the analysis of this model, we need some *matrix* concepts. We will briefly discuss some of these here, but also see Appendix A.4.

Let $A \in R^{m \times n}$ denote a rectangular array of numbers with m rows and n columns, and let a_{ij} denote the entry in the ith row and jth column (referred to as the (i, j)th entry of A). For example,

$$A = \begin{pmatrix} 1.2 & 1.0 & 0.0 \\ 3.2 & 0.2 & 6.3 \end{pmatrix} \in R^{2 \times 3}$$

denotes a 2×3 matrix and, for example, $a_{22} = 0.2$.

We can add two matrices of the same dimensions m and n by simply adding their elements componentwise. So if $A, B \in R^{m \times n}$ and $C = A + B$, then $c_{ij} = a_{ij} + b_{ij}$. Further, we can multiply a matrix by a real number c by simply multiplying every entry in the matrix by c. So if $A \in R^{m \times n}$, then $B = cA \in R^{m \times n}$ and $b_{ij} = ca_{ij}$. We will sometimes write a matrix $A \in R^{m \times n}$ in terms of its columns as $A = (\ a_1\ \cdots\ a_n\ )$ so that here $a_i \in R^m$. Finally, if $A \in R^{m \times n}$ and $b \in R^n$, then we define the product of A times b as $Ab = b_1 a_1 + \cdots + b_n a_n \in R^m$.

Suppose now that $Y \in R^n$ and that $E(Y)$ is constrained to lie in a set of the form

$$S = \left\{ \beta_1 v_1 + \cdots + \beta_k v_k : \beta_i \in R^1, i = 1, \ldots, k \right\}$$

where $v_1, \ldots, v_k$ are fixed vectors in R^n. A set such as S is called a *linear subspace* of R^n. When $\{v_1, \ldots, v_k\}$ has the *linear independence property*, namely,

$$\beta_1 v_1 + \cdots + \beta_k v_k = 0$$

if and only if $\beta_1 = \cdots = \beta_k = 0$, then we say that S has dimension k and $\{v_1, \ldots, v_k\}$ is a *basis* for S.

If we put

$$V = (v_1 \cdots v_k) = \begin{pmatrix} v_{11} & v_{12} & \cdots & v_{1k} \\ v_{21} & v_{22} & \cdots & v_{2k} \\ \vdots & \vdots & & \vdots \\ v_{n1} & v_{n2} & \cdots & v_{nk} \end{pmatrix} \in R^{n \times k},$$

then we can write

$$E(Y) = \beta_1 v_1 + \cdots + \beta_k v_k = \begin{pmatrix} \beta_1 v_{11} + \beta_2 v_{12} + \cdots + \beta_k v_{1k} \\ \beta_1 v_{21} + \beta_2 v_{22} + \cdots + \beta_k v_{2k} \\ \vdots \\ \beta_1 v_{n1} + \beta_2 v_{n2} + \cdots + \beta_k v_{nk} \end{pmatrix} = V\beta$$

for some unknown point $\beta = (\beta_1, \beta_2, \ldots, \beta_k)'$. When we observe $y \in R^n$, then the least-squares estimate of $E(Y)$ is obtained by finding the value of β that minimizes

$$\sum_{i=1}^{n} (y_i - \beta_1 v_{i1} - \beta_2 v_{i2} - \cdots - \beta_k v_{ik})^2.$$

It can be proved that a unique minimizing value for $\beta \in R^k$ exists whenever $\{v_1, \ldots, v_k\}$ is a basis. The minimizing value of β will be denoted by b and is called the *least-squares estimate* of β. The point $b_1 v_1 + \cdots + b_k v_k = Vb$ is the least-squares estimate of $E(Y)$ and is sometimes called the vector of *fitted values*. The point $y - Vb$ is called the vector of *residuals*.

We now consider how to calculate b. For this we need to understand what it means to multiply the matrix $A \in R^{m \times k}$ on the right by the matrix $B \in R^{k \times n}$. The *matrix product* AB is defined to be the $m \times n$ matrix whose (i, j)th entry is given by

$$\sum_{l=1}^{k} a_{il} b_{lj}.$$

Notice that the array A must have the same number of columns as the number of rows of B for this product to be defined. The *transpose* of a matrix $A \in R^{m \times k}$ is defined to be

$$A' = \begin{pmatrix} a_{11} & \cdots & a_{m1} \\ \vdots & & \vdots \\ a_{1k} & \cdots & a_{mk} \end{pmatrix} \in R^{k \times m},$$

namely, the ith column of A becomes the ith row of A'. For a matrix $A \in R^{k \times k}$, the *matrix inverse* of A is defined to be the matrix A^{-1} such that

$$AA^{-1} = A^{-1}A = I$$

where $I \in R^{k \times k}$ has 1's along its diagonal and 0's everywhere else and is called the $k \times k$ *identity matrix*. It is not always the case that $A \in R^{k \times k}$ has an inverse, but when it does it can be shown that the inverse is unique. Note that there are many mathematical and statistical software packages that include the facility for computing matrix products, transposes, and inverses.

We have the following fundamental result.

Theorem 10.3.7 If $E(Y) \in S = \{\beta_1 v_1 + \cdots + \beta_k v_k : \beta_i \in R^1, i = 1, \ldots, k\}$ and the columns of $V = (v_1 \cdots v_k)$ have the linear independence property, then $(V'V)^{-1}$ exists, the least-squares estimate of β is unique and is given by

$$b = \begin{pmatrix} b_1 \\ \vdots \\ b_k \end{pmatrix} = (V'V)^{-1} V'y. \tag{10.3.14}$$

Least-Squares Estimates, Predictions, and Standard Errors

For the linear regression model (10.3.13), we have that (writing X_{ij} for the jth value of X_i)

$$E\left(\left(\begin{array}{c} Y_1 \\ \vdots \\ Y_n \end{array}\right) \middle| X_{ij} = x_{ij} \text{ for all } i, j\right) = \left(\begin{array}{c} \beta_1 x_{11} + \cdots + \beta_k x_{1k} \\ \vdots \\ \beta_1 x_{n1} + \cdots + \beta_k x_{nk} \end{array}\right)$$

$$= \beta_1 v_1 + \cdots + \beta_k v_k = V\beta$$

where $\beta = (\beta_1, \ldots, \beta_k)'$ and

$$V = \left(\begin{array}{cccc} v_1 & v_2 & \cdots & v_k \end{array}\right) = \left(\begin{array}{ccc} x_{11} & \cdots & x_{1k} \\ \vdots & & \vdots \\ x_{n1} & \cdots & x_{nk} \end{array}\right) \in R^{n \times k}.$$

We will assume, hereafter, that the columns $v_1, \ldots, v_k$ of V have the linear independence property. Then (replacing expectation by conditional expectation) it is immediate that the least-squares estimate of β is given by (10.3.14).

As with the simple linear regression model, we have a number of results concerning the least-squares estimates. We state these here without proof.

Theorem 10.3.8 If the $(x_{i1}, \ldots, x_{ik}, y_i)$ are independent observations for $i = 1, \ldots, n$, and the linear regression model applies, then

$$E\left(B_i \mid X_{ij} = x_{ij} \text{ for all } i, j\right) = \beta_i$$

for $i = 1, \ldots, k$.

So Theorem 10.3.8 states that the least-squares estimates are unbiased estimates of the linear regression coefficients.

If we want to assess the accuracy of these estimates, then we need to be able to compute their standard errors.

Theorem 10.3.9 If the $(x_{i1}, \ldots, x_{ik}, y_i)$ are independent observations for $i = 1, \ldots, n$, from the linear regression model, and if $\text{Var}(Y \mid X_1 = x_1, \ldots, X_k = x_k) = \sigma^2$ for every $x_1, \ldots, x_k$, then

$$\text{Cov}\left(B_i, B_j \mid X_{ij} = x_{ij} \text{ for all } i, j\right) = \sigma^2 c_{ij} \qquad (10.3.15)$$

where c_{ij} is the (i, j)th entry in the matrix $(V'V)^{-1}$.

We have the following result concerning the estimation of the mean

$$E\left(Y \mid X_1 = x_1, \ldots, X_k = x_k\right) = \beta_1 x_1 + \cdots + \beta_k x_k$$

by the estimate $b_1 x_1 + \cdots + b_k x_k$.

Corollary 10.3.4

$$\text{Var} \left(B_1 x_1 + \cdots + B_k x_k \,|\, X_{ij} = x_{ij} \text{ for all } i, j \right)$$

$$= \sigma^2 \left(\sum_{i=1}^{k} x_i^2 c_{ii} + 2 \sum_{i<j} x_i x_j c_{ij} \right) \tag{10.3.16}$$

We also use $b_1 x_1 + \cdots + b_k x_k$ as a prediction of a new response value when $X_1 = x_1, \ldots, X_k = x_k$.

We see, from Theorem 10.3.9 and Corollary 10.3.4, that we need an estimate of σ^2 to compute standard errors. The estimate is given by

$$s^2 = \frac{1}{n-k} \sum_{i=1}^{n} \left(y_i - b_1 x_{i1} - \cdots - b_k x_{ik} \right)^2, \tag{10.3.17}$$

and we have the following result.

Theorem 10.3.10 If the $(x_{i1}, \ldots, x_{ik}, y_i)$ are independent observations for $i = 1, \ldots, n$, from the linear regression model, and if $\text{Var}(Y \,|\, X_1 = x_1, \ldots, X_k = x_k) = \sigma^2$, then

$$E \left(S^2 \,|\, X_{ij} = x_{ij} \text{ for all } i, j \right) = \sigma^2.$$

Combining (10.3.15) and (10.3.17), we deduce that the standard error of b_i is $s\sqrt{c_{ii}}$. Combining (10.3.16) and (10.3.17), we deduce that the standard error of $b_1 x_1 + \cdots + b_k x_k$ is

$$s \left(\sum_{i=1}^{k} x_i^2 c_{ii} + 2 \sum_{i<j} x_i x_j c_{ij} \right)^{1/2}.$$

The ANOVA Decomposition and F Statistics

When one of the predictors $X_1, \ldots, X_k$ is constant, then we say that the model has an *intercept term*. By convention we will always take this to be the first predictor. So when we want the model to have an intercept term, we take $X_1 \equiv 1$ and β_1 is the intercept, e.g., the simple linear regression model. In such a case, we have the following result, giving the ANOVA decomposition for this model.

Lemma 10.3.2 If, for $i = 1, \ldots, n$, the values $(x_{i1}, \ldots, x_{ik}, y_i)$ are such that the matrix V has linearly independent columns, with v_1 equal to a column of ones, then $b_1 = \bar{y} - b_2 \bar{x}_2 - \cdots - b_k \bar{x}_k$ and

$$\sum_{i=1}^{n} (y_i - \bar{y})^2 = \sum_{i=1}^{n} (b_2 (x_{i2} - \bar{x}_2) + \cdots + b_k (x_{ik} - \bar{x}_k))^2$$

$$+ \sum_{i=1}^{n} (y_i - b_1 x_{i1} - \cdots - b_k x_{ik})^2.$$

We call

$$RSS\left(X_2, \ldots, X_k\right) = \sum_{i=1}^{n} \left(b_2\left(x_{i2} - \bar{x}_2\right) + \cdots + b_k\left(x_{ik} - \bar{x}_k\right)\right)^2$$

the regression sum of squares and

$$ESS = \sum_{i=1}^{n} \left(y_i - b_1 x_{i1} + \cdots + b_k x_{ik}\right)^2$$

the error sum of squares. This leads to the following ANOVA table.

Source	Df	Sum of Squares	Mean Square
$X_2, \ldots, X_k$	$k-1$	$RSS\left(X_2, \ldots, X_k\right)$	$RSS\left(X_2, \ldots, X_k\right)/(k-1)$
Error	$n-k$	ESS	s^2
Total	$n-1$	$\sum_{i=1}^{n}\left(y_i - \bar{y}\right)^2$	

When there is an intercept term, the null hypothesis of no relationship between the response and the predictors is equivalent to $H_0 : \beta_2 = \cdots = \beta_k = 0$. As with the simple linear regression model, the mean square for regression can be shown to be an unbiased estimator of σ^2 if and only if the null hypothesis is true. Therefore, a sensible statistic to use for assessing the null hypothesis is the F-statistic

$$F = \frac{RSS\left(X_2, \ldots, X_k\right)/(k-1)}{s^2},$$

with large values being evidence against the null.

Often, we want to assess the null hypothesis $H_0 : \beta_{l+1} = \cdots = \beta_k = 0$ or, equivalently, the hypothesis that the model is given by

$$E\left(Y \mid X_1 = x_1, \ldots, X_k = x_k\right) = \beta_1 x_1 + \cdots + \beta_l x_l$$

where $l < k$. This hypothesis says that the last $k - l$ predictors $X_{l+1}, \ldots, X_k$, have no relationship with the response.

If we denote the least-squares estimates of $\beta_1, \ldots, \beta_l$, obtained by fitting the smaller model, by $b_1^*, \ldots, b_l^*$, then we have the following result.

Lemma 10.3.3 If the $(x_{i1}, \ldots, x_{ik}, y_i)$ for $i = 1, \ldots, n$ are values for which the matrix V has linearly independent columns, with v_1 equal to a column of ones, then

$$\begin{aligned} RSS\left(X_2, \ldots, X_k\right) &= \sum_{i=1}^{n} \left(b_2\left(x_{i2} - \bar{x}_2\right) + \cdots + b_k\left(x_{ik} - \bar{x}_k\right)\right)^2 \\ &\geq \sum_{i=1}^{n} \left(b_2^*\left(x_{i2} - \bar{x}_2\right) + \cdots + b_l^*\left(x_{il} - \bar{x}_l\right)\right)^2 \\ &= RSS\left(X_2, \ldots, X_l\right). \end{aligned} \qquad (10.3.18)$$

On the right of the inequality in (10.3.18) we have the regression sum of squares obtained by fitting the model based on the first l predictors. Therefore, we can interpret the difference of the left and right sides of (10.3.18), namely,

$$RSS\left(X_{l+1}, \ldots, X_k \mid X_2, \ldots, X_l\right) = RSS\left(X_2, \ldots, X_k\right) - RSS\left(X_2, \ldots, X_l\right)$$

as the contribution of the predictors $X_{l+1}, \ldots, X_k$ to the regression sum of squares when the predictors $X_1, \ldots, X_l$ are in the model. We get the following ANOVA table (actually only the first three columns of the ANOVA table) corresponding to this decomposition of the total sum of squares.

Source	Df	Sum of Squares
$X_2, \ldots, X_l$	$l-1$	$RSS\left(X_2, \ldots, X_l\right)$
$X_{l+1}, \ldots, X_k \mid X_2, \ldots, X_l$	$k-l$	$RSS\left(X_{l+1}, \ldots, X_k \mid X_2, \ldots, X_l\right)$
Error	$n-k$	ESS
Total	$n-1$	$\sum_{i=1}^{n} (y_i - \bar{y})^2$

It can be shown that the null hypothesis $H_0 : \beta_{l+1} = \cdots = \beta_k = 0$ holds if and only if

$$RSS\left(X_{l+1}, \ldots, X_k \mid X_2, \ldots, X_l\right) / (k-l)$$

is an unbiased estimator of σ^2. Therefore, a sensible statistic to use for assessing this null hypothesis is the F-statistic

$$F = \frac{RSS\left(X_{l+1}, \ldots, X_k \mid X_2, \ldots, X_l\right) / (k-l)}{s^2},$$

with large values being evidence against the null.

The Coefficient of Determination

The coefficient of determination for this model is given by

$$R^2 = \frac{RSS\left(X_2, \ldots, X_k\right)}{\sum_{i=1}^{n} (y_i - \bar{y})^2},$$

which, by Lemma 10.3.2, is always between 0 and 1. The value of R^2 gives the proportion of the observed variation in Y that is explained by the inclusion of the nonconstant predictors in the model.

It can be shown that R^2 is the square of the *multiple correlation coefficient* between Y and $X_1, \ldots, X_k$. However, we do not discuss the multiple correlation coefficient in this text.

Confidence Intervals and Testing Hypotheses

For inference we have the following result.

Theorem 10.3.11 If the conditional distribution of Y given $(X_1, \ldots, X_k) = (x_1, \ldots, x_k)$ is $N\left(\beta_1 x_1 + \cdots + \beta_k x_k, \sigma^2\right)$, and if we observe the independent values $(x_{i1}, \ldots, x_{ik}, y_i)$ for $i = 1, \ldots, n$, then the conditional distributions of the B_i and S^2, given $X_{ij} = x_{ij}$ for all i, j, are as follows.

(i) $B_i \sim N\left(\beta_1, \sigma^2 c_{ii}\right)$

(ii) $B_1 x_1 + \cdots + B_k x_k$ is distributed

$$N\left(\beta_1 x_1 + \cdots + \beta_k x_k, \sigma^2\left(\sum_{i=1}^{k} x_i^2 c_{ii} + 2\sum_{i<j} x_i x_j c_{ij}\right)\right)$$

(iii) $(n-k) S^2/\sigma^2 \sim \chi^2 (n-k)$ independent of $(B_1, \ldots, B_k)$

Corollary 10.3.5

(i) $(B_i - \beta_i)/sc_{ii}^{1/2} \sim t(n-k)$

(ii)

$$\frac{B_1 x_1 + \cdots + B_k x_k - \beta_1 x_1 - \cdots - \beta_k x_k}{S\left(\sum_{i=1}^{k} x_i^2 c_{ii} + 2\sum_{i<j} x_i x_j c_{ij}\right)^{1/2}} \sim t(n-k)$$

(iii) $H_0 : \beta_{l+1} = \cdots = \beta_k = 0$ is true if and only if

$$F = \frac{\left(RSS\left(X_2, \ldots, X_k\right) - RSS\left(X_2, \ldots, X_l\right)\right)/(k-l)}{S^2} \sim F(k-l, n-k)$$

Analysis of Residuals

In an application of the multiple regression model, we must check to make sure that the assumptions make sense. Model checking is based on the residuals $y_i - b_1 x_{i1} - \cdots - b_k x_{ik}$ (after standardization), just as discussed in Section 9.1. Note that the ith residual is simply the difference between the observed value y_i at $(x_{i1}, \ldots, x_{ik})$ and the predicted value $b_1 x_{i1} + \cdots + b_k x_{ik}$ at $(x_{i1}, \ldots, x_{ik})$.

We also have the following result (this can be proved as a Corollary of Theorem 10.3.10).

Corollary 10.3.6

(i) $E\left(Y_i - B_1 x_{i1} - \cdots - B_k x_{ik} \mid V\right) = 0$

(ii) $\text{Cov}(Y_i - B_1 x_{i1} - \cdots - B_k x_{ik}, Y_j - B_1 x_{j1} - \cdots - B_k x_{jk}, \mid V) = \sigma^2 d_{ij}$ where d_{ij} is the (i,j)th entry of the matrix $I - V\left(V'V\right)^{-1} V'$

Therefore, the standardized residuals are given by

$$\frac{y_j - b_1 x_{j1} - \cdots - b_k x_{jk}}{sd_{ii}^{1/2}}. \tag{10.3.19}$$

When s is replaced by σ in (10.3.19), Corollary 10.3.6 implies that this quantity has conditional mean 0 and conditional variance 1. Further, when the conditional distribution of the response given the predictors is normal, then it can be shown that the conditional distribution of this quantity is $N(0, 1)$. These results are also approximately true for (10.3.19) for large n. Further, it can be shown that the covariances between the standardized residuals go to 0 as $n \to \infty$, under certain reasonable conditions on distribution of the predictor variables. So one approach to model checking here is to see whether the values given by (10.3.19) look at all like a sample from the $N(0, 1)$ distribution.

What do we do if model checking leads to a failure of the model? As in Chapter 9, we can consider making various transformations of the data to see if there is a simple modification of the model that will pass. We can make transformations not only to the response variable Y, but to the predictor variables $X_1, \ldots, X_k$ as well.

An Application of Multiple Linear Regression Analysis

The computations needed to implement a multiple linear regression analysis cannot be carried out by hand. These are much too time-consuming and error-prone. It is therefore important that a statistician have a computer with suitable software available when doing a multiple linear regression analysis.

The following data are taken from *Statistical Theory and Methodology in Science and Engineering*, 2nd ed., by K. A. Brownlee (Wiley, New York, 1965). The response variable Y is stack loss (Loss), which represents 10 times the percentage of ammonia lost as unabsorbed nitric oxide, and the predictor variables are X_1 = air flow (Air), X_2 = temperature of inlet water (Temp), and X_3 = the concentration of nitric acid (Acid). Also recorded is the day (Day) on which the observation was taken.

Day	Air	Temp	Acid	Loss	Day	Air	Temp	Acid	Loss
1	80	27	89	42	12	58	17	88	13
2	80	27	88	37	13	58	18	82	11
3	75	25	90	37	14	58	19	93	12
4	62	24	87	28	15	50	18	89	8
5	62	22	87	18	16	50	18	86	7
6	62	23	87	18	17	50	19	72	8
7	62	24	93	19	18	50	19	79	8
8	62	24	93	20	19	50	20	80	9
9	58	23	87	15	20	56	20	82	15
10	58	18	80	14	21	70	20	91	15
11	58	18	89	14					

We consider the model

$$Y \mid x_1, x_2, x_3 \sim N(\beta_0 + \beta_1 x_1 + \beta_2 x_2 + \beta_3 x_3, \sigma^2).$$

Note that we have included an intercept term. Figure 10.3.9 is a normal probability plot of the standardized residuals. This looks reasonable, except for one residual, -2.63822, that diverges quite distinctively from the rest of the values, which lie close to the 45-degree line. Printing out the standardized residuals shows that this residual is associated with the observation on the 21st day. Possibly there was something unique about this day's operations, and so it is

reasonable to discard this data value and refit the model. Figure 10.3.10 is a
normal probability plot obtained by fitting the model to the first 20 observations. This looks somewhat better, but still we might be concerned about at
least one of the residuals that deviates substantially from the 45-degree line.

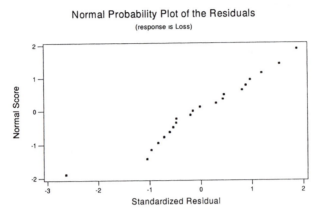

Figure 10.3.9: Normal probability plot of the standardized residuals based on all the
data.

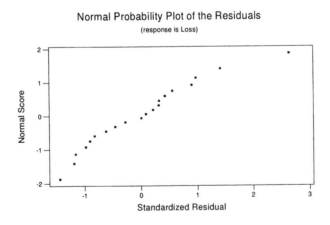

Figure 10.3.10: Normal probability plot of the standardized residuals based on the
first 20 data values.

Following the analysis of these data in *Fitting Equations to Data,* by C.
Daniel and F. S. Wood (Wiley-Interscience, New York, 1971) we consider instead
the model

$$\ln Y \mid x_1, x_2, x_3 \sim N(\beta_0 + \beta_1 x_1 + \beta_2 x_2 + \beta_3 x_3, \sigma^2), \qquad (10.3.20)$$

i.e., we transform the response variable by taking its logarithm and use all of the
data. Often, when models do not fit, simple transformations like this can lead to

major improvements. In this case, we see a much improved normal probability plot, as provided in Figure 10.3.11.

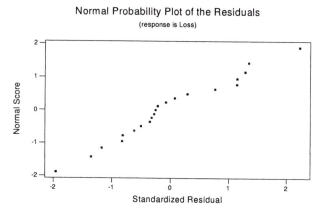

Figure 10.3.11: Normal probability plot of the standardized residuals for all the data using $\ln Y$ as the response.

We also looked at plots of the standardized residuals against the various predictors; and these looked reasonable. Figure 10.3.12 is a plot of the standardized residuals against the values of Air.

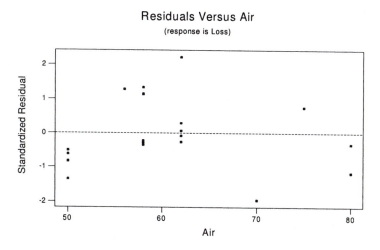

Figure 10.3.12: A plot of the standardized residuals for all the data, using $\ln Y$ as the response, against the values of the predictor Air.

Now that we have accepted the model (10.3.20), we can proceed to inferences about the unknowns of the model. The least-squares estimates of the β_i, their standard errors (Se), the corresponding t-statistics for testing the $\beta_i = 0$, and the P-values for this, are given in the following table.

Coefficient	Estimate	Se	t-statistic	P-value
β_0	-0.948700	0.647700	-1.46	0.161
β_1	0.034565	0.007343	4.71	0.000
β_2	0.063460	0.020040	3.17	0.006
β_3	0.002864	0.008510	0.34	0.742

The estimate of σ^2 is given by $s^2 = 0.0312$.

To test the null hypothesis that there is no relationship between the response and the predictors, or, equivalently, $H_0 : \beta_1 = \beta_2 = \beta_3 = 0$, we have the following ANOVA table.

Source	Df	Sum of Squares	Mean Square
X_1, X_2, X_3	3	4.9515	1.6505
Error	17	0.5302	0.0312
Total	20	5.4817	

The value of the F-statistic is given by $1.6505/0.0312 = 52.900$, and when $F \sim F(3, 17)$, we have that $P(F > 52.900) = 0.000$. So we have substantial evidence against the null hypothesis. To see how well the model explains the variation in the response, we computed the value of $R^2 = 86.9\%$. Therefore, approximately 87% of the observed variation in Y can be explained by changes in the predictors in the model.

While we have concluded that a relationship exists between the response and the predictors, it may be that some of the predictors have no relationship with the response. For example, the table of t-statistics above would seem to indicate that perhaps X_3 (Acid) is not affecting Y. We can assess this via the following ANOVA table, obtained by fitting the model $\ln Y \mid x_1, x_2, x_3 \sim N(\beta_0 + \beta_1 x_1 + \beta_2 x_2, \sigma^2)$.

Source	Df	Sum of Squares	Mean Square
X_1, X_2	2	4.9480	2.4740
$X_3 \mid X_1, X_2$	1	0.0035	0.0035
Error	17	0.5302	0.0312
Total	20	5.4817	

Note that $RSS(X_3 \mid X_1, X_2) = 4.9515 - 4.9480 = 0.0035$. The value of the F-statistic for testing $H_0 : \beta_3 = 0$ is $0.0035/0.0312 = 0.112$, and when $F \sim F(1, 17)$, we have that $P(F > 0.112) = 0.742$. So we have no evidence against the null hypothesis and can drop X_3 from the model. Actually, this is the same P-value as obtained via the t-test of this null hypothesis, as, in general, the t-test that a single regression coefficient is 0 is equivalent to the F-test. Similar tests of the need to include X_1 and X_2 do not lead us to drop these variables from the model.

So based on the above results, we decide to drop X_3 from the model and use the equation

$$E(Y \mid X_1 = x_1, X_2 = x_2) = -0.7522 + 0.035402 X_1 + 0.06346 X_2 \quad (10.3.21)$$

to describe the relationship between Y and the predictors. Note that the least-squares estimates of β_0, β_1, and β_2 in (10.3.21) are obtained by refitting the model without X_3. ∎

Summary of Section 10.3

- In this section we examined the situation in which the response variable and the predictor variables are quantitative.

- In this situation the linear regression model provides a possible description of the form of any relationship that may exist between the response and the predictors.

- Least squares is a standard method for fitting linear regression models to data.

- The ANOVA is a decomposition of the total variation observed in the response variable into a part attributable to changes in the predictor variables and a part attributable to random error.

- If we assume a normal linear regression model, then we have inference methods available such as confidence intervals and tests of significance. In particular, we have available the F-test to assess whether or not a relationship exists between the response and the predictors.

- A normal linear regression model is checked by examining the standardized residuals.

Exercises

10.3.1 Suppose that $(x_1, \ldots, x_n)$ is a sample from a Bernoulli$(0, \theta)$ distribution where $\theta \in [0, 1]$ is unknown. What is the least-squares estimate of the mean of this distribution?

10.3.2 Suppose that $(x_1, \ldots, x_n)$ is a sample from the Uniform$(0, \theta)$ where $\theta > 0$ is unknown. What is the least-squares estimate of the mean of this distribution?

10.3.3 Suppose that $(x_1, \ldots, x_n)$ is a sample from the Exponential(θ) where $\theta > 0$ is unknown. What is the least-squares estimate of the mean of this distribution?

10.3.4 Consider the $n = 11$ data values in the following table.

Observation	X	Y	Observation	X	Y
1	−5.00	−10.00	7	1.00	3.52
2	−4.00	−8.83	8	2.00	5.64
3	−3.00	−9.15	9	3.00	7.28
4	−2.00	−4.26	10	4.00	7.62
5	−1.00	−0.30	11	5.00	8.51
6	0.00	−0.04			

Suppose we consider the simple normal linear regression to describe the relationship between the response Y and the predictor X.

(a) Plot the data in a scatter plot.

(b) Calculate the least-squares line and plot this on the scatter plot in part (a).

(c) Plot the standardized residuals against X.

(d) Produce a normal probability plot of the standardized residuals.

(e) What are your conclusions based on the plots produced in parts (c) and (d)?

(f) If appropriate, calculate 0.95-confidence intervals for the intercept and slope.

(g) Construct the ANOVA table to test whether or not there is a relationship between the response and the predictors. What is your conclusion?

(h) If the model is correct, what proportion of the observed variation in the response is explained by changes in the predictor?

(i) Predict a future Y at $X = 0.0$. Is this prediction an extrapolation or an interpolation? Determine the standard error of this prediction.

(j) Predict a future Y at $X = 6.0$. Is this prediction an extrapolation or an interpolation? Determine the standard error of this prediction.

(k) Predict a future Y at $X = 20.0$. Is this prediction an extrapolation or an interpolation? Determine the standard error of this prediction. Compare this with the standard errors obtained in parts (i) and (j) and explain the differences.

10.3.5 Consider the $n = 11$ data values in the following table.

Observation	X	Y	Observation	X	Y
1	−5.00	65.00	7	1.00	6.52
2	−4.00	39.17	8	2.00	17.64
3	−3.00	17.85	9	3.00	34.28
4	−2.00	7.74	10	4.00	55.62
5	−1.00	2.70	11	5.00	83.51
6	0.00	−0.04			

Suppose we consider the simple normal linear regression to describe the relationship between the response Y and the predictor X.

(a) Plot the data in a scatter plot.

(b) Calculate the least-squares line and plot this on the scatter plot in part (a).

(c) Plot the standardized residuals against X.

(d) Produce a normal probability plot of the standardized residuals.

(e) What are your conclusions based on the plots produced in parts (c) and (d)?

(f) If appropriate, calculate 0.95-confidence intervals for the intercept and slope.

(g) Do the results of your analysis allow you to conclude that there is a relationship between Y and X? Explain why or why not.

(h) If the model is correct, what proportion of the observed variation in the response is explained by changes in the predictor?

10.3.6 Suppose that the following data record the densities of an organism in a containment vessel for 10 days.

Day	Number/Liter	Day	Number/Liter
1	1.6	6	1341.6
2	16.7	7	2042.9
3	65.2	8	7427.0
4	23.6	9	15571.8
5	345.3	10	33128.5

Suppose we consider the simple normal linear regression to describe the relationship between the response Y (density) and the predictor X (day).
(a) Plot the data in a scatter plot.
(b) Calculate the least-squares line and plot this on the scatter plot in part (a).
(c) Plot the standardized residuals against X.
(d) Produce a normal probability plot of the standardized residuals.
(e) What are your conclusions based on the plots produced in parts (c) and (d)?
(f) Can you think of a transformation of the response that might address any problems found? If so, repeat parts (a) through (e) after performing this transformation. (Hint: The scatter plot looks like exponential growth. What transformation is the inverse of exponentiation?)
(g) Calculate 0.95-confidence intervals for the appropriate intercept and slope.
(h) Construct the appropriate ANOVA table to test whether or not there is a relationship between the response and the predictors. What is your conclusion?
(i) Do the results of your analysis allow you to conclude that there is a relationship between Y and X? Explain why or why not.
(j) Compute the proportion of variation explained by the predictor for the two models you have considered. Compare the results.
(k) Predict a future Y at $X = 12$. Is this prediction an extrapolation or an interpolation?

Computer Exercises

10.3.7 Suppose we consider the simple normal linear regression to describe the relationship between the response Y (income) and the predictor X (investment) for the data in Example 10.3.9.
(a) Plot the data in a scatter plot.
(b) Calculate the least-squares line and plot this on the scatter plot in part (a).
(c) Plot the standardized residuals against X.
(d) Produce a normal probability plot of the standardized residuals.
(e) What are your conclusions based on the plots produced in parts (c) and (d)?
(f) If appropriate, calculate 0.95-confidence intervals for the intercept and slope.
(g) Do the results of your analysis allow you to conclude that there is a relationship between Y and X? Explain why or why not.
(h) If the model is correct, what proportion of the observed variation in the response is explained by changes in the predictor?

10.3.8 The following data are measurements of tensile strength (100 lb/in^2) and hardness (Rockwell E) on 20 pieces of die-cast aluminum.

Sample	Strength	Hardness	Sample	Strength	Hardness
1	293	53	11	298	60
2	349	70	12	292	51
3	340	78	13	380	95
4	340	55	14	345	88
5	340	64	15	257	51
6	354	71	16	265	54
7	322	82	17	246	52
8	334	67	18	286	64
9	247	56	19	324	83
10	348	86	20	282	56

Suppose we consider the simple normal linear regression to describe the relationship between the response Y (strength) and the predictor X (hardness).
(a) Plot the data in a scatter plot.
(b) Calculate the least-squares line and plot this on the scatter plot in part (a).
(c) Plot the standardized residuals against X.
(d) Produce a normal probability plot of the standardized residuals.
(e) What are your conclusions based on the plots produced in parts (c) and (d)?
(f) If appropriate, calculate 0.95-confidence intervals for the intercept and slope.
(g) Do the results of your analysis allow you to conclude that there is a relationship between Y and X? Explain why or why not.
(h) If the model is correct, what proportion of the observed variation in the response is explained by changes in the predictor?

10.3.9 Tests were carried out to determine the effect of gas inlet temperature (degrees Fahrenheit) and rotor speed (rpm) on the tar content (grains/cu. ft) of a gas stream, producing the following data.

Observation	Tar	Speed	Temperature
1	60.0	2400	54.5
2	65.0	2450	58.5
3	63.5	2500	58.0
4	44.0	2700	62.5
5	54.5	2700	68.0
6	26.0	2775	45.5
7	54.0	2800	63.0
8	53.5	2900	64.5
9	33.5	3075	57.0
10	44.0	3150	64.0

Suppose we consider the normal linear regression model

$$Y \mid W = w, X = x \sim N\left(\beta_1 + \beta_2 w + \beta_3 x, \sigma^2\right)$$

to describe the relationship between Y (tar content) and the predictors W (rotor speed) and X (temperature).

(a) Plot the response in scatter plots against each predictor.

(b) Calculate the least-squares equation.

(c) Plot the standardized residuals against W and X.

(d) Produce a normal probability plot of the standardized residuals.

(e) What are your conclusions based on the plots produced in parts (c) and (d)?

(f) If appropriate, calculate 0.95-confidence intervals for the regression coefficients.

(g) Construct the ANOVA table to test whether or not there is a relationship between the response and the predictors. What is your conclusion?

(h) If the model is correct, what proportion of the observed variation in the response is explained by changes in the predictors?

(i) In an ANOVA table, assess the null hypothesis that there is no effect due to W, given that X is in the model.

(j) Estimate the mean of Y when $W = 2750$ and $X = 50.0$. If we consider this value as a prediction of a future Y at these settings, is this an extrapolation or interpolation?

10.3.10 Suppose we consider the normal linear regression model

$$Y \mid X = x \sim N\left(\beta_1 + \beta_2 x + \beta_3 x^2, \sigma^2\right)$$

for the data of Exercise 10.3.5.

(a) Plot the response Y in a scatter plot against X.

(b) Calculate the least-squares equation.

(c) Plot the standardized residuals against X.

(d) Produce a normal probability plot of the standardized residuals.

(e) What are your conclusions based on the plots produced in parts (c) and (d)?

(f) If appropriate, calculate 0.95-confidence intervals for the regression coefficients.

(g) Construct the ANOVA table to test whether or not there is a relationship between the response and the predictor. What is your conclusion?

(h) If the model is correct, what proportion of the observed variation in the response is explained by changes in the predictors?

(i) In an ANOVA table, assess the null hypothesis that there is no effect due to X^2, given that X is in the model.

(j) Compare the predictions of Y at $X = 6$ using the simple linear regression model and using the linear model with a linear and quadratic term.

Problems

10.3.11 Suppose that $(x_1, \ldots, x_n)$ is a sample from the mixture distribution $0.5\text{Uniform}(0, 1) + 0.5\text{Uniform}(2, \theta)$ where $\theta > 2$ is unknown. What is the least-squares estimate of the mean of this distribution?

10.3.12 Consider the simple linear regression model and suppose that for the data collected, we have $\sum_{i=1}^{n} (x_i - \bar{x})^2 = 0$. Explain how, and for which value of x, you would estimate $E(Y \mid X = x)$.

10.3.13 For the simple linear regression model, under the assumptions of Theorem 10.3.3, establish that

$$\text{Cov}\left(Y_i - B_1 - B_2 x_i, Y_j - B_1 - B_2 x_j \mid X_1 = x_1, \ldots, X_n = x_n\right)$$

$$= \sigma^2 \delta_{ij} - \sigma^2 \left(\frac{1}{n} + \frac{(x_i - \bar{x})(x_j - \bar{x})}{\sum_{k=1}^{n}(x_k - \bar{x})^2}\right),$$

where $\delta_{ij} = 1$ when $i = j$ and is 0 otherwise. (Hint: Use Theorems 3.3.2 and 10.3.3.)

10.3.14 Establish that (10.3.11) is distributed $N(0, 1)$ when S is replaced by σ in the denominator. (Hint: Use Theorem 4.6.1 and Problem 10.3.13.)

10.3.15 (*Prediction intervals*) Under the assumptions of Theorem 10.3.6, prove that the interval

$$b_1 + b_2 x \pm s \left(1 + \frac{1}{n} + \frac{(x_i - \bar{x})^2}{\sum_{k=1}^{n}(x_k - \bar{x})^2}\right)^{1/2} t_{(1+\gamma)/2}(n-2),$$

based on independent $(x_1, y_1), \ldots, (x_n, y_n)$, will contain Y with probability equal to γ for a future independent (X, Y) with $X = x$. (Hint: Theorems 4.6.1 and 3.3.2 and Corollary 10.3.1.)

10.3.16 Consider the regression model with no intercept, given by $E(Y \mid X = x) = \beta x$ where $\beta \in R^1$ is unknown. Suppose we observe the independent values $(x_1, y_1), \ldots, (x_n, y_n)$.
(a) Determine the least-squares estimate of β.
(b) Prove that the least-squares estimate b of β is unbiased and, assuming $\text{Var}(Y \mid X = x) = \sigma^2$, prove that

$$\text{Var}(B \mid X_1 = x_1, \ldots, X_n = x_n) = \frac{\sigma^2}{\sum_{i=1}^{n} x_i^2}.$$

(c) Under the assumptions given in part (b), prove that

$$s^2 = \frac{1}{n-1} \sum_{i=1}^{n} (y_i - bx_i)^2$$

is an unbiased estimator of σ^2.
(d) Record an appropriate ANOVA decomposition for this model and a formula for R^2, measuring the proportion of the variation observed in Y due to changes in X.
(e) When $Y \mid X = x \sim N(\beta x, \sigma^2)$, and we observe the independent values $(x_1, y_1), \ldots, (x_n, y_n)$, prove that $b \sim N(\beta, \sigma^2 / \sum_{i=1}^{n} x_i^2)$.
(f) Under the assumptions of part (e), and assuming that $(n-1)S^2/\sigma^2 \sim \chi^2(n-1)$ independent of B (this can be proved), indicate how you would test the null hypothesis of no relationship between Y and X.

(g) How would you define standardized residuals for this model and use them to check model validity?

10.3.17 For data $(x_1, y_1), \ldots, (x_n, y_n)$, prove that if $\alpha_1 = \beta_1 + \beta_2 \bar{x}$ and $\alpha_2 = \beta_2$, then $\sum_{i=1}^{n} (y_i - \beta_1 - \beta_2 x_i)^2$ equals

$$\sum_{i=1}^{n} (y_i - \bar{y})^2 + n (\alpha_1 - \bar{y})^2 + \alpha_2^2 \sum_{i=1}^{n} (x_i - \bar{x})^2 - 2\alpha_2 \sum_{i=1}^{n} (x_i - \bar{x})(y_i - \bar{y}).$$

From this, deduce that $\bar{y}$ and $a = \sum_{i=1}^{n} (x_i - \bar{x})(y_i - \bar{y}) / \sum_{i=1}^{n} (x_i - \bar{x})^2$ are the least squares of α_1 and α_2, respectively.

10.3.18 For the model discussed in Section 10.3.3, prove that the prior given by $\alpha_1 \mid \alpha_2, \sigma^2 \sim N(\mu_1, \tau_1^2 \sigma^2)$, $\alpha_2 \mid \sigma^2 \sim N(\mu_2, \tau_2^2 \sigma^2)$, and $1/\sigma^2 \sim \text{Gamma}(\kappa, \nu)$. Conclude that this prior is conjugate with the posterior distribution, as specified. (Hint: The development is similar to Example 7.1.4, as detailed in Section 7.5.)

10.3.19 For the model specified in Section 10.3.3, prove that when $\tau_1 \to \infty, \tau_2 \to \infty$ and $\nu \to 0$, the posterior distribution of α_1 is given by the distribution of $\bar{y} + (2\nu_{xy}/n(2\kappa + n))^{1/2} Z$ where $Z \sim t(2\kappa + n)$ and $\nu_{xy} = (c_y^2 - a^2 c_x^2)/2$.

Challenge

10.3.20 If $X_1, \ldots, X_n$ is a sample from a distribution with finite variance, then prove that

$$\frac{X_i - \bar{X}}{\sqrt{\sum_{k=1}^{n} (X_k - \bar{X})^2}} \xrightarrow{a.s.} 0.$$

10.4 Quantitative Response and Categorical Predictors

In this section we consider the situation in which the response is quantitative and the predictors are categorical. There can be many categorical predictors, but we restrict our discussion to at most two, as this gives the most important features of the general case. The general case is left to a further course.

10.4.1 One Categorical Predictor (One-Way ANOVA)

Suppose now that the response Y is quantitative and the predictor X is categorical, taking a values or levels denoted $1, \ldots, a$. With the regression model, we assume that the only aspect of the conditional distribution of Y, given $X = x$, that changes as x changes, is the mean. We let

$$\beta_i = E(Y \mid X = i)$$

denote the mean response when the predictor X is at level i. Note that this is immediately a linear regression model.

We introduce the *dummy variables*

$$X_i = \begin{cases} 1 & X = i \\ 0 & X \neq i \end{cases}$$

for $i = 1, \ldots, a$. Notice that, whatever the value is of the response Y, only one of the dummy variables takes the value 1, and the rest take the value 0. Accordingly, we can write

$$E(Y \mid X_1 = x_1, \ldots, X_a = x_a) = \beta_1 x_1 + \cdots + \beta_a x_a,$$

because one and only one of the $x_i = 1$ whereas the rest are 0. This has exactly the same form as the model discussed in Section 10.3.4, as the X_i are quantitative. As such, all the results of Section 10.3.4 immediately apply (we will restate relevant results here).

Now suppose that we observe n_i values $(y_{i1}, \ldots, y_{in_i})$ when $X = i$, and all the response values are independent. Note that we have a independent samples. The least-squares estimates of the β_i are obtained by minimizing

$$\sum_{i=}^{a} \sum_{j=1}^{n_i} (y_{ij} - \beta_i)^2.$$

The least-squares estimates are then equal to (Problem 10.4.9)

$$b_i = \bar{y}_i = \frac{1}{n_i} \sum_{j=1}^{n_i} y_{ij}.$$

These can be shown to be unbiased estimators of the β_i.

Assuming that the conditional distributions of Y, given $X = x$, all have variance equal to σ^2, we have that the conditional variance of $\bar{Y}_i$ is given by σ^2/n_i, and the conditional covariance between $\bar{Y}_i$ and $\bar{Y}_j$, when $i \neq j$, is 0. Further, under these conditions, an unbiased estimator of σ^2 is given by

$$s^2 = \frac{1}{N-a} \sum_{i=}^{a} \sum_{j=1}^{n_i} (y_{ij} - \bar{y}_i)^2$$

where $N = n_1 + \cdots + n_k$.

If, in addition, we assume the normal linear regression model, namely,

$$Y \mid X = i \sim N\left(\beta_i, \sigma^2\right),$$

then $\bar{Y}_i \sim N\left(\beta_i, \sigma^2/n_i\right)$ independent of $(N-a) S^2/\sigma^2 \sim \chi^2(N-a)$. Therefore, by Definition 4.6.2,

$$T = \frac{\bar{Y}_i - \beta_i}{S/\sqrt{n_i}} \sim t(N-a),$$

and this leads to a γ-confidence interval of the form

$$\bar{y}_i \pm \frac{s}{\sqrt{n_i}} t_{(1+\gamma)/2} (N - a)$$

for β_i. Also we can test the null hypothesis $H_0 : \beta_i = \beta_{i0}$ by computing the P-value

$$P\left(|T| \geq \left|\frac{\bar{y}_i - \beta_{i0}}{s/\sqrt{n_i}}\right|\right) = 2\left(1 - G\left(\left|\frac{\bar{y}_i - \beta_{i0}}{s/\sqrt{n_i}}\right| ; N - a\right)\right)$$

where $G(\cdot; N - a)$ is the cdf of the $t(N - a)$ distribution. Note that these inferences are just like those derived in Section 6.3 for the location-scale normal model, except we now use a different estimator of σ^2 (with more degrees of freedom).

Often we want to make inferences about a difference of means $\beta_i - \beta_j$. Note that $E(\bar{Y}_i - \bar{Y}_j) = \beta_i - \beta_j$ and

$$\text{Var}(\bar{Y}_i - \bar{Y}_j) = \text{Var}(\bar{Y}_i) + \text{Var}(\bar{Y}_j) = \sigma^2(1/n_i + 1/n_j)$$

because $\bar{Y}_i$ and $\bar{Y}_j$ are independent. By Theorem 4.6.1,

$$\bar{Y}_i - \bar{Y}_j \sim N(\beta_i - \beta_j, \sigma^2(1/n_i + 1/n_j)).$$

Further,

$$\frac{(\bar{Y}_i - \bar{Y}_j) - (\beta_i - \beta_j)}{\sigma(1/n_i + 1/n_j)^{1/2}} \sim N(0, 1)$$

independent of $(N - a) S^2/\sigma^2 \sim \chi^2 (N - a)$. Therefore, by Definition 4.6.2,

$$
\begin{aligned}
T &= \left(\frac{(\bar{Y}_i - \bar{Y}_j) - (\beta_i - \beta_j)}{\sigma(1/n_i + 1/n_j)^{1/2}}\right) \bigg/ \sqrt{\frac{(N - a) S^2}{(N - a) \sigma^2}} \\
&= \frac{(\bar{Y}_i - \bar{Y}_j) - (\beta_i - \beta_j)}{S(1/n_i + 1/n_j)^{1/2}} \sim t(N - a).
\end{aligned}
\tag{10.4.1}
$$

This leads to the γ-confidence interval

$$\bar{y}_i - \bar{y}_j \pm s\sqrt{\frac{1}{n_i} + \frac{1}{n_j}} t_{(1+\gamma)/2} (N - a)$$

for the difference of means $\beta_i - \beta_j$. We can test the null hypothesis $H_0 : \beta_i = \beta_j$, i.e., the difference in the means equals 0, by computing the P-value

$$P\left(|T| \geq \left|\frac{\bar{y}_i - \bar{y}_j}{s\sqrt{\frac{1}{n_i} + \frac{1}{n_j}}}\right|\right) = 2\left(1 - G\left(\left|\frac{\bar{y}_i - \bar{y}_j}{s\sqrt{\frac{1}{n_i} + \frac{1}{n_j}}}\right| ; N - a\right)\right).$$

When $a = 2$, i.e., there are just two values for X, we refer to (10.4.1) as the *two-sample t-statistic*, and the corresponding inference procedures are called the *two-sample t-confidence interval* and the *two-sample t-test* for the difference of means. In this case, if we conclude that $\beta_1 \neq \beta_2$, then we are saying that a relationship exists between Y and X.

Suppose, in the general case when $a \geq 2$, we are interested in assessing whether or not there is a relationship between the response and the predictor. There is no relationship if and only if all the conditional distributions are the same; this is true, under our assumptions, if and only if $\beta_1 = \cdots = \beta_a$, i.e., if and only if all the means are equal. So testing the null hypothesis that there is no relationship between the response and the predictor is equivalent to testing the null hypothesis $H_0 : \beta_1 = \cdots = \beta_a = \beta$ for some unknown β.

If the null hypothesis is true, the least-squares estimate of β is given by $\bar{y}$, the overall average response value. In this case, we have that the total variation decomposes as (Problem 10.4.10)

$$\sum_{i=}^{a}\sum_{j=1}^{n_i}(y_{ij} - \bar{y})^2 = \sum_{i=1}^{a} n_i (\bar{y}_i - \bar{y})^2 + \sum_{i=}^{a}\sum_{j=1}^{n_i}(y_{ij} - \bar{y}_i)^2 ,$$

and so the relevant ANOVA table for testing H_0 is given below.

Source	Df	Sum of Squares	Mean Square
X	$a - 1$	$\sum_{i=1}^{a} n_i (\bar{y}_i - \bar{y})^2$	$\sum_{i=1}^{a} n_i (\bar{y}_i - \bar{y})^2 / (a - 1)$
Error	$N - a$	$\sum_{i=}^{a}\sum_{j=1}^{n_i}(y_{ij} - \bar{y}_i)^2$	s^2
Total	$N - 1$	$\sum_{i=}^{a}\sum_{j=1}^{n_i}(y_{ij} - \bar{y})^2$	

To assess H_0, we use the F-statistic

$$F = \frac{\sum_{i=1}^{a} n_i (\bar{y}_i - \bar{y})^2 / (a - 1)}{s^2},$$

because, under the null hypothesis, both the numerator and the denominator are unbiased estimators of σ^2. When the null hypothesis is false, the numerator tends to be larger than σ^2. When we add the normality assumption, we have that $F \sim F(a - 1, N - a)$, and so we compute the P-value

$$P\left(F > \frac{\sum_{i=1}^{a} n_i (\bar{y}_i - \bar{y})^2 / (a - 1)}{s^2}\right)$$

to assess whether the observed value of F is so large as to be surprising. Note that when $a = 2$, this P-value equals the P-value obtained via the two sample t-test.

If we reject the null hypothesis of no differences among the means, then we want to see where the differences exist. For this we use inference methods based on (10.4.1). Of course, we have to worry about the problem of *multiple comparisons*, as discussed in Section 9.3. Recall that this problem arises whenever we are testing many null hypotheses using a specific critical value, such as 5%, as a cut-off for a P-value, to decide whether or not a difference exists. The cut-off value for an individual P-value is referred to as the *individual error rate*. In effect, even if no differences exist, the probability of concluding that at least one difference exists, the *family error rate*, can be quite high. There are a number of procedures designed to control the family error rate when making multiple comparisons. The simplest is to lower the individual error rate, as the family error rate is typically an increasing function of this quantity. This is the approach we adopt here, and we rely on statistical software to compute and report the family error rate for us. We refer to this procedure as *Fisher's multiple comparison test*.

To check the model, we look at the standardized residuals (Problem 10.4.12) given by

$$\frac{y_{ij} - \bar{y}_i}{s\sqrt{1 - \frac{1}{n_i}}}. \tag{10.4.2}$$

We will restrict our attention to various plots of the standardized residuals for model checking.

Example 10.4.1

A study was undertaken to determine whether or not eight different types of fat are absorbed in different amounts during the cooking of donuts. The results of cooking six different donuts and then measuring the amount of fat in grams absorbed were collected. We take the variable X to be the type of fat and use the model of this section.

The collected data are presented in the following table.

Fat 1	164	177	168	156	172	195
Fat 2	172	197	167	161	180	190
Fat 3	177	184	187	169	179	197
Fat 4	178	196	177	181	184	191
Fat 5	163	177	144	165	166	178
Fat 6	163	193	176	172	176	178
Fat 7	150	179	146	141	169	183
Fat 8	164	169	155	149	170	167

A normal probability plot of the standardized residuals is provided in Figure 10.4.1. A plot of the standardized residuals against type of fat is provided in Figure 10.4.2. Neither plot gives us great grounds for concern over the validity of the model, although there is some indication of a difference in the variability of the response as the type of fat changes. Another useful plot in this situation is a side-by-side boxplot, as it shows graphically where potential differences may lie. Such a plot is provided in Figure 10.4.3.

The following table gives the mean amounts of each fat absorbed.

Fat 1	Fat 2	Fat 3	Fat 4	Fat 5	Fat 6	Fat 7	Fat 8
172.00	177.83	182.17	184.50	165.50	176.33	161.33	162.33

The grand mean response is given by 172.8.

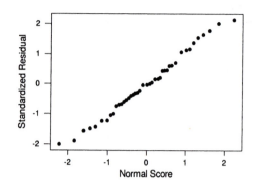

Figure 10.4.1: Normal probability plot of the standardized residuals in Example 10.4.1.

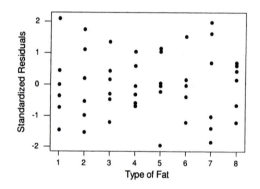

Figure 10.4.2: Standardized residuals versus type of fat in Example 10.4.1.

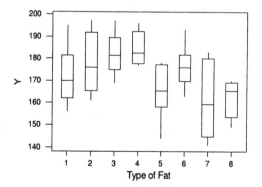

Figure 10.4.3: Side-by-side boxplots of the response versus type of fat in Example 10.4.1.

To assess the null hypothesis of no differences among the types of fat, we calculate the following ANOVA table.

Source	Df	Sum of Squares	Mean Square
X	7	3344	478
Error	40	5799	145
Total	47	9143	

Then we use the F-statistic given by $F = 478/145 = 3.3$. Because $F \sim F(7, 40)$ under H_0, we obtain the P-value $P(F > 3.3) = 0.007$. Therefore, we conclude that there is a difference among the fat types at the 0.05 level.

To ascertain where the differences exist, we look at all pairwise differences. There are $8 \cdot 7/2 = 28$ such comparisons. If we use the 0.05 level to determine whether or not a difference among means exists, then software computes the family error rate as 0.481, and this seems uncomfortably high. When we use the 0.01 level, the family error rate falls to 0.151. With the individual error rate at 0.003 the family error rate is 0.0546. Using the individual error rate of 0.003, the only differences detected among the means are those between Fat 4 and Fat 7, and Fat 4 and Fat 8. Note that Fat 4 has the highest absorption while Fats 7 and 8 have the lowest absorptions.

Overall, the results are somewhat inconclusive, as we see some evidence of differences existing, but we are left with some anomalies as well. For example, Fats 4 and 5 are not different and neither are Fats 7 and 5, while Fats 4 and 5 are deemed to be different. To resolve such conflicts requires either larger sample sizes, or a more refined experiment so that the comparisons are more accurate. ∎

10.4.2 Repeated Measures (Paired Comparisons)

Consider k quantitative variables $Y_1, \ldots, Y_k$ defined on a population Π. Suppose that our purpose is to compare the distributions of these variables. Typically, these will be similar variables, all measured in the same units.

Example 10.4.2
Suppose that Π is a set of students enrolled in a first-year program that requires students take both calculus and physics, and we want to compare the marks achieved in these subjects. If we let Y_1 denote the calculus grade and Y_2 denote the physics grade then we want to compare the distributions of these variables. ∎

Example 10.4.3
Suppose we want to compare the distributions of the duration of headaches for two treatments (A and B) in a population of migraine headache sufferers. We let Y_1 denote the duration of a headache after being administered treatment A, and let Y_2 denote the duration of a headache after being administered treatment B. ∎

The *repeated-measures* approach to this problem involves taking a random sample $\pi_1, \ldots, \pi_n$ from Π and, for each π_i, obtaining the k-dimensional value

$(Y_1(\pi_i),\ldots,Y_k(\pi_i)) = (y_{i1},\ldots,y_{ik})$. This gives a sample of n from a k-dimensional distribution. Obviously, this is called repeated measures because we are taking the measurements $Y_1(\pi_i),\ldots,Y_k(\pi_i)$ on the same π_i.

An alternative to repeated measures is to take k independent samples from Π, and, for each of these samples, obtain the values of one and only one of the variables Y_i. There is an important reason why the repeated-measures approach is preferred: We expect less variation in the values of differences, like $Y_i - Y_j$, under repeated measures sampling, than we do under independent sampling because the values $Y_1(\pi),\ldots,Y_k(\pi)$ are being taken on the *same* member of the population in repeated measures.

To see this more clearly, suppose that all of the variances and covariances exist for the joint distribution of $Y_1,\ldots,Y_k$. This implies that

$$\text{Var}(Y_i - Y_j) = \text{Var}(Y_i) + \text{Var}(Y_j) - 2\,\text{Cov}(Y_i, Y_j). \qquad (10.4.3)$$

Because Y_i and Y_j are similar variables, being measured on the same individual, we expect them to be positively correlated. Now with independent sampling, we have that $\text{Var}(Y_i - Y_j) = \text{Var}(Y_i) + \text{Var}(Y_i)$, so the variances of differences should be smaller with repeated measures than with independent sampling.

When we assume that the distributions of the Y_i differ at most in their means, then it makes sense to make inferences about the differences of the population means $\mu_i - \mu_j$, using the differences of the sample means $\bar{y}_i - \bar{y}_j$. In the repeated measures context, we can write

$$\bar{y}_i - \bar{y}_j = \frac{1}{n}\sum_{l=1}^{n}(y_{li} - y_{lj}).$$

Because the individual components of this sum are independent

$$\text{Var}(\bar{Y}_i - \bar{Y}_j) = \frac{\text{Var}(Y_i) + \text{Var}(Y_j) - 2\,\text{Cov}(Y_i, Y_j)}{n}.$$

We can consider the differences $d_1 = y_{1i} - y_{1j},\ldots, d_n = y_{ni} - y_{nj}$ to be a sample of n from a one-dimensional distribution with mean $\mu_i - \mu_j$ and variance σ^2 given by (10.4.3). Accordingly, we estimate $\mu_i - \mu_j$ by $\bar{d} = \bar{y}_i - \bar{y}_j$ and estimate σ^2 by

$$s^2 = \frac{1}{n-1}\sum_{i=1}^{n}(d_i - \bar{d})^2. \qquad (10.4.4)$$

If we assume that the joint distribution of $Y_1,\ldots,Y_k$ is multivariate normal (this means that any linear combination of these variables is normally distributed — see Problem 9.1.11), then this forces the distribution of $Y_i - Y_j$ to be $N(\mu_i - \mu_j, \sigma^2)$. Accordingly, we have all the univariate techniques discussed in Chapter 6 for inferences about $\mu_i - \mu_j$.

The discussion so far has been about whether the distributions of variables differed. Assuming these distributions differ at most in their means, this leads to a comparison of the means. We can, however, record an observation as (X, Y), where X takes values in $\{1, \ldots, k\}$ and $X = i$ means that $Y = Y_i$. Then the conditional distribution of Y given $X = i$ is the same as the distribution of Y_i. Therefore, if we conclude that the distributions of the Y_i are different, we can conclude that a relationship exists between Y and X. In Example 10.4.2, this means that a relationship exists between a student's grade and whether or not the grade was in calculus or physics. In Example 10.4.3 this means that a relationship exists between length of a headache and the treatment.

When can we assert that such a relationship is in fact a cause–effect relationship? Applying the discussion in Section 10.1.2, we know that we have to be able to assign the value of X to a randomly selected element of the population. In Example 10.4.2 we see that this is impossible, and so we cannot assert that such a relationship is a cause–effect relationship. In Example 10.4.3, however, we can indeed do this — namely, for a randomly selected individual, we randomly assign a treatment to the first headache experienced during the study period and then apply the other treatment to the second headache experienced during the study period.

A full discussion of repeated measures requires more advanced concepts in statistics. We restrict our attention now to the presentation of an example when $k = 2$. This is commonly referred to as *paired comparisons*.

Example 10.4.4 *Blood Pressure Study*
The following table came from a study of the effect of the drug captopril on blood pressure, as reported in *Applied Statistics, Principles and Examples* by D. R. Cox and E. J. Snell (Chapman and Hall, London, 1981). Each measurement is the difference in the systolic blood pressure before and after having been administered the drug.

−9	−4	−21	−3	−20
−31	−17	−26	−26	−10
−23	−33	−19	−19	−23

Figure 10.4.4 is a normal probability plot for these data and, because this looks reasonable, we conclude that the inference methods based on the assumption of normality are acceptable. Note that here we have not standardized the variable first, so we are only looking to see if the plot is reasonably straight.

The mean difference is given by $\bar{d} = -18.93$ with standard deviation $s = 9.03$. Accordingly, the standard error of the estimate of the difference in the means, using (10.4.4), is given by $s/\sqrt{15} = 2.33$. A 0.95-confidence interval for the difference in the mean systolic blood pressure, before and after being administered captopril, is then

$$\bar{d} \pm \frac{s}{\sqrt{n}} t_{0.975}(n-1) = -18.93 \pm 2.33\, t_{0.975}(14) = (-23.93, -13.93).$$

Because this does not include 0, we reject the null hypothesis of no difference in the means at the 0.05 level. The actual P-value for the two-sided test is given by

$$P\left(|T| > \left|\frac{-18.93}{2.33}\right|\right) = 0.000,$$

because $T \sim t(14)$ under the null hypothesis H_0 that the means are equal. Therefore we have strong evidence against H_0. It seems that we have strong evidence that the drug is leading to a drop in blood pressure. ∎

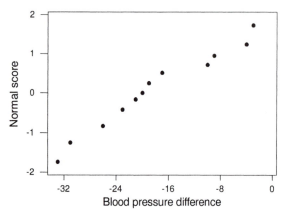

Figure 10.4.4: Normal probability plot for the data in Example 10.4.4.

10.4.3   Two Categorical Predictors (Two-Way ANOVA)

Now suppose that we have a single quantitative response Y and two categorical predictors A and B, where A takes a levels and B takes b levels. One possibility is to consider running two one-factor studies. One study will examine the relationship between Y and A, and the second study will examine the relationship between Y and B. There are several disadvantages to such an approach, however.

First, and perhaps foremost, doing two separate analyses will not allow us to determine the joint relationship A and B have with Y. This relates directly to the concept of *interaction* between predictors. We will define this concept more precisely shortly, but basically, if A and B interact, then the conditional relationship between Y and A, given $B = j$, changes in some substantive way as we change j. If the predictors A and B do not interact, then indeed we will be able to examine the relationship between the response and each of the predictors separately. But we almost never know that this is the case beforehand and must assess whether or not an interaction exists based on collected data.

A second reason for including both predictors in the analysis is that this will often lead to a reduction in the contribution of random error to the results. By this we mean that we will be able to explain some of the observed variation in Y by the inclusion of the second variable in the model. This depends, however, on the additional variable having a relationship with the response. Further, for the inclusion of a second variable to be worthwhile, this relationship must be strong enough to justify the loss in degrees of freedom available for the estimation of the contribution of random error to the experimental results. As we will see, including the second variable in the analysis results in a reduction in the degrees of freedom in the Error row of the ANOVA table. Degrees of freedom are playing the role of sample size here. The fewer the degrees of freedom in the Error row, the less accurate our estimate of σ^2 will be.

When we include both predictors in our analysis, and we have the opportunity to determine the sampling process, it is important that we *cross* the predictors. By this we mean that we observe Y at each combination

$$(A, B) = (i, j) \in \{1, \ldots, a\} \times \{1, \ldots, b\} \,.$$

Suppose then, that we have n_{ij} response values at the $(A, B) = (i, j)$ setting of the predictors. Then, letting

$$E\left(Y \mid (A, B) = (i, j)\right) = \beta_{ij}$$

be the mean response when $A = i$ and $B = j$, and introducing the dummy variables

$$X_{ij} = \begin{cases} 1 & A = i, B = j \\ 0 & A \neq i \text{ or } B \neq j, \end{cases}$$

we can write

$$\begin{aligned} E\left(Y \mid X_{ij} = x_{ij} \text{ for all } i, j\right) &= \beta_{11}x_{11} + \beta_{21}x_{21} + \cdots + \beta_{ab}x_{ab} \\ &= \sum_{i=1}^{a} \sum_{j=1}^{b} \beta_{ij}x_{ij}. \end{aligned}$$

The relationship between Y and the predictors is completely encompassed in the changes in the β_{ij} as i and j change. From this we can see that a regression model for this situation is immediately a linear regression model.

Now let y_{ijk} denote the kth response value when $X_{ij} = 1$. Then, as in Section 10.4.1, the least-squares estimate of β_{ij} is given by

$$b_{ij} = \bar{y}_{ij} = \frac{1}{n_{ij}} \sum_{k=1}^{n_{ij}} y_{ijk},$$

the mean of the observations when $X_{ij} = 1$. If in addition we assume that the conditional distributions of Y given the predictors all have variance equal to σ^2, then with $N = n_{11} + n_{21} + \cdots + n_{ab}$, we have that

$$s^2 = \frac{1}{N - ab} \sum_{i=1}^{a} \sum_{j=1}^{b} \sum_{k=1}^{n_{ij}} (y_{ijk} - \bar{y}_{ij})^2 \tag{10.4.5}$$

is an unbiased estimator of σ^2. Therefore, using (10.4.5), the standard error of $\bar{y}_{ij}$ is given by $s/\sqrt{n_{ij}}$.

With the normality assumption, we have that $\bar{Y}_{ij} \sim N\left(\beta_{ij}, \sigma^2/n_{ij}\right)$, independent of

$$\frac{(N - ab) S^2}{\sigma^2} \sim \chi^2 (N - ab) \,.$$

This leads to the γ-confidence intervals

$$\bar{y}_{ij} \pm \frac{s}{\sqrt{n_{ij}}} t_{(1+\gamma)/2} \left(N - ab\right)$$

for β_{ij} and

$$\bar{y}_{ij} - \bar{y}_{kl} \pm s\sqrt{\frac{1}{n_{ij}} + \frac{1}{n_{kl}}}\, t_{(1+\gamma)/2} \left(N - ab\right)$$

for the difference of means $\beta_{ij} - \beta_{kl}$.

We are interested in whether or not there is any relationship between Y and the predictors. There is no relationship between the response and the predictors if and only if all the β_{ij} are equal. Before testing this, however, it is customary to test the null hypothesis that there is no interaction between the predictors. The precise definition of no interaction here is that

$$\beta_{ij} = \mu_i + \upsilon_j$$

for all i and j for some constants μ_i and υ_j, i.e., the means can be expressed additively. Note that if we fix $B = j$ and let A vary, then these *response curves* (a response curve is a plot of the means of one variable while holding the value of the second variable fixed) are all parallel. This is an equivalent way of saying that there is no interaction between the predictors.

In Figure 10.4.5 we have depicted response curves in which the factors do not interact, and in Figure 10.4.6 we have depicted response curves in which they do. Note that the solid lines, for example, joining β_{11} and β_{21}, are there just to make it easier to display the parallelism (or lack thereof) and have no other significance.

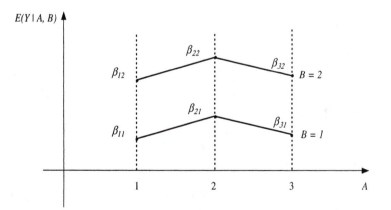

Figure 10.4.5: Response curves for expected response with two predictors, with A taking three levels and B taking two levels. The response curves are parallel, so the predictors do not interact.

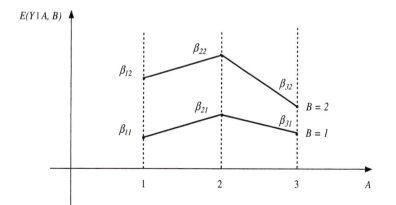

Figure 10.4.6: Response curves for expected response with two predictors, with A taking three levels and B taking two levels. The response curves are not parallel, so the predictors do interact.

To test the null hypothesis of no interaction, we must first fit the model where $\beta_{ij} = \mu_i + v_j$, i.e., find the least squares estimates of the β_{ij} under these constraints. We will not pursue the mathematics of obtaining these estimates here, but rely on software to do this for us and to compute the sum of squares relevant for testing the null hypothesis of no interaction (from the results of Section 10.3.4, we know that this is obtained by differencing the regression sum of squares obtained from the full model and the regression sums of squares obtained from the model with no interaction).

If we decide that an interaction exists, then it is immediate that both A and B have an effect on Y (if A does not have an effect, then A and B cannot interact — see Problem 10.4.11), and we must look at differences among the $\bar{y}_{ij}$ to determine the form of the relationship. If we decide that no interaction exists, then A has an effect if and only if the μ_i vary, and B has an effect if and only if the v_j vary. We can test the null hypothesis $H_0 : \mu_1 = \cdots = \mu_a$ of no effect due to A and the null hypothesis $H_0 : v_1 = \cdots = v_b$ of no effect due to V separately, once we have decided that no interaction exists.

The details for deriving the relevant sums of squares for all these hypotheses are not covered here, but many statistical packages will produce an ANOVA table, as given below.

Source	Df	Sum of Squares
A	$a-1$	$RSS(A)$
B	$b-1$	$RSS(B)$
$A \times B$	$(a-1)(b-1)$	$RSS(A \times B)$
Error	$N - ab$	$\sum_{i=1}^{a} \sum_{j=1}^{b} \sum_{k=1}^{n_{ij}} \left(y_{ijk} - \bar{y}_{ij} \right)^2$
Total	$N - 1$	$\sum_{i=1}^{a} \sum_{j=1}^{b} \sum_{k=1}^{n_{ij}} \left(y_{ijk} - \bar{y} \right)^2$

Note that if we had included only A in the model, then there would be $N - a$ degrees of freedom for the estimation of σ^2. By including B, we lose $(N - a) - (N - ab) = a(b - 1)$ degrees of freedom for the estimation of σ^2.

Using this table, we first assess the null hypothesis H_0 : no interaction between A and B, using $F \sim F((a - 1)(b - 1), N - ab)$ under H_0, via the P-value

$$P\left(F > \frac{RSS(A \times B)/(a - 1)(b - 1)}{s^2}\right)$$

where s^2 is given by (10.4.5). If we decide that no interaction exists, then we assess the null hypothesis H_0 : no effect due to A, using $F \sim F(a - 1, N - ab)$ under H_0, via the P-value

$$P\left(F > \frac{RSS(A)/(a - 1)}{s^2}\right),$$

and assess H_0 : no effect due to B, using $F \sim F(b - 1, N - ab)$ under H_0, via the P-value

$$P\left(F > \frac{RSS(B)/(b - 1)}{s^2}\right).$$

To check the model, we look at the standardized residuals given by (Problem 10.4.13)

$$\frac{y_{ijk} - \bar{y}_{ij}}{s\sqrt{1 - \frac{1}{n_{ij}}}}. \tag{10.4.6}$$

We will restrict our attention to various plots of the standardized residuals for model checking.

We consider an example of a two factor analysis.

Example 10.4.5
The data in the following table come from G. E. P. Box and D. R. Cox, "An analysis of transformations" (*Journal of the Royal Statistical Society*, Series B, 1964, p. 211) and represent survival times, in hours, of animals exposed to one of three different types of poisons and allocated four different types of treatments. We let A denote the treatments and B denote the type of poison, so we have $3 \times 4 = 12$ different (A, B) combinations. Each combination was administered to four different animals; i.e., $n_{ij} = 4$ for every i and j.

	A1	A2	A3	A4
B1	$3.1, 4.5, 4.6, 4.3$	$8.2, 11.0, 8.8, 7.2$	$4.3, 4.5, 6.3, 7.5$	$4.5, 7.1, 6.6, 6.2$
B2	$3.6, 2.9, 4.0, 2.3$	$9.2, 6.1, 4.9, 12.4$	$4.4, 3.5, 3.1, 4.0$	$5.6, 10.2, 7.1, 3.8$
B3	$2.2, 2.1, 1.8, 2.3$	$3.0, 3.7, 3.8, 2.9$	$2.3, 2.5, 2.4, 2.2$	$3.0, 3.6, 3.1, 3.3$

A normal probability plot for these data, using the standardized residuals after fitting the two-factor model, reveals a definite problem. In the above reference, a transformation of the response to the reciprocal $1/Y$ is suggested, based on a more sophisticated analysis, and this indeed leads to much more appropriate standardized residual plots. Figure 10.4.7 is a normal probability plot for the standardized residuals based on the reciprocal response. This normal probability plot looks reasonable.

Figure 10.4.8 is a plot of the standardized residuals against the various (A, B) combinations, where we have coded the combination (i, j) as $b(i - 1) + j$ with $b = 3, i = 1, 2, 3, 4$, and $j = 1, 2, 3$. This coding assigns a unique integer to each combination (i, j) and is convenient when we want to compare scatter plots of the response for each treatment. Again, this residual plot looks reasonable.

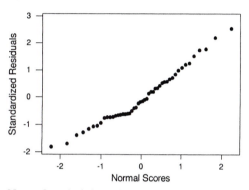

Figure 10.4.7: Normal probability plot of the standardized residuals in Example 10.4.5 using the reciprocal of the response.

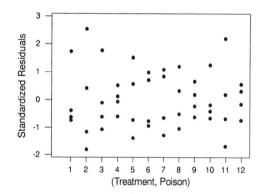

Figure 10.4.8: Scatter plot for the data in Example 10.4.4 of the standardized residuals against each value of (A, B) using the reciprocal of the response.

Below we provide the least-squares estimates of the β_{ij} for the transformed model.

	A1	A2	A3	A4
B1	0.24869	0.11635	0.18627	0.16897
B2	0.32685	0.13934	0.27139	0.17015
B3	0.48027	0.30290	0.42650	0.30918

The ANOVA table for the data, as obtained from a standard statistical package, is given below.

Source	Df	Sum of Squares	Mean Square
A	3	0.20414	0.06805
B	2	0.34877	0.17439
$A \times B$	6	0.01571	0.00262
Error	36	0.08643	0.00240
Total	47	0.65505	

From this we determine that $s = \sqrt{0.00240} = 4.89898 \times 10^{-2}$, and so the standard errors of the least-squares estimates are all equal to $s/2 = 0.0244949$.

To test the null hypothesis of no interaction between A and B, we have, using $F \sim F(6, 36)$ under H_0, the P-value

$$P\left(F > \frac{0.00262}{0.00240}\right) = P(F > 1.09) = 0.387.$$

We have no evidence against the null hypothesis.

So we can go on to test the null hypothesis of no effect due to A and we have, using $F \sim F(2, 36)$ under H_0, the P-value

$$P\left(F > \frac{0.06805}{0.00240}\right) = P(F > 28.35) = 0.000.$$

We reject this null hypothesis.

Similarly, testing the null hypothesis of no effect due to B, we have, using $F \sim F(2, 36)$ under H_0, the P-value

$$P\left(F(2, 36) > \frac{0.17439}{0.00240}\right) = P(F(2, 36) > 72.66) = 0.000.$$

We reject this null hypothesis as well.

Accordingly, we have decided that the appropriate model is the *additive model* given by $E(1/Y \mid (A, B) = (i, j)) = \mu_i + \upsilon_j$ (we are still using the transformed response $1/Y$). We can also write this as $E(1/Y \mid (A, B) = (i, j)) = (\mu_i + \alpha) + (\upsilon_j - \alpha)$ for any choice of α. Therefore, there is no unique estimate of the additive effects due to A or B. We still have, however, unique least-squares estimates of the means, and these are obtained (using software) by fitting the model with constraints on the β_{ij} corresponding to no interaction existing. These are recorded in the following table.

	A1	A2	A3	A4
B1	0.26977	0.10403	0.21255	0.13393
B2	0.31663	0.15089	0.25942	0.18080
B3	0.46941	0.30367	0.41219	0.33357

As we have decided that there is no interaction between A and B, we can assess single-factor effects by examining the response means for each factor separately. For example, the means for investigating the effect of A are given in the following table.

A1	A2	A3	A4
0.352	0.186	0.295	0.216

We can compare these means using procedures based on the t-distribution. For example, a 0.95-confidence interval for the difference in the means at levels $A1$ and $A2$ is given by

$$\bar{y}_1. - \bar{y}_2. \pm \frac{s}{\sqrt{12}} t_{0.975}(36) = (0.352 - 0.186) \pm \sqrt{\frac{0.00240}{12}} \, 2.0281$$
$$= (0.13732, 0.19468). \qquad (10.4.7)$$

This indicates that we would reject the null hypothesis of no difference between these means at the 0.05 level.

Notice that we have used the estimate of σ^2 based on the full model in (10.4.7). Logically, it would seem to make more sense to use the estimate based on fitting the additive model, because we have decided that it is appropriate. When we do so, this is referred to as *pooling*, as it can be shown that the new error estimate is calculated by adding $RSS(A \times B)$ to the original ESS and dividing by the sum of the $A \times B$ degrees of freedom and the error degrees of freedom. Not to pool is regarded as a somewhat more conservative procedure. ∎

10.4.4 Randomized Blocks

With two-factor models, we generally want to investigate whether or not both of these factors have a relationship with the response Y. Suppose, however, that we know that a factor B has a relationship with Y, and we are interested in investigating whether or not another factor A has a relationship with Y. Should we run a single-factor experiment using the predictor A, or run a two-factor experiment including the factor B?

The answer is as we have stated at the start of Section 10.4.2. Including the factor B will allow us, if B accounts for a lot of the observed variation, to make more accurate comparisons. Notice, however, that if B does not have a substantial effect on Y, then its inclusion will be a waste, as we sacrificed $a(b-1)$ degrees of freedom that would otherwise go toward the estimation of σ^2.

So it is important that we do indeed *know* that B has a substantial effect. In such a case, we refer to B as a *blocking variable*. It is important again that the blocking variable B be crossed with A. Then we can test for any effect due to A by first testing for an interaction between A and B and, if no such interaction is found, then test for an effect due to A alone, just as we have discussed in Section 10.4.3.

A special case of using a blocking variable arises when we have $n_{ij} = 1$ for all i and j. In this case, $N = ab$, so there are no degrees of freedom available for the estimation of error. In fact, we have that (Problem 10.4.14) $s^2 = 0$. Still, such a design has practical value, provided we are willing to *assume* that there is no interaction between A and B. This is called a *randomized block design*.

For a randomized block design, we have that

$$s^2 = \frac{RSS\,(A \times B)}{(a-1)\,(b-1)} \tag{10.4.8}$$

is an unbiased estimate of σ^2, and so we have $(a-1)\,(b-1)$ degrees of freedom for the estimation of error. Of course, this will not be correct if A and B do interact, but when they do not this can be a highly efficient design, as we have removed the effect of the variation due to B and require only ab observations for this. When the randomized block design is appropriate, we test for an effect due to A, using $F \sim F\,(a-1, (a-1)\,(b-1))$ under H_0, via the P-value

$$P\left(F\,(a-1, (a-1)\,(b-1)) > \frac{RSS\,(A)\,/(a-1)}{s^2}\right).$$

10.4.5 One Categorical and One Quantitative Predictor

It is also possible that the response is quantitative while some of the predictors are categorical and some are quantitative. We now consider the situation where we have one categorical predictor A, taking a values, and one quantitative predictor W. We assume that the regression model applies. Further, we will restrict our attention to the situation where we suppose that, within each level of A, the mean response varies as

$$E\,(Y \mid (A, W) = (i, w)) = \beta_{i1} + \beta_{i2}w,$$

so that we have a linear regression model.

If we introduce the dummy variables

$$X_{ij} = \begin{cases} W^{j-1} & A = i \\ 0 & A \neq i \end{cases}$$

for $i = 1, \ldots, a$ and $j = 1, 2$, then we can write the linear regression model as

$$E\,(Y \mid (X_{ij}) = (x_{ij})) = (\beta_{11}x_{11} + \beta_{12}x_{12}) + \cdots + (\beta_{a1}x_{a1} + \beta_{a2}x_{a2}).$$

Here β_{i1} is the intercept and β_{i2} is the slope specifying the relationship between Y and W when $A = i$. The methods of Section 10.3.4 are then available for inference about this model.

We also have a notion of interaction in this context, as we say that the two predictors interact if the slopes of the lines vary across the levels of A. So saying that no interaction exists is the same as saying that the response curves are parallel when graphed for each level of A. If an interaction exists, then it is definite that both A and W have an effect on Y. So the null hypothesis that no interaction exists is equivalent to $H_0 : \beta_{12} = \cdots = \beta_{a2}$.

If we decide that no interaction exists, then we can test for no effect due to W by testing the null hypothesis that the common slope is equal to 0, or we can test the null hypothesis that there is no effect due to A by testing $H_0 : \beta_{11} = \cdots = \beta_{a1}$, i.e., that the intercept terms are the same across the levels of A.

We do not pursue the analysis of this model further here. Statistical software is available, however, that will calculate the relevant ANOVA table for assessing the various null hypotheses.

Suppose that we are running an experimental design and for each experimental unit we can measure, but not control, a quantitative variable W that we believe has an effect on the response Y. If the effect of this variable is appreciable, then good statistical practice suggests that we should include this variable in the model, as we will reduce the contribution of error to our experimental results and thus make more accurate comparisons. Of course, we pay a price when we do this, as we lose degrees of freedom that would otherwise be available for the estimation of error. So we must be sure that W does have a significant effect in such a case. Also, we do not test for an effect of such a variable, as we presumably know it has an effect. This technique is referred to as the *analysis of covariance* and is obviously similar in nature to the use of blocking variables.

Summary of Section 10.4

- We considered the situation involving a quantitative response and categorical predictor variables.

- By the introduction of dummy variables for the predictor variables, we can consider this situation as a particular application of the multiple regression model of Section 10.3.4.

- If we decide that a relationship exists, then we typically try to explain what form this relationship takes by comparing means. To prevent finding too many statistically significant differences, we lower the individual error rate to ensure a sensible family error rate.

- When we have two predictors, we first check to see if the factors interact. If the two predictors interact, then both have an effect on the response.

- A special case of a two-way analysis arises when one of the predictors serves as a blocking variable. It is generally important to know that the blocking variable has an effect on the response so that we do not waste degrees of freedom by including it.

- Sometimes we can measure variables on individual experimental units that we know have an effect on the response. In such a case, we include these variables in our model, as they will reduce the contribution of random error to the analysis and make our inferences more accurate.

Exercises

10.4.1 The following values of a response Y were obtained for three settings of a categorical predictor A.

$A = 1$	2.9976	0.3606	4.7716	1.5652
$A = 2$	0.7468	1.3308	2.2167	−0.3184
$A = 3$	2.1192	2.3739	0.3335	3.3015

Suppose we assume the normal regression model for these data with one categorical predictor.
(a) Produce a side-by-side boxplot for the data.
(b) Plot the standardized residuals against A (if you are using a computer for your calculations, also produce a normal probability plot of the standardized residuals). Does this give you grounds for concern that the model assumptions are incorrect?
(c) Carry out a one-way ANOVA to test for any difference among the conditional means of Y given A.
(d) If warranted, construct 0.95-confidence intervals for the differences between the means and summarize your findings.

10.4.2 The following values of a response Y were obtained for three settings of a categorical predictor A.

$A = 1$	0.090	0.800	33.070	−1.890
$A = 2$	5.120	1.580	1.760	1.740
$A = 3$	5.080	−3.510	4.420	1.19

Suppose we assume the normal regression model for these data with one categorical predictor.
(a) Produce a side-by-side boxplot for the data.
(b) Plot the standardized residuals against A (if you are using a computer for your calculations, also produce a normal probability plot of the standardized residuals). Does this give you grounds for concern that the model assumptions are incorrect?
(c) If concerns arise about the validity of the model, can you "fix" the problem?
(d) If you have been able to fix any problems encountered with the model, carry out a one-way ANOVA to test for any differences among the conditional means of Y given A.
(e) If warranted, construct 0.95-confidence intervals for the differences between the means and summarize your findings.

10.4.3 The following table gives the percentage moisture content of two different types of cheeses determined by randomly sampling batches of cheese from the production process.

Cheese 1	39.02, 38.79, 35.74, 35.41, 37.02, 36.00
Cheese 2	38.96, 39.01, 35.58, 35.52, 35.70, 36.04

Suppose we assume the normal regression model for these data with one categorical predictor.

(a) Produce a side-by-side boxplot for the data.

(b) Plot the standardized residuals against Cheese (if you are using a computer for your calculations, also produce a normal probability plot of the standardized residuals). Does this give you grounds for concern that the model assumptions are incorrect?

(c) Carry out a one-way ANOVA to test for any differences among the conditional means of Y given Cheese. Note that this is the same as a t-test for the difference in the means.

10.4.4 In an experiment, rats were fed a stock ration for 100 days with various amounts of gossypol added. The following weight gains in grams were recorded.

0.00% Gossypol	228, 229, 218, 216, 224, 208, 235, 229, 233, 219, 224, 220, 232, 200, 208, 232
0.04% Gossypol	186, 229, 220, 208, 228, 198, 222, 273, 216, 198, 213
0.07% Gossypol	179, 193, 183, 180, 143, 204, 114, 188, 178, 134, 208, 196
0.10% Gossypol	130, 87, 135, 116, 118, 165, 151, 59, 126, 64, 78, 94, 150, 160, 122, 110, 178
0.13% Gossypol	154, 130, 118, 118, 118, 104, 112, 134, 98, 100, 104

Suppose we assume the normal regression model for these data and treat gossypol as a categorical predictor taking five levels.

(a) Create a side-by-side boxplot graph for the data. Does this give you any reason to be concerned about the assumptions that underlie an analysis based on the normal regression model?

(b) Produce a plot of the standardized residuals against the factor gossypol (if you are using a computer for your calculations, also produce a normal probability plot of the standardized residuals). What are your conclusions?

(c) Carry out a one-way ANOVA to test for any differences among the mean responses for the different amounts of gossypol.

(d) Compute 0.95-confidence intervals for all the pairwise differences of means and summarize your conclusions.

10.4.5 In an investigation into the effect of deficiencies of trace elements on a variable Y measured on sheep, the data in the following table were obtained.

Control	13.2, 13.6, 11.9, 13.0, 14.5, 13.4
Cobalt	11.9, 12.2, 13.9, 12.8, 12.7, 12.9
Copper	14.2, 14.0, 15.1, 14.9, 13.7, 15.8
Cobalt + Copper	15.0, 15.6, 14.5, 15.8, 13.9, 14.4

Suppose we assume the normal regression model for these data with one categorical predictor.

(a) Produce a side-by-side boxplot for the data.

(b) Plot the standardized residuals against the predictor (if you are using a computer for your calculations, also produce a normal probability plot of the standardized residuals). Does this give you grounds for concern that the model assumptions are incorrect?

(c) Carry out a one-way ANOVA to test for any differences among the conditional means of Y given the predictor.

(d) If warranted, construct 0.95-confidence intervals for all the pairwise differences between the means and summarize your findings.

10.4.6 Two diets were given to samples of pigs over a period of time, and the following weight gains (in lbs.) were recorded.

Diet A	$8, 4, 14, 15, 11, 10, 6, 12, 13, 7$
Diet B	$7, 13, 22, 15, 12, 14, 18, 8, 21, 23, 10, 17$

Suppose we assume the normal regression model for these data.

(a) Produce a side-by-side boxplot for the data.

(b) Plot the standardized residuals against Diet. Also produce a normal probability plot of the standardized residuals. Does this give you grounds for concern that the model assumptions are incorrect?

(c) Carry out a one-way ANOVA to test for a difference between the conditional means of Y given Diet.

(d) Construct 0.95-confidence intervals for differences between the means.

10.4.7 Ten students were randomly selected from the students in a university who took first-year calculus and first-year statistics. Their grades in these courses are recorded in the following table.

Student	1	2	3	4	5	6	7	8	9	10
Calculus	66	61	77	62	66	68	64	75	59	71
Statistics	66	63	79	63	67	70	71	80	63	74

Suppose we assume the normal regression model for these data.

(a) Produce a side-by-side boxplot for the data.

(b) Treating the calculus and statistics marks as separate samples, carry out a one-way ANOVA to test for any difference between the mean mark in calculus and the mean mark in statistics. Produce the appropriate plots to check for model assumptions.

(c) Now take into account that each student has a calculus mark and a statistics mark and test for any difference between the mean mark in calculus and the mean mark in statistics. Produce the appropriate plots to check for model assumptions. Compare your results with those obtained in part (b).

(d) Estimate the correlation between the calculus and statistics marks.

10.4.8 The following data were recorded in *Statistical Methods* by G. Snedecor and W. Cochran (Iowa State University Press, Ames, 1967) and represent the average number of florets observed on plants in seven plots. Each of the plants was either planted with high corms or low corms (a type of underground stem).

	Plot 1	Plot 2	Plot 3	Plot 4	Plot 5	Plot 6	Plot 7
Corm High	11.2	13.3	12.8	13.7	12.2	11.9	12.1
Corm Low	14.6	12.6	15.0	15.6	12.7	12.0	13.1

Suppose we assume the normal regression model for these data.
(a) Produce a side-by-side boxplot for the data.
(b) Treating the Corm High and Corm Low measurements as separate samples, carry out a one-way ANOVA to test for any difference between the population means. Produce the appropriate plots to check for model assumptions.
(c) Now take into account that each plot has a Corm High and Corm Low measurement. Compare your results with those obtained in part (b). Produce the appropriate plots to check for model assumptions.
(d) Estimate the correlation between the calculus and statistics marks.

Problems

10.4.9 Prove that $\sum_{i=}^{a} \sum_{j=1}^{n_i} (y_{ij} - \beta_i)^2$ is minimized as a function of the β_i by $\beta_i = \bar{y}_i = (y_{i1} + \cdots + y_{in_i})/n_i$ for $i = 1, \ldots, a$.

10.4.10 Prove that

$$\sum_{i=}^{a} \sum_{j=1}^{n_i} (y_{ij} - \bar{y})^2 = \sum_{i=1}^{a} n_i (\bar{y}_i - \bar{y})^2 + \sum_{i=}^{a} \sum_{j=1}^{n_i} (y_{ij} - \bar{y}_i)^2$$

where $\bar{y}_i = (y_{i1} + \cdots + y_{in_i})/n_i$ and $\bar{y}$ is the grand mean.

10.4.11 Argue that if the relationship between a quantitative response Y and two categorical predictors A and B is given by a linear regression model, then A and B both have an effect on Y whenever A and B interact. (Hint: What does it mean in terms of response curves for an interaction to exist, for an effect due to A to exist?)

10.4.12 Establish that (10.4.2) is the appropriate expression for the standardized residual for the linear regression model with one categorical predictor.

10.4.13 Establish that (10.4.6) is the appropriate expression for the standardized residual for the linear regression model with two categorical predictors.

10.4.14 Establish that $s^2 = 0$ for the linear regression model with two categorical predictors when $n_{ij} = 1$ for all i and j.

10.4.15 How would you assess whether or not the randomized block design was appropriate after collecting the data?

Computer Problems

10.4.16 Use appropriate software to carry out Fisher's multiple comparison test on the data in Exercise 10.4.5 so that the family error rate is between 0.04 and 0.05. What individual error rate is required?

10.4.17 Consider the data in Exercise 10.4.3, but now suppose we also take into account that the cheeses were made in lots where each lot corresponded to a production run. Recording the data this way, we obtain the following table.

	Lot 1	Lot 2	Lot 3
Cheese 1	39.02, 38.79	35.74, 35.41	37.02, 36.00
Cheese 2	38.96, 39.01	35.58, 35.52	35.70, 36.04

Suppose we assume the normal regression model for these data with two categorical predictors.
(a) Produce a side-by-side boxplot for the data for each treatment.
(b) Produce a table of cell means.
(c) Produce a normal probability plot of the standardized residuals and a plot of the standardized residuals against each treatment combination (code the treatment combinations so there is a unique integer corresponding to each). Comment on the validity of the model.
(d) Construct the ANOVA table testing first for no interaction between A and B and, if necessary, an effect due to A and an effect due to B.
(e) Based on the results of part (d), construct the appropriate table of means, plot the corresponding response curve, and make all pairwise comparisons among the means.
(f) Compare your results with those obtained in Exercise 10.4.4 and comment on the differences.

10.4.18 A two-factor experimental design was carried out, with factors A and B both categorical variables taking three values. Each treatment was applied four times and the following response values were obtained.

	$A = 1$		$A = 2$		$A = 3$	
$B = 1$	19.86	20.88	26.37	24.38	29.72	29.64
	20.15	25.44	24.87	30.93	30.06	35.49
$B = 2$	15.35	15.86	22.82	20.98	27.12	24.27
	21.86	26.92	29.38	34.13	34.78	40.72
$B = 3$	4.01	4.48	10.34	9.38	15.64	14.03
	21.66	25.93	30.59	40.04	36.80	42.55

Suppose we assume the normal regression model for these data with two categorical predictors.
(a) Produce a side-by-side boxplot for the data for each treatment.
(b) Produce a table of cell means.
(c) Produce a normal probability plot of the standardized residuals and a plot of the standardized residuals against each treatment combination (code the treatment combinations so there is a unique integer corresponding to each). Comment on the validity of the model.
(d) Construct the ANOVA table testing first for no interaction between A and B and, if necessary, an effect due to A and an effect due to B.

(e) Based on the results of part (d), construct the appropriate table of means, plot the corresponding response curves, and make all pairwise comparisons among the means.

10.4.19 A chemical paste is made in batches and put into casks. Ten delivery batches were randomly selected for testing; then three casks were randomly selected from each delivery and the paste strength was measured twice, based on samples drawn from each sampled cask. The response was expressed as a percentage of fill strength. The collected data are given in the following table. Suppose we assume the normal regression model for these data with two categorical predictors.

	Batch 1	Batch 2	Batch 3	Batch 4	Batch 5
Cask 1	62.8, 62.6	60.0, 61.4	58.7, 57.5	57.1, 56.4	55.1, 55.1
Cask 2	60.1, 62.3	57.5, 56.9	63.9, 63.1	56.9, 58.6	54.7, 54.2
Cask 3	62.7, 63.1	61.1, 58.9	65.4, 63.7	64.7, 64.5	58.5, 57.5
	Batch 6	Batch 7	Batch 8	Batch 9	Batch 10
Cask 1	63.4, 64.9	62.5, 62.6	59.2, 59.4	54.8, 54.8	58.3, 59.3
Cask 2	59.3, 58.1	61.0, 58.7	65.2, 66.0	64.0, 64.0	59.2, 59.2
Cask 3	60.5, 60.0	56.9, 57.7	64.8, 64.1	57.7, 56.8	58.9, 56.8

(a) Produce a side-by-side boxplot for the data for each treatment.
(b) Produce a table of cell means.
(c) Produce a normal probability plot of the standardized residuals and a plot of the standardized residuals against each treatment combination (code the treatment combinations so there is a unique integer corresponding to each). Comment on the validity of the model.
(d) Construct the ANOVA table testing first for no interaction between Batch and Cask and, if necessary, no effect due to Batch and no effect due to Cask.
(e) Based on the results of part (d), construct the appropriate table of means and plot the corresponding response curves.

10.4.20 The following data arose from a randomized block design where factor B is the blocking variable and corresponds to plot of land on which cotton is planted. Each plot was divided into five subplots, and different concentrations of fertilizer were applied to each, with the response being a strength measurement of the cotton harvested. There were three blocks and five different concentrations of fertilizer. Note that there is only one observation for each block and concentration combination. Further discussion of these data can be found in *Experimental Design, 2nd ed.* by W. G. Cochran and G. M. Cox (John Wiley and Sons, New York, 1957, pp. 107–108). Suppose we assume the normal regression model with two categorical predictors.

	$B = 1$	$B = 2$	$B = 3$
$A = 36$	7.62	8.00	7.93
$A = 54$	8.14	8.15	7.87
$A = 72$	7.70	7.73	7.74
$A = 108$	7.17	7.57	7.80
$A = 144$	7.46	7.68	7.21

(a) Construct the ANOVA table for testing for no effect due to fertilizer, and which also removes the variation due to the blocking variable.

(b) Beyond the usual assumptions that we are concerned about, what additional assumption is necessary for this analysis?

(c) Actually, the factor A is a quantitative variable. If we were to take this into account by fitting a model that had the same slope for each block but possibly different intercepts, then what benefit would be gained?

(d) Carry out the analysis suggested in part (c) and assess whether or not this model makes sense for these data.

10.5 Categorical Response and Quantitative Predictors

We now consider the situation in which the response is categorical but at least some of the predictors are quantitative. The essential difficulty in this context lies with the quantitative predictors, and so we will focus on the situation in which all the predictors are quantitative. When there are also some categorical predictors these can be handled in the same way, as we can replace each categorical predictor by a set of dummy quantitative variables, as discussed in Section 10.4.5.

For reasons of simplicity, we will restrict our attention to the situation in which the response variable Y is binary valued, and we will take these values to be 0 and 1. Suppose then that there are k quantitative predictors $X_1, \ldots, X_k$. Because $Y \in \{0, 1\}$, we have

$$E(Y \mid X_1 = x_1, \ldots, X_k = x_k) = P(Y = 1 \mid X_1 = x_1, \ldots, X_k = x_k) \in [0, 1].$$

Therefore, we cannot write $E(Y \mid x_1, \ldots, x_k) = \beta_1 x_1 + \cdots + \beta_k x_k$ without placing some unnatural restrictions on the β_i to ensure that $\beta_1 x_1 + \cdots + \beta_k x_k \in [0, 1]$.

Perhaps the simplest way around this is to use a 1–1 function $l : [0, 1] \rightarrow R^1$ and write

$$l(P(Y = 1 \mid X_1 = x_1, \ldots, X_k = x_k)) = \beta_1 x_1 + \cdots + \beta_k x_k,$$

so that

$$P(Y = 1 \mid X_1 = x_1, \ldots, X_k = x_k) = l^{-1}(\beta_1 x_1 + \cdots + \beta_k x_k).$$

We refer to l as a *link function*. There are many possible choices for l. For example, it is immediate that we can take l to be any inverse cdf function for a continuous distribution.

If we take $l = \Phi^{-1}$, i.e., the inverse cumulative distribution function (cdf) of the $N(0, 1)$ distribution, then this is called the *probit link*. A more commonly used link, due to some inherent mathematical simplicities, is the *logistic link* given by

$$l(p) = \ln\left(\frac{p}{1-p}\right). \tag{10.5.1}$$

The right-hand side of (10.5.1) is referred to as the *logit* or *log odds*. The logistic link is the inverse cdf of the logistic distribution (Exercise 10.5.1). We will restrict our discussion to the logistic link hereafter.

The logistic link implies that

$$P\left(Y = 1 \mid X_1 = x_1, \ldots, X_k = x_k\right) = \frac{\exp\left\{\beta_1 x_1 + \cdots + \beta_k x_k\right\}}{1 + \exp\left\{\beta_1 x_1 + \cdots + \beta_k x_k\right\}}, \quad (10.5.2)$$

which is a relatively simple relationship. We see immediately, however, that

$$\mathrm{Var}\left(Y \mid X_1 = x_1, \ldots, X_k = x_k\right)$$
$$= P\left(Y = 1 \mid X_1 = x_1, \ldots, X_k = x_k\right)\left(1 - P\left(Y = 1 \mid X_1 = x_1, \ldots, X_k = x_k\right)\right),$$

so the variance of the conditional distribution of Y, given the predictors, depends on the values of the predictors. Therefore, these models are not, strictly speaking, regression models as we have defined them. Still when we use the link function given by (10.5.1) we refer to this as the *logistic regression model*.

Now suppose we observe n independent observations $(x_{i1}, \ldots, x_{ik}, y_i)$ for $i = 1, \ldots, n$. We then have that, given $(x_{i1}, \ldots, x_{ik})$, the response y_i is an observation from the Bernoulli($P\left(Y = 1 \mid X_1 = x_1, \ldots, X_k = x_k\right)$) distribution. Then (10.5.2) implies that the conditional likelihood, given the values of the predictors, is

$$\prod_{i=1}^{n} \left(\frac{\exp\left\{\beta_1 x_{i1} + \cdots + \beta_k x_{ik}\right\}}{1 + \exp\left\{\beta_1 x_{i1} + \cdots + \beta_k x_{ik}\right\}}\right)^{y_i} \left(\frac{1}{1 + \exp\left\{\beta_1 x_{i1} + \cdots + \beta_k x_{ik}\right\}}\right)^{1-y_i}.$$

Inference about the β_i then proceeds via the likelihood methods discussed in Chapter 6. In fact, we need to use software to obtain the MLE's and, because the exact sampling distributions of these quantities are not available, the large sample methods discussed in Section 6.5 are used for approximate confidence intervals and P-values. Note that assessing the null hypothesis $H_0 : \beta_i = 0$ is equivalent to assessing the null hypothesis that the predictor X_i does not have a relationship with the response.

We illustrate the use of logistic regression via an example.

Example 10.5.1

The following table of data represent the

(number of failures, number of successes)

for ingots prepared for rolling under different settings of the predictor variables, $U =$ soaking time, and $V =$ heating time, as reported in *Analysis of Binary Data* by D. R. Cox (Methuen, London, 1970). A failure indicates that an ingot is not ready for rolling after the treatment. There were observations at 19 different settings of these variables.

	$V = 7$	$V = 14$	$V = 27$	$V = 51$
$U = 1.0$	$(0, 10)$	$(0, 31)$	$(1, 55)$	$(3, 10)$
$U = 1.7$	$(0, 17)$	$(0, 43)$	$(4, 40)$	$(0, 1)$
$U = 2.2$	$(0, 7)$	$(2, 31)$	$(0, 21)$	$(0, 1)$
$U = 2.8$	$(0, 12)$	$(0, 31)$	$(1, 21)$	$(0, 0)$
$U = 4.0$	$(0, 9)$	$(0, 19)$	$(1, 15)$	$(0, 1)$

Including an intercept in the model and linear terms for U and V leads to three predictor variables $X_1 \equiv 1, X_2 = U, X_3 = V$, and the model takes the form

$$P(Y = 1 \mid X_2 = x_2, X_3 = x_3) = \frac{\exp\{\beta_1 + \beta_2 x_2 + \beta_3 x_3\}}{1 + \exp\{\beta_1 + \beta_2 x_2 + \beta_3 x_3\}}.$$

Fitting the model via the method of maximum likelihood leads to the estimates given in the following table. Here z is the value of estimate divided by its standard error. Because this is approximately distributed $N(0, 1)$ when the corresponding β_i equals 0, the P-value for assessing the null hypothesis that $\beta_i = 0$ is $P(|Z| > |z|)$ with $Z \sim N(0, 1)$.

Coefficient	Estimate	Std. Error	z	P-value
β_1	5.55900	1.12000	4.96	0.000
β_2	-0.05680	0.33120	-0.17	0.864
β_3	-0.08203	0.02373	-3.46	0.001

Of course, we have to feel confident that the model is appropriate before we can go on to make formal inferences about the β_i. In this case, we note that the number of successes $s(x_2, x_3)$ in the cell of the table, corresponding to the setting $(X_2, X_3) = (x_2, x_3)$, is an observation from a

$$\text{Binomial}(m(x_2, x_3), P(Y = 1 \mid X_2 = x_2, X_3 = x_3))$$

distribution, where $m(x_2, x_3)$ is the sum of the number of successes and failures in that cell. So, for example, if $X_2 = U = 1.0$ and $X_3 = V = 7$, then $m(1.0, 7) = 10$ and $s(1.0, 7) = 10$. Denoting the estimate of $P(Y = 1 \mid X_2 = x_2, X_3 = x_3)$ by $\hat{p}(x_2, x_3)$, obtained by plugging in the MLE, we have that (Problem 10.5.5)

$$X^2 = \sum_{(x_2, x_3)} \frac{(s(x_2, x_3) - m(x_2, x_3)\hat{p}(x_2, x_3))^2}{m(x_2, x_3)\hat{p}(x_2, x_3)} \tag{10.5.3}$$

is asymptotically distributed as a $\chi^2(19 - 3) = \chi^2(16)$ distribution when the model is correct. We determine the degrees of freedom by counting the number of cells where there were observations (19 in this case, as no observations were obtained when $U = 2.8, V = 51$) and subtracting off the number of parameters estimated. For these data, $X^2 = 13.543$ and the P-value is $P(\chi^2(16) > 13.543) = 0.633$. Therefore, we have no evidence that the model is incorrect and can go to inferences about the β_i.

From the preceding table we see that the null hypothesis $H_0 : \beta_2 = 0$ is not rejected. Accordingly, we drop X_2 and fit the smaller model given by

$$P(Y = 1 \mid X_3 = x_3) = \frac{\exp\{\beta_1 + \beta_3 x_3\}}{1 + \exp\{\beta_1 + \beta_3 x_3\}}.$$

This leads to the estimates $\hat{\beta}_1 = 5.4152$ and $\hat{\beta}_3 = -0.08070$. Note that these are only marginally different than the previous estimates. In Figure 10.5.1 we present a graph of the fitted function over the range where we have observed X_3. ∎

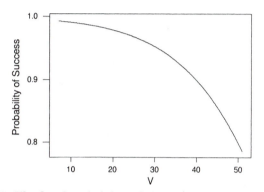

Figure 10.5.1: The fitted probability of obtaining an ingot ready to be rolled as a function of heating time in Example 10.5.1.

Summary of Section 10.5

- We have examined the situation in which we have a single binary-valued response variable and a number of quantitative predictors.

- One method of expressing a relationship between the response and predictors is via the use of a link function.

- If we use the logistic link function, then we can carry out a logistic regression analysis using likelihood methods of inference.

Exercises

10.5.1 Prove that the function $f : R^1 \to R^1$ defined by $f(x) = e^{-x} (1 + e^{-x})^{-2}$ for $x \in R^1$, is a density function with distribution function given by $F(x) = (1 + e^{-x})^{-1}$ and inverse cdf given by $F^{-1}(p) = \ln p - \ln(1 - p)$ for $p \in [0, 1]$. This is called the *logistic distribution*.

10.5.2 Suppose that a logistic regression model for a binary valued response Y is given by

$$P(Y = 1 \mid x) = \frac{\exp\{\beta_1 + \beta_2 x\}}{1 + \exp\{\beta_1 + \beta_2 x\}}.$$

Prove that the log odds at $X = x$ is given by $\beta_1 + \beta_2 x$.

Computer Exercises

10.5.3 Use software to replicate the results of Example 10.5.1.

10.5.4 Suppose that the following data were obtained for the quantitative predictor X and the binary valued response variable Y.

X	-5	-4	-3	-2	-1	0	1	2	3	4	5
Y	$0,0$	$0,0$	$0,0$	$0,0$	$1,0$	$0,0$	$1,0$	$0,1$	$1,1$	$1,1$	$1,1$

(a) Using this data, fit the logistic regression model given by

$$P\left(Y = 1 \,|\, x\right) = \frac{\exp\left\{\beta_1 + \beta_2 x + \beta_3 x^2\right\}}{1 + \exp\left\{\beta_1 + \beta_2 x + \beta_3 x^2\right\}}.$$

(b) Does the model fit the data?
(c) Test the null hypothesis $H_0 : \beta_3 = 0$.
(d) If you decide there is no quadratic effect, refit the model and test for any linear effect.
(e) Plot $P\left(Y = 1 \,|\, x\right)$ as a function of x.

Problem

10.5.5 Prove that (10.5.3) is the correct form for the chi-squared goodness-of-fit test statistic.

10.6   Further Proofs (Advanced)

Theorem 10.3.1 Suppose that $E\left(Y \,|\, X = x\right) = \beta_1 + \beta_2 x$, and we observe the independent values $(x_1, y_1), \ldots, (x_n, y_n)$ for (X, Y). Then the least-squares estimates of β_1 and β_2 are given by $b_1 = \bar{y} - b_2 \bar{x}$ and

$$b_2 = \frac{\sum_{i=1}^{n} (x_i - \bar{x})(y_i - \bar{y})}{\sum_{i=1}^{n} (x_i - \bar{x})^2},$$

whenever $\sum_{i=1}^{n} (x_i - \bar{x})^2 \neq 0$.

Proof: We need an algebraic result that will simplify our calculations.

Lemma 10.6.1 If $(x_1, y_1), \ldots, (x_n, y_n)$ are such that $\sum_{i=1}^{n} (x_i - \bar{x})^2 \neq 0$ and $q, r \in R^1$, then $\sum_{i=1}^{n} (y_i - b_1 - b_2 x_i)(q + r x_i) = 0$.

Proof: We have

$$\sum_{i=1}^{n} (y_i - b_1 - b_2 x_i) = n\bar{y} - nb_1 - nb_2\bar{x} = n\left(\bar{y} - \bar{y} + b_2\bar{x} - b_2\bar{x}\right) = 0,$$

which establishes that $\sum_{i=1}^{n} (y_i - b_1 - b_2 x_i) q = 0$ for any q. Now using this, and the formulas in Theorem 10.3.1, we obtain

$$\sum_{i=1}^{n} (y_i - b_1 - b_2 x_i) x_i = \sum_{i=1}^{n} (y_i - b_1 - b_2 x_i) (x_i - \bar{x})$$

$$= \sum_{i=1}^{n} (y_i - \bar{y} - b_2 (x_i - \bar{x})) (x_i - \bar{x})$$

$$= \sum_{i=1}^{n} (y_i - \bar{y}) (x_i - \bar{x}) - \sum_{i=1}^{n} (y_i - \bar{y}) (x_i - \bar{x}) = 0.$$

This establishes the Lemma. ∎

Returning to the proof of Theorem 10.3.1 we have

$$\sum_{i=1}^{n} (y_i - \beta_1 - \beta_2 x_i)^2$$

$$= \sum_{i=1}^{n} (y_i - b_1 - b_2 x_i - (\beta_1 - b_1) - (\beta_2 - b_2) x_i)^2$$

$$= \sum_{i=1}^{n} (y_i - b_1 - b_2 x_i)^2 - 2 \sum_{i=1}^{n} (y_i - b_1 - b_2 x_i) \{(\beta_1 - b_1) + (\beta_2 - b_2) x_i\}$$

$$+ \sum_{i=1}^{n} ((\beta_1 - b_1) + (\beta_2 - b_2) x_i)^2$$

$$= \sum_{i=1}^{n} (y_i - b_1 - b_2 x_i)^2 + \sum_{i=1}^{n} ((\beta_1 - b_1) + (\beta_2 - b_2) x_i)^2,$$

as the middle term is 0 by Lemma 10.6.1. Therefore,

$$\sum_{i=1}^{n} (y_i - \beta_1 - \beta_2 x_i)^2 \geq \sum_{i=1}^{n} (y_i - b_1 - b_2 x_i)^2$$

and $\sum_{i=1}^{n} (y_i - \beta_1 - \beta_2 x_i)^2$ takes its minimum value if and only if

$$\sum_{i=1}^{n} ((\beta_1 - b_1) + (\beta_2 - b_2) x_i)^2 = 0.$$

This occurs if and only if $(\beta_1 - b_1) + (\beta_2 - b_2) x_i = 0$ for every i. Because the x_i are not all the same value, this is true if and only if $\beta_1 = b_1$ and $\beta_2 = b_2$, which completes the proof. ∎

Theorem 10.3.2 If $E(Y \mid X = x) = \beta_1 + \beta_2 x$ and we observe the independent values $(x_1, y_1), \ldots, (x_n, y_n)$ for (X, Y), then

(i) $E(B_1 \mid X_1 = x_1, \ldots, X_n = x_n) = \beta_1$,

(ii) $E(B_2 \mid X_1 = x_1, \ldots, X_n = x_n) = \beta_2$.

Proof: From Theorem 10.3.1, and $E(\bar{Y} \mid X_1 = x_1, \ldots, X_n = x_n) = \beta_1 + \beta_2\bar{x}$, we have that

$$
\begin{aligned}
E(B_2 \mid X_1 = x_1, \ldots, X_n = x_n) &= \frac{\sum_{i=1}^{n}(x_i - \bar{x})(\beta_1 + \beta_2 x_i - \beta_1 - \beta_2\bar{x})}{\sum_{i=1}^{n}(x_i - \bar{x})^2} \\
&= \beta_2 \frac{\sum_{i=1}^{n}(x_i - \bar{x})^2}{\sum_{i=1}^{n}(x_i - \bar{x})^2} = \beta_2.
\end{aligned}
$$

Also, from Theorem 10.3.1, and what we have just proved,

$$
E(B_1 \mid X_1 = x_1, \ldots, X_n = x_n) = \beta_1 + \beta_2\bar{x} - \beta_2\bar{x} = \beta_1. \quad\blacksquare
$$

Theorem 10.3.3 If $E(Y \mid X = x) = \beta_1 + \beta_2 x$, $\text{Var}(Y \mid X = x) = \sigma^2$ for every x, and we observe the independent values $(x_1, y_1), \ldots, (x_n, y_n)$ for (X, Y), then

(i) $\text{Var}(B_1 \mid X_1 = x_1, \ldots, X_n = x_n) = \sigma^2(1/n + \bar{x}^2/\sum_{i=1}^{n}(x_i - \bar{x})^2)$,

(ii) $\text{Var}(B_2 \mid X_1 = x_1, \ldots, X_n = x_n) = \sigma^2/\sum_{i=1}^{n}(x_i - \bar{x})^2$,

(iii) $\text{Cov}(B_1, B_2 \mid X_1 = x_1, \ldots, X_n = x_n) = -\sigma^2\bar{x}/\sum_{i=1}^{n}(x_i - \bar{x})^2$.

Proof: (ii) Observe that b_2 is a linear combination of the $y_i - \bar{y}$ values, and so we can evaluate the conditional variance once we have obtained the conditional variances and covariances of the $Y_i - \bar{Y}$ values. We have that

$$
Y_i - \bar{Y} = \left(1 - \frac{1}{n}\right)Y_i - \frac{1}{n}\sum_{j \neq i} Y_j,
$$

so the conditional variance of $Y_i - \bar{Y}$ is given by

$$
\sigma^2\left(1 - \frac{1}{n}\right)^2 + \sigma^2\frac{n-1}{n^2} = \sigma^2\left(1 - \frac{1}{n}\right).
$$

When $i \neq j$ we can write

$$
Y_i - \bar{Y} = \left(1 - \frac{1}{n}\right)Y_i - \frac{1}{n}Y_j - \frac{1}{n}\sum_{k \neq i,j} Y_k,
$$

and the conditional covariance between $Y_i - \bar{Y}$ and $Y_j - \bar{Y}$ is then given by

$$
-2\sigma^2\left(1 - \frac{1}{n}\right)\frac{1}{n} + \sigma^2\frac{n-2}{n^2} = -\frac{\sigma^2}{n}
$$

(note that you can assume that the means of the expectations of the Y's are 0 for this calculation). Therefore, the conditional variance of B_2 is given by

$$
\begin{aligned}
&\text{Var}(B_2 \mid x_1, \ldots, x_n) \\
&= \sigma^2\left(1 - \frac{1}{n}\right)\frac{\sum_{i=1}^{n}(x_i - \bar{x})^2}{\left(\sum_{i=1}^{n}(x_i - \bar{x})^2\right)^2} - \frac{\sigma^2}{n}\frac{\sum_{i \neq j}(x_i - \bar{x})(x_j - \bar{x})}{\left(\sum_{i=1}^{n}(x_i - \bar{x})^2\right)^2} \\
&= \frac{\sigma^2}{\sum_{i=1}^{n}(x_i - \bar{x})^2},
\end{aligned}
$$

because

$$\sum_{i \neq j} (x_i - \bar{x})(x_j - \bar{x}) = \left(\sum_{i=1}^{n} (x_i - \bar{x}) \right)^2 - \sum_{i=1}^{n} (x_i - \bar{x})^2$$

$$= - \sum_{i=1}^{n} (x_i - \bar{x})^2.$$

(iii) We have that

$$\text{Cov}(B_1, B_2 \mid X_1 = x_1, \ldots, X_n = x_n)$$
$$= \text{Cov}(\bar{Y} - B_2 \bar{X}, B_2 \mid X_1 = x_1, \ldots, X_n = x_n)$$
$$= \text{Cov}(\bar{Y}, B_2 \mid X_1 = x_1, \ldots, X_n = x_n) - \bar{x} \, \text{Var}(B_2 \mid X_1 = x_1, \ldots, X_n = x_n)$$

and

$$\text{Cov}(\bar{Y}, B_2 \mid X_1 = x_1, \ldots, X_n = x_n)$$
$$= \frac{\sum_{i=1}^{n} (x_i - \bar{x}) \, \text{Cov}\left((Y_i - \bar{Y}), \bar{Y} \mid X_1 = x_1, \ldots, X_n = x_n\right)}{\sum_{i=1}^{n} (x_i - \bar{x})^2}$$
$$= \sigma^2 \left(1 - \frac{1}{n}\right) \frac{\sum_{i=1}^{n} (x_i - \bar{x})}{\sum_{i=1}^{n} (x_i - \bar{x})^2} = 0.$$

Therefore, $\text{Cov}(B_1, B_2 \mid X_1 = x_1, \ldots, X_n = x_n) = -\sigma^2 \bar{x} / \sum_{i=1}^{n} (x_i - \bar{x})^2$.

(i) Finally,

$$\text{Var}(B_1 \mid X_1 = x_1, \ldots, X_n = x_n) = \text{Var}(\bar{Y} - B_2 \bar{x} \mid X_1 = x_1, \ldots, X_n = x_n)$$
$$= \text{Var}(\bar{Y} \mid X_1 = x_1, \ldots, X_n = x_n) + \bar{x}^2 \, \text{Var}(B_2 \mid X_1 = x_1, \ldots, X_n = x_n)$$
$$- 2 \, \text{Cov}(\bar{Y}, B_2 \mid X_1 = x_1, \ldots, X_n = x_n)$$

where $\text{Var}(\bar{Y} \mid X_1 = x_1, \ldots, X_n = x_n) = \sigma^2 / n$, and substituting the results for (ii) and (iii) completes the proof of the theorem. ∎

Corollary 10.3.1

$$\text{Var}(B_1 + B_2 x \mid X_1 = x_1, \ldots, X_n = x_n) = \sigma^2 \left(\frac{1}{n} + \frac{(x - \bar{x})^2}{\sum_{i=1}^{n} (x_i - \bar{x})^2} \right).$$

Proof: We have that

$$\text{Var}(B_1 + B_2 x \mid X_1 = x_1, \ldots, X_n = x_n)$$
$$= \text{Var}(B_1 \mid X_1 = x_1, \ldots, X_n = x_n) + x^2 \, \text{Var}(B_2 \mid X_1 = x_1, \ldots, X_n = x_n)$$
$$+ 2x \, \text{Cov}(B_1, B_2 \mid X_1 = x_1, \ldots, X_n = x_n)$$
$$= \sigma^2 \left(\frac{1}{n} + \frac{\bar{x}^2 + x^2 - 2x\bar{x}}{\sum_{i=1}^{n} (x_i - \bar{x})^2} \right) = \sigma^2 \left(\frac{1}{n} + \frac{(x - \bar{x})^2}{\sum_{i=1}^{n} (x_i - \bar{x})^2} \right). ∎$$

Theorem 10.3.4 If $E(Y \mid X = x) = \beta_1 + \beta_2 x$, $\mathrm{Var}(Y \mid X = x) = \sigma^2$ for every x and we observe the independent values $(x_1, y_1), \ldots, (x_n, y_n)$ for (X, Y) then

$$E\left(S^2 \mid X_1 = x_1, \ldots, X_n = x_n\right) = \sigma^2.$$

Proof: We have that

$$E\left((n-2)\,S^2 \mid X_1 = x_1, \ldots, X_n = x_n\right)$$

$$= E\left(\sum_{i=1}^{n} (Y_i - B_1 - B_2 x_i)^2 \mid X_1 = x_1, \ldots, X_n = x_n\right)$$

$$= \sum_{i=1}^{n} E\left(\left(Y_i - \bar{Y} - B_2\,(x_i - \bar{x})\right)^2 \mid X_1 = x_1, \ldots, X_n = x_n\right)$$

$$= \sum_{i=1}^{n} \mathrm{Var}\left(Y_i - \bar{Y} - B_2\,(x_i - \bar{x}) \mid X_1 = x_1, \ldots, X_n = x_n\right)$$

because

$$E\left(Y_i - \bar{Y} - B_2\,(x_i - \bar{x}) \mid X_1 = x_1, \ldots, X_n = x_n\right)$$
$$= \beta_1 + \beta_2 x_i - \beta_1 - \beta_2 \bar{x} - \beta_2\,(x_i - \bar{x}) = 0.$$

Now,

$$\mathrm{Var}\left(\left(Y_i - \bar{Y} - B_2\,(x_i - \bar{x})\right) \mid X_1 = x_1, \ldots, X_n = x_n\right)$$
$$= \mathrm{Var}\left(Y_i - \bar{Y} \mid X_1 = x_1, \ldots, X_n = x_n\right)$$
$$- 2\,(x_i - \bar{x})\,\mathrm{Cov}\left(\left(Y_i - \bar{Y}\right), B_2 \mid X_1 = x_1, \ldots, X_n = x_n\right)$$
$$+ (x_i - \bar{x})^2\,\mathrm{Var}\left(B_2 \mid X_1 = x_1, \ldots, X_n = x_n\right)$$

and, using the results established about the covariances of the $Y_i - \bar{Y}$ in the proof of Theorem 10.3.3, we have that

$$\mathrm{Var}\left(Y_i - \bar{Y} \mid X_1 = x_1, \ldots, X_n = x_n\right) = \sigma^2\left(1 - \frac{1}{n}\right)$$

and

$$\mathrm{Cov}\left(Y_i - \bar{Y}, B_2 \mid X_1 = x_1, \ldots, X_n = x_n\right)$$

$$= \frac{1}{\sum_{i=1}^{n} (x_i - \bar{x})^2} \sum_{j=1}^{n} (x_j - \bar{x})\,\mathrm{Cov}\left(Y_i - \bar{Y}, Y_j - \bar{Y} \mid X_1 = x_1, \ldots, X_n = x_n\right)$$

$$= \frac{\sigma^2}{\sum_{i=1}^{n} (x_i - \bar{x})^2}\left(\left(1 - \frac{1}{n}\right)(x_i - \bar{x}) - \frac{1}{n}\sum_{j \neq i} (x_j - \bar{x})\right) = \frac{\sigma^2\,(x_i - \bar{x})}{\sum_{i=1}^{n} (x_i - \bar{x})^2},$$

because $\sum_{j \neq i} (x_j - \bar{x}) = -(x_i - \bar{x})$. Therefore,

$$\text{Var}\left(Y_i - \bar{Y} - B_2\left(x_i - \bar{x}\right) \mid X_1 = x_1, \ldots, X_n = x_n\right)$$

$$= \sigma^2 \left(1 - \frac{1}{n}\right) - 2 \frac{\sigma^2 (x_i - \bar{x})^2}{\sum_{i=1}^{n} (x_i - \bar{x})^2} + \frac{\sigma^2 (x_i - \bar{x})^2}{\sum_{i=1}^{n} (x_i - \bar{x})^2}$$

$$= \sigma^2 \left(1 - \frac{1}{n} - \frac{(x_i - \bar{x})^2}{\sum_{i=1}^{n} (x_i - \bar{x})^2}\right)$$

and

$$E\left(S^2 \mid X_1 = x_1, \ldots, X_n = x_n\right) = \frac{\sigma^2}{n-2} \sum_{i=1}^{n} \left(1 - \frac{1}{n} - \frac{(x_i - \bar{x})^2}{\sum_{i=1}^{n} (x_i - \bar{x})^2}\right) = \sigma^2$$

as was stated. ∎

Lemma 10.3.1 If $(x_1, y_1), \ldots, (x_n, y_n)$ are such that $\sum_{i=1}^{n} (x_i - \bar{x})^2 \neq 0$, then

$$\sum_{i=1}^{n} (y_i - \bar{y})^2 = b_2^2 \sum_{i=1}^{n} (x_i - \bar{x})^2 + \sum_{i=1}^{n} (y_i - b_1 - b_2 x_i)^2.$$

Proof: We have that

$$\begin{aligned}
\sum_{i=1}^{n} (y_i - \bar{y})^2 &= \sum_{i=1}^{n} y_i^2 - n\bar{y}^2 \\
&= \sum_{i=1}^{n} (y_i - b_1 - b_2 x_i + b_1 + b_2 x_i)^2 - n\bar{y}^2 \\
&= \sum_{i=1}^{n} (y_i - b_1 - b_2 x_i)^2 + \sum_{i=1}^{n} (b_1 + b_2 x_i)^2 - n\bar{y}^2,
\end{aligned}$$

because $\sum_{i=1}^{n} (y_i - b_1 - b_2 x_i)(b_1 + b_2 x_i) = 0$ by Lemma 10.6.1. Then, using Theorem 10.3.1, we have

$$\sum_{i=1}^{n} (b_1 + b_2 x_i)^2 - n\bar{y}^2 = \sum_{i=1}^{n} (\bar{y} + b_2 (x_i - \bar{x}))^2 - n\bar{y}^2 = b_2^2 \sum_{i=1}^{n} (x_i - \bar{x})^2,$$

and this completes the proof. ∎

Theorem 10.3.6 If Y given $X = x$ is distributed $N\left(\beta_1 + \beta_2 x, \sigma^2\right)$ and we observe the independent values $(x_1, y_1), \ldots, (x_n, y_n)$ for (X, Y), then the conditional distributions of B_1, B_2, and S^2 given $X_1 = x_1, \ldots, X_n = x_n$ are as follows.

(i) $B_1 \sim N(\beta_1, \sigma^2(1/n + \bar{x}^2 / \sum_{i=1}^{n} (x_i - \bar{x})^2))$

(ii) $B_2 \sim N(\beta_2, \sigma^2 / \sum_{i=1}^{n} (x_i - \bar{x})^2)$

(iii)

$$B_1 + B_2 x \sim N\left(\beta_1 + \beta_2 x, \sigma^2 \left(\frac{1}{n} + \frac{(x - \bar{x})^2}{\sum_{i=1}^{n}(x_i - \bar{x})^2}\right)\right)$$

(iv) $(n-2)\, S^2/\sigma^2 \sim \chi^2\,(n-2)$ independent of (B_1, B_2)

Proof: (i) Because B_1 can be written as a linear combination of the Y_i, Theorem 4.6.1 implies that the distribution of B_1 must be normal. The result then follows from Theorems 10.3.2 and 10.3.3. A similar proof establishes (ii) and (iii). The proof of (iv) is similar to the proof of Theorem 4.6.6, and we leave this to a further course in statistics. ∎

Corollary 10.3.2

(i) $(B_1 - \beta_1)\,/S\left(1/n + \bar{x}^2/\sum_{i=1}^{n}(x_i - \bar{x})^2\right)^{1/2} \sim t\,(n-2)$

(ii) $(B_2 - \beta_2)\left(\sum_{i=1}^{n}(x_i - \bar{x})^2\right)^{1/2}/S \sim t\,(n-2)$

(iii)

$$\frac{B_1 + B_2 x - \beta_1 - \beta_2 x}{S\left(\left(\frac{1}{n} + (x - \bar{x})^2\right)/\sum_{i=1}^{n}(x_i - \bar{x})^2\right)^{1/2}} \sim t\,(n-2),$$

(iv) If F is defined as in (10.3.8), then $H_0 : \beta_2 = 0$ is true if and only if $F \sim F\,(1, n-2)$.

Proof: (i) Because B_1 and S^2 are independent

$$\frac{B_1 - \beta_1}{\sigma\left(1/n + \bar{x}^2/\sum_{i=1}^{n}(x_i - \bar{x})^2\right)^{1/2}} \sim N(0,1)$$

independent of $(n-2)\,S^2/\sigma^2 \sim \chi^2\,(n-2)$. Therefore, applying Definition 4.6.2, we have

$$\frac{B_1 - \beta_1}{\sigma\left(1/n + \bar{x}^2/\sum_{i=1}^{n}(x_i - \bar{x})^2\right)^{1/2}\left(\frac{(n-2)S^2}{(n-2)\sigma^2}\right)^{1/2}}$$

$$= \frac{B_1 - \beta_1}{S\left(1/n + \bar{x}^2/\sum_{i=1}^{n}(x_i - \bar{x})^2\right)^{1/2}} \sim t\,(n-2).$$

(ii) The proof proceeds just as in the proof of (i).

(iii) The proof proceeds just as in the proof of (i) and also using Corollary 10.3.1.

(iv) Taking the square of the ratio in (ii) and applying Theorem 4.6.11 implies

$$G = \frac{(B_2 - \beta_2)^2}{S^2\left(\sum_{i=1}^{n}(x_i - \bar{x})^2\right)^{-1}} = \frac{(B_2 - \beta_2)^2 \sum_{i=1}^{n}(x_i - \bar{x})^2}{S^2} \sim F\,(1, n-2).$$

Now observe that F defined by (10.3.8) equals G when $\beta_2 = 0$. The converse that $F \sim F\,(1, n-2)$ only if $\beta_2 = 0$ is somewhat harder to prove and we leave this to a further course. ∎

Chapter 11

Advanced Topic — Stochastic Processes

In this chapter, we consider *stochastic processes*, which are processes that proceed randomly in *time*. That is, rather than consider fixed random variables X, Y, etc., or even sequences of independently and identically distributed random variables, we shall instead consider sequences $X_0, X_1, X_2, \ldots$ where X_n represents some random quantity *at time* n. In general, the value X_n at time n might depend on the quantity X_{n-1} at time $n-1$, or even the values X_m for other times $m < n$. Stochastic processes have a different "flavor" from ordinary random variables — because they proceed in time, they seem more "alive".

We begin with a simple but very interesting case, namely simple random walk.

11.1 Simple Random Walk

Simple random walk can be thought of as a model for repeated *gambling*. Specifically, suppose you start with \$$a$, and repeatedly make \$1 bets. At each bet, you have probability p of winning \$1, and probability q of losing \$1, where $p + q = 1$. If X_n is the amount of money you have at time n (henceforth, your *fortune* at time n), then $X_0 = a$, while X_1 could be $a + 1$ or $a - 1$, depending on whether you win or lose your first bet. Then X_2 could be $a + 2$ (if you win your first two bets), or a (if you win once and lose once), or $a - 2$ (if you lose your first two bets). Continuing in this way, we obtain a whole sequence $X_0, X_1, X_2, \ldots$ of random values, corresponding to your fortune at times $0, 1, 2, \ldots$.

We shall refer to the stochastic process $\{X_n\}$ as *simple random walk*. Another way to define this model is to start with random variables $\{Z_i\}$ that are i.i.d. with $P(Z_i = 1) = p$ and $P(Z_i = -1) = 1 - p \equiv q$, where $0 < p < 1$. (Here $Z_i = 1$ if you win the ith bet, while $Z_i = -1$ if you lose the ith bet.) We then set $X_0 = a$, and for $n \geq 1$ we set

$$X_n = a + Z_1 + Z_2 + \cdots + Z_n.$$

The following is a specific example of this.

Example 11.1.1
Consider simple random walk with $a = 8$ and $p = 1/3$, so you start with \$8 and have probability $1/3$ of winning each bet. Then the probability that you have \$9 after one bet is given by

$$P(X_1 = 9) = P(8 + Z_1 = 9) = P(Z_1 = 1) = 1/3,$$

as it should be. Also, the probability that you have \$7 after one bet is given by

$$P(X_1 = 7) = P(8 + Z_1 = 7) = P(Z_1 = -1) = 2/3.$$

On the other hand, the probability that you have \$10 after two bets is given by

$$P(X_2 = 10) = P(8 + Z_1 + Z_2 = 10) = P(Z_1 = Z_2 = 1) = (1/3)(1/3) = 1/9. \quad \blacksquare$$

Example 11.1.2
Consider again simple random walk with $a = 8$ and $p = 1/3$. Then the probability that you have \$7 after three bets is given by

$$P(X_3 = 7) = P(8 + Z_1 + Z_2 + Z_3 = 7) = P(Z_1 + Z_2 + Z_3 = -1).$$

Now, there are three different ways we could have $Z_1 + Z_2 + Z_3 = -1$. We could have either (a) $Z_1 = 1$, while $Z_2 = Z_3 = -1$, (b) $Z_2 = 1$, while $Z_1 = Z_3 = -1$, or (c) $Z_3 = 1$, while $Z_1 = Z_2 = -1$. Each of these three options has probability $(1/3)(2/3)(2/3)$. Hence,

$$P(X_3 = 7) = (1/3)(2/3)(2/3) + (1/3)(2/3)(2/3) + (1/3)(2/3)(2/3) = 4/9. \quad \blacksquare$$

If the number of bets is much larger than three, then it becomes less and less convenient to compute probabilities in the above manner. A more systematic approach is required. We turn to that next.

11.1.1   The Distribution of the Fortune

We first compute the distribution of X_n, i.e., the probability that your fortune X_n after n bets takes on various values.

Theorem 11.1.1 Let $\{X_n\}$ be simple random walk as before, and let n be a positive integer. If k is an integer such that $-n \leq k \leq n$ and $n + k$ is even, then

$$P(X_n = a + k) = \binom{n}{\frac{n+k}{2}} p^{(n+k)/2} q^{(n-k)/2}.$$

For all other values of k, we have $P(X_n = a + k) = 0$. Furthermore, $E(X_n) = a + n(2p - 1)$.

Proof: See Section 11.7. ∎

This theorem tells us the entire distribution, and expected value, of the fortune X_n at time n.

Example 11.1.3
Suppose $p = 1/3$ and $n = 8$ and $a = 1$. Then $P(X_n = 6) = 0$, because $6 = 1 + 5$, and $n + 5 = 13$ is not even. Also, $P(X_n = 13) = 0$, because $13 = 1 + 12$ and $12 > n$. On the other hand,

$$
\begin{aligned}
P(X_n = 5) &= P(X_n = 1 + 4) = \binom{n}{\frac{n+4}{2}} p^{(n+4)/2} q^{(n-4)/2} = \binom{8}{6}(1/3)^6(2/3) \\
&= \frac{8 \cdot 7}{2}(1/3)^6(2/3)^1 \approx 0.0256.
\end{aligned}
$$

Also, $E(X_n) = a + n(2p - 1) = 1 + 8(2/3 - 1) = -5/3$. ∎

Regarding $E(X_n)$, we immediately obtain the following corollary.

Corollary 11.1.1 If $p = 1/2$, then $E(X_n) = a$ for all $n \geq 0$. If $p < 1/2$, then $E(X_n) < a$ for all $n \geq 1$. If $p > 1/2$, then $E(X_n) > a$ for all $n \geq 1$.

This corollary has the following interpretation. If $p = 1/2$, then the game is *fair*, i.e., both you and your opponent have equal chance of winning each bet. Thus, the corollary says that for fair games, your expected fortune $E(X_n)$ will never change from its initial value, a.

On the other hand, if $p < 1/2$, then the game is *subfair*, i.e., your opponent's chances are better than yours. In this case, the corollary says your expected fortune will *decrease*, i.e., be less than its initial value of a. Similarly, if $p > 1/2$, then the game is *superfair*, and the corollary says your expected fortune will *increase*, i.e., be more than its initial value of a.

Of course, in a real gambling casino, the game is always subfair (which is how the casino makes its profit). Hence, in a real casino, the average amount of money with which you leave will always be less than the amount with which you entered!

Example 11.1.4
Suppose $a = 10$ and $p = 1/4$. Then $E(X_n) = 10 + n(2p - 1) = 10 - 3n/4$. Hence, we always have $E(X_n) \leq 10$, and indeed $E(X_n) < 0$ if $n \geq 14$. That is, your expected fortune is never more than your initial value of \$10, and in fact is *negative* after 14 or more bets. ∎

Finally, we note as an aside that it is possible to change your probabilities by changing your gambling *strategy*, as in the following example. Hence, the preceding analysis applies only to the strategy of betting just \$1 each time.

Example 11.1.5
Consider the "double 'til you win" gambling strategy, defined as follows. We first bet \$1. Each time we lose, we *double* our bet on the succeeding turn. As soon as we win once, we stop playing (i.e., bet zero from then on).

It is easily seen that, with this gambling strategy, we will be up $1 as soon as we win a bet (which must happen eventually because $p > 0$). Hence, with probability 1 we will gain $1 with this gambling strategy for *any* positive value of p.

This is rather surprising, because if $0 < p < 1/2$, then the odds in this game are against us. So it seems that we have "cheated fate," and indeed we have. On the other hand, we may need to lose an arbitrarily large amount of money before we win our $1, so "infinite capital" is required to follow this gambling strategy. If only finite capital is available, then it is impossible to cheat fate in this manner. For a proof of this, see more advanced probability books, e.g., page 64 of *A First Look at Rigorous Probability Theory* by J. S. Rosenthal (World Scientific Publishing, Singapore, 2000). ■

11.1.2 The Gambler's Ruin Problem

The previous subsection considered the distribution and expected value of the fortune X_n at a fixed time n. Here, we consider the *gambler's ruin* problem, which requires the consideration of many different n at once, i.e., considers the *time evolution* of the process.

Let $\{X_n\}$ be simple random walk as before, for some initial fortune a and some probability p of winning each bet. Assume a is a positive integer. Furthermore, let $c > a$ be some other integer. The gambler's ruin question is, if you repeatedly bet $1, then what is the probability that you will reach a fortune of $c *before* you lose all your money by reaching a fortune $0? In other words, will the random walk hit c before hitting 0? Informally, what is the probability that the gambler gets rich (i.e., has $c) before going broke?

More formally, let

$$\tau_0 = \min\{n \geq 0 : X_n = 0\},$$
$$\tau_c = \min\{n \geq 0 : X_n = c\}$$

be the *first hitting times* of 0 and c, respectively. That is, τ_0 is the first time your fortune reaches 0, while τ_c is the first time your fortune reaches c.

The gambler's ruin question is, what is

$$P(\tau_c < \tau_0),$$

the probability of hitting c before hitting 0? This question is not so easy to answer, because there is no limit to how long it might take until either c or 0 is hit. Hence, it is not sufficient to just compute the probabilities after 10 bets, or 20 bets, or 100 bets, or even 1,000,000 bets. Fortunately, it is possible to answer this question, as follows.

Theorem 11.1.2 Let $\{X_n\}$ be simple random walk, with some initial fortune a and probability p of winning each bet. Assume $0 < a < c$. Then the probability

$P(\tau_c < \tau_0)$ of hitting c before 0 is given by

$$P(\tau_c < \tau_0) = \begin{cases} a/c & p = 1/2 \\ \frac{1-\left(\frac{q}{p}\right)^a}{1-\left(\frac{q}{p}\right)^c} & p \neq 1/2. \end{cases}$$

Proof: See Section 11.7 for the proof. ∎

Consider some applications of this result.

Example 11.1.6
Suppose you start with \$5 (i.e., $a = 5$) and your goal is to win \$10 before going broke (i.e., $c = 10$). If $p = 0.500$, then your probability of success is $a/c = 0.500$. If $p = 0.499$, then your probability of success is given by

$$\left(1 - \left(\frac{0.501}{0.499}\right)^5\right)\left(1 - \left(\frac{0.501}{0.499}\right)^{10}\right)^{-1},$$

which is approximately 0.495. If $p = 0.501$, then your probability of success is given by

$$\left(1 - \left(\frac{0.499}{0.501}\right)^5\right)\left(1 - \left(\frac{0.499}{0.501}\right)^{10}\right)^{-1},$$

which is approximately 0.505. We thus see that in this case, small changes in p lead to small changes in the probability of winning at gambler's ruin. ∎

Example 11.1.7
Suppose now that you start with \$5000 (i.e., $a = 5000$) and your goal is to win \$10,000 before going broke (i.e., $c = 10,000$). If $p = 0.500$, then your probability of success is $a/c = 0.500$, same as before. On the other hand, if $p = 0.499$, then your probability of success is given by

$$\left(1 - \left(\frac{0.501}{0.499}\right)^{5000}\right)\left(1 - \left(\frac{0.501}{0.499}\right)^{10,000}\right)^{-1},$$

which is approximately 2×10^{-9}, i.e., two parts in a billion! Finally, if $p = 0.501$, then your probability of success is given by

$$\left(1 - \left(\frac{0.499}{0.501}\right)^{5000}\right)\left(1 - \left(\frac{0.499}{0.501}\right)^{10,000}\right)^{-1},$$

which is extremely close to 1. We thus see that in this case, small changes in p lead to extremely large changes in the probability of winning at gambler's ruin. For example, even a tiny disadvantage on each bet can lead to a very large disadvantage in the long run! The reason for this is that, to get from 5000 to 10,000, many bets must be made, so small changes in p have a huge effect overall. ∎

Finally, we note that it is also possible to use the gambler's ruin result to compute $P(\tau_0 < \infty)$, the probability that the walk will *ever* hit 0 (equivalently, that you will *ever* lose all your money), as follows.

Theorem 11.1.3 Let $\{X_n\}$ be simple random walk, with initial fortune $a > 0$, and probability p of winning each bet. Then the probability $P(\tau_0 < \infty)$ that the walk will *ever* hit 0 is given by

$$P(\tau_0 < \infty) = \begin{cases} 1 & p \le 1/2 \\ (q/p)^a & p > 1/2. \end{cases}$$

Proof: See Section 11.7 for the proof.

Example 11.1.8
Suppose $a = 2$ and $p = 2/3$. Then the probability that you will *eventually* lose all your money is given by $(q/p)^a = ((1/3)/(2/3))^2 = 1/4$. Thus, starting with just \$2, we see that $3/4$ of the time, you will be able to bet forever without ever losing all your money.

On the other hand, if $p \le 1/2$, then no matter how large a is, it is certain that you will eventually lose all your money. ∎

Summary of Section 11.1

- A simple random walk is a sequence $\{X_n\}$ of random variables, with $X_0 = 1$, and $P(X_{n+1} = X_n + 1) = p = 1 - P(X_{n+1} = X_n - 1)$.

- It follows that $P(X_n = a + k) = \binom{n}{\frac{n+k}{2}} p^{(n+k)/2} q^{(n-k)/2}$ for $k = -n, -n + 2, -n + 4, \ldots, n$, and $E(X_n) = a + n(2p - 1)$.

- If $0 < a < c$, then the gambler's ruin probability of reaching c before 0 is equal to a/c if $p = 1/2$, otherwise to $(1 - ((1-p)/p)^a)/(1 - ((1-p)/p)^c)$.

Exercises

11.1.1 Let $\{X_n\}$ be simple random walk, with initial fortune $a = 12$ and probability $p = 1/3$ of winning each bet. Compute $P(X_n = x)$ for the following values of n and x:
(a) $n = 0, x = 13$
(b) $n = 1, x = 12$
(c) $n = 1, x = 13$
(d) $n = 1, x = 11$
(e) $n = 1, x = 14$
(f) $n = 2, x = 12$
(g) $n = 2, x = 13$
(h) $n = 2, x = 14$
(i) $n = 2, x = 15$
(j) $n = 20, x = 15$
(k) $n = 20, x = 16$

(l) $n = 20, x = -18$
(m) $n = 20, x = 10$

11.1.2 Let $\{X_n\}$ be simple random walk, with initial fortune $a = 5$, and probability $p = 2/5$ of winning each bet.
(a) Compute $P(X_1 = 6, X_2 = 5)$.
(b) Compute $P(X_1 = 4, X_2 = 5)$.
(c) Compute $P(X_2 = 5)$.
(d) What is the relationship between the quantities in parts (a), (b), and (c)? Why is this so?

11.1.3 Let $\{X_n\}$ be simple random walk, with initial fortune $a = 7$, and probability $p = 1/6$ of winning each bet.
(a) Compute $P(X_1 = X_3 = 8)$.
(b) Compute $P(X_1 = 6, X_3 = 8)$.
(c) Compute $P(X_3 = 8)$.
(d) What is the relationship between the quantities in parts (a), (b), and (c)? Why is this so?

11.1.4 Suppose $a = 1000$ and $p = 0.49$.
(a) Compute $E(X_n)$ for $n = 0, 1, 2, 10, 20, 100$, and 1000.
(b) How large does n need to be before $E(X_n) < 0$?

11.1.5 Let $\{X_n\}$ be simple random walk, with initial fortune a, and probability $p = 0.499$ of winning each bet. Compute the gambler's ruin probability $P(\tau_c < \tau_0)$ for the following values of a and c. Interpret your results in words.
(a) $a = 9, c = 10$
(b) $a = 90, c = 100$
(c) $a = 900, c = 1000$
(d) $a = 9000, c = 10,000$
(e) $a = 90,000, \ c = 100,000$
(f) $a = 900,000, \ c = 1,000,000$

11.1.6 Let $\{X_n\}$ be simple random walk, with initial fortune $a = 10$ and probability p of winning each bet. Compute $P(\tau_0 < \infty)$ where $p = 0.4$, and also where $p = 0.6$. Interpret your results in words.

Problem

11.1.7 Suppose you start with $10, and repeatedly bet $2 (instead of $1), having probability p of winning each time. Suppose your goal is $100, i.e., you keep on betting until you either lose all your money, or reach $100.
(a) As a function of p, what is the probability that you will reach $100 before losing all your money? Be sure to justify your solution. (Hint: You may find yourself dividing both 10 and 100 by 2.)
(b) Suppose $p = 0.4$. Compute a numerical value for the solution in part (a).
(c) Compare the probabilities in part (b) with the corresponding probabilities if you bet just $1 each time. Which is larger?
(d) Repeat part (b) for the case where you bet $10 each time. Does the probability of success increase or decrease?

Challenge

11.1.8 Prove that the formula for the gambler's ruin probability $P(\tau_c < \tau_0)$ is a continuous function of p, by proving that it is continuous at $p = 1/2$. That is, prove that

$$\lim_{p \to 1/2} \frac{1 - ((1-p)/p)^a}{1 - ((1-p)/p)^c} = \frac{a}{c}.$$

Discussion Topics

11.1.9 Suppose you repeatedly play roulette in a real casino, betting the same amount each time, continuing forever as long as you have money to bet. Is it certain that you will eventually lose all your money? Why or why not?

11.1.10 In Problem 11.1.7 parts (c) and (d), can you explain intuitively why the probabilities change as they do, as we increase the amount we bet each time?

11.1.11 Suppose you start at a, and need to reach c, where $c > a > 0$. You must keep gambling until you reach either c or 0. Suppose you are playing a subfair game (i.e., $p < 1/2$), but you can choose how much to bet for each bet (i.e., you can bet \$1, or \$2, or more, though of course you cannot bet more than you have). What betting amounts do you think[1] will maximize your probability of success, i.e., maximize $P(\tau_0 < \tau_c)$? (Hint: The results of Problem 11.1.7 may provide a clue.)

11.2 Markov Chains

Intuitively, a *Markov chain* represents the random motion of some object. We shall write X_n for the position (or value) of the object at time n. There are then rules that give the *probabilities* for where the object will jump next.

A Markov chain requires a *state space* S, which is the set of all places the object can go. (For example, perhaps $S = \{1, 2, 3\}$, or $S = \{\text{top, bottom}\}$, or S is the set of all positive integers.)

A Markov chain also requires *transition probabilities*, which give the probabilities for where the object will jump next. Specifically, for $i, j \in S$, the number p_{ij} is the probability that, if the object is at i, it will next jump to j. Thus, the collection $\{p_{ij} : i, j \in S\}$ of transition probabilities satisfies $p_{ij} \geq 0$ for all $i, j \in S$, and

$$\sum_{j \in S} p_{ij} = 1$$

for each $i \in S$.

We shall also need to consider where the Markov chain *starts*. Often we will simply set $X_0 = s$ for some particular state $s \in S$. More generally, we could

---

[1] For more advanced results about this, see, e.g., Theorem 7.3 of *Probability and Measure* (3rd ed.) by P. Billingsley (John Wiley & Sons, New York, 1995).

have an *initial distribution* $\{\mu_i : i \in S\}$ where $\mu_i = P(X_0 = i)$. In this case, we need $\mu_i \geq 0$ for each $i \in S$, and

$$\sum_{i \in S} \mu_i = 1.$$

To summarize, here S is the state space of all places the object can go, μ_i represents the probability that the object starts at the point i, and p_{ij} represents the probability that, *if* the object is at the point i, it will then jump to the point j on the next step. In terms of the sequence of random values $X_0, X_1, X_2, \ldots$, we then have that

$$P(X_{n+1} = j \mid X_n = i) = p_{ij},$$

for any positive integer n and any $i, j \in S$. Note that we also require that this jump probability does not depend on the chain's previous history. That is, we require

$$P(X_{n+1} = j \mid X_n = i, X_{n-1} = x_{n-1}, \ldots, X_0 = x_0) = p_{ij}$$

for all n and all $i, j, x_0, \ldots, x_{n-1} \in S$.

11.2.1 Examples of Markov Chains

We present some examples of Markov chains here.

Example 11.2.1
Let $S = \{1, 2, 3\}$ consist of just three elements, and define the transition probabilities by $p_{11} = 0$, $p_{12} = 1/2$, $p_{13} = 1/2$, $p_{21} = 1/3$, $p_{22} = 1/3$, $p_{23} = 1/3$, $p_{31} = 1/4$, $p_{32} = 1/4$, and $p_{33} = 1/2$. This means that, for example, if the chain is at the state 3, then it has probability $1/4$ of jumping to state 1 on the next jump, probability $1/4$ of jumping to state 2 on the next jump, and probability $1/2$ of remaining at state 3 on the next jump.

This Markov chain jumps around on the three points $\{1, 2, 3\}$ in a random and interesting way. For example, if it starts at the point 1, then it might jump to 2 or to 3 (with probability $1/2$ each). If it jumps to (say) 3, then on the next step it might jump to 1 or 2 (probability $1/4$ each) or 3 (probability $1/2$). It continues making such random jumps forever.

Note that we can also write the transition probabilities p_{ij} in matrix form, as

$$(p_{ij}) = \begin{pmatrix} 0 & 1/2 & 1/2 \\ 1/3 & 1/3 & 1/3 \\ 1/4 & 1/4 & 1/2 \end{pmatrix}$$

(so that $p_{31} = 1/4$, etc.). The matrix (p_{ij}) is then called a *stochastic matrix*. This matrix representation is convenient sometimes. ∎

Example 11.2.2
Again, let $S = \{1, 2, 3\}$. This time define the transition probabilities $\{p_{ij}\}$ in matrix form, as

$$(p_{ij}) = \begin{pmatrix} 1/4 & 1/4 & 1/2 \\ 1/3 & 1/3 & 1/3 \\ 0.01 & 0.01 & 0.98 \end{pmatrix}.$$

This also defines a Markov chain on S. For example, from the state 3, there is probability 0.01 of jumping to state 1, probability 0.01 of jumping to state 2, and probability 0.98 of staying in state 3. ∎

Example 11.2.3
Let $S = \{\text{bedroom, kitchen, den}\}$. Define the transition probabilities $\{p_{ij}\}$ in matrix form by

$$(p_{ij}) = \begin{pmatrix} 1/4 & 1/4 & 1/2 \\ 0 & 0 & 1 \\ 0.01 & 0.01 & 0.98 \end{pmatrix}.$$

This defines a Markov chain on S. For example, from the bedroom, the chain has probability 1/4 of staying in the bedroom, probability 1/4 of jumping to the kitchen, and probability 1/2 of jumping to the den. ∎

Example 11.2.4
This time let $S = \{1, 2, 3, 4\}$, and define the transition probabilities $\{p_{ij}\}$ in matrix form, as

$$(p_{ij}) = \begin{pmatrix} 0.2 & 0.4 & 0 & 0.4 \\ 0.4 & 0.2 & 0.4 & 0 \\ 0 & 0.4 & 0.2 & 0.4 \\ 0.4 & 0 & 0.4 & 0.2 \end{pmatrix}.$$

This defines a Markov chain on S. For example, from the state 4, it has probability 0.4 of jumping to the state 1, but probability 0 of jumping to the state 2. ∎

Example 11.2.5
This time, let $S = \{1, 2, 3, 4, 5, 6, 7\}$, and define the transition probabilities $\{p_{ij}\}$ in matrix form, as

$$(p_{ij}) = \begin{pmatrix} 1 & 0 & 0 & 0 & 0 & 0 & 0 \\ 1/2 & 0 & 1/2 & 0 & 0 & 0 & 0 \\ 0 & 1/5 & 4/5 & 0 & 0 & 0 & 0 \\ 0 & 0 & 1/3 & 1/3 & 1/3 & 0 & 0 \\ 1/10 & 0 & 0 & 0 & 7/10 & 0 & 1/5 \\ 0 & 0 & 0 & 0 & 0 & 0 & 1 \\ 0 & 0 & 0 & 0 & 0 & 1 & 0 \end{pmatrix}.$$

This defines a (complicated!) Markov chain on S. ∎

Example 11.2.6 *Random Walk on the Circle*
Let $S = \{0, 1, 2, \ldots, d-1\}$, and define the transition probabilities by saying that $p_{ii} = 1/3$ for all $i \in S$, and also $p_{ij} = 1/3$ whenever i and j are "next to" each other around the circle. That is, $p_{ij} = 1/3$ whenever $j = i$, or $j = i+1$, or $j = i-1$. Also, $p_{0,d-1} = p_{d-1,0} = 1/3$. Otherwise, $p_{ij} = 0$.

If we think of the d elements of S as arranged in a circle, then our object, at each step, either stays where it is, or moves one step clockwise, or moves one step counterclockwise, each with probability $1/3$. (Note in particular that it can go around the "corner" by jumping from $d - 1$ to 0, or from 0 to $d - 1$, with probability $1/3$.) ∎

Example 11.2.7 *Ehrenfest's Urn*

Consider two urns, urn 1 and urn 2. Suppose there are d balls divided between the two urns. Suppose at each step, we choose one ball uniformly at random from among the d balls, and switch it to the opposite urn. We let X_n be the number of balls in urn 1 at time n. Thus, there are $d - X_n$ balls in urn 2 at time n.

Here the state space is $S = \{0, 1, 2, \ldots, d\}$, because these are all the possible numbers of balls in urn 1 at any time n.

Also, if there are i balls in Urn 1 at some time, then there is probability i/n that we next choose one of those i balls, in which case the number of balls in urn 1 goes down to $i - 1$. Hence,

$$p_{i,i-1} = i/d.$$

Similarly,

$$p_{i,i+1} = (d - i)/d,$$

because there is probability $(d-i)/d$ that we will instead choose one of the $d-i$ balls in Urn 2. Thus, this Markov chain moves randomly among the possible numbers $\{0, 1, \ldots, d\}$ of balls in urn 1 at each time.

One might expect that, if d is large and the Markov chain is run for a long time, there would most likely be approximately $d/2$ balls in urn 1. (We shall consider such questions in Section 11.2.4.) ∎

The above examples should convince you that Markov chains on finite state spaces come in all shapes and sizes. Markov chains on *infinite* state spaces are also important. Indeed, we have already seen one such class of Markov chains.

Example 11.2.8 *Simple Random Walk*

Let $S = \{\ldots, -2, -1, 0, 1, 2, \ldots\}$ be the set of all integers. Then S is infinite, so we cannot write the transition probabilities $\{p_{ij}\}$ in matrix form.

Fix $a \in S$, and let $X_0 = a$. Fix a real number p with $0 < p < 1$, and let $p_{i,i+1} = p$ and $p_{i,i-1} = 1 - p$ for each $i \in \mathbf{Z}$, with $p_{ij} = 0$ if $j \neq i \pm 1$. Thus, this Markov chain begins at the point a (with probability 1), and at each step either increases by 1 (with probability p) or decreases by 1 (with probability $1 - p$). It is easily seen that this Markov chain corresponds precisely to the random walk (i.e., repeated gambling) model of Section 11.1.2. ∎

Finally, we note that in a group, you can create your own Markov chain, as follows (try it — it's fun!).

Example 11.2.9

Form a group of between 5 and 50 people. Each person should secretly pick out two other people from the group, their "A person" and "B person." Also, each person should have a coin.

Take any object, such as a ball, or a pen, or a stuffed frog. Give the object to one person to start. This person should then immediately flip the coin. If the coin comes up heads, they give (or throw!) the object to their A person. If it comes up tails, they give the object to their B person. The person receiving the object should then immediately flip the coin and continue the process. (If the person also says their name, then this is a great way for everyone to meet each other!)

Continue this process for a large number of turns. What patterns do you observe? Does everyone eventually receive the object? With what frequency? How long does it take the object to return to where it started? Make as many interesting observations as you can; some of them will be related to the topics that follow. ∎

11.2.2   Computing with Markov Chains

Suppose a Markov chain $\{X_n\}$ has transition probabilities $\{p_{ij}\}$, and initial distribution $\{\mu_i\}$. Then $P(X_0 = i) = \mu_i$ for all states i. What about $P(X_1 = i)$? We have the following result.

Theorem 11.2.1 Consider a Markov chain $\{X_n\}$ with state space S, transition probabilities $\{p_{ij}\}$, and initial distribution $\{\mu_i\}$. Then for any $i \in S$,

$$P(X_1 = i) = \sum_{k \in S} \mu_k p_{ki}.$$

Proof: From the law of total probability,

$$P(X_1 = i) = \sum_{k \in S} P(X_0 = k, \ X_1 = i).$$

But $P(X_0 = k, \ X_1 = i) = P(X_0 = k)\, P(X_1 = i \mid X_0 = k) = \mu_k\, p_{ki}$ and the result follows. ∎

Consider an example of this.

Example 11.2.10
Again, let $S = \{1, 2, 3\}$, and

$$(p_{ij}) = \begin{pmatrix} 1/4 & 1/4 & 1/2 \\ 1/3 & 1/3 & 1/3 \\ 0.01 & 0.01 & 0.98 \end{pmatrix}.$$

Suppose that $P(X_0 = 1) = 1/7$, $P(X_1 = 2) = 2/7$, and $P(X_1 = 3) = 4/7$. Then

$$P(X_1 = 3) = \sum_{k \in S} \mu_k p_{k3} = (1/7)(1/2) + (2/7)(1/3) + (4/7)(0.98) \approx 0.73.$$

Thus, about 73% of the time, this chain will be in state 3 after one step. ∎

To proceed, let us write

$$P_i(A) = P(A \mid X_0 = i)$$

for the probability of the event A, *assuming* that the chain starts in the state i, that is, assuming that $\mu_i = 1$ and $\mu_j = 0$ for $j \neq i$. We then see that $P_i(X_n = j)$ is the probability that, if the chain starts in state i and is run for n steps, it will end up in state j. Can we compute this?

For $n = 0$, we must have $X_0 = i$. Hence, $P_i(X_0 = j) = 1$ if $i = j$, while $P_i(X_0 = j) = 0$ if $i \neq j$.

For $n = 1$, we see that $P_i(X_1 = j) = p_{ij}$. That is, the probability that we will be at the state j after one step is given by the transition probability p_{ij}.

What about for $n = 2$? If we start at i and end up at j after 2 steps, then we have to be at *some* state after 1 step. Let k be this state. Then we see the following.

Theorem 11.2.2 We have $P_i(X_1 = k, X_2 = j) = p_{ik} p_{kj}$.

Proof: If we start at i, then the probability of jumping first to k is equal to p_{ik}. Given that we have jumped first to k, the probability of *then* jumping to j is given by p_{kj}. Hence,

$$
\begin{aligned}
P_i(X_1 = k, X_2 = j) &= P(X_1 = k, X_2 = j \mid X_0 = i) \\
&= P(X_1 = k \mid X_0 = i)\, P(X_2 = k \mid X_1 = j, X_0 = i) \\
&= p_{ik} p_{kj}. \quad\blacksquare
\end{aligned}
$$

Using this, we obtain the following.

Theorem 11.2.3 We have $P_i(X_2 = j) = \sum_{k \in S} p_{ik} p_{kj}$.
Proof: By the law of total probability,

$$P_i(X_2 = j) = \sum_{k \in S} P_i(X_1 = k, X_2 = j),$$

so the result follows from Theorem 11.2.2. $\blacksquare$

Example 11.2.11
Consider again the chain of Example 11.2.1, with $S = \{1, 2, 3\}$ and

$$(p_{ij}) = \begin{pmatrix} 0 & 1/2 & 1/2 \\ 1/3 & 1/3 & 1/3 \\ 1/4 & 1/4 & 1/2 \end{pmatrix}.$$

Then

$$
\begin{aligned}
P_1(X_2 = 3) &= \sum_{k \in S} p_{1k} p_{k3} = p_{11} p_{13} + p_{12} p_{23} + p_{13} p_{33} \\
&= (0)(1/4) + (1/2)(1/3) + (1/2)(1/2) = 1/6 + 1/4 = 5/12. \quad\blacksquare
\end{aligned}
$$

By induction (Problem 11.2.14), we obtain the following.

Theorem 11.2.4 We have

$$P_i(X_n = j) = \sum_{i_1, i_2, \ldots, i_{n-1} \in S} p_{ii_1} p_{i_1 i_2} p_{i_2 i_3} \cdots p_{i_{n-2} i_{n-1}} p_{i_{n-1} j}.$$

Proof: See Problem 11.2.14.

Theorem 11.2.4 thus gives a complete formula for the probability, starting at a state i at time 0, that the chain will be at some other state j at time n. We see from Theorem 11.2.4 that, once we know the transition probabilities p_{ij} for all $i, j \in S$, then we can compute the values of $P_i(X_n = j)$ for all $i, j \in S$, and all positive integers n. (The computations get pretty messy, though!) The quantities $P_i(X_n = j)$ are sometimes called the *higher-order transition probabilities*.

Consider an application of this.

Example 11.2.12

Consider once again the chain with $S = \{1, 2, 3\}$ and

$$(p_{ij}) = \begin{pmatrix} 0 & 1/2 & 1/2 \\ 1/3 & 1/3 & 1/3 \\ 1/4 & 1/4 & 1/2 \end{pmatrix}.$$

Then

$$P_1(X_3 = 2) = \sum_{k \in S} \sum_{\ell \in S} p_{1k} p_{k\ell} p_{\ell 3}$$

$$= p_{11}p_{11}p_{13} + p_{11}p_{12}p_{23} + p_{11}p_{13}p_{33} + p_{12}p_{21}p_{13} + p_{12}p_{22}p_{23} + p_{12}p_{23}p_{33}$$

$$\quad + p_{13}p_{31}p_{13} + p_{13}p_{32}p_{23} + p_{13}p_{33}p_{33}$$

$$= (0)(0)(1/2) + (0)(1/2)(1/3) + (0)(1/2)(1/2) + (1/2)(1/3)(1/2)$$

$$\quad + (1/2)(1/3)(1/3) + (1/2)(1/4)(1/2) + (1/2)(1/4)(1/2)$$

$$\quad + (1/2)(1/4)(1/3) + (1/2)(1/2)(1/2)$$

$$= 31/72. \quad \blacksquare$$

Finally, we note that if we write A for the matrix (p_{ij}), and write v_0 for the row vector $(\mu_i) = (P(X_0 = i))$, and write v_1 for the row vector $(P(X_1 = i))$, then Theorem 11.2.1 can be written succinctly using matrix multiplication as $v_1 = v_0 A$. That is, the (row) vector of probabilities for the chain after one step v_1, is equal to the (row) vector of probabilities for the chain after zero steps v_0, *multiplied* by the matrix A of transition probabilities. In fact, if we write v_n for the row vector $(P(X_n = i))$, then proceeding by induction, we see that $v_{n+1} = v_n A$ for each n, so therefore $v_n = v_0 A^n$, where A^n is the nth *power* of the matrix A. In this context, Theorem 11.2.4 has a particularly nice interpretation. It says that $P_i(X_n = j)$ is equal to the (i, j) entry of the matrix A^n, i.e., the nth power of the matrix A.

11.2.3 Stationary Distributions

Suppose we have Markov chain transition probabilities $\{p_{ij}\}$ on a state space S. Let $\{\pi_i : i \in S\}$ be a probability distribution on S, so that $\pi_i \geq 0$ for all i, and $\sum_{i \in S} \pi_i = 1$. We have the following definition.

Definition 11.2.1 The distribution $\{\pi_i : i \in S\}$ is *stationary* for a Markov chain with transition probabilities $\{p_{ij}\}$ on a state space S, if $\sum_{i \in S} \pi_i p_{ij} = \pi_j$ for all $j \in S$.

The reason for the terminology "stationary" is that, if the chain begins with those probabilities, then it will *always* have those same probabilities, as the following theorem and corollary show.

Theorem 11.2.5 Suppose $\{\pi_i : i \in S\}$ is a stationary distribution for a Markov chain with transition probabilities $\{p_{ij}\}$ on a state space S. Suppose that for some integer n, we have $P(X_n = i) = \pi_i$ for all $i \in S$. Then we also have $P(X_{n+1} = i) = \pi_i$ for all $i \in S$.

Proof: If $\{\pi_i\}$ is stationary, then we compute that

$$
\begin{aligned}
P(X_{n+1} = j) &= \sum_{i \in S} P(X_n = i, X_{n+1} = j) \\
&= \sum_{i \in S} P(X_n = i)\, P(X_{n+1} = j \mid X_n = i) = \sum_{i \in S} \pi_i\, p_{ij} = \pi_j. \blacksquare
\end{aligned}
$$

By induction, we obtain the following corollary.

Corollary 11.2.1 Suppose $\{\pi_i : i \in S\}$ is a stationary distribution for a Markov chain with transition probabilities $\{p_{ij}\}$ on a state space S. Suppose that for some integer n, we have $P(X_n = i) = \pi_i$ for all $i \in S$. Then we also have $P(X_m = i) = \pi_i$ for all $i \in S$ and all integers $m > n$.

The above theorem and corollary say that, once a Markov chain is in its stationary distribution, it will remain in its stationary distribution forever more.

Example 11.2.13
Consider the Markov chain with $S = \{1, 2, 3\}$, and

$$
(p_{ij}) = \begin{pmatrix} 1/2 & 1/4 & 1/4 \\ 1/2 & 1/4 & 1/4 \\ 1/2 & 1/4 & 1/4 \end{pmatrix}.
$$

No matter where this Markov chain is, it always jumps with the same probabilities, i.e., to state 1 with probability 1/2, to state 2 with probability 1/4, or to state 3 with probability 1/4.

Indeed, if we set $\pi_1 = 1/2$, $\pi_2 = 1/4$, and $\pi_3 = 1/4$, then we see that $p_{ij} = \pi_j$ for all $i, j \in S$. Hence,

$$
\sum_{i \in S} \pi_i p_{ij} = \sum_{i \in S} \pi_i \pi_j = \pi_j \sum_{i \in S} \pi_i = \pi_j (1) = \pi_j.
$$

Thus, $\{\pi_i\}$ is a stationary distribution. Hence, once in the distribution $\{\pi_i\}$, the chain will stay in the distribution $\{\pi_i\}$ forever. ∎

Example 11.2.14
Consider a Markov chain with $S = \{0, 1\}$ and

$$(p_{ij}) = \begin{pmatrix} 0.1 & 0.9 \\ 0.6 & 0.4 \end{pmatrix}.$$

If this chain had a stationary distribution $\{\pi_i\}$, then we must have that

$$\pi_0(0.1) + \pi_1(0.6) = \pi_0,$$
$$\pi_0(0.9) + \pi_1(0.4) = \pi_1.$$

The first equation gives $\pi_1(0.6) = \pi_0(0.9)$, so $\pi_1 = (3/2)(\pi_0)$. This is also consistent with the second equation. In addition, we require that $\pi_0 + \pi_1 = 1$, i.e., that $\pi_0 + (3/2)\pi_0 = 1$, so that $\pi_0 = 2/5$. Then $\pi_1 = (3/2)(2/5) = 3/5$.

We then check that the settings $\pi_0 = 2/5$ and $\pi_1 = 3/5$ satisfy the above equations. Hence, $\{\pi_i\}$ is indeed a stationary distribution for this Markov chain. ∎

Example 11.2.15
Consider next the Markov chain with $S = \{1, 2, 3\}$, and

$$(p_{ij}) = \begin{pmatrix} 0 & 1/2 & 1/2 \\ 1/2 & 0 & 1/2 \\ 1/2 & 1/2 & 0 \end{pmatrix}.$$

We see that this Markov chain has the property that, in addition to having $\sum_{j \in S} p_{ij} = 1$, for all i, it also has $\sum_{i \in S} p_{ij} = 1$, for all j. That is, not only do the *rows* of the matrix (p_{ij}) sum to 1, but so do the *columns*. (Such a matrix is sometimes called *doubly stochastic*.)

Let $\pi_1 = \pi_2 = \pi_3 = 1/3$, so that $\{\pi_i\}$ is the uniform distribution on S. Then we compute that

$$\sum_{i \in S} \pi_i p_{ij} = \sum_{i \in S} (1/3) p_{ij} = (1/3) \sum_{i \in S} p_{ij} = (1/3)(1) = \pi_j.$$

Because this is true for all j, we see that $\{\pi_i\}$ is a stationary distribution for this Markov chain. ∎

Example 11.2.16
Consider the Markov chain with $S = \{1, 2, 3\}$, and

$$(p_{ij}) = \begin{pmatrix} 1/2 & 1/4 & 1/4 \\ 1/3 & 1/3 & 1/3 \\ 0 & 1/4 & 3/4 \end{pmatrix}.$$

Does this Markov chain have a stationary distribution?

Well, if it had a stationary distribution $\{\pi_i\}$, then the following equations would have to be satisfied:

$$\begin{aligned}
\pi_1 &= (1/2)\pi_1 + (1/3)\pi_2 + (0)\pi_3, \\
\pi_2 &= (1/4)\pi_1 + (1/3)\pi_2 + (1/4)\pi_3, \\
\pi_3 &= (1/4)\pi_1 + (1/3)\pi_2 + (3/4)\pi_3.
\end{aligned}$$

The first equation gives $\pi_1 = (2/3)\pi_2$. The second equation then gives

$$(1/4)\pi_3 = \pi_2 - (1/4)\pi_1 - (1/3)\pi_2 = \pi_2 - (1/4)(2/3)\pi_2 - (1/3)\pi_2 = (1/2)\pi_2,$$

so that $\pi_3 = 2\pi_2$.

But we also require $\pi_1 + \pi_2 + \pi_3 = 1$, i.e., $(2/3)\pi_2 + \pi_2 + 2\pi_2 = 1$, so that $\pi_2 = 3/11$. Then $\pi_1 = 2/11$, and $\pi_3 = 6/11$.

It is then easily checked that the distribution given by $\pi_1 = 2/11$, $\pi_2 = 3/11$, and $\pi_3 = 6/11$ satisfies the preceding equations, so it is indeed a stationary distribution for this Markov chain. ∎

Example 11.2.17

Consider again random walk on the circle, as in Example 11.2.6. We observe that for any state j, there are precisely three states i (namely, the state $i = j$, the state one clockwise from j, and the state one counterclockwise from j) with $p_{ij} = 1/3$. Hence, $\sum_{i \in S} p_{ij} = 1$. That is, the transition matrix (p_{ij}) is again doubly stochastic.

It then follows, just as in Example 11.2.15, that the uniform distribution, given by $\pi_i = 1/d$ for $i = 0, 1, \dots, d-1$, is a stationary distribution for this Markov chain. ∎

Example 11.2.18

For Ehrenfest's Urn (Example 11.2.7), it is not obvious what might be a stationary distribution. However, a possible solution emerges by thinking about each ball *individually*. Indeed, any given ball usually stays still, but occasionally gets flipped from one urn to the other. So it seems reasonable that in stationarity it should be equally likely to be in either urn, i.e., have probability $1/2$ of being in urn 1.

If this is so, then the total number of balls in urn 1 would have the distribution Binomial$(n, 1/2)$, since there would be n balls each having probability $1/2$ of being in urn 1.

To test this, we set $\pi_i = \binom{d}{i}/2^d$, for $i = 0, 1, \dots, d$. We then compute that if $1 \le j \le d-1$, then

$$\begin{aligned}
\sum_{i \in S} \pi_i p_{ij} &= \pi_{j-1} p_{j-1,j} + \pi_{j+1} p_{j+1,j} \\
&= \binom{d}{j-1} \frac{1}{2^d} \frac{d - (j-1)}{d} + \binom{d}{j+1} \frac{1}{2^d} \frac{j+1}{d} \\
&= \binom{d-1}{j-1} \frac{1}{2^d} + \binom{d-1}{j} \frac{1}{2^d}.
\end{aligned}$$

Next, we use the identity known as *Pascal's triangle*, which says that

$$\binom{d-1}{j-1} + \binom{d-1}{j} = \binom{d}{j}.$$

Hence, we conclude that

$$\sum_{i \in S} \pi_i p_{ij} = \binom{d}{j} \frac{1}{2^d} = \pi_j.$$

With minor modifications (Problem 11.2.15), the preceding argument works for $j = 0$ and $j = d$ as well. We therefore conclude that $\sum_{i \in S} \pi_i p_{ij} = \pi_j$, for all $j \in S$. Hence, $\{\pi_i\}$ is a stationary distribution. ∎

One easy way to check for stationarity is the following.

Definition 11.2.2 A Markov chain is said to be *reversible* with respect to a distribution $\{\pi_i\}$ if, for all $i, j \in S$, we have $\pi_i p_{ij} = \pi_j p_{ji}$.

Theorem 11.2.6 If a Markov chain is reversible with respect to $\{\pi_i\}$, then $\{\pi_i\}$ is a stationary distribution for the chain.

Proof: We compute, using reversibility, that for any $j \in S$,

$$\sum_{i \in S} \pi_i p_{ij} = \sum_{i \in S} \pi_j p_{ji} = \pi_j \sum_{i \in S} p_{ji} = \pi_j (1) = \pi_j.$$

Hence, $\{\pi_i\}$ is a stationarity distribution. ∎

Example 11.2.19
Suppose $S = \{1, 2, 3, 4, 5\}$, and the transition probabilities are given by

$$(p_{ij}) = \begin{pmatrix} 1/3 & 2/3 & 0 & 0 & 0 \\ 1/3 & 0 & 2/3 & 0 & 0 \\ 0 & 1/3 & 0 & 2/3 & 0 \\ 0 & 0 & 1/3 & 0 & 2/3 \\ 0 & 0 & 0 & 1/3 & 2/3 \end{pmatrix}.$$

It is not immediately clear what stationary distribution this chain may possess. Furthermore, to compute directly as in Example 11.2.16 would be quite messy.

On the other hand, we observe that for $1 \le i \le 4$, we always have $p_{i,i+1} = 2 p_{i+1,i}$. Hence, if we set $\pi_i = C2^i$ for some $C > 0$, then we will have

$$\pi_i p_{i,i+1} = C2^i p_{i,i+1} = C2^i 2 p_{i+1,i},$$

while

$$\pi_{i+1} p_{i+1,i} = C2^{i+1} p_{i+1,i}.$$

Hence, $\pi_i p_{i,i+1} = \pi_{i+1} p_{i+1,i}$ for each i.

Furthermore, $p_{ij} = 0$ if i and j differ by at least 2. It follows that $\pi_i p_{ij} = \pi_j p_{ji}$ for each $i, j \in S$. Hence, the chain is reversible with respect to $\{\pi_i\}$ and so $\{\pi_i\}$ is a stationary distribution for the chain.

Finally, we solve for C. We need $\sum_{i \in S} \pi_i = 1$. Hence, we must have $C = 1/\sum_{i \in S} 2^i = 1/\sum_{i=1}^{5} 2^i = 1/63$. Thus, $\pi_i = 2^i/63$, for $i \in S$. ∎

11.2.4   Markov Chain Limit Theorem

Suppose now that $\{X_n\}$ is a Markov chain, which has a stationary distribution $\{\pi_i\}$. We have already seen that, if $P(X_n = i) = \pi_i$ for all i for some n, then also $P(X_m = i) = \pi_i$ for all i for all $m > n$.

Suppose now that it is *not* the case that $P(X_n = i) = \pi_i$ for all i. One might still expect that, if the chain is run for a long time (i.e., $n \to \infty$), then the probability of being at a particular state $i \in S$ might converge to π_i, regardless of the initial state chosen. That is, one might expect that

$$\lim_{n \to \infty} P(X_n = i) = \pi_i, \tag{11.2.1}$$

for each $i \in S$, regardless of the initial distribution $\{\mu_i\}$.

This is not true in complete generality, as the following two examples show. However, we shall see in Theorem 11.2.8 that this is indeed true for most Markov chains.

Example 11.2.20
Suppose that $S = \{1, 2\}$ and that the transition probabilities are given by

$$(p_{ij}) = \begin{pmatrix} 1 & 0 \\ 0 & 1 \end{pmatrix}.$$

That is, this Markov chain never moves at all! Suppose also that $\mu_1 = 1$, i.e., that we always have $X_0 = 1$.

In this case, *any* distribution is stationary for this chain. In particular, we can take $\pi_1 = \pi_2 = 1/2$ as a stationary distribution. On the other hand, we clearly have $P_1(X_n = 1) = 1$ for all n. Because $\pi_1 = 1/2$, and $1 \neq 1/2$, we do *not* have $\lim_{n \to \infty} P(X_n = i) = \pi_i$ in this case.

We shall see later that this Markov chain is not "irreducible," which is the obstacle to convergence. ∎

Example 11.2.21
Suppose again that $S = \{1, 2\}$, but that this time the transition probabilities are given by

$$(p_{ij}) = \begin{pmatrix} 0 & 1 \\ 1 & 0 \end{pmatrix}.$$

That is, this Markov chain always moves from 1 to 2, and from 2 to 1. Suppose again that $\mu_1 = 1$, i.e., that we always have $X_0 = 1$.

We may again take $\pi_1 = \pi_2 = 1/2$ as a stationary distribution (in fact, this time the stationary distribution is unique). On the other hand, this time we clearly have $P_1(X_n = 1) = 1$ for n even, and $P_1(X_n = 1) = 0$ for n odd. Hence, again we do *not* have $\lim_{n \to \infty} P_1(X_n = 1) \to \pi_1 = 1/2$.

We shall see that here the obstacle to convergence is that the Markov chain is "periodic," with period 2. ∎

In light of these examples, we make some definitions.

Definition 11.2.3 A Markov chain is *irreducible* if it is possible for the chain to move from any state to any other state. Equivalently, the Markov chain is irreducible if for any $i, j \in S$, there is a positive integer n with $P_i(X_n = j) > 0$.

Thus, the Markov chain of Example 11.2.20 is not irreducible, because it is not possible to get from state 1 to state 2. Indeed, in that case $P_1(X_n = 2) = 0$ for all n.

Example 11.2.22
Consider the Markov chain with $S = \{1, 2, 3\}$, and

$$(p_{ij}) = \begin{pmatrix} 1/2 & 1/2 & 0 \\ 1/2 & 1/4 & 1/4 \\ 1/2 & 1/4 & 1/4 \end{pmatrix}.$$

For this chain, it is not possible to get from state 1 to state 3 in one step. On the other hand, it is possible to get from state 1 to state 2, and then from state 2 to state 3. Hence, this chain is still irreducible. ∎

Example 11.2.23
Consider the Markov chain with $S = \{1, 2, 3\}$, and

$$(p_{ij}) = \begin{pmatrix} 1/2 & 1/2 & 0 \\ 3/4 & 1/4 & 0 \\ 1/2 & 1/4 & 1/4 \end{pmatrix}.$$

For this chain, it is not possible to get from state 1 to state 3 in one step. Furthermore, it is not possible to get from state 2 to state 3, either. In fact, there is no way to *ever* get from state 1 to state 3, in any number of steps. Hence, this chain is *not* irreducible. ∎

Clearly, if a Markov chain is not irreducible, then the Markov chain convergence (11.2.1) will not always hold, because it will be impossible to ever get to certain states of the chain.

We also need the following definition.

Definition 11.2.4 Given Markov chain transitions $\{p_{ij}\}$ on a state space S, and a state $i \in S$, the *period* of i is the greatest common divisor (g.c.d.) of the set $\{n \geq 1 : p_{ii}^{(n)} > 0\}$, where $p_{ii}^{(n)} = P(X_n = i \mid X_0 = i)$.

That is, the period of i is the g.c.d. of the times at which it is possible to travel from i to i. For example, the period of i is 2 if it is only possible to travel from i to i in an even number of steps. (Such was the case for Example 11.2.21.) On the other hand, if $p_{ii} > 0$, then clearly the period of i is 1.

Clearly, if the period of some state is greater than 1, then again (11.2.1) will not always hold, because the chain will be able to reach certain states at certain times only. This prompts the following definition.

Definition 11.2.5 A Markov chain is *aperiodic* if the period of each state is equal to 1.

Example 11.2.24

Consider the Markov chain with $S = \{1, 2, 3\}$, and

$$(p_{ij}) = \begin{pmatrix} 0 & 1 & 0 \\ 0 & 0 & 1 \\ 1 & 0 & 0 \end{pmatrix}.$$

For this chain, from state 1 it is possible only to get to state 2. And from state 2 it is possible only to get to state 3. Then from state 3 it is possible only to get to state 1. Hence, it is possible only to return to state 1 after an integer multiple of 3 steps. Hence, state 1 (and, indeed, all three states) has period equal to 3, and the chain is *not* aperiodic. ∎

Example 11.2.25

Consider the Markov chain with $S = \{1, 2, 3\}$, and

$$(p_{ij}) = \begin{pmatrix} 0 & 1 & 0 \\ 0 & 0 & 1 \\ 1/2 & 0 & 1/2 \end{pmatrix}.$$

For this chain, from state 1 it is possible only to get to state 2. And from state 2 it is possible only to get to state 3. However, from state 3 it is possible to get to either state 1 or state 3. Hence, it is possible to return to state 1 after either 3 or 4 steps. Because the greatest common divisor of 3 and 4 is 1, we conclude that the period of state 1 (and, indeed, of all three states) is equal to 1, and the chain is indeed aperiodic. ∎

We note the following simple fact.

Theorem 11.2.7 If a Markov chain has $p_{ij} > 0$ for all $i, j \in S$, then the chain is irreducible and aperiodic.

Proof: If $p_{ij} > 0$ for all $i, j \in S$, then $P_i(X_1 = j) > 0$ for all $i, j \in S$. Hence, the Markov chain must be irreducible.

Also, if $p_{ij} > 0$ for all $i, j \in S$, then the set $\{n \geq 1 : p_{ii}^{(n)} > 0\}$ contains the value $n = 1$ (and, indeed, all positive integers n). Hence, its greatest common divisor must be 1. Therefore, each state i has period 1, so the chain is aperiodic. ∎

In terms of the preceding definitions, we have the following very important theorem about Markov chain convergence.

Theorem 11.2.8 Suppose a Markov chain is irreducible and aperiodic, and has a stationary distribution $\{\pi_i\}$. Then regardless of the initial distribution $\{\mu_i\}$, we have $\lim_{n \to \infty} P(X_n = i) = \pi_i$ for all states i.

Proof: For a proof of this, see more advanced probability books, e.g., pages 77–78 of *A First Look at Rigorous Probability Theory* by J. S. Rosenthal (World Scientific Publishing, Singapore, 2000).

Theorem 11.2.8 shows that stationary distributions are even more important. Not only does a Markov chain remain in a stationary distribution once it is there, but for most chains (irreducible and aperiodic ones), the probabilities *converge* to the stationary distribution in any case. Hence, the stationary distribution provides fundamental information about the long-term behavior of the Markov chain.

Example 11.2.26

Consider again the Markov chain with $S = \{1, 2, 3\}$, and

$$(p_{ij}) = \begin{pmatrix} 1/2 & 1/4 & 1/4 \\ 1/2 & 1/4 & 1/4 \\ 1/2 & 1/4 & 1/4 \end{pmatrix}.$$

We have already seen that if we set $\pi_1 = 1/2$, $\pi_2 = 1/4$, and $\pi_3 = 1/4$, then $\{\pi_i\}$ is a stationary distribution. Furthermore, we see that $p_{ij} > 0$ for all $i, j \in S$, so by Theorem 11.2.7 the Markov chain must be irreducible and aperiodic.

We conclude that $\lim_{n \to \infty} P(X_n = i) = \pi_i$ for all states i. For example, $\lim_{n \to \infty} P(X_n = 1) = 1/2$. (Also, this limit does not depend on the initial distribution, so, for example, $\lim_{n \to \infty} P_1(X_n = 1) = 1/2$ and $\lim_{n \to \infty} P_2(X_n = 1) = 1/2$, as well.)

In fact, for this example we will have $P(X_n = i) = \pi_i$ for all i provided $n \geq 1$. ∎

Example 11.2.27

Consider again the Markov chain of Example 11.2.3, with $S = \{0, 1\}$ and

$$(p_{ij}) = \begin{pmatrix} 0.1 & 0.9 \\ 0.6 & 0.4 \end{pmatrix}.$$

We have already seen that this Markov chain has a stationary distribution, given by $\pi_0 = 2/5$ and $\pi_1 = 3/5$.

Furthermore, because $p_{ij} > 0$ for all $i, j \in S$, this Markov chain is irreducible and aperiodic. Therefore, we conclude that $\lim_{n \to \infty} P(X_n = i) = \pi_i$. So, if (say) $n = 100$, then we will have $P(X_{100} = 0) \approx 2/5$, and $P(X_{100} = 1) \approx 3/5$. Once again, this conclusion does not depend on the initial distribution, so, e.g., $\lim_{n \to \infty} P_0(X_n = i) = \lim_{n \to \infty} P_1(X_n = i) = \pi_i$ as well. ∎

Example 11.2.28

Consider again the Markov chain of Example 11.2.16, with $S = \{1, 2, 3\}$, and

$$(p_{ij}) = \begin{pmatrix} 1/2 & 1/4 & 1/4 \\ 1/3 & 1/3 & 1/3 \\ 0 & 1/4 & 3/4 \end{pmatrix}.$$

We have already seen that this chain has a stationary distribution $\{\pi_i\}$ given by $\pi_1 = 2/11$, $\pi_2 = 3/11$, and $\pi_3 = 6/11$.

Now, in this case we do not have $p_{ij} > 0$ for all $i, j \in S$, because $p_{31} = 0$. On the other hand, $p_{32} > 0$ and $p_{21} > 0$, so by Theorem 11.2.3, we have

$$P_3(X_2 = 1) = \sum_{k \in S} p_{3k} p_{k1} \geq p_{32} p_{21} > 0.$$

Hence, the chain is still irreducible.

Similarly, we have $P_3(X_2 = 3) \geq p_{32}p_{23} > 0$, and $P_3(X_3 = 3) \geq p_{32}p_{21}p_{13} > 0$. Therefore, because the g.c.d. of 2 and 3 is 1, we see that the g.c.d. of the set of n with $P_3(X_n = 3) > 0$ is also 1. Hence, the chain is still aperiodic.

Because the chain is irreducible and aperiodic, it follows from Theorem 11.2.8 that $\lim_{n\to\infty} P(X_n = i) = \pi_i$, for all states i. Hence, $\lim_{n\to\infty} P(X_n = 1) = 2/11$, $\lim_{n\to\infty} P(X_n = 2) = 3/11$, and $\lim_{n\to\infty} P(X_n = 3) = 6/11$. Thus, if (say) $n = 500$, then we expect that $P(X_{500} = 1) \approx 2/11$, $P(X_{500} = 2) \approx 3/11$, and $P(X_{500} = 3) \approx 6/11$. ∎

Summary of Section 11.2

- A Markov chain is a sequence $\{X_n\}$ of random variables, having transition probabilities $\{p_{ij}\}$ such that $P(X_{n+1} = j \mid X_n = i) = p_{ij}$, and having an initial distribution $\{\mu_i\}$ such that $P(X_0 = i) = \mu_i$.

- There are many different examples of Markov chains.

- All probabilities for all the X_n can be computed in terms of $\{\mu_i\}$ and $\{p_{ij}\}$.

- A distribution $\{\pi_i\}$ is stationary for the chain if $\sum_{i \in S} \pi_i p_{ij} = \pi_j$ for all $j \in S$.

- If the Markov chain is irreducible and aperiodic, and $\{\pi_i\}$ is stationary, then $\lim_{n\to\infty} P(X_n = i) = \pi_i$ for all $i \in S$.

Exercises

11.2.1 Consider a Markov chain with $S = \{1, 2, 3\}$, $\mu_1 = 0.7$, $\mu_2 = 0.1$, $\mu_3 = 0.2$, and

$$(p_{ij}) = \begin{pmatrix} 1/4 & 1/4 & 1/2 \\ 1/6 & 1/2 & 1/3 \\ 1/8 & 3/8 & 1/2 \end{pmatrix}.$$

Compute the following quantities.
(a) $P(X_0 = 1)$
(b) $P(X_0 = 2)$
(c) $P(X_0 = 3)$
(d) $P(X_1 = 2 \mid X_0 = 1)$
(e) $P(X_3 = 2 \mid X_2 = 1)$
(f) $P(X_1 = 2 \mid X_0 = 2)$
(g) $P(X_1 = 2)$

11.2.2 Consider a Markov chain with $S = \{\text{high}, \text{low}\}$, $\mu_{\text{high}} = 1/3$, $\mu_{\text{low}} = 2/3$, and

$$(p_{ij}) = \begin{pmatrix} 1/4 & 3/4 \\ 1/6 & 5/6 \end{pmatrix}.$$

Compute the following quantities.
(a) $P(X_0 = \text{high})$
(b) $P(X_0 = \text{low})$
(c) $P(X_1 = \text{high} \,|\, X_0 = \text{high})$
(d) $P(X_3 = \text{high} \,|\, X_2 = \text{low})$
(e) $P(X_1 = \text{high})$

11.2.3 Consider a Markov chain with $S = \{0, 1\}$, and

$$(p_{ij}) = \begin{pmatrix} 0.2 & 0.8 \\ 0.3 & 0.7 \end{pmatrix}.$$

(a) Compute $P_i(X_2 = j)$ for all four combinations of $i, j \in S$.
(b) Compute $P_0(X_3 = 1)$.

11.2.4 Consider again the Markov chain with $S = \{0, 1\}$ and

$$(p_{ij}) = \begin{pmatrix} 0.2 & 0.8 \\ 0.3 & 0.7 \end{pmatrix}.$$

(a) Compute a stationary distribution $\{\pi_i\}$ for this chain.
(b) Compute $\lim_{n \to \infty} P_0(X_n = 0)$.
(c) Compute $\lim_{n \to \infty} P_1(X_n = 0)$.

11.2.5 Consider the Markov chain of Example 11.2.5, with $S = \{1, 2, 3, 4, 5, 6, 7\}$ and

$$(p_{ij}) = \begin{pmatrix} 1 & 0 & 0 & 0 & 0 & 0 & 0 \\ 1/2 & 0 & 1/2 & 0 & 0 & 0 & 0 \\ 0 & 1/5 & 4/5 & 0 & 0 & 0 & 0 \\ 0 & 0 & 1/3 & 1/3 & 1/3 & 0 & 0 \\ 1/10 & 0 & 0 & 0 & 7/10 & 0 & 1/5 \\ 0 & 0 & 0 & 0 & 0 & 0 & 1 \\ 0 & 0 & 0 & 0 & 0 & 1 & 0 \end{pmatrix}.$$

Compute the following quantities.
(a) $P_2(X_1 = 1)$
(b) $P_2(X_1 = 2)$
(c) $P_2(X_1 = 3)$
(d) $P_2(X_2 = 1)$
(e) $P_2(X_2 = 2)$
(f) $P_2(X_2 = 3)$
(g) $P_2(X_3 = 3)$
(h) $P_2(X_3 = 1)$
(i) $P_2(X_1 = 7)$
(j) $P_2(X_2 = 7)$
(k) $P_2(X_3 = 7)$
(l) $\max_n P_2(X_n = 7)$ (i.e., the largest probability of going from state 2 to state 7 in n steps, for any n)
(m) Is this Markov chain irreducible?

11.2.6 For each of the following transition probability matrices, determine (with explanation) whether it is irreducible, and whether it is aperiodic.
(a)

$$(p_{ij}) = \begin{pmatrix} 0.2 & 0.8 \\ 0.3 & 0.7 \end{pmatrix}$$

(b)

$$(p_{ij}) = \begin{pmatrix} 1/4 & 1/4 & 1/2 \\ 1/6 & 1/2 & 1/3 \\ 1/8 & 3/8 & 1/2 \end{pmatrix}$$

(c)

$$(p_{ij}) = \begin{pmatrix} 0 & 1 \\ 0.3 & 0.7 \end{pmatrix}$$

(d)

$$(p_{ij}) = \begin{pmatrix} 0 & 1 & 0 \\ 1/3 & 1/3 & 1/3 \\ 0 & 1 & 0 \end{pmatrix}$$

(e)

$$(p_{ij}) = \begin{pmatrix} 0 & 1 & 0 \\ 0 & 0 & 1 \\ 1 & 0 & 0 \end{pmatrix}$$

(f)

$$(p_{ij}) = \begin{pmatrix} 0 & 1 & 0 \\ 0 & 0 & 1 \\ 1/2 & 0 & 1/2 \end{pmatrix}$$

11.2.7 Compute a stationary distribution for the Markov chain of Example 11.2.4. (Hint: Don't forget Example 11.2.15.)

11.2.8 Show that the random walk on circle process (Example 11.2.6) is
(a) irreducible;
(b) aperiodic; and
(c) reversible with respect to its stationary distribution.

11.2.9 Show that the Ehrenfest's Urn process (Example 11.2.7) is
(a) irreducible;
(b) *not* aperiodic; and
(c) reversible with respect to its stationary distribution.

11.2.10 Consider the Markov chain with $S = \{1, 2, 3\}$, and

$$(p_{ij}) = \begin{pmatrix} 0 & 1 & 0 \\ 0 & 0 & 1 \\ 1/2 & 1/2 & 0 \end{pmatrix}.$$

(a) Determine (with explanation) whether or not the chain is irreducible.
(b) Determine (with explanation) whether or not the chain is aperiodic.

(c) Compute a stationary distribution for the chain.

(d) Compute (with explanation) a good approximation to $P_1(X_{500} = 2)$.

11.2.11 Repeat all four parts of Exercise 11.2.10 if $S = \{1, 2, 3\}$ and

$$(p_{ij}) = \begin{pmatrix} 0 & 1/2 & 1/2 \\ 0 & 0 & 1 \\ 1/2 & 1/2 & 0 \end{pmatrix}.$$

Problems

11.2.12 Consider a Markov chain with $S = \{1, 2, 3, 4, 5\}$, and

$$(p_{ij}) = \begin{pmatrix} 1/5 & 4/5 & 0 & 0 & 0 \\ 1/5 & 0 & 4/5 & 0 & 0 \\ 0 & 1/5 & 0 & 4/5 & 0 \\ 0 & 0 & 1/5 & 0 & 4/5 \\ 0 & 0 & 0 & 1/5 & 4/5 \end{pmatrix}.$$

Compute a stationary distribution $\{\pi_i\}$ for this chain. (Hint: Use reversibility, as in Example 11.2.19.)

11.2.13 Suppose 100 lily pads are arranged in a circle, numbered $0, 1, \ldots, 99$ (with pad 99 next to pad 0). Suppose a frog begins at pad 0, and each second either jumps one pad clockwise, or jumps one pad counter-clockwise, or stays where it is, each with probability $1/3$. After doing this for a month, what is the approximate probability that the frog will be at pad 55? (Hint: The frog is doing random walk on the circle, as in Example 11.2.6. Also, the results of Example 11.2.17, and of Theorem 11.2.8, may help.)

11.2.14 Prove Theorem 11.2.4. (Hint: Proceed as in the proof of Theorem 11.2.3, and use induction.)

11.2.15 In Example 11.2.18, prove that $\sum_{i \in S} \pi_i p_{ij} = \pi_j$ when $j = 0$ and when $j = d$.

Discussion Topic

11.2.16 With a group, create the "human Markov chain" of Example 12.2.9. Make as many observations as you can about the long-term behavior of the resulting Markov chain.

11.3 Markov Chain Monte Carlo

In Section 4.5, we saw that it is possible to estimate various quantities (such as properties of real objects through experimentation, or the value of complicated sums or integrals) by using *Monte Carlo techniques*, namely, by generating appropriate random variables on a computer. Furthermore, we have seen in Section 2.10 that it is quite easy to generate random variables having certain special distributions. The Monte Carlo method was used several times in Chapters 6, 7, 9, and 10 to assist in the implementation of various statistical methods.

However, for many (in fact most!) probability distributions, there is no simple, direct way to simulate (on a computer) random variables having such a distribution. We illustrate this with an example.

Example 11.3.1

Let Z be a random variable taking values on the set of all integers, with

$$P(Z = j) = C(j - 1/2)^4 e^{-3|j|} \cos^2(j) \qquad (11.3.1)$$

for $j = \ldots, -2, -1, 0, 1, 2, 3, \ldots$, where $C = 1/\sum_{j=-\infty}^{\infty}(j - 1/2)^4 e^{-3|j|} \cos^2(j)$. Now suppose that we want to compute the quantity $A = E((Z - 20)^2)$.

Well, *if* we could generate i.i.d. random variables $Y_1, Y_2, \ldots, Y_M$ with distribution given by (11.3.1), for very large M, then we could estimate A by

$$A \approx \hat{A} = \frac{1}{M} \sum_{i=1}^{M} (Y_i - 20)^2.$$

Then $\hat{A}$ would be a *Monte Carlo estimate* of A.

The problem, of course, is that it is not easy to generate random variables Y_i with this distribution. In fact, it is not even easy to compute the value of C. ∎

Surprisingly, the difficulties described in Example 11.3.1 can sometimes be solved using Markov chains. We illustrate this idea as follows.

Example 11.3.2

In the context of Example 11.3.1, suppose we could find a Markov chain on the state space $S = \{\ldots, -2, -1, 0, 1, 2, \ldots\}$ of all integers, which was irreducible and aperiodic and which had a stationary distribution given by $\pi_j = C(j - 1/2)^4 e^{-3|j|} \cos^2(j)$ for $j \in S$.

If we did, then we could run the Markov chain for a long time N, to get random values $X_0, X_1, X_2, \ldots, X_N$. For large enough N, by Theorem 11.2.8, we would have

$$P(X_N = j) \approx \pi_j = C(j - 1/2)^4 e^{-3|j|} \cos^2(j).$$

Hence, if we set $Y_1 = X_N$, then we would have $P(Y_1 = j)$ approximately equal to (11.3.1), for all integers j. That is, the value of X_N would be *approximately* as good as a true random variable Y_1 with this distribution.

Once the value of Y_1 was generated, then we could repeat the process by *again* running the Markov chain, this time to generate new random values

$$X_0^{[2]}, X_1^{[2]}, X_2^{[2]}, \ldots, X_N^{[2]}$$

(say). We would then have

$$P(X_N^{[2]} = j) \approx \pi_j = C(j - 1/2)^4 e^{-3|j|} \cos^2(j).$$

Hence, if we set $Y_2 = X_N^{[2]}$, then we would have $P(Y_2 = j)$ approximately equal to (11.3.1), for all integers j.

Continuing in this way, we could generate values $Y_1, Y_2, Y_3, \ldots, Y_M$, such that these are approximately i.i.d. from the distribution given by (11.3.1). We could then, as before, estimate A by

$$A \approx \hat{A} = \frac{1}{M} \sum_{i=1}^{M} (Y_i - 20)^2.$$

This time, the approximation has two sources of error. First, there is Monte Carlo error, because M might not be large enough. Second, there is Markov chain error, because N might not be large enough. However, if M and N are both very large, then $\hat{A}$ will be a good approximation to A. ∎

We summarize the method of the preceding example in the following theorem.

Theorem 11.3.1 (*The Markov chain Monte Carlo method*) Suppose we wish to estimate the expected value $A = E\left(h(Z)\right)$, where $P(Z = j) = \pi_j$ for $j \in S$, with $P(Z = j) = 0$ for $j \notin S$. Suppose for $i = 1, 2, \ldots, M$, we can generate values $X_0^{[i]}, X_1^{[i]}, X_2^{[i]}, \ldots, X_N^{[i]}$ from some Markov chain that is irreducible and aperiodic, and has $\{\pi_j\}$ as a stationary distribution. Let

$$\hat{A} = \frac{1}{M} \sum_{i=1}^{M} h(X_N^{[i]}).$$

If M and N are sufficiently large, then $A \approx \hat{A}$.

It is somewhat inefficient to run M different Markov chains. Instead, practitioners often just run a single Markov chain, and average over the different values of the chain. For an irreducible Markov chain run long enough, this will again converge to the right answer, as the following theorem states.

Theorem 11.3.2 (*The single-chain Markov chain Monte Carlo method*) Suppose we wish to estimate the expected value $A = E\left(h(Z)\right)$, where $P(Z = j) = \pi_j$ for $j \in S$, with $P(Z = j) = 0$ for $j \notin S$. Suppose we can generate values $X_0, X_1, X_2, \ldots, X_N$ from some Markov chain that is irreducible and aperiodic, and has $\{\pi_j\}$ as a stationary distribution. For some integer $B \geq 0$, let

$$\hat{A} = \frac{1}{N - B + 1} \sum_{i=B+1}^{N} h(X_i).$$

If $N - B$ is sufficiently large, then $A \approx \hat{A}$.

Here B is the *burn-in* time, designed to remove the influence of the chain's starting value X_0. The best choice of B remains controversial among statisticians. However, if the starting value X_0 is "reasonable," then it is okay to take

$B = 0$, provided that N is sufficiently large. This is what was done, for instance, in Example 7.3.2.

These theorems indicate that, if we can construct a Markov chain that has $\{\pi_i\}$ as a stationary distribution, then we can use that Markov chain to estimate quantities associated with $\{\pi_i\}$. This is a very helpful trick, and it has made the Markov chain Monte Carlo method into one of the most popular techniques in the entire subject of computational statistics.

However, for this technique to be useful, we need to be able to construct a Markov chain that has $\{\pi_i\}$ as a stationary distribution. This sounds like a difficult problem! Indeed, if $\{\pi_i\}$ were very *simple*, then we wouldn't need to use Markov chain Monte Carlo at all. But if $\{\pi_i\}$ is complicated, then how can we possibly construct a Markov chain that has that particular stationary distribution?

Remarkably, this problem turns out to be much easier to solve than one might expect. We now discuss one of the best solutions, the Metropolis–Hastings algorithm.

11.3.1 The Metropolis–Hastings Algorithm

Suppose we are given a probability distribution $\{\pi_i\}$ on a state space S. How can we construct a Markov chain on S that has $\{\pi_i\}$ as a stationary distribution?

One answer is given by the *Metropolis–Hastings algorithm*. It designs a Markov chain that proceeds in two stages. In the first stage, a new point is *proposed* from some *proposal distribution*. In the second stage, the proposed point is either *accepted* or *rejected*. If the proposed point is accepted, then the Markov chain moves there. If it is rejected, then the Markov chain stays where it is. By choosing the probability of accepting to be just right, we end up creating a Markov chain that has $\{\pi_i\}$ as a stationary distribution.

The details of the algorithm are as follows. We start with a state space S, and a probability distribution $\{\pi_i\}$ on S. We then choose some (simple) Markov chain transition probabilities $\{q_{ij} : i, j \in S\}$ called the *proposal distribution*. Thus, we require that $q_{ij} \geq 0$, and $\sum_{j \in S} q_{ij} = 1$ for each $i \in S$. However, we do *not* assume that $\{\pi_i\}$ is a stationary distribution for the chain $\{q_{ij}\}$; indeed, the chain $\{q_{ij}\}$ might not even *have* a stationary distribution.

Given $X_n = i$, the Metropolis–Hastings algorithm computes the value X_{n+1} as follows.

1. Choose $Y_{n+1} = j$ according to the Markov chain $\{q_{ij}\}$.

2. Set $\alpha_{ij} = \min\left\{1, \frac{\pi_j q_{ji}}{\pi_i q_{ij}}\right\}$ (the *acceptance probability*).

3. With probability α_{ij}, let $X_{n+1} = Y_{n+1} = j$ (i.e., *accepting* the proposal Y_{n+1}). Otherwise, with probability $1 - \alpha_{ij}$, let $X_{n+1} = X_n = i$ (i.e., *rejecting* the proposal Y_{n+1}).

The reason for this unusual algorithm is given by the following theorem.

Theorem 11.3.3 The preceding Metropolis–Hastings algorithm results in a Markov chain $X_0, X_1, X_2, \ldots$, which has $\{\pi_i\}$ as a stationary distribution.

Proof: See Section 11.7 for the proof.

We consider some applications of this algorithm.

Example 11.3.3

As in Example 11.3.1, suppose $S = \{\ldots, -2, -1, 0, 1, 2, \ldots\}$, and

$$\pi_j = C(j - 1/2)^4 e^{-3|j|} \cos^2(j),$$

for $j \in S$. We shall construct a Markov chain having $\{\pi_i\}$ as a stationary distribution.

We first need to choose some simple Markov chain $\{q_{ij}\}$. We let $\{q_{ij}\}$ be simple random walk with $p = 1/2$, so that $q_{ij} = 1/2$ if $j = i+1$ or $j = i-1$, and $q_{ij} = 0$ otherwise.

We then compute that if $j = i + 1$ or $j = i - 1$, then

$$
\begin{aligned}
\alpha_{ij} &= \min\left\{1, \frac{q_{ji}\pi_j}{q_{ij}\pi_i}\right\} = \min\left\{1, \frac{(1/2)C(j-1/2)^4 e^{-3|j|}\cos^2(j)}{(1/2)C(i-1/2)^4 e^{3i}\cos^2(i)}\right\} \\
&= \min\left\{1, \frac{(j-1/2)^4 e^{-3|j|}\cos^2(j)}{(i-1/2)^4 e^{3i}\cos^2(i)}\right\}. \quad\quad (11.3.2)
\end{aligned}
$$

Note that C has cancelled out, so that α_{ij} does not depend on C. (In fact, this will always be the case.) Hence, we see that α_{ij}, while somewhat messy, is still very easy for a computer to calculate.

Given $X_n = i$, the Metropolis–Hastings algorithm computes the value X_{n+1} as follows.

1. Let $Y_{n+1} = X_n + 1$ or $Y_{n+1} = X_n - 1$, with probability $1/2$ each.

2. Let $j = Y_{n+1}$, and compute α_{ij} as in (11.3.2).

3. With probability α_{ij}, let $X_{n+1} = Y_{n+1} = j$. Otherwise, with probability $1 - \alpha_{ij}$, let $X_{n+1} = X_n = i$.

These steps can all be easily performed on a computer. If we repeat this for $n = 0, 1, 2, \ldots, N-1$, for some large number N of iterations, then we will obtain a random variable X_N, where $P(X_N = j) \approx \pi_j = C(j-1/2)^4 e^{-3|j|}\cos^2(j)$, for all $j \in S$. ∎

Example 11.3.4

Again, let $S = \{\ldots, -2, -1, 0, 1, 2, \ldots\}$, and this time let $\pi_j = Ke^{-j^4}$, for $j \in S$. Let the proposal distribution $\{q_{ij}\}$ correspond to a simple random walk with $p = 1/4$, so that $Y_{n+1} = X_n + 1$ with probability $1/4$, and $Y_{n+1} = X_n - 1$ with probability $3/4$.

In this case, we compute that if $j = i + 1$, then

$$\alpha_{ij} = \min\left\{1, \frac{q_{ji}\pi_j}{q_{ij}\pi_i}\right\} = \min\left\{1, \frac{(3/4)Ke^{-j^4}}{(1/4)Ke^{-i^4}}\right\} = \min\left\{1, 3e^{j^4-i^4}\right\}. \quad (11.3.3)$$

If instead $j = i - 1$, then

$$\alpha_{ij} = \min\left\{1, \frac{q_{ji}\pi_j}{q_{ij}\pi_i}\right\} = \min\left\{1, \frac{(1/4)Ke^{-j^4}}{(3/4)Ke^{-i^4}}\right\}$$

$$= \min\left\{1, (1/3)e^{j^4 - i^4}\right\}. \tag{11.3.4}$$

(Note that the constant K has again cancelled out, as expected.) Hence, again α_{ij} is very easy for a computer to calculate.

Given $X_n = i$, the Metropolis–Hastings algorithm computes the value X_{n+1} as follows.

1. Let $Y_{n+1} = X_n + 1$ with probability 1/4, or $Y_{n+1} = X_n - 1$ with probability 3/4.

2. Let $j = Y_{n+1}$, and compute α_{ij} using (11.3.3) and (11.3.4).

3. With probability α_{ij}, let $X_{n+1} = Y_{n+1} = j$. Otherwise, with probability $1 - \alpha_{ij}$, let $X_{n+1} = X_n = i$.

Once again, these steps can all be easily performed on a computer, and if repeated for some large number N of iterations, then $P(X_N = j) \approx \pi_j = Ke^{-j^4}$ for $j \in S$. ∎

The Metropolis–Hastings algorithm can also be used for *continuous* random variables by using *densities*, as follows.

Example 11.3.5
Suppose we want to generate a sample from the distribution with density proportional to

$$f(y) = e^{-y^4}(1 + |y|)^3.$$

So the density is $C f(y)$, where $C = 1/\int_{-\infty}^{\infty} f(y)\, dy$. How can we generate a random variable Y such that Y has approximately this distribution, i.e., has probability density approximately equal to $C f(y)$?

Let us use a proposal distribution given by an $N(x, 1)$ distribution, namely, a normal distribution with mean x and variance 1. That is, given $X_n = x$, we choose Y_{n+1} by $Y_{n+1} \sim N(x, 1)$. Because the $N(x, 1)$ distribution has density $(2\pi)^{-1/2} e^{-(y-x)^2/2}$, this corresponds to a *proposal density* of $q(x, y) = (2\pi)^{-1/2} e^{-(y-x)^2/2}$.

As for the acceptance probability $\alpha(x, y)$, we again use densities, so that

$$\alpha(x, y) = \min\left\{1, \frac{C f(y)\, q(y, x)}{C f(x)\, q(x, y)}\right\}$$

$$= \min\left\{1, \frac{C e^{-y^4}\, (2\pi)^{-1/2}\, e^{-(y-x)^2/2}}{C e^{-x^4}\, (2\pi)^{-1/2}\, e^{-(x-y)^2/2}}\right\}$$

$$= \min\left\{1, e^{-y^4 + x^4}\right\}. \tag{11.3.5}$$

Given $X_n = x$, the Metropolis–Hastings algorithm computes the value X_{n+1} as follows.

1. Generate $Y_{n+1} \sim N(X_n, 1)$.

2. Let $y = Y_{n+1}$, and compute $\alpha(x, y)$ as before.

3. With probability $\alpha(x, y)$, let $X_{n+1} = Y_{n+1} = y$. Otherwise, with probability $1 - \alpha(x, y)$, let $X_{n+1} = X_n = x$.

Once again, these steps can all be easily performed on a computer, and if repeated for some large number N of iterations, then the random variable X_N will approximately have density given by $C f(y)$. ■

11.3.2   The Gibbs Sampler

In Section 8.3.3 we discussed the Gibbs sampler and its application in a Bayesian statistics problem. As we will now demonstrate, the *Gibbs sampler* is a specialized version of the Metropolis–Hastings algorithm, designed for multivariate distributions. It chooses the proposal probabilities q_{ij} just right so that we always have $\alpha_{ij} = 1$, i.e., so that no rejections are ever required.

Suppose that $S = \{\ldots, -2, -1, 0, 1, 2, \ldots\} \times \{\ldots, -2, -1, 0, 1, 2, \ldots\}$, i.e., S is the set of all ordered pairs of integers $i = (i_1, i_2)$. (Thus, $(2, 3) \in S$, and $(-6, 14) \in S$, etc.) Suppose that some distribution $\{\pi_i\}$ is defined on S. Define a proposal distribution $\{q_{ij}^{(1)}\}$ as follows.

Let $V(i) = \{j \in S : j_2 = i_2\}$. That is, $V(i)$ is the set of all states $j \in S$ such that i and j agree in their second coordinate. Thus, $V(i)$ is a vertical line in S, which passes through the point i.

In terms of this definition of $V(i)$, define $q_{ij}^{(1)} = 0$ if $j \notin V(i)$, i.e., if i and j differ in their second coordinate. If $j \in V(i)$, i.e., if i and j agree in their second coordinate, then define

$$q_{ij}^{(1)} = \frac{\pi_j}{\sum_{k \in V(i)} \pi_k}.$$

One interpretation is that, if $X_n = i$, and $P(Y_{n+1} = j) = q_{ij}^{(1)}$ for $j \in S$, then the distribution of Y_{n+1} is the *conditional* distribution of $\{\pi_i\}$, *conditional* on knowing that the second coordinate must be equal to i_2.

In terms of this choice of $q_{ij}^{(1)}$, what is α_{ij}? Well, if $j \in V(i)$, then $i \in V(j)$, and also $V(j) = V(i)$. Hence,

$$
\begin{aligned}
\alpha_{ij} &= \min\left\{1, \frac{\pi_j q_{ji}^{(1)}}{\pi_i q_{ij}^{(1)}}\right\} = \min\left\{1, \frac{\pi_j \left(\pi_i / \sum_{k \in V(j)} \pi_k\right)}{\pi_i \left(\pi_j / \sum_{l \in V(i)} \pi_l\right)}\right\} \\
&= \min\left\{1, \frac{\pi_j \pi_i}{\pi_i \pi_j}\right\} = \min\{1, 1\} = 1.
\end{aligned}
$$

That is, this algorithm accepts the proposal Y_{n+1} with probability 1, and never rejects at all!

Now, this algorithm by itself is not very useful, because it proposes only states in $V(i)$, so it never changes the value of the second coordinate at all. However, we can similarly define a horizontal line through i by $H(i) = \{j \in S : j_1 = i_1\}$, so that $H(i)$ is the set of all states j such that i and j agree in their first coordinate. That is, $H(i)$ is a horizontal line in S that passes through the point i.

We can then define $q_{ij}^{(2)} = 0$ if $j \notin H(i)$ (i.e., if i and j differ in their first coordinates), while if $j \in V(i)$ (i.e., if i and j agree in their first coordinate), then

$$q_{ij}^{(2)} = \frac{\pi_j}{\sum_{k \in H(i)} \pi_k}.$$

As before, we compute that for this proposal, we will always have $\alpha_{ij} = 1$, i.e., the Metropolis–Hastings algorithm with this proposal will never reject.

The Gibbs sampler works by *combining* these two different Metropolis–Hastings algorithms, by alternating between them. That is, given a value $X_n = i$, it produces a value X_{n+1} as follows.

1. Propose a value $Y_{n+1} \in V(i)$ according to the proposal distribution $\{q_{ij}^{(1)}\}$.

2. Always accept Y_{n+1}, and set $j = Y_{n+1}$ thus moving vertically.

3. Propose a value $Z_{n+1} \in H(j)$ according to the proposal distribution $\{q_{ij}^{(2)}\}$.

4. Always accept Z_{n+1} thus moving horizontally.

5. Set $X_{n+1} = Z_{n+1}$.

In this way, the Gibbs sampler does a "zigzag" through the state space S, alternately moving in the vertical and in the horizontal direction.

In light of Theorem 11.3.2, we immediately obtain the following.

Theorem 11.3.4 The preceding Gibbs sampler algorithm results in a Markov chain $X_0, X_1, X_2, \ldots$ that has $\{\pi_i\}$ as a stationary distribution.

The Gibbs sampler thus provides a particular way of implementing the Metropolis–Hastings algorithm in multidimensional problems, which never rejects the proposed values.

Summary of Section 11.3

- In cases that are too complicated for ordinary Monte Carlo techniques, it is possible to use Markov chain Monte Carlo techniques instead, by averaging values arising from a Markov chain.

- The Metropolis–Hastings algorithm provides a simple method of creating a Markov chain with stationary distribution $\{\pi_i\}$. Given X_n, it generates a proposal Y_{n+1} from a proposal distribution $\{q_{ij}\}$, and then either accepts this proposal (and sets $X_{n+1} = Y_{n+1}$) with probability α_{ij}, or rejects this proposal (and sets $X_{n+1} = X_n$) with probability $1 - \alpha_{ij}$.

- Alternatively, the Gibbs sampler updates the coordinates one at a time from their conditional distribution, such that we always have $\alpha_{ij} = 1$.

Exercises

11.3.1 Suppose $\pi_i = C e^{-(i-13)^4}$ for $i \in S = \{\ldots, -2, -1, 0, 1, 2, \ldots\}$, where $C = 1/\sum_{i=-\infty}^{\infty} e^{-(i-13)^4}$. Describe in detail a Metropolis–Hastings algorithm for $\{\pi_i\}$, which uses simple random walk with $p = 1/2$ for the proposals.

11.3.2 Suppose $\pi_i = C(i + 6.5)^{-8}$ for $i \in S = \{\ldots, -2, -1, 0, 1, 2, \ldots\}$, where $C = 1/\sum_{i=-\infty}^{\infty} (i + 6.5)^{-8}$. Describe in detail a Metropolis–Hastings algorithm for $\{\pi_i\}$, which uses simple random walk with $p = 5/8$ for the proposals.

11.3.3 Suppose $\pi_i = K e^{-i^4 - i^6 - i^8}$ for $i \in S = \{\ldots, -2, -1, 0, 1, 2, \ldots\}$, where $C = 1/\sum_{i=-\infty}^{\infty} e^{-i^4 - i^6 - i^8}$. Describe in detail a Metropolis–Hastings algorithm for $\{\pi_i\}$, which uses simple random walk with $p = 7/9$ for the proposals.

11.3.4 Suppose $f(x) = e^{-x^4 - x^6 - x^8}$ for $x \in R^1$. Let $K = 1/\int_{-\infty}^{\infty} e^{-x^4 - x^6 - x^8} \, dx$. Describe in detail a Metropolis–Hastings algorithm for the distribution having density $K f(x)$, which uses the proposal distribution $N(x, 1)$, i.e., a normal distribution with mean x and variance 1.

11.3.5 Let $f(x) = e^{-x^4 - x^6 - x^8}$ for $x \in R^1$, and let $K = 1/\int_{-\infty}^{\infty} e^{-x^4 - x^6 - x^8} \, dx$. Describe in detail a Metropolis–Hastings algorithm for the distribution having density $K f(x)$, which uses the proposal distribution $N(x, 10)$, i.e., a normal distribution with mean x and variance 10.

Computer Exercises

11.3.6 Run the algorithm of Exercise 11.3.1. Discuss the output.

11.3.7 Run the algorithm of Exercise 11.3.2. Discuss the output.

Problem

11.3.8 Suppose $S = \{1, 2, 3, \ldots\} \times \{1, 2, 3, \ldots\}$, i.e., S is the set of all *pairs* of positive integers. For $i = (i_1, i_2) \in S$, suppose $\pi_i = C / 2^{i_1 + i_2}$ for appropriate positive constant C. Describe in detail a Gibbs sampler algorithm for this distribution $\{\pi_i\}$.

Computer Problems

11.3.9 Run the algorithm of Exercise 11.3.4. Discuss the output.

11.3.10 Run the algorithm of Exercise 11.3.6. Discuss the output.

Discussion Topics

11.3.11 Why do you think Markov chain Monte Carlo algorithms have become so popular in so many branches of science? (List as many reasons as you can.)

11.3.12 Suppose you will be using a Markov chain Monte Carlo estimate of the form

$$\hat{A} = \frac{1}{M} \sum_{i=1}^{M} h(X_N^{[i]}).$$

Suppose also that, due to time constraints, your total number of iterations cannot be more than one million. That is, you must have $NM \leq 1,000,000$. Discuss the advantages and disadvantages of the following choices of N and M.
(a) $N = 1,000,000$, $M = 1$
(b) $N = 1$, $M = 1,000,000$
(c) $N = 100$, $M = 10,000$
(d) $N = 10,000$, $M = 100$
(e) $N = 1000$, $M = 1000$
(f) Which choice do you think would be best, under what circumstances? Why?

11.4 Martingales

In this section, we study a special class of stochastic processes called *martingales*. We shall see that these processes are characterized by "staying the same on average."

As motivation, consider again a simple random walk in the case of a fair game, i.e., with $p = 1/2$. Suppose, as in the gambler's ruin set-up, that you start at a and keep going until you hit either c or 0, where $0 < a < c$. Let Z be the value that you end up with, so that we always have either $Z = c$ or $Z = 0$. We know from Theorem 11.1.2 that in fact $P(Z = c) = a/c$, so that $P(Z = 0) = 1 - a/c$.

Let us now consider the *expected value* of Z. We have that

$$E(Z) = \sum_{z \in R^1} z \, P(Z = z) = cP(Z = c) + 0P(Z = 0) = c(a/c) = a.$$

That is, the average value of where you end up is a. But a is also the value that you started at!

This is not a coincidence. Indeed, because $p = 1/2$ (i.e., the game was fair), this means that "on average" you always stayed at a. That is, $\{X_n\}$ is a *martingale*.

11.4.1 Definition of a Martingale

We begin with the definition of a martingale. For simplicity, we assume that the martingale is a Markov chain, though this is not really necessary.

Definition 11.4.1 Let $X_0, X_1, X_2, \ldots$ be a Markov chain. The chain is a *martingale* if for all $n = 0, 1, 2, \ldots$, we have $E(X_{n+1} - X_n \mid X_n) = 0$. That is, on average the chain's value does not change, regardless of what the current value X_n actually is.

Example 11.4.1

Let $\{X_n\}$ be simple random walk with $p = 1/2$. Then $X_{n+1} - X_n$ is equal to either 1 or -1, with probability $1/2$ each. Hence,

$$E(X_{n+1} - X_n \,|\, X_n) = (1)(1/2) + (-1)(1/2) = 0,$$

so $\{X_n\}$ stays the same on average, and is a martingale. (Note that we will never actually have $X_{n+1} - X_n = 0$. However, *on average* we will have $X_{n+1} - X_n = 0$.) ∎

Example 11.4.2

Let $\{X_n\}$ be simple random walk with $p = 2/3$. Then $X_{n+1} - X_n$ is equal to either 1 or -1, with probabilities $2/3$ and $1/3$, respectively. Hence,

$$E(X_{n+1} - X_n \,|\, X_n) = (1)(2/3) + (-1)(1/3) = 1/3 \neq 0.$$

Thus, $\{X_n\}$ is *not* a martingale in this case. ∎

Example 11.4.3

Suppose we start with the number 5, and then repeatedly do the following. We either add 3 to the number (with probability $1/4$), or subtract 1 from the number (with probability $3/4$). Let X_n be the number obtained after repeating this procedure n times. Then given the value of X_n, we see that $X_{n+1} = X_n + 3$ with probability $1/4$, while $X_{n+1} = X_n - 1$ with probability $3/4$. Hence,

$$E(X_{n+1} - X_n \,|\, X_n) = (3)(1/4) + (-1)(3/4) = 3/4 - 3/4 = 0$$

and $\{X_n\}$ is a martingale. ∎

It is sometimes possible to create martingales in subtle ways, as follows.

Example 11.4.4

Let $\{X_n\}$ again be simple random walk, but this time for general p. Then $X_{n+1} - X_n$ is equal to 1 with probability p, and to -1 with probability $q = 1 - p$. Hence,

$$E(X_{n+1} - X_n \,|\, X_n) = (1)(p) + (-1)(q) = p - q = 2p - 1.$$

If $p \neq 1/2$, then this is *not* equal to 0. Hence, $\{X_n\}$ does *not* stay the same on average, so $\{X_n\}$ is *not* a martingale.

On the other hand, let

$$Z_n = \left(\frac{1-p}{p}\right)^{X_n},$$

i.e., Z_n equals the constant $(1-p)/p$ raised to the *power* of X_n. Then increasing X_n by 1 corresponds to *multiplying* Z_n by $(1-p)/p$, while decreasing X_n by 1 corresponds to *dividing* Z_n by $(1-p)/p$, i.e., multiplying by $p/(1-p)$. But

$X_{n+1} = X_n + 1$ with probability p, while $X_{n+1} = X_n - 1$ with probability $q = 1 - p$. Therefore, we see that given the value of Z_n, we have

$$
\begin{aligned}
E(Z_{n+1} - Z_n \mid Z_n) &= \left(\frac{1-p}{p} Z_n - Z_n\right) p + \left(\frac{p}{1-p} Z_n - Z_n\right)(1-p) \\
&= ((1-p)Z_n - pZ_n) + (pZ_n - (1-p)Z_n) = 0.
\end{aligned}
$$

Accordingly, $E(Z_{n+1} - Z_n \mid Z_n) = 0$, so that $\{Z_n\}$ stays the same on average, i.e., $\{Z_n\}$ is a martingale. ∎

11.4.2 Expected Values

Because martingales stay the same on average, we immediately have the following.

Theorem 11.4.1 Let $\{X_n\}$ be a martingale with $X_0 = a$. Then $E(X_n) = a$ for all n.

This theorem sometimes provides very useful information, as the following examples demonstrate.

Example 11.4.5
Let $\{X_n\}$ again be simple random walk with $p = 1/2$. Then we have already seen that $\{X_n\}$ is a martingale. Hence, if $X_0 = a$, then we will have $E(X_n) = a$ for all n. That is, for a fair game (i.e., for $p = 1/2$), no matter how long you have been gambling, your *average* fortune will always be equal to your initial fortune a. ∎

Example 11.4.6
Suppose we start with the number 10, and then repeatedly do the following. We either add 2 to the number (with probability $1/3$), or subtract 1 from the number (with probability $2/3$). Suppose we repeat this process 25 times. What is the expected value of the number we end up with?

Without martingale theory, this problem appears to be difficult, requiring lengthy computations of various possibilities for what could happen on each of the 25 steps. However, with martingale theory, it is very easy.

Indeed, let X_n be the number after n steps, so that $X_0 = 10, X_1 = 12$ (with probability $1/3$) or $X_1 = 9$ (with probability $2/3$), etc. Then, because $X_{n+1} - X_n$ equals either 2 (with probability $1/3$) or -1 (with probability $2/3$), we have

$$
E(X_{n+1} - X_n \mid X_n) = 2(1/3) + (-1)(2/3) = 2/3 - 2/3 = 0.
$$

Hence, $\{X_n\}$ is a martingale.

It then follows that $E(X_n) = X_0 = 10$, for any n. In particular, $E(X_{25}) = 10$. That is, after 25 steps, on average the number will be equal to 10. ∎

11.4.3 Stopping Times

If $\{X_n\}$ is a martingale with $X_0 = a$, then it is very helpful to know that $E(X_n) = a$ for all n. However, it is sometimes even more helpful to know that $E(X_T) = 0$, where T is a *random* time. Now, this is not always true; however, it is often true, as we shall see. We begin with another definition.

Definition 11.4.2 Let $\{X_n\}$ be a stochastic process, and let T be a random variable taking values on $0, 1, 2, \ldots$. Then T is a *stopping time* if for all $m = 0, 1, 2, \ldots$, the event $\{T = m\}$ is independent of the values $X_{m+1}, X_{m+2}, \ldots$. That is, when deciding whether or not $T = m$ (i.e., whether or not to "stop" at time m), we are not allowed to look at the future values $X_{m+1}, X_{m+2}, \ldots$.

Example 11.4.7
Let $\{X_n\}$ be simple random walk, let b be any integer, and let $\tau_b = \min\{n \geq 0 : X_n = b\}$ be the first time we hit the value b. Then τ_b is a stopping time, because the event $\tau_b = n$ depends only on $X_0, \ldots, X_n$, not on $X_{n+1}, X_{n+2}, \ldots$.

On the other hand, let $T = \tau_b - 1$, so that T corresponds to stopping just *before* we hit b. Then T is not a stopping time, because it must look at the future value X_{m+1} to decide whether or not to stop at time m. ∎

A key result about martingales and stopping times is the optional stopping theorem, as follows.

Theorem 11.4.2 (*Optional stopping theorem*) Suppose $\{X_n\}$ is a martingale with $X_0 = a$, and T is a stopping time. Suppose further that either

(a) the martingale is bounded up to time T, i.e., for some $M > 0$ we have $|X_n| \leq M$ for all $n \leq T$; or
(b) the stopping time is bounded, i.e., for some $M > 0$ we have $T \leq M$.

Then $E(X_T) = a$, i.e., on average the value of the process at the random time T is equal to the starting value a.

Proof: For a proof and further discussion, see, e.g., page 273 of *Probability: Theory and Examples (2nd ed.)* by R. Durrett (Duxbury Press, New York, 1996). ∎

Consider a simple application of this.

Example 11.4.8
Let $\{X_n\}$ be simple random walk with initial value a and with $p = 1/2$. Let $r > a > s$ be integers. Let $T = \min\{\tau_r, \tau_s\}$ be the first time the process hits either r or s. Then $r \geq X_n \geq s$ for $n \leq T$, so that condition (a) of the optional stopping theorem applies. We conclude that $E(X_T) = a$, i.e., that at time T, the walk will on average be equal to a. ∎

We shall see that the optional stopping theorem is useful in many ways.

Example 11.4.9
We can use the optional stopping theorem to find the probability that the simple random walk with $p = 1/2$ will hit r before hitting another value s.

Indeed, again let $\{X_n\}$ be simple random walk with initial value a and $p = 1/2$, with $r > a > s$ integers and $T = \min\{\tau_r, \tau_s\}$. Then as earlier, $E(X_T) = a$. We can use this to solve for $P(X_T = r)$, i.e., for the probability that the walk hits r before hitting s.

Clearly, we always have either $X_T = r$ or $X_T = s$. Let $h = P(X_T = r)$. Then $E(X_T) = hr + (1 - h)s$. Because $E(X_T) = a$, we must have $a = hr + (1 - h)s$. Solving for h, we see that

$$P(X_T = r) = \frac{a - s}{r - s}.$$

We conclude that the probability that the process will hit r before it hits s is equal to $(a - s)/(r - s)$. Note that absolutely no difficult computations were required to obtain this result. ∎

A special case of the previous example is particularly noteworthy.

Example 11.4.10
In the previous example, suppose $r = c$ and $s = 0$. Then the value $h = P(X_T = r)$ is precisely the same as the probability of success in the gambler's ruin problem. The previous example shows that $h = (a - s)/(r - s) = a/c$. This gives the same answer as Theorem 11.1.2, but with far less effort. ∎

It is impressive that, in the preceding example, martingale theory can solve the gambler's ruin problem so easily in the case $p = 1/2$. Our previous solution, without using martingale theory, was much more difficult (see Section 11.7). Even more surprisingly, martingale theory can also solve the gambler's ruin problem when $p \neq 1/2$, as follows.

Example 11.4.11
Let $\{X_n\}$ be simple random walk with initial value a and with $p \neq 1/2$. Let $0 < a < c$ be integers. Let $T = \min\{\tau_c, \tau_0\}$ be the first time the process hits either c or 0. To solve the gambler's ruin problem in this case, we are interested in $g = P(X_T = c)$. We can use the optional stopping theorem to solve for the gambler's ruin probability g, as follows.

Now, $\{X_n\}$ is not a martingale, so we cannot apply martingale theory to it. However, let

$$Z_n = \left(\frac{1 - p}{p}\right)^{X_n}.$$

Then $\{Z_n\}$ has initial value $Z_0 = ((1 - p)/p)^a$. Also, we know from Example 11.4.4 that $\{Z_n\}$ is a martingale. Furthermore,

$$0 \leq Z_n \leq \max\left\{\left(\frac{1 - p}{p}\right)^c, \left(\frac{1 - p}{p}\right)^{-c}\right\}$$

for $n \leq T$, so that condition (a) of the optional stopping theorem applies. We conclude that

$$E(Z_T) = Z_0 = \left(\frac{1 - p}{p}\right)^a.$$

Now, clearly we always have either $X_T = c$ (with probability g) or $X_T = 0$ (with probability $1 - g$). In the former case, $Z_T = ((1-p)/p)^c$, while in the latter case, $Z_T = 1$. Hence, $E(Z_T) = g((1-p)/p)^c + (1-g)(1)$. Because $E(Z_T) = ((1-p)/p)^a$, we must have

$$\left(\frac{1-p}{p}\right)^a = g\left(\frac{1-p}{p}\right)^c + (1-g)(1).$$

Solving for g, we see that

$$g = \frac{((1-p)/p)^a - 1}{((1-p)/p)^c - 1}.$$

This again gives the same answer as Theorem 11.1.2, this time for $p \neq 1/2$, but again with far less effort. ∎

Martingale theory can also tell us other surprising facts.

Example 11.4.12
Let $\{X_n\}$ be simple random walk with $p = 1/2$ and with initial value $a = 0$. Will the walk hit the value -1 some time during the first million steps? Probably yes, but not for sure. Furthermore, conditional on *not* hitting -1, it will probably be extremely large, as we now discuss.

Let $T = \min\{10^6, \tau_{-1}\}$. That is, T is the first time the process hits -1, unless that takes more than one million steps, in which case $T = 10^6$.

Now, $\{X_n\}$ is a martingale. Also T is a stopping time (because it does not look into the future when deciding whether or not to stop). Furthermore, we always have $T \leq 10^6$, so condition (b) of the optional stopping theorem applies. We conclude that $E(X_T) = a = 0$.

On the other hand, by the law of total expectation, we have

$$E(X_T) = E(X_T \,|\, X_T = -1)\, P(X_T = -1) + E(X_T \,|\, X_T \neq -1)\, P(X_T \neq -1).$$

Also, clearly $E(X_T \,|\, X_T = -1) = -1$. Let $u = P(X_T = -1)$, so that $P(X_T \neq -1) = 1 - u$. Then we conclude that $0 = (-1)u + E(X_T \,|\, X_T \neq -1)(1-u)$, so that

$$E(X_T \,|\, X_T \neq -1) = \frac{u}{1-u}.$$

Now, clearly u will be very close to 1, i.e., it is very likely that within 10^6 steps the process will have hit -1. Hence, $E(X_T \,|\, X_T \neq -1)$ is extremely large.

We may summarize this discussion as follows. Nearly always we have $X_T = -1$. However, very occasionally we will have $X_T \neq -1$. Furthermore, the average value of X_T when $X_T \neq -1$ is *so* large that overall (i.e., counting both the case $X_T = -1$ and the case $X_T \neq -1$), the average value of X_T is 0 (as it must be because $\{X_n\}$ is a martingale)! ∎

If one is not careful, then it is possible to be tricked by martingale theory, as follows.

Example 11.4.13

Suppose again that $\{X_n\}$ is simple random walk with $p = 1/2$, and with initial value $a = 0$. Let $T = \tau_{-1}$, i.e., T is the first time the process hits -1 (no matter how long that takes).

Because the process will always wait until it hits -1, we always have $X_T = -1$. Because this is true with probability 1, therefore we also have $E(X_T) = -1$.

On the other hand, again $\{X_n\}$ is a martingale, so again it appears that we should have $E(X_T) = 0$. What is going on?

The answer, of course, is that neither condition (a) nor condition (b) of the optional stopping theorem is satisfied in this case. That is, there is no limit to how large T might have to be, or how large X_n might get for some $n \leq T$. Hence, the optional stopping theorem does not apply in this case, and we cannot conclude that $E(X_T) = 0$. Instead, $E(X_T) = -1$ here. ∎

Summary of Section 11.4

- A Markov chain $\{X_n\}$ is a martingale if it stays the same on average, i.e., if $E(X_{n+1} - X_n \,|\, X_n) = 0$ for all n. There are many examples.

- A stopping time T for the chain is a nonnegative integer-valued random variable that does not look into the future of $\{X_n\}$. For example, perhaps $T = \tau_b$ is the first time the chain hits some state b.

- If $\{X_n\}$ is a martingale with stopping time T, and if either T or $\{X_n\}_{n \leq T}$ is bounded, then $E(X_T) = X_0$. This can be used to solve many problems, e.g., gambler's ruin.

Exercises

11.4.1 Suppose we define a process $\{X_n\}$ as follows. Given X_n, with probability $3/8$ we let $X_{n+1} = X_n - 4$, while with probability $5/8$ we let $X_{n+1} = X_n + C$. What value of C will make $\{X_n\}$ be a martingale?

11.4.2 Suppose we define a process $\{X_n\}$ as follows. Given X_n, with probability p we let $X_{n+1} = X_n + 7$, while with probability $1 - p$ we let $X_{n+1} = X_n - 2$. What value of p will make $\{X_n\}$ be a martingale?

11.4.3 Suppose we define a process $\{X_n\}$ as follows. Given X_n, with probability p we let $X_{n+1} = 2X_n$, while with probability $1 - p$ we let $X_{n+1} = X_n/2$. What value of p will make $\{X_n\}$ be a martingale?

11.4.4 Let $\{X_n\}$ be a martingale, with initial value $X_0 = 14$. Suppose for some n, we know that $P(X_n = 8) + P(X_n = 12) + P(X_n = 17) = 1$, i.e., X_n is always either 8, 12, or 17. Suppose further that $P(X_n = 8) = 0.1$. Compute $P(X_n = 14)$.

11.4.5 Let $\{X_n\}$ be a martingale, with initial value $X_0 = 5$. Suppose we know that $P(X_8 = 3) + P(X_8 = 4) + P(X_8 = 6) = 1$, i.e., X_8 is always either 3, 4, or 6. Suppose further that $P(X_8 = 3) = 2\,P(X_8 = 6)$. Compute $P(X_8 = 4)$.

11.4.6 Suppose you start with 175 pennies. You repeatedly flip a fair coin. Each time the coin comes up heads, you win a penny; each time the coin comes up tails, you lose a penny.
(a) After repeating this procedure 20 times, how many pennies will you have on average?
(b) Suppose you continue until you have either 100 or 200 pennies, and then you stop. What is the probability you will have 200 pennies when you stop?

11.4.7 Define a process $\{X_n\}$ by $X_0 = 27$, and $X_{n+1} = 3X_n$ with probability $1/4$, or $X_{n+1} = X_n/3$ with probability $3/4$. Let $T = \min\{\tau_1, \tau_{81}\}$ be the first time the process hits either 1 or 81.
(a) Show that $\{X_n\}$ is a martingale.
(b) Show that T is a stopping time.
(c) Compute $E(X_T)$.
(d) Compute the probability $P(X_T = 1)$ that the process hits 1 before hitting 81.

Problems

11.4.8 Let $\{X_n\}$ be a stochastic process, and let T_1 be a stopping time. Let $T_2 = T_1 + i$ and $T_3 = T_1 - i$, for some positive integer i. Which of T_2 and T_3 is necessarily a stopping time, and which is not? (Explain your reasoning.)

11.4.9 Let $\{X_n\}$ be a stochastic process, and let T_1 and T_2 be two different stopping times. Let $T_3 = \min\{T_1, T_2\}$, and $T_4 = \max\{T_1, T_2\}$.
(a) Is T_3 necessarily a stopping time? (Explain your reasoning.)
(b) Is T_4 necessarily a stopping time? (Explain your reasoning.)

11.5   Brownian Motion

The simple random walk model of Section 11.1.2 (with $p = 1/2$) can be extended to an interesting continuous-time model, called *Brownian motion*, as follows. Roughly, the idea is to speed up time faster and faster by a factor of M (for very large M), while simultaneously shrinking space smaller and smaller by a factor of $1/\sqrt{M}$. The factors of M and $1/\sqrt{M}$ are chosen just right so that, using the central limit theorem, we can derive properties of Brownian motion. Indeed, using the central limit theorem, we shall see that various distributions related to Brownian motion are in fact *normal* distributions.

   Historically, Brownian motion gets its name from Robert Brown, a botanist, who in 1828 observed the motions of tiny particles in solution, under a microscope, as they were bombarded from random directions by many unseen molecules. Brownian motion was proposed as a model for the observed chaotic, random movement of such particles. In fact, Brownian motion turns out *not* to be a very good model for such movement (for example, Brownian motion has infinite derivative, which would only make sense if the particles moved infinitely quickly!). However, Brownian motion has many useful mathematical properties and is also very important in the theory of finance because it is often used as a model of stock price fluctuations. A proper mathematical theory of Brownian motion was developed in 1923 by Norbert Wiener[2]; as a result, Brownian motion is also sometimes called the *Wiener process*.

We shall construct Brownian motion in two steps. First, we construct faster and faster random walks, to be called $\{Y_t^{(M)}\}$ where M is large. Then, we take the limit as $M \to \infty$ to get Brownian motion.

11.5.1 Faster and Faster Random Walks

To begin, we let $Z_1, Z_2, \ldots$ be i.i.d. with $P(Z_i = +1) = P(Z_i = -1) = 1/2$. For each $M \in \{1, 2, \ldots\}$, define a discrete-time random process

$$\left\{ Y_{i/M}^{(M)} : i = 0, 1, \ldots \right\},$$

by $Y_0^{(M)} = 0$, and

$$Y_{(i+1)/M}^{(M)} = Y_{\frac{i}{M}}^{(M)} + \frac{1}{\sqrt{M}} Z_{i+1},$$

for $i = 0, 1, 2, \ldots$ so that

$$Y_{i/M}^{(M)} = \frac{1}{\sqrt{M}} (Z_1 + Z_2 + \ldots + Z_i).$$

Intuitively, then, $\{Y_{i/M}^{(M)}\}$ is like ordinary (discrete-time) random walk (with $p = 1/2$), except that time has been sped up by a factor of M and space has been shrunk by a factor of $\sqrt{M}$ (each step in the new walk moves a distance $1/\sqrt{M}$). That is, this process takes lots and lots of very small steps.

To make $\{Y_{i/M}^{(M)}\}$ into a continuous-time process, we can then "fill in" the missing values by making the function linear on the intervals $[i/M, (i+1)/M]$. In this way, we obtain a continuous-time process

$$\{Y_t^{*(M)} : t \geq 0\},$$

which agrees with $\{Y_{i/M}^{(M)}\}$ whenever $t = 1/M$. In Figure 11.5.1 we have plotted

$$\{Y_{i/10}^{(10)} : i = 0, 1, \ldots, 20\}$$

(the dots) and the corresponding values of

$$\{Y_t^{*(10)} : 0 \leq t \leq 20\}$$

(the solid line), arising from the realization

$$(Z_1, \ldots, Z_{20}) = (1, -1, -1, -1, -1, 1, 1, \ldots),$$

where we have taken $1/\sqrt{10} = 0.316$.

---

[2] Wiener was such an absent-minded professor that he once got lost and couldn't find his house. In his confusion, he asked a young girl for directions, without recognizing the girl as his daughter!

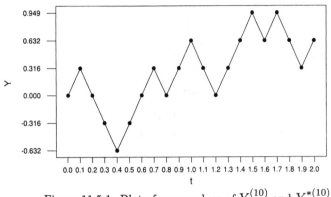

Figure 11.5.1: Plot of some values of $Y_{i/10}^{(10)}$ and $Y_t^{*(10)}$.

The collection of variables $\{Y_t^{*(M)} : t \geq 0\}$ is then a stochastic process but is now indexed by the *continuous* time parameter $t \geq 0$. This is an example of a *continuous-time stochastic process*.

Now, the factors M and $\sqrt{M}$ have been chosen carefully, as the following theorem illustrates.

Theorem 11.5.1 Let $\{Y_t^{*(M)} : t \geq 0\}$ be as defined earlier. Then for large M:
(a) For $t \geq 0$, the distribution of $Y_t^{*(M)}$ is approximately $N(0, t)$, i.e., normally distributed with mean t.
(b) For $s, t \geq 0$, the covariance

$$\text{Cov}\left(Y_t^{*(M)}, Y_t^{*(M)}\right)$$

is approximately equal to $\min\{s, t\}$.
(c) For $t \geq s \geq 0$, the distribution of the increment $Y_t^{*(M)} - Y_s^{*(M)}$ is approximately $N(0, t - s)$, i.e., normally distributed with mean 0 and variance $t - s$, and is approximately independent of $Y_s^{*(M)}$.
(d) $Y_t^{*(M)}$ is a continuous function of t.

Proof: See Section 11.7 for the proof of this result.

We shall use this limit theorem to construct Brownian motion.

11.5.2   Brownian Motion as a Limit

We have now developed the faster and faster processes $\{Y_t^{*(M)} : t \geq 0\}$, and some of their properties. Brownian motion is then defined as the limit as $M \to \infty$ of the processes $\{Y_t^{*(M)} : t \geq 0\}$. That is, we define Brownian motion $\{B_t : t \geq 0\}$ by saying that the distribution of $\{B_t : t \geq 0\}$ is equal to the limit as $M \to \infty$ of the distribution of $\{Y_t^{*(M)} : t \geq 0\}$. A graph of a typical run of Brownian motion is in Figure 11.5.2.

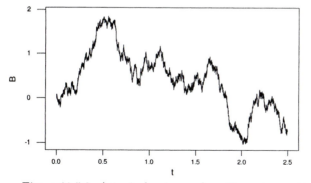

Figure 11.5.2: A typical outcome from Brownian motion.

In this way, all the properties of $Y_t^{*(M)}$ for large M, as developed in Theorem 11.5.1, will apply to Brownian motion, as follows.

Theorem 11.5.2 Let $\{B_t : t \geq 0\}$ be Brownian motion. Then
(a) B_t is normally distributed: $B_t \sim N(0, t)$ for any $t \geq 0$;
(b) $\mathrm{Cov}(B_s, B_t) = E(B_s B_t) = \min\{s, t\}$ for $s, t \geq 0$;
(c) if $0 < s < t$, then the increment $B_t - B_s$ is normally distributed: $B_t - B_s \sim N(0, t - s)$, and furthermore $B_t - B_s$ is independent of B_s;
(d) the function $\{B_t\}_{t \geq 0}$ is a continuous function.

This theorem can be used to compute many things about Brownian motion.

Example 11.5.1
Let $\{B_t\}$ be Brownian motion. What is $P(B_5 \leq 3)$?
We know that $B_5 \sim N(0, 5)$. Hence, $B_5/\sqrt{5} \sim N(0, 1)$. We conclude that

$$P(B_5 \leq 3) = P(B_5/\sqrt{5} \leq 3/\sqrt{5}) = \Phi(3/\sqrt{5}) = 0.910,$$

where

$$\Phi(x) = \int_{-\infty}^{x} \frac{1}{\sqrt{2\pi}} e^{-s^2/2} \, ds$$

is the cumulative distribution function of a standard normal distribution, and we have found the numerical value from Table D.2. Thus, about 91% of the time, Brownian motion will be less than 3 at time 5. ∎

Example 11.5.2
Let $\{B_t\}$ be Brownian motion. What is $P(B_7 \geq -4)$?
We know that $B_7 \sim N(0, 7)$. Hence, $B_7/\sqrt{7} \sim N(0, 1)$. We conclude that

$$
\begin{aligned}
P(B_7 \geq -4) &= 1 - P(B_7 \leq -4) = 1 - P(B_7/\sqrt{7} \leq -4/\sqrt{7}) \\
&= 1 - \Phi(-4/\sqrt{7}) = 1 - 0.065 = 0.935.
\end{aligned}
$$

Thus, over 93% of the time, Brownian motion will be at least -4 at time 7. ∎

Example 11.5.3

Let $\{B_t\}$ be Brownian motion. What is $P(B_8 - B_6 \leq -1.5)$?

We know that $B_8 - B_6 \sim N(0, 8 - 6) = N(0, 2)$. Hence, $(B_8 - B_6)/\sqrt{2} \sim N(0, 1)$. We conclude that

$$P(B_8 - B_6 \leq -1.5) = P((B_8 - B_6)/\sqrt{2} \leq -1.5/\sqrt{2}) = \Phi(-1.5/\sqrt{2}) = 0.144.$$

Thus, about 14% of the time, Brownian motion will decrease by at least 1.5 between time 6 and time 8. ∎

Example 11.5.4

Let $\{B_t\}$ be Brownian motion. What is $P(B_2 \leq -0.5, \ B_5 - B_2 \geq 1.5)$?

By Theorem 11.5.2, we see that $B_5 - B_2$ and B_2 are *independent*. Hence,

$$P(B_2 \leq -0.5, \ B_5 - B_2 \geq 1.5) \ = \ P(B_2 \leq -0.5) \, P(B_5 - B_2 \geq 1.5).$$

Now, we know that $B_2 \sim N(0, 2)$. Hence, $B_2/\sqrt{2} \sim N(0, 1)$, and

$$P(B_2 \leq -0.5) = P(B_2/\sqrt{2} \leq -0.5/\sqrt{2}) = \Phi(-0.5/\sqrt{2}).$$

Similarly, $B_5 - B_2 \sim N(0, 3)$, so $(B_5 - B_2)/\sqrt{3} \sim N(0, 1)$, and

$$
\begin{aligned}
P(B_5 - B_2 \geq 1.5) \ &= \ P((B_5 - B_2)/\sqrt{3} \geq 1.5/\sqrt{3}) \\
&= \ 1 - P((B_5 - B_2)/\sqrt{3} < 1.5/\sqrt{3}) = \Phi(1.5/\sqrt{3}).
\end{aligned}
$$

We conclude that

$$
\begin{aligned}
P(B_2 \leq -0.5, \ B_5 - B_2 \geq 1.5) \ &= \ P(B_2 \leq -0.5) \, P(B_5 - B_2 \geq 1.5) \\
&= \ \Phi(-0.5/\sqrt{2}) \, \Phi(1.5/\sqrt{3}) = 0.292.
\end{aligned}
$$

Thus, about 29% of the time, Brownian motion will be no more than $-1/2$ at time 2 and will then increase by at least 1.5 between time 2 and time 5. ∎

We note also that, because Brownian motion was created from simple random walks with $p = 1/2$, it follows that Brownian motion is a *martingale*. This implies that $E(B_t) = 0$ for all t, but of course, we already knew that because $B_t \sim N(0, t)$. On the other hand, we can now use the optional stopping theorem (Theorem 11.4.2) to conclude that $E(B_T) = 0$ where T is a stopping time (provided, as usual, that either T or $\{B_t : t \leq T\}$ is bounded). This allows us to compute certain probabilities, as follows.

Example 11.5.5

Let $\{B_t\}$ be Brownian motion. Let $c < 0 < b$. What is the probability the process will hit c before it hits b?

To solve this problem, we let τ_c be the first time the process hits c, and τ_b be the first time the process hits b. We then let $T = \min\{\tau_c, \tau_b\}$ be the first time the process either hits c or hits b. The question becomes, what is $P(\tau_c < \tau_b)$? Equivalently, what is $P(B_T = c)$?

To solve this, we note that we must have $E(B_T) = B_0 = 0$. But if $h = P(B_T = c)$, then $B_T = c$ with probability h, and $B_T = b$ with probability $1 - h$. Hence, we must have $0 = E(B_T) = hc + (1 - h)b$, so that $h = b/(b - c)$. We conclude that

$$P(B_T = c) = P(\tau_c < \tau_b) = \frac{b}{b - c}.$$

(Recall that $c < 0$, so that $b - c = |b| + |c|$ here.) ∎

Finally, we note that although Brownian motion is a continuous function, it turns out that, with probability one, Brownian motion is not *differentiable* anywhere at all! This is part of the reason that Brownian motion is not a good model for the movement of real particles. (See Challenge 11.5.15 for a result related to this.) However, Brownian motion has many other uses, including as a model for stock prices, which we now describe.

11.5.3 Diffusions and Stock Prices

Brownian motion is used to construct various *diffusion* processes, as follows.

Given Brownian motion $\{B_t\}$, we can let

$$X_t = a + \delta t + \sigma B_t,$$

where a and δ are any real numbers, and $\sigma \geq 0$. Then $\{X_t\}$ is a diffusion.

Here a is the initial value, δ (called the *drift*) is the average rate of increase, and σ (called the *volatility parameter*) represents the amount of randomness of the diffusion.

Intuitively, X_t is approximately equal to the linear function $a + \delta t$, but due to the randomness of Brownian motion, X_t takes on *random* values around this linear function.

The precise distribution of X_t can be computed, as follows.

Theorem 11.5.3 Let $\{B_t\}$ be Brownian motion, and let $X_t = a + \delta t + \sigma B_t$ be a diffusion. Then
(a) $E(X_t) = a + \delta t$,
(b) $\text{Var}(X_t) = \sigma^2 t$,
(c) $X_t \sim N(a + \delta t, \sigma^2 t)$.

Proof: We know $B_t \sim N(0, 1)$, so $E(B_t) = 0$ and $\text{Var}(B_t) = t$. Also, $a + \delta t$ is not random (i.e., is a *constant* from the point of view of random variables). Hence,

$$E(X_t) = E(a + \delta t + \sigma B_t) = a + \delta t + \sigma E(B_t) = a + \delta t,$$

proving part (a).

Similarly,

$$\text{Var}(X_t) = \text{Var}(a + \delta t + \sigma B_t) = \text{Var}(\sigma B_t) = \sigma^2 \text{Var}(B_t) = \sigma^2 t,$$

proving part (b).

Finally, because X_t is a linear function of the normally distributed random variable B_t, X_t must be normally distributed by Theorem 4.6.1. This proves part (c). ∎

Diffusions are often used as models for *stock prices*. That is, it is often assumed that the price X_t of a stock at time t is given by $X_t = a + \delta t + \sigma B_t$ for appropriate values of a, δ, and σ.

Example 11.5.6

Suppose a stock has initial price \$20, drift of \$3 per year, and volatility parameter 1.4. What is the probability that the stock price will be over \$30 after two and a half years?

Here the stock price after t years is given by $X_t = 20 + 3t + 1.4B_t$ and is thus a diffusion.

So, after 2.5 years, we have $X_{2.5} = 20 + 7.5 + 1.4B_{2.5} = 27.5 + 1.4B_{2.5}$. Hence,

$$
\begin{aligned}
P(X_{2.5} > 30) &= P(27.5 + 1.4B_{2.5} > 30) = P(B_{2.5} > (30 - 27.5)/1.4) \\
&= P(B_{2.5} > 1.79).
\end{aligned}
$$

But like before,

$$
\begin{aligned}
P(B_{2.5} > 1.79) &= 1 - P(B_{2.5} \le 1.79) = 1 - P(B_{2.5}/\sqrt{2.5} \le 1.79/\sqrt{2.5}) \\
&= 1 - \Phi(1.79/\sqrt{2.5}) = 0.129.
\end{aligned}
$$

We conclude that $P(X_{2.5} > 30) = 0.129$.

Hence, there is just under a 13% chance that the stock will be worth more than \$30 after two and a half years. ∎

Example 11.5.7

Suppose a stock has initial price \$100, drift of $-\$2$ per year, and volatility parameter 5.5. What is the probability that the stock price will be under \$90 after just half a year?

Well, here the stock price after t years is given by $X_t = 100 - 2t + 5.5B_t$ and is again a diffusion. So, after 0.5 years, we have $X_{0.5} = 100 - 1.0 + 5.5B_{0.5} = 99 + 5.5B_{0.5}$. Hence,

$$
\begin{aligned}
P(X_{0.5} < 90) &= P(99 + 5.5B_{0.5} < 90) = P(B_{0.5} < (90 - 99)/5.5) \\
&= P(B_{0.5} < -1.64) = P(B_{0.5}/\sqrt{0.5} \le -1.64/\sqrt{0.5}) \\
&= \Phi(-1.64/\sqrt{0.5}) = \Phi(-2.32) = 0.010.
\end{aligned}
$$

Therefore, there is about a 1% chance that the stock will be worth less than \$90 after half a year. ∎

More generally, the drift δ and volatility σ could be *functions* of the value X_t, leading to more complicated diffusions $\{X_t\}$, though we do not pursue this here.

Summary of Section 11.5

- Brownian motion $\{B_t\}_{t \geq 0}$ is created from simple random walk with $p = 1/2$, by speeding up time by a large factor M, and shrinking space by a factor $1/\sqrt{M}$.

- Hence, $B_0 = 0$, and $B_t \sim N(0, t)$, and $\{B_t\}$ has independent normal increments with $B_t - B_s \sim N(0, t - s)$ for $0 \leq s < t$, and $\text{Cov}(B_s, B_t) = \min(s, t)$, and $\{B_t\}$ is a continuous function.

- Diffusions (often used to model stock prices) are of the form $X_t = a + \delta t + \sigma B_t$.

Exercises

11.5.1 Consider the speeded-up processes $\{Y_{i/M}^{(M)}\}$ used to construct Brownian motion. Compute
(a) $P(Y_1^{(1)} = 1)$.
(b) $P(Y_1^{(2)} = 1)$.
(c) $P(Y_1^{(2)} = \sqrt{2})$. (Hint: Don't forget that $\sqrt{2} = 2/\sqrt{2}$.)
(d) $P(Y_1^{(M)} \geq 1)$ for $M = 1$, $M = 2$, $M = 3$, and $M = 4$.

11.5.2 Let $\{B_t\}$ be Brownian motion. Compute $P(B_1 \geq 1)$.

11.5.3 Let $\{B_t\}$ be Brownian motion. Compute each of the following quantities.
(a) $P(B_2 \geq 1)$
(b) $P(B_3 \leq -4)$
(c) $P(B_9 - B_5 \leq 2.4)$
(d) $P(B_{26} - B_{11} > 9.8)$
(e) $P(B_{26.3} \leq -6)$
(f) $P(B_{26.3} \leq 0)$

11.5.4 Let $\{B_t\}$ be Brownian motion. Compute each of the following quantities.
(a) $P(B_2 \geq 1, \ B_5 - B_2 \geq 2)$
(b) $P(B_5 < -2, \ B_{13} - B_5 \geq 4)$
(c) $P(B_{8.4} > 3.2, \ B_{18.6} - B_{8.4} \geq 0.9)$

11.5.5 Let $\{B_t\}$ be Brownian motion. Compute $E(B_{13}B_8)$. (Hint: Don't forget part (b) of Theorem 11.5.2.)

11.5.6 Let $\{B_t\}$ be Brownian motion. Compute $E((B_{17} - B_{14})^2)$ in two ways.
(a) Use the fact that $B_{17} - B_{14} \sim N(0, 3)$.
(b) Square it out, and compute $E(B_{17}^2) - 2\,E(B_{17}B_{14}) + E(B_{14}^2)$.

11.5.7 Let $\{B_t\}$ be Brownian motion.
(a) Compute the probability that the process hits -5 before it hits 15.
(b) Compute the probability that the process hits -15 before it hits 5.
(c) Which of the answers of parts (a) and (b) is larger? Why is this so?

(d) Compute the probability that the process hits 15 before it hits -5.

(e) What is the sum of the answers to parts (a) and (d)? Why is this so?

11.5.8 Let $X_t = 5 + 3t + 2B_t$ be a diffusion (so that $a = 5$, $\delta = 3$, and $\sigma = 2$). Compute each of the following quantities.

(a) $E(X_7)$

(b) $\text{Var}(X_{8.1})$

(c) $P(X_{2.5} < 12)$

(d) $P(X_{17} > 50)$

11.5.9 Let $X_t = 10 - 1.5t + 4B_t$. Compute $E(X_3 X_5)$.

11.5.10 Suppose a stock has initial price \$400 and has volatility parameter equal to 9. Compute the probability that the stock price will be over \$500 after 8 years, if the drift per year is equal to

(a) \$0.

(b) \$5.

(c) \$10.

(d) \$20.

11.5.11 Suppose a stock has initial price \$200 and drift of \$3 per year. Compute the probability that the stock price will be over \$250 after 10 years, if the volatility parameter is equal to

(a) 1.

(b) 4.

(c) 10.

(d) 100.

Problems

11.5.12 Let $\{B_t\}$ be Brownian motion, and let $X = 2B_3 - 7B_5$. Compute the mean and variance of X.

11.5.13 Prove that $P(B_t < x) = P(B_t > -x)$ for any $t \geq 0$ and any $x \in R^1$.

Challenges

11.5.14 Compute $P(B_s \leq x \mid B_t = y)$, where $0 < s < t$, and $x, y \in R^1$. (Hint: You will need to use conditional densities.)

11.5.15 (a) Let $f : R^1 \to R^1$ be a *Lipschitz function*, i.e., a function for which there exists $K < \infty$ such that $|f(x) - f(y)| \leq K|x-y|$ for all $x, y \in R^1$. Compute

$$\lim_{h \searrow 0} \frac{(f(t+h) - f(t))^2}{h}$$

for any $t \in R^1$.

(b) Let $\{B_t\}$ be Brownian motion. Compute

$$\lim_{h \searrow 0} E\left(\frac{(B_{t+h} - B_t)^2}{h} \right)$$

for any $t > 0$.

(c) What do parts (a) and (b) seem to imply about Brownian motion?

(d) It is a known fact that all functions that are continuously differentiable on a closed interval are Lipschitz. In light of this, what does part (c) seem to imply about Brownian motion?

Discussion Topic

11.5.16 Diffusions such as those discussed here (and more complicated, varying-coefficient versions) are very often used by major investors and stock traders to model stock prices.

(a) Do you think that diffusions provide good models for stock prices?

(b) Even if diffusions did *not* provide good models for stock prices, why might investors still need to know about them?

11.6 Poisson Processes

Finally, we turn our attention to Poisson processes. These processes are models for events that happen at random times T_n. For example, T_n could be the time of the nth fire in a city, or the detection of the nth particle by a Geiger counter, or the nth car passing a checkpoint on a road. Poisson processes provide a model for the probabilities for when these events might take place.

More formally, we let $a > 0$, and let $R_1, R_2, \ldots$ be i.i.d. random variables, each having the Exponential(a) distribution. We let $T_0 = 0$, and for $n \geq 1$,

$$T_n = R_1 + R_2 + \ldots + R_n.$$

The value T_n thus corresponds to the (random) time of the nth event.

We also define a collection of counting variables N_t, as follows. For $t \geq 0$, we let

$$N_t = \max\{n : T_n \leq t\}.$$

That is, N_t counts the number of events that have happened by time t. (In particular, $N_0 = 0$. Furthermore, $N_t = 0$ for all $t < T_1$, i.e., before the first event occurs.)

We can think of the collection of variables N_t for $t \geq 0$ as being a stochastic process, indexed by the continuous time parameter $t \geq 0$. The process $\{N_t : t \geq 0\}$ is thus another example, like Brownian motion, of a *continuous-time stochastic process*.

In fact, $\{N_t : t \geq 0\}$ is called a *Poisson process* (with *intensity a*). This name comes from the following.

Theorem 11.6.1 For any $t > 0$, the distribution of N_t is Poisson(at).

Proof: See Section 11.7 for the proof of this result. ∎

In fact, even more is true.

Theorem 11.6.2 Let $0 = t_0 \leq t_1 < t_2 < t_3 < \ldots < t_d$. Then for $i = 1, 2, \ldots, d$, the distribution of $N_{t_i} - N_{t_{i-1}}$ is Poisson($a(t_i - t_{i-1})$). Furthermore, the random variables $N_{t_i} - N_{t_{i-1}}$, for $i = 1, \ldots, d$, are independent.

Proof: See Section 11.7 for the proof of this result. ∎

Example 11.6.1
Let $\{N_t\}$ be a Poisson process with intensity $a = 5$. What is $P(N_3 = 12)$?

Here $N_3 \sim$ Poisson($3a$) $=$ Poisson(15). Hence, from the definition of the Poisson distribution, we have

$$P(N_3 = 12) = e^{-15}(15)^{12} / 12! = 0.083,$$

which is a little more than 8%. ∎

Example 11.6.2
Let $\{N_t\}$ be a Poisson process with intensity $a = 2$. What is $P(N_6 = 11)$?

Here $N_6 \sim$ Poisson($6a$) $=$ Poisson(12). Hence,

$$P(N_6 = 11) = e^{-12}(12)^{11}/11! = 0.114,$$

or just over 11%. ∎

Example 11.6.3
Let $\{N_t\}$ be a Poisson process with intensity $a = 4$. What is $P(N_2 = 3, N_5 = 4)$? (Recall that here the comma means "and" in probability statements.)

We begin by writing $P(N_2 = 3, N_5 = 4) = P(N_2 = 3, N_5 - N_2 = 1)$. This is just rewriting the question. However, it puts it into a context where we can use Theorem 11.6.2.

Indeed, by that theorem, N_2 and $N_5 - N_2$ are independent, with $N_2 \sim$ Poisson(8) and $N_5 - N_2 \sim$ Poisson(12). Hence,

$$
\begin{aligned}
P(N_2 = 3, N_5 = 4) &= P(N_2 = 3, N_5 - N_2 = 1) \\
&= P(N_2 = 3)\, P(N_5 - N_2 = 1) \\
&= e^{-8}\frac{8^3}{3!}\, e^{-12}\frac{12^1}{1!} = 0.0000021.
\end{aligned}
$$

We thus see that the event $\{N_2 = 3, N_5 = 4\}$ is very unlikely in this case. ∎

Summary of Section 11.6

- Poisson processes are models of events that happen at random times T_n.

- It is assumed that the time $R_n = T_n - T_{n-1}$ between consecutive events in Exponential(a) for some $a > 0$. Then N_t represents the total number of events by time t.

- It follows that $N_t \sim$ Poisson(at), and in fact the process $\{N_t\}_{t \geq 0}$ has independent increments, with $N_t - N_s \sim$ Poisson($a(t - s)$) for $0 \leq s < t$.

Exercises

11.6.1 Let $\{N(t)\}_{t \geq 0}$ be a Poisson process with intensity $a = 7$. Compute
(a) $P(N_2 = 13)$.
(b) $P(N_5 = 3)$.
(c) $P(N_6 = 20)$.
(d) $P(N_{50} = 340)$.
(e) $P(N_2 = 13, N_5 = 3)$.
(f) $P(N_2 = 13, N_6 = 20)$.
(g) $P(N_2 = 13, N_5 = 3, N_6 = 20)$.

11.6.2 Let $\{N(t)\}_{t \geq 0}$ be a Poisson process with intensity $a = 3$. Compute $P(N_{1/2} = 6)$ and $P(N_{0.3} = 5)$.

11.6.3 Let $\{N(t)\}_{t \geq 0}$ be a Poisson process with intensity $a = 1/3$. Compute $P(N_2 = 6)$ and $P(N_3 = 5)$.

11.6.4 Let $\{N(t)\}_{t \geq 0}$ be a Poisson process with intensity $a = 3$. Compute $P(N_2 = 6, N_3 = 5)$. Explain your answer.
11.6.5 Let $\{N(t)\}_{t \geq 0}$ be a Poisson process with intensity $a > 0$. Compute (with explanation) the conditional probability $P(N_{2.6} = 2 \,|\, N_{2.9} = 2)$.
11.6.6 Let $\{N(t)\}_{t \geq 0}$ be a Poisson process with intensity $a = 1/3$. Compute (with explanation) the following conditional probabilities.
(a) $P(N_6 = 5 \,|\, N_9 = 5)$
(b) $P(N_6 = 5 \,|\, N_9 = 7)$
(c) $P(N_9 = 5 \,|\, N_6 = 7)$
(d) $P(N_9 = 7 \,|\, N_6 = 7)$
(e) $P(N_9 = 12 \,|\, N_6 = 7)$

Problems

11.6.7 Let $\{N_t : t \geq 0\}$ be a Poisson process with intensity $a > 0$. Let $0 < s < t$, and let j be a positive integer.
(a) Compute (with explanation) the conditional probability $P(N_s = j \,|\, N_t = j)$.
(b) Does the answer in part (a) depend on the value of the intensity a? Intuitively, why or why not?

11.6.8 Let $\{N_t : t \geq 0\}$ be a Poisson process with intensity $a > 0$. Let T_1 be the time of the first event, as usual. Let $0 < s < t$.
(a) Compute $P(N_s = 1 \,|\, N_t = 1)$. (If you wish, you may use the previous problem, with $j = 1$.)
(b) Suppose t is fixed, but s is allowed to vary in the interval $(0, t)$. What does the answer to part (b) say about the "conditional distribution" of T_1, conditional on knowing that $N_t = 1$?

11.7   Further Proofs

Theorem 11.1.1 Let $\{X_n\}$ be simple random walk as above, and let n be a positive integer. Then if k is an integer such that $-n \leq k \leq n$ and $n+k$ is even,

then

$$P(X_n = a + k) = \binom{n}{\frac{n+k}{2}} p^{(n+k)/2} q^{(n-k)/2}.$$

For all other values of k, we have $P(X_n = a + k) = 0$. Furthermore,

$$E(X_n) = a + n(2p - 1).$$

Proof: Of the first n bets, let W_n be the number won, and let L_n be the number lost. Then $n = W_n + L_n$. Also, $X_n = a + W_n - L_n$.

Adding these two equations together, we conclude that $n + X_n = W_n + L_n + a + W_n - L_n = a + 2W_n$. Solving for W_n, we see that $W_n = (n + X_n - a)/2$. Because W_n must be an integer, it follows that $n + X_n - a$ must be even. We conclude that $P(X_n = a + k) = 0$ unless $n + k$ is even.

On the other hand, solving for X_n, we see that $X_n = a + 2W_n - n$, or $X_n - a = 2W_n - n$. Because $0 \le W_n \le n$, it follows that $-n \le X_n - a \le n$, i.e., that $P(X_n = a + k) = 0$ if $k < -n$ or $k > n$.

Suppose now that $k + n$ is even, and $-n \le k \le n$. Then from the above, $P(X_n = a + k) = P(W_n = (n + k)/2)$. But the distribution of W_n is clearly Binomial(n, p). We conclude that

$$P(X_n = a + k) = \binom{n}{\frac{n+k}{2}} p^{(n+k)/2} q^{(n-k)/2},$$

provided that $k + n$ is even and $-n \le k \le n$.

Finally, because $W_n \sim$ Binomial(n, p), therefore $E(W_n) = np$. Hence, because $X_n = a + 2W_n - n$, therefore $E(X_n) = a + 2E(W_n) - n = a + 2np - n = a + n(2p - 1)$, as claimed. ∎

Theorem 11.1.2 Let $\{X_n\}$ be simple random walk, with some initial fortune a and probability p of winning each bet. Assume $0 < a < c$. Then the probability $P(\tau_c < \tau_0)$ of hitting c before 0 is given by

$$P(\tau_c < \tau_0) = \begin{cases} a/c & p = 1/2 \\ \frac{1 - \left(\frac{q}{p}\right)^a}{1 - \left(\frac{q}{p}\right)^c} & p \ne 1/2. \end{cases}$$

Proof: To begin, let us write $s(b)$ for the probability $P(\tau_c < \tau_0)$ when starting at the initial fortune b, for any $0 \le b \le c$. We are interested in computing $s(a)$. However, it turns out to be easier to solve for *all* of the values $s(0), s(1), s(2), \ldots, s(c)$ simultaneously, and this is the trick we use.

We have by definition that $s(0) = 0$ (i.e., if we start with \$0, then we can never win) and $s(c) = 1$ (i.e., if we start with \$c, then we have already won). So, those two cases are easy. However, the values of $s(b)$ for $1 \le b \le c - 1$ are not obtained as easily.

Our trick will be to develop equations that relate the values $s(b)$ for *different* values of b. Indeed, suppose $1 \le b \le c-1$. It is difficult to compute $s(b)$ directly. However, it is easy to understand what will happen on the *first* bet — we will either lose \$1 with probability p, or win \$1 with probability q. That leads to the following result.

Lemma 11.7.1 For $1 \leq b \leq c - 1$, we have

$$s(b) = ps(b+1) + qs(b-1). \tag{11.7.1}$$

Proof: Suppose first that we win the first bet, i.e., that $Z_1 = 1$. *After* this first bet, we will have fortune $b+1$. We then get to "start over" in our quest to reach c before reaching 0, except this time starting with fortune $b+1$ instead of b. Hence, after winning this first bet, our chance of reaching c before reaching 0 is now $s(b+1)$. (We still do not know what $s(b+1)$ is, but at least we are making a connection between $s(b)$ and $s(b+1)$.)

Suppose instead that we lose this first bet, i.e., that $Z_1 = -1$. After this first bet, we will have fortune $b - 1$. We then get to "start over" with fortune $b - 1$ instead of b. Hence, after this first bet, our chance of reaching c before reaching 0 is now $s(b-1)$.

We can combine all of the preceding information, as follows.

$$
\begin{aligned}
s(b) &= P(\tau_c < \tau_0) \\
&= P(Z_1 = 1, \tau_c < \tau_0) + P(Z_1 = -1, \tau_c < \tau_0) \\
&= ps(b+1) + qs(b-1).
\end{aligned}
$$

That is, $s(b) = p\,s(b+1) + q\,s(b-1)$, as claimed. ∎

So, where are we? We had $c + 1$ unknowns, $s(0), s(1), \ldots, s(c)$. We now know the 2 equations $s(0) = 0$ and $s(c) = 1$, plus the $c - 1$ equations of the form $s(b) = p\,s(b+1) + q\,s(b-1)$ for $b = 1, 2, \ldots, c-1$. In other words, we have $c+1$ equations in $c+1$ unknowns, so we can now solve our problem!

The solution still requires several algebraic steps, as follows.

Lemma 11.7.2 For $1 \leq b \leq c - 1$, we have

$$s(b+1) - s(b) = \frac{q}{p}(s(b) - s(b-1)).$$

Proof: Recalling that $p + q = 1$ We rearrange (11.7.1) as follows,

$$
\begin{aligned}
s(b) &= p\,s(b+1) + q\,s(b-1), \\
(p+q)s(b) &= p\,s(b+1) + q\,s(b-1), \\
q(s(b) - s(b-1)) &= p(s(b+1) - s(b)),
\end{aligned}
$$

and finally

$$s(b+1) - s(b) = \frac{q}{p}(s(b) - s(b-1)),$$

which gives the result. ∎

Lemma 11.7.3 For $0 \le b \le c$, we have

$$s(b) = \sum_{i=0}^{b-1} \left(\frac{q}{p}\right)^i s(1). \tag{11.7.2}$$

Proof: Applying the equation of Lemma 11.7.2 with $b = 1$, we obtain

$$s(2) - s(1) = \frac{q}{p}(s(1) - s(0)) = \frac{q}{p}s(1)$$

(because $s(0) = 0$). Applying it *again* with $b = 2$, we obtain

$$s(3) - s(2) = \frac{q}{p}(s(2) - s(1)) = \left(\frac{q}{p}\right)^2 (s(1) - s(0)) = \left(\frac{q}{p}\right)^2 s(1).$$

By induction, we see that

$$s(b+1) - s(b) = \left(\frac{q}{p}\right)^b s(1),$$

for $b = 0, 1, 2, \ldots, c - 1$. Hence, we compute that for $b = 0, 1, 2, \ldots, c$,

$$
\begin{aligned}
&s(b) \\
&= (s(b) - s(b-1)) + (s(b-1) - s(b-2)) + (s(b-2) - s(b-3)) + \cdots \\
&\quad + (s(1) - s(0)) \\
&= \sum_{i=0}^{b-1}(s(i+1) - s(i)) = \sum_{i=0}^{b-1} \left(\frac{q}{p}\right)^i s(1).
\end{aligned}
$$

This gives the result. ∎

We are now able to finish the proof of Theorem 11.1.2.

If $p = 1/2$, then $q/p = 1$, so (11.7.2) becomes $s(b) = bs(1)$. But $s(c) = 1$, so we must have $cs(1) = 1$, i.e., $s(1) = 1/c$. Then $s(b) = bs(1) = b/c$. Hence, $s(a) = a/c$ in this case.

If $p \ne 1/2$, then $q/p \ne 1$, so (11.7.2) is a geometric series, and becomes

$$s(b) = \frac{(q/p)^b - 1}{(q/p) - 1}s(1).$$

Because $s(c) = 1$, we must have

$$1 = \frac{(q/p)^c - 1}{(q/p) - 1}s(1),$$

so

$$s(1) = \frac{(q/p) - 1}{(q/p)^c - 1}.$$

Then

$$s(b) = \frac{(q/p)^b - 1}{(q/p) - 1} s(1) = \frac{(q/p)^b - 1}{(q/p) - 1} \frac{(q/p) - 1}{(q/p)^c - 1} = \frac{(q/p)^b - 1}{(q/p)^c - 1}. \cdots$$

Hence,

$$s(a) = \frac{(q/p)^a - 1}{(q/p)^c - 1}$$

in this case. ∎

Theorem 11.1.3 Let $\{X_n\}$ be simple random walk, with initial fortune $a > 0$ and probability p of winning each bet. Then the probability $P(\tau_0 < \infty)$ that the walk will *ever* hit 0 is given by

$$P(\tau_0 < \infty) = \begin{cases} 1 & p \leq 1/2 \\ (q/p)^a & p > 1/2. \end{cases}$$

Proof: By continuity of probabilities, we see that

$$P(\tau_0 < \infty) = \lim_{c \to \infty} P(\tau_0 < \tau_c) = \lim_{c \to \infty} (1 - P(\tau_c < \tau_0)).$$

Hence, if $p = 1/2$, then $P(\tau_0 < \infty) = \lim_{c \to \infty} (1 - a/c) = 1$.

Now, if $p \neq 1/2$, then

$$P(\tau_0 < \infty) = \lim_{c \to \infty} \left(1 - \frac{1 - (q/p)^a}{1 - (q/p)^c} \right).$$

If $p < 1/2$, then $q/p > 1$, so $\lim_{c \to \infty}(q/p)^c = \infty$, and $P(\tau_0 < \infty) = 1$. If $p > 1/2$, then $q/p < 1$, so $\lim_{c \to \infty}(q/p)^c = 0$, and $P(\tau_0 < \infty) = (q/p)^a$. ∎

Theorem 11.3.3 The preceding Metropolis–Hastings algorithm results in a Markov chain $X_0, X_1, X_2, \ldots$, which has $\{\pi_i\}$ as a stationary distribution.

Proof: We shall prove that the resulting Markov chain is *reversible* with respect to $\{\pi_i\}$, i.e., that

$$\pi_i P(X_{n+1} = j \mid X_n = i) = \pi_j P(X_{n+1} = i \mid X_n = j), \qquad (11.7.3)$$

for $i, j \in S$. It will then follow from Theorem 11.2.6 that $\{\pi_i\}$ is a stationary distribution for the chain.

We thus have to prove (11.7.3). Now, (11.7.3) is clearly true if $i = j$, so we can assume that $i \neq j$.

But if $i \neq j$, and $X_n = i$, then the only way we can have $X_{n+1} = j$ is if $Y_{n+1} = j$ (i.e., we *propose* the state j, which we will do with probability p_{ij}). Also we *accept* this proposal (which we will do with probability α_{ij}). Hence,

$$P(X_{n+1} = j \mid X_n = i) = q_{ij}\alpha_{ij} = q_{ij} \min\left\{1, \frac{\pi_j q_{ji}}{\pi_i q_{ij}}\right\} = \min\left\{q_{ij}, \frac{\pi_j q_{ji}}{\pi_i}\right\}.$$

It follows that $\pi_i P(X_{n+1} = j \mid X_n = i) = \min\{\pi_i q_{ij}, \pi_j q_{ji}\}$.

Similarly, we compute that $\pi_j P(X_{n+1} = i \mid X_n = j) = \min\{\pi_j q_{ji}, \pi_i q_{ij}\}$. It follows that (11.7.3) is true. ∎

Theorem 11.5.1 Let $\{Y_t^{*(M)} : t \geq 0\}$ be as defined earlier. Then for large M:
(a) For $t \geq 0$, the distribution of $Y_t^{*(M)}$ is approximately $N(0, t)$, i.e., normally distributed with mean t.
(b) For $s, t \geq 0$, the covariance

$$\text{Cov}\left(Y_t^{*(M)}, Y_t^{*(M)}\right)$$

is approximately equal to $\min\{s, t\}$.
(c) For $t \geq s \geq 0$, the distribution of the increment

$$Y_t^{*(M)} - Y_s^{*(M)}$$

is approximately $N(0, t-s)$, i.e., normally distributed with mean 0 and variance $t - s$, and is approximately independent of $Y_s^{*(M)}$.
(d) $Y_t^{*(M)}$ is a continuous function of t.

Proof: Write $\lfloor r \rfloor$ for the greatest integer not exceeding r, so that, e.g., $\lfloor 7.6 \rfloor = 7$. Then we see that for large M, t is very close to $\lfloor tM \rfloor / M$, so that $Y_t^{*(M)}$ is very close (formally, within $O(1/M)$ in probability) to

$$A = Y_{\lfloor tM \rfloor / M}^{(M)} = \frac{1}{\sqrt{M}}(Z_1 + Z_2 + \cdots + Z_{\lfloor tM \rfloor}).$$

Now, A is equal to $1/\sqrt{M}$ times the sum of $\lfloor tM \rfloor$ different i.i.d. random variables, each having mean 0 and variance 1. It follows from the central limit theorem, that A converges in distribution to the distribution $N(0, t)$ as $M \to \infty$. This proves part (a).

For part (b), note that also $Y_s^{*(M)}$ is very close to

$$B = Y_{\lfloor sM \rfloor / M}^{(M)} = \frac{1}{\sqrt{M}}(Z_1 + Z_2 + \cdots + Z_{\lfloor sM \rfloor}).$$

Because $E(Z_i) = 0$, we must have $E(A) = E(B) = 0$, so that $\text{Cov}(A, B) = E(AB)$.

For simplicity, assume $s \leq t$; the case $s > t$ is similar. Then we have

$$
\begin{aligned}
\text{Cov}(A, B) &= E(AB) \\
&= \frac{1}{M} E\left((Z_1 + Z_2 + \cdots + Z_{\lfloor sM \rfloor})(Z_1 + Z_2 + \cdots + Z_{\lfloor tM \rfloor})\right) \\
&= \frac{1}{M} E\left(\sum_{i=1}^{\lfloor sM \rfloor} \sum_{j=1}^{\lfloor tM \rfloor} Z_i Z_j\right) = \frac{1}{M} \sum_{i=1}^{\lfloor sM \rfloor} \sum_{j=1}^{\lfloor tM \rfloor} E(Z_i Z_j).
\end{aligned}
$$

Now, we have $E(Z_i Z_j) = 0$ unless $i = j$, in which case $E(Z_i Z_j) = 1$. There will be precisely $\lfloor sM \rfloor$ terms in the sum for which $i = j$, namely, one for each value of i (since $t \geq s$). Hence,

$$\mathrm{Cov}(A, B) = \frac{\lfloor sM \rfloor}{M},$$

which converges to s as $M \to \infty$. This proves part (b).

Part (c) follows very similarly to part (a). Finally, part (d) follows because the function $Y_t^{(M)}$ was constructed in a continuous manner (as in Figure 11.5.1). ∎

Theorem 11.6.1 For any $t > 0$, the distribution of N_t is Poisson(at).

Proof: We first require a technical lemma.

Lemma 11.7.4 Let $g_n(t) = e^{-at} a^n t^{n-1}/(n-1)!$ be the density of the Gamma(n, a) distribution. Then for $n \geq 1$,

$$\int_0^t g_n(s)\, ds = \sum_{i=n}^{\infty} e^{-at}(at)^i/i!. \tag{11.7.4}$$

Proof: If $t = 0$, then both sides are 0. For other t, differentiating with respect to t, we see (setting $j = i - 1$) that

$$\frac{\partial}{\partial t} \sum_{i=n}^{\infty} e^{-at}(at)^i = \sum_{i=n}^{\infty}(-ae^{-at}(at)^i/i!) + e^{-at}a^i t^{i-1}/(i-1)!)$$

$$= \sum_{i=n}^{\infty}(-e^{-at}a^{i+1}t^i/i!) + \sum_{j=n-1}^{\infty} e^{-at}a^{j+1}t^j/j!$$

$$= e^{-at}a^{(n-1)+1}t^{n-1}/(n-1)!$$

$$= g_n(t) = \frac{\partial}{\partial t}\int_0^t g_n(s)\, ds.$$

Because this is true for all $t \geq 0$, we see that (11.7.4) is satisfied for any $n \geq 0$. ∎

Recall (Example 2.4.8) that the Exponential(λ) distribution is the same as the Gamma($1, \lambda$) distribution. Furthermore, recall (Problem 2.9.12) that if $X \sim$ Gamma(α_1, λ) and $Y \sim$ Gamma(α_2, λ) are independent, then $X + Y \sim$ Gamma($\alpha_1 + \alpha_2, \lambda$).

Now, in our case we have $T_n = R_1 + R_2 + \cdots + R_n$, where $R_i \sim$ Exponential(a) = Gamma($1, a$). It follows that $T_n \sim$ Gamma(n, a). Hence, the density of T_n is $g_n(t) = e^{-at} a^n t^{n-1}/(n-1)!$.

Now, the event that $N_t \geq n$ (i.e., that the number of events by time t is at least n) is the same as the event that $T_n \leq t$ (i.e., that the nth event occurs before time n). Hence,

$$P(N_t \geq n) = P(T_n \leq t) = \int_0^t g_n(s)\, ds\,.$$

Then by Lemma 11.7.4,

$$P(N_t \geq n) = \sum_{i=n}^{\infty} e^{-at} \frac{(at)^i}{i!} \tag{11.7.5}$$

for any $n \geq 1$. If $n = 0$, then both sides are 1, so in fact (11.7.5) holds for any $n \geq 0$.

Using this, we see that

$$P(N_t = j) = P(N_t \geq j) - P(N_t \geq j+1)$$

$$= \left(\sum_{i=j}^{\infty} e^{-at}(at)^i/i! \right) - \left(\sum_{i=j+1}^{\infty} e^{-at}(at)^i/i! \right) = e^{-at}(at)^j/j!.$$

It follows that $N_t \sim \text{Poisson}(at)$, as claimed. ∎

Theorem 11.6.2 Let $0 = t_0 \leq t_1 < t_2 < t_3 < \cdots < t_d$. Then for $i = 1, 2, \ldots, d$, the distribution of $N_{t_i} - N_{t_{i-1}}$ is $\text{Poisson}(a(t_i - t_{i-1}))$. Furthermore, the random variables $N_{t_i} - N_{t_{i-1}}$, for $i = 1, \ldots, d$, are independent.

Proof: From the memoryless property of the exponential distributions (Problem 2.4.4), it follows that regardless of the values of N_s for $s \leq t_{i-1}$, this will have no effect on the distribution of the increments $N_t - N_{t_{i-1}}$ for $t > t_{i-1}$. That is, the process $\{N_t\}$ starts fresh at each time t_{i-1}, except from a different initial value $N_{t_{i-1}}$ instead of from $N_0 = 0$.

Hence, the distribution of $N_{t_{i-1}+u} - N_{t_{i-1}}$ for $u \geq 0$ is identical to the distribution of $N_u - N_0 = N_u$ and is independent of the values of N_s for $s \leq t_{i-1}$. Because we already know that $N_u \sim \text{Poisson}(au)$, it follows that $N_{t_{i-1}+u} - N_{t_{i-1}} \sim \text{Poisson}(au)$ as well. In particular, $N_{t_i} - N_{t_{i-1}} \sim \text{Poisson}(a(t_i - t_{i-1}))$ as well, with $N_{t_i} - N_{t_{i-1}}$ independent of $\{N_s : s \leq t_{i-1}\}$. The result follows. ∎

Appendix A

Mathematical Background

To understand this book, it is necessary to know certain mathematical subjects listed below. Because it is assumed the student has already taken a course in calculus, such things as derivatives, integrals, and infinite series are treated quite briefly below. Multivariable integrals are treated in somewhat more detail.

A.1   Derivatives

From calculus, we know that the *derivative* of a function f is its instantaneous rate of change:

$$f'(x) = \frac{d}{dx}f(x) = \lim_{h \to 0} \frac{f(x+h) - f(x)}{h}.$$

In particular, the reader should recall from calculus that

$$\tfrac{d}{dx}5 = 0, \qquad \tfrac{d}{dx}x^3 = 3x^2, \qquad \tfrac{d}{dx}x^n = nx^{n-1},$$

$$\tfrac{d}{dx}e^x = e^x, \qquad \tfrac{d}{dx}\sin x = \cos x, \qquad \tfrac{d}{dx}\cos x = -\sin x,$$

etc. Hence, if $f(x) = x^3$, then $f'(x) = 3x^2$, and, e.g., $f'(7) = 3\,7^2 = 147$.

Derivatives respect addition and scalar multiplication, so if f and g are functions and C is a constant, then

$$\frac{d}{dx}(C\,f(x) + g(x)) = C\,f'(x) + g'(x).$$

Thus,

$$\frac{d}{dx}(5x^3 - 3x^2 + 7x + 12) = 15x^2 - 6x + 7,$$

etc.

Finally, derivatives satisfy a *chain rule*, that if a function can be written as a *composition* of two other functions, as in $f(x) = g(h(x))$, then $f'(x) =$

$g'(h(x)) \, h'(x)$. Thus,

$$\tfrac{d}{dx} e^{5x} = 5e^{5x},$$

$$\tfrac{d}{dx} \sin(x^2) = 2x \cos(x^2),$$

$$\tfrac{d}{dx} (x^2 + x^3)^5 = 5(x^2 + x^3)^4 (2x + 3x^2),$$

etc.

Higher-order derivatives are defined by

$$f''(x) = \frac{d}{dx} f'(x), \; f'''(x) = \frac{d}{dx} f''(x),$$

etc. In general, the rth-order derivative $f^{(r)}(x)$ can be defined inductively by $f^{(0)}(x) = f(x)$ and

$$f^{(r)}(x) = \frac{d}{dx} f^{(r-1)}(x)$$

for $r \geq 1$. Thus, if $f(x) = x^4$, then $f'(x) = 4x^3$, $f''(x) = f^{(2)}(x) = 12x^2$, $f^{(3)}(x) = 24x$, $f^{(4)}(x) = 24$, etc.

Derivatives are used often in this text.

A.2   Integrals

If f is a function, and $a < b$ are constants, then the *integral* of f over the interval $[a, b]$, written

$$\int_a^b f(x) \, dx,$$

represents adding up the values $f(x)$, multiplied by the widths of small intervals around x. That is, $\int_a^b f(x) \, dx \approx \sum_{i=1}^d f(x_i)(x_i - x_{i-1})$, where $a = x_0 < x_1 < \ldots < x_d = b$, and where $x_i - x_{i-1}$ is small.

More formally, we can set $x_i = a + (i/d)(b - a)$ and let $d \to \infty$, to get a formal definition of integral as

$$\int_a^b f(x) \, dx = \lim_{d \to \infty} \sum_{i=1}^d f\left(a + (i/d)(b - a)\right)(1/d).$$

To compute $\int_a^b f(x) \, dx$ in this manner each time would be tedious. Fortunately, the *fundamental theorem of calculus* provides a much easier way to compute integrals. It says that if $F(x)$ is any function with $F'(x) = f(x)$, then $\int_a^b f(x) \, dx = F(b) - F(a)$. Hence,

$$\int_a^b 3x^2 \, dx = b^3 - a^3,$$

$$\int_a^b x^2 \, dx = \tfrac{1}{3}(b^3 - a^3),$$

$$\int_a^b x^n \, dx = \tfrac{1}{n+1}(b^{n+1} - a^{n+1}),$$

and

$$\int_a^b \cos x \, dx = \sin b - \sin a,$$

$$\int_a^b \sin x \, dx = -(\cos b - \cos a),$$

$$\int_a^b e^{5x} \, dx = \tfrac{1}{5}(e^{5b} - e^{5a}).$$

A.3 Infinite Series

If $a_1, a_2, a_3, \ldots$ is an infinite sequence of numbers, we can consider the infinite sum (or *series*)

$$\sum_{i=1}^{\infty} a_i = a_1 + a_2 + a_3 + \ldots$$

Formally, $\sum_{i=1}^{\infty} a_i = \lim_{N \to \infty} \sum_{i=1}^{N} a_i$. This sum may be finite or infinite.

For example, clearly $\sum_{i=1}^{\infty} 1 = 1 + 1 + 1 + 1 + \ldots = \infty$. On the other hand, because

$$\frac{1}{2} + \frac{1}{4} + \frac{1}{8} + \frac{1}{16} + \ldots + \frac{1}{2^n} = \frac{2^n - 1}{2^n},$$

we see that

$$\frac{1}{2} + \frac{1}{4} + \frac{1}{8} + \frac{1}{16} + \ldots = \sum_{i=1}^{\infty} \frac{1}{2^i} = \lim_{N \to \infty} \sum_{i=1}^{N} \frac{1}{2^i} = \lim_{N \to \infty} \frac{2^N - 1}{2^N} = 1.$$

More generally, we compute that

$$\sum_{i=1}^{\infty} a^i = \frac{a}{1 - a}$$

whenever $|a| < 1$.

One particularly important kind of infinite series is a *Taylor series*. If f is a function, then its Taylor series is given by

$$f(0) + x f'(0) + \frac{1}{2!} x^2 f''(0) + \frac{1}{3!} x^3 f^{(3)}(0) + \ldots = \sum_{i=0}^{\infty} \frac{1}{i!} x^i f^{(i)}(0).$$

(Here $i! = i(i-1)(i-2) \ldots (2)(1)$ stands for i *factorial*, with $0! = 1! = 1$, $2! = 2$, $3! = 6$, $4! = 24$, etc.) Usually, $f(x)$ will be exactly equal to its Taylor series expansion, thus,

$$
\begin{aligned}
\sin x &= x - x^3/3 + x^5/5 - x^7/7 + \ldots, \\
\cos x &= 1 - x^2/2 + x^4/4 - x^5/5 + \ldots, \\
e^x &= 1 + x + x^2/2! + x^3/3! + x^4/4! + \ldots, \\
e^{5x} &= 1 + 5x + (5x)^2/2! + (5x)^3/3! + (5x)^4/4! + \ldots,
\end{aligned}
$$

etc. If $f(x)$ is a polynomial (e.g., $f(x) = x^3 - 3x^2 + 2x - 6$), then the Taylor series of $f(x)$ is precisely the same function as $f(x)$ itself.

A.4   Matrix Multiplication

A *matrix* is any $r \times s$ collection of numbers, e.g.,

$$A = \begin{pmatrix} 8 & 6 \\ 5 & 2 \end{pmatrix}, \quad B = \begin{pmatrix} 3 & 6 & 2 \\ -7 & 6 & 0 \end{pmatrix}, \quad C = \begin{pmatrix} 2 & -1 \\ 3/5 & 2/5 \\ -0.6 & -17.9 \end{pmatrix},$$

etc.

Matrices can be *multiplied*, as follows. If A is an $r \times s$ matrix, and B is an $s \times u$ matrix, then the product AB is an $r \times u$ matrix whose i, j entry is given by $\sum_{k=1}^{s} A_{ik}B_{kj}$, a sum of products. For example, with A and B as above, if $M = AB$, then

$$\begin{aligned} M &= \begin{pmatrix} 8 & 6 \\ 5 & 2 \end{pmatrix} \begin{pmatrix} 3 & 6 & 2 \\ -7 & 6 & 0 \end{pmatrix} \\ &= \begin{pmatrix} 8\,(3) + 6\,(-7) & 8\,(6) + 6\,(6) & 8\,(2) + 6\,(0) \\ 5\,(3) + 2\,(-7) & 5\,(6) + 2\,(6) & 5\,(2) + 2\,(0) \end{pmatrix} = \begin{pmatrix} -18 & 84 & 16 \\ 1 & 42 & 10 \end{pmatrix}, \end{aligned}$$

as, for example, the $(2, 1)$ entry of M equals $5\,(3) + 2\,(-7) = 1$.

Matrix multiplication turns out to be surprisingly useful, and it is used in various places in this book.

A.5   Partial Derivatives

Suppose f is a function of *two* variables, as in $f(x, y) = 3x^2y^3$. Then we can take a *partial derivative* of f with respect to x, writing

$$\frac{\partial}{\partial x} f(x, y),$$

by varying x while keeping y fixed. That is,

$$\frac{\partial}{\partial x} f(x, y) = \lim_{h \to 0} \frac{f(x + h, y) - f(x, y)}{h}.$$

This can be computed simply by regarding y as a constant value. For the example above,

$$\frac{\partial}{\partial x}(3x^2y^3) = 6xy^3.$$

Similarly, by regarding x as constant and varying y, we see that

$$\frac{\partial}{\partial y}(3x^2y^3) = 9x^2y^2.$$

Other examples include

$$\frac{\partial}{\partial x}(18e^{xy} + x^6y^8 - \sin(y^3)) = 18ye^{xy} + 6x^5y^8,$$

$$\frac{\partial}{\partial y}(18e^{xy} + x^6y^8 - \sin(y^3)) = 18xe^{xy} + 8x^6y^7 - 3y^2\sin(y^3),$$

etc.

If f is a function of three or more variables, then partial derivatives may similarly be taken. Thus,

$$\frac{\partial}{\partial x}(x^2y^4z^6)) = 2xy^4z^6, \quad \frac{\partial}{\partial y}(x^2y^4z^6)) = 4x^2y^3z^6, \quad \frac{\partial}{\partial z}(x^2y^4z^6)) = 6x^2y^4z^5,$$

etc.

A.6   Multivariable Integrals

If f is a function of two or more variables, we can still compute integrals of f. However, instead of taking integrals over an interval $[a, b]$, we must take integrals over higher-dimensional *regions*.

For example, let $f(x, y) = x^2y^3$, and let R be the rectangular region given by $R = \{0 \le x \le 1, 5 \le y \le 7\} = [0, 1] \times [5, 7]$. What is

$$\int_R \int f(x, y)\, dx\, dy,$$

the integral of f over the region R? In geometrical terms, it is the volume under the graph of f (and this is a surface) over the region R. But how do we compute this?

Well, if y is constant, we know that

$$\int_0^1 f(x, y)\, dx = \int_0^1 x^2y^3\, dx = \frac{1}{3}y^3. \tag{A.6.1}$$

This corresponds to adding up the values of f along one "strip" of the region R, where y is constant. In Figure A.6.1 we show the region on integration $R = [0, 1] \times [5, 7]$. The value of (A.6.1), when $y = 6.2$, is $(6.2)^3/3 = 79.443$, and this is the area under the curve $x^2 (6.2)^3$ over the line $[0, 1] \times \{6.2\}$.

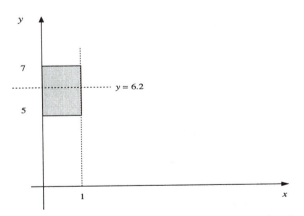

Figure A.6.1: Plot of the region of integration (shaded) $R = [0, 1] \times [5, 7]$ together with the line at $y = 6.2$.

If we then add up the values of the areas over these strips along all different possible y values, then we obtain the overall integral or volume, as follows:

$$\int_R \int f(x,y)\, dx\, dy \;=\; \int_5^7 \left(\int_0^1 f(x,y)\, dx\right) dy = \int_5^7 \left(\int_0^1 x^2 y^3\, dx\right) dy$$

$$=\; \int_5^7 \left(\frac{1}{3} y^3\right) dy = \frac{1}{3}\frac{1}{4}(7^4 - 5^4) = 148\,.$$

So the volume under the the graph of f and over the region R is given by 148.

Note that we can also compute this integral by integrating first y and then x, and we get the same answer:

$$\int_R \int f(x,y)\, dx\, dy \;=\; \int_0^1 \left(\int_5^7 f(x,y)\, dy\right) dx = \int_0^1 \left(\int_5^7 x^2 y^3\, dy\right) dx$$

$$=\; \int_0^1 \left(\frac{1}{4} x^2 (7^4 - 5^4)\right) dx = \frac{1}{3}\frac{1}{4}(7^4 - 5^4) = 148\,.$$

A.6.1 Nonrectangular Regions

If the region R is not a rectangle, then the computation is more complicated. The idea is that, for each value of x, we integrate y over only those values for which the point (x,y) is inside R.

For example, suppose that R is the triangle given by $R = \{(x,y) : 0 \le x \le 2y \le 6\}$. In Figure A.6.2 we have plotted this region together with the slices at $x = 3$ and $y = 3/2$. We use the x-slices to determine the limits on y for fixed x when we integrate out y first, and we use the y-slices to determine the limits on x for fixed y when we integrate out x first.

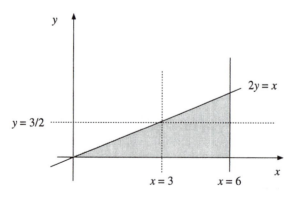

Figure A.6.2: The integration region (shaded) $R = \{(x,y) : \; 0 \le x \le 2y \le 6\}$ together with the slices at $x = 3$ and $y = 3/2$.

Then x can take any value between 0 and 6. However, once we know x, then

y can only take values between $x/2$ and 3. Hence, if $f(x, y) = xy + x^6 y^8$, then

$$\int_R \int f(x, y) \, dx \, dy$$

$$= \int_0^6 \left(\int_{x/2}^3 f(x, y) \, dy \right) dx = \int_0^6 \left(\int_{x/2}^3 (xy + x^6 y^8) \, dy \right) dx$$

$$= \int_0^6 \left(x \frac{1}{2} (3^2 - (x/2)^2) + (x^6 \frac{1}{9} (3^9 - (x/2)^9)) \right) dx$$

$$= \int_0^6 \left(\frac{9}{2} x - \frac{9}{8} x^3 + 2187 x^6 - x^1 5/4608 \right) dx$$

$$= \frac{9}{2} \frac{1}{2} (6^2 - 0^2) - \frac{9}{8} \frac{1}{4} (6^4 - 0^4) + 2187 \frac{1}{7} (6^7 - 0^7) - \frac{1}{16} (6^1 5 - 0^1 5)/4608$$

$$= 81 - 364.5 + 87460005 - 6377292 = 81082429.5.$$

Once again, we can compute the same integral in the opposite order, by integrating first x and then y. In this case, y can take any value between 0 and 3. Then, for a given value of y, we see that x can take values between 0 and $2y$. Hence,

$$\int_R \int f(x, y) \, dx \, dy = \int_0^3 \left(\int_0^{2y} f(x, y) \, dx \right) dy = \int_0^3 \left(\int_0^{2y} (xy + x^6 y^8) \, dx \right) dy.$$

We leave it as an exercise for the reader to finish this integral, and see that the same answer as above is obtained.

Functions of three or more variables can also be integrated over regions of the corresponding dimension three or higher. For simplicity, we do not emphasize such higher-order integrals in this book.

Appendix B

Computations

All the computations found in this text were carried out using Minitab. This statistical software package is very easy to learn and use. Other packages such as SAS or S-Plus could also be used for this purpose.

Most of the computations were performed using Minitab like a calculator, i.e., data were entered and then a number of Minitab commands were accessed to obtain the quantities desired. No programming is required for these computations.

There were a few computations, however, that did involve a bit of programming. Typically, this was a computation in which a number of operations were required to be performed many times, and so looping was desirable. In each such case, we have recorded here the Minitab code that we used for these computations. As the following examples show, these programs were never very involved.

Students can use these programs as templates for writing their own Minitab programs. Actually, the language is so simple that we feel that anyone using another language for programming can read these programs and use them as templates in the same way. Simply think of the symbols $c1$, $c2$, etc. as arrays where we address the ith element in the array $c1$ by $c1(i)$. Further, there are constants $k1$, $k2$, etc.

A Minitab program is called a *macro* and must start with the statement `gmacro` and end with the statement `endmacro`. The first statement after `gmacro` gives a name to the program. Comments in a program, put there for explanatory purposes, start with `note`.

If the file containing the program is called `prog.txt` and this is stored in the root directory of a disk drive called `c`, then the Minitab command

```
MTB> %c:/prog.txt
```

will run the program. Any output will either be printed in the Session window (if you have used a `print` command) or stored in the Minitab worksheet.

More details on Minitab can be found using `Help` in the program or from a manual that explains the language, such as *Minitab Manual for Moore and*

McCabe's Introduction to the Practice of Statistics, 4th ed. by M. J. Evans (W. H. Freeman, New York, 2002.)

Bootstrapping in Example 6.4.2

The following Minitab code generates 1000 bootstrap samples from the data in c1, calculates the median of each of these samples, and then calculates the sample variance of these medians.

```
gmacro
bootstrapping
base 34256734
note - original sample is stored in c1
note - bootstrap sample is placed in c2 (each one overwritten)
note - medians of bootstrap samples are stored in c3
note - k1 = size of data set (and bootstrap samples)
let k1=15
do k2=1:1000
sample 15 c1 c2;
replace.
let c3(k2)=median(c2)
enddo
note - k3 equals (6.4.5)
let k3=(stdev(c3))**2
print k3
endmacro
```

Sampling from the Posterior in Example 7.3.1

The following Minitab code generates a sample of 10^4 from the joint posterior in this example. Note that in Minitab the Gamma(α, β) density takes the form

$$\frac{\beta^{-\alpha}}{\Gamma(\alpha)} x^{\alpha-1} e^{-x/\beta}.$$

So to generate from a Gamma(α, β) distribution, as defined in this book, we must put the second shape parameter equal to $1/\beta$ in Minitab.

```
gmacro
normalpost
note - the base command sets the seed for the random numbers
base 34256734
note - the parameters of the posterior
note - k1 = first parameter of the gamma distribution
note      = (alpha_0 + n/2)
let k1=9.5
note - k2 = 1 / beta
let k2=1/77.578
```

```
note - k3 = posterior mean of mu
let k3=5.161
note - k4 = (n + 1/(tau_0 squared) )^(-1)
let k4=1/15.5
note - main loop
note - c3 contains generated value of sigma**2
note - c4 contains generated value of mu
note - c5 contains generated value of coefficient of variation
do k5=1:10000
random 1 c1;
gamma k1 k2.
let c3(k5)=1/c1(1)
let k6=sqrt(k4/c1(1))
random 1 c2;
normal k3 k6.
let c4(k5)=c2(1)
let c5(k5)=sqrt(c3(k5))/c4(k5)
enddo
endmacro
```

Calculating the Estimates and Standard Errors in Example 7.3.1

We have a sample of 10^4 values from the posterior distribution of ψ stored in C5. The following computations use this sample to calculate an estimate of the posterior probability that $\psi \leq 0.5$ (k1), and to calculate the standard error of this estimate (k2), the estimate minus three times its standard error (k3), and the estimate plus three times its standard error (k4).

```
MTB > let c6=c5 le .5
MTB > let k1=mean(c6)
MTB > let k2=sqrt(k1*(1-k1))/sqrt(10000)
MTB > let k3=k1-3*k2
MTB > let k4=k1+3*k2
MTB > print k1 k2 k3 k4
Data Display
K1 0.289300
K2 0.004533
K3 0.275411
K4 0.302609
```

Using the Gibbs Sampler in Example 7.3.2

The following Minitab code generates a chain of length 10^4 values using the Gibbs sampler described there.

```
gmacro
gibbs
```

```
base 34256734
note - data sample is stored in c1
note - starting value for mu.
let k1=mean(c1)
note - starting value for sigma**2
let k2=stdev(c1)
let k2=k2**2
note - lambda
let k3=3
note - sample size
let k4=15
note - n/2 + alpha_0 + 1/2
let k5=k4/2 +2+.5
note - mu_0
let k6=4
note - tau_0**2
let k7=2
note - beta_0
let k8=1
let k9=(k3/2+.5)
note - main loop
do k100=1:10000
note - generate the nu_i in c10
do k111=1:15
let k10=.5*(((c1(k111)-k1)**2)/(k2*k3) +1)
let k10=1/k10
random 1 c2;
gamma k9 k10.
let c10(k111)=c2(1)
enddo
note - generate sigma**2 in c20
let c11=c10*((c1-k1)**2)
let k11=.5*sum(c11)/k3+.5*((k1-k6)**2)/k7 +k8
let k11=1/k11
random 1 c2;
gamma k5 k11.
let c20(k100)=1/c2(1)
let k2=1/c2(1)
note - generate mu in c21
let k13=1/(sum(c10)/k3 +1/k7)
let c11=c1*c10/k3
let k14=sum(c11)+k6/k7
let k14=k13*k14
let k13=sqrt(k13*k2)
random 1 c2;
normal k14 k13.
```

```
let c21(k100)=c2(1)
let k1=c2(1)
enddo
endmacro
```

Batching in Example 7.3.2

The following Minitab code divides the generated sample, obtained via the preceding Gibbs sampling code, into batches, and calculates the batch means.

```
gmacro
batching
note - k2= batch size
let k2=40
note - k4 holds the batch sums
note - c1 contains the data to be batched (10000 data values)
note - c2 will contain the batch means (250 batch means)
do k10=1:10000/40
let k4=0
do k20=0:39
let k3=c1(k10+k20)
let k4=k4+k3
enddo
let k11=floor(k10/k2) +1
let c2(k11)=k4/k2
enddo
endmacro
```

Simulating a Sample from the Distribution of the Discrepancy Statistic in Example 9.1.2

The following code generates a sample from the discrepancy statistic specified when in this example.

```
gmacro
goodnessoffit
base 34256734
note - generated sample is stored in c1
note - residuals are placed in c2
note - value of D(r) are placed in c3
note - k1 = size of data set
let k1=5
do k2=1:10000
random k1 c1
let k3=mean(c1)
let k4=sqrt(k1-1)*stdev(c1)
let c2=((c1-k3)/k4)**2
```

```
let c2=loge(c2)
let k5=-sum(c2)/k1
let c3(k2)=k5
enddo
endmacro
```

Generating from a Dirichlet Distribution in Example 10.2.3

The following code generates a sample from a Dirichlet$(\alpha_1, \alpha_2, \alpha_3, \alpha_4,)$ distribution.

```
gmacro
dirichlet
note - the base command sets the seed for the random number
note    generator (so you can repeat a simulation).
base 34256734
note - here we provide the algorithm for generating from a
note    Dirichlet(k1,k2,k3,k4) distribution.
note - assign the values of the parameters.
let k1=2
let k2=3
let k3=1
let k4=1.5
let k5=K2+k3+k4
let k6=k3+k4
note - generate the sample with i-th sample in i-th row of
note    c2, c3, c4, c5, ....
do k10=1:5
random 1 c1;
beta k1 k5.
let c2(k10)=c1(1)
random 1 c1;
beta k2 k6.
let c3(k10)=(1-c2(k10))*c1(1)
random 1 c1;
beta k3 k4.
let c4(k10)=(1-c2(k10)-c3(k10))*c1(1)
let c5(k10)= 1-c2(k10)-c3(k10)-c4(k10)
enddo
endmacro
```

Appendix C

Common Distributions

We record here the most commonly used distributions in probability and statistics and some of their basic characteristics.

C.1 Discrete Distributions

1. *Bernoulli*(θ), $\theta \in [0, 1]$ (same as Binomial$(1, \theta)$).

probability function: $p(x) = \theta^x (1 - \theta)^{1-x}$ for $x = 0, 1$.
mean: θ.
variance: $\theta(1 - \theta)$.
moment-generating function: $m(t) = (1 - \theta + \theta e^t)$ for $t \in R^1$.

2. *Binomial*(n, θ), $n > 0$ an integer, $\theta \in [0, 1]$.

probability function: $p(x) = \binom{n}{x} \theta^x (1 - \theta)^{n-x}$ for $x = 0, 1, \ldots, n$.
mean: $n\theta$.
variance: $n\theta(1 - \theta)$.
moment-generating function: $m(t) = (1 - \theta + \theta e^t)^n$ for $t \in R^1$.

3. *Geometric*(θ), $\theta \in [0, 1]$ (same as Negative-Binomial$(1, \theta)$).

probability function: $p(x) = (1 - \theta)^x \theta$ for $x = 0, 1, 2, \ldots$.
mean: $(1 - \theta)/\theta$.
variance: $(1 - \theta)/\theta^2$.
moment-generating function: $m(t) = \theta(1 - (1 - \theta) e^t)^{-1}$ for $t < -\ln(1 - \theta)$.

4. *Hypergeometric*(N, M, n), $M \leq N, n \leq N$ all positive integers.

probability function:

$$p(x) = \binom{M}{x} \binom{N - M}{n - x} / \binom{N}{n} \text{ for } \max(0, n + M - N) \leq x \leq \min(n, M).$$

mean: $n\frac{M}{N}$.
variance: $n\frac{M}{N} \left(1 - \frac{M}{N}\right) \frac{N-n}{N-1}$.

5. *Multinomial*$(n, \theta_1, \ldots, \theta_k)$, $n > 0$ an integer, each $\theta_i \in [0,1]$, $\theta_1 + \cdots + \theta_k = 1$.
probability function:

$$p(x_1, \ldots, x_k) = \binom{n}{x_1 \ldots x_k} \theta_1^{x_1} \cdots \theta_k^{x_k} \text{ where each } x_i \in \{0, 1, \ldots, n\}$$
$$\text{and } x_1 + \cdots + x_k = n.$$

mean: $E(X_i) = n\theta_i$.
variance: $\text{Var}(X_i) = n\theta_i (1 - \theta_i)$.
covariance: $\text{Cov}(X_i, X_j) = -n\theta_i\theta_j$ when $i \neq j$.

6. *Negative-Binomial*(r, θ), $r > 0$ an integer, $\theta \in [0,1]$.
probability function: $p(x) = \binom{r-1+x}{x}\theta^r (1-\theta)^x$ for $x = 0, 1, 2, 3, \ldots$.
mean: $r(1-\theta)/\theta$.
variance: $r(1-\theta)/\theta^2$.
moment-generating function: $m(t) = \theta^r (1 - (1-\theta)e^t)^{-r}$ for $t < -\ln(1-\theta)$.

7. *Poisson*(λ), $\lambda > 0$.
probability function: $p(x) = \frac{\lambda^x}{x!}e^{-\lambda}$ for $x = 0, 1, 2, 3, \ldots$.
mean: λ.
variance: λ.
moment-generating function: $m(t) = \exp\{\lambda(e^t - 1)\}$ for $t \in R^1$.

C.2   Absolutely Continuous Distributions

1. *Beta*(a, b), $a > 0, b > 0$ (same as Dirichlet(a, b)).
density function: $f(x) = \frac{\Gamma(a+b)}{\Gamma(a)\Gamma(b)}x^{a-1}(1-x)^{b-1}$ for $x > 0$.
mean: $a/(a+b)$.
variance: $ab/(a+b+1)(a+b)^2$.

2. *Bivariate Normal*$\left(\mu_1, \mu_2, \sigma_1^2, \sigma_2^2, \rho\right)$ for $\mu_1, \mu_2 \in R^1$, $\sigma_1^2, \sigma_2^2 > 0$, $\rho \in [-1, 1]$.
density function:

$$f_{X_1, X_2}(x_1, x_2)$$
$$= \frac{1}{2\pi\sigma_1\sigma_2\sqrt{1-\rho^2}} \exp\left\{-\frac{1}{2(1-\rho^2)}\left[\left(\frac{x_1-\mu_1}{\sigma_1}\right)^2 + \left(\frac{x_2-\mu_2}{\sigma_2}\right)^2 - 2\rho\left(\frac{x_1-\mu_1}{\sigma_1}\right)\left(\frac{x_2-\mu_2}{\sigma_2}\right)\right]\right\}$$
$$\text{for } x_1 \in R^1, x_2 \in R^1.$$

mean: $E(X_i) = \mu_i$.
variance: $\text{Var}(X_i) = \sigma_i^2$.
covariance: $\text{Cov}(X_1, X_2) = \rho\sigma_1\sigma_2$.

3. $\chi^2(\alpha)$ or *Chi-squared*(α), $\alpha > 0$ (same as Gamma$(\alpha/2, 1/2)$).

density function: $f(x) = 2^{-\alpha/2} \left(\Gamma(\frac{\alpha}{2})\right)^{-1} x^{(\alpha/2)-1} e^{-x/2}$ for $x > 0$.

mean: α.

variance: 2α.

moment-generating function: $m(t) = (1 - 2t)^{-\alpha/2}$ for $t < 1/2$.

4. *Dirichlet*$(\alpha_1, \ldots, \alpha_{k+1})$, $\alpha_i > 0$ for each i.

density function:

$$f_{X_1,\ldots,X_k}(x_1, \ldots, x_k)$$
$$= \frac{\Gamma(\alpha_1 + \cdots + \alpha_{k+1})}{\Gamma(\alpha_1) \cdots \Gamma(\alpha_{k+1})} x_1^{\alpha_1 - 1} \cdots x_k^{\alpha_k - 1} (1 - x_1 - \cdots - x_k)^{\alpha_{k+1} - 1}$$

for $x_i \geq 0, i = 1, \ldots, k$ and $0 \leq x_1 + \cdots + x_k \leq 1$.

mean:
$$E(X_i) = \frac{\alpha_i}{\alpha_1 + \cdots + \alpha_{k+1}}.$$

variance:
$$\text{Var}(X_i) = \frac{\alpha_i (\alpha_1 + \cdots + \alpha_{k+1} - \alpha_i)}{(\alpha_1 + \cdots + \alpha_{k+1})^2 (1 + \alpha_1 + \cdots + \alpha_{k+1})}.$$

covariance: when $i \neq j$
$$\text{Cov}(X_i, X_j) = \frac{-\alpha_i \alpha_j}{(\alpha_1 + \cdots + \alpha_{k+1})^2 (1 + \alpha_1 + \cdots + \alpha_{k+1})}.$$

5. *Exponential*(λ), $\lambda > 0$ (same as Gamma$(1, \lambda)$).

density function: $f(x) = \lambda e^{-\lambda x}$ for $x > 0$.

mean: λ^{-1}.

variance: λ^{-2}.

moment-generating function: $m(t) = \lambda (\lambda - t)^{-1}$ for $t < \lambda$.

Note that some books and software packages instead replace λ by $1/\lambda$ in the definition of the Exponential(λ) distribution — always check this when using another book or when using software to generate from this distribution.

6. $F(\alpha, \beta)$, $\alpha > 0, \beta > 0$.

density function:

$$f(u) = \frac{\Gamma\left(\frac{\alpha+\beta}{2}\right)}{\Gamma\left(\frac{\alpha}{2}\right)\Gamma\left(\frac{\beta}{2}\right)} \left(\frac{\alpha}{\beta} u\right)^{(\alpha+1)-1} \left(1 + \frac{\alpha}{\beta} u\right)^{-(\alpha+\beta)/2} \frac{\alpha}{\beta}$$

for $x > 0$.

mean: $\beta/(\beta - 2)$ when $\beta > 2$.

variance: $2\beta^2 (\alpha + \beta - 2)/\alpha (\beta - 2)^2 (\beta - 4)$ when $\beta > 4$.

7. $Gamma(\alpha, \lambda)$, $\alpha > 0, \lambda > 0$.

density function: $f(x) = \frac{\lambda^\alpha x^{\alpha-1}}{\Gamma(\alpha)} e^{-\lambda x}$ for $x > 0$.
mean: α/λ.
variance: α/λ^2.
moment-generating function: $m(t) = \lambda^\alpha (\lambda - t)^{-\alpha}$ for $t < \lambda$.

Note that some books and software packages instead replace λ by $1/\lambda$ in the definition of the $Gamma(\alpha, \lambda)$ distribution — always check this when using another book or when using software to generate from this distribution.

8. $N(\mu, \sigma^2)$, $\mu \in R^1, \sigma^2 > 0$.

density function: $f(x) = \left(2\pi\sigma^2\right)^{-1/2} \exp\left(-\frac{1}{2\sigma^2}(x - \mu)^2\right)$ for $x \in R^1$.
mean: μ.
variance: σ^2.
moment-generating function: $m(t) = \exp\left(\mu t + \sigma^2 t^2/2\right)$ for $t \in R^1$.

9. $t(\alpha)$ or $Student(\alpha)$, $\alpha > 0$.

density function:

$$f(x) = \frac{\Gamma\left(\frac{\alpha+1}{2}\right)}{\Gamma\left(\frac{1}{2}\right)\Gamma\left(\frac{\alpha}{2}\right)} \left(1 + \frac{x^2}{\alpha}\right)^{-(\alpha+1)/2} \frac{1}{\sqrt{\alpha}}$$
$$\text{for } x \in R^1.$$

mean: 0 when $\alpha > 1$.
variance: $\alpha/(\alpha - 2)$ when $\alpha > 2$.

10. $Uniform[L, R]$, $R > L$.

density function: $f(x) = 1/(R - L)$ for $L < x < R$.
mean: $(L + R)/2$.
variance: $(R - L)^2/12$.
moment-generating function: $m(t) = \left(e^{Rt} - e^{Lt}\right)/t(R - L)$.

Appendix D

Tables

The following tables can be used for various computations. It is recommended, however, that the reader become familiar with the use of a statistical software package instead of relying on the tables. Computations of a much greater variety and accuracy can be carried out using the software, and in the end, it is much more convenient.

D.1   Random Numbers

Each line in Table D.1 is a sample of 40 random digits, i.e., 40 independent and identically distributed (i.i.d.) values from the uniform distribution on the set $\{0, 1, 2, 3, 4, 5, 6, 7, 8, 9\}$.

Suppose we want a sample of five i.i.d. values from the uniform distribution on $S = \{1, 2, \ldots, 25\}$, i.e., a random sample of five, with replacement, from S. To do this, pick a starting point in the table and start reading off successive (nonoverlapping) two-digit numbers, treating a pair such as 07 as 7, and discarding any pairs that are not in the range 1 to 25, until you have five values. For example, if we start at line 110, we read the pairs (* indicates a sample element) 38, 44, 84, 87, 89, 18*, 33, 82, 46, 97, 39, 36, 44, 20*, 06*, 76, 68, 80, 87, 08*, 81, 48, 66, 94, 87, 60, 51, 30, 92, 97, 00, 41, 27, 12*. We can see at this point that we have a sample of five given by 18, 20, 6, 8, 12.

If we want a random sample of five, without replacement, from S, then we proceed as above but now ignore any repeats in the generated sample until we get the five numbers. In this preceding case, we did not get any repeats, so this is also a simple random sample of size five without replacement.

Table D.1 Random Numbers							
Line							
101	19223	95034	05756	28713	96409	12531	42544	82853
102	73676	47150	99400	01927	27754	42648	82425	36290
103	45467	71709	77558	00095	32863	29485	82226	90056
104	52711	38889	93074	60227	40011	85848	48767	52573
105	95592	94007	69971	91481	60779	53791	17297	59335
106	68417	35013	15529	72765	85089	57067	50211	47487
107	82739	57890	20807	47511	81676	55300	94383	14893
108	60940	72024	17868	24943	61790	90656	87964	18883
109	36009	19365	15412	39638	85453	46816	83485	41979
110	38448	48789	18338	24697	39364	42006	76688	08708
111	81486	69487	60513	09297	00412	71238	27649	39950
112	59636	88804	04634	71197	19352	73089	84898	45785
113	62568	70206	40325	03699	71080	22553	11486	11776
114	45149	32992	75730	66280	03819	56202	02938	70915
115	61041	77684	94322	24709	73698	14526	31893	32592
116	14459	26056	31424	80371	65103	62253	50490	61181
117	38167	98532	62183	70632	23417	26185	41448	75532
118	73190	32533	04470	29669	84407	90785	65956	86382
119	95857	07118	87664	92099	58806	66979	98624	84826
120	35476	55972	39421	65850	04266	35435	43742	11937
121	71487	09984	29077	14863	61683	47052	62224	51025
122	13873	81598	95052	90908	73592	75186	87136	95761
123	54580	81507	27102	56027	55892	33063	41842	81868
124	71035	09001	43367	49497	72719	96758	27611	91596
125	96746	12149	37823	71868	18442	35119	62103	39244

Table D.1 Random Numbers *(Continued)*

Line								
126	96927	19931	36809	74192	77567	88741	48409	41903
127	43909	99477	25330	64359	40085	16925	85117	36071
128	15689	14227	06565	14374	13352	49367	81982	87209
129	36759	58984	68288	22913	18638	54303	00795	08727
130	69051	64817	87174	09517	84534	06489	87201	97245
131	05007	16632	81194	14873	04197	85576	45195	96565
132	68732	55259	84292	08796	43165	93739	31685	97150
133	45740	41807	65561	33302	07051	93623	18132	09547
134	27816	78416	18329	21337	35213	37741	04312	68508
135	66925	55658	39100	78458	11206	19876	87151	31260
136	08421	44753	77377	28744	75592	08563	79140	92454
137	53645	66812	61421	47836	12609	15373	98481	14592
138	66831	68908	40772	21558	47781	33586	79177	06928
139	55588	99404	70708	41098	43563	56934	48394	51719
140	12975	13258	13048	45144	72321	81940	00360	02428
141	96767	35964	23822	96012	94591	65194	50842	53372
142	72829	50232	97892	63408	77919	44575	24870	04178
143	88565	42628	17797	49376	61762	16953	88604	12724
144	62964	88145	83083	69453	46109	59505	69680	00900
145	19687	12633	57857	95806	09931	02150	43163	58636
146	37609	59057	66967	83401	60705	02384	90597	93600
147	54973	86278	88737	74351	47500	84552	19909	67181
148	00694	05977	19664	65441	20903	62371	22725	53340
149	71546	05233	53946	68743	72460	27601	45403	88692
150	07511	88915	41267	16853	84569	79367	32337	03316
151	03802	29341	29264	80198	12371	13121	54969	43912
152	77320	35030	77519	41109	98296	18984	60869	12349
153	07886	56866	39648	69290	03600	05376	58958	22720
154	87065	74133	21117	70595	22791	67306	28420	52067
155	42090	09628	54035	93879	98441	04606	27381	82637
156	55494	67690	88131	81800	11188	28552	25752	21953
157	16698	30406	96587	65985	07165	50148	16201	86792
158	16297	07626	68683	45335	34377	72941	41764	77038
159	22897	17467	17638	70043	36243	13008	83993	22869
160	98163	45944	34210	64158	76971	27689	82926	75957
161	43400	25831	06283	22138	16043	15706	73345	26238
162	97341	46254	88153	62336	21112	35574	99271	45297
163	64578	67197	28310	90341	37531	63890	52630	76315
164	11022	79124	49525	63078	17229	32165	01343	21394
165	81232	43939	23840	05995	84589	06788	76358	26622

D.2 Standard Normal Cdf

If $Z \sim N(0,1)$, then we can use Table D.2 to compute the cumulative distribution function (cdf) Φ for Z. For example, suppose we want to compute $\Phi(z) = P(Z < 1.03)$. The symmetry of the $N(0,1)$ distribution about 0 implies that $\Phi(z) = 1 - \Phi(-z)$, so using Table D.2, we have that $P(Z < 1.03) = P(Z < 1.03) = 1 - P(Z < -1.03) = 1 - 0.1515 = 0.8485$.

Table D.2 Standard Normal Cdf										
z	.00	.01	.02	.03	.04	.05	.06	.07	.08	.09
−3.4	.0003	.0003	.0003	.0003	.0003	.0003	.0003	.0003	.0003	.0002
−3.3	.0005	.0005	.0005	.0004	.0004	.0004	.0004	.0004	.0004	.0003
−3.2	.0007	.0007	.0006	.0006	.0006	.0006	.0006	.0005	.0005	.0005
−3.1	.0010	.0009	.0009	.0009	.0008	.0008	.0008	.0008	.0007	.0007
−3.0	.0013	.0013	.0013	.0012	.0012	.0011	.0011	.0011	.0010	.0010
−2.9	.0019	.0018	.0018	.0017	.0016	.0016	.0015	.0015	.0014	.0014
−2.8	.0026	.0025	.0024	.0023	.0023	.0022	.0021	.0021	.0020	.0019
−2.7	.0035	.0034	.0033	.0032	.0031	.0030	.0029	.0028	.0027	.0026
−2.6	.0047	.0045	.0044	.0043	.0041	.0040	.0039	.0038	.0037	.0036
−2.5	.0062	.0060	.0059	.0057	.0055	.0054	.0052	.0051	.0049	.0048
−2.4	.0082	.0080	.0078	.0075	.0073	.0071	.0069	.0068	.0066	.0064
−2.3	.0107	.0104	.0102	.0099	.0096	.0094	.0091	.0089	.0087	.0084
−2.2	.0139	.0136	.0132	.0129	.0125	.0122	.0119	.0116	.0113	.0110
−2.1	.0179	.0174	.0170	.0166	.0162	.0158	.0154	.0150	.0146	.0143
−2.0	.0228	.0222	.0217	.0212	.0207	.0202	.0197	.0192	.0188	.0183
−1.9	.0287	.0281	.0274	.0268	.0262	.0256	.0250	.0244	.0239	.0233
−1.8	.0359	.0351	.0344	.0336	.0329	.0322	.0314	.0307	.0301	.0294
−1.7	.0446	.0436	.0427	.0418	.0409	.0401	.0392	.0384	.0375	.0367
−1.6	.0548	.0537	.0526	.0516	.0505	.0495	.0485	.0475	.0465	.0455
−1.5	.0668	.0655	.0643	.0630	.0618	.0606	.0594	.0582	.0571	.0559
−1.4	.0808	.0793	.0778	.0764	.0749	.0735	.0721	.0708	.0694	.0681
−1.3	.0968	.0951	.0934	.0918	.0901	.0885	.0869	.0853	.0838	.0823
−1.2	.1151	.1131	.1112	.1093	.1075	.1056	.1038	.1020	.1003	.0985
−1.1	.1357	.1335	.1314	.1292	.1271	.1251	.1230	.1210	.1190	.1170
−1.0	.1587	.1562	.1539	.1515	.1492	.1469	.1446	.1423	.1401	.1379
−0.9	.1841	.1814	.1788	.1762	.1736	.1711	.1685	.1660	.1635	.1611
−0.8	.2119	.2090	.2061	.2033	.2005	.1977	.1949	.1922	.1894	.1867
−0.7	.2420	.2389	.2358	.2327	.2296	.2266	.2236	.2206	.2177	.2148
−0.6	.2743	.2709	.2676	.2643	.2611	.2578	.2546	.2514	.2483	.2451
−0.5	.3085	.3050	.3015	.2981	.2946	.2912	.2877	.2843	.2810	.2776
−0.4	.3446	.3409	.3372	.3336	.3300	.3264	.3228	.3192	.3156	.3121
−0.3	.3821	.3783	.3745	.3707	.3669	.3632	.3594	.3557	.3520	.3483
−0.2	.4207	.4168	.4129	.4090	.4052	.4013	.3974	.3936	.3897	.3859
−0.1	.4602	.4562	.4522	.4483	.4443	.4404	.4364	.4325	.4286	.4247
−0.0	.5000	.4960	.4920	.4880	.4840	.4801	.4761	.4721	.4681	.4641

D.3 Chi-Squared Distribution Quantiles

If $X \sim \chi^2(df)$, then we can use Table D.3 to obtain some quantiles for this distribution. For example, if $df = 10$ and $P = 0.98$, then $x_{0.98} = 21.16$ is the 0.98-quantile of this distribution.

Table D.3 $\chi^2(df)$ Quantiles									
				P					
df	0.75	0.85	0.90	0.95	0.975	0.98	0.99	0.995	0.9975	0.999
1	1.32	2.07	2.71	3.84	5.02	5.41	6.63	7.88	9.14	10.83
2	2.77	3.79	4.61	5.99	7.38	7.82	9.21	10.60	11.98	13.82
3	4.11	5.32	6.25	7.81	9.35	9.84	11.34	12.84	14.32	16.27
4	5.39	6.74	7.78	9.49	11.14	11.67	13.28	14.86	16.42	18.47
5	6.63	8.12	9.24	11.07	12.83	13.39	15.09	16.75	18.39	20.51
6	7.84	9.45	10.64	12.59	14.45	15.03	16.81	18.55	20.25	22.46
7	9.04	10.75	12.02	14.07	16.01	16.62	18.48	20.28	22.04	24.32
8	10.22	12.03	13.36	15.51	17.53	18.17	20.09	21.95	23.77	26.12
9	11.39	13.29	14.68	16.92	19.02	19.68	21.67	23.59	25.46	27.88
10	12.55	14.53	15.99	18.31	20.48	21.16	23.21	25.19	27.11	29.59
11	13.70	15.77	17.28	19.68	21.92	22.62	24.72	26.76	28.73	31.26
12	14.85	16.99	18.55	21.03	23.34	24.05	26.22	28.30	30.32	32.91
13	15.98	18.20	19.81	22.36	24.74	25.47	27.69	29.82	31.88	34.53
14	17.12	19.41	21.06	23.68	26.12	26.87	29.14	31.32	33.43	36.12
15	18.25	20.60	22.31	25.00	27.49	28.26	30.58	32.80	34.95	37.70
16	19.37	21.79	23.54	26.30	28.85	29.63	32.00	34.27	36.46	39.25
17	20.49	22.98	24.77	27.59	30.19	31.00	33.41	35.72	37.95	40.79
18	21.60	24.16	25.99	28.87	31.53	32.35	34.81	37.16	39.42	42.31
19	22.72	25.33	27.20	30.14	32.85	33.69	36.19	38.58	40.88	43.82
20	23.83	26.50	28.41	31.41	34.17	35.02	37.57	40.00	42.34	45.31
21	24.93	27.66	29.62	32.67	35.48	36.34	38.93	41.40	43.78	46.80
22	26.04	28.82	30.81	33.92	36.78	37.66	40.29	42.80	45.20	48.27
23	27.14	29.98	32.01	35.17	38.08	38.97	41.64	44.18	46.62	49.73
24	28.24	31.13	33.20	36.42	39.36	40.27	42.98	45.56	48.03	51.18
25	29.34	32.28	34.38	37.65	40.65	41.57	44.31	46.93	49.44	52.62
26	30.43	33.43	35.56	38.89	41.92	42.86	45.64	48.29	50.83	54.05
27	31.53	34.57	36.74	40.11	43.19	44.14	46.96	49.64	52.22	55.48
28	32.62	35.71	37.92	41.34	44.46	45.42	48.28	50.99	53.59	56.89
29	33.71	36.85	39.09	42.56	45.72	46.69	49.59	52.34	54.97	58.30
30	34.80	37.99	40.26	43.77	46.98	47.96	50.89	53.67	56.33	59.70
40	45.62	49.24	51.81	55.76	59.34	60.44	63.69	66.77	69.70	73.40
50	56.33	60.35	63.17	67.50	71.42	72.61	76.15	79.49	82.66	86.66
60	66.98	71.34	74.40	79.08	83.30	84.58	88.38	91.95	95.34	99.61
80	88.13	93.11	96.58	101.9	106.6	108.1	112.3	116.3	120.1	124.8
100	109.1	114.7	118.5	124.3	129.6	131.1	135.8	140.2	144.3	149.4

D.4 t Distribution Quantiles

Table D.4 contains some quantiles for t or Student distributions. For example, if $X \sim t(df)$, with $df = 10$ and $P = 0.98$, then $x_{0.98} = 2.359$ is the 0.98-quantile of the $t(10)$ distribution. Recall that the $t(df)$ distribution is symmetric about 0 so, for example, $x_{0.25} = -x_{0.75}$.

Table D.4 $t(df)$ Quantiles										
	P									
df	0.75	0.85	0.90	0.95	0.975	0.98	0.99	0.995	0.9975	0.999
1	1.000	1.963	3.078	6.314	12.71	15.89	31.82	63.66	127.3	318.3
2	0.816	1.386	1.886	2.920	4.303	4.849	6.965	9.925	14.09	22.33
3	0.765	1.250	1.638	2.353	3.182	3.482	4.541	5.841	7.453	10.21
4	0.741	1.190	1.533	2.132	2.776	2.999	3.747	4.604	5.598	7.173
5	0.727	1.156	1.476	2.015	2.571	2.757	3.365	4.032	4.773	5.893
6	0.718	1.134	1.440	1.943	2.447	2.612	3.143	3.707	4.317	5.208
7	0.711	1.119	1.415	1.895	2.365	2.517	2.998	3.499	4.029	4.785
8	0.706	1.108	1.397	1.860	2.306	2.449	2.896	3.355	3.833	4.501
9	0.703	1.100	1.383	1.833	2.262	2.398	2.821	3.250	3.690	4.297
10	0.700	1.093	1.372	1.812	2.228	2.359	2.764	3.169	3.581	4.144
11	0.697	1.088	1.363	1.796	2.201	2.328	2.718	3.106	3.497	4.025
12	0.695	1.083	1.356	1.782	2.179	2.303	2.681	3.055	3.428	3.930
13	0.694	1.079	1.350	1.771	2.160	2.282	2.650	3.012	3.372	3.852
14	0.692	1.076	1.345	1.761	2.145	2.264	2.624	2.977	3.326	3.787
15	0.691	1.074	1.341	1.753	2.131	2.249	2.602	2.947	3.286	3.733
16	0.690	1.071	1.337	1.746	2.120	2.235	2.583	2.921	3.252	3.686
17	0.689	1.069	1.333	1.740	2.110	2.224	2.567	2.898	3.222	3.646
18	0.688	1.067	1.330	1.734	2.101	2.214	2.552	2.878	3.197	3.611
19	0.688	1.066	1.328	1.729	2.093	2.205	2.539	2.861	3.174	3.579
20	0.687	1.064	1.325	1.725	2.086	2.197	2.528	2.845	3.153	3.552
21	0.686	1.063	1.323	1.721	2.080	2.189	2.518	2.831	3.135	3.527
22	0.686	1.061	1.321	1.717	2.074	2.183	2.508	2.819	3.119	3.505
23	0.685	1.060	1.319	1.714	2.069	2.177	2.500	2.807	3.104	3.485
24	0.685	1.059	1.318	1.711	2.064	2.172	2.492	2.797	3.091	3.467
25	0.684	1.058	1.316	1.708	2.060	2.167	2.485	2.787	3.078	3.450
26	0.684	1.058	1.315	1.706	2.056	2.162	2.479	2.779	3.067	3.435
27	0.684	1.057	1.314	1.703	2.052	2.158	2.473	2.771	3.057	3.421
28	0.683	1.056	1.313	1.701	2.048	2.154	2.467	2.763	3.047	3.408
29	0.683	1.055	1.311	1.699	2.045	2.150	2.462	2.756	3.038	3.396
30	0.683	1.055	1.310	1.697	2.042	2.147	2.457	2.750	3.030	3.385
40	0.681	1.050	1.303	1.684	2.021	2.123	2.423	2.704	2.971	3.307
50	0.679	1.047	1.299	1.676	2.009	2.109	2.403	2.678	2.937	3.261
60	0.679	1.045	1.296	1.671	2.000	2.099	2.390	2.660	2.915	3.232
80	0.678	1.043	1.292	1.664	1.990	2.088	2.374	2.639	2.887	3.195
100	0.677	1.042	1.290	1.660	1.984	2.081	2.364	2.626	2.871	3.174
1000	0.675	1.037	1.282	1.646	1.962	2.056	2.330	2.581	2.813	3.098
∞	0.674	1.036	1.282	1.645	1.960	2.054	2.326	2.576	2.807	3.091
	50%	70%	80%	90%	95%	96%	98%	99%	99.5%	99.8%
	Confidence level									

D.5 *F* Distribution Quantiles

If $X \sim F(ndf, ddf)$, then we can use Table D.5 to obtain some quantiles for this distribution. For example, if $ndf = 3, ddf = 4$, and $P = 0.975$, then $x_{.975} = 9.98$ is the 0.975-quantile of the $F(3, 4)$ distribution.

Note that, if $X \sim F(ndf, ddf)$, then $Y = 1/X \sim F(ddf, ndf)$ and $P(X \leq x) = P(Y \geq 1/x)$.

		\multicolumn{6}{c}{**Table D.5 $F(ndf, ddf)$ Quantiles**}				
		\multicolumn{6}{c}{*ndf*}				
ddf	*P*	1	2	3	4	5	6
1	0.900	39.86	49.50	53.59	55.83	57.24	58.20
	0.950	161.45	199.50	215.71	224.58	230.16	233.99
	0.975	647.79	799.50	864.16	899.58	921.85	937.11
	0.990	4052.18	4999.50	5403.35	5624.58	5763.65	5858.99
	0.999	405284.07	499999.50	540379.20	562499.58	576404.56	585937.11
2	0.900	8.53	9.00	9.16	9.24	9.29	9.33
	0.950	18.51	19.00	19.16	19.25	19.30	19.33
	0.975	38.51	39.00	39.17	39.25	39.30	39.33
	0.990	98.50	99.00	99.17	99.25	99.30	99.33
	0.999	998.50	999.00	999.17	999.25	999.30	999.33
3	0.900	5.54	5.46	5.39	5.34	5.31	5.28
	0.950	10.13	9.55	9.28	9.12	9.01	8.94
	0.975	17.44	16.04	15.44	15.10	14.88	14.73
	0.990	34.12	30.82	29.46	28.71	28.24	27.91
	0.999	167.03	148.50	141.11	137.10	134.58	132.85
4	0.900	4.54	4.32	4.19	4.11	4.05	4.01
	0.950	7.71	6.94	6.59	6.39	6.26	6.16
	0.975	12.22	10.65	9.98	9.60	9.36	9.20
	0.990	21.20	18.00	16.69	15.98	15.52	15.21
	0.999	74.14	61.25	56.18	53.44	51.71	50.53
5	0.900	4.06	3.78	3.62	3.52	3.45	3.40
	0.950	6.61	5.79	5.41	5.19	5.05	4.95
	0.975	10.01	8.43	7.76	7.39	7.15	6.98
	0.990	16.26	13.27	12.06	11.39	10.97	10.67
	0.999	47.18	37.12	33.20	31.09	29.75	28.83
6	0.900	3.78	3.46	3.29	3.18	3.11	3.05
	0.950	5.99	5.14	4.76	4.53	4.39	4.28
	0.975	8.81	7.26	6.60	6.23	5.99	5.82
	0.990	13.75	10.92	9.78	9.15	8.75	8.47
	0.999	35.51	27.00	23.70	21.92	20.80	20.03
7	0.900	3.59	3.26	3.07	2.96	2.88	2.83
	0.950	5.59	4.74	4.35	4.12	3.97	3.87
	0.975	8.07	6.54	5.89	5.52	5.29	5.12
	0.990	12.25	9.55	8.45	7.85	7.46	7.19
	0.999	29.25	21.69	18.77	17.20	16.21	15.52

Table D.5 $F(ndf, ddf)$ Quantiles *(continued)*							
				ndf			
ddf	P	7	8	9	10	11	12
1	0.900	58.91	59.44	59.86	60.19	60.47	60.71
	0.950	236.77	238.88	240.54	241.88	242.98	243.91
	0.975	948.22	956.66	963.28	968.63	973.03	976.71
	0.990	5928.36	5981.07	6022.47	6055.85	6083.32	6106.32
	0.999	592873.29	598144.16	602283.99	605620.97	608367.68	610667.82
2	0.900	9.35	9.37	9.38	9.39	9.40	9.41
	0.950	19.35	19.37	19.38	19.40	19.40	19.41
	0.975	39.36	39.37	39.39	39.40	39.41	39.41
	0.990	99.36	99.37	99.39	99.40	99.41	99.42
	0.999	999.36	999.37	999.39	999.40	999.41	999.42
3	0.900	5.27	5.25	5.24	5.23	5.22	5.22
	0.950	8.89	8.85	8.81	8.79	8.76	8.74
	0.975	14.62	14.54	14.47	14.42	14.37	14.34
	0.990	27.67	27.49	27.35	27.23	27.13	27.05
	0.999	131.58	130.62	129.86	129.25	128.74	128.32
4	0.900	3.98	3.95	3.94	3.92	3.91	3.90
	0.950	6.09	6.04	6.00	5.96	5.94	5.91
	0.975	9.07	8.98	8.90	8.84	8.79	8.75
	0.990	14.98	14.80	14.66	14.55	14.45	14.37
	0.999	49.66	49.00	48.47	48.05	47.70	47.41
5	0.900	3.37	3.34	3.32	3.30	3.28	3.27
	0.950	4.88	4.82	4.77	4.74	4.70	4.68
	0.975	6.85	6.76	6.68	6.62	6.57	6.52
	0.990	10.46	10.29	10.16	10.05	9.96	9.89
	0.999	28.16	27.65	27.24	26.92	26.65	26.42
6	0.900	3.01	2.98	2.96	2.94	2.92	2.90
	0.950	4.21	4.15	4.10	4.06	4.03	4.00
	0.975	5.70	5.60	5.52	5.46	5.41	5.37
	0.990	8.26	8.10	7.98	7.87	7.79	7.72
	0.999	19.46	19.03	18.69	18.41	18.18	17.99
7	0.900	2.78	2.75	2.72	2.70	2.68	2.67
	0.950	3.79	3.73	3.68	3.64	3.60	3.57
	0.975	4.99	4.90	4.82	4.76	4.71	4.67
	0.990	6.99	6.84	6.72	6.62	6.54	6.47
	0.999	15.02	14.63	14.33	14.08	13.88	13.71

	**Table D.5** $F(ndf, ddf)$ **Quantiles** *(continued)*
		colspan="6"	ndf
ddf	P	15	20	30	60	120	10000
1	0.900	61.22	61.74	62.26	62.79	63.06	63.32
	0.950	245.95	248.01	250.10	252.20	253.25	254.30
	0.975	984.87	993.10	1001.41	1009.80	1014.02	1018.21
	0.990	6157.28	6208.73	6260.65	6313.03	6339.39	6365.55
	0.999	615763.66	620907.67	626098.96	631336.56	633972.40	636587.61
2	0.900	9.42	9.44	9.46	9.47	9.48	9.49
	0.950	19.43	19.45	19.46	19.48	19.49	19.50
	0.975	39.43	39.45	39.46	39.48	39.49	39.50
	0.990	99.43	99.45	99.47	99.48	99.49	99.50
	0.999	999.43	999.45	999.47	999.48	999.49	999.50
3	0.900	5.20	5.18	5.17	5.15	5.14	5.13
	0.950	8.70	8.66	8.62	8.57	8.55	8.53
	0.975	14.25	14.17	14.08	13.99	13.95	13.90
	0.990	26.87	26.69	26.50	26.32	26.22	26.13
	0.999	127.37	126.42	125.45	124.47	123.97	123.48
4	0.900	3.87	3.84	3.82	3.79	3.78	3.76
	0.950	5.86	5.80	5.75	5.69	5.66	5.63
	0.975	8.66	8.56	8.46	8.36	8.31	8.26
	0.990	14.20	14.02	13.84	13.65	13.56	13.46
	0.999	46.76	46.10	45.43	44.75	44.40	44.06
5	0.900	3.24	3.21	3.17	3.14	3.12	3.11
	0.950	4.62	4.56	4.50	4.43	4.40	4.37
	0.975	6.43	6.33	6.23	6.12	6.07	6.02
	0.990	9.72	9.55	9.38	9.20	9.11	9.02
	0.999	25.91	25.39	24.87	24.33	24.06	23.79
6	0.900	2.87	2.84	2.80	2.76	2.74	2.72
	0.950	3.94	3.87	3.81	3.74	3.70	3.67
	0.975	5.27	5.17	5.07	4.96	4.90	4.85
	0.990	7.56	7.40	7.23	7.06	6.97	6.88
	0.999	17.56	17.12	16.67	16.21	15.98	15.75
7	0.900	2.63	2.59	2.56	2.51	2.49	2.47
	0.950	3.51	3.44	3.38	3.30	3.27	3.23
	0.975	4.57	4.47	4.36	4.25	4.20	4.14
	0.990	6.31	6.16	5.99	5.82	5.74	5.65
	0.999	13.32	12.93	12.53	12.12	11.91	11.70

Table D.5 $F(ndf, ddf)$ **Quantiles** (*continued*)

ddf	P	ndf 1	2	3	4	5	6
8	0.900	3.46	3.11	2.92	2.81	2.73	2.67
	0.950	5.32	4.46	4.07	3.84	3.69	3.58
	0.975	7.57	6.06	5.42	5.05	4.82	4.65
	0.990	11.26	8.65	7.59	7.01	6.63	6.37
	0.999	25.41	18.49	15.83	14.39	13.48	12.86
9	0.900	3.36	3.01	2.81	2.69	2.61	2.55
	0.950	5.12	4.26	3.86	3.63	3.48	3.37
	0.975	7.21	5.71	5.08	4.72	4.48	4.32
	0.990	10.56	8.02	6.99	6.42	6.06	5.80
	0.999	22.86	16.39	13.90	12.56	11.71	11.13
10	0.900	3.29	2.92	2.73	2.61	2.52	2.46
	0.950	4.96	4.10	3.71	3.48	3.33	3.22
	0.975	6.94	5.46	4.83	4.47	4.24	4.07
	0.990	10.04	7.56	6.55	5.99	5.64	5.39
	0.999	21.04	14.91	12.55	11.28	10.48	9.93
11	0.900	3.23	2.86	2.66	2.54	2.45	2.39
	0.950	4.84	3.98	3.59	3.36	3.20	3.09
	0.975	6.72	5.26	4.63	4.28	4.04	3.88
	0.990	9.65	7.21	6.22	5.67	5.32	5.07
	0.999	19.69	13.81	11.56	10.35	9.58	9.05
12	0.900	3.18	2.81	2.61	2.48	2.39	2.33
	0.950	4.75	3.89	3.49	3.26	3.11	3.00
	0.975	6.55	5.10	4.47	4.12	3.89	3.73
	0.990	9.33	6.93	5.95	5.41	5.06	4.82
	0.999	18.64	12.97	10.80	9.63	8.89	8.38
13	0.900	3.14	2.76	2.56	2.43	2.35	2.28
	0.950	4.67	3.81	3.41	3.18	3.03	2.92
	0.975	6.41	4.97	4.35	4.00	3.77	3.60
	0.990	9.07	6.70	5.74	5.21	4.86	4.62
	0.999	17.82	12.31	10.21	9.07	8.35	7.86
14	0.900	3.10	2.73	2.52	2.39	2.31	2.24
	0.950	4.60	3.74	3.34	3.11	2.96	2.85
	0.975	6.30	4.86	4.24	3.89	3.66	3.50
	0.990	8.86	6.51	5.56	5.04	4.69	4.46
	0.999	17.14	11.78	9.73	8.62	7.92	7.44

		ndf					
ddf	P	7	8	9	10	11	12
8	0.900	2.62	2.59	2.56	2.54	2.52	2.50
	0.950	3.50	3.44	3.39	3.35	3.31	3.28
	0.975	4.53	4.43	4.36	4.30	4.24	4.20
	0.990	6.18	6.03	5.91	5.81	5.73	5.67
	0.999	12.40	12.05	11.77	11.54	11.35	11.19
9	0.900	2.51	2.47	2.44	2.42	2.40	2.38
	0.950	3.29	3.23	3.18	3.14	3.10	3.07
	0.975	4.20	4.10	4.03	3.96	3.91	3.87
	0.990	5.61	5.47	5.35	5.26	5.18	5.11
	0.999	10.70	10.37	10.11	9.89	9.72	9.57
10	0.900	2.41	2.38	2.35	2.32	2.30	2.28
	0.950	3.14	3.07	3.02	2.98	2.94	2.91
	0.975	3.95	3.85	3.78	3.72	3.66	3.62
	0.990	5.20	5.06	4.94	4.85	4.77	4.71
	0.999	9.52	9.20	8.96	8.75	8.59	8.45
11	0.900	2.34	2.30	2.27	2.25	2.23	2.21
	0.950	3.01	2.95	2.90	2.85	2.82	2.79
	0.975	3.76	3.66	3.59	3.53	3.47	3.43
	0.990	4.89	4.74	4.63	4.54	4.46	4.40
	0.999	8.66	8.35	8.12	7.92	7.76	7.63
12	0.900	2.28	2.24	2.21	2.19	2.17	2.15
	0.950	2.91	2.85	2.80	2.75	2.72	2.69
	0.975	3.61	3.51	3.44	3.37	3.32	3.28
	0.990	4.64	4.50	4.39	4.30	4.22	4.16
	0.999	8.00	7.71	7.48	7.29	7.14	7.00
13	0.900	2.23	2.20	2.16	2.14	2.12	2.10
	0.950	2.83	2.77	2.71	2.67	2.63	2.60
	0.975	3.48	3.39	3.31	3.25	3.20	3.15
	0.990	4.44	4.30	4.19	4.10	4.02	3.96
	0.999	7.49	7.21	6.98	6.80	6.65	6.52
14	0.900	2.19	2.15	2.12	2.10	2.07	2.05
	0.950	2.76	2.70	2.65	2.60	2.57	2.53
	0.975	3.38	3.29	3.21	3.15	3.09	3.05
	0.990	4.28	4.14	4.03	3.94	3.86	3.80
	0.999	7.08	6.80	6.58	6.40	6.26	6.13

Table D.5 $F(ndf, ddf)$ **Quantiles** (*continued*)

ddf	P	15	20	30	60	120	10000
Table D.5 $F(ndf, ddf)$ **Quantiles** *(continued)*							
				ndf			
8	0.900	2.46	2.42	2.38	2.34	2.32	2.29
	0.950	3.22	3.15	3.08	3.01	2.97	2.93
	0.975	4.10	4.00	3.89	3.78	3.73	3.67
	0.990	5.52	5.36	5.20	5.03	4.95	4.86
	0.999	10.84	10.48	10.11	9.73	9.53	9.34
9	0.900	2.34	2.30	2.25	2.21	2.18	2.16
	0.950	3.01	2.94	2.86	2.79	2.75	2.71
	0.975	3.77	3.67	3.56	3.45	3.39	3.33
	0.990	4.96	4.81	4.65	4.48	4.40	4.31
	0.999	9.24	8.90	8.55	8.19	8.00	7.82
10	0.900	2.24	2.20	2.16	2.11	2.08	2.06
	0.950	2.85	2.77	2.70	2.62	2.58	2.54
	0.975	3.52	3.42	3.31	3.20	3.14	3.08
	0.990	4.56	4.41	4.25	4.08	4.00	3.91
	0.999	8.13	7.80	7.47	7.12	6.94	6.76
11	0.900	2.17	2.12	2.08	2.03	2.00	1.97
	0.950	2.72	2.65	2.57	2.49	2.45	2.41
	0.975	3.33	3.23	3.12	3.00	2.94	2.88
	0.990	4.25	4.10	3.94	3.78	3.69	3.60
	0.999	7.32	7.01	6.68	6.35	6.18	6.00
12	0.900	2.10	2.06	2.01	1.96	1.93	1.90
	0.950	2.62	2.54	2.47	2.38	2.34	2.30
	0.975	3.18	3.07	2.96	2.85	2.79	2.73
	0.990	4.01	3.86	3.70	3.54	3.45	3.36
	0.999	6.71	6.40	6.09	5.76	5.59	5.42
13	0.900	2.05	2.01	1.96	1.90	1.88	1.85
	0.950	2.53	2.46	2.38	2.30	2.25	2.21
	0.975	3.05	2.95	2.84	2.72	2.66	2.60
	0.990	3.82	3.66	3.51	3.34	3.25	3.17
	0.999	6.23	5.93	5.63	5.30	5.14	4.97
14	0.900	2.01	1.96	1.91	1.86	1.83	1.80
	0.950	2.46	2.39	2.31	2.22	2.18	2.13
	0.975	2.95	2.84	2.73	2.61	2.55	2.49
	0.990	3.66	3.51	3.35	3.18	3.09	3.01
	0.999	5.85	5.56	5.25	4.94	4.77	4.61

		Table D.5 $F(ndf, ddf)$ **Quantiles** *(continued)*					
				ndf			
ddf	P	1	2	3	4	5	6
15	0.900	3.07	2.70	2.49	2.36	2.27	2.21
	0.950	4.54	3.68	3.29	3.06	2.90	2.79
	0.975	6.20	4.77	4.15	3.80	3.58	3.41
	0.990	8.68	6.36	5.42	4.89	4.56	4.32
	0.999	16.59	11.34	9.34	8.25	7.57	7.09
20	0.900	2.97	2.59	2.38	2.25	2.16	2.09
	0.950	4.35	3.49	3.10	2.87	2.71	2.60
	0.975	5.87	4.46	3.86	3.51	3.29	3.13
	0.990	8.10	5.85	4.94	4.43	4.10	3.87
	0.999	14.82	9.95	8.10	7.10	6.46	6.02
30	0.900	2.88	2.49	2.28	2.14	2.05	1.98
	0.950	4.17	3.32	2.92	2.69	2.53	2.42
	0.975	5.57	4.18	3.59	3.25	3.03	2.87
	0.990	7.56	5.39	4.51	4.02	3.70	3.47
	0.999	13.29	8.77	7.05	6.12	5.53	5.12
60	0.900	2.79	2.39	2.18	2.04	1.95	1.87
	0.950	4.00	3.15	2.76	2.53	2.37	2.25
	0.975	5.29	3.93	3.34	3.01	2.79	2.63
	0.990	7.08	4.98	4.13	3.65	3.34	3.12
	0.999	11.97	7.77	6.17	5.31	4.76	4.37
120	0.900	2.75	2.35	2.13	1.99	1.90	1.82
	0.950	3.92	3.07	2.68	2.45	2.29	2.18
	0.975	5.15	3.80	3.23	2.89	2.67	2.52
	0.990	6.85	4.79	3.95	3.48	3.17	2.96
	0.999	11.38	7.32	5.78	4.95	4.42	4.04
10000	0.900	2.71	2.30	2.08	1.95	1.85	1.77
	0.950	3.84	3.00	2.61	2.37	2.21	2.10
	0.975	5.03	3.69	3.12	2.79	2.57	2.41
	0.990	6.64	4.61	3.78	3.32	3.02	2.80
	0.999	10.83	6.91	5.43	4.62	4.11	3.75

Table D.5 $F(ndf, ddf)$ **Quantiles** *(continued)*

ddf	P	7	8	9	10	11	12
15	0.900	2.16	2.12	2.09	2.06	2.04	2.02
	0.950	2.71	2.64	2.59	2.54	2.51	2.48
	0.975	3.29	3.20	3.12	3.06	3.01	2.96
	0.990	4.14	4.00	3.89	3.80	3.73	3.67
	0.999	6.74	6.47	6.26	6.08	5.94	5.81
20	0.900	2.04	2.00	1.96	1.94	1.91	1.89
	0.950	2.51	2.45	2.39	2.35	2.31	2.28
	0.975	3.01	2.91	2.84	2.77	2.72	2.68
	0.990	3.70	3.56	3.46	3.37	3.29	3.23
	0.999	5.69	5.44	5.24	5.08	4.94	4.82
30	0.900	1.93	1.88	1.85	1.82	1.79	1.77
	0.950	2.33	2.27	2.21	2.16	2.13	2.09
	0.975	2.75	2.65	2.57	2.51	2.46	2.41
	0.990	3.30	3.17	3.07	2.98	2.91	2.84
	0.999	4.82	4.58	4.39	4.24	4.11	4.00
60	0.900	1.82	1.77	1.74	1.71	1.68	1.66
	0.950	2.17	2.10	2.04	1.99	1.95	1.92
	0.975	2.51	2.41	2.33	2.27	2.22	2.17
	0.990	2.95	2.82	2.72	2.63	2.56	2.50
	0.999	4.09	3.86	3.69	3.54	3.42	3.32
120	0.900	1.77	1.72	1.68	1.65	1.63	1.60
	0.950	2.09	2.02	1.96	1.91	1.87	1.83
	0.975	2.39	2.30	2.22	2.16	2.10	2.05
	0.990	2.79	2.66	2.56	2.47	2.40	2.34
	0.999	3.77	3.55	3.38	3.24	3.12	3.02
10000	0.900	1.72	1.67	1.63	1.60	1.57	1.55
	0.950	2.01	1.94	1.88	1.83	1.79	1.75
	0.975	2.29	2.19	2.11	2.05	1.99	1.95
	0.990	2.64	2.51	2.41	2.32	2.25	2.19
	0.999	3.48	3.27	3.10	2.96	2.85	2.75

Table D.5 $F(ndf, ddf)$ **Quantiles** *(continued)*

ddf	P	ndf 15	20	30	60	120	10000
15	0.900	1.97	1.92	1.87	1.82	1.79	1.76
	0.950	2.40	2.33	2.25	2.16	2.11	2.07
	0.975	2.86	2.76	2.64	2.52	2.46	2.40
	0.990	3.52	3.37	3.21	3.05	2.96	2.87
	0.999	5.54	5.25	4.95	4.64	4.47	4.31
20	0.900	1.84	1.79	1.74	1.68	1.64	1.61
	0.950	2.20	2.12	2.04	1.95	1.90	1.84
	0.975	2.57	2.46	2.35	2.22	2.16	2.09
	0.990	3.09	2.94	2.78	2.61	2.52	2.42
	0.999	4.56	4.29	4.00	3.70	3.54	3.38
30	0.900	1.72	1.67	1.61	1.54	1.50	1.46
	0.950	2.01	1.93	1.84	1.74	1.68	1.62
	0.975	2.31	2.20	2.07	1.94	1.87	1.79
	0.990	2.70	2.55	2.39	2.21	2.11	2.01
	0.999	3.75	3.49	3.22	2.92	2.76	2.59
60	0.900	1.60	1.54	1.48	1.40	1.35	1.29
	0.950	1.84	1.75	1.65	1.53	1.47	1.39
	0.975	2.06	1.94	1.82	1.67	1.58	1.48
	0.990	2.35	2.20	2.03	1.84	1.73	1.60
	0.999	3.08	2.83	2.55	2.25	2.08	1.89
120	0.900	1.55	1.48	1.41	1.32	1.26	1.19
	0.950	1.75	1.66	1.55	1.43	1.35	1.26
	0.975	1.94	1.82	1.69	1.53	1.43	1.31
	0.990	2.19	2.03	1.86	1.66	1.53	1.38
	0.999	2.78	2.53	2.26	1.95	1.77	1.55
10000	0.900	1.49	1.42	1.34	1.24	1.17	1.03
	0.950	1.67	1.57	1.46	1.32	1.22	1.03
	0.975	1.83	1.71	1.57	1.39	1.27	1.04
	0.990	2.04	1.88	1.70	1.48	1.33	1.05
	0.999	2.52	2.27	1.99	1.66	1.45	1.06

D.6   Binomial Distribution Probabilities

If $X \sim \text{Binomial}(n, p)$, then Table D.6 contains entries computing

$$P(X = k) = \left(\begin{array}{c} n \\ k \end{array} \right) p^k (1-p)^{n-k}$$

for various values of n, k, and p.

Note if $X \sim \text{Binomial}(n, p)$, then $P(X = k) = P(Y = n - k)$ where $Y = n - X \sim \text{Binomial}(n, 1 - p)$.

Table D.6 Binomial Probabilities

n	k	.01	.02	.03	.04	.05	.06	.07	.08	.09
2	0	.9801	.9604	.9409	.9216	.9025	.8836	.8649	.8464	.8281
	1	.0198	.0392	.0582	.0768	.0950	.1128	.1302	.1472	.1638
	2	.0001	.0004	.0009	.0016	.0025	.0036	.0049	.0064	.0081
3	0	.9703	.9412	.9127	.8847	.8574	.8306	.8044	.7787	.7536
	1	.0294	.0576	.0847	.1106	.1354	.1590	.1816	.2031	.2236
	2	.0003	.0012	.0026	.0046	.0071	.0102	.0137	.0177	.0221
	3				.0001	.0001	.0002	.0003	.0005	.0007
4	0	.9606	.9224	.8853	.8493	.8145	.7807	.7481	.7164	.6857
	1	.0388	.0753	.1095	.1416	.1715	.1993	.2252	.2492	.2713
	2	.0006	.0023	.0051	.0088	.0135	.0191	.0254	.0325	.0402
	3			.0001	.0002	.0005	.0008	.0013	.0019	.0027
	4									.0001
5	0	.9510	.9039	.8587	.8154	.7738	.7339	.6957	.6591	.6240
	1	.0480	.0922	.1328	.1699	.2036	.2342	.2618	.2866	.3086
	2	.0010	.0038	.0082	.0142	.0214	.0299	.0394	.0498	.0610
	3		.0001	.0003	.0006	.0011	.0019	.0030	.0043	.0060
	4						.0001	.0001	.0002	.0003
	5									
6	0	.9415	.8858	.8330	.7828	.7351	.6899	.6470	.6064	.5679
	1	.0571	.1085	.1546	.1957	.2321	.2642	.2922	.3164	.3370
	2	.0014	.0055	.0120	.0204	.0305	.0422	.0550	.0688	.0833
	3		.0002	.0005	.0011	.0021	.0036	.0055	.0080	.0110
	4					.0001	.0002	.0003	.0005	.0008
	5									
	6									
7	0	.9321	.8681	.8080	.7514	.6983	.6485	.6017	.5578	.5168
	1	.0659	.1240	.1749	.2192	.2573	.2897	.3170	.3396	.3578
	2	.0020	.0076	.0162	.0274	.0406	.0555	.0716	.0886	.1061
	3		.0003	.0008	.0019	.0036	.0059	.0090	.0128	.0175
	4				.0001	.0002	.0004	.0007	.0011	.0017
	5								.0001	.0001
	6									
	7									
8	0	.9227	.8508	.7837	.7214	.6634	.6096	.5596	.5132	.4703
	1	.0746	.1389	.1939	.2405	.2793	.3113	.3370	.3570	.3721
	2	.0026	.0099	.0210	.0351	.0515	.0695	.0888	.1087	.1288
	3	.0001	.0004	.0013	.0029	.0054	.0089	.0134	.0189	.0255
	4			.0001	.0002	.0004	.0007	.0013	.0021	.0031
	5							.0001	.0001	.0002
	6									
	7									
	8									

Table D.6 Binomial Probabilities *(continued)*

n	k	.10	.15	.20	.25	.30	.35	.40	.45	.50
2	0	.8100	.7225	.6400	.5625	.4900	.4225	.3600	.3025	.2500
	1	.1800	.2550	.3200	.3750	.4200	.4550	.4800	.4950	.5000
	2	.0100	.0225	.0400	.0625	.0900	.1225	.1600	.2025	.2500
3	0	.7290	.6141	.5120	.4219	.3430	.2746	.2160	.1664	.1250
	1	.2430	.3251	.3840	.4219	.4410	.4436	.4320	.4084	.3750
	2	.0270	.0574	.0960	.1406	.1890	.2389	.2880	.3341	.3750
	3	.0010	.0034	.0080	.0156	.0270	.0429	.0640	.0911	.1250
4	0	.6561	.5220	.4096	.3164	.2401	.1785	.1296	.0915	.0625
	1	.2916	.3685	.4096	.4219	.4116	.3845	.3456	.2995	.2500
	2	.0486	.0975	.1536	.2109	.2646	.3105	.3456	.3675	.3750
	3	.0036	.0115	.0256	.0469	.0756	.1115	.1536	.2005	.2500
	4	.0001	.0005	.0016	.0039	.0081	.0150	.0256	.0410	.0625
5	0	.5905	.4437	.3277	.2373	.1681	.1160	.0778	.0503	.0313
	1	.3280	.3915	.4096	.3955	.3602	.3124	.2592	.2059	.1563
	2	.0729	.1382	.2048	.2637	.3087	.3364	.3456	.3369	.3125
	3	.0081	.0244	.0512	.0879	.1323	.1811	.2304	.2757	.3125
	4	.0004	.0022	.0064	.0146	.0284	.0488	.0768	.1128	.1562
	5		.0001	.0003	.0010	.0024	.0053	.0102	.0185	.0312
6	0	.5314	.3771	.2621	.1780	.1176	.0754	.0467	.0277	.0156
	1	.3543	.3993	.3932	.3560	.3025	.2437	.1866	.1359	.0938
	2	.0984	.1762	.2458	.2966	.3241	.3280	.3110	.2780	.2344
	3	.0146	.0415	.0819	.1318	.1852	.2355	.2765	.3032	.3125
	4	.0012	.0055	.0154	.0330	.0595	.0951	.1382	.1861	.2344
	5	.0001	.0004	.0015	.0044	.0102	.0205	.0369	.0609	.0937
	6			.0001	.0002	.0007	.0018	.0041	.0083	.0156
7	0	.4783	.3206	.2097	.1335	.0824	.0490	.0280	.0152	.0078
	1	.3720	.3960	.3670	.3115	.2471	.1848	.1306	.0872	.0547
	2	.1240	.2097	.2753	.3115	.3177	.2985	.2613	.2140	.1641
	3	.0230	.0617	.1147	.1730	.2269	.2679	.2903	.2918	.2734
	4	.0026	.0109	.0287	.0577	.0972	.1442	.1935	.2388	.2734
	5	.0002	.0012	.0043	.0115	.0250	.0466	.0774	.1172	.1641
	6		.0001	.0004	.0013	.0036	.0084	.0172	.0320	.0547
	7				.0001	.0002	.0006	.0016	.0037	.0078
8	0	.4305	.2725	.1678	.1001	.0576	.0319	.0168	.0084	.0039
	1	.3826	.3847	.3355	.2670	.1977	.1373	.0896	.0548	.0313
	2	.1488	.2376	.2936	.3115	.2965	.2587	.2090	.1569	.1094
	3	.0331	.0839	.1468	.2076	.2541	.2786	.2787	.2568	.2188
	4	.0046	.0185	.0459	.0865	.1361	.1875	.2322	.2627	.2734
	5	.0004	.0026	.0092	.0231	.0467	.0808	.1239	.1719	.2188
	6		.0002	.0011	.0038	.0100	.0217	.0413	.0703	.1094
	7			.0001	.0004	.0012	.0033	.0079	.0164	.0312
	8					.0001	.0002	.0007	.0017	.0039

Table D.6 Binomial Probabilities *(continued)*

n	k	.01	.02	.03	.04	.05	.06	.07	.08	.09
9	0	.9135	.8337	.7602	.6925	.6302	.5730	.5204	.4722	.4279
	1	.0830	.1531	.2116	.2597	.2985	.3292	.3525	.3695	.3809
	2	.0034	.0125	.0262	.0433	.0629	.0840	.1061	.1285	.1507
	3	.0001	.0006	.0019	.0042	.0077	.0125	.0186	.0261	.0348
	4			.0001	.0003	.0006	.0012	.0021	.0034	.0052
	5						.0001	.0002	.0003	.0005
	6									
	7									
	8									
	9									
10	0	.9044	.8171	.7374	.6648	.5987	.5386	.4840	.4344	.3894
	1	.0914	.1667	.2281	.2770	.3151	.3438	.3643	.3777	.3851
	2	.0042	.0153	.0317	.0519	.0746	.0988	.1234	.1478	.1714
	3	.0001	.0008	.0026	.0058	.0105	.0168	.0248	.0343	.0452
	4			.0001	.0004	.0010	.0019	.0033	.0052	.0078
	5					.0001	.0001	.0003	.0005	.0009
	6									.0001
	7									
	8									
	9									
	10									
12	0	.8864	.7847	.6938	.6127	.5404	.4759	.4186	.3677	.3225
	1	.1074	.1922	.2575	.3064	.3413	.3645	.3781	.3837	.3827
	2	.0060	.0216	.0438	.0702	.0988	.1280	.1565	.1835	.2082
	3	.0002	.0015	.0045	.0098	.0173	.0272	.0393	.0532	.0686
	4		.0001	.0003	.0009	.0021	.0039	.0067	.0104	.0153
	5				.0001	.0002	.0004	.0008	.0014	.0024
	6							.0001	.0001	.0003
	7									
	8									
	9									
	10									
	11									
	12									
15	0	.8601	.7386	.6333	.5421	.4633	.3953	.3367	.2863	.2430
	1	.1303	.2261	.2938	.3388	.3658	.3785	.3801	.3734	.3605
	2	.0092	.0323	.0636	.0988	.1348	.1691	.2003	.2273	.2496
	3	.0004	.0029	.0085	.0178	.0307	.0468	.0653	.0857	.1070
	4		.0002	.0008	.0022	.0049	.0090	.0148	.0223	.0317
	5			.0001	.0002	.0006	.0013	.0024	.0043	.0069
	6						.0001	.0003	.0006	.0011
	7								.0001	.0001
	8									
	9									
	10									
	11									
	12									
	13									
	14									
	15									

Table D.6   Binomial Probabilities *(continued)*

n	k	.10	.15	.20	.25	.30	.35	.40	.45	.50
9	0	.3874	.2316	.1342	.0751	.0404	.0207	.0101	.0046	.0020
	1	.3874	.3679	.3020	.2253	.1556	.1004	.0605	.0339	.0176
	2	.1722	.2597	.3020	.3003	.2668	.2162	.1612	.1110	.0703
	3	.0446	.1069	.1762	.2336	.2668	.2716	.2508	.2119	.1641
	4	.0074	.0283	.0661	.1168	.1715	.2194	.2508	.2600	.2461
	5	.0008	.0050	.0165	.0389	.0735	.1181	.1672	.2128	.2461
	6	.0001	.0006	.0028	.0087	.0210	.0424	.0743	.1160	.1641
	7			.0003	.0012	.0039	.0098	.0212	.0407	.0703
	8				.0001	.0004	.0013	.0035	.0083	.0176
	9						.0001	.0003	.0008	.0020
10	0	.3487	.1969	.1074	.0563	.0282	.0135	.0060	.0025	.0010
	1	.3874	.3474	.2684	.1211	.0725	.0403	.0207	.0098	
	1	.3874	.3474	.2684	.1877	.0725	.0403	.0207	.0207	.0098
	2	.1937	.2759	.3020	.2816	.2335	.1757	.1209	.0763	.0439
	3	.0574	.1298	.2013	.2503	.2668	.2522	.2150	.1665	.1172
	4	.0112	.0401	.0881	.1460	.2001	.2377	.2508	.2384	.2051
	5	.0015	.0085	.0264	.0584	.1029	.1536	.2007	.2340	.2461
	6	.0001	.0012	.0055	.0162	.0368	.0689	.1115	.1596	.2051
	7		.0001	.0008	.0031	.0090	.0212	.0425	.0746	.1172
	8			.0001	.0004	.0014	.0043	.0106	.0229	.0439
	9					.0001	.0005	.0016	.0042	.0098
	10							.0001	.0003	.0010
12	0	.2824	.1422	.0687	.0317	.0138	.0057	.0022	.0008	.0002
	1	.3766	.3012	.2062	.1267	.0712	.0368	.0174	.0075	.0029
	2	.2301	.2924	.2835	.2323	.1678	.1088	.0639	.0339	.0161
	3	.0852	.1720	.2362	.2581	.2397	.1954	.1419	.0923	.0537
	4	.0213	.0683	.1329	.1936	.2311	.2367	.2128	.1700	.1208
	5	.0038	.0193	.0532	.1032	.1585	.2039	.2270	.2225	.1934
	6	.0005	.0040	.0155	.0401	.0792	.1281	.1766	.2124	.2256
	7		.0006	.0033	.0115	.0291	.0591	.1009	.1489	.1934
	8		.0001	.0005	.0024	.0078	.0199	.0420	.0762	.1208
	9			.0001	.0004	.0015	.0048	.0125	.0277	.0537
	10					.0002	.0008	.0025	.0068	.0161
	11						.0001	.0003	.0010	.0029
	12								.0001	.0002
15	0	.2059	.0874	.0352	.0134	.0047	.0016	.0005	.0001	
	1	.3432	.2312	.1319	.0668	.0305	.0126	.0047	.0016	.0005
	2	.2669	.2856	.2309	.1559	.0916	.0476	.0219	.0090	.0032
	3	.1285	.2184	.2501	.2252	.1700	.1110	.0634	.0318	.0139
	4	.0428	.1156	.1876	.2252	.2186	.1792	.1268	.0780	.0417
	5	.0105	.0449	.1032	.1651	.2061	.2123	.1859	.1404	.0916
	6	.0019	.0132	.0430	.0917	.1472	.1906	.2066	.1914	.1527
	7	.0003	.0030	.0138	.0393	.0811	.1319	.1771	.2013	.1964
	8		.0005	.0035	.0131	.0348	.0710	.1181	.1647	.1964
	9		.0001	.0007	.0034	.0116	.0298	.0612	.1048	.1527
	10			.0001	.0007	.0030	.0096	.0245	.0515	.0916
	11				.0001	.0006	.0024	.0074	.0191	.0417
	12					.0001	.0004	.0016	.0052	.0139
	13						.0001	.0003	.0010	.0032
	14								.0001	.0005
	15									

Table D.6 Binomial Probabilities *(continued)*

n	k	.01	.02	.03	.04	.05	.06	.07	.08	.09
20	0	.8179	.6676	.5438	.4420	.3585	.2901	.2342	.1887	.1516
	1	.1652	.2725	.3364	.3683	.3774	.3703	.3526	.3282	.3000
	2	.0159	.0528	.0988	.1458	.1887	.2246	.2521	.2711	.2818
	3	.0010	.0065	.0183	.0364	.0596	.0860	.1139	.1414	.1672
	4		.0006	.0024	.0065	.0133	.0233	.0364	.0523	.0703
	5			.0002	.0009	.0022	.0048	.0088	.0145	.0222
	6				.0001	.0003	.0008	.0017	.0032	.0055
	7						.0001	.0002	.0005	.0011
	8								.0001	.0002
	9									
	10									
	11									
	12									
	13									
	14									
	15									
	16									
	17									
	18									
	19									
	20									

Table D.6 Binomial Probabilities *(continued)*

n	k	.10	.15	.20	.25	.30	.35	.40	.45	.50
20	0	.1216	.0388	.0115	.0032	.0008	.0002			
	1	.2702	.1368	.0576	.0211	.0068	.0020	.0005	.0001	
	2	.2852	.2293	.1369	.0669	.0278	.0100	.0031	.0008	.0002
	3	.1901	.2428	.2054	.1339	.0716	.0323	.0123	.0040	.0011
	4	.0898	.1821	.2182	.1897	.1304	.0738	.0350	.0139	.0046
	5	.0319	.1028	.1746	.2023	.1789	.1272	.0746	.0365	.0148
	6	.0089	.0454	.1091	.1686	.1916	.1712	.1244	.0746	.0370
	7	.0020	.0160	.0545	.1643	.1643	.1844	.1659	.1221	.0739
	8	.0004	.0046	.0222	.0609	.1144	.1614	.1797	.1623	.1201
	9	.0001	.0011	.0074	.0271	.0654	.1158	.1597	.1771	.1602
	10		.0002	.0020	.0099	.0308	.0686	.1171	.1593	.1762
	11			.0005	.0030	.0120	.0336	.0710	.1185	.1602
	12			.0001	.0008	.0039	.0136	.0355	.0727	.1201
	13				.0002	.0010	.0045	.0146	.0366	.0739
	14					.0002	.0012	.0049	.0150	.0370
	15						.0003	.0013	.0049	.0148
	16							.0003	.0013	.0046
	17								.0002	.0011
	18									.0002
	19									
	20									

Index

684